INTERNATIONAL SERIES
OF
MONOGRAPHS ON PHYSICS

GENERAL EDITORS
R. K. ADAIR   R. J. ELLIOTT
W. C. MARSHALL   D. H. WILKINSON

THE PRINCIPLES OF
NUCLEAR
MAGNETISM

BY

A. ABRAGAM

PROFESSEUR AU COLLÈGE DE FRANCE
CHEF DE DÉPARTEMENT AU
COMMISSARIAT À L'ÉNERGIE ATOMIQUE

CLARENDON PRESS · OXFORD

Oxford University Press, Walton Street, Oxford OX2 6DP
Oxford New York Toronto
Delhi Bombay Calcutta Madras Karachi
Petaling Jaya Singapore Hong Kong Tokyo
Nairobi Dar es Salaam Cape Town
Melbourne Auckland
and associated companies in
Beirut Berlin Ibadan Nicosia

Oxford is a trade mark of Oxford University Press

Published in the United States
by Oxford University Press, New York

© Oxford University Press 1961

First published 1961
Reprinted (with corrections) 1962, 1967, 1970, 1973, 1978
First published in paperback 1983
Reprinted 1985, 1986

ISBN 0–19–851236–8
ISBN 0–19–852014 X Pbk.

All rights reserved. No part of this publication may be reproduced, stored in a retrieval system, or transmitted, in any form or by any means, electronic, mechanical, photocopying, recording, or otherwise, without the prior permission of Oxford University Press

This book is sold subject to the condition that it shall not, by way of trade or otherwise, be lent, re-sold, hired out or otherwise circulated without the publisher's prior consent in any form of binding or cover other than that in which it is published and without a similar condition including this condition being imposed on the subsequent purchaser

Printed in Hong Kong

PREFACE

People have now-a-days got a strange opinion that everything should be taught by lectures. Now, I cannot see that lectures can do so much good as reading the books from which the lectures are taken.

DR. JOHNSON (1766)

SINCE the first successful detection of nuclear resonance signals late in 1945, nuclear magnetism has developed at a pace which after fifteen years still shows no sign of slackening. Besides its first and obvious application to the measurement of nuclear moments, it has become a major tool in the study of the finer properties of matter in bulk. Structure of molecules, reaction rates and chemical equilibria, chemical bonding, crystal structures, internal motions in solids and liquids, electronic densities in metals, alloys, and semiconductors, internal fields in ferromagnetic and antiferromagnetic substances, density of states in superconductors, properties of quantum liquids, are some of the topics where nuclear magnetism has so far provided specific and detailed information.

Detailed studies of nuclear relaxation processes combined with electron spin resonance have led to original methods of nuclear alignment which may in the near future open up new experimental possibilities to nuclear physicists as well as to the specialists of very low temperatures.

On a more fundamental level, systems of nuclear spins provide examples of problems in statistical mechanics and in the theory of irreversible processes that are sufficiently simple to allow a clean and unambiguous solution both theoretically and experimentally, but are still sufficiently complex to be far from trivial. Thus the hitherto purely academic concept of negative temperature has found its full physical significance in nuclear magnetism. Thus also the trend toward thermal equilibrium of a system (the spin system) coupled to a thermal bath (the lattice) can be investigated, theoretically and experimentally, more fully than is usually possible in other fields of physics.

One of the more remarkable features of nuclear magnetism and, at least for the author of this book, the most deeply satisfying, is the very close connexion between theory and experiment that leaves little room for a theory that could not be tested by a suitable experiment or for an experiment that does not admit of a well-defined theoretical interpretation; this perhaps to a greater extent than in any other branch of modern physics.

Such at any rate was the view that has presided over the writing of this book. It is an attempt to provide an account of the principles and methods of nuclear magnetism that would be acceptable to the theorist and of use to the experimentalist; a wish often expressed, seldom fulfilled, that stands a better chance, it is hoped, in a field where theorist and experimentalist as often as not are the same person.

On the contents of this book the following comments can be made.

The experimental methods are discussed very briefly and in very general terms. The lack of space and the lack of competence of the author have precluded any detailed description of the experimental techniques currently in use. One can only hope that the definite need for a monograph dealing with these problems will be satisfied in the future.

There are no tables of important constants such as nuclear moments, quadrupole couplings, chemical shifts, relaxation times, and spin-spin couplings. The amount of labour involved in the compiling of such data, the space requirements, and also the danger of rapid obsolescence have led to the conclusion that their presentation in review articles in a form comparable to that currently used for electron resonance data is much more suitable. For the same reason it did not seem possible nor indeed desirable to include an extensive bibliography of the published work in the field of nuclear magnetism, which at the time of writing contains well over 2,000 titles and is continually increasing.

The selection of the references included in this book was made on the following basis. The viewpoint adopted is that in a book, in contrast to an article, no references need be given for credit and that it is the privilege and indeed the role of the author to blend together the contributions of the various workers in the field and to modify them as he sees fit, for the benefit of the reader. In fairness to all concerned it should therefore be made perfectly clear that the references given in each case are not necessarily considered by the author as the most original or the most important contributions in the field but rather as convenient sources of information to supplement that given in the book. To give one example among many, the important work of Kubo and Tomita, which has influenced the author very strongly, is not quoted because it was felt that the gist of its contents was dealt with in sufficient detail to make further reference to the original paper less useful for the reader than that to some other, perhaps less important, work. Under such circumstances it was clear that a name index would serve no useful purpose and it has accordingly been omitted.

PREFACE

To conclude, one may ask what category of readers I had in mind in writing this book. This is a difficult question which may be answered in part by saying what kind of book I would have liked to have written. What I had in mind was to attempt for nuclear magnetism what Van Vleck has done for the theory of electric and magnetic susceptibilities in a book which thirty years after its publication is still *the* book on the subject. A *demi-succès* in this undertaking would be good enough.

It is not possible to acknowledge in detail my great indebtedness for illuminating discussions and constructive criticism to so many friends and colleagues and especially to the members of our group at Saclay and to our visitors from abroad. I would like, however, to thank particularly 'the long legg'd spinner', my good friend Ionel Solomon who has patiently coached me in the mysteries of the lock-in detector and who has kindly prepared some beautiful photographs with the express purpose of illustrating this book.

The planning of the book and the writing of many chapters have benefited from the help and advice of Professor R. V. Pound more than I can say. Although I must take full responsibility for the entire book, those familiar with his work will have no difficulty in discerning his influence. Let him be assured of my deep gratitude.

I am very grateful to many colleagues and to the editors of the following journals for permission to reproduce a number of figures from these journals: *Acta Metallurgica—Comptes Rendus de l'Academie des Sciences—Journal of the Physical Society of Japan—Molecular Physics—Nature—Physica—Physical Review—Physical Review Letters—Proceedings of the Physical Society—Proceedings of the Royal Society—Science—The Journal of Chemical Physics—The Physics and Chemistry of Solids*. My special thanks go to those authors who kindly took the trouble of sending original photographs.

I am greatly indebted to Madame Houzé and Madame Gugenberger for their help with the preparation of the manuscript and to Madame Gugenberger for her competent handling of the proof. Mr. Timothy Hall has kindly prepared the subject index.

A. A.

Saclay
October 1960

PREPACE FOR THE PAPERBACK EDITION

THIS book is the first (and possibly the last) attempt to give in a single volume a comprehensive account of all the concepts, methods, theories and results of the discipline called Nuclear Magnetism, that is the collective behaviour of nuclear spins in bulk matter, as it was perceived at the time of writing more than twenty years ago.

The subject did not stand still to say the least and the book could not possibly contain now *all* the useful information, even if it ever did, which I doubt. However, although not *all* of Nuclear Magnetism is in there, whatever is, is both important and *not wrong*. The secret for this in my view is the following: never put in a book (as opposed to a Research paper) anything that you do not understand thoroughly. There is, I believe (I quote from memory), a statement by Freeman Dyson: "any book published at time t and up to date to time $(t - \tau)$ will be obsolete at time $(t + \tau)$". This may be true if you gather uncritically in your book all the information available up to time $(t - \tau)$ but not if you filter it the way I have indicated.

Much of what one does not understand at the time of writing turns out later to follow from faulty experiment or faulty theory and becomes useless ballast in a book. The fact that there is very little of this in the present book, may be a justification for issuing it as a paperback.

Collège de France
July 82

CONTENTS

I. GENERAL INTRODUCTION — 1
  A. Nuclear paramagnetism — 1
  B. Special features of radio-frequency spectroscopy — 3
  C. The phenomenon of resonance; 'resonant' and 'non-resonant' methods — 4
    (a) Beam measurements of atomic, molecular, and nuclear magnetic moments — 5
    (b) Measurement of the magnetic moment of the neutron — 6
    (c) The fine structure of the hydrogen atom — 7
    (d) The fine structure of positronium — 8
    (e) The magnetic moment of the μ-meson — 9
    (f) Resonance detected by optical means — 10
    (g) Perturbed angular correlations — 11
  D. Nuclear magnetic resonance — 13
    (a) The principle — 13
    (b) Various extensions and generalizations — 15
REFERENCES — 17

II. MOTION OF FREE SPINS — 19
  A. Classical treatment — 19
  B. Quantum mechanical treatment — 22
  C. Quantum mechanical description of a statistical ensemble of free spins. Density matrix — 24
  D. Relation with the perturbation method — 27
  E. Transient effects — 32
    (a) Free precession — 32
    (b) Spin echoes — 33
    (c) The adiabatic theorem, adiabatic passage — 34
  F. The general problem of two levels coupled by an r.f. field. The fictitious spin $\frac{1}{2}$ — 36
REFERENCES — 38

III. MACROSCOPIC ASPECTS OF NUCLEAR MAGNETISM — 39
  I. INTRODUCTION — 39
    A. Static susceptibility — 39
    B. Resonant absorption of r.f. energy — 40
  II. THE PHENOMENOLOGICAL EQUATIONS OF BLOCH — 44
    A. Steady-state solutions—saturation — 45
    B. Steady-state solutions in an inhomogeneous field — 49
    C. Modified Bloch equations in low fields — 53

CONTENTS

III. TRANSIENT METHODS IN NUCLEAR MAGNETISM	57
A. The method of the spin echoes	58
(a) Spin diffusion	59
(b) Coherent and incoherent pulses	62
B. Free precession	63
(a) Free precession in the earth's field	64
C. Adiabatic passage	65
Applications of the fast-passage methods	66
(a) Measurement of the relaxation time T_1	66
(b) Measurement of T_2 in liquids	67
(c) Other applications of adiabatic fast passage	68
D. The method of transient nutation	68
IV. DETECTION METHODS	71
A. General	71
B. Detection of steady-state nuclear signals	75
(a) Q-meter detection	75
(b) Bridge and crossed coils methods	76
(c) Marginal oscillator	77
(d) Audio-modulation, narrow band amplification, phase sensitive detection, signal-to-noise ratio	78
(1) Lock-in detection and signal-to-noise ratio	79
(2) Signal-to-noise ratio	82
(e) Transient effects in steady-state detection	85
C. Transient methods of detection	86
(a) Adiabatic fast passage	86
(b) Pulse methods, coherent and incoherent pulses	87
D. Negative absorption—masers	89
APPENDIX. Proof of the Kramers–Krönig relations	93
REFERENCES	96
IV. DIPOLAR LINE WIDTH IN A RIGID LATTICE	97
I. INTRODUCTION	97
A. The local field	97
B. General theory of magnetic absorption	98
II. BROADENING BY LIKE SPINS	103
A. Dipole–dipole interaction	103
B. Definition of the moments	106
C. Principle of the calculation of moments	108
D. Calculation of the second and fourth moments	111
E. Relationship between the line shape and the free precession signal	114
F. A comparison between theory and experiment	115
III. DIPOLAR BROADENING BY UNLIKE SPINS	122

CONTENTS

IV. Dipolar Broadening in Magnetically Diluted Substances ... 125
   A. The method of moments ... 125
   B. The statistical theory ... 126

V. Modifications in the Dipolar Broadening caused by the Existence of Quadrupole Couplings ... 128

REFERENCES ... 132

V. SPIN TEMPERATURE ... 133
   A. Non-interacting spins ... 133
   B. Interacting spins in high field ... 136
   C. Interacting spins in low fields ... 144
   D. Zeeman system with more than one spin species ... 150
   E. Dynamics of thermal spin-spin processes ... 154

REFERENCES ... 158

VI. ELECTRON–NUCLEUS INTERACTIONS ... 159

I. Electrostatic Couplings ... 159
   A. The Hamiltonian ... 159
   B. Ionic crystals ... 166
   C. Molecular crystals ... 169

II. Magnetic Interactions ... 170
   A. The coupling Hamiltonian ... 170
   B. The effect of electron-nucleus coupling in diamagnetic substances ... 173
     (a) General ... 173
     (b) Calculation of the chemical shift ... 175
     (c) Indirect interaction between nuclear spins in diamagnetic substances ... 183
        (1) The orbital coupling ... 184
        (2) The Heitler–London approximation ... 186
        (3) The method of molecular orbitals ... 190
   C. The effect of electron-nucleus coupling in paramagnetic substances ... 191
     (a) Non-metals ... 191
        (1) Nature of the coupling ... 191
        (2) Observability of nuclear resonance ... 193
     (b) Metals ... 199
        (1) The frequency shift in metals ... 199
        (2) The indirect interactions in metals ... 206
   D. Nuclear resonance in antiferromagnetic and ferromagnetic substances ... 210

REFERENCES ... 214

VII. FINE STRUCTURE OF RESONANCE LINES—QUADRUPOLE EFFECTS   216

  I. FINE STRUCTURE CAUSED BY DIPOLAR COUPLING   216

    A. Rigid lattice   216
      (a) Two identical spins (two protons) I^1 and I^2 216
      (b) Systems of more than two spins   222

    B. Nuclear resonance in solid hydrogen   223
      (a) Introduction. System of two interacting protons   223
      (b) Solid hydrogen   225
      (c) *Ortho-* and *para-*hydrogen   226
      (d) Crystalline potential   227
      (e) Magnetic resonance in a strong field   228
      (f) Magnetic resonance in zero field   231
      (g) Magnetic resonance in HD and D_2 231

  II. ENERGY LEVELS OF NUCLEAR SPINS IN THE PRESENCE OF QUADRUPOLE INTERACTIONS   232

    A. High magnetic fields   233
      (a) Energy levels in single crystals   233
      (b) Imperfect cubic crystals   237
        (1) Powder pattern   237
        (2) First-order broadening in imperfect crystals   237
        (3) Transient methods, multiple echoes   241
        (4) Second-order quadrupole broadening in imperfect crystals   246

    B. Low magnetic fields   249
      (a) Zero field spectra   249
        (1) Integer spins   250
        (2) Half-integer spins   251
      (b) Zeeman splittings of quadrupole levels   253
        (1) Integer spins   253
        (2) Half-integer spins   254
      (c) Transient methods   257
        (1) Transient magnetization in zero field   257
        (2) Transient magnetization in a small magnetic field H_0 260

  APPENDIX. Sign of the quadrupole coupling   261

REFERENCES   263

VIII. THERMAL RELAXATION IN LIQUIDS AND GASES   264

  I. INTRODUCTION   264
    A. Coupling of the nuclear spins with the radiation field   264
    B. Coupling of the spin system with the lattice   267

  II. RELAXATION IN LIQUIDS AND GASES   268
    A. General   268
    B. Definitions   270

C. Motion of a system subject to a perturbation which is a random function of time — 272
  (a) Transition probability — 272
  (b) The master equation for populations — 274
  (c) The master equation for the density matrix — 276
  (d) The master equation in operator form — 278
  (e) Macroscopic differential equations — 280
  (f) Summary of the notation introduced in this section — 281
  (g) Justification of the four assumptions leading to the generalized master equation — 282
D. Quantum mechanical formulation of the problem — 283
E. Relaxation by dipolar coupling — 289
  (a) Like spins — 290
  (b) Unlike spins — 294
  (c) Correlation functions resulting from random molecular rotation or translation — 297
    (1) Rotation — 298
    (2) Translation — 300
F. Other mechanisms of relaxation in liquids — 305
  (a) General — 305
  (b) (1) Scalar spin-spin coupling — 306
    (2) Scalar relaxation of the first kind — 308
    (3) Scalar relaxation of the second kind — 309
  (c) Quadrupole relaxation in liquids through molecular reorientation — 313
  (d) Relaxation through anisotropic chemical shift combined with molecular reorientation — 315
G. Nuclear relaxation in gases — 316
  (a) The H_2 molecule—diatomic molecules — 316
  (b) Relaxation in monatomic gases — 322

III. COMPARISON BETWEEN THEORY AND EXPERIMENT — 323
A. Dipolar coupling between like spins — 324
  (a) Short correlation times, relative values of T_1 — 324
  (b) Absolute values of T_1 — 326
  (c) Long correlation times — 327
B. Coupling between unlike spins — 328
  (a) Single irradiation methods — 328
  (b) Double irradiation methods — 333
    (1) The HF molecule — 333
    (2) Coupling between a nuclear spin and an electronic spin — 338
C. Electric quadrupole relaxation in liquids — 346
D. Nuclear relaxation in gases — 349
  (a) Nuclear relaxation in hydrogen gas — 349

(b) Nuclear relaxation in liquid hydrogen 350
(c) Relaxation in monatomic gases 352

REFERENCES 353

IX. THERMAL RELAXATION AND DYNAMIC POLARIZATION IN SOLIDS 354

I. CONDUCTION ELECTRONS AND SPIN-LATTICE RELAXATION IN METALS 355
  A. An elementary calculation of the relaxation time . . 356
  B. Nuclear relaxation time and spin temperature . . . 359
  C. Dynamic nuclear polarization in metals (the Overhauser effect) . . 364
    (a) Fermi statistics and non-equilibrium electron spin distribution 364
    (b) Dynamic polarization 367
    (c) Coupled equations for nuclear and electron spin polarization 368
  D. Comparison with experiment 370
    (a) Measurements of T_1 370
    (b) Dynamic polarization experiments 373
    (c) Dynamic nuclear polarization in metals at the temperatures of liquid helium 375

II. NUCLEAR RELAXATION CAUSED BY FIXED PARAMAGNETIC IMPURITIES 378
  A. Theory 379
  B. Comparison with experiment 386

III. MAGNETIC RELAXATION AND DYNAMIC POLARIZATION IN SEMICONDUCTORS AND INSULATORS . . . 389
  A. Relaxation by conduction electrons in semiconductors . 389
  B. Dynamic polarization by fixed paramagnetic impurities—solid state effect 392

IV. RELAXATION BY THERMAL VIBRATIONS IN A CRYSTALLINE LATTICE 401
  A. Lattice vibrations and phonons 402
  B. Transition probabilities induced by the spin-phonon coupling 404
  C. Magnetic and quadrupole relaxation by spin-phonon coupling 409
    (a) Magnetic relaxation 409
    (b) Quadrupole relaxation 411
  D. Ultrasonic experiments 417
    (a) Quadrupole transitions 419
    (b) Magnetic transitions 421

REFERENCES 423

X. THEORY OF LINE WIDTH IN THE PRESENCE OF MOTION OF THE SPINS 424

I. INTRODUCTION 424

II. THE ADIABATIC CASE 427
  A. General theory 427
  B. Exchange narrowing 435
  C. Brownian motion narrowing 439
III. THE NON-ADIABATIC LINE WIDTH 441
  A. Line width and transverse relaxation time 441
  B. General case 442
IV. DESTRUCTION OF FINE STRUCTURES THROUGH MOTION 447
V. INFLUENCE OF INTERNAL MOTIONS IN SOLIDS ON THE WIDTH AND RELAXATION PROPERTIES OF ZEEMAN RESONANCE LINES 451
  A. Rotational motions 451
  B. Translational diffusion in solids 458
VI. INFLUENCE OF INTERNAL MOTIONS IN SOLIDS ON THE WIDTH AND RELAXATION OF QUADRUPOLE RESONANCE LINES 467
  A. Torsion oscillations 468
    (a) The spin Hamiltonian 468
    (b) The line width 470
    (c) Relaxation time 472
    (d) The spectral densities 473
  B. Hindered rotations 474
    (a) Fast motion 474
    (b) Slow motion 477
REFERENCES 479

XI. MULTIPLET STRUCTURE OF RESONANCE LINES IN LIQUIDS 480
  I. ENERGY LEVELS OBSERVED BY CONTINUOUS WAVE METHODS 480
    A. $J \ll \delta$ 482
    B. J and δ comparable for two spins $\tfrac{1}{2}$ 484
    C. J and δ comparable, for two groups G and G' of p equivalent spins i and p' equivalent spins i', respectively 488
    D. Perturbation method 489
    E. Isochronous non-equivalent spins 491
  II. MULTIPLET SPECTRA OBSERVED BY TRANSIENT METHODS 495
    A. The method of free precession 495
    B. The method of spin echoes 497
  III. LINE WIDTH PROBLEMS IN MULTIPLET SPECTRA 501
    A. Effects of quadrupole relaxation and chemical exchange 501
    B. Effects of magnetic relaxation 506
REFERENCES 510

XII. THE EFFECTS OF STRONG RADIO-FREQUENCY FIELDS 511

I. Strong Radio-frequency Fields in Liquids 511
A. 'Non-viscous liquids' 511
B. 'Viscous liquids' 517
C. Bloch equations for a 'simple' line 522
D. Decoupling of spins through 'stirring' by a radio-frequency field 527
    (a) Introduction 527
    (b) The intermediate pattern (elementary theory) 530
    (c) The intermediate pattern (detailed theory) 533

II. Strong Radio-frequency Fields in Solids 539
A. Introduction 539
B. Spin temperature in the rotating frame, reversible fast passage 545
C. Spin temperature in the rotating frame, steady-state solutions 555
D. Spin-lattice relaxation in the rotating frame 560
    (a) Relaxation for a single spin species 560
    (b) Relaxation in the presence of two spin species 562
E. Double irradiation methods 566
    (a) Rotary saturation 566
        (1) Rotary saturation in liquids 566
        (2) Rotary saturation in solids 569
    (b) Line narrowing by double frequency irradiation 570
    (c) Transient methods of double irradiation 578

REFERENCES 580

INDEX OF NUCLEAR SPECIES 583

SUBJECT INDEX 591

PLATES

The following figures appear as plates at the places shown:

Figs. III, 2–III, 10	between pages 64 and 65
Fig. III, 11	facing page 76
III, 15	85
III, 18	96
III, 19	97
VII, 10	245
VIII, 6 and 7	338
VIII, 8	339
VIII, 10	341
IX, 1B	378
IX, 4 and 5	396
XI, 2	483
XI, 7	496
XI, 8 and 9	497
XII, 1	522

I
GENERAL INTRODUCTION

. . . pale, deluding beams.
PURCELL (*Dido and Aeneas*)

A. Nuclear paramagnetism

THE subject matter of this book is the magnetic behaviour of assemblies of large numbers of atomic nuclei.

We denote their collective macroscopic magnetic properties by the term 'nuclear magnetism' in analogy with the term electronic magnetism for assemblies of electrons.

It is essentially this collective aspect of the problem that distinguishes both theory and experiments in the field of nuclear magnetism proper, from other research in the field of nuclear moments.

Many atomic nuclei in their ground state have a non-zero spin angular momentum $I\hbar$ (integer or half integer in units of $\hbar$) and a dipolar magnetic moment $\mu = \gamma\hbar I$ collinear with it. With few exceptions, the order of magnitude of these moments is between 10^{-3} and 10^{-4} Bohr magnetons. It is these moments that give rise to nuclear magnetism. Without attempting a detailed parallel between nuclear and electronic magnetism, a few differences can be pointed out. Of the three usual aspects of magnetism, namely ferromagnetism (or antiferromagnetism), diamagnetism, and paramagnetism, only the last is of interest in nuclear magnetism. It will be remembered that ferromagnetism may arise when the temperature, T of the sample, times the Boltzmann constant k (i.e. kT), becomes comparable to the couplings between the spins. The strong exchange coupling of electrostatic origin that gives rise to electronic ferromagnetism is absent in nuclear magnetism, and, because of the smallness of nuclear moments, the magnetic coupling between nuclear spins is such that temperatures of the order of 10^{-7} °K or less would be required for a possible observation of nuclear ferromagnetism (or antiferromagnetism). This makes nuclear ferromagnetism a subject beyond experimental possibilities for the time being (although possibly not for ever). A nuclear analogue of electronic diamagnetism, that is of magnetism arising from the Larmor precession of the electronic charges in an applied magnetic field, is not easy to visualize, but at least in bulk matter it can reasonably be expected to be entirely negligible.

We are thus left with nuclear paramagnetism. While for electronic paramagnetism there is an appreciable contribution from the orbital motion of the bound electrons, the nuclear paramagnetism of orbital origin is entirely negligible in bulk matter (for reasons that will appear in Chapter VI) and only that due to the nuclear spins will be considered.

The existence of spin paramagnetism, that is, the appearance in a sample containing a large number of elementary spin-moments of a net macroscopic magnetization when placed in a magnetic field H_0, is due to the fact that different orientations of the spins with respect to the field, described by different values of the quantum number $I_z = m$ of the spin quantized along the field, correspond to different magnetic energies E_m. According to the fundamental Boltzmann law of statistical mechanics, the populations P_m of the energy levels are proportional to $\exp(-E_m/kT) = \exp(\gamma \hbar m H_0/kT)$. The net magnetization of a sample containing N spins will thus be

$$M = N\gamma\hbar \frac{\sum_{m=-I}^{I} m \exp(\gamma \hbar m H_0/kT)}{\sum_{m=-I}^{I} \exp(\gamma \hbar m H_0/kT)}. \qquad (1)$$

In nuclear magnetism, where the ratio $\gamma \hbar H_0/kT$ is almost always a very small number, it is permissible to make a linear expansion of the Boltzmann exponential, thus obtaining

$$M = \frac{N\gamma^2\hbar^2 H_0}{kT} \frac{\sum_{m=-I}^{I} m^2}{2I+1} = \frac{N\gamma^2\hbar^2 I(I+1)}{3kT} H_0 = \chi_0 H_0, \qquad (2)$$

where χ_0 is the static nuclear susceptibility. The proportionality of χ_0 to $1/T$ is the well-known Curie law.

Since the static nuclear susceptibility is, according to (2), proportional to $\gamma^2\hbar^2 I(I+1)$, that is, to the square of the magnitude of the elementary nuclear moment, it will be smaller than the electronic paramagnetic susceptibility by a factor of the order of 10^{-6} to 10^{-8}. Because of this smallness it is very difficult to observe by the conventional magnetostatic methods. To give an example, the static proton susceptibility of a cubic centimetre of water is at room temperature of the order of 3×10^{-10} c.g.s. units. Although a static measurement of nuclear magnetic susceptibility had been made as early as 1937 (**1**), on solid hydrogen at $2°$ K (to take advantage of the large factor $1/T$ in (2)), nuclear magnetism would probably have remained a simple curiosity if much more sensitive methods using the principle of resonance had not been developed.

The paramount importance of these methods in nuclear magnetism is such that the phenomenon itself has often been described as nuclear magnetic resonance. It is worth emphasizing, however, that the phenomenon of nuclear magnetic resonance, to be described in great detail in this book, although the main if not practically the only experimental means of studying nuclear magnetism, does not cover all the aspects of it. For that reason a somewhat more general title has been given to this book. Finally, it should be mentioned that electronic paramagnetism, although much more accessible to conventional static methods than nuclear paramagnetism, has also greatly benefited from the use of resonance techniques. The vast subject of electron magnetic resonance is outside of the scope of this book, although because of the close connexion between nuclear and electron paramagnetism certain topics may be considered to belong to either subject.

B. Special features of radio-frequency spectroscopy

Nuclear magnetic resonance is a branch of Spectroscopy, which is here understood in its broadest sense to encompass all studies of the nature of the energy levels of material systems and of the transitions induced between them through absorption or emission of electromagnetic radiation. More specifically, nuclear magnetic resonance is a branch of radio-frequency spectroscopy with a domain of frequencies extending at present from, say, a thousand Mc/s, down to 2 kc/s; the latter is the separation in frequency units between the two Zeeman levels of a proton spin in the earth magnetic field.

In spite of the extent of this frequency scale and of the great variety of the phenomena studied, r.f. spectroscopy owes its unity to a few characteristics which have important consequences for the theory. The electromagnetic radiations of interest can be produced by means of electronic generators with frequencies defined and measurable with considerable precision. As a consequence, and in contrast with the situation prevailing in infra-red, optical, and γ-ray spectroscopy, the uncertainty $\Delta \nu$ in the frequency of the radiation will as a rule be smaller than the width $\Delta E/h$ (in frequency units) of the levels between which the transition is induced. Because of the low-energy $h\nu$ of each photon and of the narrowness $\Delta \nu$ of their frequency range, very low power suffices to create an astronomically large number of photons (say 10^{17}) per unit frequency range. As is well known from the quantum theory of radiation, in the presence of such a large number of photons, induced emission and absorption are overwhelmingly larger than spontaneous emission, which in r.f. spectroscopy is a negligible phenomenon. The description of induced emission, or absorption (in contrast with spontaneous emission) does not require a quantum mechanical description of the radiation field, and this is a considerable simplification.

Another important feature of the radiation produced by an electronic generator is its coherence. In a classical description of the electromagnetic field this is expressed very simply, by saying that each component of the electromagnetic field is a sine (or cosine) function of time, with a well-defined phase. In photon language a well-defined phase of the field entails an uncertainty in the total number of photons, since according to the quantum theory of radiation there is an uncertainty relation of the form $\Delta n\,\Delta\phi \sim 1$ between the phase ϕ of a component of the electromagnetic field and the number n of photons of that frequency (per mode of oscillation of the radiation field). In r.f. spectroscopy, where n is a very large number, it is possible to define simultaneously with great accuracy the number of photons and thus the amplitude, and the phase, of the r.f. field, without contradicting the uncertainty relation above.

The conclusion of what has been said is that in r.f. spectroscopy it is convenient and legitimate to describe the r.f. electromagnetic field as a classical quantity, a description that will be used henceforth.

C. The phenomenon of resonance; 'resonant' and 'non-resonant' methods

In order to measure the separation Δ between two energy levels of a system, which is the fundamental spectroscopic problem, a natural procedure is to find a measurable quantity G, which is a known function of the separation Δ, and to measure it. The precision in such a method, which we shall call 'static' or 'non-resonant', is often poor, first because it is at most the precision with which G can be measured and secondly because the function that relates the measured G to the quantity of interest Δ may contain other parameters not always well known.

In a 'resonant' method the system under consideration is irradiated by an r.f. field, the frequency of which can be changed continuously. As long as the resonance condition $\Delta = h\nu$ is not fulfilled, the probability for a transition of the system to be induced by the field is very weak, but it increases considerably when the value $\nu_0 = \Delta/h$ is reached. If the increase of the rate of transitions causes a *detectable* (rather than measurable) change in the system, the fact that a resonance has occurred can be ascertained and the measurement of Δ is reduced to the measurement of a frequency. Thus each 'resonant' experiment involves two steps: (a) induce, or 'drive' the resonance, (b) detect its occurrence. The problem of detection, by far the more difficult of the two, has been solved in a large (and still growing) number of ways, each best adapted to the special features of the system studied.

In order to place nuclear magnetic resonance in the much wider field of r.f. spectroscopy, it is instructive to consider the principle of a few 'resonant' experiments exhibiting a large variety in the means chosen for detection. In some fields a 'static' method was used prior to a 'resonant' one, in others a 'resonant' method, possible in principle, has not yet been realized in practice. A comparison between the two methods is illuminating. Such being the purpose of this review, the description of the various experiments will be extremely brief and no attention will be paid to the historical order.

(a) *Beam measurements of atomic, molecular, and nuclear magnetic moments*

A static method was first used in the historical experiment of Gerlach and Stern (**2**). An atom with magnetic moment μ moves along a direction OX in an inhomogeneous field H_0 which, as well as its gradient $\partial H_0/\partial z$, is perpendicular to OX. The atom experiences a force $\mu_z(\partial H_0/\partial z)$ that deflects it through an amount proportional to μ_z. For a spin, of say $\tfrac{1}{2}$, μ_z may have two values $\pm\mu$, and the atoms after crossing the inhomogeneous field produce two symmetrical spots where they impinge on a screen perpendicular to OX. From the distance between the spots, μ can be obtained if the field gradient and the time spent by the atoms in the inhomogeneous field are known. In spite of many improvements on the simple scheme described, the accuracy is poor.

In the first version of the 'resonant' method (**3**) the atoms cross two regions A and B of inhomogeneous field with rigorously opposite gradients $\partial H_0/\partial z$ separated by a region C of homogeneous field. The construction of the apparatus is such that atoms or molecules emerging from a point source O, converge after crossing the three regions A, B, C to an image D, where a particle detector can be placed. Although the trajectories of particles with different orientations of their magnetic moments are different inside the beam, they all converge on to D if no reorientation of these moments occur on the way. A small r.f. field placed in the region C of homogeneous field H_0, will, if it has the correct frequency, induce transitions between the different magnetic states of the particle, causing it to change its trajectory in the second inhomogeneous region B and to miss the detector. A measurable change in the flux of particles striking the detector determines the frequency at which the resonance occurs.

The accuracy is considerably higher than in the static measurement. Forgoing all details, the following important features of the method should be stressed.

(α) The physical phenomenon leading to detection of the resonance is a change in the trajectory of the particle whose spin undergoes a transition, resulting in a change of the flux of particles at the detector. Each particle that hits the detector releases an energy, considerably higher than that involved in the electromagnetic transition that would cause it to miss the detector. We shall call this type of detection 'trigger' detection. It can be very sensitive (most strikingly so, for radioactive particles), and beams of small intensities can be used.

(β) Consequently the collisions between particles can be made very rare and the interactions very weak. Each atom or molecule behaves practically as an isolated system.

(γ) The transition of a particle is detected irrespective of whether it gains or loses magnetic energy. The effects of absorption and induced emission add up and the existence of differences among the populations of the magnetic states of the particle is not required. Unpolarized beams can be used. This is an exceptional feature for a 'resonant' experiment as will appear in the following.

(δ) The transitions induced by the r.f. field are magnetic dipole transitions.

Beam experiments based on electric rather than magnetic resonance have also been performed on molecules. Their principle is the same as outlined above, provided the word magnetic is replaced everywhere by electric.

(b) *Measurement of the magnetic moment of the neutron*

This measurement is based on the fact that the absorption coefficient of a beam of polarized neutrons by a ferromagnetic substance depends on the angle between that polarization and the magnetization of the ferromagnetic. A neutron beam is directed across two slabs of saturated iron magnetized along a direction OZ perpendicular to the direction OX

of the beam. As it leaves the first, or polarizing, slab, because of the spin-dependent absorption, the neutron beam is partially polarized. In a later version of the experiment greater polarization is achieved through reflection of an unpolarized beam from a magnetized single crystal mirror. If between the two slabs a field H_0 is applied at right angles to the magnetization and the direction of the beam, the spin of a neutron, spending a time t in that field, will have precessed around the field through an angle $\alpha = \omega t$ proportional to both the magnetic moment of the neutron and the magnitude of the field. The intensity transmitted through the second, or analysing, slab will be a periodic function of the field H_0 with a period $2\pi/\omega$, from which the magnetic moment of the neutron can in principle be obtained. The accuracy is very poor.

In the much more accurate 'resonant' version of the experiment (4), a field H_0 parallel to the magnetization of the iron, rather than perpendicular to it, placed between the polarizer and the analyser, does not affect the transmission of the neutrons through the analyser. A small r.f. field H_1 perpendicular to H_0 changes the polarization of the beam by inducing transitions between the states $I_z = \pm\frac{1}{2}$ of the neutron spin. The resonance is detected by a change in the intensity of the transmitted beam. Of the four characteristics (α)–(δ) of the previous method, only the third, (γ), is modified. A difference of populations between the states $I_z = \pm\frac{1}{2}$ of the neutron spin is necessary in order for the resonance to be detectable.

(c) *The fine structure of the hydrogen atom* (5)

The Dirac one-particle theory predicts the same energy for the $2s_{\frac{1}{2}}$ and $2p_{\frac{1}{2}}$ levels of the hydrogen atom. The experiment to be described measures the separation between these levels predicted as due to the interaction between the bound electron and the radiation field.

A beam of hydrogen atoms is excited by electronic collisions into the states $2s_{\frac{1}{2}}$ and $2p_{\frac{1}{2}}$. The latter return very rapidly (10^{-9} sec), to the ground state $1s_{\frac{1}{2}}$ with emission of light, so that the beam practically contains only atoms in states $1s_{\frac{1}{2}}$ (ground) and $2s_{\frac{1}{2}}$ (metastable). The beam then strikes a detector, sensitive to excited atoms $2s_{\frac{1}{2}}$ (but not to atoms in the ground state), which, losing their excitation energy $E(2s_{\frac{1}{2}}) - E(1s_{\frac{1}{2}})$, cause secondary electrons to be emitted from the detector. An r.f. field of frequency $\Delta E/h$, superposed on the beam, will transfer the atoms from the metastable state $2s_{\frac{1}{2}}$ into $2p_{\frac{1}{2}}$, from which they immediately return to the ground state before reaching the detector. The resonance is detected through the decrease in the current of secondary electrons. Of the four characteristics of this experiment the following differ from Rabi's experiment.

(γ) An inequality of populations between the two states linked by the r.f. transitions is required. It is produced by the very short lifetime of one of them, $2p_{\frac{1}{2}}$.

(δ) The transition is electric dipole.

An experiment, bearing to this one the same relation that the Stern and Gerlach experiment bears to that of Rabi, is easily imagined. An applied static electric field would mix the states $2s_{\frac{1}{2}}$ and $2p_{\frac{1}{2}}$ by an amount that would depend in a known way on their distance Δ. Each state $2s_{\frac{1}{2}}$, now partially $2p_{\frac{1}{2}}$, will have a finite probability of returning to the ground state with emission of light before reaching the detector. Knowing the time spent by each atom in the beam, the lifetime of the state $2p_{\frac{1}{2}}$, and the strength of the applied electric field, it is possible in principle, although rather hopeless in practice, to obtain Δ from the decrease in the current of secondary electrons as a function of the applied field.

This method of Lamb for detecting the resonance is not applicable to the measurement of the fine structure interval $E(3s_{\frac{1}{2}})-E(3p_{\frac{1}{2}})$ of hydrogen, for in contrast to the $2s_{\frac{1}{2}}$ lifetime of $\frac{1}{7}$ of a second, $3s_{\frac{1}{2}}$ has a lifetime of $1 \cdot 6 \times 10^{-7}$ seconds against emission of the H_α line to $2p$ states. The $3p$ states with a lifetime of $5 \cdot 4 \times 10^{-9}$, and thus less populated than $3s_{\frac{1}{2}}$, may decay either with an H_α quantum to $2s$ or to $1s$ with emission of L_β line (branching ratio $1:7\cdot5$). Resonant transitions equalizing the populations of $3s$ and $3p$ will produce a decrease in the amount of H_α light emitted, which can thus be used as a detector of the r.f. resonance (6).

(d) *The fine structure of positronium* (7)

The ground state of positronium, a bound system made of an electron and a positron, has a fine structure, where a distance Δ separates a lower singlet state ^1S from a higher triplet state ^3S. Either state has a finite lifetime against annihilation of the electron and the positron into γ-radiation. In all experiments aiming to measure Δ, the basic fact is that the singlet state can decay into two photons with an inverse lifetime $1/\tau_s = \lambda_s \simeq 8 \times 10^9$, while a selection rule causes the triplet to decay into three photons at a rate $1/\tau_t = \lambda_t \simeq 7 \times 10^6$, a thousand times slower.

In a 'non-resonant' method a static field H_0 mixes, with an admixture coefficient proportional to $1/\Delta$, the singlet state with the substate $S_z = 0$

of the triplet, which, becoming then partly singlet, acquires a finite probability of decaying into two photons. From the increase in the rate of decay into two photons as a function of the applied field Δ can be, and has been, obtained.

A straightforward procedure to measure Δ by a resonance method would be to induce transitions between the singlet and the triplet state by means of an r.f. field of frequency Δ/h, and to detect the resonance through an increase in the number of two photon decays. In practice, Δ/h being of the order of 10^5 Mc/s, it is difficult to produce appreciable r.f power at this frequency. For that reason, a magnetic field H_0 is applied to the system, and transitions at a much lower frequency are induced between the substate $S_z = 0$ of the triplet and the two substates $S_z = \pm 1$. The applied field H_0 introduces an admixture from the singlet state into the triplet substate $S_z = 0$ which then has a finite probability of decaying into two photons. On the other hand, the triplet substates $S_z = \pm 1$ remain pure and can decay into three photons only. The resonance is detected through the increase in the number of the two photon decays and since the resonance frequency is a known function of Δ and H_0, Δ can be computed from it. The accuracy is much higher than in the 'non-resonant' method.

All characteristics (α)–(γ) are similar to those of Lamb's experiment; (δ) is different—the transitions in positronium being magnetic rather than electric dipole.

One will notice that the resonant experiment on positronium bears the same relationship to Lamb's experiment as the non-resonant one to the hypothetical Lamb's experiment in an applied static electric field.

(e) *The magnetic moment of the μ-meson*

The measurement of the magnetic moment of the μ meson (spin $\frac{1}{2}$) has been made possible by the discovery that the non-conservation of parity in the $\pi \to \mu + \nu$ decay leads to a high degree of polarization in μ-meson beams emerging from cyclotrons. As a consequence of this polarization the angular distribution of the electrons following the decay $\mu = e + \nu + \bar{\nu}$ is highly anisotropic, the electron direction being strongly correlated with the μ spin direction.

In the original static experiment the counting rate of electrons in a given direction was measured as a periodic function of the strength of an applied field (perpendicular to the initial polarization of the beam) which caused the spin of the μ-meson to precess around it (8). From the period of the precession the magnetic moment of the μ-meson could be obtained.

In a resonance experiment a d.c. magnetic field is applied parallel to the initial polarization of the μ-meson and a reorientation of the spin of the μ-meson by an r.f. field is detected by a change in the electron-counting rate in a given direction.

This measurement was sufficiently precise for the main limitation in the evaluation of the anomalous magnetic moment of the μ-meson to be caused by the uncertainty in the value of its mass.

'Trigger' detection, absence of interactions between the very few μ-mesons existing in the sample at a given time, differences of populations due to the non-conservation of parity, and magnetic dipole transition are the standard features of this experiment. A difference from the experiments described so far is that this one is performed in bulk matter, where the μ-meson is stopped, and the interactions of the μ-spin with its surroundings are not necessarily negligible as appears from the considerable depolarization of μ-spins (shown by the reduced anisotropy of the decay electrons) that occurs in certain media.

(*f*) *Resonance detected by optical means* (9)

In this method, polarized light derived from a gas-discharge lamp is shone on an absorption cell filled with the same gas at a low pressure. It is easily seen that if the excited state of the atoms of the cell, formed by absorption of the incident light, has a non-zero angular momentum, its various magnetic substates will be unequally populated, and the re-emitted light will be polarized. The Zeeman splittings that exist between those substates in an applied magnetic field can be measured by a resonance experiment where r.f. transitions between those substates equalize their populations, the resonance being detected through the ensuing change in the polarization of the re-emitted light. This detection is still of the 'trigger' type.

If the ground state of the atoms of the cell also has a finite total angular momentum including nuclear spin, and since the selection rules are different for excitation by polarized light and de-excitation through spontaneous emission, it is easily seen that the magnetic substates of the ground state also become unequally populated. This method of polarization of the ground state has received the name of optical pumping (9). The destruction of these inequalities of populations by a resonant r.f. field can also be detected by a change in the re-emitted polarized light or even more simply by a change in the absorption coefficient for the incident polarized light. The pure nuclear magnetic resonance of Hg^{201} which has only nuclear angular momentum in its ground state, could be detected by that method for the first time (**10**).

An interesting development of these experiments is the polarization of atoms of a different species or even of free electrons, contained in the cell, through exchange collisions with atoms polarized by optical pumping. Thus, hydrogen atoms have been polarized by collisions with optically 'pumped' sodium atoms. This could be detected by a resonant r.f. field disorienting the hydrogen atoms, which in turn through exchange collisions disoriented the sodium atoms. A resonant transition of the hydrogen atoms could thus be detected through a change in the polarization of light re-emitted by excited sodium atoms.

(*g*) *Perturbed angular correlations*

The principle of measurements of nuclear magnetic moments in excited nuclear states is similar to the one just described for atoms. The angular correlation existing between two cascade radiations emitted by a nucleus may be interpreted as an anisotropic distribution of the second radiation caused by the inequality of populations among the magnetic substates of the intermediate nucleus. This inequality is due to the fact that the direction of emission of the first radiation, chosen as direction of quantization for the definition of the magnetic substates of the intermediate nucleus, is for the latter a privileged direction in space. In a static field H_0 the magnetic moment of the intermediate nucleus precesses around it with a frequency proportional to μH_0 and so does the anisotropy pattern of the angular distribution of the second radiation.

Several nuclear moments (excited states of Cd^{111} and Pb^{204}) have been measured in that fashion.

In a resonance method the field H_0 applied along the direction of the first radiation does not perturb the correlation, but the addition to it of a small r.f. field H_1 at right angles to H_0 does perturb it if the resonance frequency is correct. This experiment has not yet been successfully performed at the time of writing this book, for reasons that will appear in the next section.

The resemblance between these measurements of nuclear moments and those of the magnetic moment of the μ-meson is evident. In particular, although the interactions between the various radioactive nuclei, few and far apart, are negligible, the effects on the nuclear spins of their surroundings may be important.

The static method has also been applied to the measurements of moments of excited nuclear states, formed by Coulomb excitation, and decaying through γ-emission. The direction of reference for the anisotropy of the γ-emission is that of the incoming charged particle.

We shall stop here this review of resonant experiments using 'trigger' detection. Quite a few examples could be added even now and it is likely that more will be available in the future. Of those gathered here, the following remarks can be made.

Because of the great sensitivity of the 'trigger' detection methods, the number of elementary systems (atoms, molecules, or nuclei) studied is usually very small in comparison with normal atomic densities in bulk matter, and the interactions between them are negligible.

With the exception of the experiments on the μ-meson and on perturbed angular correlations, the influence of the surroundings is also negligible and each atom, molecule, or nucleus is practically an isolated system.

In each of the experiments outlined the difference in populations, necessary for the detectability of the resonance (with the exception of the Rabi method, which does not require any), is due either to a geometrical feature of the experiment such as the existence of a privileged direction of propagation or polarization in the way the system under study has been 'prepared', or to a dynamical feature such as very different lifetimes of the two states between which the resonant transition occurs.

Another characteristic common to most of the systems reviewed is the existence of a specific time τ available for the 'driving' of the resonant transition, which must have a probability per unit time W not negligible with respect to $1/\tau$ in order for the transition to be detectable.

This time τ is the transit time in the r.f. region of the beam of molecules, neutrons, or atoms, in the experiments of Rabi, Alvarez and Bloch, and Lamb (on the $2s$ levels for the latter). It is the lifetime of the triplet state of positronium, of the μ-meson, of the excited atomic states, in some of the other experiments. The characteristic time is the lifetime of the intermediate nuclear state for perturbed angular correlations and the reason for the lack of 'resonant' experiments with these is due to the lack of intermediate nuclear states with lifetimes sufficiently long to allow an appreciable r.f. 'driving' of transitions between their substates, and sufficiently short to permit the use of standard coincidence techniques.

Finally, one may point to the great diversity of techniques to 'prepare' the system for resonance and to detect the latter, each technique taking advantage of a special feature of the system to be studied. This must be contrasted with the universality of electromagnetic detection methods to be considered now.

D. Nuclear magnetic resonance

(a) *The principle*

In this method, developed independently by two research groups headed respectively by F. Bloch (**11**) and E. M. Purcell (**12**), the detection of the passage through the resonance is based on a modification occurring at resonance in the electromagnetic device itself that 'drives' the resonant transition of interest. These effects have been described in several ways; none is the best for every application. In connexion with the viewpoint adopted so far the simplest description of the phenomenon (mainly used by Purcell and his co-workers) is that of an absorption by the nuclear spin system of electromagnetic energy provided by an r.f. generator. Since, as will be shown in Chapters II and III, the absorption is proportional to the electromagnetic energy localized in the resonator, coil, or cavity that produces the 'driving' field, nuclear magnetic absorption can be described and detected as an additional load or a change in the quality factor Q of the resonant circuit of the 'driving system'. More (although not very much more) will be said about the detection problems in Chapter III. The electromagnetic detection of resonance has important consequences.

The lack of sensitivity of the electromagnetic means of detection, due to the fact that the energies involved are precisely those of the transitions induced, not amplified by any 'trigger' effect, requires the presence of nuclear moments in large numbers (10^{18} or more). This in turn makes it necessary to use bulk matter samples: solids, liquids, or gases under appreciable pressure. Then the nuclear spins prior to the 'driving' of the resonance will in general be in thermal equilibrium with their surroundings or 'lattice'. The unequal populations of the magnetic-energy levels E_m of the nuclear spins are then proportional to $\exp(-E_m/kT)$, where T is the temperature of the lattice. It is these inequalities among populations that, making absorption larger than induced emission, lead to a net absorption by the spin system of electromagnetic energy supplied by the r.f. generator. Thus, in contrast to other resonance experiments in bulk matter such as, say, the measurement of the magnetic moment of the μ-meson, the coupling of the spins with the lattice is an essential feature of the whole scheme of electromagnetic detection, rather than a small and more or less inconvenient perturbation.

A very important problem is then that of spin lattice relaxation, that is, mechanisms whereby the nuclear spins are 'informed', as it were, of the temperature of the lattice and come into thermal equilibrium with it.

Although a single time constant does not always describe adequately the trend of nuclear spins towards equilibrium with the lattice, such a constant is usually introduced to give at least the time scale of the process, and is given the name of spin-lattice relaxation time T_1.

The study of relaxation mechanisms and the calculation of relaxation times is one of the major problems of nuclear magnetism and an important part of this book will be devoted to it.

A steady-state nuclear resonance experiment can thus be interpreted as the result of two competing processes, the relaxation which tends to establish between the populations the Boltzmann inequalities, and the driving electromagnetic field that tends to destroy them. Another way of looking at the nuclear resonance phenomenon, mainly due to F. Bloch, and more illuminating for certain experiments, is to describe it as a forced precession of the nuclear magnetization in the applied r.f. field, the effect of the precession being to induce a detectable electromotive force in a receiving coil.

This description will be developed in more detail in Chapter III.

The large nuclear concentrations and the small distances that exist between nuclear spins in bulk matter have another important consequence, which is the existence of relatively strong spin interactions. These are particularly important in solids, for in liquids their effectiveness is considerably reduced, as will appear in more detail later, by the rapid relative motions of the nuclei.

These interactions in solids have two consequences. First, a broadening of the resonance lines occurs since each spin 'sees' now besides the applied field a small local field due to its neighbours, which has different values for different spins of the sample, whence a spread in the observed resonance frequencies. The calculation of the resulting line shapes is an important problem of nuclear magnetism. Secondly, the tight coupling between the spins allows rapid energy transfer from one spin to another, leading under circumstances to be discussed in more detail later, to the establishment of a thermal equilibrium inside the nuclear spin system itself in a time much shorter than T_1, and at a temperature that may be quite different from that of the lattice. This time, called spin-spin relaxation time, is represented in the literature by the symbol T_2.

(b) Various extensions and generalizations

The simple scheme of a magnetic-resonance experiment outlined above has now received appreciable extension.

Transient methods of various kinds where the spins are not in equilibrium with the lattice have been widely utilized. Dynamical polarization methods whereby it is possible to increase the difference in populations between two levels many times above its thermal equilibrium value have been discovered and applied. Double irradiation methods have appeared where a transition is being driven at a certain frequency and detected through its effect (for instance through spin-spin coupling) on another transition, which, detected in the normal way, is used as an indicator for the first one.

It is also worth pointing out that the association as a rule of electromagnetic detection of resonance with the use of bulk samples, where the nuclear polarization is related to the temperature of the lattice, may suffer some exceptions. Thus, for electronic magnetic resonance at least, it has been possible to use electromagnetic detection for the observation in a gas cell of the resonance of rubidium[87] polarized by optical pumping (13).

Conversely, it is at least conceivable to use other means of detection for the resonance of nuclear spins embedded in bulk matter. Long ago it had been suggested (14), although never realized, that it might be possible to detect the passage through resonance by observing a rise in the temperature of the sample, following a sudden increase at resonance in the absorption of electromagnetic energy by the nuclear spins. More recently it was suggested (15), and unsuccessfully attempted, to detect the nuclear resonance of radioactive nuclei aligned at very low temperatures reached by adiabatic demagnetization, by destroying the anisotropy of emitted γ-radiation with an r.f. field that equalized the populations of the aligned nuclei.

The first successful experiment of this kind (16), based on dynamic polarization, was performed at a temperature of $1 \cdot 5°$ K, where the electronic if not the nuclear splittings are comparable to kT. Driving a forbidden transition of the spectrum of radioactive Co[60] enhanced considerably, according to a mechanism to be explained later, the nuclear alignment of Co[60], which was observed by the appearance (rather than destruction) of anisotropy in the emitted γ-radiation. The value of the driving frequency at which this effect appeared gave the value of the splitting for the transition driven. Other experiments of this kind have since been performed on radioactive P^{32}, As^{76}, and Sb^{122} embedded in silicon. Although in these experiments the spins certainly cannot be said to be in equilibrium with the lattice, the fact that there is during the experiment, or was prior to its start, a thermal contact between the energy levels of the spin system and the lattice is essential, and in this respect dynamical polarization differs deeply from other means of polarization described in Section C.

No mention has been made so far of another important nuclear property connected with the spatial orientation of the nuclear spin, namely the nuclear quadrupole moment, which gives a measure of the lack of sphericity in the distribution of electric charge inside the nucleus. A more precise mathematical definition of the nuclear quadrupole moment will be given in Chapter VI. In an inhomogeneous electric field of symmetry lower than cubic, such as exists in certain crystals or molecules, a nuclear quadrupole moment has energy states that correspond to various orientations of the nuclear spin with respect to the axes of symmetry of the local electric field. The splittings between these levels range from a fraction of a megacycle up to a few hundreds or in some cases a few thousands of Mc/s, depending on the nucleus and the nature of the sample.

An r.f. magnetic field may induce transitions between these levels, detected by standard electromagnetic methods (17). The name currently given to that phenomenon, pure quadrupole resonance, is unfortunate, for the character of the transition induced is magnetic dipole and it is due to the coupling of an r.f. magnetic field with the magnetic moment of the nucleus. In fact, when a magnetic field is applied to a sample containing nuclei with quadrupole moments, there is a continuum of situations between the pure Zeeman resonance (when the symmetry of the local electric field is cubic or higher) and the so-called pure quadrupole resonance (when the applied magnetic field is vanishingly small). There is nothing special about the latter situation except the possibility in that case of obtaining sharp lines in polycrystalline samples.

The existence of quadrupole moments has made it possible to induce electric quadrupole transitions between energy levels of nuclear spins (18). Although a simple order of magnitude calculation shows that fields produced by conductors external to the sample are much too homogeneous for that purpose, adequate field gradients can be produced by the periodic motion of ionic charges inside the sample, induced by acoustical vibrations driven by an external generator. The detection of the resonance could be made either by examining, with straight magnetic-resonance methods, the reduced inequalities of populations following an acoustical irradiation, or even, in later experiments, through the direct absorption of acoustical energy by the nuclear spins, appearing as an additional load on the acoustical generator. We terminate this review of various means of detection of nuclear resonance in bulk matter by the remark that even the static effects of nuclear magnetism, however small, might be put to good use for that purpose. The torque exerted on the nuclear magnetization of a sample suspended in an inhomogeneous magnetic field could be changed by the destruction of that magnetization through application of a saturating r.f. field. The detection of that change, thus separated from the much larger torque exerted on the moment due to the diamagnetic magnetization of the electrons, provides a means for detecting the nuclear resonance.

Variations of such a method could have interest for other combinations, including the presence of large electric quadrupole splittings in single crystals. In this last case saturation of one transition of the spectrum could lead to a large *increase* of a magnetic moment of the sample along a crystalline axis rather than along the applied field, permitting a deflexion even in a *homogeneous* field, a favourable circumstance for a resonance experiment.

REFERENCES

1. B. LAZAREW and L. SCHUBNIKOW, *Phys. Z. Sowjet.* **11**, 445, 1937.
2. W. GERLACH and O. STERN, *Ann. Physik*, **74**, 673, 1924.
3. I. I. RABI, J. R. ZACHARIAS, S. MILLMAN, and P. KUSCH, *Phys. Rev.* **53**, 318, 1938.
4. L. W. ALVAREZ and F. BLOCH, ibid. **57**, 111, 1940.
5. W. E. LAMB and R. C. RETHERFORD, ibid. **72**, 241, 1947.
6. W. E. LAMB and T. M. SANDERS, ibid. **103**, 313, 1956.
7. M. DEUTSCH and E. DULIT, ibid. **84**, 601, 1951.
   M. DEUTSCH and S. C. BROWN, ibid. **85**, 1047, 1952.
8. R. L. GARWIN, L. M. LEDERMAN, and M. WEINRICH, ibid. **105**, 1415, 1957.
9. J. BROSSEL and A. KASTLER, *C.R. Acad. Sci.* **229**, 1213, 1949.
   A. KASTLER, *J. Physique*, **11**, 255, 1950.
10. B. CAGNAC, J. BROSSEL, and A. KASTLER, *C.R. Acad. Sci.* **246**, 1827, 1958.
11. F. BLOCH, W. W. HANSEN, and M. PACKARD, *Phys. Rev.* **70**, 474, 1946.
12. E. M. PURCELL, H. C. TORREY, and R. V. POUND, ibid. **69**, 37, 1946.
13. T. R. CARVER, *J. Phys. Rad.* **19**, 872, 1958.
14. C. J. GORTER, *Physica*, **3**, 995, 1936.

15. N. Bloembergen and G. M. Temmer, *Phys. Rev.* **89**, 883, 1953.
16. C. D. Jeffries, ibid. **106**, 164, 1957.
    M. Abraham, R. W. Kedzie, and C. D. Jeffries, ibid. **106**, 165, 1957.
17. H. G. Dehmelt and H. Krüger, *Naturwiss.* **37**, 111, 1950.
18. W. G. Proctor and W. H. Tanttila, *Phys. Rev.* **98**, 1854, 1955.

II

MOTION OF FREE SPINS

Liberté, égalité, réversibilité

A BASIC problem in nuclear magnetism is the description of the behaviour of a free spin in a uniform magnetic field. A free spin is a system with an angular momentum $\mathbf{I}\hbar$ and a magnetic moment $\mathbf{M} = \gamma\hbar\mathbf{I}$. This problem will be treated classically first, and then quantum-mechanically.

A. Classical treatment

According to the classical theory of electromagnetism, a magnetic moment $\mathbf{M}$ experiences, in a field $\mathbf{H}$, a torque $\mathbf{C} = \mathbf{M} \wedge \mathbf{H}$, equal to the rate of change $\hbar(d\mathbf{I}/dt)$ of its angular momentum. Since $\mathbf{M} = \gamma\hbar\mathbf{I}$, the motion of the magnetic moment is described by the equation

$$\frac{d\mathbf{M}}{dt} = \gamma \mathbf{M} \wedge \mathbf{H}. \tag{1}$$

In order to solve (1) it is useful to transform to rotating coordinates, a technique that has proved of great assistance in the description of magnetic resonance. Let S' be a frame of reference rotating with respect to the laboratory frame S with an angular velocity represented by a vector $\boldsymbol{\omega}$. According to the general law of relative motion, the time derivative $d\mathbf{A}/dt$ of any time-dependent vector $\mathbf{A}(t)$, computed in the laboratory frame S, and its derivative $\partial\mathbf{A}/\partial t$ computed in the moving frame S', are related through

$$\frac{d\mathbf{A}}{dt} = \frac{\partial\mathbf{A}}{\partial t} + \boldsymbol{\omega} \wedge \mathbf{A}. \tag{2}$$

Combining (1) and (2), the motion of the magnetic moment in the rotating frame is given by the equation

$$\frac{\partial\mathbf{M}}{\partial t} = \gamma \mathbf{M} \wedge \left(\mathbf{H} + \frac{\boldsymbol{\omega}}{\gamma}\right). \tag{3}$$

This has the same form as equation (1) provided the magnetic field $\mathbf{H}$ is replaced by an effective field $\mathbf{H}_e = \mathbf{H} + (\boldsymbol{\omega}/\gamma)$, the sum of the laboratory field $\mathbf{H}$ and a fictitious field $\mathbf{H}_f = +(\boldsymbol{\omega}/\gamma)$.

This result will be applied first for a field $\mathbf{H} = \mathbf{H}_0$, constant in time. By choosing a rotating frame with $\boldsymbol{\omega} = -\gamma\mathbf{H}_0$, the effective field $\mathbf{H}_e$ vanishes. In this frame $\partial \mathbf{M}/\partial t = 0$ and the magnetic moment is a fixed vector. Therefore, with respect to the laboratory frame it precesses with an angular velocity ω_0, where $\omega_0 = -\gamma H_0$ is called the Larmor frequency of the spin in the applied field $\mathbf{H}_0$. The magnitude H_0 of the applied field is positive but γ can be positive or negative. From the measurement of such a Larmor frequency γ can be obtained if H_0 is known or vice versa.

The unit vector $\mathbf{k}$ of the z-axis in the laboratory frame having been chosen parallel to $\mathbf{H}_0$, suppose next that the total field $\mathbf{H}$ is the sum of the constant field $\mathbf{H}_0 = H_0\mathbf{k} = -(\omega_0/\gamma)\mathbf{k}$ and of a field $\mathbf{H}_1$ perpendicular to $\mathbf{H}_0$ and rotating around it with an angular velocity ω.

The unit vector $\mathbf{i}$ of the x-axis in the rotating frame S' being taken along the field $\mathbf{H}_1$, the effective field $\mathbf{H}_e$ is static in S' and is given by

$$\mathbf{H}_e = \left(H_0 + \frac{\omega}{\gamma}\right)\mathbf{k} + H_1\mathbf{i}. \tag{4}$$

Let ω_1 be defined as $\omega_1 = -\gamma H_1$, where the magnitude H_1 of the rotating field is positive and ω_1 has the sign of $-\gamma$.

The magnitude of the effective field $\mathbf{H}_e$ is

$$H_e = \left[\left(H_0 + \frac{\omega}{\gamma}\right)^2 + H_1^2\right]^{\frac{1}{2}} = -\frac{a}{\gamma}, \tag{5}$$

where
$$a = -[(\omega_0 - \omega)^2 + \omega_1^2]^{\frac{1}{2}}\frac{\gamma}{|\gamma|}. \tag{5'}$$

The angle θ between $\mathbf{H}_e$ and the applied field $\mathbf{H}_0$ which goes from 0 to π is given unambiguously by

$$\tan\theta = \frac{H_1}{H_0 + (\omega/\gamma)} = \frac{\omega_1}{\omega_0 - \omega}, \tag{6}$$

$$\sin\theta = \frac{\omega_1}{a}, \quad \cos\theta = \frac{\omega_0 - \omega}{a}.$$

In the rotating frame S', the motion of the magnetic moment $\mathbf{M}$ is a Larmor precession around the effective field $\mathbf{H}_e$, of angular velocity $a = -\gamma H_e$. Its motion in the laboratory frame is the combination of this with the rotation ω of S' with respect to S, around $\mathbf{H}_0$.

If at time $t = 0$ the magnetic moment is aligned along H_0, at time t the angle between them will be

$$\cos\alpha = \cos^2\theta + \sin^2\theta \cos at = 1 - 2\sin^2\theta \sin^2 \tfrac{1}{2}at. \tag{7}$$

So far no assumptions have been made about the relative magnitudes of H_0 and H_1. In practice H_1 will often be much smaller than H_0. Then, from equations (6) and (7) it is seen that θ and hence α remain very small unless the difference $|\omega-\omega_0|$ becomes comparable to $|\omega_1|$. This is a resonance phenomenon. A rotating field H_1, small compared with the constant field H_0, can reorient appreciably a magnetic moment only if its frequency of rotation, ω, is in the neighbourhood of the Larmor frequency ω_0. The width of the resonance, that is, the value of the difference $|\omega-\omega_0|$ below which the effect is appreciable, is of the order of ω_1.

Before going to the quantum mechanical treatment one should remark that in practice oscillating rather than rotating fields are used. A linearly polarized field $2H_1 \cos \omega t$ can be considered as a superposition of two fields of amplitude H_1, rotating in opposite directions with angular velocities $\pm\omega$. If $H_1 \ll H_0$ the effect of a rotating field on a magnetic moment is negligible unless its frequency of rotation ω is in the neighbourhood of the Larmor frequency ω_0. But then the effect of the component rotating at the frequency $-\omega$, off resonance by 2ω, is so small as to be, usually, completely negligible. The effect of the counter-rotating field can be computed to the lowest order through the following elementary considerations.

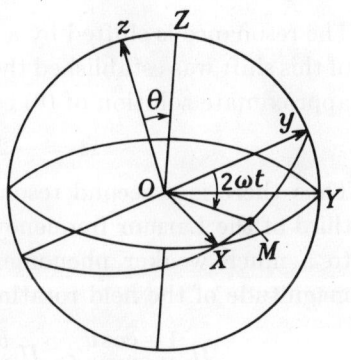

Fig. II, 1

The component $-\omega$ of the oscillating field is assumed to be so different from ω_0 that
$$|\omega_0-(-\omega)| = |\omega_0+\omega| \gg |\omega_1|.$$
In the frame rotating with the velocity $-\omega$ the applied field H_0 and the then static component of the r.f. field can be replaced by a large effective static field (Fig. II, 1)
$$H_e = \frac{1}{|\gamma|}[(\omega+\omega_0)^2+\omega_1^2]^{\frac{1}{2}} = -\frac{\Omega}{\gamma},$$
making with H_0 the small angle θ, where $\tan\theta = \omega_1/(\omega_0+\omega)$. In this frame the remaining rotating component $+\omega$ is the sum of three fields: an oscillating field parallel to $\mathbf{H}_e$, of frequency 2ω and amplitude $H_1 \sin\theta$, and in the plane perpendicular to $\mathbf{H}_e$ a field of amplitude $\frac{1}{2}H_1(1+\cos\theta)$ rotating with the frequency 2ω, and a very small field of amplitude $\frac{1}{2}H_1(1-\cos\theta)$ rotating with a frequency -2ω. If we neglect the effect

of the parallel field, the resonance condition for these two rotating components is $\Omega^2 = 4\omega^2$ or

$$(\omega-\omega_0)(3\omega+\omega_0) = \omega_1^2. \qquad (8)$$

One solution of (8) for $\Omega = +2\omega$ is to a first approximation

$$\omega = \omega_0 + \frac{\omega_1^2}{4\omega_0} = \omega_0\left\{1 + \left(\frac{H_1}{2H_0}\right)^2\right\}. \qquad (9)$$

The resonance is shifted by a relative amount $(H_1/2H_0)^2$. The existence of this shift was established theoretically at an early date (1). The second approximate solution of (8) corresponds to

$$\Omega = -2\omega \quad \text{or} \quad \omega \simeq -\tfrac{1}{3}\omega_0. \qquad (10)$$

Thus there is a second resonance for a frequency approximately one-third of the Larmor frequency of the spin. This resonance corresponds to a much weaker phenomenon than the principal one because the magnitude of the field rotating at the frequency -2ω is

$$H_1\frac{1-\cos\theta}{2} \sim H_1\frac{\theta^2}{4} \sim \frac{H_1}{4}\left(\frac{\omega_1}{\omega_0+\omega}\right)^2 \sim H_1\left(\frac{3H_1}{4H_0}\right)^2.$$

A more elaborate calculation shows that higher order resonances exist for $|\omega_0/\omega| = 2n+1$, where n is an integer. The existence of these subsidiary resonances was demonstrated theoretically (2) and experimentally (3) in connexion with optical pumping experiments. For the ratios (H_1/H_0) usually met in nuclear resonance these effects are very small and will be neglected from now on in this discussion.

B. Quantum mechanical treatment

In the Heisenberg representation the equations of motion of the angular momentum operator $\hbar\mathbf{I}$ are

$$\frac{\hbar}{i}\frac{d\mathbf{I}}{dt} = [\mathscr{H}, \mathbf{I}] = [-\gamma\hbar\mathbf{H}.\mathbf{I}, \mathbf{I}], \qquad (11)$$

where $\mathscr{H} = -\gamma\hbar(\mathbf{H}.\mathbf{I})$ is the Hamiltonian describing the coupling of the spin with the applied field.

For the z component

$$\frac{\hbar}{i}\frac{dI_z}{dt} = -\gamma\hbar\{H_x[I_x, I_z]+H_y[I_y, I_z]\} = \frac{\gamma\hbar}{i}\{I_x H_y - I_y H_x\} = \frac{\gamma\hbar}{i}[\mathbf{I}\wedge\mathbf{H}]_z. \qquad (12)$$

This equation has the same form as the classical equation (1).

This identity has the following consequences: the expectation value $\langle I_z\rangle$ taken over the wave function of a free spin $\mathbf{I}$ also obeys the classical equation and can therefore be calculated classically. This result will first be applied to the problem of a spin $I = \frac{1}{2}$ in a magnetic field, sum of a d.c. field H_0 along the z-axis and a field H_1 rotating in the xy plane. Let p_+ and p_- be the probabilities of finding the spin in the states $I_z = +\frac{1}{2}$ and $I_z = -\frac{1}{2}$ respectively: $\langle I_z \rangle = \frac{1}{2}(p_+ - p_-) = \frac{1}{2}(1 - 2p_-)$. If at time $t = 0$, $p_+ = 1$, $p_- = 0$, at time t, $\langle I_z(t) \rangle$ will be $\frac{1}{2}\cos\alpha$, where α is the angle calculated classically by the formula (7), and hence

$$p_- = P(t) = \tfrac{1}{2}(1-\cos\alpha) = \sin^2\theta \sin^2 \tfrac{1}{2}at = \frac{\omega_1^2}{a^2}\sin^2 \tfrac{1}{2}at. \qquad (13)$$

This is the well-known formula of Rabi for a spin $\frac{1}{2}$ in a rotating field.

Since equations (12) are linear, it follows that they are also obeyed by the total moment of a large number of identical, non-interacting spins. Furthermore, when the components of angular momentum, measured in units of $\hbar$, are large numbers, it is permissible to treat them as classical quantities. It follows that the classical equation (1) describes the observable behaviour of a sample containing a large number of identical spins provided interactions among them can be neglected.

An explicit solution of the Schrödinger equation describing the motion of a spin I in a rotating field will now be derived:

$$i\dot\psi = \frac{\mathscr{H}}{\hbar}\psi = -\gamma\{H_0 I_z + H_1(I_x \cos\omega t + I_y \sin\omega t)\}\psi$$

$$= \{\omega_0 I_z + \tfrac{1}{2}\omega_1(I_+ e^{-i\omega t} + I_- e^{i\omega t})\}\psi, \qquad (14)$$

where $I_\pm = I_x \pm i I_y$. Transformation to the rotating frame of coordinates is equivalent to the substitution $\psi = U\psi_e = \exp(-i\omega I_z t)\psi_e$. The Schrödinger equation in the new frame is

$$i\dot\psi_e = \left\{ U^{-1}\frac{\mathscr{H}}{\hbar}U - iU^{-1}\dot U \right\}\psi_e. \qquad (15)$$

From the commutation relations one obtains

$$\left.\begin{aligned}
U^{-1}I_z U &= e^{i\omega t I_z} I_z e^{-i\omega t I_z} = I_z \\
e^{i\omega I_z t} I_+ e^{-i\omega I_z t} &= e^{i\omega t} I_+ \\
e^{i\omega I_z t} I_- e^{-i\omega I_z t} &= e^{-i\omega t} I_- \\
U^{-1}\mathscr{H}U &= \hbar(\omega_0 I_z + \omega_1 I_x)
\end{aligned}\right\}, \qquad (15')$$

$$i\dot\psi_e = \{(\omega_0 - \omega)I_z + \omega_1 I_x\}\psi_e. \qquad (16)$$

This integrates immediately to

$$\psi_e = \exp[-i\{(\omega_0 - \omega)I_z + \omega_1 I_x\}t]\psi_e(0),$$

where $\psi_e(0) = \psi(0)$,

$$\psi = \exp(-i\omega I_z t)\exp\{-ia(\mathbf{n}.\mathbf{I})t\}\psi(0), \qquad (17)$$

where a is given by (5') and $\mathbf{n}$ is a unit vector of components $n_z = \cos\theta$, $n_x = \sin\theta$, $n_y = 0$.

From (17) it is easy to compute the probability amplitude $A_{m'm}$ and the probability $P_{m'm} = |A_{m'm}|^2$ for finding the spin in the state m' at time t, knowing that it was at time $t = 0$ in the state m:

$$A_{m'm} = (m'|\psi(t)) = (m'|\exp\{-(i\omega I_z t)\}\exp\{-ia(\mathbf{n}.\mathbf{I})t\}|m), \qquad (18)$$

$$P_{m'm} = |(m'|\exp\{-ia(\mathbf{n}.\mathbf{I})t\}|m)|^2. \qquad (18')$$

(18') is easily computed in the case of a spin $\tfrac{1}{2}$. In that case $\mathbf{I} = \tfrac{1}{2}\boldsymbol{\sigma}$ where σ_x, σ_y, σ_z are the Pauli matrices which have the following properties:

$$(\mathbf{n}.\boldsymbol{\sigma})^{2p} = |\mathbf{n}|^{2p} = 1, \qquad (\mathbf{n}.\boldsymbol{\sigma})^{2p+1} = (\mathbf{n}.\boldsymbol{\sigma}),$$

$$\exp\{-ia(\mathbf{n}.\mathbf{I})t\} = \exp\{-\tfrac{1}{2}ia(\mathbf{n}.\boldsymbol{\sigma})t\} = \cos\tfrac{1}{2}at - i(\mathbf{n}.\boldsymbol{\sigma})\sin\tfrac{1}{2}at,$$

and therefore

$$P_{-\tfrac{1}{2},\tfrac{1}{2}} = |(-\tfrac{1}{2}|\cos\tfrac{1}{2}at - i\sin\tfrac{1}{2}at(\cos\theta\,\sigma_z + \sin\theta\,\sigma_x)|\tfrac{1}{2})|^2 = \sin^2\theta\sin^2\tfrac{1}{2}at, \qquad (19)$$

which is the same as (13) already proved by a semi-classical argument. In the general case $I > \tfrac{1}{2}$, $A_{m'm}$ can be obtained explicitly by using the theory of the irreducible representations of the group of rotations. Rather than reproduce the long derivation here only the formula for $P_{m'm}$ is given (4):

$$P_{mm'} = (\cos\tfrac{1}{2}\alpha)^{4I}(I+m)!(I+m')!(I-m)!(I-m')! \times$$

$$\times \left[\sum_{\lambda=0}^{2I}(-1)^\lambda \frac{(\tan\tfrac{1}{2}\alpha)^{2\lambda-m+m'}}{\lambda!(\lambda-m+m')!(I+m-\lambda)!(I-m'-\lambda)!}\right]^2, \qquad (19')$$

where $\sin^2\tfrac{1}{2}\alpha = \sin^2\theta\sin^2(\tfrac{1}{2}at)$.

C. Quantum mechanical description of a statistical ensemble of free spins. Density matrix

Consider a sample containing a large number N of identical nuclear spins, for simplicity spins $i = \tfrac{1}{2}$, which has been placed in a static field H_0 for a long time. A measurement of the macroscopic nuclear magnetization will, as is well known, yield a value $M_z = M_0 = \chi_0 H_0$, where χ_0 is the static nuclear susceptibility, in a direction parallel to the field, and values $M_x = M_y = 0$ at right angles to the field. One may investigate whether this behaviour is compatible with a description of all individual nuclear spins by the same wave function:

$$\psi = a_1\xi + a_2\eta, \quad \text{where} \quad |a_1|^2 + |a_2|^2 = 1 \qquad (20)$$

and ξ and η are eigenstates of the operator i_z with eigenvalues $+\frac{1}{2}$ and $-\frac{1}{2}$. One finds easily

$$\left.\begin{aligned}\frac{M_z}{\gamma\hbar} &= \frac{M_0}{\gamma\hbar} = N\langle i_z\rangle = N(\psi|i_z|\psi) = \tfrac{1}{2}N\{|a_1|^2-|a_2|^2\} \\ \frac{M_+}{\gamma\hbar} &= N(\psi|i_+|\psi) = Na_1^* a_2 \\ \frac{M_-}{\gamma\hbar} &= \phantom{N(\psi|i_+|\psi) = {}} Na_1 a_2^* \end{aligned}\right\} \quad (21)$$

The first equation (21) is self-evident and has been used already in the present discussion: $|a_1|^2$ and $|a_2|^2$ are the probabilities p_+ and p_- for finding each spin in the state $\pm\frac{1}{2}$ or, as one often says, the relative populations of these states. However, the second equation (21) shows clearly that unless a_1 or $a_2 = 0$ (which is a case of complete polarization for all the spins of the sample), the magnetization of the sample should have a component in the equatorial plane in contradiction with experiments.

Thus the hypothesis that the coefficients a_1 and a_2 are the same for all the spins of the sample is untenable. Instead one must assume that they are different for different spins and that the average of the product $a_1^* a_2$ taken over the whole sample

$$\overline{a_1^* a_2} = \frac{1}{N}\sum_i a_1^{i*} a_2^i = 0.$$

In the same way one must use the averages $\overline{|a_1|^2} = p_+$ and $\overline{|a_2|^2} = p_-$ for the relative populations of the two levels. If now a rotating field H_1 is introduced suddenly at time $t = 0$, for a duration t, the coefficients $a_1(t)$ and $a_2(t)$ can be calculated as linear combinations of $a_1(0)$ and $a_2(0)$ using the probability amplitude (18), and the average $\overline{a_1^*(t)a_2(t)}$ can be computed. Its value can actually be predicted very simply since we know from the equations (1) the motion of the macroscopic magnetization of the sample. Thus at resonance, for instance, for $\omega = \omega_0$, the magnetic moment will have precessed around Oz by an angle $\omega_0 t$ and at the same time around the axis Ox of the rotating frame by an angle $\omega_1 t$. The result is

$$M_z(t) = M_0\cos\omega_1 t, \qquad M_+(t) = M_0\sin(\omega_1 t)e^{i\omega t}$$

and, comparing this with the equations (21), one obtains

$$\overline{a_1^*(t)a_2(t)} = \tfrac{1}{2}\{\overline{|a_1(0)|^2}-\overline{|a_2(0)|^2}\}\sin(\omega_1 t)e^{i\omega t},$$
$$\overline{|a_1(t)|^2}-\overline{|a_2(t)|^2} = \{\overline{|a_1(0)|^2}-\overline{|a_2(0)|^2}\}\cos\omega_1 t. \qquad (22)$$

From this very simple example one can conclude that the description of an ensemble of spins i by the relative populations $p_m = \overline{|a_m|^2}$ of their energy levels E_m is in general incomplete and should be supplemented by the knowledge of cross-products such as $\overline{a_m a_{m'}^*}$.

This result is well known in quantum statistics where one introduces a statistical operator ρ represented by a so-called density matrix with matrix elements

$$\rho_{mm'} = (m|\rho|m') = \overline{a_{m'}^* a_m}. \tag{23}$$

ρ contains all the information necessary for the description of a statistical ensemble of identical systems. In particular, the predicted value of any observable Q of the system is given by the equation

$$\langle Q \rangle = \text{trace}\{\rho Q\}, \tag{24}$$

of which the equations (21) are a special case.

The variation of ρ with time is given by the equation

$$\frac{\hbar}{i}\frac{d\rho}{dt} = -[\mathcal{H}, \rho], \tag{25}$$

where $\mathcal{H}$ is the Hamiltonian of an individual system. This equation is very similar to Heisenberg's equations of motion (note, however, the change in sign) and $\rho(t)$ is related to $\rho(0)$ through

$$\rho(t) = U(t)\rho(0)U^{-1}(t), \tag{26}$$

where $U(t)$ is a unitary operator. In the case of a time-independent Hamiltonian $\mathcal{H}$, U can be written

$$U = \exp\left\{-i\frac{\mathcal{H}}{\hbar}t\right\}. \tag{26'}$$

A complication, often encountered in practice, results from the fact that the systems (spins here) of the ensemble sometimes differ statistically in their Hamiltonians, as well as in their wave functions (coefficients a_m) at $t = 0$. For instance, because of the inevitable inhomogeneity of the applied field over the dimensions of the ensemble, the Larmor frequencies of all the spins will be distributed over a corresponding range. In that case one should subdivide the statistical ensemble into sub-ensembles sufficiently large for each of them to contain a large number of individual systems (spins) but at the same time sufficiently small for the variations of the individual Hamiltonians to be negligible inside each of them.

For each of the sub-ensembles one can then define density matrices $\rho_i(t)$ and $\rho_i(0)$ related by

$$\rho_i(t) = U_i(t)\rho_i(0)U_i^{-1}(t), \tag{27}$$

and then define an average density matrix ρ for the whole ensemble.

It is quite clear that the relation between $\rho(t)$ and $\rho(0)$, thus defined, will *not* in general be unitary.

A more detailed discussion and a generalization of the properties of the density matrix will be deferred until later.

In the next section the results obtained so far will be compared with those of the conventional theory of transition probabilities which uses the perturbation method.

D. Relation with the perturbation method

In the perturbation method, the unperturbed system is the spin in the d.c. magnetic field H_0, described by the unperturbed Hamiltonian

$$\mathscr{H}_0 = -\gamma\hbar H_0 I_z = \hbar\omega_0 I_z, \tag{28}$$

and the perturbation is the coupling of this spin with the rotating field, $H_x = H_1 \cos\omega t$, $H_y = H_1 \sin\omega t$, described by a perturbation Hamiltonian

$$\mathscr{H}_1 = -\gamma\hbar(H_x I_x + H_y I_y) = -\tfrac{1}{2}\gamma\hbar H_1(I_+ e^{-i\omega t} + I_- e^{i\omega t}). \tag{28'}$$

The eigenvalues of $\mathscr{H}_0$ are of the form $m\hbar\omega_0$ where the integer or half integer m takes $2I+1$ values from $-I$ to $+I$. Because of the properties of the operators $I_\pm$, $\mathscr{H}_1$ has only matrix elements between states for which m differs by one unit. It is well known in quantum mechanics that a time-dependent perturbation of frequency ω can only induce transitions between states separated by an energy interval $\Delta E = \hbar\omega_0$, if ω is in the neighbourhood of ω_0. Physically this result expresses the conservation of energy, the change in energy $\hbar\omega_0$ of the system being compensated by the energy $\hbar\omega$ of the photon which is absorbed or emitted. If H_1 is a true rotating field the equality $\omega = \omega_0$ has to be satisfied with respect to the sign as well. For an oscillating field $2H_1 \cos\omega t$, if $|\omega| = |\omega_0|$, one of the two rotating components of the oscillating field has the right sense of rotation.

In practice, the energy levels E_a and E_b will have a certain width and the energy difference $\Delta E = E_a - E_b = \hbar\omega_0$ will be described by a distribution function $\rho(\Delta E)$ around a central value $\Delta E^0 = \hbar\omega_0^0$, with $\int \rho(\Delta E) d(\Delta E) = 1$. Instead of $\rho(\Delta E)$ one can introduce distribution functions $g(\nu_0) = h\rho(h\nu_0)$ or $f(\omega_0) = \hbar\rho(\hbar\omega_0)$ also normalized through $\int g(\nu_0) d\nu_0 = \int f(\omega_0) d\omega_0 = 1$.

The width δ of the distribution is then of the order of $[f(\omega_0)]^{-1}$ on the ω_0 scale. The transition probability per unit time induced by the r.f. field of frequency ω between two states $|a\rangle$ and $|b\rangle$ is given by the well-known formula

$$W_{ab} = \frac{2\pi}{\hbar}|(b|\mathcal{H}_1|a)|^2 \rho(\Delta E), \qquad (29)$$

where we make $\Delta E = \hbar\omega$, which for the Hamiltonians (28) and (28') yields

$$W_{m,m-1} = \frac{2\pi}{\hbar}\frac{\gamma^2\hbar^2}{4}H_1^2|(m|I_+|m-1)|^2\frac{g(\nu)}{h} = \tfrac{1}{4}\omega_1^2(I+m)(I-m+1)g(\nu). \qquad (30)$$

In particular, for spin $\tfrac{1}{2}$,

$$W_{\frac{1}{2}\to-\frac{1}{2}} = W_{-\frac{1}{2}\to\frac{1}{2}} = W = \tfrac{1}{4}\omega_1^2 g(\nu) = \tfrac{1}{2}\omega_1^2 \pi f(\omega). \qquad (30')$$

It is interesting to compare the exact solution (19), (19') with the approximate solutions (30), (30'). For simplicity the comparison will be made only for spin $\tfrac{1}{2}$. According to (30') a spin in the state $+\tfrac{1}{2}$ at time $t=0$ should, at time t, have a probability $P = Wt$ of being found in the state $-\tfrac{1}{2}$, at least for such small values of t that $Wt \ll 1$, since this result has been obtained by a first-order perturbation theory. On the other hand, the exact formula (19) gives, at resonance, for small values of t, the entirely different result $P = \tfrac{1}{4}\omega_1^2 t^2$. It is therefore very important to reconcile these results in order to understand why the perturbation method, widely used in the theory of nuclear resonance, does give results in agreement with experiment. Since formula (30') was established under the assumption of a distribution $f(\omega_0)$ of the Larmor frequencies, the same distribution should be introduced in the exact treatment by multiplying the expression (19) by the shape function $f(\omega_0)$ and integrating over ω_0. This procedure gives

$$P(t) = \omega_1^2 \int \frac{\sin^2 \tfrac{1}{2}at}{a^2} f(\omega_0)\, d\omega_0, \qquad (31)$$

and a transition probability per unit time

$$\frac{dP}{dt} = \tfrac{1}{2}\omega_1^2 \int \frac{\sin at}{a} f(\omega_0)\, d\omega_0 = \tfrac{1}{2}\omega_1^2 \int_{-\infty}^{\infty} \frac{\sin t[u^2+\omega_1^2]^{\frac{1}{2}}}{(u^2+\omega_1^2)^{\frac{1}{2}}} f(\omega+u)\, du, \qquad (31')$$

where $u = \omega_0 - \omega$.

If it is assumed that the width δ of the distribution is much larger than ω_1, for values of t such that $\delta t \gg 1$, $f(\omega+u)$ will be practically constant over the range of values of u which contribute significantly to the integral and can be taken out of the integral. One then gets

$$\frac{dP}{dt} = \tfrac{1}{2}\pi\omega_1^2 f(\omega) J_0(\omega_1 t), \tag{32}$$

where J_0 is a Bessel function. For small values of the argument $J_0 \sim 1$ and

$$\frac{dP}{dt} \simeq \tfrac{1}{2}\pi\omega_1^2 f(\omega), \quad \text{as in formula (30').} \tag{32'}$$

One must remember, however, that (32') is valid only under the assumption $\omega_1 t \ll 1 \ll \delta t$ and, since $\delta \simeq 1/f(\omega) \gg \omega_1$ was assumed, one also has

$$Wt \simeq \frac{\omega_1^2 t}{\delta} = \frac{\omega_1}{\delta}\omega_1 t \ll 1.$$

Although the equivalence of the exact treatment and of the perturbation theory for short times such that $Wt \ll 1$ has thus been established, the validity of the method of the transition probabilities over long times $(Wt \gg 1)$ is still not established.

Consider a statistical ensemble of spins with relative populations $p_+(t)$ and $p_-(t)$, yielding

$$M_z(t) = \tfrac{1}{2}N\gamma\hbar\{p_+(t)-p_-(t)\}.$$

If an r.f. field is applied at time zero the customary theory of transition probabilities leads to the following equations:

$$\frac{dp_+}{dt} = -W(p_+-p_-),$$
$$\frac{dp_-}{dt} = -W(p_--p_+), \tag{33}$$

which integrates to

$$p_+-p_- = \{p_+(0)-p_-(0)\}\exp(-2Wt),$$
$$M_z(t) = M_z(0)\exp(-2Wt), \tag{34}$$

or

$$\frac{dM_z}{dt} = -2WM_z(0)\exp(-2Wt).$$

On the other hand, from (32) one easily obtains

$$\frac{dM_z}{dt} = -2WM_z(0)J_0(\omega_1 t). \tag{35}$$

The two expressions (34) and (35) coincide for small t such that $Wt \ll 1$ (although not too small, since (35) has been established under the assumption $\delta t \gg 1$) and they both vanish for t very large. However, in the intermediate region the two results behave quite differently and, since (35) is the exact formula, one is inclined to question the validity of equations (33).

It is not really surprising that the transition-probability method could lead to an incorrect, or at least an incomplete, description since it deliberately overlooks the coherence of the applied r.f. field and deals only with populations, that is, with diagonal elements of the density matrix ρ ignoring the off-diagonal ones. On the other hand, the comparison made above between the predictions of the formulae (34) and (35) relative to the destruction of the magnetization existing at time $t = 0$ by an r.f. field applied at that time did not correspond to a usual physical situation. It was assumed that at time $t = 0$ the sample had a magnetization $M_0 = \chi_0 H_0$ parallel to the applied field H_0 which, as was shown earlier, corresponded to the absence of off-diagonal matrix elements ρ_{12} and ρ_{21} of ρ and to the inequality of the diagonal elements

$$\rho_{11} = \overline{|a_1|^2} = p_+ \quad \text{and} \quad \rho_{22} = \overline{|a_2|^2} = p_-.$$

This situation is created by a relaxation mechanism which does two things: it tends to destroy the off-diagonal elements, if any, of ρ and to give to the diagonal elements, or populations, the values corresponding to thermal equilibrium. The two processes of which the first is usually called the spin-spin relaxation and the second the spin-lattice relaxation need not, and in fact usually do not, proceed with the same speed, but can be associated with time constants T_2 and T_1 with $T_1 \geqslant T_2$. Under the combined effect of the relaxation and of the r.f. field a steady state is reached where the magnetization is different from zero and where the departure $M_z - M_0$, from the equilibrium value M_0, can be measured. In particular, if the r.f. field is sufficiently weak, this departure, proportional to the r.f. transition probability W, gives a direct measurement of the shape function $f(\omega)$ which describes the distribution of the energy levels of the spin system. It will be shown for a simple model that the transition probability method and the exact formula (19) lead in that case to the same result.

Let us assume that each spin undergoes at random intervals with a mean value T_1, collisions of such a nature that there is no correlation between the values of the wave function $\begin{pmatrix} a_1 \\ a_2 \end{pmatrix}$ of the spin before and after the collision (strong collisions). Then, immediately after a collision,

$$\overline{a_2^* a_1} = \rho_{12} = \rho_{21}^* = 0,$$
$$\overline{|a_1|^2} = \rho_{11} = (p_+)_0,$$
$$\overline{|a_2|^2} = \rho_{22} = (p_-)_0,$$

where $(p_+)_0$ and $(p_-)_0$ are the equilibrium populations.

These hypotheses, applied to the transition probability method, lead to equations

$$\frac{dp_+}{dt} = -W(p_+ - p_-) - \frac{1}{T_1}\{p_+ - (p_+)_0\},$$
$$\frac{dp_-}{dt} = -W(p_- - p_+) - \frac{1}{T_1}\{p_- - (p_-)_0\}, \quad (36)$$

which have a steady-state solution

$$M_z = \frac{M_0}{1 + 2WT_1}.$$

Then
$$\frac{M_0 - M_z}{M_0} = \frac{2WT_1}{1 + 2WT_1}. \quad (37)$$

For $WT_1 \ll 1$,

$$\frac{M_0 - M_z}{M_0} \simeq 2WT_1 \quad \text{(condition of no saturation)}.$$

On the other hand, to apply the exact formula (19) for the transition probability $P(t) = (\omega_1^2/a^2)\sin^2 \tfrac{1}{2}at$ one must be careful to choose the starting time as immediately after a collision, since it is only then that the off-diagonal elements ρ_{12} vanish and (19) is valid. The equation giving $p_+(t)$ can then be written

$$p_+(t) = \exp\!\left(\frac{-t}{T_1}\right)\{p_-(0)P(t) + p_+(0)[1 - P(t)]\} + $$
$$+ \int_0^t \frac{dt'}{T_1} \exp\!\left\{-\left(\frac{t-t'}{T_1}\right)\right\}\{(p_-)_0 P(t-t') + (p_+)_0[1 - P(t-t')]\}. \quad (38)$$

The first term is the contribution to $p_+(t)$ from the spins which underwent no collision between the switching on of the r.f. field and the time t, and the integral is the sum of the contributions from the spins which had their *last* collision at time t'.

A similar relation can be written for $p_-(t)$. A steady-state solution can be obtained by making t go to infinity. It gives

$$M_z = M_0 \int_0^\infty \frac{dt'}{T_1} \exp\!\left(\frac{-t'}{T_1}\right)\{1 - 2P(t')\}$$
$$= M_0\!\left\{1 - \frac{\omega_1^2}{\omega_1^2 + (1/T_1^2) + (\omega - \omega_0)^2}\right\}. \quad (39)$$

This may be written

$$\frac{M_0 - M_z}{M_0} = \frac{\omega_1^2 T_1^2}{\omega_1^2 T_1^2 + (\omega - \omega_0)^2 T_1^2 + 1}.$$

To be equivalent to (37) and to establish thereby the correctness of the perturbation theory method for this 'strong-collision' model, one must show that
$$2W = (\omega_1^2 T_1)/[(\omega-\omega_0)^2 T_1^2+1].$$
This is indeed correct for the model chosen and in fact the well-known Lorentzian shape factor
$$f(\omega) = \frac{1}{\pi} \frac{T_1}{1+(\omega-\omega_0)^2 T_1^2}, \qquad (39')$$
characteristic of spectroscopic lines emitted with mean life T_1, can be recognized. The model is far too special to be characteristic of the general situation in magnetic resonance spectroscopy. For example, such a situation as a distribution of Larmor frequencies ω_0 in the ensemble must be treated by multiplying both sides of (39) by the shape function $f(\omega_0)$ and integrating over ω_0. If it is assumed that the width δ of this function is much larger than both ω_1 and $1/T_1$, such integration gives
$$M_z = M_0\left\{1 - \frac{\pi \omega_1^2 f(\omega)}{\sqrt{\{\omega_1^2+(1/T_1^2)\}}}\right\} = M_0\left\{1 - \frac{2W}{\sqrt{\{\omega_1^2+(1/T_1^2)\}}}\right\}.$$
For strengths of the r.f. field so small that $\omega_1 T_1 \ll 1$,
$$\frac{M_0-M_z}{M_0} = 2WT_1.$$
This is the same result as by the perturbation theory for the condition $WT_1 \ll 1$.

In addition, under restriction allowing the application of the 'strong collision' model, that is, under the assumption that the shape function $f(\omega)$ is given by (39'), it is found that both methods give
$$\frac{M_0-M_z}{M_0} = \frac{2WT_1}{1+2WT_1}$$
without restriction on WT_1.

E. Transient effects

(a) Free precession

If the relaxation phenomena are neglected, the solutions of the equations of motion in an applied d.c. field $H_0 = -\omega_0/\gamma$ are
$$M_z(t) = M_z(0),$$
$$M_+(t) = M_+(0)\exp(i\omega_0 t). \qquad (40)$$
For a sample in thermal equilibrium at time $t = 0$, $M_z(0) = \chi_0 H_0$, $M_+(0) = 0$, and no transverse magnetization exists. Such a magnetization can be created by applying a rotating field H_1 for a time τ. If its

frequency ω is equal to the Larmor frequency ω_0, at the end of the r.f. pulse, one will have

$$M_z(\tau) = M_0 \cos \omega_1 \tau,$$
$$M_+(\tau) = M_0 \sin \omega_1 \tau \exp(i\omega_0 \tau). \qquad (41)$$

After the r.f. field is cut off, taking (41) as initial condition for (40), there results a precessing magnetization $M_+(t) = M_0 \sin \omega_1 \tau \exp(i\omega_0 t)$ which keeps its amplitude permanently. This amplitude is a maximum following what is called a 90° pulse ($\omega_1 \tau = \tfrac{1}{2}\pi$). It vanishes after a 180° pulse ($\omega_1 \tau = \pi$). In practice there may be a scatter in the Larmor frequencies described by the shape function $f(\omega_0)$. At the end of the r.f. pulse the magnetization corresponding to each value of ω_0 can be computed using the rotating coordinates device. The result becomes very simple if one assumes that the amplitude ω_1 of the r.f. field is much larger than the width δ of the shape function and that the duration of the pulse is of the order of $1/\omega_1$, for then the resonance condition is approximately fulfilled for all the spins of the sample, the effective field H_e has the same direction OX perpendicular to the applied field, the angle of precession around H_e has the same value ($\omega_1 \tau$), and (41) is still valid. Once the pulse is over the precessing magnetization will be

$$M_+(t) = M_0 \sin \omega_1 \tau \int f(\omega_0) \exp(i\omega_0 t) \, d\omega_0$$
$$= M_0 \sin(\omega_1 \tau) \exp(i\omega_0^0 t) \int_{-\infty}^{\infty} f(\omega_0^0 + u) \exp(iut) \, du, \qquad (42)$$

where ω_0^0 is the central Larmor frequency. The transverse magnetization still precesses at the mean Larmor frequency but with a time-dependent amplitude $G_1(t)$ which is the Fourier transform of the shape function.

As t goes to infinity, because of the destructive interference among the contributions of the different parts of the sample to the transverse magnetization, $G_1(t)$ goes to zero, and the observation of its decay gives the same information on the shape function as the observation of the resonance with a vanishingly small r.f. field. For instance, if the shape function is a Lorentz curve $f(\omega_0^0 + u) = (b/\pi)[1/(b^2 + u^2)]$, the decay shape will be the exponential $G(t) = G(0)\exp(-bt)$. It will be shown later how this result, established under very special assumptions, can be generalized to apply to real systems.

(b) Spin echoes

In spite of the fact that the precessing magnetization has disappeared after a time θ, which, because of the general properties of the Fourier transform is of the order of $1/\delta$ (δ = width of the shape function), it is possible to restore it to its initial value by application of a second appropriately dimensioned r.f. pulse. Studies based on this technique are known as spin echoes. Consider what happens at time t in the frame S' rotating at the central frequency $\omega_0^0 = \omega$ of the spectral distribution. With respect to that frame, the spins which have a Larmor frequency $\omega_0 = \omega_0^0 + u$ will have precessed by an angle ut. Suppose now that at time t a 180° pulse is applied, corresponding to a rotation of π around an effective field of direction X' in the equatorial plane, of the frame S' (which may or may not coincide with the direction X of the effective field for the first pulse). At the end of the pulse the spins of frequency $\omega_0^0 + u$ will have with respect to X' the phase $-u(t+\tau)$ (apart from a constant) and therefore at time $2(t+\tau)$ they will all be aligned again. Thus the transverse magnetization of all parts of the sample will add constructively to a value equal to the one immediately following the first pulse. Again a more detailed study of this phenomenon and its possible applications will be deferred until later.

(c) *The adiabatic theorem, adiabatic passage*

From the equation of motion (1) we may immediately deduce

$$\frac{d}{dt}(M^2) = 2\mathbf{M}\frac{d\mathbf{M}}{dt} = 0. \tag{43}$$

The magnitude of the magnetization M is a constant of the motion, whatever the variation of H with time. It will now be shown that if this variation is sufficiently slow, the angle of the magnetization with the instantaneous direction of the field is also a constant of the motion.

The variation with time of the vector $\mathbf{H}$ can be described quite generally by the vector equation

$$\frac{d\mathbf{H}}{dt} = \mathbf{\Omega} \wedge \mathbf{H} + \Omega_1 \mathbf{H}, \tag{44}$$

where the vector $\mathbf{\Omega}$ and the scalar Ω_1 have the dimensions of a frequency. Consider a moving frame S' where the z-axis is continuously aligned along the instantaneous direction of the field $\mathbf{H}$. According to (44) the relative motion of S' with respect to the laboratory will be a rotation about an instantaneous axis $\mathbf{\Omega}$. In that frame the magnetization will change in time according to

$$\frac{\partial \mathbf{M}}{\partial t} = \gamma \mathbf{M} \wedge \left(\mathbf{H} + \frac{\mathbf{\Omega}}{\gamma}\right). \tag{45}$$

By definition, in this frame $H_x = H_y = 0$ and

$$\frac{\partial M_z}{\partial t} = M_x \Omega_y - M_y \Omega_x.$$

If $|\Omega| \ll |\gamma H|$ then, approximately,

$$\frac{\partial M_x}{\partial t} \cong \gamma H M_y, \qquad \frac{\partial M_y}{\partial t} \cong -\gamma H M_x.$$

M_x and M_y are approximately sinusoidal functions with instantaneous frequency $\omega_0(t) = -\gamma H(t)$.

After a long time t, the change in M_z will be

$$\Delta M_z = M_z(t) - M_z(0) = \int_0^t [M_x(t')\Omega_y(t') - M_y(t')\Omega_x(t')] \, dt.$$

If the variation of $\mathbf{\Omega}$ with time is sufficiently slow, or to be precise, if its Fourier expansion has negligible components at frequencies of the order of $|\gamma H(t)|$, then, for any t,

$$|\Delta M_z| \sim \left| \frac{M\Omega}{\gamma H} \right| \ll M,$$

and M_z, that is, the component of $\mathbf{M}$ along the field, will remain constant. This is the adiabatic theorem. This result may be applied to the case of a magnetic moment in a rotating field $\mathbf{H}_0 + \mathbf{H}_1$ of the type previously defined where the d.c. field H_0 is being slowly varied. From the definition of the effective field H_e we have

$$\frac{d\mathbf{H}_e}{dt} = \cos\theta \frac{\dot{H}_0}{H_e} \mathbf{H}_e + \sin\theta \frac{\dot{H}_0}{H_e} (\mathbf{n} \wedge \mathbf{H}_e), \qquad (46)$$

where $\mathbf{n}$ is a unit vector orthogonal to $\mathbf{H}_0$ and $\mathbf{H}_1$. Comparing this with equation (44) gives

$$\Omega = \sin\theta \frac{\dot{H}_0}{H_e} = H_1 \frac{\dot{H}_0}{H_e^2}. \qquad (47)$$

The quantity Ω is the smaller, the farther from resonance. From the adiabatic condition $|\Omega| \ll |\gamma H_e|$ we have

$$\dot{H}_0 \ll \frac{\gamma H_e^2}{\sin\theta}.$$

This condition is strongest at resonance and gives

$$\dot{H}_0 \ll \gamma H_1^2. \qquad (48)$$

For the applicability of the adiabatic theorem we also require that the spectrum of $\Omega(t)$ contains no frequencies comparable to γH_e. If the time variation of H_0 is a modulation $H_0 = H_0^* + H_m \cos pt$ this requires

$\rho \ll |\gamma H_1|$, whereas (48) becomes $\rho H_m \ll \gamma H_1^2$. If we superimpose the condition $H_m \gg H_1$, which is not incompatible with the previous ones, the duration $\tau = |H_1/\dot{H}_0|$ of the passage through the resonance will be a small fraction of the modulation period.

If we start from a value of H_0 say far above resonance where the effective field is practically parallel to $\mathbf{H}_0$, and go through the resonance to the other side, far below the resonance, the magnetic moment $\mathbf{M}$, initially parallel to $\mathbf{H}_0$, will remain continuously parallel to $\mathbf{H}_e$ and, thus, end up antiparallel to $\mathbf{H}_0$. At the passage through the resonance there will be a transverse magnetization equal to the initial value M_0. If there is a distribution of Larmor frequencies, the magnetic moment of the sample can still be reversed by adiabatic passage, since the condition (48) is independent of the width δ of the shape function. However, the maximum value of the transverse magnetization during the passage is reduced in a ratio of the order of ω_1/δ if $\delta \gg |\omega_1|$. The modification caused in the theory of adiabatic passage by the existence of relaxation and the practical applications of this method will be discussed later.

F. The general problem of two levels coupled by an r.f. field. The fictitious spin ½

If we apply to a system a perturbation which is a sinusoidal function of time with a frequency ω in the neighbourhood of the frequency $\omega_0 = (E_a - E_b)/h$, where E_a and E_b are the energy levels of two states $|a\rangle$ and $|b\rangle$ of the system, it is permissible to a good approximation to disregard the other levels of the system. We then have a system with two degrees of freedom where all the relevant physical observables can be represented by 2×2 matrices. Each operator Q can then be written as $Q = \tfrac{1}{2}q_0 + \mathbf{q}\cdot\mathbf{s}$, where $q_0 = \mathrm{tr}\{Q\}$ inside the manifold of the two levels, and $\mathbf{s} = \tfrac{1}{2}\boldsymbol{\sigma}$, where σ_ξ, σ_η, σ_ζ are the usual Pauli matrices. Thus the total Hamiltonian of the system including the r.f. perturbation can be written as
$$\mathcal{H} = \tfrac{1}{2}E_0 - \gamma'\hbar(\mathbf{H}'\cdot\mathbf{s}),$$
where the four quantities E_0, $\gamma'H'_\xi$, $\gamma'H'_\eta$, $\gamma'H'_\zeta$ are easily determined from the matrix elements of $\mathcal{H}$, and H' and γ' are a fictitious magnetic field and a fictitious gyromagnetic ratio. Similarly, the density matrix ρ can be written as $\rho = \tfrac{1}{2} + \mathbf{m}\cdot\mathbf{s}$.

The equation of motion $(\hbar/i)(d\rho/dt) = -[\mathcal{H}, \rho]$ leads immediately to
$$\frac{d\mathbf{m}}{dt} = \gamma'\mathbf{m}\wedge\mathbf{H}', \tag{49}$$

which enables us to treat **m** as a fictitious magnetic moment. The expectation value of any operator Q will be

$$\langle Q \rangle = \mathrm{tr}(\rho Q) = \tfrac{1}{2}\{q_0 + (\mathbf{q} \cdot \mathbf{m})\}. \tag{50}$$

As an example consider the case of a spin I in a magnetic field where the levels $I_z = m$ are not equidistant because of a small quadrupole interaction and therefore the various transitions $|\Delta m| = 1$ can be induced separately. For instance, for the transition $-\tfrac{1}{2} \to \tfrac{1}{2}$,

$$\gamma' = \gamma, \qquad H'_z = H_0, \qquad H'_\xi + iH'_\eta = [I(I+1) + \tfrac{1}{4}]^{\frac{1}{2}} H_1 e^{i\omega t},$$

$$H'_1 e^{i\omega t} = [I + \tfrac{1}{2}] H_1 e^{i\omega t}, \qquad I_z = s_\zeta, \qquad I_{x,y} = [I + \tfrac{1}{2}] s_{\xi,\eta},$$

$$\langle I_x + i I_y \rangle = [I + \tfrac{1}{2}](m_\xi + i m_\eta).$$

Since the transverse magnetization $\langle I_x + iI_y \rangle$ is proportional to the transverse component of **m**, in order to give it its maximum value after a pulse one must have a pulse of a duration such that $\gamma H'_1 \tau = \tfrac{1}{2}\pi$ or $\gamma H_1 \tau = \tfrac{1}{2}\pi/(I+\tfrac{1}{2})$. For instance, for $I = \tfrac{5}{2}$ the pulse which will give the maximum decay tail will have a duration τ such that $\gamma H_1 \tau = \tfrac{1}{6}\pi$.

As another example, consider the problem of interchanging the populations of the two levels $M_F = 0$ of a hydrogen atom in a d.c. magnetic field H_0, by means of an oscillating magnetic field $H_1 \cos \omega t$, parallel to H_0, where $M_F = I_z + S_z$ is the component of the total atomic spin (electronic plus nuclear) along the field.

This problem occurs in a device for producing polarized protons in a cyclotron (5).

Defining the two states $|a\rangle$ and $|b\rangle$ with $M_F = 0$ as eigenstates of a fictitious spin $s_\zeta = \pm\tfrac{1}{2}$, the Hamiltonian of the system reduced to these two states can be written

$$\mathcal{H} = -\hbar \gamma' H'_0 s_\zeta + \hbar \gamma' H'_1 \{ s_\xi \cos \omega t + s_\eta \sin \omega t \}, \tag{51}$$

where

$$-\gamma' H'_0 = (E_a - E_b)/\hbar; \qquad \gamma' H'_1 = H_1 \langle a | \gamma_e S_z + \gamma_I I_z | b \rangle \tag{52}$$

(γ_e and γ_I are the gyromagnetic ratios of the electron and the proton).

Then, with the density matrix of the reduced system written as $\tfrac{1}{2} + \mathbf{m} \cdot \mathbf{s}$, the interchange of the populations of $|a\rangle$ and $|b\rangle$ corresponds to the reversal of the fictitious magnetization **m** and from the criterion (48), $\dot{H}'_0 \ll \gamma' H'^2_1$, the conditions for such a reversal are obtained from (52).

More examples of this formalism will be described later.

REFERENCES

1. F. BLOCH and A. SIEGERT, *Phys. Rev.* **57**, 522, 1940.
2. J. WINTER, *C.R. Acad. Sci.* **241**, 375, 1955.
3. J. MARGERIE and J. BROSSEL, ibid. **241**, 373, 1955.
4. E. MAJORANA, *Nuovo Cim.* **9**, 43, 1932.
5. A. ABRAGAM and J. WINTER, *Phys. Rev. Letters*, **1**, 374, 1958.

III

MACROSCOPIC ASPECTS OF NUCLEAR MAGNETISM

Felix qui potuit rerum cognoscere causas

I. Introduction

A. Static susceptibility

We have already stressed in the two previous chapters the paramount importance of relaxation phenomena in the study of nuclear magnetism. It is because of the existence of relaxation mechanisms that the nuclear spins can 'feel' the temperature of the lattice, and that differences of populations can appear between the various energy levels of the nuclear spin system, leading to a net absorption by the spin system of r.f. power supplied by an external generator. It is as a consequence of these differences that a net nuclear magnetization

$$M = \chi_0 H_0 = \frac{N\gamma^2\hbar^2 I(I+1)H_0}{3kT} \qquad (1)$$

appears in an applied field H_0.

It is worth pointing out parenthetically that the Curie law expressed by formula (1), derived in Chapter I under the assumption of isolated nuclear spins in an applied d.c. field $H_z = H_0$, is in fact a good deal more general, as can be demonstrated easily using the formalism of the density matrix introduced in Chapter II.

Suppose that the Hamiltonian of a system of N nuclear spins $\mathbf{I}^j$ contains besides their Zeeman energy $\mathcal{H}_0 = -\gamma\hbar H_0 \sum_j I_z^j = -\gamma\hbar H_0 J_z$, where $J_z = \sum I_z^j$, other terms such as dipolar couplings between the spins or quadrupolar couplings with local gradients, which we shall describe as an extra term, $\mathcal{H}_1$, in the Hamiltonian. The assumption of thermal equilibrium between the spin system and the lattice is expressed by a statistical spin operator

$$\rho = \exp\left\{-\frac{\mathcal{H}_0+\mathcal{H}_1}{kT}\right\}\Big/\mathrm{tr}\left[\exp\left\{-\frac{\mathcal{H}_0+\mathcal{H}_1}{kT}\right\}\right]. \qquad (2)$$

The macroscopic nuclear magnetic moment of the sample will be

$M = \text{tr}\{\gamma\hbar J_z \rho\}$ or, if we assume that the ratio $(\mathcal{H}_0+\mathcal{H}_1)/kT$ is small,

$$M \cong \frac{-\text{tr}\{\gamma\hbar J_z(\mathcal{H}_0+\mathcal{H}_1)\}}{\text{tr}\{1\}} \frac{1}{kT}. \tag{3}$$

If $\mathcal{H}_1$ is such that $\text{tr}\{\mathcal{H}_1 J_z\} = 0$, which happens to be the case for both spin-spin couplings and quadrupole couplings, it will be seen that

$$M \cong \frac{\gamma^2\hbar^2 H_0 \text{tr}\{J_z^2\}}{\text{tr}\{1\}} \frac{1}{kT} = \frac{N\gamma^2\hbar^2 I(I+1)H_0}{3kT},$$

in accordance with (1) and irrespective of the relative magnitude of $\mathcal{H}_0$ and $\mathcal{H}_1$, provided that both are much smaller than kT.

The conditions under which a nuclear magnetic resonance is observed are manifold. A first distinction can be drawn between the use of very small r.f. fields which can be assumed to leave the populations of the spin states practically unperturbed, and that of large r.f. fields which decrease appreciably the differences between the populations of the spin states, a phenomenon known as saturation. Another distinction can be made between so-called slow-passage methods where at each stage of the experiment a quasi steady-state is reached, and transient methods where non-equilibrium situations occur.

Yet another can be drawn between the behaviour of solid samples and fluid (liquid or gaseous) samples. In the solids there is usually a tight coupling between the nuclear spins which complicates considerably the phenomenon of nuclear resonance. The effects of this coupling will be examined at some length in the next two chapters.

In fluids this coupling is quenched to a considerable extent by the rapid relative motions of the spins, as will be explained in Chapters VIII and X, and as a consequence the dynamics of nuclear magnetism are a good deal simpler there.

B. Resonant absorption of r.f. energy

The amount of r.f. energy absorbed per unit time by a sample containing per unit volume N spins I of magnetic moment $\gamma\hbar I$ is easily computed using the formula (30) of Chapter II, which gives the transition probability per unit time induced by a rotating r.f. field of frequency $\omega = 2\pi\nu$ and amplitude $H_1 = -\omega_1/\gamma$.

If the assumption of negligible saturation can be made, the difference in populations between the states $I_z = m$ and $I_z = m-1$ is for each spin:

$$P_{m-1} - P_m = P_m\left\{\exp\left(\frac{\hbar\omega_0}{kT}\right) - 1\right\}$$

$$\cong \frac{1}{(2I+1)}\left\{\exp\left(\frac{\hbar\omega_0}{kT}\right) - 1\right\} \cong \frac{1}{(2I+1)}\frac{\hbar\omega_0}{kT}.$$

The total energy absorbed per unit time will thus be

$$P = \hbar\omega \frac{\hbar\omega_0}{kT} \frac{1}{(2I+1)} \frac{\pi\omega_1^2}{2} \sum_{m=I}^{-I+1} |(m|I_+|m-1)|^2 Nf(\omega)$$

$$= \frac{\hbar^2\omega\omega_0 \pi\omega_1^2}{2(2I+1)kT} \text{tr}\{I_+ I_-\} Nf(\omega)$$

$$= \frac{\hbar^2\omega\omega_0}{kT} \frac{\gamma^2 H_1^2}{6} I(I+1) Nf(\omega) 2\pi. \tag{4}$$

The origin of the finite width of spin levels described by the shape function $f(\omega)$ occurring in (4) need not be specified in those formulae. It may be caused by dipolar interactions between the spins, by the inhomogeneity of the applied field, by fluctuating local magnetic fields such as those that exist in metals and are due to the spins of conduction electrons, etc. It is sufficient to know that a relaxation mechanism of unspecified nature maintains the spins at the temperature of the lattice and thus the populations P_m of the spin levels at their Boltzmann values.

It is important to realize, however, that the very fact that r.f. energy is being absorbed by the spin system requires a transverse component of nuclear magnetization, incompatible with a rigorous description of the spin system by the populations of its levels. As explained in Chapter II, such a description would imply the absence of off-diagonal matrix elements for the statistical operator and thus no transverse magnetization.

Let us assume that the rotating field of amplitude H_1 is actually produced by a linearly polarized field $H_x = 2H_1 \cos\omega t$, the counter-rotating component having a negligible effect as explained previously.

The r.f. power absorbed by the spin system is

$$P = -\overline{\mathbf{M} \cdot \frac{d\mathbf{H}}{dt}} = -\overline{M_x \frac{dH_x}{dt}}. \tag{5}$$

If the r.f. excitation $2H_1 \cos\omega t$ of the spin system is sufficiently small, the response $M_x(t)$ of the spin system may be assumed proportional to it and can be written

$$M_x(t) = 2H_1\{\chi'(\omega)\cos\omega t + \chi''(\omega)\sin\omega t\}, \tag{6}$$

where $\chi'(\omega)$ and $\chi''(\omega)$, independent of H_1, are the so-called real and imaginary parts of the r.f. susceptibility $\chi = \chi' - i\chi''$, defined by the relations
$$H_x = 2H_1 \text{re}\{e^{i\omega t}\}, \qquad M_x = 2H_1 \text{re}\{\chi e^{i\omega t}\}, \tag{7}$$
where 're' means real part.

The method of calculating χ' and χ'' from the microscopic structure of the spin system will be explained in Chapter IV.

Carrying (6) over into (5), we find $P = 2H_1^2 \chi'' \omega$ which, by comparison with (4) and using the formula (1) for χ_0, gives

$$\chi''(\omega) = \tfrac{1}{2}\pi\chi_0\omega_0 f(\omega). \qquad (8)$$

The fact that, with our sign conventions, $\omega_0 = -\gamma H_0$, and thus also $\chi''(\omega)$ may take negative values, should not worry the reader since the absorbed power $P = 2H_1^2\chi''\omega$ will have the sign of $\chi_0\omega\omega_0$ or, since $\omega \simeq \omega_0$, the sign of $\chi_0\omega_0^2$.

It will be noticed that from the relation (8) between $\chi''(\omega)$ and the static susceptibility χ_0, all quantum-mechanical quantities have disappeared. This is a consequence of the so-called Kramers–Krönig relations, which exist for linear systems between the real and imaginary parts of their response to a sinusoidal excitation and will be derived at the end of this chapter. These relations are

$$\chi'(\omega) - \chi_\infty = \frac{1}{\pi}\mathscr{P}\int_{-\infty}^{\infty}\frac{\chi''(\omega')\,d\omega'}{\omega'-\omega},$$

$$\chi''(\omega) = -\frac{1}{\pi}\mathscr{P}\int_{-\infty}^{\infty}\frac{\chi'(\omega')-\chi_\infty}{\omega'-\omega}\,d\omega'. \qquad (8')$$

In (8') $\quad \mathscr{P}\int_{-\infty}^{\infty}\dfrac{g(\omega')\,d\omega'}{\omega'-\omega} \quad$ means $\quad \lim\limits_{\epsilon\to 0}\left(\int_{-\infty}^{\omega-\epsilon} + \int_{\omega+\epsilon}^{\infty}\right).$

Some caution should be exercised in applying these formulae to nuclear magnetism. According to the definition (6) of χ' and χ'', χ' is clearly an even and χ'' an odd function of ω. In nuclear magnetism we often calculate the response to rotating rather than oscillating fields and the precessing magnetization thus calculated can be considered as a response to an oscillating field only because the effect of the counter-rotating component is neglected. Let

$$M_x(t) = 2H_1\{\chi'_R(\omega)\cos\omega t + \chi''_R(\omega)\sin\omega t\}$$

be the response to a field rotating at a frequency ω:

$$H_x = H_1\cos\omega t, \qquad H_y = H_1\sin\omega t.$$

The response to a linearly polarized field: $H_x = 2H_1\cos\omega t$, $H_y = 0$, sum of two rotating fields with opposite frequencies, will be if the system is linear:

$$M_x(t) = 2H_1\{\chi'(\omega)\cos\omega t + \chi''(\omega)\sin\omega t\},$$

with
$$\chi'(\omega) = \chi'_R(\omega)+\chi'_R(-\omega),$$
$$\chi''(\omega) = \chi''_R(\omega)-\chi''_R(-\omega). \tag{8''}$$

Since $\chi'_R(\omega)$ and $\chi''_R(\omega)$ are very small unless $\omega \cong \omega_0$, which is the resonance frequency, replacing χ' and χ'' in (8') by their values (8''), writing $\omega = \omega_0+y$, $\omega' = \omega_0+y'$ and neglecting small terms, we get the K.–K. relations into a form better adapted for later purposes:

$$\chi'_R(\omega_0+y) \cong \frac{1}{\pi}\mathscr{P}\int_{-\infty}^{\infty}\frac{\chi''_R(\omega_0+y')\,dy'}{y'-y},$$

$$\chi''_R(\omega_0+y) \cong -\frac{1}{\pi}\mathscr{P}\int_{-\infty}^{\infty}\frac{\chi'_R(\omega_0+y')\,dy'}{y'-y}. \tag{8'''}$$

If $\chi''_R(\omega_0+y')$ is an even function of y' (symmetrical resonance curve) it is clear from the first equation (8''') that $\chi'_R(\omega_0+y)$ is an odd function of y, and $\chi'_R(\omega_0) = 0$. In order to demonstrate (8), we make $\omega = 0$ in the first relation (8'), getting

$$\chi_0 = \frac{1}{\pi}\mathscr{P}\int_{-\infty}^{\infty}\frac{\chi''_R(\omega')-\chi''_R(-\omega')}{\omega'}d\omega'$$

$$= \frac{2}{\pi}\mathscr{P}\int_{-\infty}^{\infty}\frac{\chi''_R(\omega')\,d\omega'}{\omega'} \cong \frac{2}{\pi\omega_0}\int_{-\infty}^{\infty}\chi''_R(\omega')\,d\omega'.$$

Writing $\chi''_R(\omega) = af(\omega)$ where a is a constant and $f(\omega)$ is a shape function normalized to unity, we get $a = \frac{1}{2}\pi(\omega_0\chi_0)$ whence (8).

The shape function $f(\omega)$ will in general be a bell-shaped narrow curve with a maximum at the Larmor frequency ω_0 of the spin system and, since $f(\omega)$ is normalized to unity, it will have a width Δ such that $\Delta f(\omega_0) \sim 1$. For instance, the Lorentz shape already met with in Chapter II is described by

$$f(\omega) = \frac{T_2}{\pi}\frac{1}{1+(\omega-\omega_0)^2 T_2^2}. \tag{9}$$

The half width at half intensity is equal to $\Delta = 1/T_2$ and $f(\omega_0) = T_2/\pi$. For such a shape, according to (8),

$$\chi''(\omega_0) = \tfrac{1}{2}\omega_0 T_2 \chi_0 = \tfrac{1}{2}\chi_0\frac{\omega_0}{\Delta}. \tag{10}$$

The same relation is valid with a factor of order unity for other line shapes. It is because resonance lines can be very narrow in nuclear magnetism, and thus ω_0/Δ can be a large number, that the resonant r.f. susceptibility is much larger than the static one, and resonant methods can be so much more sensitive than static methods.

The simplicity of the results thus obtained for vanishingly small r.f. fields has to be contrasted with the complications that arise as soon as the r.f. field is sufficiently large for the saturation to set in. Definite assumptions have to be made then about the internal structure of the spin system, the origin of the line width and the relaxation mechanisms, in order to be able to predict the behaviour of the spin system submitted to large r.f. perturbations. For a very special model (no interactions between the spins, and strong collisions) a calculation of such behaviour was given in Chapter II.

II. The Phenomenological Equations of Bloch

In 1946 Felix Bloch proposed for the description of magnetic properties of ensembles of nuclei in external magnetic fields a set of very simple equations derived from phenomenological arguments, that have proved exceedingly fruitful, and, for liquid samples at least, have provided in most cases a correct quantitative description of the detailed behaviour of the phenomena. The heuristic argument for obtaining these equations is as follows.

First, in an arbitrary *homogeneous* field the equation of motion of the nuclear magnetization for an ensemble of *free* spins was shown in Chapter II to be $d\mathbf{M}/dt = \gamma \mathbf{M} \wedge \mathbf{H}$.

Second, in a *static* field $H_z = H_0$, the trend of the magnetization towards its equilibrium value $M_z = M_0 = \chi_0 H_0$ can often be described with good accuracy by the equation $dM_z/dt = -\{(M_z - M_0)/T_1\}$. T_1 is called the longitudinal relaxation time. Third, if by any means such as an r.f. pulse, the nuclear magnetization is given a component at right angles to the applied field H_0, the various local fields, owing to the fact that the spins are actually not free but interact with each other and with their surroundings, cause the transverse magnetization to decay at a rate which can often be represented by the equations

$$\frac{dM_x}{dt} = -\frac{M_x}{T_2}, \qquad \frac{dM_y}{dt} = -\frac{M_y}{T_2},$$

where T_2 is called the transverse relaxation time.

Fourth, and this is a new assumption rather than a consequence of the three previous points, in the presence of an applied field, the sum of a d.c. field and a much smaller r.f. field, the motion due to relaxation

Ch. III MACROSCOPIC ASPECTS OF NUCLEAR MAGNETISM   45

can be superposed on the motion of the free spins, leading to the equation

$$\frac{d\mathbf{M}}{dt} = \gamma \mathbf{M} \wedge \mathbf{H} - \frac{M_x \mathbf{i}' + M_y \mathbf{j}'}{T_2} - \frac{M_z - M_0}{T_1}\mathbf{k}', \qquad (11)$$

where $\mathbf{i}', \mathbf{j}', \mathbf{k}'$ are the unit vectors of the laboratory frame of reference. Postponing until later the question of the validity of the equations (11), we shall examine some of their consequences.

A. Steady-state solutions—saturation

Assume that the applied field is the sum of a d.c. field

$$H_z = H_0 = -\frac{\omega_0}{\gamma}$$

and of an r.f. field $\mathbf{H}_1$ of amplitude $H_1 = -\omega_1/\gamma$ rotating at a frequency ω in the neighbourhood of ω_0. This field will usually be one of the rotating components of an applied field $H_x = 2H_1 \cos \omega t$, linearly polarized along the OX axis of the laboratory frame, the effect of the counter-rotating component being neglected. In the frame rotating around H_0 at the frequency ω there is an effective static field

$$\mathbf{H}_{\text{eff}} = \left(H_0 + \frac{\omega}{\gamma}\right)\mathbf{k} + H_1 \mathbf{i} = \frac{(\omega - \omega_0)\mathbf{k} - \omega_1 \mathbf{i}}{\gamma} = \frac{\Delta \omega \, \mathbf{k} - \omega_1 \mathbf{i}}{\gamma}, \qquad (12)$$

where $\mathbf{i}, \mathbf{j}, \mathbf{k} = \mathbf{k}'$ are the unit vectors of the rotating frame and $\omega_0 = -\gamma H_0$, $\omega_1 = -\gamma H_1$. In the rotating frame the equation of motion is

$$\frac{d\mathbf{M}}{dt} = \gamma(\mathbf{M} \wedge \mathbf{H}_{\text{eff}}) - \frac{\tilde{M}_x \mathbf{i} + \tilde{M}_y \mathbf{j}}{T_2} - \frac{M_z - M_0}{T_1}\mathbf{k}, \qquad (13)$$

where $\tilde{M}_x$ and $\tilde{M}_y$ are the transverse components of $\mathbf{M}$ in that frame. Equation (13) can be rewritten as

$$\frac{d\tilde{M}_x}{dt}\mathbf{i} + \frac{d\tilde{M}_y}{dt}\mathbf{j} + \frac{d\tilde{M}_z}{dt}\mathbf{k}$$
$$= (\tilde{M}_x \mathbf{i} + \tilde{M}_y \mathbf{j} + M_z \mathbf{k}) \wedge (\Delta\omega \, \mathbf{k} - \omega_1 \mathbf{i}) - \frac{\mathbf{i}\tilde{M}_x + \mathbf{j}\tilde{M}_y}{T_2} - \frac{M_z - M_0}{T_1}\mathbf{k}, \quad (13')$$

or

$$\frac{d\tilde{M}_x}{dt} = -\frac{\tilde{M}_x}{T_2} + \Delta\omega \, \tilde{M}_y,$$

$$\frac{d\tilde{M}_y}{dt} = -\Delta\omega \, \tilde{M}_x - \frac{\tilde{M}_y}{T_2} - \omega_1 M_z,$$

$$\frac{d\tilde{M}_z}{dt} = \omega_1 \tilde{M}_y - \frac{M_z - M_0}{T_1}. \qquad (14)$$

The general solution of (14), for fixed values of the parameters, is a sum of decreasing exponential terms and of a steady-state solution obtained by setting

$$\frac{d\tilde{M}_x}{dt} = \frac{d\tilde{M}_y}{dt} = \frac{dM_z}{dt} = 0.$$

If a sufficiently long time has elapsed for the transient exponentials to have decayed, the steady-state solution can be written

$$\tilde{M}_x = \frac{\Delta\omega\,\gamma H_1 T_2^2}{1+(T_2\Delta\omega)^2+\gamma^2 H_1^2 T_1 T_2} M_0,$$

$$\tilde{M}_y = \frac{\gamma H_1 T_2}{1+(T_2\Delta\omega)^2+\gamma^2 H_1^2 T_1 T_2} M_0,$$

$$M_z = \frac{1+(\Delta\omega T_2)^2}{1+(T_2\Delta\omega)^2+\gamma^2 H_1^2 T_1 T_2} M_0. \tag{15}$$

The fact that the three components of **M** are proportional to M_0 is not surprising: in the absence of initial polarization, that is, of inequalities of populations among the magnetic energy levels, the nuclear resonance cannot be observed.

From the values (15) of the components $\tilde{M}_x$ and $\tilde{M}_y$, the transverse components M_x and M_y in the laboratory frame can be calculated through

$$M_x+iM_y = (\tilde{M}_x+i\tilde{M}_y)e^{i\omega t},$$
$$M_x = \tilde{M}_x \cos\omega t - \tilde{M}_y \sin\omega t,$$
$$M_y = \tilde{M}_x \sin\omega t + \tilde{M}_y \cos\omega t. \tag{16}$$

A notation introduced by Felix Bloch and widely used in the literature is

$$u = \tilde{M}_x, \qquad v = -\frac{\gamma}{|\gamma|}\tilde{M}_y = \frac{\omega_0}{|\omega_0|}\tilde{M}_y = \pm\tilde{M}_y. \tag{16'}$$

The components of the magnetization in the laboratory frame are functions of time and can induce in a coil a detectable voltage at the frequency ω.

Experimental devices to be described briefly later in this chapter permit observation of u and v separately or of various combinations of these two quantities.

For negligible saturation, that is, for $\gamma^2 H_1^2 T_1 T_2 \ll 1$, v, which for reasons to appear shortly is called the absorption, can be written

$$v = -\pi|\gamma|H_1 M_0 f_{T_2}(\Delta\omega), \tag{17}$$

where $f_{T_2}(\Delta\omega)$ is the normalized Lorentz shape function (9) with half width at half intensity $1/T_2$.

For appreciable saturation, v can be written

$$v = \frac{-\pi|\gamma|H_1 M_0}{(1+\gamma^2 H_1^2 T_1 T_2)^{\frac{1}{2}}} f_{T_2'}(\Delta\omega), \tag{18}$$

where the half width T_2' is

$$\frac{1}{T_2'} = \frac{1}{T_2}(1+\gamma^2 H_1^2 T_1 T_2)^{\frac{1}{2}}. \tag{18'}$$

In presence of saturation the resonance curve representing v still has the Lorentz shape but becomes broader in the ratio

$$\frac{T_2}{T_2'} = (1+\gamma^2 H_1^2 T_1 T_2)^{\frac{1}{2}}.$$

The value of $|v|$ is maximum at resonance and equal to

$$|v|_{\max} = \frac{|\gamma|H_1 T_2}{1+\gamma^2 H_1^2 T_1 T_2} M_0. \tag{18''}$$

For small values of H_1, $|v|_{\max}$ is proportional to H_1; then it passes through a maximum reached when $\gamma^2 H_1^2 T_1 T_2 = 1$ and equal to

$$|v|_{\max,\max} = \tfrac{1}{2} M_0 \sqrt{\left(\frac{T_2}{T_1}\right)}, \tag{19}$$

and decreases towards zero as H_1 is increased further.

In practice T_2 is sometimes shorter than T_1 so that the steady state transverse magnetization is smaller than M_0. The behaviour of the function u is different from that of the function v. It is an odd rather than even function of $\Delta\omega$, vanishes at resonance, and has a maximum and a minimum on either side of the resonance for

$$\Delta\omega = \pm \frac{(1+\gamma^2 H_1^2 T_1 T_2)^{\frac{1}{2}}}{T_2} = \pm \frac{1}{T_2'}$$

with $\quad u_{\max} = -u_{\min} = \tfrac{1}{2}|\gamma|H_1 T_2' M_0.$

As H_1 increases $|u_{\max}| = |u_{\min}|$ grows steadily to an asymptotic value

$$|u_{\max}|_{\text{asymp}} = |v|_{\max,\max} = \frac{M_0}{2}\sqrt{\left(\frac{T_2}{T_1}\right)}. \tag{20}$$

Finally M_z, which at resonance and for small values of H_1 differs from M_0 only to the second order in H_1, decreases steadily towards zero as H_1 increases.

A convenient geometrical representation of the solutions (15) is obtained if, through elimination of $\Delta\omega$ and H_1 from (15), a relation is derived between M_z, $\tilde{M}_x$, and $\tilde{M}_y$. It shows with a little algebra that the locus of the vector **M** as a function of H_1 and $\Delta\omega$ is an ellipsoid with axes $2a_x = 2a_y = M_0\sqrt{(T_2/T_1)}$, $2a_z = M_0$, centred at the point $x = y = 0$, $z = \tfrac{1}{2}M_0$.

All the preceding results for u and v are easily visualized in this manner.

The fact that for large H_1 the dispersion remains finite, whereas the absorption goes down to zero, makes it preferable, while searching for signals in samples with unknown relaxation times, to use strong r.f. fields and detection devices to be described later, sensitive to the dispersion rather than to the absorption.

The r.f. power absorbed by the spin system is easily calculated from the formula
$$\frac{dE}{dt} = -P = -\mathbf{M}\frac{d\mathbf{H}}{dt}.$$

The vector $d\mathbf{H}/dt$ has in the rotating frame components:
$$\tilde{X} = \tilde{Z} = 0, \qquad \tilde{Y} = \omega H_1,$$
and
$$P = -\mathbf{M}\frac{d\mathbf{H}}{dt} = -\omega \tilde{M}_y H_1,$$
or, by (15),
$$P = \frac{-\omega \gamma H_1^2 T_2}{1+(T_2\Delta\omega)^2+\gamma^2 H_1^2 T_1 T_2} M_0 = \frac{\chi_0 H_1^2 \omega \omega_0 T_2}{1+(T_2\Delta\omega)^2+\gamma^2 H_1^2 T_1 T_2}. \tag{21}$$

Since $v = (\omega_0/|\omega_0|)\tilde{M}_y$, P can be written as $P = -|\omega|H_1 v$, whence the name of absorption component given to the quantity v. For large values of H_1, P approaches asymptotically at resonance the maximum value $\chi_0 H_0^2/T_1 = M_0 H_0/T_1$. This can be interpreted by saying that the maximum energy that can be transferred to the lattice in the time T_1 is the energy $\chi_0 H_0^2$, twice the paramagnetic energy $\tfrac{1}{2}\chi_0 H_0^2$ stored in the sample at thermal equilibrium. It must be pointed out, however, and will be shown later that the 'signal' available for the detection of the absorption is proportional to v rather than to the absorbed power P. From the definition (6) of the r.f. susceptibilities χ' and χ'' and the equation (16) we obtain
$$\chi'(\omega) = \frac{\tilde{M}_x}{2H_1}, \qquad \chi''(\omega) = \frac{\tilde{M}_y}{2H_1},$$
or, replacing in (15) M_0 by $\chi_0 H_0 = -\chi_0 \omega_0/\gamma$,
$$\chi'_R = -\frac{\tfrac{1}{2}\omega_0 \Delta\omega T_2^2}{1+(T_2\Delta\omega)^2+\gamma^2 H_1^2 T_1 T_2}\chi_0,$$
$$\chi''_R = \frac{\tfrac{1}{2}\omega_0 T_2}{1+(T_2\Delta\omega)^2+\gamma^2 H_1^2 T_1 T_2}\chi_0. \tag{22}$$

χ'_R and χ''_R can be rewritten as

$$\chi'_R = -\tfrac{1}{2}\pi\chi_0(\Delta\omega T'_2)f_{T'_2}(\Delta\omega)\omega_0,$$
$$\chi''_R = \tfrac{1}{2}\pi\chi_0(1+\gamma^2 H_1^2 T_1 T_2)^{-\tfrac{1}{2}}\omega_0 f_{T'_2}(\Delta\omega), \quad (23)$$

where $T'_2 = T_2[1+\gamma^2 H_1^2 T_1 T_2]^{-\tfrac{1}{2}}$ and $f_{T'_2}(\Delta\omega)$ is the normalized Lorentz shape function

$$\frac{T'_2}{\pi}\frac{1}{1+(T'_2\Delta\omega)^2}.$$

For negligible saturation,

$$\chi'_R = -\tfrac{1}{2}\pi\chi_0\,\omega_0(\Delta\omega\,T_2)f_{T_2}(\Delta\omega),$$
$$\chi''_R = \tfrac{1}{2}\pi\chi_0\,\omega_0 f_{T_2}(\Delta\omega). \quad (24)$$

The reader can easily satisfy himself that the susceptibilities $\chi'_R(\omega)$ and $\chi''_R(\omega)$ defined by (24) do verify the K.–K. relations (8'''). On the other hand, it is important to notice that the susceptibilities (23) describing the existence of saturation *do not* verify the K.–K. relations as indeed they should not, since the response of the spin system does not depend linearly on H_1. It is actually easy to verify that in that case relations of the type (8''') exist between χ'_R and $\chi''_R[1+\gamma^2 H_1^2 T_1 T_2]^{\tfrac{1}{2}}$ as defined by (23).

A well-known, but sometimes forgotten, point is that the susceptibility of an assembly of spins is a tensor, correlating the magnetization with the applied field by a relation of the form

$$M_i = \sum_j \chi_{ij} H_j.$$

Thus χ'_R and χ''_R given by (22) are the real and imaginary part of the component χ_{xx} of the susceptibility tensor. In some schemes of detection, to be described later, we are interested in the component χ_{yx} that gives the values of M_y when the r.f. field is applied along ox. Since (neglecting the effect of the counter-rotating component of the r.f. field) the magnetization precesses with the velocity ω around the axis oz, it is clear that $M_y(t) = M_x(t-\pi/2\omega)$, whence

$$\chi'_{yx}\cos\omega t + \chi''_{yx}\sin\omega t = \chi'_{xx}\cos(\omega t - \tfrac{1}{2}\pi) + \chi''_{xx}\sin(\omega t - \tfrac{1}{2}\pi),$$
$$\chi'_{yx} = -\chi''_{xx}, \quad \chi''_{yx} = \chi'_{xx}.$$

B. Steady-state solutions in an inhomogeneous field

It often happens, especially for liquid samples, that the spread $\gamma\Delta H$ in Larmor frequencies due to the inhomogeneity of the applied field across the sample results for the observed absorption curve in a width much greater than that predicted by the equations (17) and (18), namely $1/T_2'$ for negligible saturation and $1/T_2'$ if saturation is present. Suppose that $h(x)\,dx$ (normalized to $\int h(x)\,dx = 1$) represents the relative weight of spins with Larmor frequencies ω_0 between ω_0^0+x and ω_0^0+x+dx (ω_0^0 being the value of the Larmor frequency with the maximum weight). For those spins the steady-state solutions (15) are valid with $-\Delta\omega = \omega_0-\omega = \omega_0^0+x-\omega$ and the total response of the sample will be obtained by multiplying these solutions by $h(x)\,dx$ and integrating over x. Thus for v as given by (18), we obtain

$$v = \frac{-\pi|\gamma|H_1M_0}{(1+\gamma^2H_1^2T_1T_2)^{\frac{1}{2}}} \int_{-\infty}^{\infty} f_{T_2'}(\omega_0^0-\omega+x)h(x)\,dx. \qquad (25)$$

If the width is entirely determined by the inhomogeneity of the field for all obtainable values of $1/T_2' = (1/T_2)[1+\gamma^2H_1^2T_1T_2]^{\frac{1}{2}}$, $h(x)$ has a much slower dependence on x than $f_{T_2'}(\omega_0^0-\omega+x)$, which has a relatively sharp maximum for $\omega_0^0-\omega+x = 0$, and v can be written

$$v = \frac{-\pi|\gamma|H_1M_0}{(1+\gamma^2H_1^2T_1T_2)^{\frac{1}{2}}} h(\omega-\omega_0^0). \qquad (26)$$

The dependence of $|v_{\max}| = \pi|\gamma|H_1M_0h(0)[1+\gamma^2H_1^2T_1T_2]^{-\frac{1}{2}}$ on H_1, that is, the saturation, is quite different from that predicted previously from equation (18″) since $|v_{\max}|$ increases steadily towards a maximum value

$$|v_{\max}|_{\text{asymp}} = \frac{\pi h(0)}{\sqrt{(T_1T_2)}} M_0 \qquad (26')$$

quite different from (19). At the same time, according to (26), the shape of $v(\omega)$, entirely contained in $h(\omega-\omega_0^0)$, does not change with H_1.

To pursue the discussion farther, it is helpful for the sake of simplicity to assume that the shape function $h(x)$ of the field inhomogeneity also has the Lorentz form:

$$h(x) = \frac{1}{1+(xT_2^\dagger)^2}\frac{T_2^\dagger}{\pi}, \qquad (26'')$$

where $1/T_2^\dagger = \gamma\Delta H$ is the half width at half intensity, due to the field inhomogeneity. The integral (25) can then be performed exactly. It is the folding (*Faltung*) of two Lorentz distributions, well known to result into a Lorentz distribution with a width that is the sum of the widths of the two components.

If we define $T_2'^*$ through

$$\frac{1}{T_2'^*} = \frac{1}{T_2'} + \frac{1}{T_2^\dagger}, \qquad (27)$$

or, for negligible H_1, $$\frac{1}{T_2^*} = \frac{1}{T_2} + \frac{1}{T_2^\dagger},$$

(25) can be rewritten

$$v = \frac{-\pi|\gamma|H_1 M_0}{(1+\gamma^2 H_1^2 T_1 T_2)^{\frac{1}{2}}} f_{T_2'^*}(\omega-\omega_0^0), \qquad (28)$$

or, for small H_1 and negligible saturation,

$$v = -\pi|\gamma|H_1 M_0 f_{T_2^*}(\omega-\omega_0^0). \qquad (28')$$

If $$\frac{1}{T_2^\dagger} \gg \frac{1}{T_1}, \frac{1}{T_2} \qquad (28'')$$

it is possible to give to H_1 values, large enough to have $\gamma^2 H_1^2 T_1 T_2 \gg 1$, but at the same time small enough to have

$$\frac{1}{T_2'^*} = \frac{1}{T_2'} + \frac{1}{T_2^\dagger} = \frac{(1+\gamma^2 H_1^2 T_1 T_2)^{\frac{1}{2}}}{T_2} + \frac{1}{T_2^\dagger} \cong \frac{1}{T_2^\dagger}.$$

The quantity $|v_{\max}|$ then, according to (28), reaches an asymptotic value

$$|v_{\max}|_{\text{asymp}} \cong \frac{M_0 T_2^\dagger}{\sqrt{(T_1 T_2)}} \cong \frac{M_0}{2}\sqrt{\left(\frac{T_2^\dagger}{T_1}\right)} 2\sqrt{\left(\frac{T_2^\dagger}{T_2}\right)} \ll \frac{M_0}{2}\sqrt{\left(\frac{T_2^\dagger}{T_1}\right)}. \qquad (29)$$

If H_1 were further increased to such an extent that $1/T_2'$ ceased to be small compared with $1/T_2^\dagger$, v would decrease further and eventually tend to zero. The conclusion of this discussion is that it is not correct, even qualitatively, as is sometimes done, to take into account the inhomogeneity of the applied field by introducing into the Bloch equations a generalized transverse relaxation time T_2^*.

It is interesting to notice that the assumption of the Lorentz shape (26″) for the field inhomogeneity does in fact, according to a theorem given in Chapter II, lead to an exponential decay rate,

$$\frac{dM_{\text{transv}}}{dt} = -\frac{M_{\text{transv}}}{T_2^*},$$

for the transverse magnetization in a free precession experiment. The discussion above shows that the heuristic assumption four, made earlier and stating that the contributions to $d\mathbf{M}/dt$ from the torque due to the applied field (d.c. plus r.f.) and from the observed free decay could be simply added, was indeed an extra assumption on the conditions for the validity of the Bloch equations, since there are cases when it is clearly invalid.

The fact that there is a fundamental difference in nature between the broadening expressed by $1/T_2$ in the Bloch equations and that caused by the inhomogeneity of the field, will appear even more clearly later in this chapter in connexion with the spin echo methods.

To calculate the so-called dispersion component u we make use of the remark made at the end of Section II A, that in presence of saturation

$$\chi'_R(\omega_0+y) = [1+\gamma^2 H_1^2 T_1 T_2]^{\frac{1}{2}} \frac{1}{\pi} \mathscr{P} \int_{-\infty}^{\infty} \frac{\chi''_R(\omega_0+y')}{y'-y} dy', \qquad (30)$$

or, integrating over all the Larmor frequencies $\omega_0 = \omega_0^0 + x$ of the sample,

$$(\chi'_R)_{\text{total}} = [1+\gamma^2 H_1^2 T_1 T_2]^{\frac{1}{2}} \frac{1}{\pi} \int_{-\infty}^{\infty} h(x) dx \, \mathscr{P} \int_{-\infty}^{\infty} \frac{\chi''_R(\omega_0^0+x+y')}{y'-y} dy'. \qquad (31)$$

Replacing χ''_R in (31) by its value (23) and reversing the order of the integrations,

$$(\chi'_R)_{\text{total}} \cong \tfrac{1}{2}\pi \chi_0 \omega_0^0 \frac{1}{\pi} \mathscr{P} \int_{-\infty}^{\infty} \frac{dy'}{y'-y} \int_{-\infty}^{\infty} h(x) f_{T_2'}(\omega_0^0+x+y') dx. \qquad (32)$$

Since the variation of $f_{T_2'}$ is much faster than that of $h(x)$ and has a sharp maximum for $\omega_0^0+x+y' = 0$, (32) can be rewritten

$$(\chi'_R)_{\text{total}} = \tfrac{1}{2}\pi \chi_0 \omega_0^0 \frac{1}{\pi} \mathscr{P} \int_{-\infty}^{\infty} \frac{dy'}{y'-y} h(-\omega_0^0-y'). \qquad (33)$$

The integrated real part of the susceptibility, $(\chi'_R)_{\text{tot}}$, has thus the same value, and the dispersion curve has the same shape as if there were no saturation at all and the shape of the absorption curve were given by $h(x)$.

To sum up these results, the following simple rules should be retained. For large inhomogeneous broadening, the shape of the absorption v does not change with saturation and its absolute value varies as $H_1[1+\gamma^2 H_1^2 T_1 T_2]^{-\frac{1}{2}}$, that is, it tends to a limit. The dispersion u has a shape derived from that of the absorption by the K.–K. relations and varies linearly with H_1. These simple rules break down only when H_1 reaches values such that the saturated width $1/T_2'$ becomes comparable to the width of the distribution h. It appears from the discussion above that a field inhomogeneity $\Delta H \gg |1/\gamma T_2|$, due for instance to a poor magnet, causes a considerable decrease in the steady-state r.f. magnetization.

Suppose that by some means the real T_2 of the sample is reduced well below the value $1/|\gamma\Delta H|$. The broadening due to the field inhomogeneity is then a small perturbation on the real width and the maximum absorption and dispersion signals become $\frac{1}{2}M_0\sqrt{(T_2/T_1)}$. If in the process of shortening T_2, T_1 is also shortened and becomes comparable to T_2, the transverse magnetization becomes $\simeq M_0$, a considerable improvement. It will be shown in Chapter VIII that dissolving paramagnetic impurities in an otherwise diamagnetic liquid results precisely in short and comparable values of T_1 and T_2. This procedure known as 'doping' is widely used to improve the steady-state signals.

C. Modified Bloch equations in low fields

The validity of the Bloch equations (11) and of their steady-state solutions (15) becomes questionable when the d.c. applied field H_0 is comparable either to the line width in gauss $1/\gamma T_2$ or to the amplitude H_1 of the r.f. field.

It has been suggested on theoretical grounds that the assumption of a magnetization relaxing towards the d.c. equilibrium value $\mathbf{M}_0 = \chi_0 \mathbf{H}_0$ should be replaced, when an r.f. field $H_1(t)$ is present, by that of relaxation towards the instantaneous value $\chi_0(\mathbf{H}_0+\mathbf{H}_1(t))$. The distinction, academic in high fields, becomes important when H_0 and H_1 are comparable.

The assumption of the relaxation of $\mathbf{M}$ towards $\chi_0(\mathbf{H}_0+\mathbf{H}_1(t))$ permits us, as will appear presently, to get rid of an unsatisfactory feature of the Bloch equations. The r.f. power absorbed, given by formula (21), goes to zero with $H_0 = -\omega_0/\gamma$, even if the r.f. frequency ω has a finite value. This result is difficult to accept on theoretical grounds and is in conflict with the low-frequency relaxation measurements of the Dutch school.

The justification from first principles of the validity of modified Bloch equations will be given in Chapter XII and we shall be content for the time being to explore some of their consequences. It will be shown in Chapter XII that if H_1 is not small compared with H_0, the conditions for the validity of the simple assumption of relaxation towards $\chi_0(\mathbf{H}_0+\mathbf{H}_1(t))$ are such that they also result in the equality $T_1 = T_2 = T$.

The modified equations adopted are thus

$$\frac{d\mathbf{M}}{dt} = \gamma \mathbf{M} \wedge \mathbf{H} - \frac{\mathbf{M}-\chi_0\mathbf{H}}{T}. \qquad (34)$$

It becomes necessary now to discriminate more carefully between the response to an oscillating and to a rotating field, the effect of the counter-rotating component in the former case being appreciable if H_1 is comparable with H_0.

We first look for steady-state solutions in a rotating field. With the same notations as in (14) except for the simplification $T_1 = T_2$, (34) can be written in the rotating frame

$$\frac{d\tilde{M}_x}{dt} = -\frac{(\tilde{M}_x - \chi_0 H_1)}{T} + \Delta\omega \tilde{M}_y,$$

$$\frac{d\tilde{M}_y}{dt} = -\Delta\omega \tilde{M}_x - \frac{\tilde{M}_y}{T} - \omega_1 M_z,$$

$$\frac{dM_z}{dt} = \omega_1 \tilde{M}_y - \frac{M_z - M_0}{T}. \tag{35}$$

We express the transverse components $\tilde{M}_x$ and $\tilde{M}_y$, the steady-state solution of (35), by means of the r.f. susceptibilities $\chi_R'^M(\omega)$ and $\chi_R''^M(\omega)$. The subscript R stands for rotating and the superscript M is to remind us that (35) are the modified Bloch equations. We find

$$\chi_R'^M = \frac{\tilde{M}_x}{2H_1} = -\frac{1}{2} \frac{\omega_0 \Delta\omega T^2 - [1 + (\gamma H_1 T)^2]}{1 + (T\Delta\omega)^2 + (\gamma H_1 T)^2} \chi_0,$$

$$\chi_R''^M = -\frac{\tilde{M}_y}{2H_1} = \frac{1}{2} \frac{\omega T}{1 + (T\Delta\omega)^2 + (\gamma H_1 T)^2} \chi_0. \tag{36}$$

For negligible saturation ($|\gamma| H_1 T \ll 1$), (36) can be rewritten

$$\chi_R'^M = \chi_R' + \frac{1}{\omega_0 T} \chi_R'', \qquad \chi_R''^M = \chi_R'' - \frac{1}{\omega_0 T} \chi_R', \tag{37}$$

where
$$\chi_R' = -\frac{\chi_0}{2} \frac{\omega_0 \Delta\omega T^2}{1 + (T\Delta\omega)^2}, \qquad \chi_R'' = \frac{\chi_0}{2} \frac{\omega_0 T}{1 + (T\Delta\omega)^2} \tag{37'}$$

are the rotating field susceptibilities (24) for the unmodified Bloch equations (11) (with $T_1 = T_2$). The susceptibilities $\chi_R'^M$ and $\chi_R''^M$ (37) satisfy the K.–K. relations. This follows immediately from the fact that χ_R' and χ_R'' (37') verify these relations. For M_z we obtain

$$M_z = \chi_0 \left[H_0 + \frac{(\omega/\gamma)(\gamma H_1 T)^2}{1 + (T\Delta\omega)^2 + (\gamma H_1 T)^2} \right], \tag{38}$$

which is to be contrasted with the value (15) previously obtained which can be rewritten as

$$M_z = \chi_0 \left[H_0 + \frac{(\omega_0/\gamma)(\gamma H_1 T)^2}{1 + (T\Delta\omega)^2 + (\gamma H_1 T)^2} \right]. \tag{38'}$$

(Remember the definition $\omega_0 = -\gamma H_0$.)

From a comparison between (36) and (22) on the one hand, (38) and (38') on the other, interesting features of the modified Bloch equations appear. The most striking is the fact that for $H_0 = -\omega_0/\gamma = 0$, (38), in contradistinction to (38'), predicts a non-vanishing magnetization M_z. This is plausible physically. In the rotating frame the spins 'see' a static field H_{eff} with components $H_z = -\omega/\gamma$, $H_x = H_1$, around which

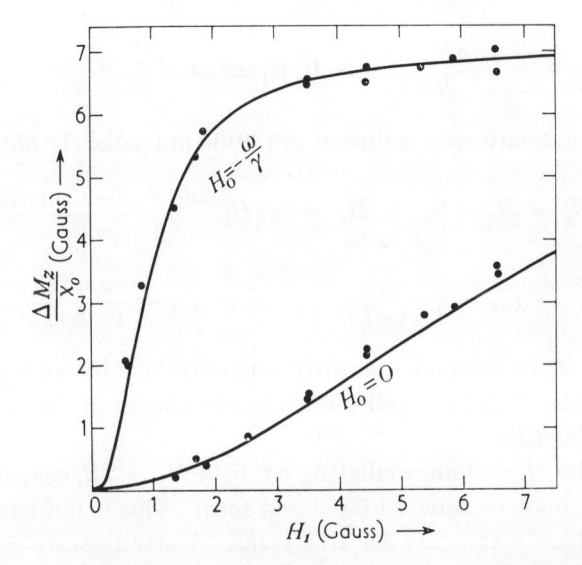

Fig. III, 1. Observed signal in a d.c. field H_0 as a function of r.f. amplitude in solid diphenyl, picril hydrazil. The full curves give the theoretical prediction of eqn. (38), assuming $T_1 = T_2 = T = 6.2 \times 10^{-8}$ sec, measured independently.

they precess, while relaxing towards $M_x^0 = \chi_0 H_1$. There is clearly a lack of symmetry between positive and negative values of M_z, resulting in a non-zero steady value of M_z. The predictions of formula (38) have been tested experimentally (1).

Fig. III, 1 shows $\Delta M_z = M_z - \chi_0 H_0$ predicted by (38), as a function of the strength H_1 of the rotating field, for (a) $H_0 = -\omega/\gamma$, (b) $H_0 = 0$, together with the experimental points. The agreement with the theory is gratifying and demonstrates clearly that the unmodified Bloch equations, which predict $M_z = 0$ for $H_0 = 0$, are inadequate.

The other interesting feature of the modified Bloch equations is that the absorbed power $P = 2\chi'' H_1^2 \omega$ does not vanish with H_0 since $\chi_R''^M$ given by (36) is proportional to ω rather than to ω_0.

Referring for further details to reference (1), we pass on to the steady-state solution of (34) for a linearly polarized r.f. field.

Consider first for the sake of comparison with the response to a rotating field, the case $H_0 = 0$. The equations (34) then read

$$\frac{dM_x}{dt} = -\frac{M_x - \chi_0 2H_1 \cos \omega t}{T},$$

$$\frac{dM_y}{dt} = 2\gamma M_z H_1 \cos \omega t - \frac{M_y}{T},$$

$$\frac{dM_z}{dt} = -2\gamma M_y H_1 \cos \omega t - \frac{M_z}{T}. \tag{39}$$

The exact steady-state solution of (39) is immediately obtained; it is

$$M_y = M_z = 0, \qquad M_x = 2\chi_0 H_1 \frac{\cos \omega t + (\omega T)\sin \omega t}{1+(\omega T)^2}, \tag{39'}$$

whence $\quad \chi' = \chi'_{xx} = \dfrac{\chi_0}{1+(\omega T)^2}, \qquad \chi'' = \chi''_{xx} = \dfrac{\chi_0(\omega T)}{1+(\omega T)^2}, \tag{40}$

and all the other components of the susceptibility tensor vanish. This is the formula of Debye, well known in the theory of dielectric dispersion and relaxation.

For a finite H_0 and an oscillating r.f. field $H_x = 2H_1 \cos \omega t$, the equations (34) cannot be solved in a closed form. This is not surprising for such was already the case for free spins. A steady-state solution can be sought in the form of a series

$$M_i = \sum_{n=0}^{\infty} Ain \cos(n\omega t) + Bin \sin(n\omega t), \tag{41}$$

where i stands for x, y, z.

The appearance in (41) of harmonics with $n > 1$ could be interpreted as an absorption of more than one quantum by the spin system, a phenomenon already mentioned in Chapter II in connexion with the response of an assembly of *free* spins to an oscillating rather than rotating r.f. field. The behaviour of the solutions is determined by the two dimensionless parameters

$$q_0 = |\gamma H_0 T| = |\omega_0 T| \quad \text{and} \quad q_1 = |\gamma H_1 T| = |\omega_1 T|,$$

of which the first measures the sharpness of the resonance, and the second the amount of saturation.

For negligible saturation χ'_M and χ''_M are immediately obtained from the solution (37) for a rotating field, since a linear field is a superposition

Ch. III MACROSCOPIC ASPECTS OF NUCLEAR MAGNETISM

of two such fields:

$$\chi'^{M}(\omega) = \chi'^{M}_{R}(\omega) + \chi'^{M}_{R}(-\omega)$$
$$= \chi'_{R}(\omega) + \chi'_{R}(-\omega) + \frac{1}{\omega_0 T}\{\chi''_{R}(\omega) + \chi''_{R}(-\omega)\},$$
$$\chi''^{M}(\omega) = \chi''^{M}_{R}(\omega) - \chi''^{M}_{R}(-\omega)$$
$$= \chi''_{R}(\omega) - \chi''_{R}(-\omega) - \frac{1}{\omega_0 T}\{\chi'_{R}(\omega) - \chi'_{R}(-\omega)\}. \quad (42)$$

From (37) and (37') it is easily verified that when $|H_0| = |\omega_0/\gamma| \to 0$, $\chi'^{M}(\omega)$ and $\chi''^{M}(\omega)$ tend towards the Debye values (40).

If the nuclear transverse magnetization is detected through the e.m.f. it induces in a coil at right angles to the coil that produces the field H_1, it is necessary to know $\chi'_{yx}(\omega)$ and $\chi''_{yx}(\omega)$. These are easily computed as sums of the contributions from the response to the rotating and counter-rotating components of the applied field $H_x = 2H_1 \cos\omega t$, and yield

$$\chi'_{yx}(\omega) = -[\chi''^{M}_{R}(\omega) + \chi''^{M}_{R}(-\omega)],$$
$$\chi''_{yx}(\omega) = +[\chi'^{M}_{R}(\omega) - \chi'^{M}_{R}(-\omega)], \quad (43)$$

where χ'^{M}_{R} and χ''^{M}_{R} are given by (37). We shall not discuss the case of a linearly polarized r.f. field in the presence of saturation, that is for $|\gamma H_1 T|$ comparable to unity or larger. If H_1/H_0 is small the distinction between rotating and linear r.f. field and between modified and unmodified Bloch equations is negligible and the problem has already been solved and discussed. If, on the other hand, one has simultaneously $H_1/H_0 \sim 1$ and $|\gamma H_1 T| \sim 1$, the problem is exceedingly complicated, for it requires the solution of a system of equations with an infinite number of unknown quantities Ain and Bin and its practical importance in nuclear magnetism hardly justifies the effort.

III. Transient Methods in Nuclear Magnetism

The Bloch equations are obtained from the simpler equations $d\mathbf{M}/dt = \gamma \mathbf{M} \wedge \mathbf{H}$ valid for an assembly of free spins through addition of the relaxation terms that contain the relaxation times T_1 and T_2. The physical meaning of T_1, as the time constant measuring the rate of progress of the spin system towards thermal equilibrium with the lattice, is perfectly clear and it can be shown experimentally in many cases, and theoretically in some, that such a progress can indeed be described by a single exponential with a time constant T_1. An unambiguous definition of T_2 is much more difficult to give and conflicting interpretations of its meaning are found in the literature.

From the macroscopic viewpoint used in this chapter the following operational definition of T_2 could be given: it has been shown experimentally that for wide categories of samples of bulk matter (mainly liquids) placed in sufficiently homogeneous fields, the motion of the macroscopic nuclear magnetization, including its saturation behaviour, is accurately described by Bloch equations with appropriate values of the constants T_1 and T_2. Then it follows from the Bloch equations that both the inverse of the unsaturated line width and the time constant for the free decay of the transverse magnetization are equal to T_2. The field can be said to be sufficiently homogeneous if a reduction in the size of the sample does not lead to a reduction in the line width. In practice, however, for many liquid samples, such a state can never be reached, the inhomogeneity of the applied field still being the main cause of width for the smallest samples that give an observable signal. A different operational definition must then be sought for T_2.

A. The method of the spin echoes

It has already been mentioned in Chapter II that the decay of the precessing transverse magnetization, following a pulse of 90°, and caused by destructive interference between the contributions from moments in different parts of the sample, precessing at different Larmor frequencies, is not an irreversible phenomenon.

In a field of total inhomogeneity ΔH across the sample, where the lifetime of the precessing magnetization is of the order of $(\gamma \Delta H)^{-1}$, a pulse of 180°, separated from the first 90° pulse by a time τ that could be much longer than $(\gamma \Delta H)^{-1}$, is followed at time 2τ by a refocusing in phase of all the elementary moments and the restoring of the full transverse magnetization that was created by the first pulse. This phenomenon is called a spin echo (2). This magnetization that disappears again in a time $\sim |\gamma \Delta H|^{-1}$ can be made to reappear *ad infinitum* at times $4\tau, 6\tau,..., 2n\tau,...$ by means of further 180° pulses applied at times $3\tau, 5\tau,..., (2n-1)\tau,...$. It can be shown that it is not necessary that the refocusing pulse be of 180° for the echo to occur and, historically, 90° refocusing pulses were first used (2). The techniques of refocusing pulses of 180° are, however, the easiest to interpret and the most widely used at present. We shall limit ourselves to these.

The completely reversible behaviour thus described for the magnetization is connected with the assumption of free spins, made in Chapter II. In a real sample the spins are submitted to internal magnetic fields resulting from couplings with their neighbours or with electronic spins if the substance is not perfectly diamagnetic. Electric local fields also act on their quadrupole moments. In liquids all these fields are modulated at a fast rate and in a random fashion, by the Brownian motion of the molecules. It will be shown in Chapter VIII under what fairly general conditions the effect of these fields is to cause an irreversible exponential decay of the transverse magnetization with a time constant T_2. If these conditions are satisfied, consider a sequence of experiments, each started from a nuclear magnetization M_0 in thermal equilibrium by a 90° pulse, followed at a time τ, different in each experiment, by a 180° pulse. The height of the echo observed at time 2τ should be proportional to $\exp(-2\tau/T_2)$, thus providing an operational definition and a measure of the transverse relaxation time T_2. This procedure, or method A (3), is time consuming, since an interval several times T_1 must elapse between each measurement to let the nuclear magnetization reach its equilibrium value M_0. A variant, or method B (3), consists in observing the heights $f(n)$ of the echoes at times $2\tau, 4\tau,...$, $2n\tau$ following a 90° pulse at time 0, and 180° pulses at times $\tau, 3\tau,...$, $(2n-1)\tau$. In this method $f(n) = \exp(-2n\tau/T_2)$. Apart from the saving of time, method B has another advantage to be described now.

(a) *Spin diffusion*

Fig. III, 2 shows a sequence of echoes obtained by method A, the echoes corresponding to different values of τ being reproduced on the same photograph by multiple exposure. It is quite clear that their dependence on τ is not exponential. This behaviour is due to the self diffusion of the molecules containing the nuclear spins, inside the sample, and thus across the inhomogeneous applied field (2). The basic assumption, made in Chapter II to explain the formation of the echo, is that the Larmor frequency $\omega_0 = \omega_0^0 + u$ of each spin (where ω_0^0 is the central Larmor frequency) remains the same between the first 90° pulse and the echo. There is then exact compensation between the angle $2u\tau$ through which each spin precesses (in the frame rotating at the central frequency ω_0^0) during that period and the angle $2u\tau$, through which it is set back at time τ, by the 180° pulse. It is clear that if after the first interval the spin has drifted into a region where the applied field and thus the Larmor frequency $\omega_0^0 + u$ is different, the compensation is not exact and an extra damping of the transverse magnetization will occur. This damping can be evaluated very simply by adding a diffusion term to the Bloch equation (4).

The time-dependent macroscopic nuclear magnetization is now a function $\mathbf{M}(\mathbf{r}, t)$ of space coordinates as well, obeying the equations

$$\frac{\partial \mathbf{M}(\mathbf{r}, t)}{\partial t} = \gamma \mathbf{M} \wedge \mathbf{H}(\mathbf{r}, t) - \frac{M_x \mathbf{i}' + M_y \mathbf{j}'}{T_2} - \frac{M_z - M_0}{T_1} \mathbf{k}' + D\nabla^2 \mathbf{M}. \quad (44)$$

The extra term $D\nabla^2\mathbf{M}$ represents the contribution of the diffusion to the rate of change of the magnetization considered as a macroscopic fluid. Strictly speaking $M_0 = \chi_0 H_0(\mathbf{r})$ is also space dependent and this dependence can be shown (4) to result in the addition of an extra term $-D\nabla^2(M_0)$ to (44). Its influence is entirely negligible for the small field gradients found in nuclear magnetic resonance experiments and we shall overlook it. Let $H_x = H_y = 0$, $H_z = H_0$ be the average value of the applied field over the sample. In each point of the sample the magnetic field vector $\mathbf{H}(\mathbf{r})$ will be slightly different because of the imperfections of the magnet. Small changes in the components x and y of the field will only affect the Larmor frequency in second order and can be disregarded. If the sample is sufficiently small the spatial dependence of H_z across the sample can be written to a first approximation as

$$H_z = H_0 + (\mathbf{G} \cdot \mathbf{r}),$$

where the vector $\mathbf{G}$ may be assumed constant through the sample. With these assumptions and in the absence of r.f. field, the equation of motion for the transverse precessing magnetization $m = M_x + iM_y$ follows from (44):

$$\frac{\partial m}{\partial t} = i\omega_0 m - \frac{m}{T_2} - i\gamma(\mathbf{G} \cdot \mathbf{r})m + D\nabla^2 m, \quad (45)$$

or, introducing $\psi(\mathbf{r}, t)$ through $m = \psi e^{(i\omega_0 t - t/T_2)}$,

$$\frac{\partial \psi}{\partial t} = -i\gamma(\mathbf{G} \cdot \mathbf{r})\psi + D\nabla^2 \psi. \quad (46)$$

The free precession described by the equation (46) is actually perturbed by the refocusing 180° pulses. To understand what happens, assume first that the diffusion term is absent and suppose that the sequence of pulses is as in method B. (The amplitude of the echo measured in method A is given naturally by the same formula as the first echo of method B and the effects of the diffusion on the results of method A can be computed by a specialization of the formula giving these effects for method B.)

In the frame rotating with the velocity $\omega = \omega_0^0$ a first 90° pulse is applied along an axis OY, thus bringing the magnetization along the axis OX. A sequence of 180° pulses is then applied at times $\tau, 3\tau,...,$ $(2n-1)\tau$ according to the method B above, and we suppose (a simplifying restriction to be lifted later) that the 180° pulses are applied along the axis OX (rather than OY). The effect of each pulse is to set back the phase of the precessing magnetization at each point by twice the angle $\alpha = -\gamma(\mathbf{G}\cdot\mathbf{r})\tau$ through which it has precessed in the rotating frame during the time τ. It is thus clear that at each point the magnetization will be along OX at times $2\tau, 4\tau,\ldots, 2n\tau$ and that between the times $(2n-1)\tau$ and $(2n+1)\tau$ the solution ψ of (46) in the absence of the diffusion term is

$$\psi = A \exp\{-i\gamma(\mathbf{G}\cdot\mathbf{r})(t-2n\tau)\}. \tag{47}$$

If we attempt to take into account the term $D\nabla^2\psi$ by assuming that A is actually a function of time (but not of space) $A(t)$, (46) leads for A to the following equation, valid between the times $(2n-1)\tau$ and $(2n+1)\tau$:

$$\frac{dA}{dt} = -AD\gamma^2 G^2(t-2n\tau)^2. \tag{48}$$

The equation (48) integrated between $t = (2n+1)\tau$ and $(2n-1)\tau$ gives

$$A\{(2n+1)\tau\} = A\{(2n-1)\tau\}\exp\{-\tfrac{2}{3}D\gamma^2 G^2\tau^3\},$$

whence by induction

$$A\{(2n+1)\tau\} = A(\tau)\exp\{-\tfrac{2}{3}D\gamma^2 G^2\tau^3 n\}. \tag{48'}$$

The same equation integrated between $2n\tau$ and $(2n+1)\tau$ gives

$$A(2n\tau) = A\{(2n+1)\tau\}\exp\{\tfrac{1}{3}D\gamma^2 G^2\tau^3\} = A(\tau)\exp\{-\tfrac{1}{3}D\gamma^2 G^2\tau^3(2n-1)\}. \tag{48''}$$

Making $n = 0$ in (48''), we find

$$A(0) = A(\tau)\exp\{\tfrac{1}{3}D\gamma^2 G^2\tau^3\}, \tag{48'''}$$

whence finally from (48'') and (48''')

$$A(2n\tau) = A(0)\exp\{-\tfrac{2}{3}D\gamma^2 G^2\tau^3 n\}.$$

Since at time $t = 0$ the diffusion had no time to act, $A(0) = 1$, and the total attenuation due to diffusion on the nth echo occurring at the time $t = 2n\tau$ is

$$\exp\{-\tfrac{2}{3}D\gamma^2 G^2\tau^3 n\} = \exp\{-\tfrac{1}{3}D\gamma^2 G^2\tau^2 t\}. \tag{49}$$

This is an exponential decay with a rate constant

$$\frac{1}{T_2^\dagger} = \tfrac{1}{3}D\gamma^2 G^2\tau^2, \tag{50}$$

which can in principle, if not in practice, be made arbitrarily small by reducing the intervals 2τ between consecutive 180° pulses.

The decay (49) has to be contrasted with the much faster and non-exponential decay

$$\exp\left\{-\frac{D\gamma^2 G^2 t^3}{12}\right\} \tag{51}$$

of the first echo observed at time $t = 2\tau$ for variable τ in the method A. Fig. III, 2 shows the truly exponential decay of the echoes obtained by method B. It should be remembered, however, that what is measured there is $1/T_2 + 1/T_2'$, where $1/T_2'$ given by (50) is an instrumental correction to be made as small as possible. The observation of the decay (51) of the first echo in method A becomes, if the field gradient is known, an excellent method for the measurement of the self-diffusion coefficient D, and the most accurate results on self-diffusion in water have been obtained by this method.

The main limitation on the accuracy of the method B is the impossibility of reducing τ, that is of increasing the number n of pulses beyond a certain value determined by the experimental conditions, because the inevitable imperfections in the refocusing 180° pulses have a cumulative effect when applied in large numbers. Techniques have been developed to circumvent these drawbacks but their description is outside the scope of this book.

(b) Coherent and incoherent pulses

The assumption made in the previous section that for the refocusing pulse of 180° the rotating r.f. field was along the axis OX of the rotating frame implied a definite phase relationship between the pulses actually produced by a linearly polarized r.f. field of which the rotating field is one component. Suppose, on the other hand, that for each 180° pulse the rotating field makes each time an arbitrary angle $\psi_1, \psi_2, ..., \psi_n, ...$ (incoherent pulses) with the direction OX which was that of the precessing magnetization immediately after the first 90° pulse. It is clear that the effect of the incoherence is to change the phase of the refocused magnetization at the nth echo. If at the time of formation of the nth echo, the echo $(n-1)$ has decayed completely, which implies $\tau \gg |\gamma \Delta H|^{-1}$, there will be no interference between them, and the amplitude of the signal will be unchanged. It is thus clear that the simplifying assumptions made about the phase relationships between the various pulses do not affect the conclusions about the effects of diffusion.

As an elementary illustration of the formation of spin echoes a parallel is sometimes drawn between the precessing nuclear moments and runners at a race track. These runners, starting together at time $t = 0$, will progressively spread out on the race track because of their different speeds, and if the race goes on for a sufficiently long time they will eventually be uniformly distributed all round the track. If, then, at time τ they reverse their steps and run in the opposite direction but with the same speeds, they will end up bunched together at time $t = 2\tau$. The experiment of nuclear magnetism that would parallel this situation is not, however, the one described previously but rather one where the applied d.c. field would be reversed at time $t = \tau$ from its value H_0 to an equal and opposite value $-H_0$, in a very short time causing a reversal of the Larmor precessions. The rapid reversal of fields of the order of several thousand gauss currently used in nuclear magnetism is hardly feasible, but a very similar experiment is easily performed if the d.c. field is the sum of a large and very homogeneous field H_0 and of a much smaller and inhomogeneous field H'_0 produced by an auxiliary coil and capable of damping the precessing magnetization in a short time. It is clear that the reversal of H'_0 only, is sufficient to refocus the nuclear spins and produce an echo. The point of spoiling a very homogeneous field H_0, where free precession can be observed for a long time, by superposing on it a much poorer field H'_0, where this precession can only be observed by means of echoes, may appear questionable. It will appear later that experiments very similar in principle to the one above may have their use. For brevity we shall call spin echoes of this type 'racetrack' echoes.

A better analogue for the current type of spin echoes would be ants crawling at different speeds on the edge of a pancake. A reversal of the pancake, will, if the ants keep crawling in the same direction in space, bunch them together. By analogy the usual spin echoes might be termed 'pancake echoes'.

B. Free precession

It was shown in Chapter II that for free spins the shape of the decay of the free precession was the Fourier transform of the distribution of Larmor frequencies, and this result was generalized in the present chapter for a Lorentz line shape caused by relaxation and corresponding to an exponential decay of free precession. A further extension showing quite generally that the amplitude of the free decay is the Fourier transform of the unsaturated line shape will be given in Chapter IV. Thus, and in contradistinction to spin-echo experiments, it is seldom that free precession methods will provide information not available from steady-state observation of the resonance line.

From the mechanism of formation of the echo it is evident that the shape of the echo is the same as that of two decay tails put together symmetrically, and that the more inhomogeneous the field, the shorter the decay and the narrower the echo. Fig. III, 3 shows an example of this situation.

Since the free precession signal following a 90° pulse is proportional to the d.c. nuclear magnetization that existed just before the pulse, it may be conveniently used for relatively long T_1 as a method for measuring the growth of the d.c. magnetization towards its equilibrium value M_0. The experimental procedure is the following (3). After the nuclear magnetization has reached its equilibrium value M_0 a 180° pulse gives to it the value $M_z = -M_0$. From then on the time-dependent value of M_z, resulting from the equation $dM_z/dt = (M_0 - M_z)/T_1$, is given by $M_z(t) = M_0(1 - 2e^{-t/T_1})$ and can be measured by the size of the signal following a 90° pulse applied a time t after the first 180° pulse. To obtain the curve $M_z(t)$, one must wait a time several times T_1 after each 90° pulse, before again applying a 180° pulse, and a 90° pulse a time t later. This is a method of the A type. Fig. III, 4 shows the variation of $M_z(t)$ proportional to the height of the signals following 90° pulses for various values of t. In particular, the value $M_z = 0$ is obtained for

$$t_1 = T_1 \log 2 = 0{\cdot}69 T_1.$$

(a) *Free precession in the earth's field*

There is at least one situation where, in spite of the duality expressed by the Fourier relationship between the responses of the two methods, the free precession gives results beyond those obtained by steady-state observation: usage of samples with long relaxation times T_1 in very low fields such as the earth field. Assuming that the relative inhomogeneity $\Delta H/H_0$ of the d.c. field is roughly independent of H_0, very small absolute inhomogeneities ΔH result in low fields and in particular in the earth's field $H_0 \sim 0{\cdot}5$ gauss. The observation there of extremely narrow lines, or conversely of very long free decays, is hampered by the smallness of the signal owing to the smallness of the equilibrium magnetization $M_0 = \chi_0 H_0$. The long relaxation time T_1 makes it possible, however, to polarize the sample in a large field $\mathbf{H}_p + \mathbf{H}_0$, with $H_p \gg H_0$, where it acquires the magnetization $M_p \cong \chi_0 H_p$, then, suppressing H_p in a time short compared to T_1, to observe in field H_0 the free precession of the large magnetization M_p, at the frequency $\omega_0 = -\gamma H_0$. The details of the experiment depend somewhat on the rate at which H is cut down to the value H_0. If the condition of adiabatic passage established in Chapter II, $dH/dt \ll \gamma H^2$, is satisfied continuously, the magnetization 'follows' the instantaneous field H and ends up aligned

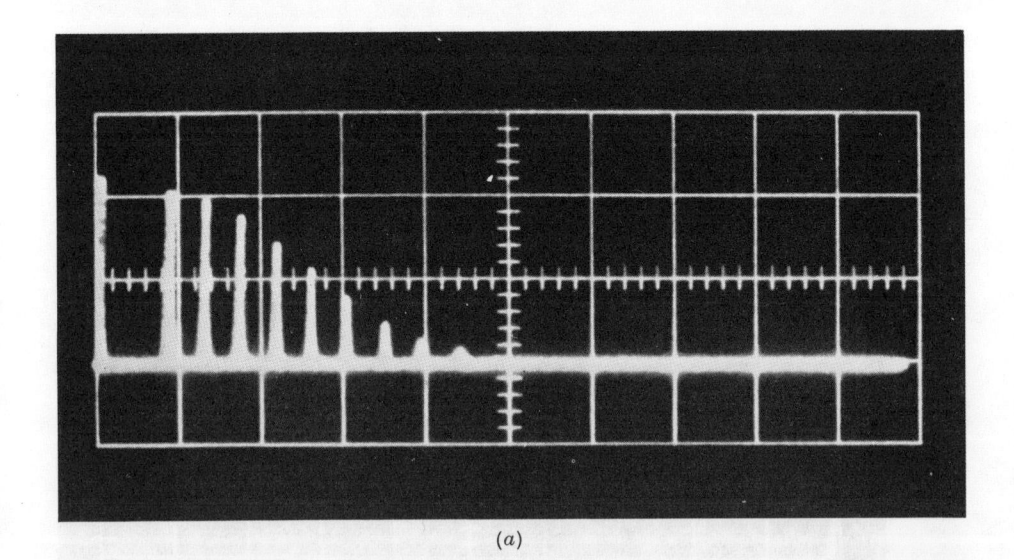

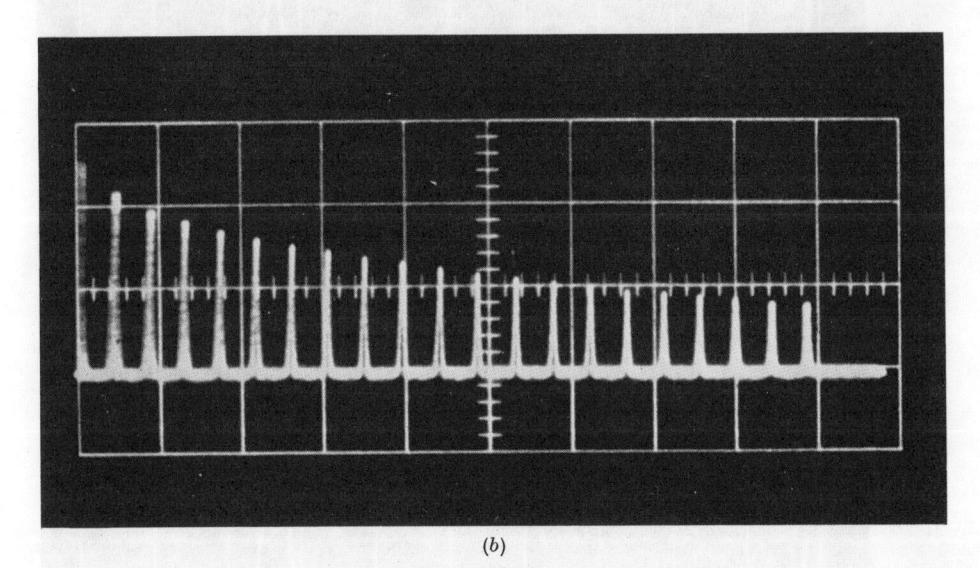

FIG. III, 2. Sequence of echoes obtained from protons in ordinary water, using respectively method A (top) and method B (bottom). The non-exponential decay of the echo envelope in method A, proportional to $\exp(-kt^3)$, is to be contrasted with its exponential decay in method B.

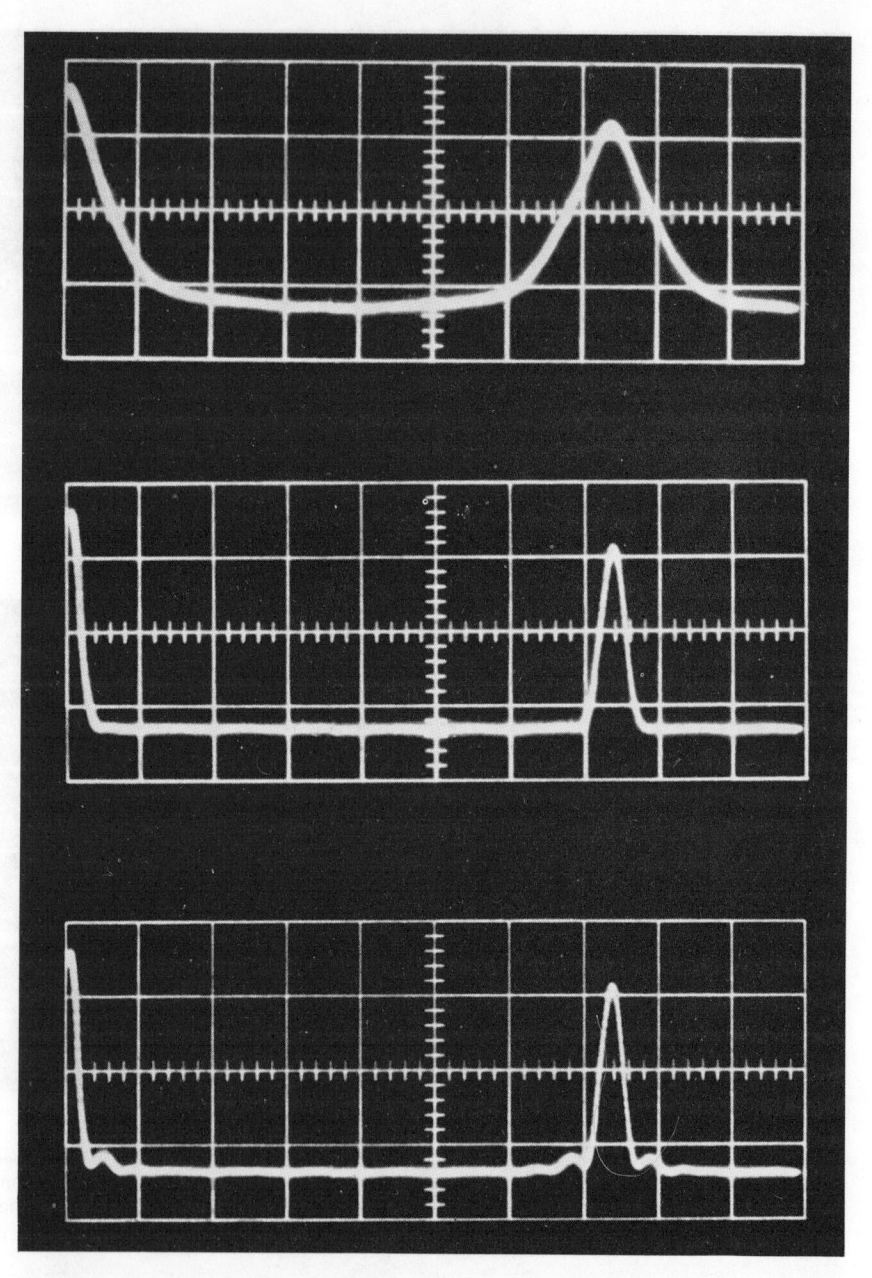

FIG. III, 3. The decay tail and echo, obtained from protons in water, corresponding to positions of the sample in the magnet with decreasing field homogeneity: at centre of the magnet (top); shifted radially by 2 cm (middle); by 5 cm (bottom).

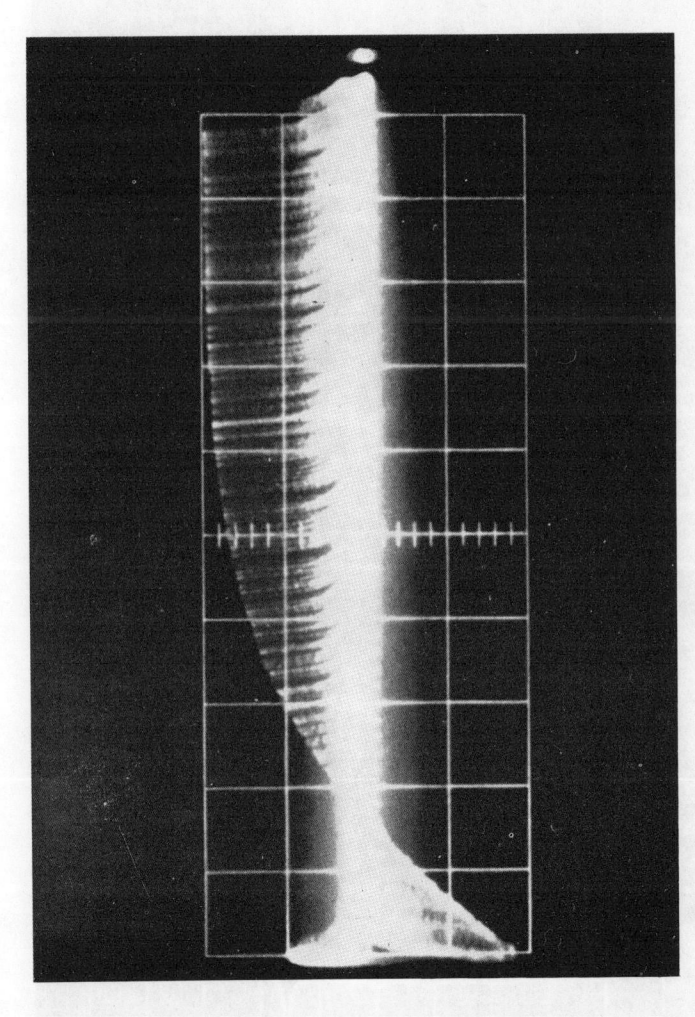

FIG. III, 4. Measurement of T_1. A 180° pulse at $t = 0$ is followed at a time t by a 90° pulse. Signals from doped water following 90° pulses, produced at variable times t, are superposed on the photograph. Their envelope is the curve $S = S_0(2 - e^{-t/T_1})$. Linear sweep: 5 millisec per division.

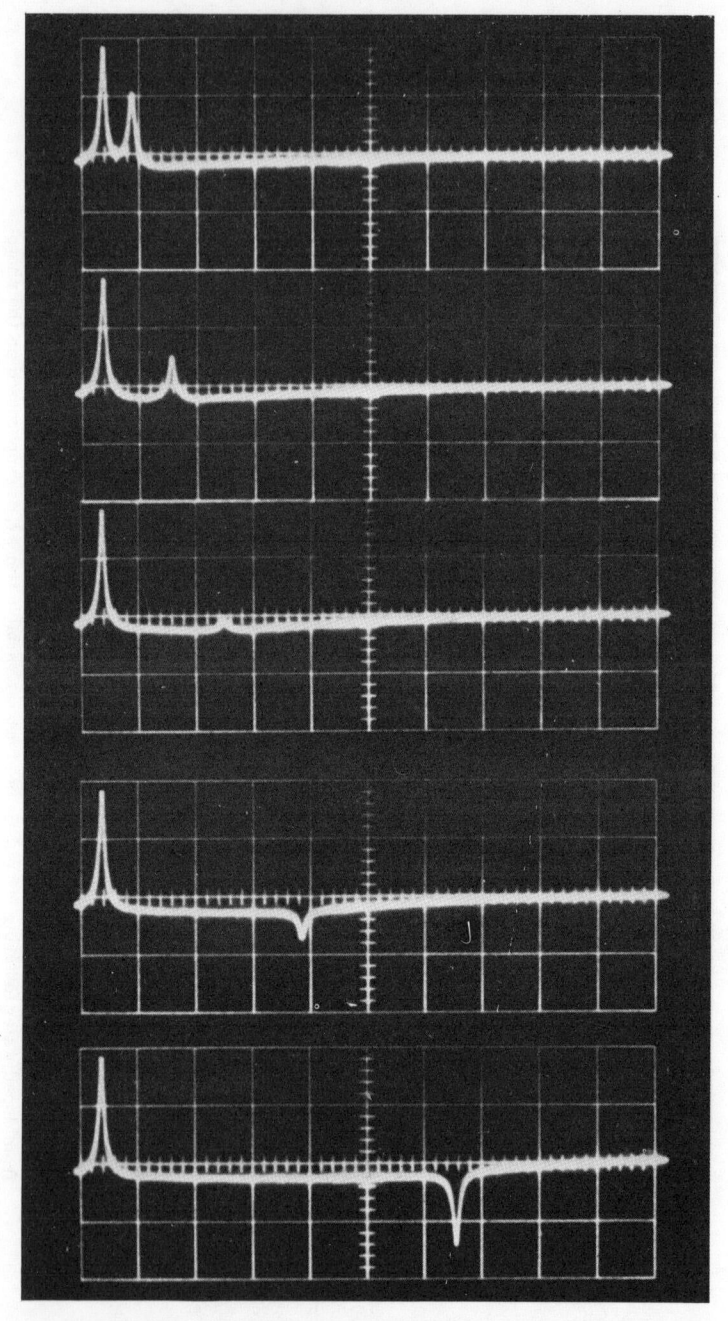

FIG. III, 5. Measurement of T_1 by rapid passage with the one-shot method. The sample is doped water and the sweep 20 millisec per division.

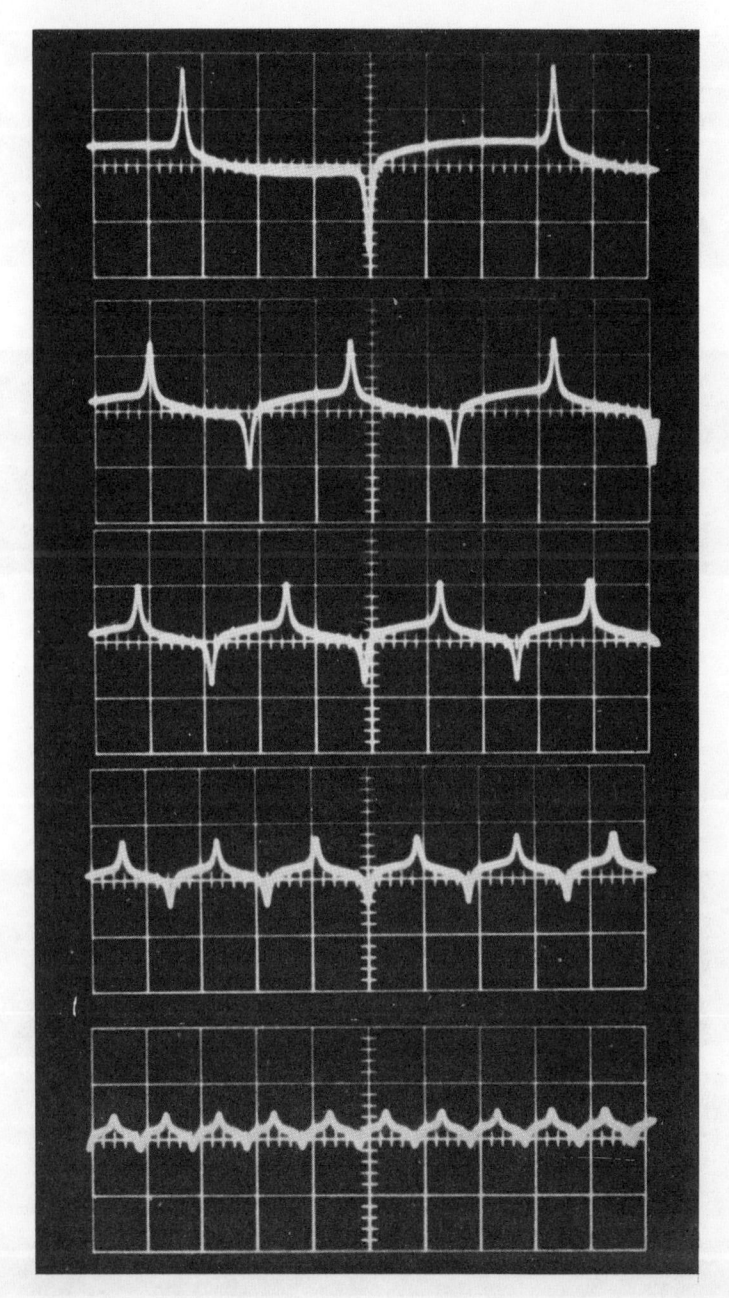

FIG. III, 6. Measurement of relaxation time in doped water by rapid passage. The sweep is 20 millisec per division. The interval τ between two consecutive passages is decreased from top to bottom and can be read on the figure. The size of the signal is proportional to $(1-e^{-\tau/T_1})/(1+e^{-\tau/T_1})$.

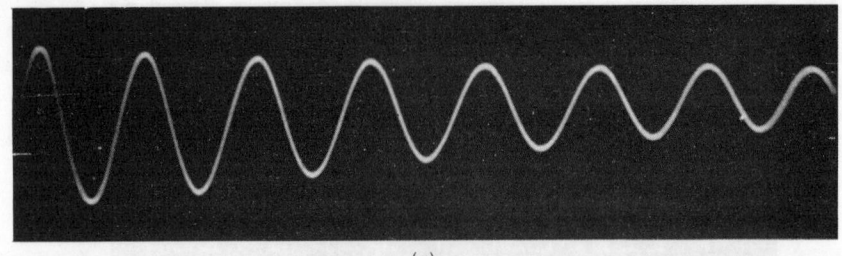

(a)

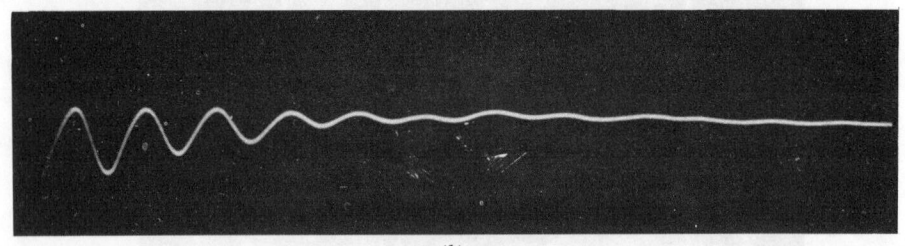

(b)

Fig. III, 7. Proton resonance in glycerine.
Sweep time = 0·01 sec; $H_1 = 0·17$ gauss.
(a) Exact resonance.
(b) H_0 displaced off resonance by 0·21 gauss.

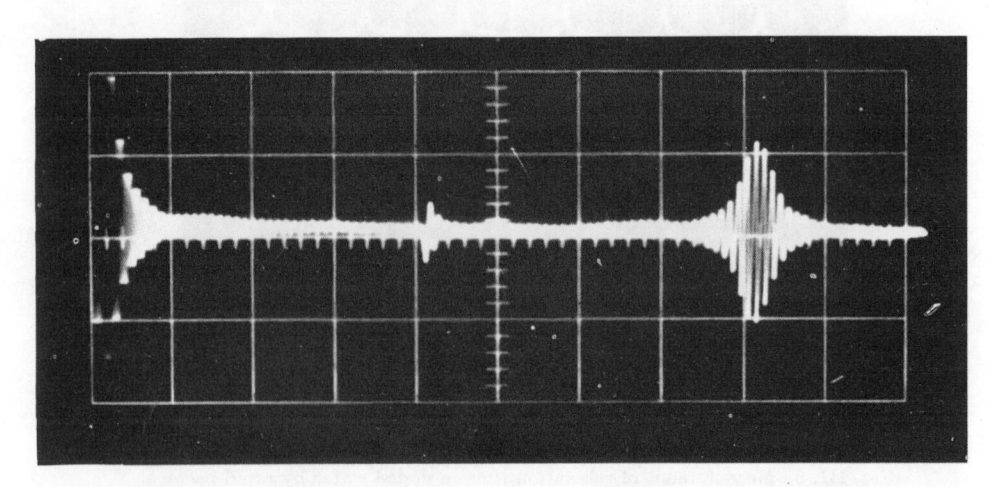

Fig. III, 8. A 'rotary echo' in doped water. The total trace is 100 millisec long.

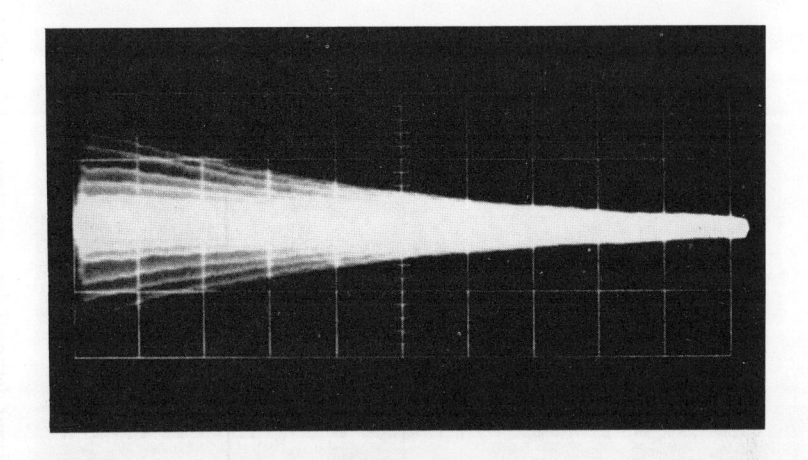

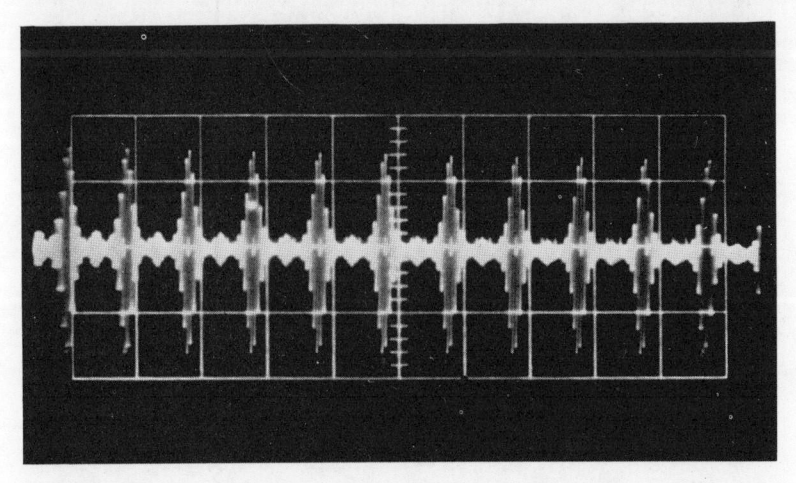

FIG. III, 9. Rotary echoes of the type B in oxygen-free benzene. *Top*: the whole pattern. The pulse rate is 50 per sec and the trace is 40 sec long. *Bottom*: part of the pattern enlarged 100 times to show the detail of the echoes.

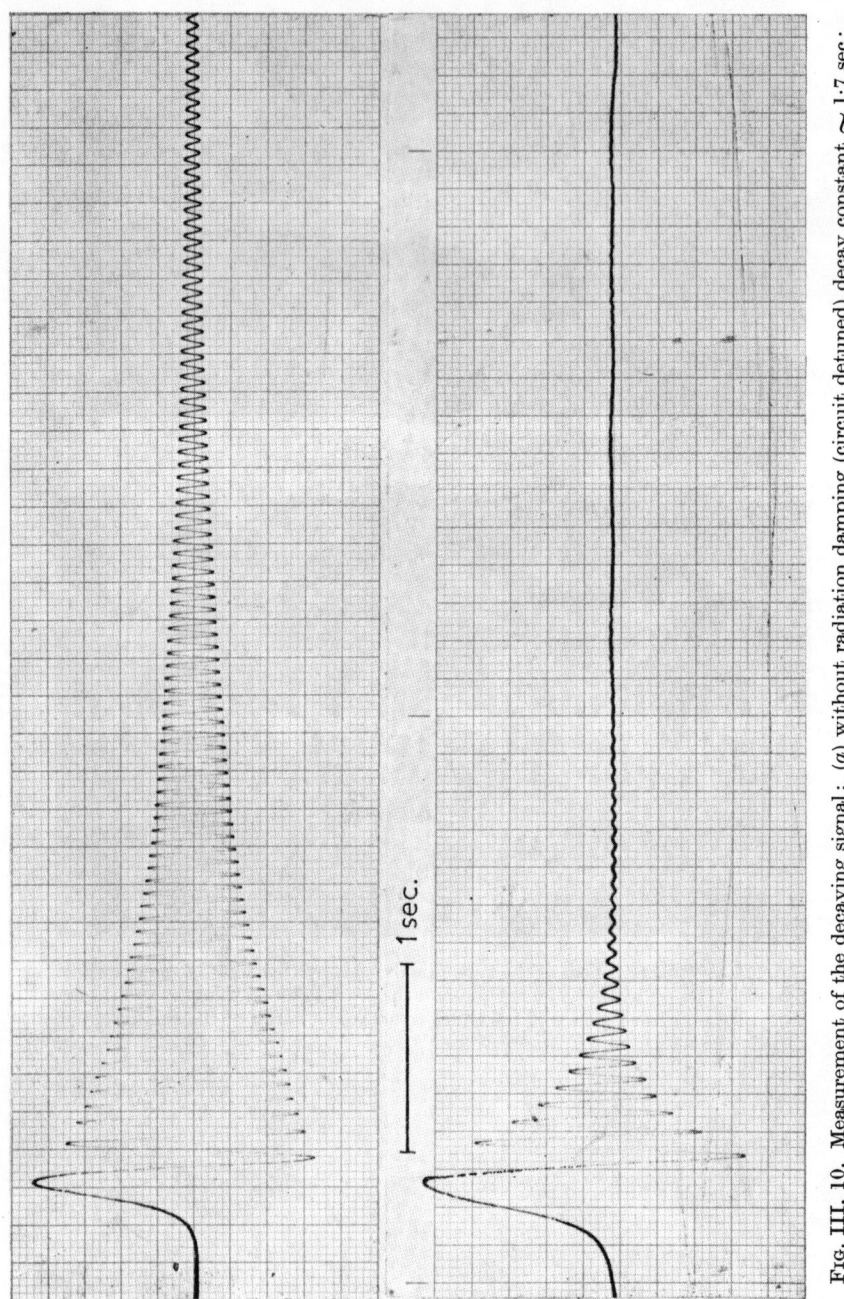

FIG. III, 10. Measurement of the decaying signal: (a) without radiation damping (circuit detuned) decay constant ~ 1.7 sec; (b) with radiation damping (circuit tuned) decay constant 0·3 sec. With the parameters used in the experiment τ given by (66') is 0·36 sec.

along the earth's field H_0. A 90° pulse is then necessary for the observation of free precession.

An elegant alternative (5) is to reduce the polarizing field so rapidly that the magnetization does not have time to follow it and ends up at right angles to the earth's field, giving a detectable precession signal. In practice the experiment is performed in two stages: first the polarizing field H_p, produced by a solenoid and perpendicular to the earth field, is reduced adiabatically (although in a time short compared with $T_1 \sim 3$ sec for water) from, say, 100 gauss down to a value H'_p of 2 or 3 gauss, then the remaining field H'_p is cut off non-adiabatically in a time short compared with $(\gamma H'_p)^{-1}$. A magnetometer for the measurement of the earth's field based on this experiment has been constructed.

C. Adiabatic passage

The principle of the adiabatic passage as a means of reversing the nuclear magnetization by sweeping, in the presence of an r.f. field H_1 of frequency ω, the d.c. field H_0 through the resonance value $H_0^* = -\omega/\gamma$, has been explained in Chapter II.

It was shown there that for free spins the condition for a complete reversal of the magnetization was

$$\frac{dH_0}{dt} \ll \gamma H_1^2. \qquad (52)$$

It was also shown there that a spread $|\Delta\omega_0| = |\gamma \Delta H|$ in the Larmor frequencies, caused by an inhomogeneity ΔH of the applied field, did not affect the reversal of the magnetization. If $\Delta H \gg H_1$, the maximum value of the transverse magnetization during the passage is reduced in the ratio $H_1/\Delta H$, because the various spins of the sample do not pass through resonance at the same time. On the other hand, if $H_1 \gg \Delta H$, all the spins are practically at resonance at the same time and the inhomogeneity of the field has no effect on the maximum transverse magnetization which is then equal to M_0. If we take as the time origin the passage through resonance, the transverse magnetization will be at time t (t positive after the passage, and negative before):

$$M_{tr} = \pm M_0 \sin\theta = \frac{\pm M_0 H_1}{[H_1^2 + (H_0 - H_0^*)^2]^{\frac{1}{2}}} = \frac{\pm M_0}{[1 + (t(\dot{H}_0/H_1))^2]^{\frac{1}{2}}}. \qquad (53)$$

The sign in (53) is positive or negative depending on whether at resonance the transverse magnetization is parallel or antiparallel to H_1 (in the rotating frame). This in turn depends on whether the rapid passage was started from above resonance ($H_0 > H_0^*$) or from below

($H_0 < H_0^*$). Since the principle of the adiabatic passage is that the angle between **M** and the effective field $\mathbf{H}_{\text{eff}} = \mathbf{H}_0 - \mathbf{H}_0^* + \mathbf{H}_1$ remains constant, this angle is practically zero if one starts from far above resonance and π if one starts from far below, whence the double sign in (53). (It should be stressed that this double sign has nothing to do with the double sign in the definition (16') of the absorption component $v = \pm \tilde{M}_y$.) The fact that the spins are not free is easily taken into account if the nuclear spin system is known to verify the Bloch equations. To the condition (52) it is sufficient to add that the effects of relaxation must be negligible during the time of passage τ through resonance

$$\tau \cong H_1 \bigg/ \left|\frac{dH_0}{dt}\right| \ll T_1, T_2 \tag{54}$$

or, since in practice $T_2 \ll T_1$, rewriting (52) and (54) together,

$$\frac{1}{T_2} \ll \frac{1}{H_1}\left|\frac{dH_0}{dt}\right| \ll |\gamma H_1|. \tag{55}$$

The inequality (54) has caused the adiabatic passage to be called in the literature adiabatic fast passage or simply fast passage. In solids, where Bloch equations are known to be invalid, a constant T_2 is sometimes introduced as a measure of the inverse line width, or of the time required for the establishment of internal equilibrium among the spins. It is important to realize that the extension of (55) to solids is completely unwarranted and, in view of the very short T_2 encountered there, much too stringent. The conditions replacing (54) in solids will be discussed in Chapter XII, and we shall be content to state here that the adiabatic fast passage has been observed in solids under conditions compatible with the much weaker requirement

$$\frac{1}{T_1} \ll \frac{1}{H_1}\left|\frac{dH_0}{dt}\right| \ll |\gamma H_1| \simeq \frac{1}{T_2}. \tag{56}$$

APPLICATIONS OF THE FAST-PASSAGE METHODS

(a) *Measurement of the relaxation time T_1*

For long relaxation times it is sufficient to measure the ratio of the signals corresponding to two consecutive fast passages separated by a time t. The signal observed during the first passage corresponds to a transverse magnetization which under the right conditions is equal to M_0. At the end of this passage $M_z = -M_0$ and t seconds later it is $M_z = M_0(1 - 2e^{-t/T_1})$. The signal of the second fast passage going back through the resonance is proportional to $-M_z$. The negative sign is due to the fact that the first and the second signals are observed starting from opposite sides of the resonance. In principle this method is similar to that using free precession signals following a sequence of a 180° and a 90° pulse. If the time t separating two consecutive fast passages is short compared with T_1 the two signals have the same sign, and opposite signs if $t \gg T_1$ (Fig. III, 5). Similar results are obtained if the steady state, resulting from a periodic sweep through the resonance, is observed. For a symmetrical sweep, where a time τ separates two fast passages in opposite directions, two equal and opposite signals are observed. They correspond to a steady state value M_1 of the nuclear magnetization M_z given by

$$M_1 = M_0 \frac{1-\exp(-\tau/T_1)}{1+\exp(-\tau/T_1)}. \tag{57}$$

(57) is easily obtained by writing the steady-state condition: just before a passage the magnetization is M_1, just after a passage it is $-M_1$, and τ seconds later it is given by

$$M_z(\tau) - M_0 = (-M_1 - M_0)\exp(-\tau/T_1).$$

Writing that $M_z(\tau) = M_1$, (57) follows. For $\tau \gg T_1$, $M_1 \cong M_0$ as resulted from the qualitative discussion above. For $\tau \ll T_1$, M_1 is very small.

Fig. III, 6 illustrates this behaviour. The same method can be applied to solids and has been successfully used in metals (6).

(b) *Measurement of T_2 in liquids*

The inhomogeneity of the applied fields H_0 that in liquids prevents the measurement of T_2 through the observation of the line width or of the free decay, and is circumvented by the spin-echo techniques, can also be overcome by the application of a resonant r.f. field H_1 of amplitude much larger than the total inhomogeneity ΔH of the applied field. Suppose that by some means the equilibrium nuclear magnetization M_0 is brought in the frame rotating at the frequency $\omega = \omega_0^0$ (average Larmor frequency of the imperfectly homogeneous field) along $H_{\text{eff}} = H_1$. If $|\gamma H_1| \gg 1/T_1, 1/T_2$ the steady-state solution (15) of the Bloch equations gives approximately $\tilde{M}_x(\infty) = \tilde{M}_y(\infty) = M_z(\infty) = 0$. Since the initial conditions are

$$\tilde{M}_x = M_0, \qquad \tilde{M}_y = M_z = 0, \tag{58}$$

it is quite reasonable to expect and easy to show from the equations (14) that the transient solution will be

$$\tilde{M}_x \cong M_0 e^{-t/T_2}, \qquad \tilde{M}_y \cong M_z = 0,$$

and the decay of the transverse magnetization will give T_2. Since the magnetization is constantly aligned along H_1 the influence of the inhomogeneity $\delta \mathbf{H}(\mathbf{r})$ of the applied field will be negligible if $H_1 \gg |\delta \mathbf{H}(\mathbf{r})|$. The magnetization can be aligned along H_1 by means of a 90° pulse which brings M_0 into the XY plane but at right angles to H_1, followed immediately by a phase shift of 90° of the rotating field that brings it along $\tilde{M}_x$ (7).

Another way of bringing M along H_1 is a fast passage. Starting from a value cf H_0, say, far above resonance, the passage must be stopped at resonance rather than far below, as in the measurement of T_1. The main difficulty in this technique is to stop at exact resonance or to stay there because of possible drifts of H_0. Failure to achieve exact resonance results according to (15) in a large (and unknown) change in $\tilde{M}_x(\infty) \cong \{(H_0 - H_0^*)/H_1\} M_0(T_2/T_1)$ and in an appreciable uncertainty in the measured value of T_2.

(c) *Other applications of adiabatic fast passage*

The adiabatic fast passage has some of the features of the 180° pulse, since it reverses the nuclear magnetization, and also of the 90° pulse, since it produces a transverse magnetization equal to M_0 and detectable by the flux it induces in a coil. This fact makes it a method of detection of the nuclear resonance, which under favourable circumstances (long T_1) can make it superior to the steady-state or slow-passage method. Indeed, fast passage was the method used by the Stanford group in the first observation of proton resonance in water. This point will be discussed further in Section IV on detection methods.

The fast passage is also the ideal method for measuring the nuclear magnetization 'on the fly' when M_z does not have the equilibrium value $\chi_0 H_0$; for instance, polarizing the sample in a field very different (much smaller or much higher) than the one at which the resonance occurs. Examples of this method will be found in Chapter V.

D. The method of transient nutation

In this method (8) also the effect of the inhomogeneity ΔH of the applied field H_0 on the measurement of T_2 is overcome by the use of a strong r.f. field $H_1 \gg \Delta H$, but in a manner which is very different from that used in fast passage. The r.f. field H_1 is introduced suddenly (non-adiabatically) at time $t = 0$, and the ensuing motion of the nuclear magnetization is observed in the presence of the r.f. field, differing in this respect from the methods using free precession and spin echoes. An interesting analogy can be drawn between this experiment and the free precession in the earth's field (5), described in Section III B. In both experiments the nuclear magnetization precesses around a weak and thus very homogeneous field (in absolute value) and the precession can be observed for many cycles. In one of the experiments the weak field is the earth's field and the precession takes place in the laboratory frame; in the other it takes place in the rotating frame, around the effective field, which at resonance is H_1. In both experiments a large nuclear magnetization is acquired by polarizing the sample in a large d.c. field which is then suppressed by a sudden change in the experimental conditions. In the first experiment, this change is a rapid cut off of the current producing the polarizing field; in the second, this field is cancelled in the rotating frame by the fictitious field $\omega/\gamma = -H_0$, introduced suddenly by the pulse giving rise to the rotating field H_1. Since, however, the applied field $H_0 = -\omega_0/\gamma$ is not perfectly homogeneous, this cancellation is not perfect and the effective field around which the precessing takes place is $\mathbf{H}_{\text{eff}}(\mathbf{r}) = \delta H_0(\mathbf{r})\mathbf{k} + H_1 \mathbf{i}$ at resonance and $\{H_0 - H_0^* + \delta H_0(\mathbf{r})\}\mathbf{k} + H_1 \mathbf{i}$ off resonance. As explained previously, inhomogeneities of H_0 at right angles to it have a negligible effect. If $H_1 \gg \Delta H$, the effect of ΔH in damping the precession is of second order and thus negligible at exact resonance, but important off resonance. Fig. III, 7 shows the precession of the magnetization around the effective field (a) at exact resonance, (b) off resonance for $|H_0 - H_0^*| \sim H_1$. The damping due to the inhomogeneity is much more severe off resonance, and the condition of minimum damping is actually a sensitive test for exact resonance in this experiment.

Immediately before the application of the r.f. field the initial conditions for the magnetization are

$$\tilde{M}_x = \tilde{M}_y = 0, \qquad M_z = M_0, \tag{58'}$$

and its behaviour after the r.f. field is applied is obtained from the transient solution of the Bloch equations (14) with the initial conditions (58'). The whole difference between the method outlined in the last section and the present one is contained in the difference between the initial conditions (58) and (58'). The calculations are straightforward but tedious: the detail is given in (8). In practice the only case of interest is one where $|\gamma H_1| \gg 1/T_2, 1/T_1$, which can be seen as follows: unless $1/T_2 \ll \gamma \Delta H$ the inhomogeneity of the field does not hamper the measurement of the real line width $1/T_2$ in a steady-state experiment and transient methods are not necessary; unless $H_1 \gg \Delta H$ the effect of inhomogeneity is not eliminated by the r.f. field. The calculations are greatly simplified by the assumption $|\gamma H_1| \gg 1/T_2$ and the result for the absorption component $\tilde{M}_y = \pm v$ is

$$\tilde{M}_y = -M\sin\theta\exp\left\{-\left[\frac{1}{T_2} - \frac{1}{2}\left(\frac{1}{T_2} - \frac{1}{T_1}\right)\sin^2\theta\right]t\right\}\times\sin(\Omega t), \qquad (59)$$

where
$$\tan\theta = \frac{H_1}{H_0 - H_0^*}, \qquad \Omega = \frac{\omega_1}{\sin\theta} = \frac{-\gamma H_1}{\sin\theta}. \qquad (59')$$

At resonance (59) becomes

$$\tilde{M}_y \simeq -M_0\exp\left\{-\tfrac{1}{2}t\left(\frac{1}{T_1} + \frac{1}{T_2}\right)\right\}\sin\omega_1 t. \qquad (60)$$

The motion is a damped precession around H_1 with a Larmor frequency $\omega_1 = -\gamma H_1$ and a damping constant $\tfrac{1}{2}(1/T_1 + 1/T_2)$.

Formulae (59) and (60) are valid for a perfectly homogeneous field H_0. Actually, to obtain the response at resonance one should write $H_0 = H_0^* + \delta H$ in (59') and integrate (59) over the shape function $g(\delta H)$ of the field inhomogeneity. According to the qualitative arguments, above, if $H_1 \gg \Delta H \simeq 1/g(0)$ the result should differ from (60) only in second order in ΔH. Assuming, for example, a gaussian shape $g(\delta H)$ with full width at half intensity $\Delta H_0 \ll H_1$, it is found (8) that

$$\tilde{M}_y \simeq \frac{-M_0}{[1+(t/T_H)^2]^{\frac{1}{4}}}\exp\left\{-\frac{1}{2}\left(\frac{1}{T_1}+\frac{1}{T_2}\right)t\right\}\sin(\gamma H_1 t + \phi), \qquad (61)$$

where ϕ is a slowly varying function of time, and

$$T_H = \frac{1}{|\gamma|\Delta H_0}\times\frac{5\cdot 6 H_1}{\Delta H_0}. \qquad (61')$$

As predicted, the inhomogeneity damping due to ΔH_0 is decreased in a ratio $\Delta H_0/H_1$ and can be made arbitrarily small if H_1 is large enough. Since an independent measurement of T_1 is not hampered by the inhomogeneity of H_0, from the knowledge of $\tfrac{1}{2}(1/T_1+1/T_2)$, T_2 can be obtained. The limitation in this direction is the inhomogeneity of H_1 which in practice increases much faster than H_1 itself, for, in order to produce a larger H_1, smaller r.f. coils are required if the r.f. power is to be kept within reasonable limits.

This difficulty is overcome in an elegant manner in the 'rotary echoes' method (9). In this method the rotating r.f. field H_1 is reversed at times $\tau, 3\tau,\ldots, (2n-1)\tau$, and echoes (of the race-track type) are observed at times $2\tau, 4\tau,\ldots, 2n\tau$. The inhomogeneities of H_0 are eliminated by a strong H_1, and those of H_1 by the echo method. Figs. III, 8 and III, 9 show rotary echoes in water where the relaxation time has been shortened by dissolving paramagnetic ions (doped water) and in oxygen-free benzene, respectively.

IV. Detection Methods
A. General

A thorough discussion of the experimental techniques of nuclear magnetism would probably require a book of a size comparable to the present volume. The only purpose of the present section is thus to give some very simple ideas about the possibilities and the limitations of electromagnetic detection of nuclear magnetic signals. The reader interested in actually building or operating experimental equipment will have to look elsewhere for information.

First a reminder of some very simple features of r.f. coils is in order. Consider a coil of n turns of diameter d, total turn area $A = \frac{1}{4}n\pi d^2$, inductance L, and volume V_c. Assume for simplicity that the field H produced by a current i circulating in the coil is uniform inside it and vanishes outside, which is correct for a long solenoid and not very wrong for other shapes.

The magnetic flux across the coil is $\Phi = HA = Li$. The magnetic energy stored in the coil is $\frac{1}{2}Li^2 = V_c H^2/8\pi$, from which the approximate relation is obtained:
$$L = 4\pi A^2/V_c. \tag{62}$$

If the current i has the time dependence $i = I\cos\omega t$, the quality factor Q of the coil is defined by $P = \omega W/Q$, where $P = \frac{1}{2}rI^2$ is the power dissipated in the resistance r of the coil and $W = \frac{1}{2}LI^2$ is the maximum energy stored in the coil. From this definition we get $Q = L\omega/r$. $Q = 100$ is a reasonable order of magnitude for an r.f. coil.

If the coil is tuned in parallel with a condenser C at a frequency ω such that $L\omega^2 C = 1$, it is easily seen that an e.m.f. $\mathscr{E}$ of that frequency induced in the coil will result in a voltage $v = Q\mathscr{E}$ across its terminals. Finally, to calculate the parallel impedance of such a circuit, the series resistance $r = L\omega/Q$ can with good accuracy be replaced by a shunt resistance $R = QL\omega$.

To get back to nuclear magnetism, consider a test-tube full of water and a coil wrapped around it. Assume for the coil the following parameters: $d = 1$ cm, $n = 10$, $Q = 100$. The coil is resonated at 42·6 Mc/sec, which is the Larmor frequency $\nu_p = |\omega_p|/2\pi = |\gamma_p|H_0/2\pi$ of the proton in a field of 10,000 gauss where the sample is supposed to be placed; assume that an r.f. pulse of angle θ has created a transverse magnetization $M = M_0 \sin\theta = \chi_0 H_0 \sin\theta$, and consider the problem of detecting its free precession.

First, one may remark that there is a difference in energy per c.c. equal to $M_0 H_0(1-\cos\theta) = \chi_0 H_0^2(1-\cos\theta)$, that is, for $\theta = \frac{1}{2}\pi$, of the order of $3\times 10^{-10}\times 10^8$ erg $\simeq 1\cdot 8\times 10^{10}$ eV between the state created by the r.f. pulse and the state where the spins are in thermal equilibrium. The number 3×10^{-10} is the susceptibility of the protons in water at room temperature as given by (1).

This energy available from the 7×10^{22} protons of one c.c. of water is equal to the energy of about 10^4 γ-rays emitted by the nuclei of radioactive Co^{60}. This helps us to understand why electromagnetic methods of detection require so much larger numbers of nuclear spins than the 'trigger' detection methods, outlined in Chapter I.

Secondly, compute the voltage $v = Q\mathscr{E}$ available at the terminals of the coil, where $\mathscr{E}$, the amplitude of the e.m.f. induced in the coil by the precessing magnetization, is given by the Faraday law:

$$\mathscr{E} = 4\pi\omega_0 MA, \qquad (63)$$

where $A = \tfrac{1}{4}\pi d^2 n$ is the total coil area crossed by the magnetic flux.

$$\mathscr{E}_{\text{volts}} = 10^{-8}\,4\pi\omega_0 M\tfrac{1}{4}\pi d^2 n = 10^{-8}\pi^2\gamma_p H_0^2 \chi_0 d^2 n \sin\theta, \qquad (63')$$

whence, for $\theta = \tfrac{1}{2}\pi$, $\qquad v = Q\mathscr{E}$,

$$v_{\text{volts}} = 10^{-8}\times 2\pi^3 \nu_p H_0 \chi_0 d^2 n Q$$
$$= 10^{-8}\times 2\pi^3 \times 42\cdot 6\times 10^6 \times 10^4 \times 3\times 10^{-10}\times 10\times 100$$
$$\simeq 0\cdot 08 \text{ volt}, \qquad (63'')$$

which is certainly a voltage of macroscopic magnitude.

Finally one should inquire about the duration of the transient voltage (63). A first limitation is the so-called inhomogeneity damping, that is, the disappearance of the transverse magnetization because of destructive interference between contributions from various parts of the sample with different Larmor frequencies. Its time constant is of the order of $(\gamma_p \Delta H)^{-1}$ and in an exceedingly good magnet where the inhomogeneity over the sample would be smaller than, say, 10^{-4} gauss (that is, in relative magnitude smaller than 10^{-8}), $|\gamma_p \Delta H|^{-1} \simeq 0\cdot 4$ sec. This kind of decay is not irreversible, however, since, as explained previously, in the absence of other causes of damping the same transverse magnetization and thus the same voltage can be recovered as many times as desired by refocusing 180° pulses. The name of inhomogeneity damping is thus somewhat improper. A second damping, real, because irreversible, is the decay, described by the transverse relaxation time T_2, and due to the perturbations that the nuclear spins experience from their surroundings on the microscopic scale. This time is of the order of 3 sec in water.

Finally, a third limitation is the so-called (also improperly) radiation damping. It is clear that the current induced in the coil by the precessing magnetization dissipates some power in the form of Joule heat. This power can only be provided at the expense of the nuclear magnetic energy $-\mathbf{M}\cdot\mathbf{H}$ of the sample and must necessarily result in a tilting of the vector magnetization towards its equilibrium position M_0, parallel to the applied field, where the nuclear magnetic energy is lowest. Although this process results eventually in the vanishing of the transverse magnetization, it hardly deserves the name of damping, since the length of the magnetization vector is unchanged. Furthermore, if one starts with a pulse of angle $\theta > 90°$, the transverse magnetization returning from its value $M_0 \sin \theta$ to the final value zero will pass through a maximum value M_0, in the plane perpendicular to H_0. The rate of change of M_z is easily computed. The amplitude $2H_1$ of the linearly polarized magnetic field originating from the current circulating in the coil, of amplitude I, is, by the definition of the inductance L, given by $2H_1 A = LI$.

If the circuit is tuned,

$$2H_1 = LI/A = \frac{L\mathscr{E}}{rA} = \frac{L}{rA} 4\pi\omega M A = 4\pi Q M = 4\pi Q M_0 \sin \theta. \quad (64)$$

Choose the z-axis along the applied field H_0 and the x-axis along the axis of the r.f. field. The rate of change of the component M_z is given by

$$\frac{dM_z}{dt} = -\gamma M_y H_x \quad (65)$$

or, since $M_x = M \sin \omega t$, $M_y = M \cos \omega t$, $H_x = -2H_1 \cos \omega t$,

$$M_z = M_0 \cos \theta, \quad M = M_0 \sin \theta,$$

$$\frac{d}{dt}(M_0 \cos \theta) = 4\pi Q \gamma M_0^2 \sin^2\theta \cos^2 \omega t. \quad (65')$$

Taking an average over a Larmor period,

$$\frac{d\theta}{dt} = -2\pi Q \gamma M_0 \sin \theta. \quad (66)$$

which integrates to

$$\tan \tfrac{1}{2}\theta = (\tan \tfrac{1}{2}\theta)_0 \exp(-2\pi Q \gamma M_0 t).$$

The time constant

$$\tau = (2\pi Q \gamma M_0)^{-1} = (2\pi Q \chi_0 \omega_0)^{-1} = (4\pi^2 Q \chi_0 \nu_0)^{-1} \quad (66')$$

is with our values of the parameters $\simeq 0.02$ sec, which turns out to be shorter than the other two time constants, $|\gamma_p \Delta H|^{-1}$ and T_2, and should suppress the free precession rather sharply. Fig. III, 10 (**10**) shows the free precession tail (*a*) with the resonating circuit untuned, which increases τ and reduces the radiation damping by a factor Q, (*b*) with the same circuit tuned at $\omega = \omega_0$. It is easy to show by a direct calculation or to deduce from the general Fourier relationship between free precession and steady-state experiments, that in the latter the attenuation of the transverse magnetization due to radiation damping manifests itself through an extra broadening of the resonance, of the order of $\delta H \sim 1/|\gamma\tau|$ in gauss.

In practice, the magnets used in nuclear magnetism, unless specially built for high resolution, are more likely to produce over the sample field inhomogeneities of the order of 10^{-2} rather than 10^{-4} gauss, thus preventing the observation of the radiation damping because of a stronger inhomogeneity damping. It might be thought that the radiation damping could still be observed then in the envelope of the echoes obtained by a sequence of a large number of 180° pulses. Such a view is actually mistaken for, if in a single echo, because of radiation damping, the magnetization swings towards M_0 by an angle ϵ, the following 180° pulse will swing it in the opposite direction by the same angle and no cumulative effect on the echo envelope will occur.

The foregoing might give the impression that nuclear signals correspond to voltages of many millivolts and that the radiation damping may be a severe broadening mechanism. Neither is generally true. According to (63′) the voltage is proportional to $\gamma\chi_0$, that is, by the relation (1) to $\gamma^3 I(I+1)N$. This last quantity, because of smaller gyromagnetic ratio, and smaller nuclear densities N, may be for many nuclei (especially for rare isotopes) smaller than the value calculated for the protons of water by several orders of magnitude. Furthermore, much greater line widths, in particular in solids, cause a much faster damping of the precession. The time available for the observation of the signal is reduced accordingly, which, as will appear shortly, leads to poor signal-to-noise ratios. Steady-state or continuous wave (C.W.) methods, where the response of the nuclear-spin system to a continuously applied r.f. field can be observed for a long time, are generally used to improve these ratios. Finally, it has been assumed that the sample filled the inside of the coil entirely. It is easy to see that if the geometry is such that the r.f. field H_1 is approximately uniform inside the coil the voltages (63) and (63′) are reduced by a filling factor

$$\eta \cong \frac{\text{volume of sample}}{\text{volume of coil}}$$

Ch. III MACROSCOPIC ASPECTS OF NUCLEAR MAGNETISM 75

which may be as small as 1/10 or less. It is possible to take into account the lack of uniformity of H_1 inside the coil by a more general definition of the filling factor η, a complication disregarded here.

B. Detection of steady-state nuclear signals

(a) Q-meter detection

A current $i = I\cos\omega t = \mathrm{re}(Ie^{j\omega t})$ circulating in a cylindrical coil of volume V_c and inductance L produces an r.f. field $2H_1\cos\omega t$ linearly polarized along its axis. Departing slightly from notations used elsewhere in this book, we shall assume in this section that $\omega = 2\pi\nu$ is a positive quantity. The angular precession of the nuclear magnetization around H_0 forced by the r.f. field will then be represented by $\epsilon\omega = \pm\omega$, the sign being $+$ or $-$ depending on whether $\omega_0 = -\gamma H_0$ is positive or negative. The flux of the r.f. field across the coil is $Li = L\,\mathrm{re}(Ie^{j\omega t})$. The nuclear magnetization of a sample of volume $V_s = \eta V_c$ placed inside the coil, has a component along the r.f. field, equal to $M_x = \mathrm{re}(2\chi H_1 e^{j\omega t})$ inside the sample and zero outside. The complex r.f. susceptibility $\chi = \chi'-j\chi''$ is vanishingly small except at resonance. The flux of the induction vector $\mathbf{B} = \mathbf{H}+4\pi\mathbf{M}$ across the coil is clearly $\mathrm{re}\{L(1+4\pi\eta\chi)Ie^{j\omega t}\}$, which can be expressed by saying that the inductance of the coil, in the presence of nuclear magnetization, takes the complex value $L(1+4\pi\eta\chi)$.

A tuned circuit with a coil containing the sample, supplied from a constant-current generator, will develop across its terminals a voltage proportional to its parallel impedance equal to

$$Z = \left[\frac{1}{r+jL\omega(1+4\pi\chi\eta)}+jC\omega\right]^{-1}, \qquad (67)$$

or, if the circuit is tuned at the frequency ω, so that $LC\omega^2 = 1$,

$$Z = R\left[\frac{1+j\,4\pi Q\eta\chi}{1+4\pi\eta\chi-j/Q}\right]^{-1}, \qquad (68)$$

where $R = L^2\omega^2/r$ is the shunt resistance of the circuit. The r.f. susceptibility χ, although larger than the d.c. susceptibility χ_0 by the ratio $\omega/\Delta\omega$, will still in general be a small number and except for very strong signals, and with the assumption $Q \gg 1$, (68) can be rewritten

$$Z \cong R[1-j\,4\pi\eta Q\chi] = R[1-4\pi\eta Q\chi''-j\,4\pi Q\eta\chi']. \qquad (69)$$

When the d.c. field H_0 is swept through the resonance there is a relative change in Z and thus also a relative change $\delta v/v$ in the output voltage $\delta v/v = -j\,4\pi\eta Q\chi$. The relative change in the amplitude of Z, proportional to the change in the detected signal, is $|1-jQ4\pi\eta\chi|-1$, which in first order is $\simeq -4\pi\eta Q\chi''$. The dispersion is not observed in this device which may be inconvenient when searching for unknown resonances as explained in Section II A.

(b) Bridge and crossed coils methods

The previous device also has another drawback. The nuclear signal δv appears as a very small modulation on the voltage v_0 that exists in its absence. Amplifying δv to a value $A\,\delta v$ acceptable for detection would lead to forbiddingly high values for Av_0. It is then preferable to compensate v_0 by adding to it, before r.f. amplification, a voltage v_1 almost equal and opposite. The voltage to be amplified and detected is thus
$$\mathscr{V} = v_0 - v_1 - v_0\,4\pi jQ\eta\chi.$$

In practice it is preferable to avoid complete compensation and to keep $|v_0-v_1| \gg \delta v$. It is possible to give to the difference v_0-v_1 any phase with respect to v_0. Writing

$$v_0 - v_1 = \alpha v_0 e^{j\phi}, \quad \text{where } \alpha \text{ is real,}$$

we get

$$\begin{aligned}\mathscr{V} &= \alpha v_0 e^{j\phi}\left\{1 - \frac{4\pi jQ\eta\chi}{\alpha}e^{-j\phi}\right\} \\ &= \alpha v_0 e^{j\phi}\left\{1 - \frac{4\pi Q\eta}{\alpha}(\chi''\cos\phi + \chi'\sin\phi) - j\frac{4\pi Q\eta}{\alpha}(\chi'\cos\phi - \chi''\sin\phi)\right\}.\end{aligned}$$
(70)

For $\phi = 0$, only χ'' contributes in first order to the amplitude of $\mathscr{V}$, and only χ' for $\phi = \tfrac{1}{2}\pi$. For intermediate values of ϕ this contribution is a mixture of χ' and χ''. Fig. III, 11 shows the shape of the detected signals proportional to the amplitude of $\mathscr{V}$, displayed on an oscilloscope: (a) pure absorption, (b) pure dispersion, (c) mixture.

The compensation between v_0 and v_1 can be made by standard bridge techniques.

This compensation is achieved in a different manner in the crossed coil system. There, the driving r.f. field $2H_1\cos\omega t$ is produced in a transmitter coil of axis ox whilst the voltage δv induced by the precessing magnetization is observed at the terminals of a receiver coil of axis oy orthogonal to the transmitter coil. If the orthogonality is perfect so that no magnetic flux produced by the transmitter coil crosses the receiver coil, the voltage induced in the latter by a magnetization

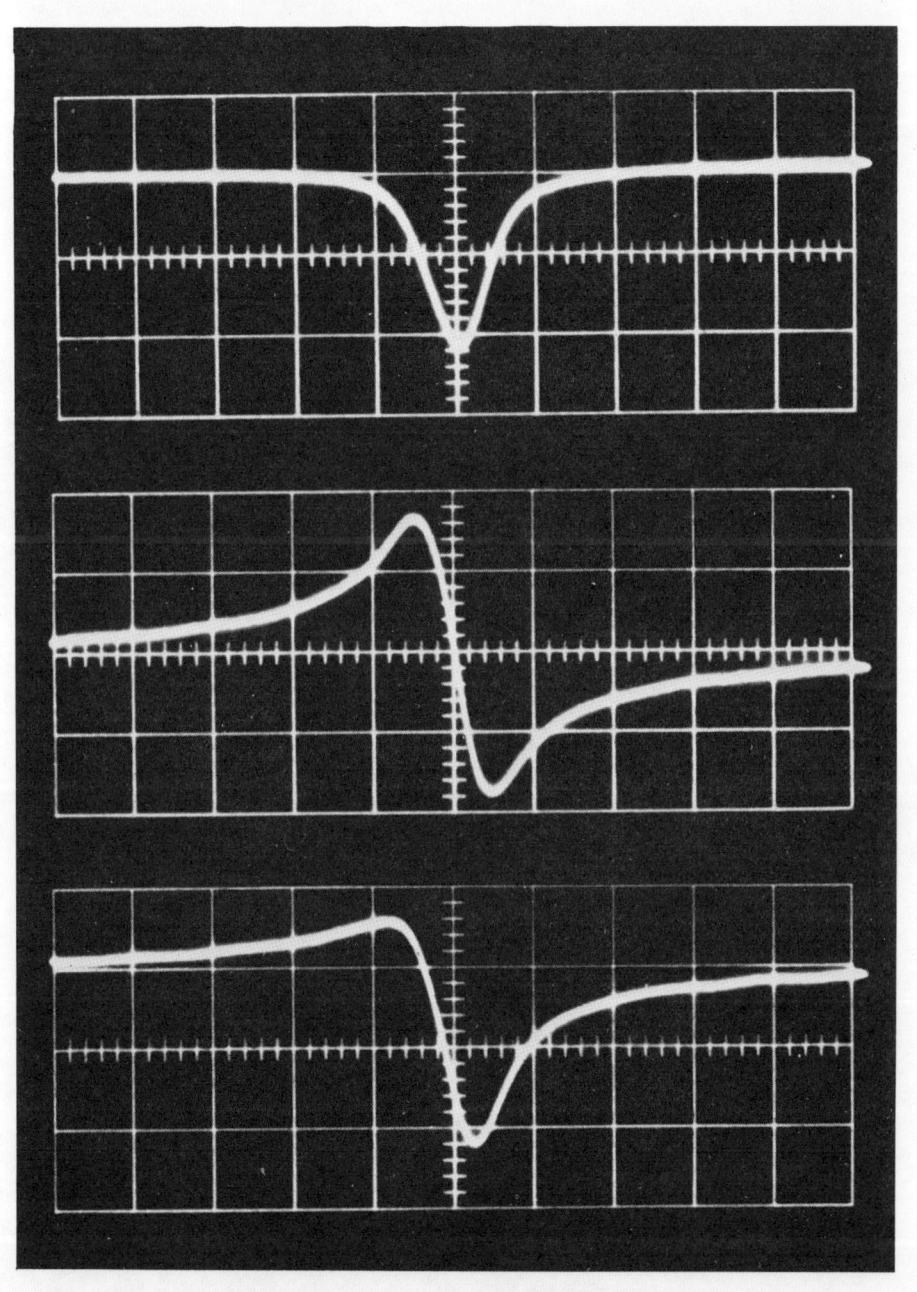

FIG. III, 11. Oscilloscope slow passage signals obtained from heavily doped water.
Top: absorption; *middle*: dispersion; *bottom*: mixture of the two modes. Sweep: 100 millisec per division; 0·2 gauss per division.

precessing with an angular velocity $\epsilon\omega$ will be proportional to

$$\frac{d}{dt}M_y(t) = \frac{d}{dt}\left\{M_x\left(t-\frac{\epsilon\pi}{2\omega}\right)\right\}$$

$$= \frac{d}{dt}\{\chi'(\omega)\cos(\omega t-\tfrac{1}{2}\epsilon\pi)+\chi''(\omega)\sin(\omega t-\tfrac{1}{2}\epsilon\pi)\}$$

$$= \epsilon\omega\{\chi'(\omega)\cos\omega t+\chi''(\omega)\sin\omega t\}, \tag{71}$$

and its amplitude will be proportional to $[\chi'^2(\omega)+\chi''^2(\omega)]^{\frac{1}{2}}$. If the orthogonality of the two coils is not perfect, and the angle between them is $\tfrac{1}{2}\pi-\theta$, there is a leakage flux into the receiver coil, proportional to $\sin\theta\cos\omega t$, that induces a leakage voltage

$$\mathscr{V}_L = A\omega\sin\theta\sin\omega t \tag{71'}$$

to be added to δv. If $\mathscr{V}_L \gg \delta v$, the change at resonance in the amplitude of the voltage in the receiver coil will be

$$|\mathscr{V}_L+\delta v|-|\mathscr{V}_L| \propto \epsilon\omega\chi''(\omega). \tag{72}$$

Thus the existence of leakage flux permits observation of $\chi''(\omega)$ rather than of $[\chi'^2+\chi''^2]^{\frac{1}{2}}$ with the crossed coils device. This device has another interesting feature apparent in formula (72): the detectable change of the voltage amplitude has a sign which, all things being equal, changes with the sign of ω_0, that is, with the sign of γ. The relative signs of two nuclear moments $\gamma\hbar I$ and $\gamma'\hbar I'$ can be obtained by comparing their signals under the conditions just described, with the same frequency and the same setting of the leakage flux (necessarily in different d.c. fields).

Thus the crossed-coil system provides as an extra piece of information, the relative sign of nuclear moments, unobtainable with the single-coil devices.

Various means have been developed for injecting into the receiver coil, voltages proportional to $\cos\omega t$ rather than to $\sin\omega t$, permitting the observation of χ' rather than of χ''. We shall not describe them here.

(c) Marginal oscillator

The principle of this device, denoted familiarly as the 'Pound box', is the following: the level of oscillation of an r.f. oscillator decreases when its load (conveniently represented by a shunt conductance in its output tuned circuit) is increased. This decrease is the sharper the lower the initial level of oscillation. A sample containing nuclear spins placed in the coil of the tuned circuit of the oscillator will at resonance absorb r.f. energy and may be considered as an additional load. The decrease in the level of oscillation is used to detect the resonance. Only the absorption part χ'' of the nuclear r.f. susceptibility can be detected in this manner. Simplicity and easy variation of the frequency within large limits (in contradistinction to bridge circuits) are among the main advantages of the Pound box. Its main weakness is the difficulty of producing very small r.f. fields H_1 sometimes necessary for long relaxation times T_1 to avoid saturation, or large H_1, useful for very short relaxation times. The former comes from the existence of a minimum level of oscillation, below which the oscillator becomes unstable, the latter from excessive noise that appears for large amplitudes of oscillation.

(d) *Audio-modulation, narrow band amplification, phase sensitive detection, signal-to-noise ratio*

In all the devices described in this section, the phenomenon of magnetic resonance manifests itself through a small change in the r.f. voltage available at the terminals of an apparatus. A linear detector, that is, an apparatus the output of which is a d.c. voltage proportional to the amplitude of an r.f. input, would thus exhibit the phenomenon of resonance as a small change in its d.c. output, proportional to the change in the amplitude of the r.f. input. This change is distinguishable from a change in the input voltage occurring in the absence of resonance and due to an instability of the apparatus, by its dependence on the applied field H_0. Modulating this field at an audio-frequency Ω, say in and out of resonance, that is by an amplitude larger than the line width, will make the change in the detected amplitude a periodic function of time $S(t)$ of period $2\pi/\Omega$, that can be separated by means of a high-pass filter from the d.c. part of the detected voltage and thus made insensitive to accidental variations of the latter.

For very weak signals the main problem is that of signal-to-noise ratio. Since the mean noise power existing in a frequency interval $\Delta\nu$ is equal to $4kT\Delta\nu$, a narrow band amplifier following the detector will decrease the noise considerably. In order that the signal $S(t)$ should not be distorted by a device of narrow band width $\Delta\nu$, field modulation of amplitude much smaller than the line width is often used. The signal $S(t)$ will then be with a good approximation proportional to

$$\frac{d\chi'}{d\omega}\cos(\Omega t+\phi) \quad \text{or} \quad \frac{d\chi''}{d\omega}\cos(\Omega t+\phi),$$

depending on whether one has chosen to observe χ' or χ'', as explained in (b). The detected amplitude of the audio-frequency signal is thus directly proportional to the derivative of χ' or χ''. Simultaneously with the modulation of the d.c. field according to the law

$$H = H_0 + H_m \cos \Omega t,$$

a continuous slow motion ('scanning') of the central value H_0 enables this value of the derivative to be measured in various points of the resonance curve. Fig. III, 12 shows signals observed in this fashion, proportional to the derivatives of χ' and χ''.

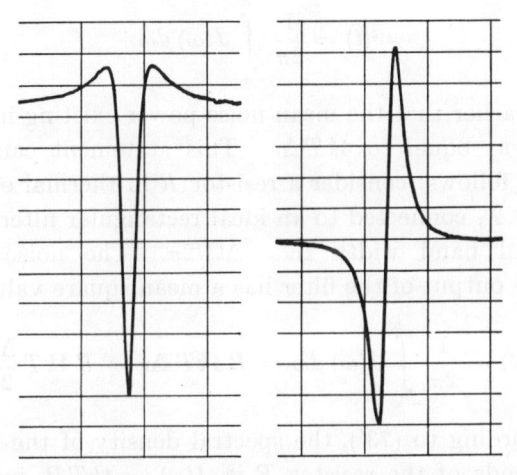

Fig. III, 12. Derivatives of absorption and dispersion in heavily doped water (half-intensity half width 0·45 gauss). Amplitude of peak to peak modulation: 0·1 gauss.

On the other hand, for strong signals with large signal-to-noise ratios, wide band amplifiers can be used and the whole curve, χ' or χ'', can be displayed without distortion on an oscilloscope, by a field modulation of amplitude several times the line width.

(1) *Lock-in detection and signal-to-noise ratio*

We now discuss briefly the problem of the signal-to-noise ratio and the principle of a device known as the lock-in amplifier whose purpose is to increase this ratio for very weak signals.

Since noise is essentially a random phenomenon describable by means of random functions the reader is referred for a brief outline of the main features of such functions to Section II B of Chapter VIII.

The noise voltage $v(t)$ existing at the output of a radioelectric circuit is a random function of time. No prediction can be made about its value at any time t, but precise statements can be made about ensemble averages, taken over statistical ensembles of such circuits, of quantities related to $v(t)$, such as the average values $\overline{v(t)}$ and $\overline{v^2(t)}$, the noise correlation function $G(\tau) = \overline{v(t)v(t+\tau)}$, and its spectral density

$$J(\omega) = \int_{-\infty}^{\infty} G(\tau)e^{-i\omega\tau}\,d\tau.$$

The mean value $\overline{v(t)}$ is zero and according to equation (17) of Chapter VIII the mean square value $\overline{v^2(t)}$ is

$$\overline{v^2(t)} = \frac{1}{2\pi} \int_{-\infty}^{\infty} J(\omega)\,d\omega. \tag{73}$$

We stated earlier that the mean noise power existing in a frequency interval $\Delta\nu$ was equal to $4kT\,\Delta\nu$. This statement can now be reformulated as follows: consider a resistor R in thermal equilibrium at a temperature T, connected to an ideal rectangular filter of mean frequency Ω and band width $\Delta\nu = \Delta\Omega/2\pi$. The noise voltage $v(t)$ existing at the output of the filter has a mean square value

$$\overline{v^2(t)} = \frac{1}{2\pi} \int_{-\infty}^{\infty} J(\omega)\,d\omega = R\,4kT\,\Delta\nu = R\,4kT\frac{\Delta\omega}{2\pi} \tag{73'}$$

and thus, according to (73'), the spectral density of the noise voltage between the ends of the resistor R is $J(\omega) = 4kTR$, independent of the frequency (white thermal noise).

If a signal at frequency Ω, $S = S_0 \sin(\Omega t)$ is sent through the filter at the same time, we shall define the signal-to-noise ratio as the ratio

$$S_0/[\overline{v^2(t)}]^{\frac{1}{2}}.$$

According to (73') this will be equal to $S_0(4kTR\,\Delta\nu)^{-\frac{1}{2}}$ and will increase as $\Delta\nu$ decreases.

For the weakest signals, band widths narrower than 1/10 of a cycle per second are required. It is extremely difficult to make audio amplifiers with such narrow bands and also to stabilize the modulation frequency Ω within a small fraction of such width. In the combination known as the lock-in amplifier or phase sensitive detector an alternative method is used.

The audio frequency signal is first amplified together with the noise voltage, the band width $\Delta\Omega$ of the amplifier satisfying simply the modest requirement $\Delta\Omega \ll \Omega$, easily achieved if $\Delta\Omega$ is of the order of a few cycles. The spectral density $J(\omega) = \int_{-\infty}^{\infty} \overline{v(t)v(t+\tau)}e^{-i\omega\tau}\,d\tau$ of the amplified

Ch. III MACROSCOPIC ASPECTS OF NUCLEAR MAGNETISM   81

noise voltage $v(t)$ vanishes outside the band width $\Delta\Omega$ of the filter, and the correlation time τ_e of the correlation function

$$G(\tau) = \overline{v(t)v(t+\tau)}$$

is of the order of $1/\Delta\Omega$.

In a lock-in amplifier, the output

$$y(t) = S_0 \sin\Omega t + v(t)$$

is 'mixed' with, that is, is multiplied by, a periodic function $F(t)$ of period $2\pi/\Omega$ and the result is integrated over a time $\theta \gg 1/\Delta\Omega$.

Assume first that the periodic function $F(t)$ is simply equal to $\sin\Omega t$. The integrated output after mixing will be

$$\int_0^\theta y(t)\sin\Omega t\, dt = \tfrac{1}{2} S_0 \theta\left[1 - \frac{\sin\Omega\theta}{\Omega\theta}\right] + V(\theta),$$

with $$V(\theta) = \int_0^\theta v(t')\sin\Omega t'\, dt' = \frac{1}{2i}\int_0^\theta v(t')\{e^{i\Omega t'} - e^{-i\Omega t'}\}\, dt'. \quad (74)$$

Since $\Omega\theta \gg \Delta\Omega\theta \gg 1$, the first term of (74) or signal term is very nearly $\tfrac{1}{2} S_0 \theta$. The second or noise term $V(\theta)$ is a random quantity and a definite statement can only be made about its mean square value $\overline{V^2(\theta)}$,

$$\overline{V^2(\theta)} = \tfrac{1}{4}\int_0^\theta dt' \int_0^\theta dt''\, G(t'-t'')[e^{i\Omega(t'-t'')} + e^{-i\Omega(t'-t'')} - e^{i\Omega(t'+t'')} - e^{-i\Omega(t'+t'')}].$$

From the assumption $\theta \gg 1/\Delta\Omega \sim \tau_c$, it follows that with good accuracy

$$\overline{V^2(\theta)} = \tfrac{1}{2}\theta \int_{-\infty}^{\infty} G(\tau)e^{i\Omega\tau}\, d\tau = \tfrac{1}{2}\theta J(\Omega).$$

The signal-to-noise ratio will be the ratio

$$\tfrac{1}{2} S_0 \theta [\overline{V^2(\theta)}]^{-\tfrac{1}{2}} = S_0 \left[\frac{4kTR}{\tfrac{1}{2}\theta}\right]^{-\tfrac{1}{2}}.$$

It is the same as if a filter with a band width $\Delta\nu = 2/\theta$ were used. In practice, the periodic 'mixing' function $F(t)$ will contain other terms besides $\sin\Omega t$. The simplest way of realizing the multiplication of the amplifier output $y(t)$ by $F(t)$ is to use a chopper that reverses the sign of $y(t)$ every half period, thus multiplying $y(t)$ by a periodic function $F(t)$ equal to -1 for $-\pi/\Omega < t < 0$ and to $+1$ for $0 < t < \pi/\Omega$.

The Fourier expansion of such a function is

$$F(t) = \frac{4}{\pi} \sum_{n=0}^{\infty} \frac{\sin[(2n+1)\Omega t]}{(2n+1)}.$$

The reader will easily convince himself that the contribution to the integrated output of terms with $n \neq 0$ is negligible if the frequency Ω is much larger than the bandwidth $\Delta\Omega$ of the amplifier.

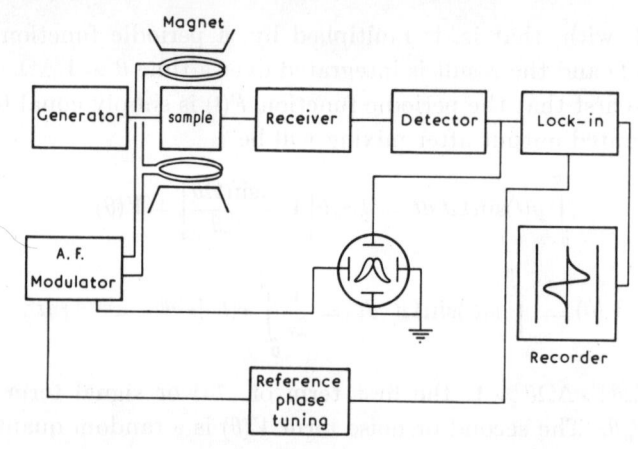

FIG. III, 13. Block diagram of a C.W. detection scheme.

In order for the integration over a long time to give a finite value for the signal, a phase coherence must exist between the signal $S_0 \sin(\Omega t)$ and $F(t)$. In practice the same generator supplies the coil that modulates the applied field H_0 and thus the signal at the frequency Ω, and the mixing signal represented by $F(t)$.

Fig. III, 13 represents a block diagram for steady state detection of nuclear magnetic signals.

(2) *Signal-to-noise ratio*

The order of magnitude of the signal-to-noise ratio obtainable in magnetic resonance will be computed under the following optimum conditions: negligible inhomogeneity broadening and $T_1 = T_2$. This situation will occur in good magnets for liquid samples with fast relaxation. The condition $1/T_2 \gg \gamma \Delta H$ will in general require a 'doping' of the sample with paramagnetic impurities for spins $\frac{1}{2}$, devoid of quadrupole moments (see Chapter VIII). The r.f. field H_1 will be of the order of $1/\gamma T_2$ for observation of the absorption and may be $\gg 1/\gamma T_2$ if the dispersion is observed. The maximum value of the transverse magnetization is then $\frac{1}{2}M_0 = \frac{1}{2}\chi_0 H_0$ for either component.

The amplitude of the r.f. voltage available at the terminals of the coil is, by a slight change of (63),

$$v = \tfrac{1}{2}\eta 4\pi\omega_0 \chi_0 H_0 A Q$$

or, according to (62),

$$v = \tfrac{1}{2}\eta 4\pi\omega_0 \chi_0 H_0 Q \left(\frac{LV_c}{4\pi}\right)^{\frac{1}{2}}. \qquad (75)$$

V_c is the volume of the coil and the filling factor

$$\eta \cong V_s/V_c = \frac{\text{volume of sample}}{\text{volume of coil}}.$$

The noise voltage existing at the same terminals in a frequency interval $\Delta\nu$ is $[4kTR\Delta\nu]^{\frac{1}{2}}$, where $R = QL\omega$ is the shunt resistance of the tuned circuit. If $\Delta\nu$ is the width of the narrow band audio-frequency amplifier, the signal-to-noise ratio $(v/f)[4kTR\Delta\nu]^{-\frac{1}{2}}$, where $f > 1$ is a factor that takes into account all additional noise of the apparatus, can be written conveniently

$$\psi = \frac{\pi}{f}\left[\tfrac{1}{2}\eta\left(\frac{\nu}{\Delta\nu}\right)Q\chi_0\left(\frac{\chi_0 H_0^2 V_s}{kT}\right)\right]^{\frac{1}{2}}. \qquad (76)$$

ψ appears thus as a product of dimensionless quantities. In particular, $\chi_0 H_0^2 V_s = V_s M_0 H_0$ is the magnetic energy of the nuclear spins of the sample in thermal equilibrium. With the same assumptions as before

$$\eta = 1, \quad \nu = 42\cdot 6 \text{ Mc/s}, \quad \Delta\nu = 1 \text{ c/s}, \quad Q = 100,$$

$$\chi_0 = 3\times 10^{-10}, \quad H_0 = 10^4 \text{ gauss}, \quad V_s = 1 \text{ c.c.}, \quad T = 300°\text{ K},$$

we find
$$\psi \cong \frac{1}{f} 2 \times 10^6. \qquad (76')$$

Little significance should be attached to the factor 2 in (76') since the above calculation gives only an order of magnitude and disregards among other things the reduction in signal due to audio-frequency modulation, which for a sinusoidal modulation is by a factor $(8)^{\frac{1}{2}}$. In practice, with great care, values of noise figures as low as 2 can be achieved. The enormous signal-to-noise ratio (76') would only be reduced by a factor 10^2 if a band width of 10 kc/s were used, making display on an oscilloscope possible.

Using (76'), we can rewrite (76) as

$$\psi = \frac{10^6}{f} V_c^{-\frac{1}{2}} \tfrac{4}{3} I(I+1) \left(\frac{\gamma}{\gamma_p}\right)^{\frac{5}{2}} \left(\frac{N}{N_p}\right) V_s \left(\frac{H}{10^4}\right)^{\frac{3}{2}} \left(\frac{Q}{100}\right)^{\frac{1}{2}} \left(\frac{T}{300}\right)^{-\frac{3}{2}} (\Delta\nu)^{-\frac{1}{2}}, \qquad (77)$$

where V_c = volume of the coil in c.c., V_s = volume of the sample in c.c.; γ and N are the gyromagnetic factor and the number of nuclear spins per c.c., respectively, and γ_p and N_p the corresponding quantities for the protons in water. The factor $V_s N$ in (77) shows that for a given coil the signal-to-noise ratio is proportional to the total number of spins inside the coil.

As an example consider the signal from deuterons in ordinary water under the same conditions:

$$H_0 = 10^4 \text{ gauss}, \quad Q = 100, \quad T = 300° \text{ K}, \quad N/N_p = 1\cdot56 \times 10^{-4},$$
$$\gamma/\gamma_p = 0\cdot152, \quad \tfrac{4}{3}I(I+1) = \tfrac{8}{3},$$

whence $\psi \sim 4/f$. Fig. III, 14 shows a signal of deuterons obtained for $H_0 = 12{,}000$ gauss, $V_s = 2$ c.c., $V_c = 6$ c.c., $Q \simeq 50$, $\Delta\nu \simeq 0\cdot1$ c/s. The observed signal-to-noise ratio is in qualitative agreement with (77).

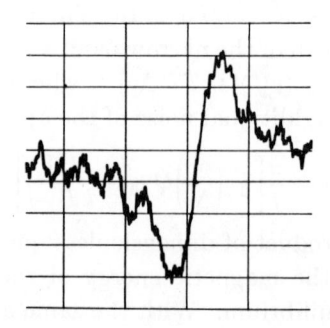

FIG. III, 14. Deuteron signal in natural (doped) water.
$H_0 = 12$ kilogauss; lock-in time constant $\tau = 20$ sec.
Volume of the sample $V_s = 2$ c.c.; volume of the coil $V_c = 6$ c.c.

For solid samples where the lines are broad the signals are much weaker. It is difficult to make quantitative predictions about signal-to-noise ratios, for the Bloch equations, which give the maximum values obtainable under steady-state conditions for the components of the transverse magnetization, are not valid in solids.

For solids, where the line width is caused by a dipolar coupling between spins of the same species, it is not unreasonable to extrapolate for the absorption (but not the dispersion!) the predictions of the Bloch equations and assume that (77) should be multiplied by $(\gamma\Delta H T_1)^{-\frac{1}{2}}$, where ΔH is the half width at half intensity (see Chapter XII). For very long T_1 such as exist in very pure solids, especially at low temperatures, this factor and thus the signal-to-noise ratio become exceedingly small and non-steady-state methods to be described shortly must be used.

Finally, one should restate that in the absence of saturation, that is for very small applied r.f. fields, the signal is always proportional to the

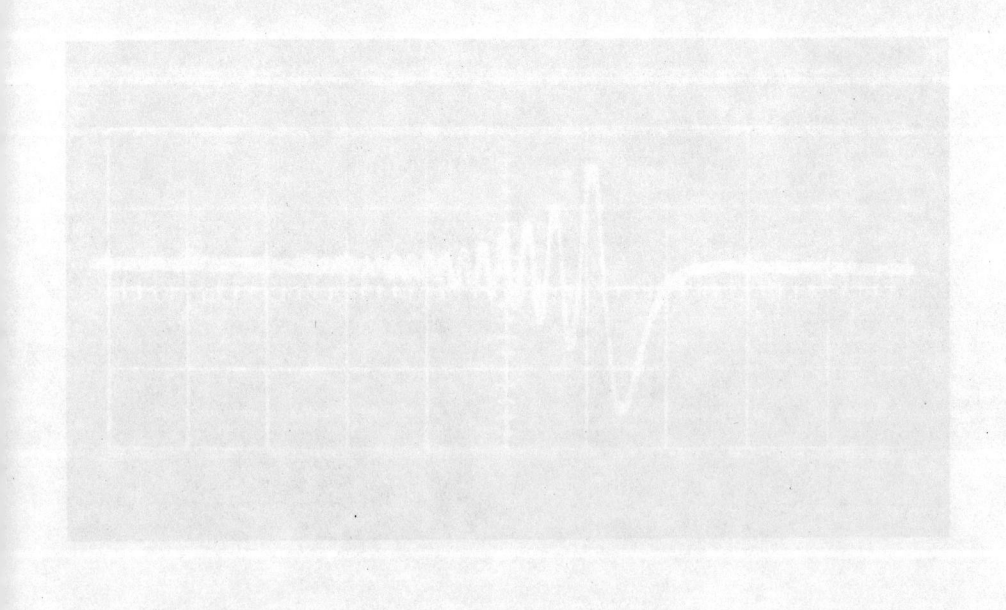

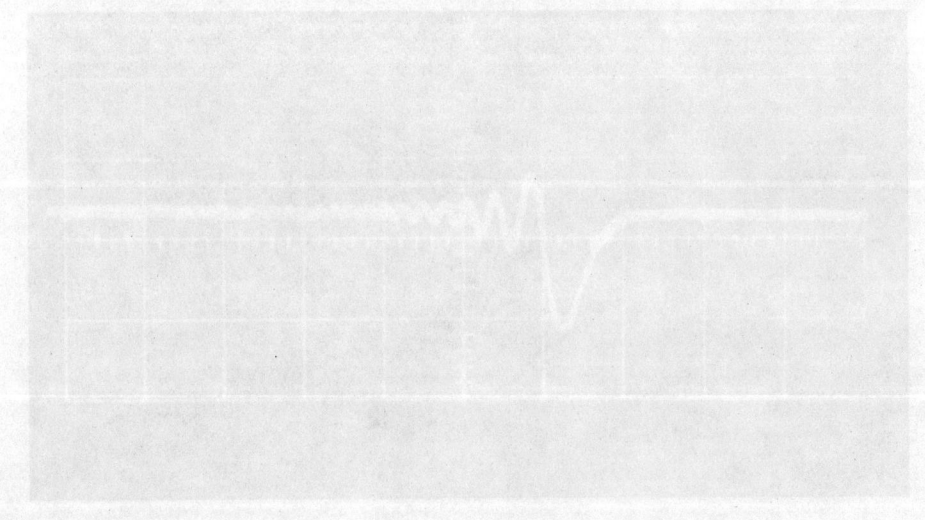

FIG. 111-13. Wiggles obtained with a homogeneous (top) and an inhomogeneous (bottom) field. Sweep factor: 0.1 sec and 1 milligauss per division.

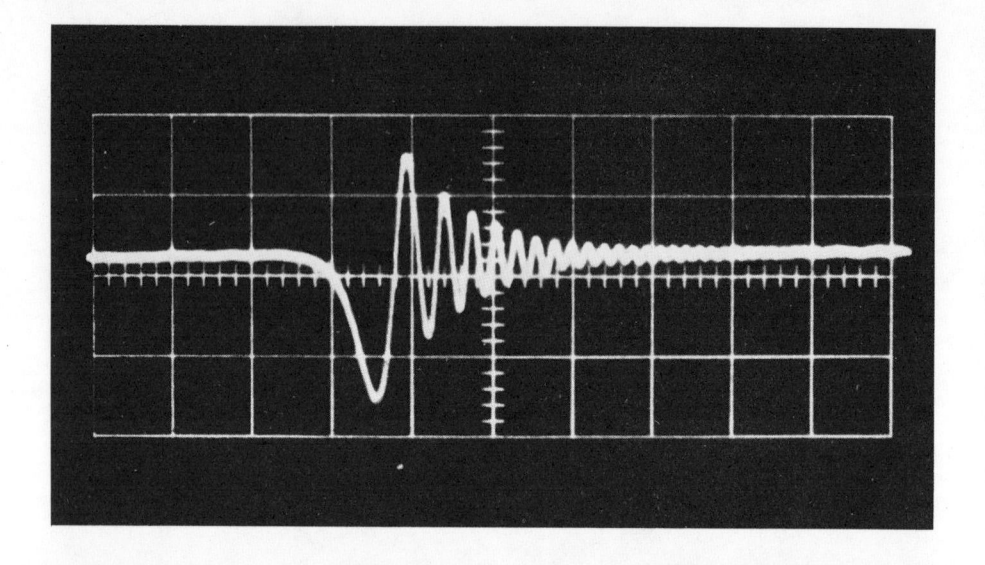

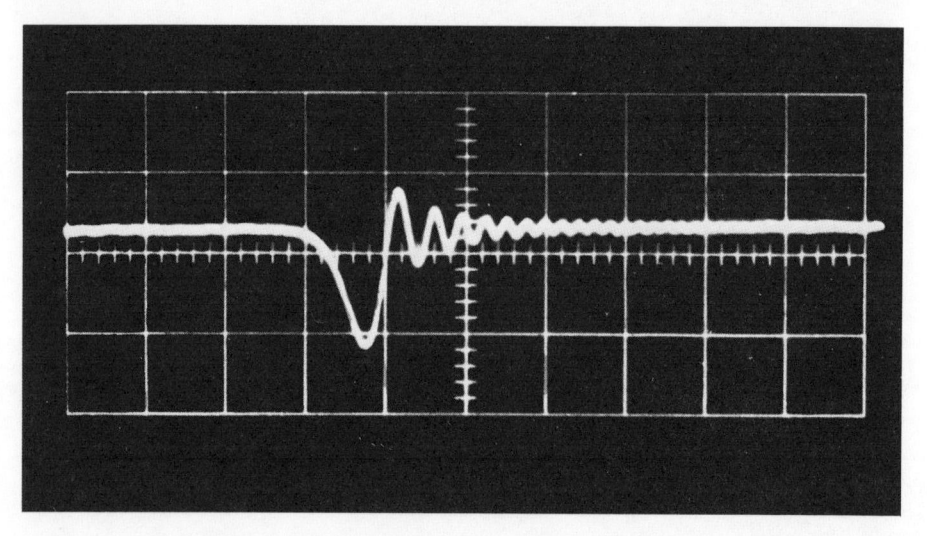

FIG. III, 15. Wiggles in doped water in a homogeneous (top) and an inhomogeneous (bottom) field. Linear sweep: 0·1 sec and 5 milligauss per division.

inverse of the line width whatever the origin of the latter. It is only for the evaluation of the optimum signal-to-noise ratio that may involve appreciable saturation, that the physical origin of the broadening is important.

(e) *Transient effects in steady-state detection*

The above analysis of the shapes and magnitudes of the signals in the presence of audio-frequency modulation was based on the use of steady-state solutions of the Bloch equations, and assumed implicitly that the modulation period $2\pi/\Omega$ and *a fortiori* the 'scanning time' spent in crossing the line, were very long in comparison with T_1 and T_2, a condition that may be difficult to realize in practice. If the line is inhomogeneously broadened with a width ΔH, and the modulation and the scanning are defined by a law $H = H_0(t) + H_m \cos\Omega t$, no less than seven parameters having the dimension of a frequency may be introduced to discuss the shape and the size of the signals observed, namely,

$$\frac{1}{T_1}, \quad \frac{1}{T_2}, \quad \gamma\Delta H, \quad \gamma H_1, \quad \gamma H_m, \quad \Omega, \quad \frac{1}{\tau} = \frac{1}{H_1}\frac{dH_0}{dt}.$$

The shapes of the signals depend considerably on the relative values of these parameters and may become very complicated. They can in principle be analysed mathematically, using Bloch equations, and a considerable literature exists on this subject. However, there is nothing fundamental in the complications of these shapes and the rather negative conclusion (by no means restricted to nuclear magnetism) to be drawn from their analysis is the following: whenever the methods of observation are such that a complicated mathematical treatment is required to establish a relationship between the data (signals) and the physical nature of the system studied, the methods of observation are inadequate and must be changed, whenever possible.

In particular, the early measurements of relaxation times (11) from the saturation behaviour of the signals, in the presence of inhomogeneity broadening and field modulation, are difficult to interpret, and have gradually been replaced by direct methods using pulses, spin echoes, or adiabatic fast passage.

Because it occurs often and is explained easily we shall, however, describe the transient phenomenon of 'wiggles' shown in Fig. III, 15. The wiggles are the oscillations that appear on the side of the resonance curve, *following* the passage through the resonance. If the r.f. field H_1 of frequency ω is not sufficiently strong for the adiabatic fast passage condition $dH_0/dt \ll \gamma H_1^2$ to be fulfilled, the magnetization does not 'follow' completely the effective field H_{eff}. When the d.c. field $H_0(t)$ has passed beyond the resonant value $H_0^* = -\omega/\gamma$, there still remains a transverse component of the magnetization, precessing freely at the frequency $\omega_0(t) = -\gamma H_0(t) \neq \omega = -\gamma H_0^*$. The wiggles represent the beats at a frequency $\omega_0(t)-\omega$ of the voltage induced in the coil by that magnetization, with the r.f. voltage of frequency ω that produces the field H_1. If the sweep of the field is linear, the shape of the beats is represented approximately by

$$\cos\left[\int \{\omega_0(t')-\omega\}\,dt'\right] \simeq \cos[\tfrac{1}{2}\gamma \dot{H}_0 t^2],$$

where t is the time counted from the passage through resonance. The damping of the wiggles is the normal inhomogeneity damping of free precession. The better the field, the longer the wiggles.

C. Transient methods of detection

The principles and applications of these methods have already been described in Section III and only a few comments of a more experimental nature will be added here.

(a) Adiabatic fast passage

For very long T_1 of the order of several hours or more that exist in some solids, the adiabatic fast passage is sometimes the only method that gives tolerable signal-to-noise ratios. If the r.f. field H_1 is sufficiently large, transient transverse magnetizations equal to M_0 can be obtained. Furthermore, many passages back and forth can be made during a time short compared with T_1 without destroying the magnetization (see Chapter XII), and the contributions from the transient signals, thus induced, can be integrated in a lock-in amplifier, improving considerably the signal-to-noise ratio. Fig. III, 16 shows a recorded signal of Si[29] (isotopic abundance 4·7 per cent., $T_1 \simeq 2$ hours at room temperature) obtained by this method. The peculiar shape of the curve (which is not the derivative of a bell-shaped absorption curve) is a consequence of the nature of the lock-in amplifier. Let

$$H = H_0(t) + H_m \cos\Omega t$$

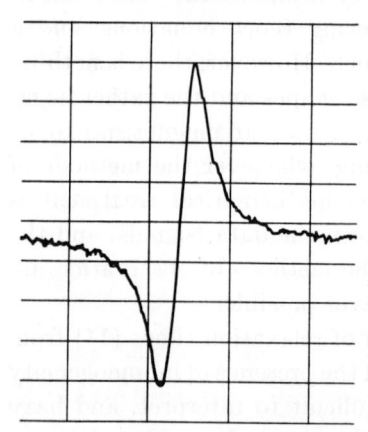

FIG. III, 16. Lock-in adiabatic fast passage signal of Si[29] in fused silica. $T_1 = 2$ hours, $H_1 = 0\cdot6$ gauss, amplitude of peak-to-peak modulation 1·6 gauss, modulation frequency 20 c/s.

be the time-dependence of the applied field, where the variation of $H_0(t)$ is slow and a linear function of time (scanning). During a cycle of modulation the effective field, and thus also the nuclear magnetization that 'follows' it, swings back and forth around the average position reached for $\cos \Omega t = 0$ that makes with oz the angle $\theta(t)$ given by $\tan \theta = H_1/(H_0+\omega/\gamma) = H_1/(H_0(t)-H_0^*)$. For $H_0(t) > H_0^*$, $\theta < \frac{1}{2}\pi$ and the first half of the modulation cycle corresponds to a smaller transverse magnetization than the second; the situation is reversed for $H_0(t) < H_0^*$ and $\theta > \frac{1}{2}\pi$. Since the lock-in always reverses the sign of the induced voltage during the same half period of the cycle, the sign of the signal, the difference of the contributions from the two half-periods, changes upon crossing the resonance, for $H_0(t) = H_0^*$. This method can measure the relative magnitude and sign of two nuclear magnetizations (not to be confused with the relative signs of two gyromagnetic moments!), but provides no information about the shape of the resonance curve, since the width of the signal is clearly of the order of H_1.

(b) Pulse methods, coherent and incoherent pulses

Pulse methods are generally used when strong nuclear signals are available and little attention need be paid to the signal-to-noise problems. Since these methods can create transverse nuclear magnetizations comparable to the d.c. magnetization M_0, it might appear at first sight that at least for broad lines, better signal-to-noise ratios might be obtained than in C.W. methods. That it is actually not so, results from the fact that the duration of the order of $(\gamma \Delta H)^{-1}$ of such signals is accordingly small, and the necessity of using broad band amplifiers in order to observe them, thus increasing the noise considerably, more than offsets the effect of dealing with large transverse magnetizations.

When more than one pulse is used, as in the echo method, there are essentially two techniques: incoherent and coherent pulses. The oscillating r.f. field produced in a coil by a sequence of pulses occurring at times $t_1, t_2, ..., t_k$ and of durations $\tau_1, \tau_2, ..., \tau_k$ can be represented by the function
$$H(t) = \sum_k A_k(t) \cos(\omega t + \psi_k),$$
where $A_k(t) = 0$ outside of the interval $t_k \leqslant t \leqslant t_k+\tau_k$ and is approximately constant inside. The pulses are called incoherent if the phases ψ_k are distributed at random, and coherent if their values can be controlled (in particular, given the same value ψ). The orientation of the rotating field H_1 with respect to the rotating frame varies at random between pulses in the first method, but is well defined (in particular,

can be made fixed) in the second. The block diagrams for the two types of apparatus are given in Fig. III, 17.

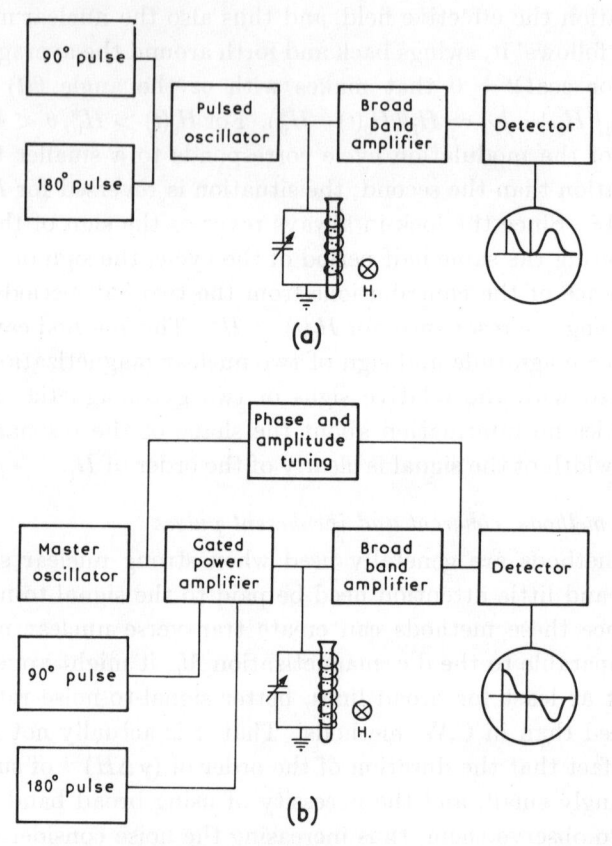

FIG. III, 17. Block diagrams of: (a) Incoherent spin echo apparatus.
(b) Coherent spin echo apparatus.

The main advantages of the technique of incoherent pulses are simplicity and the ease with which the frequency can be changed. The pulsed oscillator is started at times t_1, t_2,..., t_k and stopped at times $t_1+\tau_1$,..., $t_k+\tau_k$ by a timer. Incoherent pulses can be used when the interval between two consecutive echoes is sufficiently large for the first of them to have decayed completely, when the second appears. In the so-called method B (3) the distance τ between echoes is related to the diffusion damping constant by the formula (50) $1/T_2^\dagger = \tfrac{1}{3}D\gamma G^2\tau^2$. For long relaxation times T_2, in order to keep $(1/T_2^\dagger)/(1/T_2)$ small, short intervals τ and good magnets and thus small field gradients G must be used. Successive echoes then almost overlap as shown in Fig.

III, 3, and must necessarily be coherent. Incoherent pulses can thus only be used for short relaxation times $T_2 \leqslant 0\cdot 1$ sec. A different type of echo for which coherent pulses are necessary, the so-called multiple echoes that occur in certain solids, will be described in Chapter VII. In the technique of coherent pulses a master oscillator that generates the radio-frequency runs continuously and the role of the timer is to connect or disconnect at prescribed intervals, or 'gate', the power amplifier that produces the large r.f. H_1. Coherent pulses have another advantage besides controlling the phase of H_1: they provide a signal of reference for detection to which the nuclear signal can be added. The phase as well as the amplitude of the nuclear signal can then be measured and, furthermore, for the weaker signals the signal-to-noise ratio is considerably improved.

D. Negative absorption—masers

The static nuclear susceptibility $\chi_0 = \gamma^2\hbar^2 I(I+1)N/3kT$ is always a positive number and the magnetization $M_0 = \chi_0 H_0$ is always parallel to the applied field, an essential feature of paramagnetic substances. It is possible, however, as we saw earlier, to create by a 180° pulse or by a fast passage, situations where the magnetization is antiparallel to the applied field. The higher Zeeman energy levels are then more populated than the lower ones and a small external r.f. field will receive r.f. energy from the system, a situation that could be described as a negative nuclear magnetic absorption. It is clear that no steady state can result under those conditions since both the r.f. transition probability and the relaxation tend to reduce the magnetization as long as it is negative.

It will be shown in Chapters VIII and IX that if spins S of another species than the species I are present in the sample, it is possible under certain conditions by applying a strong r.f. field at the Larmor frequency ω_s (or sometimes $\omega_s \pm \omega_I$) to create for the spins I a steady state where their magnetization is antiparallel to d.c. field H_0 and much larger than $\chi_0 H_0$.

Postponing until then the explanation of this remarkable phenomenon, which we shall call negative dynamic polarization and which can be described very simply in a phenomenological manner by assigning to the spins I a large and negative d.c. susceptibility χ_0^d, we shall review some of its consequences.

First, as already mentioned, in the presence of negative dynamic polarization, energy will pass from the nuclear spin system to an applied r.f. field and the absorption signal will change its sign (and its magnitude). Fig. III, 18 shows absorption signals from the protons of water, obtained by Q-meter detection at 3,000 gauss. The normal signal and the strong negative signal that corresponds to $\chi_0^d \cong -50\chi_0$ have been superposed on the photograph.

As long as the voltage δv induced by the nuclear magnetization is small compared with the voltage v that produces the driving r.f. field H_1, the symmetry between positive and negative polarization is complete. For instance, a spin system with negative polarization placed in the coil of a marginal oscillator will act as a small negative load, raising rather than lowering its level of oscillation. However, as is well known from elementary network theory, when negative resistances become comparable in magnitude to normal positive resistances in a circuit, auto-oscillations may set in. Thus a marginal oscillator slightly below the threshold of oscillation may start an oscillation if a sample containing negatively polarized nuclear spins is placed in its tank circuit, the extra negative resistance provided by the nuclear spin system being sufficient to get the system above the threshold.

The condition for the establishment of a steady auto-oscillation of the system: nuclear spins plus a passive tuned circuit, is that the electromagnetic energy provided by the spin system compensates exactly the losses in the circuit.

This condition can be obtained from equation (68), which gives the shunt impedance Z of such a system. We write that for an auto-oscillator Z becomes infinite, that is, Z^{-1} vanishes. Since Z is complex we obtain the two equations

$$\chi'(\omega) = 0, \qquad 4\pi\eta Q\chi''(\omega) = -1. \tag{77'}$$

As is always the case in the theory of auto-oscillators, a linear theory is sufficient to give the threshold for the oscillation and its frequency but a non-linear theory is required to predict the actual level of oscillation. In a linear theory where the response of the nuclear spin system is assumed to be proportional to the r.f. field H_1 that exists inside the sample, χ'' is given quite generally by the formula (8):

$$\chi'' = \tfrac{1}{2}\pi\chi_0\,\omega f(\omega),$$

where $f(\omega)$ is the shape function of the nuclear resonance line, and $\chi'(\omega)$ follows from the K.-K. relations (8'). If the resonance shape is symmetrical around the resonance frequency ω_0, $\chi'(\omega_0) = 0$ and the first relation (77') shows that the oscillation frequency ω is equal to the absolute value of the Larmor frequency $|\omega_0|$ (remember the convention $\omega > 0$, of this section). If $\chi''(\omega)$ has the Lorentz shape (9),

$$\chi''(|\omega_0|) = |\omega_0|\tfrac{1}{2}T_2\chi_0.$$

We shall write generally $\chi''(|\omega_0|)$ as $\tfrac{1}{2}(|\omega_0|)T_2^*\chi_0$, it being understood that the meaning of T_2^* is $\pi f(|\omega_0|)$. The condition $4\pi\eta Q\chi'' = -1$ can be rewritten
$$2\pi|\omega_0|\eta Q_0 \chi_0 T_2^* = -1, \tag{78}$$

which is only possible if χ_0 is a dynamic negative d.c. susceptibility χ_0^d. We write Q_0 to indicate that it is the threshold value for which an oscillation may start. The condition (78) can be rewritten

$$2\pi\eta Q_0|\gamma|\,|M_0^d| = \frac{1}{T_2^*}, \tag{79}$$

where $|M_0^d| = |\chi_0^d H_0|$ is the absolute value of the d.c. nuclear magnetization of the sample. The left-hand side of (79) is the constant $1/\tau$ for the radiation damping given by formula (66'), which in a field of 10,000 gauss for $Q = 100$, $\eta = 1$, $|\dot M_0| = 3\times 10^{-10}\times 10^4$ was found to be approximately 50 sec^{-1}. Thus if a negative d.c. magnetization $M_0^d = -3\times 10^{-6}$ can be given to the protons of water, a proton oscillator of the so-called 'maser' type may operate if $1/T_2^* \leqslant 50$ sec^{-1}. The line width $\gamma\Delta H = 1/T_2^*$ expressed in gauss will be $\simeq 2$ milligauss for protons, which sets a severe but by no means impossible requirement on the homogeneity of the magnet. As shown in Fig. III, 18 a dynamical d.c. susceptibility χ_0^d of the order of $-50\chi_0$ was obtained in a field of 3,000 gauss, resulting in a d.c. magnetization $|M_d|$ of the order of 5×10^{-5}. A proton maser has actually been operated in this field (12) and also in the earth's field where it can be used as a magnetometer (13) (see Chapter VIII).

To proceed further and estimate the level at which the maser will oscillate if the Q of the circuit is larger than the threshold value Q_0 given by (78), we shall assume that the nuclear magnetization obeys Bloch equations and write the unsaturated line width as $1/\gamma T_2$ rather than $1/\gamma T_2^*$. If $Q > Q_0$ and the maser oscillates at a non-negligible level, the r.f. field in the coil will induce a certain amount of saturation, decreasing χ'' in the ratio Q_0/Q in order for (77') to remain valid. Since it follows from the Bloch equations that the saturation reduces the r.f. susceptibility in the ratio $[1+\gamma^2 H_1^2 T_1 T_2]$, the value of the r.f. field H_1 created by the current induced in the coil by the precessing magnetization is given by
$$\gamma^2 H_1^2 T_1 T_2 = (Q/Q_0)-1. \tag{80}$$

The power $P = 2\omega H_1^2 \chi''$ dissipated in the coil and provided by the nuclear spin system is calculated from (79) and (80) and found to be

$$P = |M_d|\frac{H_0}{T_1}\left(1 - \frac{Q_0}{Q}\right). \tag{81}$$

By using the formula (68) rather than the more general formula (67) for the impedance of the tuned circuit containing the nuclear spins, we implicitly assumed that the frequency $\omega_c = (LC)^{-\frac{1}{2}}$ of the tuned circuit was equal to the frequency ω of the oscillation, that is, to $|\omega_0|$. If $\omega_c \neq |\omega_0|$, the frequency of oscillation ω will not be equal to ω_0 either, a phenomenon known as 'pulling'. Let us write in (67):

$$C\omega = \frac{1}{L\omega(1-\delta)}, \quad \text{whence} \quad \delta = \frac{\omega^2 - \omega_c^2}{\omega^2} \simeq 2\frac{\omega - \omega_c}{\omega} \simeq 2\frac{(\omega - \omega_c)}{\omega_c}.$$

Eqn. (67) can be rewritten

$$Z = L\omega Q(1-\delta)\left[\frac{1 + j4\pi Q\eta\chi + jQ\delta}{1 + 4\pi\eta\chi - j/Q}\right]^{-1}. \tag{82}$$

The condition $Z^{-1} = 0$ becomes

$$4\pi Q\eta\chi''(\omega) = -1, \quad 4\pi\eta\chi'(\omega) = -\delta. \tag{83}$$

From (22) we obtain

$$\chi'(\omega) = -T_2(\omega - |\omega_0|)\chi''(\omega) \simeq -T_2(\omega - |\omega_0|)\chi''(\omega_0)$$

and, from (83),

$$T_2(\omega - |\omega_0|) = -2Q\frac{\omega - \omega_c}{\omega_c} \simeq 2Q\frac{\omega_c - |\omega_0|}{\omega_c}. \tag{84}$$

From (84) we see that $|\omega - |\omega_0||$ is smaller than $|\omega - \omega_c|$ by the ratio $2Q/\omega_c T_2$ which except for very low frequencies is very small. The 'pulling' is thus almost always negligible.

We shall conclude by remarking that the nuclear spin oscillator is simpler than most oscillators from a theoretical point of view, since its behaviour, independent of the characteristics of electronic tubes, is entirely calculable from first principles even in the non-linear region.

Further study (14) of properties of nuclear spin maser oscillators such as band width, noise figure, dynamic behaviour in reaching the steady-state oscillation when the conditions for it have been created, that could be pursued by an extension of the methods outlined here, are outside the scope of this book.

As already stated at the beginning of this section, this review of experimental methods in nuclear magnetism is far from complete. Thus no mention has been made of super-regenerative methods, nor of some interesting but rather special methods such as observation of the resonance in low fields through a change in the longitudinal component M_z of the nuclear magnetization (1).

We conclude with a few words about the production of the magnetic fields of several thousands gauss, currently used in nuclear magnetism. The two main requirements to be met by the magnets producing these fields are homogeneity in space and stability in time. We shall not discuss the engineering problems connected with these requirements and will simply mention two rather elegant devices that have been used in that connexion. The first, aimed at solving the inhomogeneity problem, is that of the 'spinning' sample. Suppose that the inhomogeneity ΔH of the field over the sample is smaller than, say, one milligauss, so that the Larmor frequencies of any two nuclei differ by less than $\Delta \nu = \gamma/2\pi \times 10^{-3}$ c/s (4 c/s for protons). A macroscopic motion ('spinning') of the whole sample at a frequency much higher than $\Delta \nu$, will cause each nuclear spin to 'see' all the values of the applied field within the interval ΔH and will produce an appreciable narrowing of the resonance line, the effective field 'seen' by each proton being the average of all those 'seen' during the motion. Fig. III, 19 shows two signals obtained with and without spinning.

In connexion with the stability problem it can be remarked that in order to keep the resonance condition $\omega = |\gamma H_0|$ with a relative accuracy of, say, 10^{-8} over long periods of time it may not be necessary that H_0 and ω be separately stable to the same extent, if a device locking them together can be imagined. Several schemes have been tried with varying success. One of the simplest in principle, if possibly not in practice, would be the use of a maser oscillator operating in the same magnet since its frequency is proportional to the applied field.

APPENDIX
Proof of the Kramers–Krönig relations

The relations (8′) between the real and the imaginary parts of the complex susceptibility $\chi = \chi' - i\chi''$ are very general and apply to many systems besides assemblies of nuclear spins. Consider a physical system S, with an input at which a time-dependent excitation $\mathscr{E}(t)$ can be applied, and an output where a response $R(t)$ can be observed. $\mathscr{E}$ and R may be physical quantities of the same nature or of different nature. In nuclear magnetism $\mathscr{E}$ is a magnetic field and R a magnetization but many other examples can be imagined: $\mathscr{E}$ electric field and R electric polarization, $\mathscr{E}$ input and R output voltages, $\mathscr{E}$ incoming flux of particles on a scatterer and R outgoing flux, etc.

The very definition of χ implies that the response to a monochromatic excitation $\mathscr{E}$ is monochromatic. In order to prove the K.–K. relations (8′), we make a few general assumptions to be satisfied by the systems S which we consider.

(i) *The systems S are linear*

We mean thereby that if R_1 and R_2 are responses to excitations $\mathscr{E}_1$ and $\mathscr{E}_2$ the response to the excitation $c_1\mathscr{E}_1+c_2\mathscr{E}_2$ will be $c_1 R_1+c_2 R_2$. Saturation is thus explicitly excluded.

(ii) *The systems S are stationary*

If $R(t)$ is the response to $\mathscr{E}(t)$, the response to $\mathscr{E}(t-t_0)$ will be $R(t-t_0)$. The condition of a monochromatic response to a monochromatic excitation follows from (i) and (ii). Let $f(t)$ be the response to the unit impulse excitation $\delta(t)$ and, according to (ii), $f(t-t')$ be the response to $\delta(t-t')$. A monochromatic excitation $e^{i\omega t}$ can be rewritten:

$$\mathscr{E}(t) = e^{i\omega t} = \int_{-\infty}^{\infty} e^{i\omega t'}\delta(t-t')\,dt'$$

with, because of (i) and (ii), a response

$$R(t) = \int_{-\infty}^{\infty} e^{i\omega t'}f(t-t')\,dt' = e^{i\omega t}\int_{-\infty}^{\infty} e^{-i\omega t'}f(t')\,dt' \tag{85}$$

which is indeed monochromatic with the same frequency.

(iii) *The systems S obey the principle of causality*

If $\mathscr{E}(t) = 0$ for $t < t_0$, $R(t) = 0$ for $t < t_0$. This assumption is the key to the proof of the K.-K. relations. It implies in particular that $f(t)$ (which incidentally is real, being the response of a physical system to a real excitation $\delta(t)$) vanishes for $t < 0$. Equation (85) can thus be written:

$$e^{i\omega t}\int_0^{\infty} f(t')e^{-i\omega t'}\,dt',$$

whence, from the definition of χ,

$$\mathscr{E} = e^{i\omega t}, \qquad R = \chi(\omega)e^{i\omega t},$$

$$\chi = \int_0^{\infty} f(t)e^{-i\omega t}\,dt, \qquad \chi' = \int_0^{\infty} f(t)\cos\omega t\,dt, \qquad \chi'' = \int_0^{\infty} f(t)\sin\omega t\,dt. \tag{86}$$

(iv) *Finite total response to a finite total excitation*

We define the total excitation and the total response by the relations:

$$E_T(t) = \int_{-\infty}^{t} |\mathscr{E}(t')|\,dt',$$

$$R_T(t) = \int_{-\infty}^{t} |R(t')|\,dt'.$$

The condition (iv) requires $R_T(t)$ to be finite if $\mathscr{E}_T(t)$ is finite. It follows immediately from (iv) that $\int_0^{\infty}|f(t)|\,dt$ is finite. One should not, however, conclude from it that $f(t)$ has no singularities, for in the simplest case where $R = \mathscr{E}$, $f(t) = \delta(t)$. It is still possible, however, to state that $|tf(t)| \to 0$ when $t \to 0$, for otherwise the integral $\int_0^{\infty}|f(t)|\,dt$ would diverge.

Ch. III MACROSCOPIC ASPECTS OF NUCLEAR MAGNETISM 95

The imaginary part $\chi''(\omega)$ is expected on physical grounds to vanish when ω tends to infinity, for otherwise an infinite absorption of energy by the system would occur. That this is actually so is easily verified from:

$$\lim_{\omega \to \infty} \chi''(\omega) = \lim_{\omega \to \infty} \int_0^\infty f(t) \sin \omega t \, dt = \lim_{\omega \to \infty} \int_0^\infty dt\, f(t) \int_0^\omega t \cos(\omega' t)\, d\omega'$$

$$= \lim_{\omega \to \infty} \tfrac{1}{2} \int_0^\infty tf(t) \int_{-\omega}^{\omega} e^{i\omega' t}\, d\omega' \to \pi \int_0^\infty tf(t)\delta(t)\, dt = 0.$$

A similar calculation shows that $\chi'_\infty = \lim_{\omega \to \infty} \chi'(\omega)$ does not necessarily vanish. This was to be expected, for in the simplest case when $\mathscr{E} = R$, $\chi'(\omega) = 1$ for all excitation frequencies. The function $\psi(\omega) = \chi(\omega) - \chi_\infty$, where $\chi(\omega)$ is defined by

$$\chi(\omega) = \int_0^\infty f(t) e^{-i\omega t}\, dt, \tag{86}$$

is a complex function of the real variable ω, which vanishes at both ends of the real axis. According to a general theorem in the theory of analytical functions of a complex variable, the function $\psi(z)$, where the complex variable z replaces ω

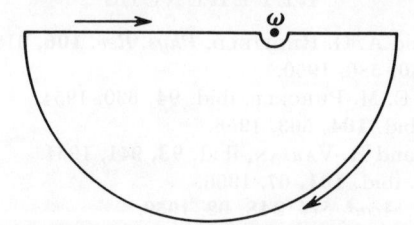

FIG. III, 20. Integration contour.

in (86), is in half the plane im$(z) \leqslant 0$ an analytical regular function of the variable z. The important point for the validity of this statement is that the integral in (86) is over *positive* values of t only, which in turn is a consequence of the causality principle (iii). Then, applying the theorem of residues to the function

$$\xi(z) = \psi(z)/(z-\omega)$$

for the contour shown in Fig. III, 20, we find

$$\mathscr{P} \int_{-\infty}^{\infty} \frac{\chi(\omega') - \chi_\infty}{\omega' - \omega}\, d\omega' + \pi i \{\chi(\omega) - \chi_\infty\} = 0. \tag{87}$$

The real and imaginary parts of (87) give respectively

$$\chi'(\omega) - \chi_\infty = \frac{1}{\pi} \mathscr{P} \int_{-\infty}^{\infty} \frac{\chi''(\omega')\, d\omega'}{\omega' - \omega},$$

$$\chi''(\omega) = -\frac{1}{\pi} \mathscr{P} \int_{-\infty}^{\infty} \frac{\chi'(\omega') - \chi_\infty}{\omega' - \omega}\, d\omega'.$$

These are the K.-K. formulae in the form (8') most suitable for the study of nuclear magnetism. Using the fact that χ' is even and χ'' is odd they are sometimes rewritten as

$$\chi'(\omega)-\chi_\infty = \frac{2}{\pi}\mathscr{P}\int_0^\infty \frac{\chi''(\omega')\omega'\,d\omega'}{\omega'^2-\omega^2},$$

$$\chi''(\omega) = -\frac{2\omega}{\pi}\mathscr{P}\int_0^\infty \frac{\chi'(\omega')-\chi_\infty}{\omega'^2-\omega^2}\,d\omega'. \tag{88}$$

As a special case consider a system with a monochromatic absorption response:

$$\chi''(\omega) = \delta(\omega-\omega_0)-\delta(-\omega-\omega_0) = \delta(\omega-\omega_0)-\delta(\omega+\omega_0). \tag{89}$$

The relations (88) give $\quad \chi'(\omega)-\chi_\infty = \dfrac{2}{\pi}\dfrac{\omega_0}{\omega_0^2-\omega^2},$ (89')

which is the response of an undamped harmonic oscillator. Conversely, it would have been possible to start from (89) and (89') obtained as a limiting case of a slightly damped oscillator, and use these relations as a starting-point to demonstrate the K.-K. relations.

REFERENCES

1. G. WHITFIELD and A. G. REDFIELD, *Phys. Rev.* **106**, 918, 1957.
2. E. HAHN, ibid. **80**, 580, 1950.
3. H. Y. CARR and E. M. PURCELL, ibid. **94**, 630, 1954.
4. H. C. TORREY, ibid. **104**, 563, 1956.
5. M. E. PACKARD and R. VARIAN, ibid. **93**, 941, 1954.
6. A. G. REDFIELD, ibid. **101**, 67, 1956.
7. I. SOLOMON, *C.R. Acad. Sci.* **248**, 92, 1959.
8. H. C. TORREY, *Phys. Rev.* **76**, 1059, 1949.
9. I. SOLOMON, *Phys. Rev. Letters*, **2**, 301, 1959.
10. A. SZÖKE and S. MEIBOOM, *Phys. Rev.* **113**, 585, 1959.
11. N. BLOEMBERGEN, E. M. PURCELL, and R. V. POUND, ibid. **73**, 679, 1948.
12. E. ALLAIS, *C.R. Acad. Sci.* **246**, 2123, 1958.
13. A. ABRAGAM, J. COMBRISSON, and I. SOLOMON, ibid. **245**, 157, 1957.
14. I. SOLOMON and J. FREYCENON, *L'Onde Électrique*, 1960.

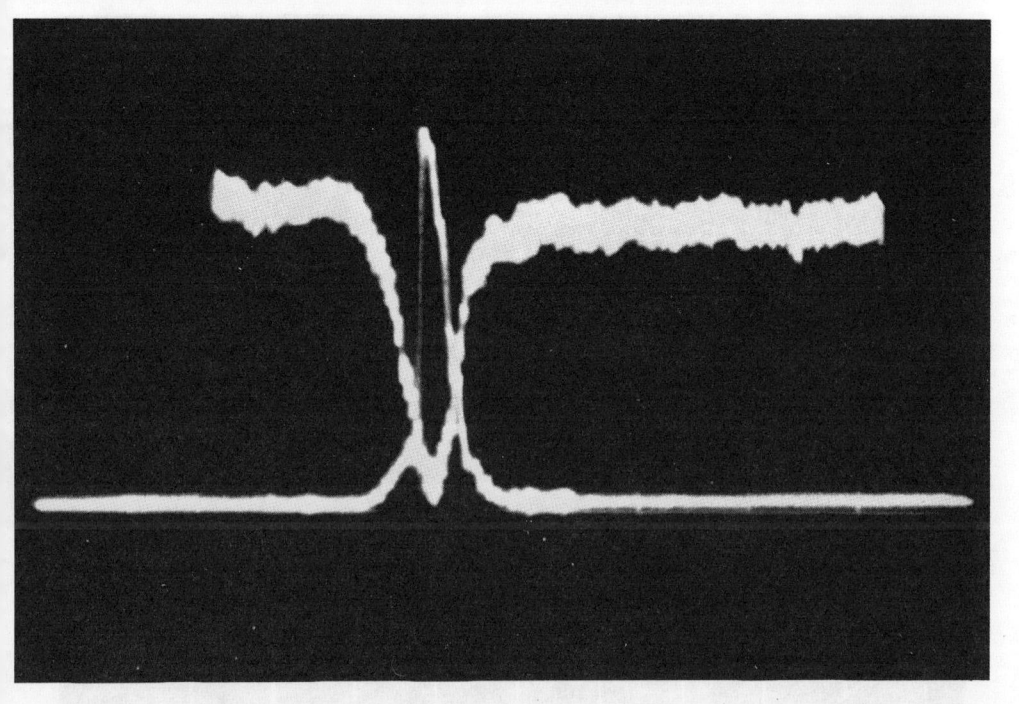

FIG. III, 18. Normal and enhanced signal of protons in water at 3000 gauss. The enhancement is of the order of −50. The scale is different for the two signals as can be seen from the reduced noise on the lower trace.

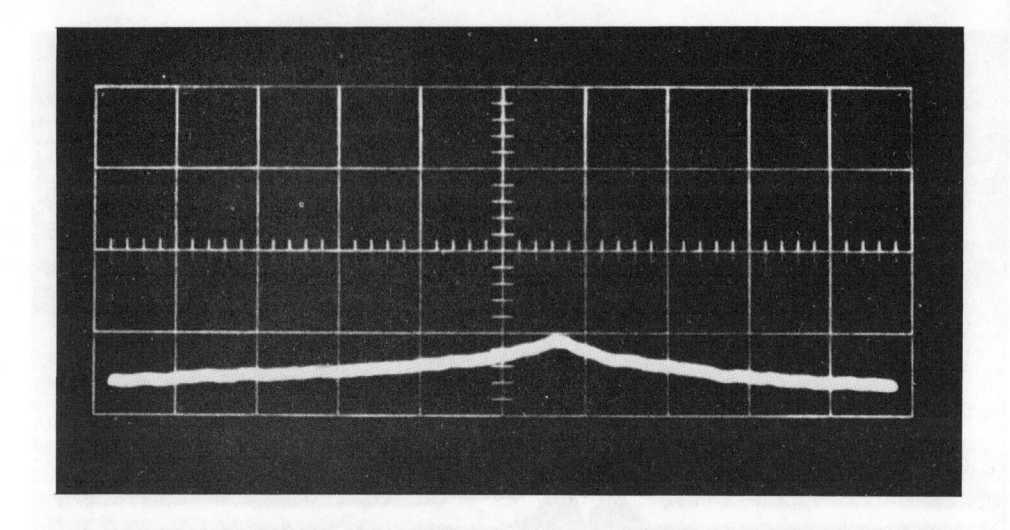

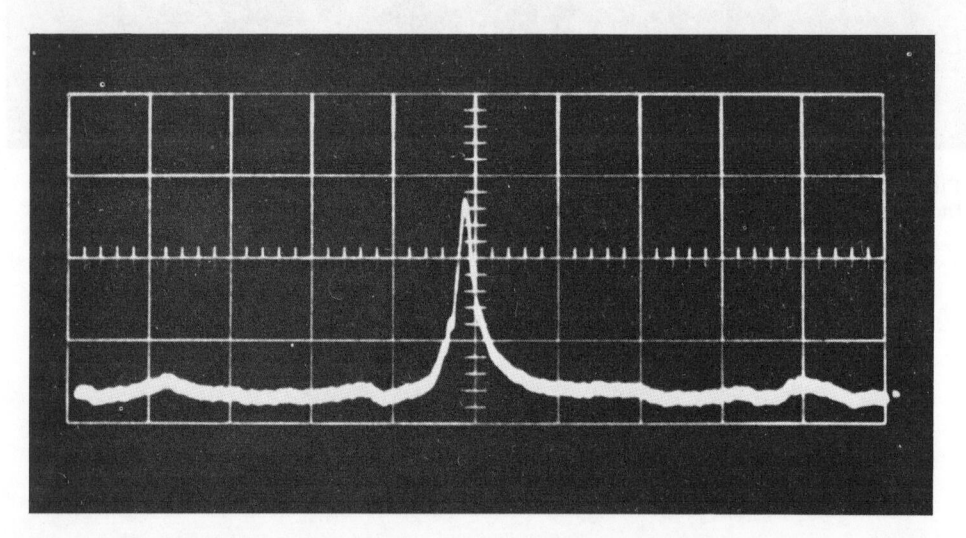

FIG. III, 19. Signal from ordinary water. *Top*: without spinning; *bottom*: spinning at 20 c/s. Linear sweep, 12 milligauss and 5 sec per division. Note the sidebands at 4·8 milligauss in the bottom photograph, caused by the 20 c/s rotation.

IV

DIPOLAR LINE WIDTH IN A RIGID LATTICE

United we fall, divided we stand

I. Introduction

It has already been stated in Chapter II that a magnetic resonance line of a system of spins in an inhomogeneous magnetic field has a certain width owing to the spread of their Larmor frequencies. A broadening of a similar nature can be produced in imperfect crystals by the coupling of nuclear quadrupole moments with small electric field gradients having values which differ from one lattice site to another in a random fashion. In both cases the line width is due to the differences among the resonance frequencies of the individual spins rather than to interactions among them, and the corresponding broadening of the line is called inhomogeneous broadening.

The quadrupolar inhomogeneous broadening will be considered in Chapter VII.

The situation is very different if the line width is due to the existence of couplings between neighbouring spins. This problem will be considered now.

A. The local field

The interaction between two nuclear spins depends on the magnitude and orientation of their magnetic moments and also on the length and orientation of the vector describing their relative positions. The effects of this interaction depend strongly on whether this vector is fixed in space or changes rapidly because of relative motion in the nuclei.

The latter case, which is the rule in liquids and gases, will be studied later and we shall limit ourselves for the present to a rigid lattice where the nuclei can be considered as fixed, a not unreasonable approximation in many solids at room temperature, particularly in ionic crystals.

The interaction of two magnetic moments $\mathbf{\mu}_1 = \gamma_1 \hbar \mathbf{I}_1$ and $\mathbf{\mu}_2 = \gamma_2 \hbar \mathbf{I}_2$ is the well-known dipole-dipole coupling

$$W_{12} = \frac{\gamma_1 \gamma_2 \hbar^2}{r_{12}^3} \left\{ \mathbf{I}_1 \cdot \mathbf{I}_2 - 3 \frac{(\mathbf{I}_1 \cdot \mathbf{r}_{12})(\mathbf{I}_2 \cdot \mathbf{r}_{12})}{r_{12}^2} \right\}. \tag{1}$$

(1) can be written as

$$W_{12} = -\mathbf{\mu}_2 \cdot \mathbf{H}_{12} = -\gamma_2 \hbar \mathbf{I}_2 \cdot \mathbf{H}_{12},$$

where H_{12} is called the local field produced by spin 1 at the site of spin 2, a very useful concept. Since nuclear magnetic moments are of the order of 10^{-3} Bohr magnetons or 10^{-23} c.g.s. and the internuclear distances of the order of Angströms, local fields in a rigid lattice will in general be of the order of a few gauss.

The interaction of two identical dipoles in a large field H_0 can be described in classical terms as follows. The first dipole μ_1 makes a Larmor precession around the field H_0 and thus has a static component along the field and a rotating component in a plane perpendicular to the field. The static component of μ_1 produces at the site of the dipole μ_2 a small static field of orientation relative to H_0 that depends on the relative position of the spins. For large H_0, only the component parallel or antiparallel to H_0 significantly changes the net static field. Since each spin in a lattice has several neighbours, with various relative positions and orientations, the static component of the local field has different values at different sites, and this causes a spread in the Larmor frequencies and a broadening of the line.

The rotating component of μ_1 produces at the site of μ_2 a local magnetic field rotating at the Larmor frequency of μ_1, which is also the Larmor frequency of μ_2. It has a component in the plane perpendicular to H_0 and can therefore change appreciably the orientation of μ_2 by a resonance phenomenon described in Chapter II. The corresponding line width was there shown to be of the order of magnitude of the rotating field, which is here of the same order of magnitude as the local static field and therefore contributes a comparable amount to the broadening.

It is necessary to realize that these two contributions are actually distinct. If the two spins are unlike, the rotating field produced by μ_1 is off resonance for μ_2 and exerts on it a negligible torque, whereas the static field produced by μ_1 at μ_2 is as effective as for like spins. Other things being equal, like neighbours are more effective than unlike ones in broadening the resonance line.

B. General theory of magnetic absorption

In order to obtain a quantitative description of the line shape determined by dipolar broadening it is necessary to introduce a formalism more general than the one used in Chapter II for inhomogeneous broadening. There, two things were shown:

(i) An r.f. field of frequency ω, sufficiently small to preclude saturation, applied to a sample containing nuclear spins, induced in the component of the macroscopic magnetization of the sample along the applied d.c. field a steady change $\Delta M_z = M_z - M_0$ proportional to the shape function $f(\omega)$ which described the spread of the Larmor frequencies among the various spins.

(ii) The time-dependence of the amplitude of the precessing magnetization obtained after a 90° pulse was described by the Fourier transform of the shape function.

These results will now be reformulated and derived again under more general assumptions.

When the spins of the sample are all coupled together by dipolar interaction, the concept of individual independent spins with stationary states becomes inadequate, as exemplified by the fact that the rotating local field produced by one spin reorients its neighbours.

It becomes necessary then to consider the sample as a single large spin system and the transitions induced by the r.f. field as transitions between the various energy levels of this system. The statistical description by a density matrix changes accordingly. Instead of being a statistical ensemble of spins described by a $(2I+1) \times (2I+1)$ density matrix, the *whole* sample of N spins now becomes a single element of a statistical ensemble and is described by a $(2I+1)^N \times (2I+1)^N$ density matrix. Such a change is by no means restricted to nuclear magnetism but occurs quite generally in statistical mechanics whenever one goes from a description of loosely coupled individual systems such as the molecules of a gas at low pressure, to the description of strongly coupled systems such as the atoms of crystal. The first description corresponds to the Maxwell–Boltzmann approach and the second to the Gibbs approach.

Using the Gibbs approach, the description of the steady state can be made as follows. If a linearly polarized r.f. field $H_1 \cos \omega t$ is applied to a system of spins along an axis Ox, the system acquires under steady state conditions a magnetization with a component along the same axis:

$$M_x = H_1\{\chi'(\omega)\cos \omega t + \chi''(\omega)\sin \omega t\}. \tag{1'}$$

The condition of linearity or non-saturation implies that χ' and χ'' are independent of H_1. It was shown in Chapter III that it is possible to measure χ' and χ'' separately and that χ'' is proportional to the rate of absorption of the r.f. energy by the sample.

We are going to derive a general formula for $\chi''(\omega)$. It has been shown that in the linear domain there exist general relations between $\chi'(\omega)$ and $\chi''(\omega)$ due to Kramers and Krönig, independent of the nature of the system considered and enabling one to calculate either quantity when the other one is known for all values of the frequency.

In the following, to avoid confusion we shall represent by the letter M the macroscopic value of the magnetization of the sample and by the script letter $\mathscr{M}$ the corresponding quantum-mechanical operator. We have the relation

$$M = \langle \mathscr{M} \rangle = \mathrm{tr}\{\rho \mathscr{M}\}, \qquad (2)$$

where ρ is the statistical operator or density matrix describing the spin system. Let $\hbar \mathscr{H}$ be the total Hamiltonian of the system in the absence of an applied r.f. field. If the system is in thermal equilibrium at a temperature T prior to the application of the r.f. field its statistical operator will be given by

$$\rho_0 = \exp(-\mathscr{H}\hbar/kT)/\mathrm{tr}\{\exp(-\mathscr{H}\hbar/kT)\}, \qquad (3)$$

which simply means that the statistical behaviour of the sample can be described by assigning to its energy levels $\hbar E_n$ populations proportional to $\exp(-\hbar E_n/kT)$.

In the presence of the r.f. field the equation of motion of ρ is

$$\frac{\hbar}{i}\frac{d\rho}{dt} = [\rho, \hbar\mathscr{H} - V\mathscr{M}_x H_1 \cos\omega t], \qquad (4)$$

where V is the volume of the sample. To solve (4) we make the substitution

$$\rho^* = e^{i\mathscr{H}t}\rho e^{-i\mathscr{H}t}, \qquad (5)$$

which transforms equation (4) into

$$\frac{\hbar}{i}\frac{d\rho^*}{dt} = -VH_1[\rho^*(t), e^{i\mathscr{H}t}\mathscr{M}_x e^{-i\mathscr{H}t}]\cos\omega t. \qquad (6)$$

We suppose that the r.f. field has been applied when the sample was in thermal equilibrium and we had

$$\rho(-\infty) = \rho_0 = \rho^*(-\infty).$$

At time t the solution of (6) to the first order in H_1 will be

$$\rho^*(t) = \rho^*(-\infty) - \frac{iH_1 V}{\hbar}\int_{-\infty}^{t}[\rho^*(-\infty), e^{i\mathscr{H}t'}\mathscr{M}_x e^{-i\mathscr{H}t'}]\cos\omega t'\,dt' \qquad (7)$$

or, going back to ρ by formula (5),

$$\rho(t) = \rho_0 - \frac{iH_1 V}{\hbar}\int_{-\infty}^{t}[\rho_0, e^{-i\mathscr{H}(t-t')}\mathscr{M}_x e^{i\mathscr{H}(t-t')}]\cos\omega t'\,dt'$$

$$= \rho_0 - \frac{iH_1 V}{\hbar}\int_{0}^{\infty}[\rho_0, e^{-i\mathscr{H}t'}\mathscr{M}_x e^{i\mathscr{H}t'}]\cos\omega(t-t')\,dt'. \qquad (8)$$

If we assume that before the application of the r.f. field there was no magnetization along the x-axis, so that
$$M_x(-\infty) = \text{tr}\{\rho_0 \mathcal{M}_x\} = 0,$$
we obtain
$$\langle \mathcal{M}_x(t) \rangle = -\frac{iH_1 V}{\hbar} \text{tr}\left\{\int_0^\infty [\rho_0, e^{-i\mathcal{H}t'} \mathcal{M}_x e^{i\mathcal{H}t'}] \mathcal{M}_x \cos \omega(t-t')\, dt'\right\} \quad (9)$$

and, according to the definition (1'),
$$\chi''(\omega) = -\frac{iV}{\hbar} \int_0^\infty \sin(\omega t') \text{tr}\{[\rho_0, e^{-i\mathcal{H}t'} \mathcal{M}_x e^{i\mathcal{H}t'}] \mathcal{M}_x\}\, dt'. \quad (10)$$

If we use the fact that the temperatures are almost always sufficiently high to allow a linear expansion of the equilibrium density matrix (3) into
$$\rho_0 \cong \left\{\mathcal{E} - \frac{\mathcal{H}\hbar}{kT}\right\} \frac{1}{\text{tr}\,\mathcal{E}},$$
where $\mathcal{E}$ is the unit operator, the susceptibility $\chi''(\omega)$ becomes
$$\chi''(\omega) = -\frac{i}{kT} \frac{V}{\text{tr}\{\mathcal{E}\}} \sum_{n,n'} \int_0^\infty \sin(\omega t') e^{-i(E_n - E_{n'})t'} |\langle n|\mathcal{M}_x|n'\rangle|^2 (E_{n'} - E_n)\, dt', \quad (11)$$

which, integrated by parts, gives
$$\chi''(\omega) = \frac{V}{\text{tr}\{\mathcal{E}\}} \frac{\omega}{kT} \int_0^\infty \cos(\omega t') \sum_{n,n'} |\langle n|\mathcal{M}_x|n'\rangle|^2 e^{-i(E_n - E_{n'})t'}\, dt'. \quad (12)$$

The equation (12) can be rewritten in a compact form in two ways, both very useful.

First, introducing the Heisenberg operator $\mathcal{M}_x(t)$ by the formula
$$\mathcal{M}_x(t) = e^{i\mathcal{H}t} \mathcal{M}_x e^{-i\mathcal{H}t}, \quad (12')$$
(12) can be written
$$\chi''(\omega) = \frac{\omega V}{kT} \frac{1}{\text{tr}(\mathcal{E})} \int_0^\infty \cos \omega t'\, G(t')\, dt', \quad (13)$$
where $G(t)$ is defined as
$$G(t) = \text{tr}(\mathcal{M}_x(t) \mathcal{M}_x). \quad (13')$$
We will call $G(t)$ the correlation function or the relaxation function of the magnetization of the system.

Secondly, (12) can be rewritten
$$\chi''(\omega) = \frac{1}{\text{tr}(\mathcal{E})} \frac{\omega V}{4kT} \int_{-\infty}^{+\infty} (e^{i\omega t'} + e^{-i\omega t'}) \sum_{n,n'} |\langle n|\mathcal{M}_x|n'\rangle|^2 e^{-i(E_n - E_{n'})t'}\, dt'$$

which, using the well-known formula for the δ function

$$\delta(x) = \frac{1}{2\pi} \int_{-\infty}^{+\infty} e^{ixt}\, dt,$$

becomes
$$\chi''(\omega) = \frac{2\pi}{\mathrm{tr}(\mathscr{E})} \frac{\omega V}{4kT} \sum{}' |\langle n|\mathscr{M}_x|n'\rangle|^2, \tag{14}$$

where the summation $\sum'$ is to be taken only over those levels where $|E_n - E_{n'}| = \omega$. Actually, taking the transition probability method for granted, (14) is often used as a starting-point to derive (13) through the integral representation of the δ function. Equation (14) states quite generally that the shape function $f(\omega)$ which determines the shape of the line is proportional to the restricted sum $\sum' |\langle n|\mathscr{M}_x|n'\rangle|^2$, where the dependence on ω comes precisely from the restriction on the summation by the condition $|E_n - E_{n'}| = \omega$. The formulae (13) and (14) are extremely general and independent of the assumption that the magnetic absorption spectrum of the system contains one or several sharp resonance lines. However, in nuclear magnetic resonance this assumption is valid and will be made henceforth. Mathematically it can be formulated as follows:

The Hamiltonian $\hbar\mathscr{H}$ of the system is the sum of a main part $\hbar\mathscr{H}_0$ and a small perturbing part which it is convenient to write as $\hbar\epsilon\mathscr{H}_1$, where ϵ is a measure of the smallness of the perturbation.

In the absence of $\mathscr{H}_1$ the absorption spectrum of the system consists of one or several infinitely sharp lines of frequencies ω_α and the susceptibility $\chi''(\omega)$ can be written

$$\chi_0''(\omega) = \sum A_\alpha \delta(\omega - \omega_\alpha), \tag{15}$$

and conversely the relaxation function $G(t)$ proportional to its Fourier transform can be written as

$$G_0(t) = \sum G_\alpha^0 e^{i\omega_\alpha t}. \tag{15'}$$

If the perturbation $\hbar\epsilon\mathscr{H}_1$ is introduced, the relaxation function becomes $G(\epsilon, t)$, which can in principle be calculated to any order in ϵ by a perturbation method, and then the susceptibility $\chi''(\omega, \epsilon)$ obtained as its Fourier transform.

Before developing this calculation in some detail we shall consider briefly the relationship between $\chi''(\omega)$ and the behaviour of the magnetization after an r.f. pulse. It is a well-known and rather obvious fact that for linear systems the steady-state response to an excitation $\cos \omega t$ is the Fourier transform of the transient response to an infinitely sharp pulse $\delta(t)$.

In practice, however, to approximate such a pulse we should have to apply to the spin system a transient magnetic field considerably larger than the d.c. field H_0.

For a system of interacting nuclear spins in a magnetic field with a sharp resonance line at a frequency ω_0 it is possible to approximate the effect of an infinitely sharp d.c. pulse by applying an r.f. pulse of much longer duration τ and smaller amplitude H_1 at a frequency $\omega \simeq \omega_0$. For, in the frame rotating at the frequency ω there is only a d.c. field H_1 and for it to approximate an infinitely sharp pulse of finite strength it is sufficient that H_1 be much larger than the local field, a much less stringent condition.

We shall come back to this point in a quantitative fashion after a more detailed study of the dipolar broadening.

II. Broadening by Like Spins

A. Dipole-dipole interaction

The total Hamiltonian of a system of identical interacting spins in a large external field can be written

$$\hbar \mathcal{H} = \hbar(\mathcal{H}_0 + \mathcal{H}_1), \tag{16}$$

where the main Hamiltonian is

$$\hbar \mathcal{H}_0 = \sum_j Z^j = -\gamma \hbar H_0 \sum_j I_z^j, \tag{16'}$$

with the energy levels given by $\hbar E_M^0 = -\gamma \hbar H_0 M$, where M is an eigenvalue of

$$I_z = \sum_j I_z^j.$$

The perturbing Hamiltonian $\hbar \mathcal{H}_1$ responsible for the broadening is

$$\hbar \mathcal{H}_1 = \sum_{j<k} W_{jk} = \sum_{j<k} \frac{\hbar^2 \gamma^2}{r_{jk}^3} \left\{ \mathbf{I}^j \cdot \mathbf{I}^k - 3 \frac{(\mathbf{I}^j \cdot \mathbf{r}_{jk})(\mathbf{I}^k \cdot \mathbf{r}_{jk})}{r_{jk}^2} \right\}. \tag{16''}$$

Let us first examine in some detail the interaction between two spins represented for brevity by $\mathbf{i}$ and $\mathbf{i'}$. Let θ and ϕ be the polar coordinates of the vector $\mathbf{r}$ describing their relative positions, the z-axis being parallel to the applied field. $W_{ii'}$ can be transcribed as

$$W_{ii'} = \frac{\gamma^2 \hbar^2}{r^3} \{ \mathbf{i} \cdot \mathbf{i'} - 3[i_z \cos\theta + \sin\theta(i_x \cos\phi + i_y \sin\phi)] \times$$
$$\times [i'_z \cos\theta + \sin\theta(i'_x \cos\phi + i'_y \sin\phi)] \}$$
$$= \frac{\gamma^2 \hbar^2}{r^3} \{ \mathbf{i} \cdot \mathbf{i'} - 3[i_z \cos\theta + \tfrac{1}{2}\sin\theta(i_+ e^{-i\phi} + i_- e^{i\phi})] \times$$
$$\times [i'_z \cos\theta + \tfrac{1}{2}\sin\theta(i'_+ e^{-i\phi} + i'_- e^{i\phi})] \}$$
$$= \frac{\gamma^2 \hbar^2}{r^3} (A+B+C+D+E+F), \tag{17}$$

where

$$\left.\begin{aligned}
A &= i_z i'_z (1 - 3\cos^2\theta) \\
B &= -\tfrac{1}{4}(1 - 3\cos^2\theta)(i_+ i'_- + i_- i'_+) = \tfrac{1}{2}(1 - 3\cos^2\theta)(i_z i'_z - \mathbf{i}.\mathbf{i}') \\
C &= -\tfrac{3}{2} \sin\theta \cos\theta\, e^{-i\phi}(i_z i'_+ + i'_z i_+) \\
D &= C^* = -\tfrac{3}{2} \sin\theta \cos\theta\, e^{i\phi}(i_z i'_- + i'_z i_-) \\
E &= -\tfrac{3}{4} \sin^2\theta\, e^{-2i\phi} i_+ i'_+ \\
F &= E^* = -\tfrac{3}{4} \sin^2\theta\, e^{2i\phi} i_- i'_-
\end{aligned}\right\}. \quad (18)$$

The reason for decomposing W in this fashion is the following. According to formula (14),

$$\chi''(\omega) \propto \sum_{n,n'}{}' |\langle n|\mathscr{M}_x|n'\rangle|^2,$$

which makes it necessary to determine the change in the energy levels of $\hbar\mathscr{H}_0$, caused by the presence of $\hbar\mathscr{H}_1$, and in that respect the various operators A through F make contributions which are qualitatively different. These operators acting on a state of the unperturbed Hamiltonian, characterized by the values $i_z = m$, $i'_z = m'$, affect this state as follows:

$$\left.\begin{array}{llll}
A & \Delta m = 0 & \Delta m' = 0 & \Delta(m+m') = 0 \\
B & \Delta m = \pm 1 & \Delta m' = \mp 1 & \Delta(m+m') = 0 \\
C & \Delta m = \begin{cases} 0 \\ 1 \end{cases} & \Delta m' = \begin{cases} 1 \\ 0 \end{cases} & \Delta(m+m') = 1 \\
D & \Delta m = \begin{cases} 0 \\ -1 \end{cases} & \Delta m' = \begin{cases} -1 \\ 0 \end{cases} & \Delta(m+m') = -1 \\
E & \Delta m = 1 & \Delta m' = 1 & \Delta(m+m') = 2 \\
F & \Delta m = -1 & \Delta m' = -1 & \Delta(m+m') = -2
\end{array}\right\} \quad (19)$$

Let us now consider an energy level $\hbar E^0_M = -\gamma\hbar H_0 M$ of the Hamiltonian (16′). This level is highly degenerate since there are many ways in which the individual values $I^j_z = m^j$ can be added to give a value $M = \sum m^j$. The level $\hbar E^0_M$ thus corresponds to a degenerate manifold of states $|M\rangle$, the degeneracy being lifted, at least partially, by the perturbing Hamiltonian $\hbar\mathscr{H}_1$ which splits the level $\hbar E^0_M$ into many sub-levels. According to first-order perturbation theory, only those parts of the perturbing Hamiltonian which have matrix elements, inside the manifold $|M\rangle$, that is, such that acting on a state $|M\rangle$ they induce no change in the value of M, contribute in first order to the splitting of the level $\hbar E^0_M$. Looking back at the formulae (19), we see that only the parts A and B of W satisfy this condition and should be retained for a perturbation calculation of the energy levels of $\hbar\mathscr{H}$.

The term A has the same form as for two classical interacting dipoles and describes the effect of the static local field mentioned in the introduction. The term B allows a simultaneous reversal of two neighbouring spins in opposite directions. Called somewhat familiarly the 'flip-flop' term it corresponds to the resonance effects of the rotating local field described in the introduction. The effect of a term such as C is to admix into a state $|M\rangle$ with unperturbed energy $\hbar E_M^0 = -\gamma\hbar H_0 M$, a small amount of a state $|M-1\rangle$. A correct eigenstate of $\hbar\mathcal{H}$ will thus be $|M\rangle + \alpha|M-1\rangle + ...$, where α is a small quantity. The coupling of the spin system with an r.f. field applied along ox is proportional to $I_x = \sum I_x^j$ and can only induce transitions with $\Delta M = \pm 1$. Weak transitions of approximate energy $2\hbar\omega_0$ between a state, say, $|M-2\rangle$ + small admixture, of approximate energy $-\gamma\hbar H_0(M-2)$, and a state $|M\rangle + \alpha|M-1\rangle + ...$ become possible with a probability of the order of α^2. A line, too weak usually to be observed, thus appears at a frequency $2\omega_0$ and also, as can easily be seen, lines of comparable intensities at the frequencies 0 and $3\omega_0$.

The justification for keeping in the perturbing Hamiltonian $\hbar\mathcal{H}_1$ only the terms A and B that commute with $\mathcal{H}_0$ and are usually called the adiabatic or secular part of $\hbar\mathcal{H}_1$, denoted henceforth as $\hbar\mathcal{H}'_1$, can be presented in a slightly different manner. Since $\chi''(\omega)$ is proportional to the Fourier transform of $G(t) = \text{tr}\{\mathcal{M}_x(t)\mathcal{M}_x\}$, it can be calculated if $\mathcal{M}_x(t) = e^{i\mathcal{H}t}\mathcal{M}_x e^{-i\mathcal{H}t}$ is known. Now $\mathcal{M}_x(t)$ obeys the equation

$$\frac{1}{i}\frac{d\mathcal{M}_x}{dt} = [\mathcal{H}_0 + \mathcal{H}_1, \mathcal{M}_x(t)]. \tag{20}$$

This equation is usually solved by using the so-called interaction representation where an operator $\tilde{\mathcal{M}}_x(t)$ is defined through

$$\tilde{\mathcal{M}}_x(t) = e^{-i\mathcal{H}_0 t}\mathcal{M}_x(t)e^{i\mathcal{H}_0 t}. \tag{21}$$

If the perturbation $\hbar\mathcal{H}_1$ is absent, $\tilde{\mathcal{M}}_x(t)$ is just $\mathcal{M}_x$ and is time-independent. Therefore, if $\hbar\mathcal{H}_1$ is small, it is reasonable to expect that the variation of $\tilde{\mathcal{M}}_x(t)$ will be slow. From (20) and (21) the equation of motion of $\tilde{\mathcal{M}}_x(t)$ becomes

$$\frac{1}{i}\frac{d\tilde{\mathcal{M}}_x}{dt} = [e^{-i\mathcal{H}_0 t}\mathcal{H}_1 e^{i\mathcal{H}_0 t}, \tilde{\mathcal{M}}_x]. \tag{22}$$

As a basic set of states we choose eigenstates of $\hbar\mathcal{H}_0$ labelled $|E^0, s\rangle$ where the second quantum number s is introduced because of degeneracy to discriminate between eigenstates of $\hbar\mathcal{H}_0$ of equal unperturbed energy $\hbar E^0$.

Equation (22) can then be written in matrix form:

$$\frac{1}{i}\frac{d}{dt}\langle E^0 s|\tilde{\mathscr{M}}_x|E^{0\prime}s'\rangle = \sum_{E''s''} e^{-i(E^0-E^{0\prime\prime})t}\langle E^0 s|\mathscr{H}_1|E^{0\prime\prime}s''\rangle\langle E^{0\prime\prime}s''|\tilde{\mathscr{M}}_x|E^{0\prime}s'\rangle -$$
$$- e^{-i(E^{0\prime\prime}-E^{0\prime})t}\langle E^0 s|\tilde{\mathscr{M}}_x|E^{0\prime\prime}s''\rangle\langle E^{0\prime\prime}s''|\mathscr{H}_1|E^{0\prime}s'\rangle. \tag{23}$$

Since the variation of $\tilde{\mathscr{M}}_x$ is slow it is reasonable to assume that on the right-hand side of (23) the contribution of terms with rapidly varying exponentials $e^{-i(E^0-E^{0\prime\prime})t}$ averages approximately to zero, and to neglect them in comparison with the secular terms for which $E^{0\prime\prime} = E^{0\prime}$ or $E^{0\prime\prime} = E^0$. But this is precisely equivalent to retaining in the perturbing Hamiltonian $\hbar\mathscr{H}_1$ the part that commutes with $\hbar\mathscr{H}_0$.

The equation (23) then reduces to

$$\frac{1}{i}\frac{d}{dt}\langle E^0 s|\tilde{\mathscr{M}}_x|E^{0\prime}s'\rangle = \sum_{s''}\langle E^0 s|\mathscr{H}_1|E^0 s''\rangle\langle E^0 s''|\tilde{\mathscr{M}}_x|E^{0\prime}s'\rangle -$$
$$-\langle E^0 s|\tilde{\mathscr{M}}_x|E^{0\prime}s''\rangle\langle E^{0\prime}s''|\mathscr{H}_1|E^{0\prime}s'\rangle. \tag{23$'$}$$

The task of finding the eigenstates even of the reduced or, as it is sometimes called, the 'truncated' Hamiltonian:

$$\hbar(\mathscr{H}_0+\mathscr{H}'_1) = \sum_j Z^j + \sum_{j<k}\gamma^2\hbar^2\frac{A_{jk}+B_{jk}}{r_{jk}^3}$$

in order to calculate the susceptibility, given by equation (14), or alternatively of calculating $G(t)$ by solving the reduced system (23$'$), is a formidable one and no exact solution of this problem exists. It is, however, possible to go beyond the qualitative statements of the introduction by using the method of the moments of the resonance curve which will be described now.

B. Definition of the moments

For a resonance curve described by a normalized shape function $f(\omega)$ with a maximum at a frequency ω_0, the nth moment M_n with respect to the point ω_0 will be defined as

$$M_n = \int (\omega-\omega_0)^n f(\omega)\, d\omega.$$

If $f(\omega)$ is symmetrical with respect to ω_0 all odd moments vanish. The knowledge of the moments gives some information on the shape of the resonance curve and in particular on the rate at which it falls down to zero in the wings, far away from ω_0.

The utility of the moments resides in the fact that they can be calculated from first principles without having to find the eigenstates of the total Hamiltonian $\hbar\mathscr{H}$. Before describing the principle of that calculation, consider two examples of resonance curves of different shapes. The gaussian curve is described by the normalized function

$$f(\omega) = \frac{1}{\Delta\sqrt{(2\pi)}} \exp\left(\frac{-(\omega-\omega_0)^2}{2\Delta^2}\right), \qquad (24)$$

from which one computes easily

$$M_2 = \Delta^2, \qquad M_4 = 3\Delta^4,$$
$$M_{2n} = 1.3.5\ldots(2n-1)\Delta^{2n},$$

the odd moments vanishing. The half width at half intensity δ, given by $f(\omega_0+\delta) = \tfrac{1}{2}f(\omega_0)$ or $\exp(-\delta^2/2\Delta^2) = \tfrac{1}{2}$, comes out to be

$$\delta = \Delta\sqrt{(2\log 2)} = 1\cdot 18\Delta.$$

It is seen that for a gaussian curve the knowledge of the second moment $M_2 = \Delta^2$ provides a fair approximation for the line width δ. Another shape often encountered in magnetic resonance and already met in Chapter II for lines broadened by strong collisions, is the Lorentz shape described by the normalized function

$$f(\omega) = \frac{\delta}{\pi} \frac{1}{\delta^2+(\omega-\omega_0)^2}, \qquad (25)$$

where δ is the half width at half intensity.

No second or higher moment can be defined for this shape, for the corresponding integrals diverge. It happens sometimes that the theory provides finite values for the second and fourth moment of curves which experimentally fit the Lorentz shape in the region of observation. Very far in the wings where the absorption is too weak to be measured accurately the line must cut off more rapidly than the Lorentz shape, to fit finite values of M_2 and M_4.

A crude but convenient trial model is the description of the curve by the formula (25) within the interval $|\omega-\omega_0| \leqslant \alpha$ with $\alpha \gg \delta$, and the assumption that it is zero outside this interval. We then find, neglecting terms of the order of δ/α,

$$M_2 = \Delta^2 = \frac{2\alpha\delta}{\pi}, \qquad M_4 = \frac{2\alpha^3\delta}{3\pi}, \qquad (25')$$

whence δ and α are computed if M_2 and M_4 are known. Since

$$M_4/(M_2)^2 = \frac{\pi\alpha}{6\delta}$$

this approximation can only be tried when the theoretical ratio $M_4/(M_2)^2$

turns out to be a large number. We then find

$$\frac{\delta}{\Delta} = \frac{\pi}{2\sqrt{3}} \left(\frac{M_2^2}{M_4}\right)^{\frac{1}{2}}. \qquad (25'')$$

The half intensity width is much smaller than the r.m.s. width. On the other hand, the assumption of a gaussian shape may appear reasonable whenever the ratio $M_4/(M_2)^2$ is of the order of 3.

C. Principle of the calculation of moments

The main weakness of the method of moments is that an important contribution to the value of a moment, which is more important the higher the moment, comes from the wings of the curve which are not actually observed. In particular, care must be exercised to exclude from the calculated moments of the magnetic resonance line centred at the Larmor frequency $\omega = \omega_0$, the contributions of the satellite lines at frequencies $\omega = 0$, $2\omega_0$, $3\omega_0$ previously mentioned. It is easy to see that in spite of their weak intensity, caused by their remoteness from the central frequency ω_0, their contribution to the second moment would be comparable to that of the main line, and would be even higher for higher moments. The way to get rid of them is to restrict the perturbing Hamiltonian $\hbar\mathcal{H}_1$, responsible for the broadening, to its secular part $\hbar\mathcal{H}'_1$, which commutes with $\mathcal{H}_0$ and thus is unable to mix together states of different total M: such mixing is the origin of the troublesome satellite lines. The reason for reducing the dipolar Hamiltonian to its secular part

$$\hbar\mathcal{H}'_1 = \sum_{j<k} \frac{\gamma^2 \hbar^2}{r_{jk}^3} (A_{jk} + B_{jk})$$

is therefore not only to make the calculation of the moments simpler but to make it correct.

Before starting the calculation let us remark that the magnetic resonance line is symmetrical with respect to the central frequency ω_0. This can be seen as follows. If $|a\rangle$ and $|b\rangle$ are two eigenstates of $\hbar(\mathcal{H}_0 + \mathcal{H}'_1)$ separated in energy by $\hbar(E_a - E_b) = \hbar\omega_0 + \delta_{ab}$, the two states $|\tilde{a}\rangle$ and $|\tilde{b}\rangle$ obtained from $|a\rangle$ and $|b\rangle$ respectively by reversing all the spins, will also be eigenstates of $\hbar(\mathcal{H}_0 + \mathcal{H}'_1)$ with $\hbar(E_{\tilde{b}} - E_{\tilde{a}}) = \hbar\omega_0 - \delta_{ab}$; and thus to each transition at a frequency $\omega_0 + u$ there corresponds a transition of equal intensity at a frequency $\omega_0 - u$. If $f(\omega)$ is the shape function, $h(u) = f(\omega_0 + u)$ is an *even* function of u. Since moments of a curve are proportional to the derivatives at the origin of its Fourier transform we shall use for their calculation the formula (13). Because the nuclear magnetic resonance line is narrow we can neglect the variation of the factor ω over its width and assume that the shape of the line is described by $\chi''(\omega)/\omega$ as well as by $\chi''(\omega)$. Then, $f(\omega)$ being the normalized shape function, (13) can be rewritten

$$f(\omega) = \mathscr{A} \int_0^\infty G(t)\cos\omega t\, dt, \qquad (26)$$

where the constant $\mathscr{A}$ is determined by the normalization of $f(\omega)$ and the even function $G(t)$ previously defined is $\mathrm{tr}\{\mathscr{M}_x(t)\mathscr{M}_x\}$. Conversely,

$$G(t) = \frac{2}{\pi\mathscr{A}} \int_0^\infty f(\omega)\cos\omega t\, d\omega. \qquad (27)$$

According to the previous considerations, in the definition of

$$\mathscr{M}_x(t) = e^{i\mathscr{H}t}\mathscr{M}_x e^{-i\mathscr{H}t}$$

we must take $\mathscr{H} = \mathscr{H}_0 + \mathscr{H}'_1$ rather than $\mathscr{H} = \mathscr{H}_0 + \mathscr{H}_1$. This leads to an important simplification. Since $\mathscr{H}_0$ and $\mathscr{H}'_1$ commute it is permissible to write

$$\exp i(\mathscr{H}_0 + \mathscr{H}'_1)t = \exp(i\mathscr{H}_0 t)\exp(i\mathscr{H}'_1 t).$$

Because the Zeeman Hamiltonian $\hbar\mathscr{H}_0$ is equal to $\hbar\omega_0 I_z$, $G(t)$ can be rewritten as

$$G(t) = \mathrm{tr}\{e^{i\omega_0 I_z t}e^{i\mathscr{H}'_1 t}\mathscr{M}_x e^{-i\mathscr{H}'_1 t}e^{-i\omega_0 I_z t}\mathscr{M}_x\}. \qquad (28)$$

The trace of a product of operators is invariant through a cyclic permutation, whence

$$G(t) = \mathrm{tr}\{e^{i\mathscr{H}'_1 t}\mathscr{M}_x e^{-i\mathscr{H}'_1 t}e^{-i\omega_0 I_z t}\mathscr{M}_x e^{i\omega_0 I_z t}\}. \qquad (28')$$

The operator $e^{i\omega_0 I_z t}$ represents a rotation of angle $\omega_0 t$ along the z-axis, and so we have

$$e^{-i\omega_0 I_z t}\mathscr{M}_x e^{i\omega_0 I_z t} = \mathscr{M}_x \cos\omega_0 t + \mathscr{M}_y \sin\omega_0 t,$$

$$G(t) = \cos\omega_0 t\, \mathrm{tr}\{e^{i\mathscr{H}'_1 t}\mathscr{M}_x e^{-i\mathscr{H}'_1 t}\mathscr{M}_x\} + \sin\omega_0 t\, \mathrm{tr}\{e^{i\mathscr{H}'_1 t}\mathscr{M}_x e^{-i\mathscr{H}'_1 t}\mathscr{M}_y\}. \qquad (29)$$

It is easy to see that the second term of (29) vanishes since a rotation of the spins through 180° around, say, ox, leaves $\mathscr{H}'_1$ and $\mathscr{M}_x$ unchanged but turns $\mathscr{M}_y$ into $-\mathscr{M}_y$.

In (27) replacing $G(t)$ by $G_1(t)\cos\omega_0 t$, where

$$G_1(t) = \mathrm{tr}\{e^{i\mathscr{H}'_1 t}\mathscr{M}_x e^{-i\mathscr{H}'_1 t}\mathscr{M}_x\}$$

will be called the reduced auto-correlation function, and writing

$$h(u) = f(\omega_0 + u),$$

one gets

$$G_1(t)\cos\omega_0 t = \frac{2}{\pi\mathscr{A}} \int_{-\omega_0}^{+\infty} f(\omega_0 + u)\cos(\omega_0 + u)t\, du$$

or, extending the lower limit to $-\infty$, which is legitimate if the line is narrow,

$$G_1(t)\cos\omega_0 t = \frac{2}{\pi\mathscr{A}}\left\{\cos\omega_0 t \int_{-\infty}^{+\infty} h(u)\cos ut\, du - \sin\omega_0 t \int_{-\infty}^{+\infty} h(u)\sin ut\, du\right\}.$$

Since $h(u)$ is an even function the second integral vanishes and one gets

$$G_1(t) = \mathrm{tr}\{e^{i\mathscr{H}'_1 t}\mathscr{M}_x e^{-i\mathscr{H}'_1 t}\mathscr{M}_x\},$$

$$G_1(t) = \frac{2}{\pi\mathscr{A}} \int_{-\infty}^{+\infty} h(u)\cos ut\, du. \tag{30}$$

The various moments of the distribution curve $h(u)$ around the resonance frequency $\omega = \omega_0$ are given by

$$M_n = \int_{-\infty}^{+\infty} h(u)u^n\, du.$$

The odd moments vanish and the even ones are given by

$$M_{2n} = (-1)^n \tfrac{1}{2}\pi\mathscr{A}\left(\frac{d^{2n}G_1(t)}{dt^{2n}}\right)_{t=0} = (-1)^n\left(\frac{d^{2n}G_1(t)}{dt^{2n}}\right)_{t=0}\!\!\Big/G_1(0). \tag{31}$$

In order to obtain the moments of the resonance curve it is then sufficient to expand $G_1(t)$ in (30) in powers of t, the coefficients of the expansion being traces of operators which are polynomial functions of $\mathscr{H}'_1$ and $\mathscr{M}_x$.

The gist of the method is that these traces are independent of the basic states chosen for their evaluation and can be computed, for instance, in a representation where the values $m^j = I_z^j$ of the individual spins (called from this the m^j representation) are good quantum numbers, thus circumventing the insoluble problem of finding the eigenstates $|n\rangle$ of the total Hamiltonian. From the definition (30) of $G_1(t)$, we find that the value of its pth derivative for $t = 0$ is given by

$$\left(\frac{d^p G_1(t)}{dt^p}\right)_{t=0} = (i)^p \,\mathrm{tr}\Big\{\underbrace{[\mathscr{H}'_1,[\mathscr{H}'_1,[...,[\mathscr{H}'_1,\mathscr{M}_x]...]\,]\mathscr{M}_x}_{p\text{ times}}\Big\}. \tag{32}$$

The formula (32) is derived simply by starting from the differential equation satisfied by the time-dependent operator $\mathscr{M}'_x(t) = e^{i\mathscr{H}'_1 t}\mathscr{M}_x e^{-i\mathscr{H}'_1 t}$, which reads

$$\frac{1}{i}\frac{d\mathscr{M}'_x}{dt} = [\mathscr{H}'_1, \mathscr{M}'_x]. \tag{33}$$

The solution of this equation can be represented by a series

$$\mathscr{M}'_x(t) = \mathscr{M}_x + \mathscr{M}'^{(1)}_x(t) + ... + \mathscr{M}'^{(n)}_x(t)$$

where the various terms are obtained by induction through the relation

$$\mathscr{M}_x'^{(n)}(t) = i\int_0^t [\mathscr{H}'_1, \mathscr{M}_x'^{(n-1)}(t')]\,dt',$$

whence (32) follows immediately.

From (31) and (32) we obtain for the first two even moments

$$M_2 = -\frac{\mathrm{tr}\{[\mathscr{H}'_1, I_x]^2\}}{\mathrm{tr}\{I_x^2\}}, \tag{34}$$

$$M_4 = \frac{\mathrm{tr}\{[\mathscr{H}'_1, [\mathscr{H}'_1, I_x]]^2\}}{\mathrm{tr}\{I_x^2\}}. \tag{34'}$$

In (34) we have replaced $\mathscr{M}_x$ by the total spin I_x, proportional to it. It should be remembered that since we have defined the Hamiltonian as $\hbar\mathscr{H}$, these moments correspond to widths measured on the $\omega = 2\pi\nu$ scale.

D. Calculation of the second and fourth moments (1)

The commutator $[\mathscr{H}'_1, I_x]$ contained in the expressions (34) for M_2 and M_4 is given by the formula

$$[\mathscr{H}'_1, I_x] = \gamma^2\hbar \sum_{j<k} \frac{1}{r_{jk}^3}\Big[A_{jk} + B_{jk}, \sum_l I_x^l\Big]$$

$$= \gamma^2\hbar \sum_{j<k} \frac{1}{r_{jk}^3}\tfrac{3}{2}(1 - 3\cos^2\theta_{jk})[I_z^j I_z^k, I_x^j + I_x^k], \tag{35}$$

since $I_x^j + I_x^k$ commutes with $\mathbf{I}^j \cdot \mathbf{I}^k$. The fact that the commutator $[\mathscr{H}'_1, I_x]$ and thus also the second moment is unaffected by any interaction among the spins that commutes with I_x, and in particular by any scalar interaction of the type $\mathbf{I}^j \cdot \mathbf{I}^k$, will prove important later on. From the elementary properties of the components of the angular momentum we get

$$[\mathscr{H}'_1, I_x] = \tfrac{3}{2}i\gamma^2\hbar \sum_{j<k} \frac{1 - 3\cos^2\theta_{jk}}{r_{jk}^3}(I_z^j I_y^k + I_z^k I_y^j), \tag{36}$$

$$-\mathrm{tr}\{[\mathscr{H}'_1, I_x]^2\} = \tfrac{2}{9}\gamma^4\hbar^2 I^2(I+1)^2(2I+1)^N \sum_{j<k} b_{jk}^2, \tag{37}$$

where
$$b_{jk} = \frac{3}{2}\frac{1 - 3\cos^2\theta_{jk}}{r_{jk}^3},$$

N is the number of the spins, and θ_{jk} the angle of the vector $\mathbf{r}_{jk}$ with the applied magnetic field H_0.

The sum $\sum_{j<k} b_{jk}^2$ is absolutely convergent so that we can neglect surface effects and write

$$\sum_{j<k} b_{jk}^2 = \tfrac{1}{2}\sum_{j\neq k} b_{jk}^2 = \tfrac{1}{2}N\sum_{k} b_{jk}^2,$$

where the last single sum is independent of the index j. Similarly,

$$\mathrm{tr}\{I_x^2\} = \sum_j \mathrm{tr}\{I_x^j\}^2 = \tfrac{1}{3}NI(I+1)(2I+1)^N,$$

whence the expression for the second moment (Van Vleck formula):

$$M_2 = \overline{\Delta\omega^2} = \tfrac{1}{3}\gamma^4\hbar^2 I(I+1)\sum_k b_{jk}^2$$

$$= \tfrac{3}{4}\gamma^4\hbar^2 I(I+1)\sum_k \frac{(1-3\cos^2\theta_{jk})^2}{r_{jk}^6}. \tag{38}$$

For a powder made of crystallites of random orientations it is permissible to average $(1-3\cos^2\theta_{jk})^2$ over all directions, leading to

$$M_2 = \tfrac{3}{5}\gamma^4\hbar^2 I(I+1)\sum_k \frac{1}{r_{jk}^6}, \tag{39}$$

which for a simple cubic lattice of lattice constant d, where

$$\sum_k \frac{1}{r_{jk}^6} = \frac{8\cdot 5}{d^6},$$

gives
$$M_2 = 5\cdot 1\gamma^4\hbar^2 I(I+1)\frac{1}{d^6}. \tag{39'}$$

In the case of a single crystal for a simple cubic lattice it is found (1) that
$$M_2 = \overline{\Delta\omega^2} = 12\cdot 3\gamma^4\hbar^2 I(I+1)\frac{1}{d^6}(\lambda_1^4+\lambda_2^4+\lambda_3^4-0\cdot 187), \tag{39''}$$

where $\lambda_1, \lambda_2, \lambda_3$ are the cosines of the direction of the applied field with respect to the crystal axes.

In connexion with these results we may remark that the second moment M_2 is $\tfrac{9}{4} = (\tfrac{3}{2})^2$ times larger than what would be obtained by naïvely taking for the perturbing Hamiltonian $\hbar\mathscr{H}'_1$ the static part

$$\gamma^2\hbar \sum_{j<k} \frac{A_{jk}}{r_{jk}^3}$$

only, neglecting the flip-flop part represented by the terms B. Conversely, a larger value would have been obtained if instead of 'truncating' $\mathscr{H}_1$ all the terms from A to F had been included. A simple calculation shows that for a powder, replacing $\mathscr{H}'_1$ by $\mathscr{H}_1$ in the commutator $[\mathscr{H}'_1, I_x]$ results in an increase of M_2 in the ratio 10/3.

It is not always possible from the knowledge of the second moment only, to draw even qualitative conclusions as to the width of the resonance curve since, as was pointed out before, any interaction that commutes with I_x, however strongly it affects the line shape, makes no contribution to the second moment. It is therefore advisable to calculate at least the fourth moment using the second of the equations (34). This calculation although straightforward is fairly tedious and for that reason, the final result only will be given:

$$M_4 = \overline{\Delta\omega^4}$$
$$= \gamma^8 \hbar^4 \left\{ 3\left(\sum_k b_{jk}^2\right)^2 - \frac{1}{3N}\sum_{jkl\neq} b_{jk}^2(b_{jl}-b_{kl})^2 - \tfrac{1}{5}\sum_k b_{jk}^4\left(8 + \frac{3}{2I(I+1)}\right) \right\} \times$$
$$\times \left[\frac{I(I+1)}{3}\right]^2. \quad (40)$$

The symbol $\sum_{jkl\neq}$ means that no two indices should be equal in the triple summation. Even for a simple cubic lattice the evaluation of (40) is extremely cumbersome if the direction of the field with respect to the crystal axes is arbitrary.

If only the first term were present inside the curly bracket one would have exactly $M_4 = 3(M_2)^2$ suggesting a Gaussian shape. An approximate calculation of (40) for a simple cubic lattice and for $I = \tfrac{1}{2}$ leads to the following values for the ratio $M_4/(M_2)^2$, taking the first twenty-six nearest neighbours for M_4:

Gaussian | Direction of the fields . 100 110 111
3 | $M_4/(M_2)^2$ . . . 2·07 2·22 2·30

Finally, we may mention that the sixth moment has also been computed for a simple cubic lattice with the field along 100 and gives for spin $I = \tfrac{1}{2}$, $M_6/(M_2)^3 \simeq 5$ (instead of 15 for a gaussian). Both the ratios $M_4/(M_2)^2$ and $M_6/(M_2)^3$ are significantly smaller than for a gaussian and would indicate a curve with a flatter top than that of a gaussian, but this departure is not so important as to make a gaussian model, which has the merit of simplicity, grossly incorrect.

Instead of $M_4/(M_2)^2$ the fourth root of this quantity is usually given in the literature, attenuating somewhat artificially the difference between the real curve and a gaussian

	Field parallel to		
Gaussian	100	110	111
1·32	1·20	1·22	1·23

E. Relationship between the line shape and the free precession signal

Before comparing these results with experiment it will be shown that the reduced auto-correlation function $G_1(t)$, related to the shape function $f(\omega) = h(\omega-\omega_0)$ by the Fourier transformation of equation (30), has a simple physical interpretation. It is proportional to the amplitude of the free precession signal after a 90° r.f. pulse, and may be measured directly. In special cases this measurement may be more convenient than the continuous wave study of the resonance curve with a very weak r.f. field.

To prove this statement let us assume that before the r.f. pulse the spin system is in thermal equilibrium and is described by a statistical operator

$$\rho_{\text{equil}} \propto \exp\left(-\frac{\hbar\mathcal{H}}{kT}\right) \cong \mathcal{E} - \frac{\hbar\mathcal{H}}{kT} \simeq \mathcal{E} - \frac{\omega_0 V}{\gamma kT} M_z, \tag{41}$$

where the small differences in populations due to the perturbing dipolar Hamiltonian $\hbar\mathcal{H}_1$ are neglected. At time $t = 0$ a rotating field H_1 is applied along the axis oy of a frame rotating with an angular velocity $\omega = \omega_0$ for a duration τ such that $|\gamma|H_1\tau = \frac{1}{2}\pi$. If the field H_1 is much larger than the local field, and therefore $|\mathcal{H}'_1\tau| \ll 1$, we can disregard the existence of the dipolar broadening for the duration of the pulse. The net effect of the 90° r.f. pulse is then to transform the operator $\mathcal{M}_z$ into $\mathcal{M}_x$ in the rotating frame and into

$$e^{i\omega I_z\tau}\mathcal{M}_x e^{-i\omega I_z\tau} = \mathcal{M}_x\cos\omega\tau - \mathcal{M}_y\sin\omega\tau$$

in the laboratory frame. At the end of the pulse, the statistical operator describing the spin system is then

$$\rho(\tau) = \mathcal{E} - \frac{\omega_0 V}{\gamma kT}[\mathcal{M}_x\cos\omega_0\tau - \mathcal{M}_y\sin\omega_0\tau], \tag{42}$$

since $\omega = \omega_0$. From then on the evolution of the statistical operator is governed by the Hamiltonian $\hbar\mathcal{H} = \hbar(\mathcal{H}_0+\mathcal{H}_1)$ where, as was demonstrated previously, it is permissible to replace $\mathcal{H}_1$ by the secular part $\mathcal{H}'_1$. Thus, at time t,

$$\rho(t) = \mathcal{E} - \frac{\omega_0 V}{\gamma kT} e^{i(\mathcal{H}_0+\mathcal{H}'_1)(t-\tau)}[\mathcal{M}_x\cos\omega_0\tau - \mathcal{M}_y\sin\omega_0\tau]e^{-i(\mathcal{H}_0+\mathcal{H}'_1)(t-\tau)}$$

which, since $\mathcal{H}_0 = \omega_0 I_z$ and $\mathcal{H}'_1$ commute, can be rewritten as

$$\rho(t) = \mathcal{E} - \frac{\omega_0 V}{\gamma kT}\{e^{i\mathcal{H}'_1(t-\tau)}\mathcal{M}_x e^{-i\mathcal{H}'_1(t-\tau)}\cos\omega_0 t - e^{i\mathcal{H}'_1(t-\tau)}\mathcal{M}_y e^{-i\mathcal{H}'_1(t-\tau)}\sin\omega_0 t\}.$$

$$\tag{43}$$

The component M_x of the magnetization is given by

$$\langle \mathscr{M}_x(t) \rangle = \text{tr}\{\rho(t)\mathscr{M}_x\} = -\frac{\omega_0 V}{\gamma k T}\cos\omega_0 t\, \text{tr}\{e^{i\mathscr{H}'_1 t}\mathscr{M}_x e^{-i\mathscr{H}'_1 t}\mathscr{M}_x\}. \quad (44)$$

In writing (44) use has been made of the fact already established that

$$\text{tr}\{e^{i\mathscr{H}'_1 t}\mathscr{M}_x e^{-i\mathscr{H}'_1 t}\mathscr{M}_y\} = 0$$

and of the assumption $|\mathscr{H}'_1 \tau| \ll 1$. The equation (44) demonstrates that the reduced auto-correlation function

$$G_1(t) = \text{tr}\{e^{i\mathscr{H}'_1 t}\mathscr{M}_x e^{-i\mathscr{H}'_1 t}\mathscr{M}_x\} \quad (44')$$

is indeed proportional to the time-dependent amplitude of the precessing magnetization of the sample. To conclude, whether the experimental problem of the determination of the line shape is approached by the continuous wave method with a vanishingly small r.f. field, or by a transient method utilizing strong r.f. pulses, the theoretical problem is solved in principle by the determination of $G_1(t)$ and is reduced to its expansion in powers of t through the formulae (32). In practice, however, such a calculation is excruciatingly tedious for all but the first few terms and one is led to utilize more or less crude models, that is, to make guesses about the general forms of $G_1(t)$ (or its Fourier transform $f(\omega)$), justified by more or less plausible physical arguments, or simply by good agreement with experiment, and to introduce into the assumed analytical expression for $G_1(t)$ enough parameters to fit the values of the moments calculated from first principles. We saw two examples of this, when first a gaussian shape was assumed with an r.m.s. width given by $(M_2)^{\frac{1}{2}}$ and then a truncated Lorentz curve when the ratio $M_4/(M_2)^2$ turned out to be a large number. A more elaborate example will be presented shortly.

F. A comparison between theory and experiment

Calcium fluoride CaF_2 is an excellent substance for testing the theory of pure dipolar broadening. It is easily obtainable in single crystals and contains only one species of nuclei with spins different from zero, those of F^{19}. These spins are $\frac{1}{2}$, so that quadrupole effects are absent. They have large magnetic moments and form a simple cubic lattice. In an early comparison with theory of the various aspects of line width a satisfactory agreement was found (2).

Some more detailed work will be discussed here (3), (4). Using the continuous-wave method one obtains the three resonance curves of Fig. IV, 1 normalized to the same area (3). Actually, as explained in

Chapter III, it is the derivatives of the resonance curves that are observed experimentally and the curves of Fig. IV, 1 were obtained through integration of those with averaging over several experimental runs. Their tops are distinctly flatter than those of gaussian curves

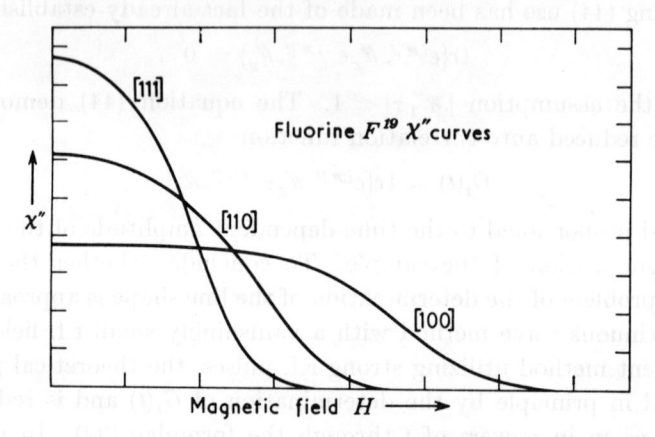

FIG. IV, 1. F^{19} absorption lines in CaF$_2$ with the magnetic field in the [100], [110], and [111] crystal directions.

with the same area and same peak heights. The values of the corresponding moments are listed together with some earlier results of (2) in Tables I and II. The agreement with theory is quite gratifying.

TABLE I. *Root mean second moments of the* F^{19} *absorption line in* CaF$_2$ *for the magnetic field H along* [100], [110], *and* [111] *crystal directions*

$\langle \Delta \nu^2 \rangle^{\frac{1}{2}}$ in gauss

Direction of H	Data 1	Data 2	Data 3	Pake and Purcell (2)	Theory
[100]	3.49±0.20	3.51±0.30	3.47±0.17	3.68±0.20	3.60
[110]	2.17±0.07	2.20±0.07	2.26±0.07	2.25±0.20	2.24
[111]	1.52±0.03	1.59±0.06	..	1.77±0.20	1.53

TABLE II. *Root mean fourth moments of the* F^{19} *absorption line in* CaF$_2$ *for the magnetic field H along* [100], [110], *and* [111] *crystal directions*

$\langle \Delta \nu^4 \rangle^{\frac{1}{4}}$ in gauss

Direction of H	Data 1	Data 2	Data 3	Theory
[100]	4.17±0.15	4.19±0.21	4.13±0.11	4.31
[110]	2.65±0.05	2.68±0.08	2.76±0.06	2.73
[111]	1.88±0.03	1.96±0.05	..	1.88

On the other hand, using transient methods, one can measure the signal of the free precession after a 90° pulse, that is $G_1(t)$ of (44′) (**4**).

The experimental points are supplemented by those obtained by taking the Fourier transforms of the C.W. curves of reference (**3**). The results are shown in Fig. IV, 2 for three orientations of the magnetic field with respect to the crystal axes. The fact that both types of experimental points fall remarkably well on a single curve demonstrates the excellence of the experimental work performed using either method (and possibly the quality of the differential analyser used for the computation of the Fourier transform) but does *not* constitute a check of the theory of dipolar broadening since the relationship between $\chi''(\omega)$ and $G_1(t)$ is the consequence of a purely mathematical theorem. Because of the general properties of the Fourier transform the two experimental methods are complementary to a large extent. In particular the behaviour of $G_1(t)$ for large t is very sensitive to the detailed shape of $\chi''(\omega)$ for small values of ω where all bell-shaped curves obtained by the C.W. methods look very much alike. A very remarkable and rather unexpected feature of the decay curves in (**4**) is their oscillatory character, which demonstrates much more strikingly than any calculation of moments the marked departure of $\chi''(\omega)$ from the gaussian shape.

It is apparent that fourth order or even sixth order polynomial expansion of $G_1(t)$, using (32), is quite inadequate to describe the behaviour of this function in the range of observation. The calculation made in (**4**) is based on the following remark:

$\mathcal{H}'_1$ is a sum of two terms α and β

$$\alpha = -\frac{\gamma^2 \hbar}{3} \sum_{j<k} b_{jk} \mathbf{I}^j \cdot \mathbf{I}^k,$$

$$\beta = \gamma^2 \hbar \sum_{j<k} b_{jk} I_z^j I_z^k,$$

$$b_{jk} = \tfrac{3}{2}(1 - 3\cos^2\theta_{jk})/r_{jk}^3, \tag{45}$$

where α and β do not commute with each other, and α commutes with $\mathcal{M}_x$ but β does not. Now, if α and β did commute $G_1(t)$ could easily be evaluated in a closed form, since then one would have

$$G_1(t) = \text{tr}\{e^{i\beta t} e^{i\alpha t} \mathcal{M}_x e^{-i\alpha t} e^{-i\beta t} \mathcal{M}_x\}$$
$$= \text{tr}\{e^{i\beta t} \mathcal{M}_x e^{-i\beta t} \mathcal{M}_x\}. \tag{46}$$

The operator β is diagonal in the $\mathcal{M}^j$ representation and the evaluation of (46) is straightforward and leads, for spins $\tfrac{1}{2}$, to the infinite product

$$G_1(t) = \text{tr}\{\mathcal{M}_x^2\} \prod_k{}' \cos(\gamma^2 \hbar b_{jk} t) \tag{47}$$

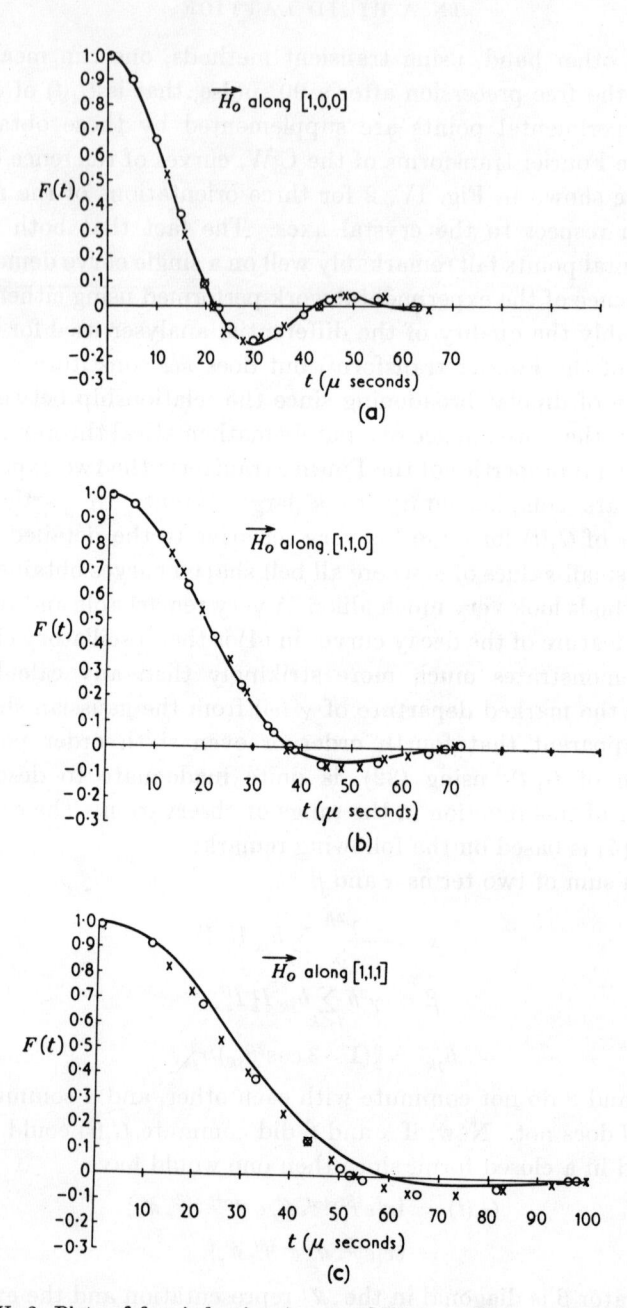

FIG. IV, 2. Plots of free induction decays of F^{19} in a single crystal of CaF$_2$. The crosses are the observed decay amplitudes, the circles are the Fourier transform data from the C.W. curves of Fig. IV, 1. The full curves are the theoretical shapes calculated using the expansion (49′).

(the sign ′ meaning that the value $k = j$ is excluded). Since, however, α and β do not commute, the authors rewrite

$$e^{it(\alpha+\beta)} \quad \text{as} \quad e^{i\beta t}X(t)e^{i\alpha t},$$

where
$$X(t) = e^{-i\beta t}e^{i(\alpha+\beta)t}e^{-i\alpha t},$$

and then
$$G_1(t) = \mathrm{tr}\{e^{i\beta t}X(t)\mathcal{M}_x X^\dagger(t)e^{-i\beta t}\mathcal{M}_x\},$$

a rigorous expression so far.

They then proceed to expand the operator $X(t)$ only, as a power series:

$$X(t) = 1 + \sum_n \frac{C_n}{n!}t^n, \tag{48}$$

where the C_n are operators.

This leads for $G_1(t)$ to a series expansion

$$G_1(t) = \sum_n F_n(t)\frac{t^n}{n!}, \tag{49}$$

where
$$F_n(t) = \sum_{p+q=n} \frac{n!}{p!\,q!} \mathrm{tr}\{e^{i\beta t}C_p \mathcal{M}_x C_q^\dagger e^{-i\beta t}\mathcal{M}_x\}.$$

In spite of the fact that the extreme complication of the calculations prevents progress beyond $n = 4$, the theoretical curves C (Fig. IV, 2) obtained by using

$$G_1^{(4)}(t) = 1 + F_2(t)\frac{t^2}{2!} + F_4(t)\frac{t^4}{4!} \tag{49′}$$

(the odd terms vanish) are in fairly good agreement with experiment and exhibit the observed oscillatory pattern. It is clear that the series expansion (49) converges much faster than the usual power series:

$$G_1(t) = \sum_n \frac{M_n t^n}{n!}. \tag{50}$$

It should be pointed out, however, that the series expansion (49) is obviously not unique, that no mathematical proof is given of the fact that it should converge faster than the power series (50), and that the real argument in its favour is the agreement with experiment, which is good as appears on Fig. IV, 2. From the theoretical point of view all that can be stated with certainty is that the expansion of $G_1^{(4)}(t)$ in power series is correct up to the term in t^4. It can then be argued that any even function $F(t)$ with $F(0) = 1$, $F''(0) = M_2$, $F^{\mathrm{iv}}(0) = M_4$ has the same *a priori* claim to represent $G_1(t)$ as (49), agreement with the experimental $G_1(t)$ over a wide range, and simplicity being the main criteria for its choice. The experimental curves bear a strong resemblance

to the analytical curve
$$F(t) = \exp\left[-\frac{a^2 t^2}{2}\right]\frac{\sin bt}{bt}, \tag{51}$$
which has a series expansion
$$F(t) = 1 - \frac{t^2}{2}\left(a^2 + \frac{b^2}{3}\right) + \frac{t^4}{4!}\left(3a^4 + 2a^2b^2 + \frac{b^4}{5}\right) -$$
$$-\frac{t^6}{6!}\left(15a^6 + 15a^4b^2 + 3a^2b^4 + \frac{b^6}{7}\right). \tag{51'}$$

Using F as a trial representation of $G_1(t)$, we choose a and b to satisfy $a^2 + \tfrac{1}{3}b^2 = M_2$; $3a^4 + 2a^2b^2 + \tfrac{1}{5}b^4 = M_4$, where M_2 and M_4 are theoretical values taken from Tables I and II and converted from the H_0 scale into the $\omega_0 = \gamma H_0$ scale. The function F thus determined is plotted in Fig. IV, 3 and its agreement with experiment is quite remarkable.

In that connexion the following remarks can be made. In the direction [100], for which the sixth moment has been calculated, we can compare the theoretical ratio $M_6/(M_2)^3 \simeq 5$ with the ratio deduced from $F(t)$:
$$\frac{15a^6 + 15a^4b^2 + 3a^2b^4 + \tfrac{1}{7}b^6}{(a^2 + \tfrac{1}{3}b^2)^3}, \tag{52}$$
which depends only on b/a determined from
$$\frac{M_4}{(M_2)^2} = \frac{3a^4 + 2a^2b^2 + \tfrac{1}{5}b^4}{(a^2 + \tfrac{1}{3}b^2)^2} = 2\cdot 07. \tag{53}$$

The ratio (52) comes out to be 5·67 in remarkable agreement with the theoretical value 5, if one remembers that for a Gaussian this ratio is 15. A further test of the trial shape (51) can be made as follows. Using a digital computer to extract from the curves of Fig. IV, 1 the values of the sixth and eighth moments, the experimental ratios thus obtained are compared with those predicted by (51), where a and b were chosen to fit the theoretical values of M_2 and M_4.

The results are shown in Table III together with the predictions for a gaussian and a rectangular line shape. (Theoretical means extracted from (51).)

TABLE III

	$M_6/(M_2)^3$	$M_8/(M_2)^4$
Experimental [100]	5·49	17·2
Theoretical [100]	5·67	19·8
Experimental [111]	7·40	29·2
Theoretical [111]	7·37	33·5
Gaussian	15	105
Rectangular	3·9	9·0

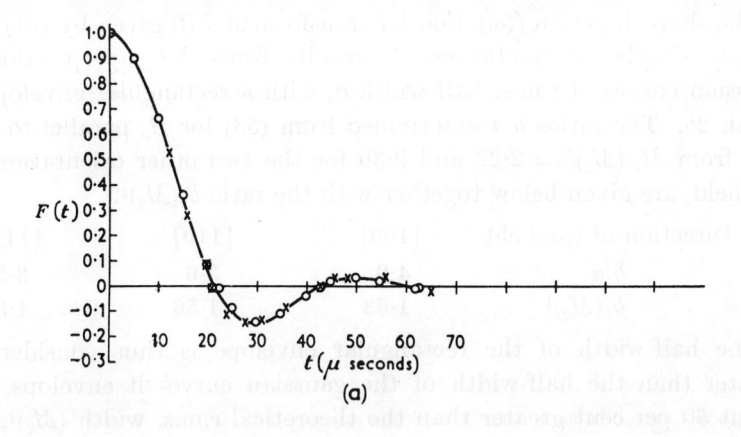

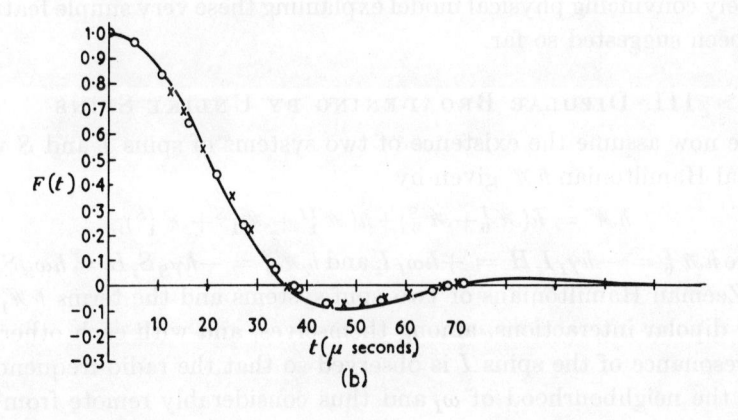

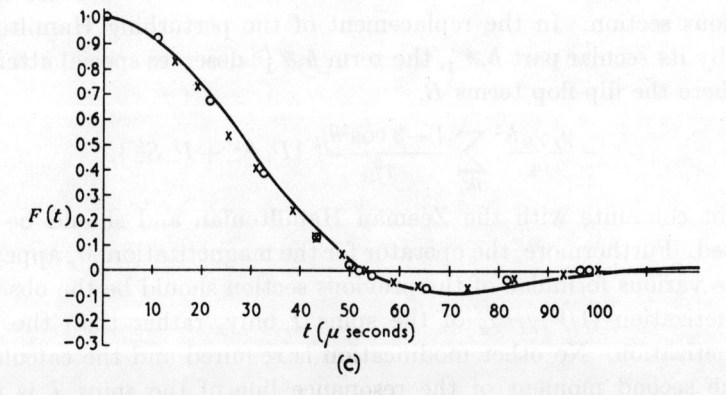

FIG. IV, 3. The experimental points, circles and crosses, are the same as in Fig. IV, 2. The full curve is the theoretical curve (51), where the parameters a and b have been adjusted to fit the theoretical values of the second and fourth moments.

The shape function $f(\omega)$, Fourier transform of $F(t)$ given by (51), has a very simple interpretation. It results from the superposition of gaussian curves of r.m.s. half width a, with a rectangular envelope of width $2b$. The ratios b/a determined from (53) for H_0 parallel to 100, and from $M_4/(M_2)^2 = 2\cdot 22$ and $2\cdot 30$ for the two other orientations of the field, are given below together with the ratio $b/(M_2)^{\frac{1}{2}}$.

Direction of the field	[100]	[110]	[111]
b/a	4·9	3·6	3·2
$b/(M_2)^{\frac{1}{2}}$	1·63	1·56	1·52

The half-width of the rectangular envelope is thus considerably greater than the half-width of the gaussian curves it envelops, and about 50 per cent greater than the theoretical r.m.s. width $(M_2)^{\frac{1}{2}}$. No entirely convincing physical model explaining these very simple features has been suggested so far.

III. Dipolar Broadening by Unlike Spins

We now assume the existence of two systems of spins I and S with a total Hamiltonian $\hbar \mathcal{H}$ given by

$$\hbar \mathcal{H} = \hbar(\mathcal{H}_0^I + \mathcal{H}_0^S) + \hbar(\mathcal{H}_1^{II} + \mathcal{H}_1^{SS} + \mathcal{H}_1^{IS}), \tag{54}$$

where $\hbar \mathcal{H}_0^I = -\hbar \gamma_I I_z H = +\hbar \omega_I I_z$ and $\hbar \mathcal{H}_0^S = -\hbar \gamma_S S_z H = \hbar \omega_S S_z$ are the Zeeman Hamiltonians of two spin systems and the terms $\hbar \mathcal{H}_1$ are their dipolar interactions, among themselves, and with each other. If the resonance of the spins I is observed so that the radio-frequency ω is in the neighbourhood of ω_I and thus considerably remote from ω_S, the following modifications have to be made to the treatment of the previous section. In the replacement of the perturbing Hamiltonian $\hbar \mathcal{H}_1$ by its secular part $\hbar \mathcal{H}_1'$, the term $\hbar \mathcal{H}_1^{IS}$ deserves special attention for there the flip-flop terms B,

$$-\frac{\gamma_I \gamma_S \hbar^2}{4} \sum_{jk'} \frac{1 - 3\cos^2 \theta_{jk'}}{r_{jk'}^3} \{I_+^j S_-^{k'} + I_-^j S_+^{k'}\}, \tag{54'}$$

do not commute with the Zeeman Hamiltonian and should be suppressed. Furthermore, the operator for the magnetization M_x appearing in the various formulae of the previous section should be the observed magnetization $(1/V)\gamma_I \hbar I_x$ of the spins I only, rather than the total magnetization. No other modification is required and the calculation of the second moment of the resonance line of the spins I is quite straightforward and gives

$$M_2^I = \overline{\Delta \omega_I^2} = \overline{(\Delta \omega_I^2)_{II}} + \overline{(\Delta \omega_I^2)_{IS}},$$

where the first term is identical with (38) whereas the second is given by

$$(\overline{\Delta\omega_I^2})_{IS} = \tfrac{1}{3}\gamma_I^2\gamma_S^2 S(S+1)\hbar^2 \sum_{k'} \frac{(1-3\cos^2\theta_{jk'})^2}{r_{jk'}^6}, \tag{55}$$

and the sum $\sum_{k'}$ is to be taken over all the sites of the spins $S^{k'}$ surrounding a spin I^j. The only comments to be made on (55) are first that the contributions to the second moment from the dipolar couplings II and IS are additive and second that the numerical factor $\tfrac{1}{3}$ in front of (55) is $\tfrac{4}{9}$ of the factor $\tfrac{3}{4}$ in front of the expression (38) for the second moment of a line broadened by like spins. This reduction due to the truncation of the flip-flop terms (54') between unlike spins, had already been predicted, although not quantitatively, by a classical argument in the introduction.

As for the broadening by like spins only, the fourth moment should be calculated to appreciate whether the r.m.s. width, given by the second moment, and the half intensity width are significantly different. Rather than give the straightforward but tedious calculation and the complicated final result a short qualitative discussion of its main physical features will be made. The fourth moment can be written symbolically as

$$M_4 \propto (\mathcal{H}_1^{II})^4 + (\mathcal{H}_1^{IS})^4 + (\mathcal{H}_1^{II})^2(\mathcal{H}_1^{IS})^2 + (\mathcal{H}_1^{SS})^2(\mathcal{H}_1^{IS})^2, \tag{56}$$

where the various terms, describing contributions from the different parts of $\mathcal{H}_1$ and interferences between them, are respectively proportional to

$$\gamma_I^8 \quad \gamma_I^4\gamma_S^4 \quad \gamma_I^6\gamma_S^2 \quad \gamma_I^2\gamma_S^6.$$

If $\gamma_S \gg \gamma_I$ the last term can be the largest in M_4, and make $M_4/(M_2)^2 \propto (\gamma_S/\gamma_I)^2$ a large number. Then, according to the discussion of Section II B of this chapter, we are led to expect a quasi-Lorentzian shape and a half intensity width much smaller than the r.m.s. width. This effect can be interpreted as a first instance of a phenomenon to be studied in great detail, later on: the motion narrowing. The physical explanation of this narrowing in the present case is that a strong coupling $\mathcal{H}_{SS}$ between the spins S leads to frequent flip-flops between them and thus to a random but rapid modulation of the local field 'seen' by a spin I, which diminishes its efficiency in broadening the resonance line of the spins I. A detailed discussion of motion narrowing will be presented in Chapter X.

A striking example of this situation is provided by the resonances of K^{39} in solid FK and Ag^{109} in solid FAg (5). The ratios $\gamma(F^{19})/\gamma(K^{39})$ and $\gamma(F^{19})/\gamma(Ag^{109})$ are very nearly equal and of the order of 20. The main contribution to the second moment M_2 is then proportional to $\gamma_I^2 \gamma_S^2$ and that to the fourth moment M_4 is proportional to $\gamma_I^2 \gamma_S^6$, where S represents the spin $\frac{1}{2}$ of F^{19}, and I the spin of either K^{39} ($\frac{3}{2}$) or Ag^{109} ($\frac{1}{2}$).

Either contribution is independent of the value I of the 'resonant' spin. Since FK and FAg have the same structure although different unit cell dimensions, the ratio $M_4/(M_2)^2$, proportional to $(\gamma_S/\gamma_I)^2$ (and incidentally independent of the value S of the 'non-resonant' spin), has very nearly the same value for Ag^{109} and K^{39} and is computed to be 57.

Since this is a large number, a trial representation of the resonance curve by a truncated Lorentzian curve leads according to eqn. (25″) to a half intensity half width δ smaller than the r.m.s. half width $\Delta = M_2^{\frac{1}{2}}$, by a factor:

$$\frac{\delta}{\Delta} = \frac{\pi}{2\sqrt{3}}\left(\frac{M_2^2}{M_4}\right)^{\frac{1}{2}} = \frac{\pi}{2\sqrt{3}\sqrt{57}} = 0 \cdot 12,$$

a very appreciable narrowing.

The computed r.m.s. half widths are:

$$Ag^{109} \quad \Delta = 1 \cdot 95 \text{ gauss},$$
$$K^{39} \quad \Delta = 1 \cdot 6 \text{ gauss},$$

and thus the expected half intensity half widths are:

$$Ag^{109} \quad \delta = 1 \cdot 95 \times 0 \cdot 12 = 0 \cdot 235 \text{ gauss},$$
$$K^{39} \quad \delta = 1 \cdot 6 \times 0 \cdot 12 = 0 \cdot 192 \text{ gauss}.$$

The observed values are:

$$Ag^{109} \quad \delta = 0 \cdot 3 \text{ gauss} \pm 0 \cdot 15,$$
$$K^{39} \quad \delta = 0 \cdot 5 \text{ gauss} \pm 0 \cdot 05.$$

The large uncertainty in $\delta(Ag^{109})$ is due to very poor signal-to-noise ratio.

Because of the weakness of the resonance it was necessary to observe the absorption signal with a large modulation amplitude and appreciable r.f. field amplitude. The real unsaturated line width is likely to be smaller than the observed value, possibly by as much as a factor 2, the exact amount of spurious broadening being difficult to estimate. Under these conditions the agreement with the theory may be considered as satisfactory. It is worth pointing out that at least for Ag^{109} if the narrowing did not occur and the real width was comparable to the r.m.s. width, the signal could not have been detected at all.

IV. Dipolar Broadening in Magnetically Diluted Substances

A. The method of moments

Another situation where the half intensity width can be significantly smaller than the r.m.s. width is that of magnetically diluted substances where the magnetic moments are distributed at random. The physical explanation is that in a magnetically diluted substance most nuclei 'see' a very weak local field and so we have a narrow line and a small half intensity width, whereas an appreciable contribution to the calculated second moment comes from a few clusters of two or more nuclei which have Larmor frequencies so different from the central frequency γH_0 that their contributions to the line intensity are too far in the wings to be observed. The usual procedure of studying the ratio $M_4/(M_2)^2$ has been applied (6) to the case of a regular cubic crystal lattice, when only a fraction f of the crystal sites is occupied by nuclear spins.

In the formulae (38) and (40) for M_2 and M_4, the summations have to be carried out over the occupied crystal sites only. It is more convenient, however, to use sums over all crystal sites. For the second moment the restricted sum $\widetilde{\sum_k} b_{jk}^2$ taken over occupied sites only is equal to $f \sum_k b_{jk}^2$ taken over all sites. If with increasing dilution the shape of the line did not change, the half intensity width would be proportional to the r.m.s. width and therefore to $\sqrt{f}$.

On the other hand, for the fourth moment (40) restricted double sums are equal to the corresponding sums over all sites, times f^2, whereas the single restricted sum $\widetilde{\sum_k} b_{jk}^4$ is equal to the corresponding full sum, times f. For a simple cubic lattice and the field along the 100 axis the formula giving the fourth moment becomes

$$\overline{\Delta\omega^4} = 3(\overline{\Delta\omega^2})^2 \left\{ 0\cdot 63 + \frac{1}{f}\left(0\cdot 098 - \frac{0\cdot 021}{I(I+1)}\right)\right\}. \tag{57}$$

(A slightly different value, 0·74 instead of 0·63, is given in (6) for the first term in the curly bracket, a discrepancy immaterial in the present discussion.)

For $f > 0\cdot 1$ the curly bracket is of the order of unity and accordingly the r.m.s. width Δ and the half intensity width δ should be of the same order of magnitude. On the other hand, for $f < 0\cdot 01$, $M_4/(M_2)^2 \gg 3$ and the model of the truncated Lorentz curve which leads to $\delta \ll \Delta$

can be tried. For $f \ll 0{\cdot}01$ and $I = \tfrac{1}{2}$ we find, according to (25″) and (39″),

$$\delta = \frac{\pi}{2\sqrt{3}}\left(\frac{M_2^2}{M_4}\right)^{\frac{1}{2}}\Delta = 5{\cdot}3 f \gamma \frac{\gamma\hbar}{d^3}. \tag{57′}$$

For very large dilutions the line width is proportional to the concentration f rather than to $\sqrt{f}$.

B. The statistical theory

In view of the arbitrariness of the model using a truncated Lorentz curve it is desirable to check the previous results through a totally different, if not more rigorous, approach, the statistical theory (7). In this theory, where we assume for simplicity spins $\tfrac{1}{2}$, the resonance frequency of a given spin, or rather its departure from the mean Larmor frequency $\omega_0 = \gamma H_0$, will be a function of the positions $\mathbf{r}_1,..., \mathbf{r}_N$ and orientations $\epsilon_1,..., \epsilon_N$ of all the other spins where $\epsilon_i = \pm 1$. It will also be assumed for simplicity that this function $\omega(\mathbf{r}_1\epsilon_1, \mathbf{r}_2\epsilon_2,..., \mathbf{r}_N\epsilon_N)$ is a sum $\omega(\mathbf{r}_1,\epsilon_1)+\omega(\mathbf{r}_2,\epsilon_2)+...+\omega(\mathbf{r}_N,\epsilon_N)$ of contributions from the various neighbours of the spin under consideration or reference spin. This amounts to assuming that the orientation of a perturbing spin A is independent of that of another perturbing spin B; not an unreasonable assumption for very diluted substances where a spin is unlikely to be appreciably influenced by more than one neighbour. The 'truncated coupling' between two spins will then be taken as

$$\frac{3}{2}\frac{\gamma^2\hbar^2}{r^3}(1-3\cos^2\theta)s_z s_z' \tag{58}$$

neglecting the scalar term, $-\tfrac{1}{2}(\gamma^2\hbar^2/r^3)(1-3\cos^2\theta)\mathbf{s}\cdot\mathbf{s}'$. This procedure is legitimate if two spins only are interacting and finds its justification here in the assumption of great dilution. If the orientation of the reference spin is, say, $\epsilon = +1$, we shall have

$$\omega(\mathbf{r}_i, \epsilon_i) = \epsilon_i \frac{3}{4}\frac{\gamma^2\hbar}{r_i^3}(1-3\cos^2\theta_i).$$

One can get rid of the ϵ's by allowing the variables r_i to have positive as well as negative values and doubling accordingly the volume over which the position of each spin can be chosen at random. $\omega(\mathbf{r}, \epsilon)$ then becomes just $\omega(\mathbf{r})$. The intensity of absorption $I(\omega)\,d\omega$ will be proportional to the volume of the $3N$ dimensional phase space where the following condition is satisfied:

$$\omega \leqslant \omega(\mathbf{r}_1,..., \mathbf{r}_N) = \omega(\mathbf{r}_1)+...+\omega(\mathbf{r}_N) \leqslant \omega+d\omega, \tag{58′}$$

$$I(\omega)\,d\omega = \frac{1}{\mathscr{V}}\int_{\mathscr{E}(\omega,d\omega)} d\tau_1,..., d\tau_N, \tag{59}$$

where $\mathscr{V} = (2V)^N$ is the total volume of the phase space, V is the volume of the sample, and $\mathscr{E}(\omega, d\omega)$ the region of the phase space where (58′) is satisfied. (59) can be rewritten as an integral over the whole phase space:

$$I(\omega)\,d\omega = \frac{d\omega}{\mathscr{V}} \int_{\mathscr{V}} \delta[\omega - \omega(\mathbf{r}_1, \mathbf{r}_2, ..., \mathbf{r}_N)]\,d\tau_1\,d\tau_2\ldots d\tau_N$$

or, using the relation $\delta(x) = (1/2\pi) \int\limits_{-\infty}^{+\infty} e^{itx}\,dt$,

$$I(\omega) = \frac{1}{2\pi}\int\limits_{-\infty}^{+\infty} e^{i\omega t}\,dt\left[\frac{1}{2V}\int\limits_{2V} e^{-itB}\,d\tau\right]^N,$$

where
$$B = \frac{3}{4}\frac{\gamma^2\hbar}{r^3}(1 - 3\cos^2\theta)$$

and
$$\int\limits_{2V} = 2\pi\int\limits_0^\pi \sin\theta\,d\theta \int\limits_{-\infty}^{+\infty} e^{-itB}\,r^2\,dr.$$

If N and V tend to infinity, leaving the density $n = N/V$ finite, the integral $(1/2V)\int\limits_{2V} e^{-itB}\,d\tau$, which is the ratio of two infinite quantities, can be rewritten as

$$\frac{1}{2V}\int\limits_{2V} e^{-itB}\,d\tau = 1 - \frac{1}{2V}\int\limits_{2V}(1 - e^{-itB})\,d\tau = 1 - \frac{n}{2N}\int\limits_{2V}(1 - e^{-itB})\,d\tau.$$

We shall see shortly that

$$V' = \lim_{V\to\infty} \int\limits_{2V} (1 - e^{-itB})\,d\tau = \lambda|t|. \tag{60}$$

This gives

$$I(\omega) = \lim_{N\to\infty}\frac{1}{2\pi}\int\limits_{-\infty}^{+\infty} e^{i\omega t}\,dt\left[1 - \frac{n\lambda|t|}{2N}\right]^N = \frac{1}{2\pi}\int\limits_{-\infty}^{+\infty} e^{i\omega t}e^{-\frac{1}{2}n\lambda|t|}\,dt. \tag{61}$$

From (61) it is clear that the absorption curve has a Lorentz shape with a width $\frac{1}{2}n\lambda$. To calculate λ we have to compute V' in (60):

$$V' = 2\pi\int\limits_0^\pi \sin\theta\,d\theta\int\limits_{-\infty}^{+\infty} r^2\,dr\{1 - e^{-(it/r^3)\frac{3}{4}\gamma^2\hbar(3\cos^2\theta - 1)}\}. \tag{61′}$$

Putting $\frac{3}{4}t\gamma^2\hbar(3\cos^2\theta - 1) = b$, $x = 1/r^3$, the integral over r becomes

$$I_1 = \frac{1}{3}\int\limits_{-\infty}^{+\infty}\frac{1 - e^{-ibx}}{x^2}\,dx = \frac{\pi|b|}{3} = \frac{1}{4}\pi\gamma^2\hbar|t||3\cos^2\theta - 1|.$$

Then

$$V' = \tfrac{1}{2}\pi^2\gamma^2\hbar|t| \int_0^\pi |3\cos^2\theta - 1|\sin\theta\,d\theta = \tfrac{1}{2}\pi^2\gamma^2\hbar|t|\frac{8}{3\sqrt{3}} = \lambda|t| \quad (62)$$

and the width $\delta = \tfrac{1}{2}n\lambda$ becomes

$$\delta = \frac{2\pi^2}{3\sqrt{3}}\gamma^2\hbar n \simeq 3\cdot 8\,\gamma^2\hbar n. \quad (63)$$

The formula (63) is to be compared with (57') obtained by the method of moments for the direction (100) of a simple cubic lattice. Since in that case the density $n = f/d^3$ the agreement between the two results is quite good.

The statistical theory leads to an absorption curve which has a Lorentz shape over the whole frequency range without cut-off. This results from the fact that we allowed the spins to come arbitrarily near each other, integrating over all values of r. Physically there will evidently be a distance of closest approach and accordingly a cut-off of the intensity for very large frequencies.

V. Modifications in the Dipolar Broadening caused by the Existence of Quadrupole Couplings

When a nucleus possesses a quadrupole moment and its environment has a symmetry lower than cubic, its energy spectrum is different from the simple Zeeman pattern and more than one resonance frequency can be observed. The dipolar broadening of the resonance line corresponding to each of these frequencies can be studied by the methods of the previous sections and, in particular, the moments of the resonance lines can be calculated. The calculations made for the pure Zeeman case have to be amended in two respects:

First, in the 'truncation' which reduces the dipolar Hamiltonian $\hbar\mathcal{H}_1$ to the part $\hbar\mathcal{H}'_1$ that commutes with the main Hamiltonian $\hbar\mathcal{H}_0$, a change in the latter due to the quadrupole interaction causes a corresponding change in the definition of $\hbar\mathcal{H}'_1$.

Second, the operator I_x representing the coupling of the spin system with the r.f. field has to be truncated as well through removal of its matrix elements linking states other than those between which the observed transition occurs. A step in that direction had already been made in Section III on the broadening by unlike spins where the magnetic moment $\gamma\hbar I_x$ of the 'resonant' nuclei only was used in the calculation, rather than the total magnetic moment.

Only a fraction of the many situations which can arise in practice have received a theoretical treatment. The results of calculations of second moments will be listed for the following cases:

(α) First we consider the $\tfrac{1}{2} \to -\tfrac{1}{2}$ transition for nuclei of half integral spin, when the quadrupole coupling is small compared with the Zeeman energy (8). The reason for selecting this transition among the $2I$ that are observable in principle, is that, its frequency being unaffected to the first order by the quadrupole coupling and thus also by its local changes caused by the imperfections of the crystal, the dipolar coupling is likely to be the main broadening agency for that transition. In particular, in imperfect cubic crystals, local imperfections can broaden all the other lines to the extent of making them unobservable, leaving the $\tfrac{1}{2} \to -\tfrac{1}{2}$ transition only (see Chapter VII).

(i) *Interaction between like spins.* By like spins we understand those that have not only the same gyromagnetic ratio γ but also are situated at crystal sites where the field gradient has the same magnitude and orientation. Since the dipolar coupling is a two-body interaction, for the calculation of the second moment by the formula (34) it is sufficient to consider two such spins with an unperturbed Hamiltonian

$$\hbar \mathscr{H}_0 = -\gamma \hbar H_0 (I_z + I'_z) + \delta (I_z^2 + I'^2_z)$$

and a dipolar coupling

$$\hbar \mathscr{H}_1 = \frac{\gamma^2 \hbar^2}{r^3} \left\{ \mathbf{I} \cdot \mathbf{I}' - 3 \frac{(\mathbf{I} \cdot \mathbf{r})(\mathbf{I}' \cdot \mathbf{r})}{r^2} \right\}.$$

The only matrix elements of $\hbar \mathscr{H}_1$ that should be retained to constitute the 'truncated' $\hbar \mathscr{H}'_1$ are all the diagonal ones $\langle m, m' | \hbar \mathscr{H}_1 | m, m' \rangle$ (terms A of equation (18)) and among the off-diagonal ones those of the form $(m, m \mp 1 | \hbar \mathscr{H}_1 | m \pm 1, m)$. It should be remarked that these are not all the non-vanishing matrix elements of the terms B of (18), which are given more generally by $\langle m, m' | \hbar \mathscr{H}_1 | m \pm 1, m' \mp 1 \rangle$. Furthermore, the only matrix elements of I_x that should be retained are the following: $\langle \pm \tfrac{1}{2} | I_x | \mp \tfrac{1}{2} \rangle$. With these modifications the calculation of the second moment is quite straightforward and leads to the result

$$\overline{\Delta \omega^2} = F_L(I) \gamma^4 \hbar^2 \sum_k b_{jk}^2, \qquad (64)$$

where $F_L(I)$, which was $\tfrac{1}{3} I(I+1)$ for the pure Zeeman case, becomes

$$F_L = \tfrac{4}{27} I(I+1) + \frac{2 I^2 (I+1)^2 + 3 I(I+1) + \tfrac{13}{8}}{18(2I+1)}. \qquad (64')$$

(ii) *Broadening by unlike spins.* It is quite clear that only the diagonal part A of the dipolar coupling should be retained, and only the matrix element $(\pm\tfrac{1}{2}|I_x|\mp\tfrac{1}{2})$ for the spin of the 'resonant' nucleus. The second moment comes out to be the same as in formula (55) for the broadening by unlike spins in the pure Zeeman case, the change due to the truncation of I_x in the numerator and the denominator of (34) cancelling out.

(iii) *Broadening by semi-like spins.* By this we mean spins with the same gyromagnetic ratio and thus the same resonance frequency for the $\tfrac{1}{2} \to -\tfrac{1}{2}$ transition, but situated at different sites and therefore having different quadrupole couplings. In that case

$$\hbar\mathcal{H}_0 = -\gamma\hbar H_0(I_z+I'_z)+\delta I_z^2+\delta' I'^2_z.$$

The truncation of I_x is the same as in (i). In the truncated $\hbar\mathcal{H}'_1$ all the diagonal matrix elements of $\hbar\mathcal{H}_1$ should be retained, but among the off-diagonal ones only the following:

$$\langle \pm\tfrac{1}{2}, \mp\tfrac{1}{2}|\hbar\mathcal{H}_1|\mp\tfrac{1}{2}, \pm\tfrac{1}{2}\rangle.$$

The second moment then comes out to be

$$\overline{\Delta\omega^2} = F_{SL}(I)\gamma^4\hbar^2 \sum_k b_{jk}^2$$

with $\qquad F_{SL}(I) = \tfrac{1}{9}\{\tfrac{4}{3}I(I+1)+\tfrac{1}{2}(2I+1)+\tfrac{1}{32}(2I+1)^3\}.$ (65)

The ratios $\qquad r_L = \dfrac{F_L(I)}{\tfrac{1}{3}I(I+1)} \quad$ and $\quad r_{SL} = \dfrac{F_{SL}(I)}{\tfrac{1}{3}I(I+1)}$

of the second moments in cases (i) and (iii) to that obtained in the absence of quadrupole interaction are given below:

I	3/2	5/2	7/2
r_L	9/10	107/105	881/756
r_{SL}	4/5	257/315	164/189

It is not quite clear whether $F_L(I)$ or $F_{SL}(I)$ should be used for the second moment of the central line $\tfrac{1}{2} \to -\tfrac{1}{2}$ in an imperfect cubic crystal where the satellite lines are washed out by the imperfections, for although in principle the nuclei 'see' different field gradients (which accounts for the disappearance of the satellites), these gradients are least different for nearest neighbours which are the most effective for the dipolar broadening. A compromise would be to introduce a coherence radius r_c inside which the neighbours are like, and semi-like outside, writing for the second moment

$$\overline{\Delta\omega^2} = \gamma^4\hbar^2\Big\{F_L\Big(\sum_{r_{jk}<r_c}b_{jk}^2\Big)+F_{SL}\Big(\sum_{r_{jk}>r_c}b_{jk}^2\Big)\Big\}.$$

(β) The second moment has also been calculated for the lines corresponding to the transitions $1 \to 0$ and $0 \to -1$, separated by a small quadrupole interaction and broadened by like neighbours, in the case of spins $I = 1$. The calculation is analogous to those outlined in (α) and leads to

$$\overline{\Delta\omega^2} = \tfrac{5}{9}\gamma^4\hbar^2 \sum_k b_{jk}^2 \tag{66}$$

(instead of $\tfrac{1}{3}I(I+1) = \tfrac{2}{3}$ for the straight Zeeman case).

(γ) The case of the pure quadrupole resonance has also been considered (9). The principle of the calculation is again very similar and the only point worth mentioning is that the truncated Hamiltonian $\hbar\mathcal{H}_1'$ may contain some E and F terms (equation (18)) since for two like spins two states such as, say, $(1, 0)$ and $(0, -1)$ between which E and F have non-vanishing matrix elements, have the same unperturbed energy. As a consequence, terms with an angular dependence other than $b_{jk}^2 = [\tfrac{3}{2}(1-3\cos^2\theta_{jk})]^2$ appear in the formula for the second moment. The following results have been obtained for spins 1 and $\tfrac{3}{2}$ with the assumption of a broadening due to like neighbours and of axial symmetry for the electric field gradient.

$I = 1$:

$$\overline{\Delta\omega^2} = \tfrac{1}{4}\gamma^4\hbar^2 \sum_k r_{jk}^{-6}[5(1-3\gamma_{jk}^2)^2+9(1-\gamma_{jk}^2)^2-2(1-3\gamma_{jk}^2)(\alpha_{jk}^2-\beta_{jk}^2)], \tag{67}$$

where α_{jk}, β_{jk}, γ_{jk} are the cosines of $\mathbf{r}_{jk}$ with respect to a frame where the axis of symmetry is the z-axis and the applied r.f. field is along the x-axis.

$I = \tfrac{3}{2}$:

$$\overline{\Delta\omega^2} = \tfrac{1}{96}\gamma^4\hbar^2 \sum_k r_{jk}^{-6}\{207(1-3\gamma_{jk}^2)^2+1512\gamma_{jk}^2(1-\gamma_{jk}^2)+$$
$$+459(1-\gamma_{jk}^2)^2+108(1-3\gamma_{jk}^2)(\alpha_{jk}^2-\beta_{jk}^2)\}. \tag{68}$$

If we assume for argument's sake that the nuclear sites are those of a cubic lattice (although the environment of each nucleus is, of course, not cubic), with the axis of the gradient parallel to one of the cubic axes, we find

$$I = 1: \qquad \overline{\Delta\omega^2} = 28\cdot 4\frac{\gamma^4\hbar^2}{d^6},$$

$$I = \tfrac{3}{2}: \qquad \overline{\Delta\omega^2} = 60\frac{\gamma^4\hbar^2}{d^6}.$$

By comparison, in the pure Zeeman case, with the field along one of the cubic axes, the formula (39″) gives

$$I = 1: \qquad \overline{\Delta\omega^2} = 20\frac{\gamma^4\hbar^2}{d^6},$$

$$I = \tfrac{3}{2}: \qquad \overline{\Delta\omega^2} = 37\cdot 5\frac{\gamma^4\hbar^2}{d^6}.$$

The same calculation can be made when the broadening is due to unlike neighbours S.

The following cases have been treated:

	Resonant spin I	Non-resonant spin S
A_1	1	Arbitrary, no quadrupole split
A_2	1	Half-integer with split
A_3	1	Integer with split
B_1	$\tfrac{3}{2}$	Arbitrary, no quadrupole split
B_2	$\tfrac{3}{2}$	Half-integer with split
B_3	$\tfrac{3}{2}$	Integer with split

All results can be summarized in the following formula:

$$\overline{\Delta\omega^2} = \tfrac{1}{3}\gamma^2\gamma'^2\hbar^2 S(S+1) \sum_{k'} r_{jk'}^{-6}\{[1+F(I)G(S)](1-3\gamma_{jk'}^2)^2+ \\ +9[G(S)+2F(I)]\gamma_{jk'}^2(1-\gamma_{jk'}^2)+9F(I)G(S)(1-\gamma_{jk'}^2)^2\}$$

where $F(I)$ and $G(S)$ are defined as follows:

$$F(1) = \tfrac{3}{2}, \qquad F(\tfrac{3}{2}) = \tfrac{1}{2};$$

$$G(S) = 1 \quad \text{for cases } A_1 \text{ and } B_1,$$
$$= \tfrac{3}{8}(2S+1)/S(S+1) \quad \text{for cases } A_2 \text{ and } B_2,$$
$$= 0 \quad \text{for } A_3 \text{ and } B_3.$$

REFERENCES

1. J. H. VAN VLECK, *Phys. Rev.* **74**, 1168, 1948.
2. G. E. PAKE and E. M. PURCELL, ibid., p. 1184, 1948.
3. C. R. BRUCE, ibid. **107**, 43, 1957.
4. I. J. LOWE and R. E. NORBERG, ibid., p. 46, 1957.
5. A. ABRAGAM and J. WINTER, *C.R. Acad. Sci.* **249**, 1633, 1959.
6. C. KITTEL and E. ABRAHAMS, ibid. **90**, 238, 1953.
7. P. W. ANDERSON, ibid. **82**, 342, 1951.
8. K. KAMBE and J. F. OLLOM, *J. Phys. Soc. Japan*, **11**, 50, 1956.
9. A. ABRAGAM and K. KAMBE, *Phys. Rev.* **91**, 894, 1953.

V

SPIN TEMPERATURE

Some like it hot

A. Non-interacting spins

THE concept of a spin temperature distinct from the lattice temperature is often used in nuclear magnetism for a pictorial description of a saturation experiment. An analogy can be drawn between a system of nuclear spins in a large magnetic field coupled to a lattice and exposed to an r.f. field inducing transitions between their Zeeman levels, and a wire with an electric current flowing through it, immersed in a bath. The establishment of an equilibrium temperature of the wire, higher than that of the bath, represents a balance between two competing processes: the input of heat dissipated in the wire by the electric current and the heat transfer from the wire to the bath. Similarly, the heat dissipated inside the spin system (the wire) by the r.f. field (the electric current) is transferred to the lattice (the bath) by a spin-lattice relaxation mechanism and the idea of a spin temperature higher than the lattice temperature follows naturally from this analogy.

To put the comparison on a more quantitative basis consider first a system of non-interacting spins $I = \frac{1}{2}$ and assign relative populations p_+ and p_- with $p_+ + p_- = 1$, to the two Zeeman levels. Since the spins are in equilibrium with the lattice prior to the application of the r.f. field, the thermal equilibrium ratio (p_+/p_-) is $\exp(+\gamma\hbar H/kT)$, and in the high-temperature approximation

$$p_+ - p_- = \frac{\gamma\hbar H}{2kT}.$$

Because the probability of a transition induced by the r.f. field is equal to that of the opposite transition, the r.f. field tends to decrease the difference between the populations below its thermal equilibrium value. If p_+ and p_- are the new steady-state values of the populations, established as the result of a competition between the r.f. irradiation and spin-lattice relaxation, we can tentatively define a spin 'temperature' T_S higher than the lattice temperature, by the relation

$$\frac{p_+}{p_-} = \exp\left(+\frac{\gamma\hbar H}{kT_S}\right). \tag{1}$$

In particular, an infinite spin 'temperature' T_S will describe a situation where p_+ and p_- have been made equal.

The definition of a spin 'temperature' for a spin I higher than $\frac{1}{2}$ is less obvious, for the existence of a temperature implies that the ratio of the populations of two adjacent Zeeman levels is the same for all levels and it is not clear whether such a situation does exist in the presence of an r.f. field. It can be easily verified that for high lattice temperature this is indeed so if the transition probability between two levels $|m\rangle$ and $|m-1\rangle$ induced by the relaxation mechanism depends on m in the same way as the transition probability due to the r.f. field, that is proportionally to $|(m|I_x|m-1)|^2 \propto (I+m)(I-m+1)$ (magnetic relaxation) and also in some other cases, to which it is thus possible to extend the definition of the spin 'temperature', introduced for spin $\frac{1}{2}$, through the relation

$$\frac{p_m}{p_{m-1}} = \exp\left(\frac{\gamma\hbar H}{kT_S}\right). \tag{2}$$

A more serious objection to the definition of a spin 'temperature' in the presence of an r.f. field is the existence of a transverse nuclear magnetization, as demonstrated for instance by Bloch's equations, incompatible with a complete description of the statistical behaviour of the spin system by the populations of its energy levels and *a fortiori* by a temperature. Therefore, while admitting the possibility of using the r.f. field to 'prepare' the spin system in a given state it is preferable not to attempt to describe its behaviour by a 'temperature' while the r.f. field is on (see, however, Chapter XII). Immediately after the suppression of the r.f. field the nuclear magnetization still has a transverse component and therefore, as explained in Chapter II, the density matrix of the spin system has off-diagonal elements. As long as these off-diagonal elements exist no completely meaningful description of the state of the spin system by a 'temperature' is possible and a time of the order of the lifetime of these off-diagonal elements, defined somewhat loosely as T_2, should elapse before attempting such a description. One interesting exception occurs when the spin system has been prepared by an 180° pulse or a rapid passage at the end of which there is no transverse nuclear magnetization and the longitudinal magnetization is antiparallel to the applied field. This corresponds to a situation where the higher energy levels are more populated than the lower ones and according to definition (2) to a negative spin 'temperature'. It should be realized that a negative temperature is 'hotter', not 'colder', than any positive temperature since to a system at an infinite temperature, extra energy should be supplied to bring it to a negative temperature. It is clear that only systems with an upper limit to their energy spectrum can have negative temperatures.

Since the time T_1 has been defined as the time necessary for the spin system to come into equilibrium with the lattice, it is clear that the concept of a spin 'temperature' different from the lattice temperature will thus be of interest when $T_2 \ll T_1$. This in turn makes it necessary for us to include in the picture the interactions between the nuclear spins which are strong in solids.

Before doing so consider a spin system describable by a spin 'temperature' according to formula (2), which can be positive or negative depending on the manner in which the spin system has been prepared, and assume that the applied magnetic field is being varied from the initial value H_0 to a different value H_1. This variation is supposed to be sufficiently slow for the conditions of the quantum mechanical adiabatic passage to be realized so that the populations P_m of the various levels $I_z = m$ are unchanged, but at the same time sufficiently fast for its total duration τ to be much shorter than T_1, permitting us to neglect the coupling of the spins with the lattice. Because of the equidistance of the Zeeman levels it is still possible to define a spin 'temperature' $T_S(H)$ proportional to the applied field by the formula (2) and to write for the magnetization of the spins Curie's formula $M_Z = CH/T_S$, which, since T_S is proportional to the field, simply expresses the fact that the magnetization $M_Z = N\gamma\hbar \sum_m p_m m$ is an adiabatic invariant.

It may be as well at this stage to point to a rather obvious but none the less frequent confusion originating from the use of the word adiabatic in two different senses. First, adiabatic in the quantum mechanical sense, or as it is sometimes called in the Ehrenfest sense, describes the evolution of a statistical ensemble when some external parameter is changed in such a way that no transitions are induced and the populations of the various energy levels remain unchanged. Secondly, adiabatic in the thermodynamic sense describes the reversible change of a system in thermal equilibrium when no heat is allowed to flow into or out of the system. It is obvious that except for very special cases such as the one of equidistant levels, if thermal equilibrium, that is a Boltzmann distribution of populations, exists at time $t = 0$, it will *not* remain of the Boltzmann form in an Ehrenfest adiabatic transformation when the energy levels change but not the populations. The two definitions are thus clearly incompatible in general. In the following, adiabatic will mean adiabatic in the Ehrenfest sense, the other type of transformation being called isentropic.

Although within the limitations described, an unambiguous and self-consistent definition of a spin 'temperature' can be made, the general usefulness of this concept, and more particularly of negative temperature, simply as a way of assigning larger populations to the higher energy states of a system than to the lower ones may not seem very impressive.

The real usefulness and deep physical meaning of the concept of spin temperature will appear later when the interactions between the nuclear spins will be taken into account, but even for non-interacting spins the concept of negative temperature can be of interest as is shown by the following example taken from electronic resonance. In the process of spin-lattice relaxation described by the usual equation

$$\frac{dM_z}{dt} = \frac{M_0 - M_z}{T_1}$$

if in the initial state $M_z = -M_0$, the time necessary for M_z to become zero, before reaching the equilibrium value $+M_0$, is $T_1 \ln 2$, of the same order of magnitude as, say, the time T_1 required to reach the value $M_0(e-1)/e \simeq \frac{2}{3} M_0$ starting from $M_z = 0$. At very low temperatures, however, and for electronic spins which then have a much greater heat capacity than the lattice, it is found that the recovery of the magnetization to its equilibrium value after a complete saturation is a much slower process than the disappearance of a negative magnetization after a 180° pulse. This asymmetry is easily understood in terms of spin temperature.

After saturation the spins are at an infinite temperature, and since the heat capacity of the lattice is very small its temperature becomes practically infinite too and it is unable to 'cool' the spin system by receiving energy from it. On the other hand, after the reversal of the magnetization the spin system at a negative temperature is 'hotter' than the lattice and energy will flow from the spin system into the lattice, leading to a disappearance of the magnetization, and this even if the lattice is at a practically infinite temperature. Although this prediction could be made without speaking of spin temperature by using the language of populations only, the concept of negative temperature is helpful in making the prediction obvious.

B. Interacting spins in high field

From the existence of a tight coupling between nuclear spins in a solid, it follows that the components $I_z^j = m^j$ of the individual spins are not good quantum numbers and it has been mentioned before that the correct approach to this problem, previously used in the calculation of the moments of the resonance curve, is a collective one due to Gibbs, where the whole sample is treated as a single spin-system.

At the expense of some rigour, it is still possible, however, at least in fields much higher than the local field, to use the much simpler Maxwell–Boltzmann picture of individual spins with well-defined energy levels, between which the dipolar interaction induces transitions at a certain rate W. As appeared in Chapter IV, a dipolar interaction between two spins can, in principle, induce several kinds of transitions: flip of one spin only, flips of both spins in the same direction, opposite flips of the two spins, but only the last one can take place with a non-negligible probability in a rigid lattice if the applied field is much higher than the local field, since it is the only one that conserves the total Zeeman energy.

An approximate evaluation of its probability W can be made as follows. Consider two identical spins ($\frac{1}{2}$ for simplicity) in a high magnetic field; at time zero, the spins are in the $+$ and $-$ states respectively and we seek the probability per unit time, W, that these spins exchange their orientations, passing from the state $|+, -)$ to the state $|-, +)$, by means of their dipolar interaction.

Imagine first that these two spins are isolated from the effects of all the other spins of the crystal. The states $|+, -)$ and $|-, +)$ are not eigenstates of the total Hamiltonian, Zeeman plus dipolar, but rather linear combinations of these states; singlet $\psi_S = \{|+-) - |-+)\}/\sqrt{2}$ and triplet $\psi_T = \{|+-) + |-+)\}/\sqrt{2}$. If at time zero the two-spin system is found in the state $|+, -)$, it goes periodically from this state to the state $|-, +)$ and returns to the initial state with a frequency

$$\omega_1 = \frac{E_S - E_T}{\hbar} = \frac{\gamma^2 \hbar}{2r^3}(1 - 3\cos^2\theta),$$

where r is the distance PP' between the two spins and θ the angle PP' makes with the direction of the applied field H_0. This problem is formally identical to the one of an r.f. field of amplitude $H_1 = \omega_1/\gamma$ inducing transitions between the two states $|+, -)$ and $|-, +)$. The fact that in the absence of dipolar interactions these two states have the same energy, that is, that the frequency of the imaginary radio field is zero, need not trouble us, since this is precisely the situation to which one is reduced by using a rotating coordinate system.

We introduce the other spins of the crystal by supposing that their effect is to give a certain width to the two levels $|+,-)$ and $|-,+)$ of the pair considered. If this width is large with respect to ω_1, which is reasonable in view of the fact that each spin has many neighbours, the treatment used in Chapter II to pass from equation (13) to equation (30') is still valid. We get $W = \frac{1}{2}\pi\omega_1^2 f(0)$, where the function $f(\omega)$ normalized to unity represents the shape of the levels of the two-spin system under the action of all the other spins. We make the crude assumption that this function has the same, approximately gaussian, shape as the magnetic resonance line of the entire spin system, whence $f(0) = 1/\sqrt{(2\pi\overline{\Delta\omega^2})}$ where $\overline{\Delta\omega^2}$ is given by the Van Vleck formula. To be specific, suppose that we are dealing with a simple cubic lattice. Then it is readily found by using equations (39') and (39'') of Chapter IV that the probability of a flip-flop between nearest neighbours is of the order of:

$$W \sim \frac{(\overline{\Delta\omega^2})^{\frac{1}{2}}}{50} \quad \text{if } H_0 \text{ is applied parallel to a crystalline axis and perpendicular to } PP',$$

$$W \sim \frac{4(\overline{\Delta\omega^2})^{\frac{1}{2}}}{50} \quad \text{if } H_0 \text{ is parallel to } PP', \tag{3}$$

$$W \sim \frac{(\overline{\Delta\omega^2})^{\frac{1}{2}}}{30} \quad \text{if the substance is a powder and } W \text{ is averaged over all directions.}$$

For spins $\frac{1}{2}$, the spin flips do not change the populations but a spin 'temperature' can always be defined through the relation (1). The mutual flips still have the interesting effect of suppressing local differences in spin 'temperatures' by a transport process which has received the name of spin diffusion. To see that this process really obeys a diffusion equation consider first a linear chain of spins $\frac{1}{2}$ separated by an interval a, and assume for simplicity that a mutual flip of probability per unit time W can take place between nearest neighbours only. For a spin of abscissa x the rate of change of, say, p_+ is given by

$$\frac{1}{W}\frac{\partial p_+(x)}{\partial t} = p_-(x)[p_+(x+a)+p_+(x-a)]-p_+(x)[p_-(x+a)+p_-(x-a)] \tag{4}$$

or, writing $p_+-p_- = p$, $p_++p_- = 1$ and assuming p so small that terms quadratic in p are negligible,

$$\frac{\partial p(x)}{\partial t} = W[p(x+a)+p(x-a)-2p(x)] \tag{5}$$

which, if $p(x)$ does not vary appreciably over a distance a, can be approximated by the uni-dimensional diffusion equation

$$\frac{\partial p}{\partial t} = Wa^2 \frac{\partial^2 p}{\partial x^2}. \tag{6}$$

If we neglect the anisotropy of W, (6) can be extended to the three-dimensional form

$$\frac{\partial p}{\partial t} = D\Delta p \quad \text{with } D = Wa^2. \tag{7}$$

The constant D is very small. With $W \sim 10^3 \sec^{-1}$, $a \sim 10^{-8}$, D is 10^{-13}, whilst it is of the order of 10^{-5} for water molecules. The time t, required to transport p over a distance r is of the order of r^2/D, which for macroscopic values of r is exceedingly long but is only about a second for a transport of p over 30 atomic distances. Spin diffusion has been described here for the case of spins $\frac{1}{2}$ but clearly exists for higher spins too. We shall see in a later chapter the important role it plays in the nuclear spin-lattice relaxation by paramagnetic impurities.

For spins larger than $\frac{1}{2}$ where the definition of a spin 'temperature' by equation (2) imposes a condition on the populations, the mutual flip which, thanks to the equidistance of the Zeeman levels, can modify the values of the populations, is a powerful mechanism for the establishment of a spin 'temperature'.

This can be seen quite generally as follows: it is a well-known result of statistical mechanics that if a given energy E has to be distributed between N identical systems having individual energy levels, the most probable distribution of populations among these states will be the Boltzmann one. For this distribution to be reached starting from any other initial distribution, a coupling mechanism is required which can transfer energy from one individual system to another and change the populations, keeping the total energy E of the N systems constant. The spin-spin interactions, through the simultaneous opposite flip of two spins, thanks to the equidistance of the levels of the individual spins, do precisely this.

To demonstrate experimentally the role of spin-spin interactions in establishing a spin 'temperature', a non-Boltzmann distribution of populations must first be created, and this is, in general, not achieved by partial saturation by an r.f. field. On the other hand, this result can be obtained by irradiating nuclei with a quadrupole moment in a cubic crystal, by an ultrasonic wave at twice the Larmor frequency. This experiment (1) was performed on Na^{23} and Cl^{35} in a single crystal of NaCl. The transition probability A induced by the ultrasonic wave could be made so large as to saturate the resonance $\Delta m = 2$ in a time that was short compared with the spin-lattice relaxation time. The populations of the four levels $I_z = -\frac{3}{2}$ to $I_z = \frac{3}{2}$, of a nuclear spin $I = \frac{3}{2}$ prior to the ultrasonic irradiation, are respectively $\frac{1}{4}(1-3\epsilon)$, $\frac{1}{4}(1-\epsilon)$, $\frac{1}{4}(1+\epsilon)$, $\frac{1}{4}(1+3\epsilon)$ with $\epsilon = \gamma\hbar H_0/2kT$, which leads to a magnetization per nucleus:

$$M_z = \frac{\gamma^2\hbar I(I+1)H}{3kT} = \frac{5\gamma^2\hbar H}{4kT}.$$

Immediately after such an irradiation lasting a time large compared with $1/A$ but short compared with T_1, for an isolated nucleus, the populations of the levels $\frac{3}{2}$ and $-\frac{1}{2}$ are equal, as are also those of the levels $\frac{1}{2}$ and $-\frac{3}{2}$. The four populations are thus $\frac{1}{4}(1-\epsilon)$, $\frac{1}{4}(1+\epsilon)$, $\frac{1}{4}(1-\epsilon)$, $\frac{1}{4}(1+\epsilon)$, leading to a total magnetic moment

$$\langle M_z \rangle = \frac{\gamma^2\hbar H}{4kT}, \tag{8}$$

five times smaller than the thermal equilibrium value. Experimentally, after a strong ultrasonic irradiation vanishingly small values of M_z, much smaller than (8), were found. This is clearly to be expected if a Boltzmann distribution of populations is continuously maintained by the spin-spin interactions, and a simple calculation shows that the decay of the magnetization during the irradiation is an exponential one, with a time constant τ given by $1/\tau = \frac{8}{5}A$, and that in the presence of spin-lattice relaxation a steady-state value $M_z = M_0[1+8AT_1/5]^{-1}$ is reached for M_z, that is arbitrary small if A is sufficiently large. The proof is as follows.

Let $p_{\frac{3}{2}}$, $p_{\frac{1}{2}}$, $p_{-\frac{1}{2}}$, $p_{-\frac{3}{2}}$ be the populations of the levels $I_z = m$ of Na^{23} irradiated by ultrasonic waves at twice the Larmor frequency and A the corresponding transition probability per unit time, between the levels $\frac{3}{2} \leftrightarrow -\frac{1}{2}$ and $\frac{1}{2} \leftrightarrow -\frac{3}{2}$.

The rates of change of these populations are given by the following equations:

$$\left.\begin{aligned}\frac{dp_{\frac{3}{2}}}{dt} &= -A(p_{\frac{3}{2}}-p_{-\frac{1}{2}})+F_{\frac{3}{2}}(p_m)+G_{\frac{3}{2}}(p_m) \\ \frac{dp_{\frac{1}{2}}}{dt} &= -A(p_{\frac{1}{2}}-p_{-\frac{3}{2}})+F_{\frac{1}{2}}(p_m)+G_{\frac{1}{2}}(p_m) \\ \frac{dp_{-\frac{1}{2}}}{dt} &= -A(p_{-\frac{1}{2}}-p_{\frac{3}{2}})+F_{-\frac{1}{2}}(p_m)+G_{-\frac{1}{2}}(p_m) \\ \frac{dp_{-\frac{3}{2}}}{dt} &= -A(p_{-\frac{3}{2}}-p_{\frac{1}{2}})+F_{-\frac{3}{2}}(p_m)+G_{-\frac{3}{2}}(p_m)\end{aligned}\right\}. \tag{9}$$

The terms F_m and G_m represent the contributions to the rates of change of the populations from the spin-lattice relaxation and the spin-spin coupling, respectively. We shall not require their exact expressions, which depend in particular on the type of spin-lattice relaxation mechanism assumed (magnetic dipole or electric quadrupole). Multiplying the first equation by $\frac{3}{2}$, the second by $\frac{1}{2}$,..., and adding them together we get on the left-hand side $d\langle I_z\rangle/dt$. On the right-hand side the sum $\frac{3}{2}F_{\frac{3}{2}}(\ )+...-\frac{3}{2}F_{-\frac{3}{2}}(\ )$ represents the contribution to the rate of change of $\langle I_z\rangle$ from the spin-lattice relaxation mechanism, which, if we assume a single spin-lattice relaxation time, an assumption to be justified in Chapter IX, can be written as $-[\langle I_z\rangle-I_0]/T_1$. Similarly, $\frac{3}{2}G_{\frac{3}{2}}$ to $-\frac{3}{2}G_{-\frac{3}{2}}$ represents the contribution to $d\langle I_z\rangle/dt$ from the spin-spin interaction, and this should be zero since, because of conservation of energy, the spin-spin interaction is unable to change $\langle I_z\rangle$. We are thus left with the equation

$$\frac{d\langle I_z\rangle}{dt} = -\frac{\langle I_z\rangle-I_0}{T_1} - 2A\{(p_{\frac{3}{2}}-p_{-\frac{3}{2}})+(p_{\frac{1}{2}}-p_{-\frac{1}{2}})\}.$$

Since the spin-spin interactions constantly maintain a Boltzmann distribution between the various populations,

$$p_{\frac{3}{2}}-p_{-\frac{3}{2}} = 3(p_{\frac{1}{2}}-p_{-\frac{1}{2}}),$$

and

$$\langle I_z\rangle = \tfrac{3}{2}(p_{\frac{3}{2}}-p_{-\frac{3}{2}})+\tfrac{1}{2}(p_{\frac{1}{2}}-p_{-\frac{1}{2}}) = \tfrac{10}{2}(p_{\frac{1}{2}}-p_{-\frac{1}{2}}) = \tfrac{10}{6}(p_{\frac{3}{2}}-p_{-\frac{3}{2}}),$$

$$\frac{d\langle I_z\rangle}{dt} = -\frac{\langle I_z\rangle-I_0}{T_1} - \tfrac{8}{5}A\langle I_z\rangle.$$

The limiting value of $\langle I_z\rangle$ obtained when $d\langle I_z\rangle/dt = 0$ is

$$\langle I_z\rangle = \frac{I_0}{1+8AT_1/5}.$$

The necessity of the equidistance of the Zeeman levels for the establishment of a spin 'temperature' through mutual spin 'flips' starting from a non-Boltzmann distribution of populations must be compared with a similar requirement for the conservation of a Boltzmann distribution when a parameter of the system (the magnetic field) is varied adiabatically. Both are a consequence of a general requirement of statistical mechanics for the existence of a temperature in a large system, namely ergodicity, that is absence of any constant of motion (good quantum number in quantum statistics) apart from the total energy. It is clear that in a system S composed of a large number N of identical elementary systems S^i in interaction with each other, the non-equidistance of their energy levels leads to the existence for the total system of extra constants of motion, the populations P_m of the individual energy levels. Indeed these populations can be expressed as the expectation values taken over the whole system S of the operators:

$$\mathscr{P}_m = \frac{1}{N} \sum_i \mathscr{P}_m^i, \tag{10}$$

where $\mathscr{P}_m^i$ is the projection operator on the state $|m\rangle$ for the system S^i.

The ergodicity of a spin system can be quenched, for instance, by the appearance of quadrupole splittings such as occur in non-cubic, or imperfect cubic crystals. Indeed it was this quenching of the spin flips by the quadrupole broadening that was partly responsible for the changes in the dipolar broadening in formula (64) to (66) of Chapter IV. As another example of a spin system where the spin-spin interactions are unable to establish a temperature, consider a substance such as LiF where the nuclear spins of Li[7] and F[19] have different Larmor frequencies. It is usual to look at it as two different systems, the system Li[7] and the system F[19] (or three systems if the small admixture of Li[6] is not neglected). For each of these systems the levels in a high field are equidistant and a spin temperature can be defined, which, however, need not be the same for both systems since energy cannot be transferred from one to the other. Thus, raising the temperature of the fluorine system by saturating the resonance of F[19], or even making it negative by a rapid passage, will not affect the temperature of the lithium system.

For the purpose of the present discussion it is enlightening to consider a crystal of Li[7]F[19] (neglecting for simplicity Li[6]) as a statistical ensemble of identical systems of a single species. The individual system is a Li[7] spin *plus* a F[19] spin associated with it by a constant but otherwise arbitrary lattice vector. The eight energy levels of such a system are given by the formulae

$$\hbar E(m, m') = \hbar H(\gamma_\text{F} m + \gamma_\text{Li} m'),$$

where $m = \pm \tfrac{1}{2}$ and $m' = \pm \tfrac{1}{2}, \pm \tfrac{3}{2}$.

Although not equidistant, they vary linearly with the field and therefore have properties intermediate between those of the Zeeman levels of a single species and those of, say, atoms with hyperfine structure levels given by the Breit–Rabi formulae. If a Boltzmann distribution exists for a given value of the applied field for the eight levels of our individual systems, it keeps that form as the field varies. On the other hand, if it is *not* of the Boltzmann form, spin-spin interactions cannot bring it to that form. All this is, of course, also evident by the consideration of Li[7] and F[19] (and also Li[6]) as different systems.

A type of phenomenon for which the concept of spin temperature provides a convenient, if not indispensable, description, is the so-called 'cross-over' which occurs when two normally distinct frequencies ω_{ab} and $\omega_{a'b'}$, belonging to the same spin or atom or two different ones, coincide for a special value of an external parameter, such as the applied d.c. field.

Let N_a and N_b be the populations of the two levels separated by $\hbar\omega_{ab}$, and $N_{a'}$ and $N_{b'}$ those of interval $\hbar\omega_{a'b'}$. If a dipolar coupling exists between the two spins or atoms, one of which has in its spectrum the frequency ω_{ab} and the other the frequency $\omega_{a'b'}$ (in particular, the spectra can be identical and both contain each frequency), and, if by a change of the external parameter, ω_{ab} is made equal to $\omega_{a'b'}$, mutual flips occur and a steady state is reached where $N_a/N_b = N_{a'}/N_{b'}$. This is conveniently described by the assignment of the same spin 'temperature' to the two pairs of levels. If the external parameter is being varied continuously through the value for which $\omega_{ab} = \omega_{a'b'}$, the criterion for the equalization of the two spin temperatures during the crossing is that the 'cross-over' time, spent in the region where the difference $|\omega_{ab}-\omega_{a'b'}|$ is smaller than the corresponding line width, will be long compared with the inverse $1/W$ of the probability for the mutual flip.

An interesting situation arises when the spin-lattice relaxation time corresponding to one of the transitions, say ω_{ab}, is very long and the other one very short. Then if the cross-over time is long compared not only with the mutual flip probability $1/W$ but also to the fast relaxation time T_1, immediately after the cross-over we have

$$N_a/N_b = N_{a'}/N_{b'} = \exp(-\hbar\omega_{a'b'}/kT),$$

where T is the temperature of the lattice. This effect has been observed (2) in a single crystal of *para*dichlorobenzene, where for certain values of the applied d.c. field, the Zeeman splitting of the protons coincides with one of the frequencies of Cl[35] or Cl[37], which have quadrupole splittings of the order of 30 Mc/s in zero field. The protons, which have a spin-lattice relaxation time of several minutes, could be polarized in less than a second by thermal contact with the chlorines. The various frequencies of the spectrum of the chlorine nuclei could thus be determined by deducing the values of the fields at which the cross-over took place from the observation of the proton polarization.

Another case of interest is that when both spin-lattice relaxation times are very long. To the relation $N_a/N_b = N_{a'}/N_{b'}$ after the cross-over, the conservation of energy for the combined spin system that is practically isolated from the lattice during the cross-over adds the supplementary condition:

$(N_a + N_{a'})$ before cross-over $= (N_a + N_{a'})$ after cross-over.

An example (3) of this situation is provided by the spectrum of phosphorus impurity atoms with nuclear and electronic spin $\frac{1}{2}$, embedded in a silicon lattice. At liquid helium temperatures the spin-lattice relaxation times of those atoms are of the order of minutes. The Hamiltonian of such an atom is

$$\mathscr{H} = -\gamma_e \hbar H S_z - \gamma_n \hbar H I_z + A \mathbf{I} \cdot \mathbf{S}.$$

Its four energy levels are given by the well-known Breit–Rabi formulae and it can be shown that three of those become equidistant for a value $H^* = A/\hbar(\gamma_e + \gamma_n)$ of the field H, which for phosphorus is 42 gauss. If the field is changed from a large value H_0 to any value above 42 gauss and back at a rate fast compared with T_1, the populations are unchanged and the process is reversible. If, however, the critical value $H^* = 42$ gauss is crossed, an irreversible change in the populations occurs, which is completely accounted for by the establishment through the crossing of a spin temperature between the three, then equidistant levels.

C. Interacting spins in low fields

The introduction of interactions between the nuclear spins, although strengthening the concept of spin temperature, still does not permit us in the light of the evidence gathered so far to conclude that it is more than a convenient description of results that could be easily formulated without making use of it. The situations examined so far belonged to the high field case where the energies of the individual spins are much larger than the interactions between them, and one may well ask what happens when this restriction is lifted.

The first experiment of that kind (4) (referred to in the following as experiment A) was performed on a crystal of LiF, where the spin-lattice relaxation times in a high field were of the order of 2 minutes for F[19] and 5 minutes for Li[7], so that experiments of short duration could be performed during which the spin system was practically isolated from the lattice. After the crystal had reached its equilibrium nuclear polarization (as measured by the nuclear resonance signal of Li[7]) in a high field of the order of 7000 gauss, it was rapidly taken out of the magnet gap into the earth's field for a second or so, then returned into the gap. The same nuclear signal was observed as before its removal, showing that the change of the field from 7000 gauss down to the earth's field value was a reversible process for the spin system, a result that is surprising at first sight. The reversibility of the reduction of the applied field down to values much larger than the local field was obviously to be expected independently of any thermodynamic considerations: as long as $H \gg H_{\text{local}}$, because of conservation of energy, the only transitions induced by the spin-spin interactions are opposite flip-flops of neighbouring spins which do not change the total magnetic moment M_0. On the other hand, when the applied field falls well below the local field, one is tempted to say that two neighbouring spins can undergo all the transitions for which matrix elements exist in their dipole-dipole interactions (flip of a single spin, flip of both spins in the same direction, not prevented now by conservation of energy) and to expect a complete disorientation of all the spins and no nuclear resonance signal upon coming back into the high field. The passage through zero field would be irreversible.

The weakness of the argument which leads to this wrong conclusion is in an incorrect description of the spin system in low fields. When the interaction between the spins becomes comparable to their Zeeman energy, the concept of energy levels of individual spins becomes meaningless and one must use the Gibbs approach and speak in terms of energy levels and eigenstates of the whole sample. The explanation proposed (4) for the reversibility of the demagnetization down to practically zero field, is the now non-trivial assumption of the existence for all values of the applied field of a thermal equilibrium inside the spin system with a spin temperature $T_S(H)$. Mathematically it can be formulated by saying that for all values of H, the density operator of the spin system is

$$\rho = \exp\{-\mathscr{H}(H)/kT_S(H)\}/\text{tr}[\exp\{-\mathscr{H}(H)/kT_S(H)\}], \quad (11)$$

where the total Hamiltonian $\mathscr{H}(H)$,

$$\mathscr{H} = \mathscr{H}_0 + \mathscr{H}_1 = \sum_i \mathscr{H}_0^i + \sum_{i<k} \mathscr{H}_1^{ik} = -HM_z + \sum_{i<k} \mathscr{H}_1^{ik}, \quad (12)$$

is the sum of the Zeeman energies and the dipolar interactions, and the spin temperature $T_S(H)$ is a parameter, whose dependence on H is determined by the isentropic character of the demagnetization. Another argument in favour of that hypothesis was the existence in very low fields of a single and much shorter (15 sec) spin-lattice relaxation time T_1 for Li^7 and F^{19}, suggesting strong thermal contact between the two nuclear species.

This interpretation has been further strengthened by another experiment (5) (experiment B), where prior to the demagnetization the nuclear magnetization was made antiparallel to the field by a very fast (fraction of a microsecond) reversal of the applied field. The situation in a high field was then describable by a negative temperature, and again it was found that the subsequent demagnetization was a reversible process and that after return into the gap the sample still exhibited a nuclear magnetization antiparallel to the field. The interpretation of this experiment parallels the previous one except for the change of a negative sign for the spin temperature.

Although the assumption of a spin temperature, positive or negative, gives a satisfactory interpretation of the reversibility of the demagnetization in the experiments A and B, it is not certain that it is unique and indeed some other had been proposed. It is possible to give to this assumption a further test and demonstrate the identity of spin temperature and thermodynamic temperature by an experiment C (6) based on the following considerations.

It is clear that when a spin system which has been allowed to come into equilibrium with a lattice at a temperature T_0 in a large field H_0, is demagnetized into a weak field $H \ll H_{\text{local}}$, some kind of order must exist in this system since, according to experiment A, upon being brought back into H_0 it has the same magnetic moment as before demagnetization. If we are to believe the assumption of a spin temperature this order is adequately described by a certain spin temperature $T_S \ll T_0$ which is derived simply from the initial conditions H_0, T_0 by writing that for a change dH of the field, the work dW done on the system by the applied field is equal to the change dU of the internal energy of the system. Since $U = \langle \mathscr{H} \rangle = \text{tr}\{\rho \mathscr{H}\}$ and $dW = -\langle M_z \rangle dH$, this can be written

$$\frac{d}{dH}\text{tr}\{\rho \mathscr{H}\} = -\text{tr}\{\rho M_z\},$$

where
$$\rho = \exp\{-\mathscr{H}(H)/kT(H)\}/\text{tr}[\exp\{-\mathscr{H}(H)/kT(H)\}],$$

$$\mathscr{H}(H) = \mathscr{H}^0(H) + \mathscr{H}^1 = -HM_z + \mathscr{H}^1,$$

$$M_z = \hbar \sum_i \gamma_i I_z^i,$$

$$\mathscr{H}^1 = \hbar^2 \sum_{i<k} \gamma_i \gamma_k r_{ik}^{-3} \left[(\mathbf{I}^i \cdot \mathbf{I}^k) - \frac{3(\mathbf{I}^i \cdot \mathbf{r}_{ik})(\mathbf{I}^k \cdot \mathbf{r}_{ik})}{r_{ik}^2} \right]. \quad (13)$$

The solution of (13) is easily obtained in the case of high temperatures when it is permissible to use a linear expansion of the exponential, which is legitimate in the three experiments A, B, C. Making use of the relations:

$$\mathrm{tr}\{\mathcal{H}^0\} = \mathrm{tr}\{\mathcal{H}^1\} = \mathrm{tr}\{\mathcal{H}^1\mathcal{H}^0\} = 0,$$

we get $\quad \dfrac{d}{dH}\left\{\dfrac{1}{kT(H)}\mathrm{tr}\{H^2 M_z^2 + (\mathcal{H}^1)^2\}\right\} = \dfrac{H}{kT(H)}\mathrm{tr}\{M_z^2\}$

or
$$\frac{dT}{T} = \frac{H\,dH}{H^2 + H_L^2}, \tag{14}$$

where H_L is the local field defined by

$$H_L^2 = \mathrm{tr}\{(\mathcal{H}^1)^2\}/\mathrm{tr}\{M_z^2\}. \tag{15}$$

The calculation of H_L^2 by the trace (15) is very similar to the evaluations of second moments in Chapter IV, with the difference that here the total, rather than the truncated dipolar Hamiltonian $\mathcal{H}^1$ is being used.

For LiF it is found $H_L^2 = 60\cdot 6$ gauss^2, $H_L = 7\cdot 77$ gauss. The equation (14) is easily integrated to

$$\frac{T}{T_0} = \left[\frac{H^2 + H_L^2}{H_0^2 + H_L^2}\right]^{\frac{1}{2}}. \tag{16}$$

In particular, if the initial field $H_0 \gg H_L$ and the final field $H \ll H_L$,

$$\frac{T}{T_0} = \frac{H_L}{H_0}. \tag{17}$$

Conversely, (16) also permits the calculation of the magnetic moment reached in a large field H^* where it can be measured by a resonance experiment, after the spin system has been polarized in a field H_0 by thermal contact with a lattice at a temperature T_0. The spin temperature T^* in the field H^* is related to T_0 and H_0 by

$$\frac{T^*}{T_0} = \frac{[H^{*2} + H_L^2]^{\frac{1}{2}}}{[H_0^2 + H_L^2]^{\frac{1}{2}}},$$

whence $\quad \langle M_z(H^*)\rangle = \dfrac{CH^*}{T^*} = \dfrac{CH^*}{T_0}\dfrac{[H_0^2 + H_L^2]^{\frac{1}{2}}}{[H^{*2} + H_L^2]^{\frac{1}{2}}}$

and, since $H^* \gg H_L$,

$$\langle M_z(H^*)\rangle = C\frac{[H_0^2 + H_L^2]^{\frac{1}{2}}}{T_0}, \tag{18}$$

which is independent of H^* and entirely determined by the initial conditions H_0 and T_0. The nuclear resonance signal observed at a frequency $\Omega^* = -\gamma H^*$ will be proportional to $\langle M_z(H^*)\rangle$ and therefore a function $S(H_0, T_0)$ of the field and temperature at which the sample was 'prepared'. The principle of the experiment C performed in liquid helium at 2° K on a crystal of LiF is then the following. Suppose that a spin system prepared at $T_0 = 300°$ K in a field

$$H_0 = H_L \times \frac{300}{2} = 7{\cdot}77 \times \frac{300}{2} = 1165 \text{ gauss}$$

is demagnetized into zero field. According to (17), if the assumption of the spin temperature is correct, its temperature in zero field is $T_S = 2°$ K.

It is easy to produce a situation where the spin system is indeed describable by a genuine thermodynamic temperature of 2° K by letting it come into equilibrium in zero field with a cold lattice at 2° K. The identity of spin temperature and thermodynamic temperature would then be demonstrated by the identity of the two signals observed in a large field H^*:

$$S(2°, 0) = S\left(300°, \frac{300}{2} H_L\right). \tag{19}$$

Actually, because of the unavoidable change of gain of the apparatus performing experiments at two different lattice temperatures, instead of (19) one has (19′):

$$S(2°, 0) = \lambda S\left(300°, \frac{300}{2} H_L\right), \tag{19′}$$

where λ is the change in the gain of the apparatus on going from 300° K to 2° K. λ is easily eliminated by making an extra measurement at 2° K, polarizing in a field $H_1 \gg H_L$, say $H_1 \sim 50$ gauss. Since both H_1 and H_0 are in the high field region where the magnetic moment is an adiabatic invariant, we know that

$$S(2°, H_1) = \lambda\left[\left(\frac{300}{2} H_1\right) \middle/ \left(\frac{300}{2} H_L\right)\right] S\left(300°, \frac{300}{2} H_L\right)$$

$$= \lambda S\left(300°, \frac{300}{2} H_L\right) H_1/H_L.$$

In order to prove (19′) experimentally it is then sufficient to verify that

$$S(2°, 0) = (H_L/H_1) S(2°, H_1). \tag{20}$$

In this way no measurements have to be made at room temperature at all. It is still better to check the general relation (18),

$$S(2°, H_0) \propto [H_0^2 + H_L^2]^{\frac{1}{2}}, \tag{21}$$

of which (20) is a special case, by plotting the curve $S(H_0)$.

To check the predicted behaviour (21), laboratory polarizing fields H_0 from zero to several times H_L were used. The fixed frequency selected for the observation of the resonance of Li^7 was 8 Mc and H^* was

4840 gauss, chosen simply as a matter of convenience. To make the observations at this field, several hundred times larger than the polarization fields, the measurements had to be made quickly. The method of fast passage is ideally suited for this. The experimental method, simplified, was consequently to permit the sample to reach equilibrium at the chosen weak field, the lattice temperature being 2° K, and then to raise the magnetic field suddenly to a value greater than 4840 gauss, making a fast passage observation on Li[7], 'on the fly'.

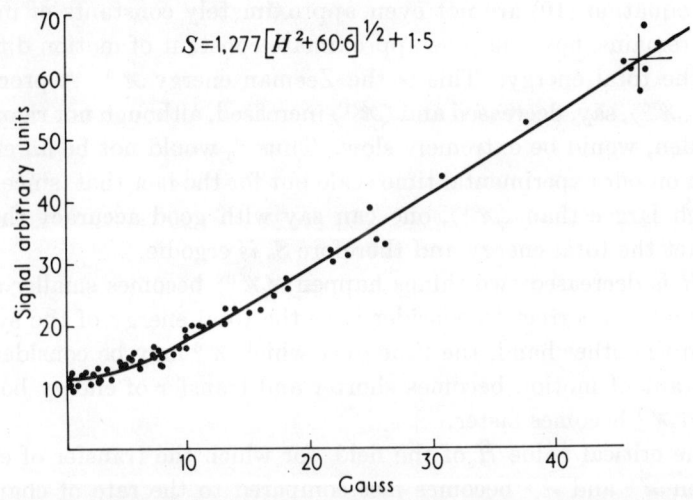

FIG. V, 1. The fast-passage signal amplitude of Li[7] measured in arbitrary units as a function of the magnetic field in which the LiF sample reached equilibrium at 2° K. The full curve is the theoretical expression (21) normalized and corrected for finite spin-lattice relaxation time, as explained in the text.

The experimental results are summarized by the points of Fig. V, 1, where the amplitude of the signals observed, in arbitrary units, is given as a function of the polarizing field H_0. The full curve is the theoretical curve (21), normalized to pass through the average of six points taken around $H_0 \sim 50$ gauss and corrected for a small increase of 1·5 units due to the growth of the signals during the operation caused by a finite, although very long, spin-lattice relaxation time. The agreement between theory and experiment is satisfactory and gives confidence in the reality of spin temperature.

Since the assumption of a spin temperature correctly describes the behaviour of a Zeeman spin system, when the applied field is varied all the way down to zero, it is interesting to inquire whether it was to be expected and why. This is a difficult problem and only a qualitative discussion will be given.

It has already been mentioned that a necessary condition for a large system to be able to come into thermal equilibrium starting from a non-equilibrium situation or to remain in thermal equilibrium when some external parameter is slowly varied, is that the system be ergodic.

Now consider a Zeeman spin system S_0 with a single species of spins. In the low field region it is reasonable to assume that it is an ergodic system without any extra constants of motion. In the high field range, because of the nature of the Zeeman splittings, the population operators $\mathscr{P}_m$ of equation (10) are not even approximately constants of motion. There remains, however, one approximate constant of motion different from the total energy. This is the Zeeman energy $\mathscr{H}^0$. A process in which $\langle\mathscr{H}^0\rangle$, say, decreased and $\langle\mathscr{H}^1\rangle$ increased, although not rigorously forbidden, would be extremely slow. Thus S_0 would not be an ergodic system on our experimental time scale but for the fact that, since $\langle\mathscr{H}^0\rangle$ is much larger than $\langle\mathscr{H}^1\rangle$, one can say with good accuracy that $\mathscr{H}^0$ is in fact the total energy and therefore S_0 *is* ergodic.

As H is decreased two things happen: $\langle\mathscr{H}^0\rangle$ becomes smaller and it is less and less correct to consider it as the total energy of the system, but, on the other hand, the time over which $\mathscr{H}^0$ may be considered as a constant of motion becomes shorter and transfer of energy between $\mathscr{H}^0$ and $\mathscr{H}^1$ becomes faster.

If the critical value $\tilde{H}$ of the field, for which the transfer of energy between $\mathscr{H}^0$ and $\mathscr{H}^1$ becomes fast compared to the rate of change of the field, is still large compared to H_L, the system is approximately ergodic through the whole range of H and the demagnetization is a reversible isentropic process, its small irreversibility being of the order of $H_L^2/\tilde{H}^2$. Since within experimental error the process is reversible, $\tilde{H} \gg H_L$. It should finally be remarked that whilst in high fields there is an obvious symmetry between positive and negative temperature since a reversal of all the spins simply changes the sign of the Zeeman energy, the situation is different in zero field where a reversal of the spins does *not* change the energy. Positive and negative temperatures in zero field clearly correspond to different relative arrangements of the spins.

D. Zeeman system with more than one spin species

The behaviour of a Zeeman spin system with two species of spins (such as LiF) provides an example for the previous discussion. In the low-field range it can be assumed that the total energy is the only constant of motion and that the system is ergodic: as evidence to that effect we may quote the identity of the spin-lattice relaxation times of Li^7 and F^{19}, and the audio-frequency experiments (7) whereby in a field of 42 gauss the application of an audio-frequency field at the Larmor frequency of F^{19} affected the nuclear resonance signal of Li^7, observed subsequently in a high field.

On the contrary, in a high field there are two distinct approximate constants of motion, the Zeeman energies $\mathcal{H}^{0'} = -\gamma'\hbar H_0 I'_z$ and $\mathcal{H}^{0''} = -\gamma''\hbar H_0 I''_z$ of both species of spins (there are three species if Li^6 is taken into account). Therefore the total spin system is *not* ergodic and the demagnetization should be an irreversible process except for special initial conditions. Just as in the case of one species, it is permissible, in a high field, to consider Li^7 and F^{19} as two distinct ergodic systems each with a temperature of its own. The fact that we do not know how to handle the lithium-fluorine interaction and what temperature, if any, should be assigned to it, is not disturbing, for this interaction is very small compared with the Zeeman energies of both systems.

By the mixing field we shall mean the field $\tilde{H}$ at which the rate of exchange of energy between Zeeman energy and spin-spin interaction (and thus also between the Zeeman energies of the two species) becomes fast compared with the rate of change of the applied field. If $\tilde{H}$ is large compared with the local field the whole process of demagnetization can be described simply by assuming that above the mixing field we have two distinct spin systems with energies

$$\mathcal{H}^{0'} = -\gamma'\hbar H_0 I'_z \quad \text{and} \quad \mathcal{H}^{0''} = -\gamma''\hbar H_0 I''_z$$

describable by temperatures T' and T'' and that below $\tilde{H}$ we have a simple Boltzmann system with a Hamiltonian

$$\mathcal{H} = \mathcal{H}^{0'} + \mathcal{H}^{0''} + \mathcal{H}^{1'} + \mathcal{H}^{1''} + \mathcal{H}^2,$$

where $\mathcal{H}^{1'}$, $\mathcal{H}^{1''}$, and $\mathcal{H}^2$ are respectively the Li^7–Li^7, F^{19}–F^{19}, and Li^7–F^{19} interactions. Once the mixing is accomplished the subsequent behaviour of the spin system is reversible, as if it were a single ergodic system at a temperature T. If one includes Li^6 the generalization is obvious.

Starting the demagnetization process at a field H_0 with initial temperature T'_i and T''_i for both systems, we arrive at the field $\tilde{H}$ with temperatures

$$T' = T'_i \frac{\tilde{H}}{H_0} \quad \text{and} \quad T'' = T''_i \frac{\tilde{H}}{H_0}.$$

After the mixing, the new temperature T is obtained by writing that the total energy, or, since $\tilde{H} \gg H_L$, the expectation value of the Zeeman energy, is conserved.

$$\langle \mathcal{H}^{0\prime} \rangle + \langle \mathcal{H}^{0\prime\prime} \rangle \simeq \langle \mathcal{H} \rangle,$$

$$\frac{1}{T'}\mathrm{tr}\{(\mathcal{H}^{0\prime})^2\} + \frac{1}{T''}\mathrm{tr}\{(\mathcal{H}^{0\prime\prime})^2\} \simeq \frac{1}{T}\mathrm{tr}\{(\mathcal{H}^{0\prime})^2 + (\mathcal{H}^{0\prime\prime})^2\},$$

$$\frac{1}{T} = \frac{(N'/T')\gamma'^2 I'(I'+1) + (N''/T'')\gamma''^2 I''(I''+1)}{N'\gamma'^2 I'(I'+1) + N''\gamma''^2 I''(I''+1)}, \quad (22)$$

where N' and N'' are the numbers of spins of each species. N' and N'' may be replaced by the isotopic abundances p' and p'' respectively. Defining μ by $\mu = p''\gamma'' I''(I''+1)/p'\gamma' I'(I'+1)$,

$$\frac{1}{T} = \frac{1/T' + \mu/T''}{1+\mu} = \frac{H_0}{\tilde{H}} \cdot \frac{1/T'_i + \mu/T''_i}{1+\mu}. \quad (23)$$

If we raise the field back to H_0, the common temperature of the total spin system will be $T_1 = T(H_0/\tilde{H})$, given by

$$\frac{1}{T_1} = \left(\frac{\tilde{H}}{H_0}\right)\frac{1}{T} = \frac{1/T'_i + \mu/T''_i}{1+\mu}. \quad (24)$$

If before demagnetization the magnetic moments of the two systems were

$$M'_i = C'H_0/T'_i, \qquad M''_i = C''H_0/T''_i \quad \text{with } C''/C' = \mu,$$

after demagnetization they become

$$M' = C'H_0/T_1, \qquad M'' = C''H_0/T_1$$

or

$$M' = \frac{M'_i + M''_i}{1+\mu}, \qquad M'' = \frac{\mu(M'_i + M''_i)}{1+\mu}. \quad (25)$$

The formulae (24) and (25) call for the following comments.

If at the start the two spin systems are at the same temperature $T'_i = T''_i = T_i$, the demagnetization process is reversible just as for a single species and the final temperature T_1 is equal to T_i as demonstrated by (24).

The two spin systems can both have negative temperatures since their energies have an upper bound and it is possible to make calorimetry experiments where positive and negative temperatures are exactly on the same footing. Thus it is possible for the equilibrium temperature to be negative after mixing, which is impossible in the case of thermal contact between a spin system and a lattice. This has actually been verified (experiment D) (6) on the sample of LiF already used for experiment C. The two species, Li[7] and F[19], ignoring Li[6] for the time being, could be prepared in any one of three well-known states, namely those characterized by magnetization of M_0, 0, and $-M_0$. M_0 was obtained by simply allowing the species in question to rest several relaxation times in a strong field, $-M_0$ by reversing the polarization by fast passage, and a polarization of zero was obtained by saturating the species in question by modulating the field over the resonance value.

With two species in the desired state the process of mixing was performed by removing the crystal from the gap for about one second. The two spin systems, isolated from each other by the different spacing of their energy levels in a strong field, lose their identities in weak or zero field, and find, in a time T_2, a common temperature. Subsequent examination shows polarizations characteristic of a common spin temperature after mixing.

If we define the relative magnetization m of either species as $m = M/M_0$, where M_0 is its equilibrium magnetization, it is easily found from formulae (23) and (25) that after the mixing we should have

$$m_{\text{final}}^{\text{Li,F}} = 0{\cdot}56 m_{\text{initial}}^{\text{F}} + 0{\cdot}44 m_{\text{initial}}^{\text{Li}}. \tag{26}$$

Table I gives the experimental results of experiment D.

The departures from the theoretical results (26) are completely accounted for by the finite relaxation time T_1 of both F^{19} and Li^7. The mixing time τ is a quickly varying function of the residual field in which the mixing takes place. Longer than T_1, above 100 gauss, it is 6 seconds at 70 gauss, and unobservably short at 30 gauss.

TABLE I

Expt.	(M/M_0) before mixing		(M/M_0) after mixing	
	Fluorine	Lithium	Fluorine	Lithium
(a)	1	1	0·95	0·95
(b)	1	0	0·42	0·51
(c)	0	1	0·42	0·43
(d)	1	−1	0·27	0·20
(e)	−1	1	0·05	0·00
(f)	0	−1	−0·16	−0·17
(g)	−1	0	−0·29	−0·34
(h)	−1	−1	−0·71	−0·73

[Experiments (a) and (h) will be recognized as the experiments A and B previously described.]

Experiment (b) of Table I could have been performed with the same outcome, however long the thermal relaxation time of Li^7 in a high field. This suggests a possibly useful method of polarizing a nuclear system with a very long thermal relaxation time, that is, it could be 'pumped' into a polarized state by cooling it at regular intervals by thermal contact with a system with a shorter T_1. This was strikingly demonstrated by using a powdered sample of CsCl for which $T_1(\text{Cl}^{35}) = 3 \cdot 5$ sec, $T_1(\text{Cs}) = 9$ min, but T_1 (common) $= 20$ sec in the earth's field. Commencing with both systems unpolarized, the sample was quickly removed from and restored to the magnetic field at 6-second intervals for a total time of 2 minutes, after which the Cs showed a polarization of $0 \cdot 7 M_0$, which otherwise would have taken about 10 minutes to achieve.

Similarly Li[6], which has an isotopic abundance of approximately 7 per cent, could be polarized by thermal mixing with F[19] and Li[7]. Because of its small magnetic moment and abundance a single mixing was sufficient to give to Li[6] its full equilibrium polarization. It could as well be prepared in a negative temperature state by reversing the magnetization of Li[7] and F[19] by rapid passage before mixing. Experiment showed that once polarized in a high field (12,000 gauss) at a lattice temperature of 77° K, it could spend a whole day at 300° K in a field of a few hundred gauss without losing its polarization, thus demonstrating a very long T_1, and the practical difficulty of polarizing it otherwise than through mixing.

E. Dynamics of thermal spin-spin processes

From the experiments A, B, C, D the conclusion can be drawn that spin temperature is a concept with a real physical meaning which within its domain of validity can be very useful.

We have not discussed the difficult problem of the dynamics of isentropic demagnetization which should, for instance, provide from first principles quantitative predictions for the variation of the mixing time with the residual field. As an admittedly crude criterion for the maximum rate at which the applied d.c. field should be varied in order that the spin system should remain in internal thermal equilibrium we could propose:

$$\frac{dH}{dt} \ll \gamma H_L^2, \qquad (27)$$

which expresses that the Larmor frequency γH_L of a spin in a local field is large compared with the inverse of the time $\theta = H_L/(dH/dt)$ required to sweep through the width of the spins energy levels. All one can say is that condition (27) was satisfied in experiments C and D where the demagnetization was indeed isentropic and violated in the very fast field reversal through which a negative temperature was created in experiment B.

Much more theoretical work is required in order to give an account of the dynamics of spin calorimetry, although some interesting attempts have already been made.

As an example we mention an order of magnitude estimate of the mixing time τ in LiF (8), which in a field of 75 gauss was measured to be 6 seconds (6), a surprisingly short value, at first glance, for so high a field.

The most likely mechanism for the establishment of thermal equilibrium between nuclear spins of F^{19} and Li^7 would seem to be a flip-flop in opposite directions between two such spins, a process that does not conserve Zeeman energy. It could thus only occur because of the finite width of the resonance of either spin and of the existence of an overlap between the tails of their resonance lines. The probability $W \simeq 1/\tau$ of such a flip-flop would be equal to that of a flip-flop between like spins that does conserve Zeeman energy and is of the order of $1/T_2$, times an overlap factor of the order of

$$\frac{\int G_F(\nu_F+u)G_{Li}(\nu_{Li}+u)\,du}{G_F(\nu_F)}. \tag{28}$$

The functions G_F and G_{Li} are the normalized shape functions for the resonance of either spin, centred around ν_F and ν_{Li} respectively.

If a gaussian shape is assumed for either resonance, (28) is clearly the folding of two gaussian curves and yields (neglecting factors of order unity):

$$\exp\{-[(\nu_F-\nu_{Li})^2/2[\langle\Delta\nu_F^2\rangle+\langle\Delta\nu_{Li}^2\rangle]]\}, \tag{28'}$$

where $\langle\Delta\nu_F^2\rangle$ and $\langle\Delta\nu_{Li}^2\rangle$ are the second moments of either resonance.

In a field of 75 gauss, the difference $\nu_F-\nu_{Li}$ is several times the combined line width in (28'). Because of the very sharp fall of the gaussian shape, away from resonance, the overlap between the resonance curves of F^{19} and Li^7 is extremely small and the value found for τ is many orders of magnitude too large.

There is, however, a higher-order process to be considered, the simultaneous flip of two Li^7 and one F^{19} spin. The energy $h(\nu_F-2\nu_{Li})$ required for this three-spin process is more than three times smaller than $h(\nu_F-\nu_{Li})$ required by the former one.

This three-spin process can be thought of as caused by the overlap of the resonance curve of F^{19} with the satellite line existing at frequency $2\nu_{Li}$ in the spectrum of Li^7.

We explained in Chapter IV the origin of these satellites, smaller than the main line by a factor of the order of $\langle\Delta\nu^2\rangle/\nu^2$, while describing the prescription for truncating the dipolar Hamiltonian in order to discard the contribution of these satellites to the second moment of the main line. Although the satellite is smaller than the main line by

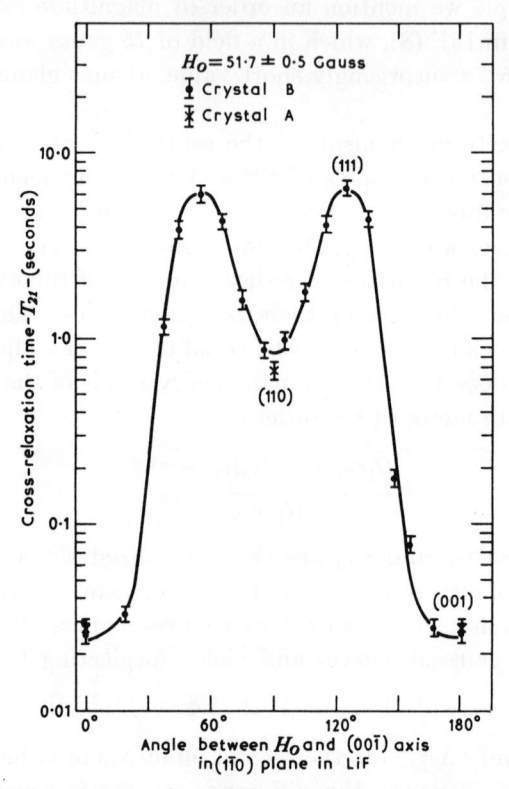

FIG. V, 2. Cross relaxation time T_{21} plotted against angle in the $(1\bar{1}0)$ plane of LiF, crystal B, at 51·7 gauss. The cross marks in the (110) and (001) directions were taken from crystal A.

a factor of the order of $\langle\Delta\nu^2\rangle/\nu^2$, this is more than offset by a tremendous increase in the exponential due to the replacement of $(\nu_F-\nu_{Li})^2$ by $(\nu_F-2\nu_{Li})^2$. If we take

$$\frac{1}{\tau} \simeq \frac{1}{T_2}\frac{\langle\Delta\nu^2\rangle_{Li}}{\nu_{Li}^2}\exp\left\{-\frac{(\nu_F-2\nu_{Li})^2}{2[\langle\Delta\nu_F^2\rangle+\langle\Delta\nu_{Li}^2\rangle]}\right\}, \qquad (29)$$

a value of τ of 6 seconds for 75 gauss, decreasing according to (29) with the applied field, is correct in order of magnitude. In view of the crudeness of the model such agreement with experiment is gratifying and it seems very likely that the basic idea of the role played by three-spin processes is essentially correct.

Apart from the very steep variation of the mixing time τ with the d.c. field H_0, careful measurements, where attention was paid to the orientation of the crystal with respect to the field, showed a very

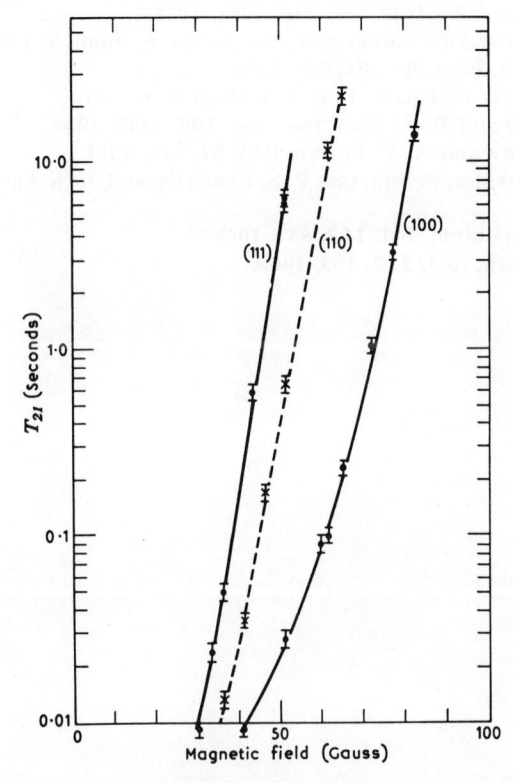

FIG. V, 3. Cross relaxation time T_{21} plotted against H_0 for three crystal orientations. The (111) direction was taken on crystal B, the (110) and (100) were from crystal A.

strong anisotropy of the mixing time τ (**10**) (or as it is also called, the cross relaxation time T_{21} (**8**)). Fig. V, 2 shows the dependence of T_{21} on the orientation of the field, exhibiting an overall change by a factor greater than 500. The largest value of T_{21} is obtained as expected for the field along the (111) direction, where resonance lines of either spin are narrowest and their overlap smallest. Fig. V, 3 shows the dependence of T_{21} on the field for three orientations of the crystal. This dependence can be represented with good accuracy by a gaussian law: $T_{21} \propto \exp(\alpha H_0^2)$ for the directions (111) and (110), but not for (100).

[In these measurements, two different single crystals of LiF were used (called A and B in the captions of Fig. V, 2 and V, 3).]

REFERENCES

1. W. G. PROCTOR and W. A. ROBINSON, *Phys. Rev.* **104**, 1344, 1956.
2. M. GOLDMAN, *C.R. Acad. Sci.* **246**, 1058, 1958.
3. A. ABRAGAM and J. COMBRISSON, *Nuovo Cim.* **6**, suppl. 3, 1197, 1957.
4. R. V. POUND, *Phys. Rev.* **81**, 156, 1951.
5. E. M. PURCELL and R. V. POUND, ibid., p. 279, 1951.
6. A. ABRAGAM and W. G. PROCTOR, ibid. **109**, 1441, 1958.
7. N. F. RAMSEY and R. V. POUND, ibid. **81**, 278, 1951.
8. N. BLOEMBERGEN, S. SHAPIRO, P. S. PERSHAN, and J. O. ARTMAN, ibid. **114**, 445, 1959.
9. R. T. SCHUMACHER, ibid. **112**, 837, 1958.
10. P. S. PERSHAN, ibid. **117**, 109, 1960.

VI

ELECTRON-NUCLEUS INTERACTIONS

On a souvent besoin d'un plus petit que soi
LA FONTAINE

LITTLE or no attention has been paid so far in this book to the magnetic interactions of the nuclear spins with electron currents and to their electrostatic interactions with electron charges. Since the nuclei have magnetic moments they are sensitive to the magnetic fields originating in the spin and orbital currents of the electrons. Although they have no electric dipole moment for reasons to be considered presently, and thus are insensitive to homogeneous electric fields, the nuclei do have quadrupole electric moments on which sufficiently inhomogeneous electric fields, such as produced by electron clouds, exert detectable torques.

Coupling as they do the electron system and the nuclear spin system, these interactions can manifest themselves in the study of either system.

As early as 1924 the magnetic coupling of the electrons with the nucleus had been proposed to explain the hyperfine structure observed in optical spectra.

Later, departures from the simple interval rule which would result from a purely magnetic interaction led to the assumption of an additional electrostatic coupling between the electronic charge and the nuclear quadrupole moment. Effects of this interaction are also perceptible in the study of the fine structure of rotational spectra of molecules. Finally, the discovery of a hyperfine structure of resonance lines in paramagnetic electron resonance gave considerable information on the nature of these couplings in bulk matter. Within the scope of this volume, it is from the other end, that is from the viewpoint of their influence on the nuclear transitions, that these couplings will be considered.

I. ELECTROSTATIC COUPLINGS

A. The Hamiltonian

Atomic energy levels are usually calculated under the simplifying assumption of a point nucleus with a charge Ze. If the finite nuclear dimensions are taken into account the changes in the atomic energy levels are exceedingly small and, for nuclear spins different from zero, are best observed through the removal of the degeneracy of the atomic levels associated with the various orientations of the nuclear spin. The Hamiltonian responsible for the splittings between those levels can be determined as follows, using the correspondence principle as a starting point. If we describe the nucleus and the electron cloud as two classical charge distributions $\rho_n(r_n)$ and $\rho_e(r_e)$, their mutual electrostatic energy is

$$W_E = \iint \frac{\rho_e(r_e)\rho_n(r_n)\,dr_e\,dr_n}{|\mathbf{r}_n-\mathbf{r}_e|}, \tag{1}$$

which can be expanded through the classical formula

$$\frac{1}{|\mathbf{r}_n-\mathbf{r}_e|} = 4\pi \sum_{l=0}^{\infty} \sum_{m=-l}^{l} \frac{1}{2l+1} \frac{r_<^l}{r_>^{l+1}} Y_l^{m*}(\theta_n,\phi_n) Y_l^m(\theta_e,\phi_e), \tag{2}$$

where the symbols $r_<$ and $r_>$ mean that the larger of the two numbers r_e and r_n is in the denominator, and the smaller in the numerator. If the small penetration of the electron inside the nucleus is neglected we may assume $r_e > r_n$ and write

$$W_E = \sum_{l,m} A_l^m B_l^{m*},$$

where
$$A_l^m = \sqrt{\left(\frac{4\pi}{2l+1}\right)} \int \rho_n(r_n) r_n^l\, Y_l^m(\theta_n,\phi_n)\,dr_n,$$

$$B_l^m = \sqrt{\left(\frac{4\pi}{2l+1}\right)} \int \rho_e(r_e) r_e^{-(l+1)} Y_l^m(\theta_e,\phi_e)\,dr_e. \tag{3}$$

If the state of the nucleus is described by a wave function $\psi_n(R_1,...,R_A)$ of the coordinates of its A nucleons, the nuclear charge density can be written as the expectation value of the operator density of charge at the point r_n:

$$\rho_n(r_n) = \left(\psi_n \middle| \sum_{i=1}^{A} e_i \delta(r_n - R_i) \middle| \psi_n\right), \tag{4}$$

where $e_i = e$ for a proton and zero for a neutron. From (3) and (4), A_l^m can be written as an expectation value $A_l^m = \langle \mathscr{A}_l^m \rangle$ where the nuclear operator $\mathscr{A}^m$ is defined by

$$\mathscr{A}_l^m = \sqrt{\left(\frac{4\pi}{2l+1}\right)} \sum_i e_i R_i^l\, Y_l^m(\Theta_i,\Phi_i), \tag{5}$$

R_i, Θ_i, Φ_i being the polar coordinates of the A nucleons. Similarly, B_l^m is the expectation value of the electron operator $\mathscr{B}_l^m$:

$$\mathscr{B}_l^m = -e\sqrt{\left(\frac{4\pi}{2l+1}\right)} \sum_{i=1}^{N} r_i^{-(l+1)} Y_l^m(\theta_i,\phi_i) \tag{6}$$

where r_i, θ_i, ϕ_i are the coordinates of the electrons. The energy of electrostatic interaction between the electrons and the nucleus is then the expectation value of a Hamiltonian

$$\mathscr{H}_E = \sum_{l,m} \mathscr{A}_l^m \mathscr{B}_l^{m*}. \qquad (7)$$

From the definitions (5) and (6) it is clear that the operators $\mathscr{A}_l^m$ and $\mathscr{B}_l^m$ transform under rotation of coordinate axes in the same way as spherical harmonics of order l. This is the definition of tensor operators of order l. The tensor operator $\mathscr{A}_l$ with $2l+1$ components $\mathscr{A}_l^m$ is called the multipole moment of order l, of the nucleus. Although for the calculation of the splittings of the atomic energy levels, only the expectation value $A_l^m = \langle \mathscr{A}_l^m \rangle$ of the nuclear multipole operator, taken over the wave function of a nuclear eigenstate, is required, this operator does in fact have off-diagonal matrix elements between some nuclear states of different energy. The study of these matrix elements, which are responsible for γ-ray transitions between those states, is outside the scope of this book.

For the diagonal matrix elements $A_l^m = \langle \mathscr{A}_l^m \rangle$ odd values of l are forbidden if we assume, as seems well established experimentally, that stationary nuclear states have well-defined parities. In particular, nuclei should have no permanent electric dipole moments ($l = 1$), in agreement with experimental evidence (**1**). Off-diagonal matrix elements of the electric dipole nuclear moment between nuclear states of different parity, may of course exist. Further information and limitations on the values of the matrix elements of the nuclear multipole operators result from their tensor character and are based on the following fundamental theorem (**2**).

Consider two manifolds $|\alpha, J)$ and $|\alpha', J')$ where α and α' are eigenvalues of an operator or of a set of operators commuting with each other and with the operator angular momentum J. (In particular one may have $\alpha = \alpha'$, $J = J'$.) For given α, α', J, J' it is possible to define $(2J+1) \times (2J'+1) \times (2l+1)$ numbers which are the various matrix elements of the $2l+1$ components of a given tensor operator $\mathscr{A}_l$. The theorem states that this set of matrix elements for any tensor operator $\mathscr{A}_l$ differs from the set for any other tensor operator $\mathscr{A}'_l$ only by a constant factor represented by the notation $(\alpha J \| \mathscr{A}_l \| \alpha' J')$. More precisely, it can be shown that a matrix element such as $(\alpha' J' M' | \mathscr{A}_l^m | \alpha J M)$, where $M = J_z$, $M' = J'_z$, is equal to

$$(\alpha' J' M' | \mathscr{A}_l^m | \alpha J M) = (JlMm | JlJ'M')(\alpha' J' \| \mathscr{A}_l \| \alpha J), \qquad (8)$$

where the first factor is the well-known Clebsch–Gordan coefficient for the coupling of angular momenta. In particular, it follows from well-known properties of this coupling that for the matrix element (8) to be different from zero, it is necessary that

$$|J-J'| \leqslant l \leqslant |J+J'|.$$

It follows that for a nucleus of spin I, $A_l^m = \langle \mathscr{A}_l^m \rangle$ will only be different from zero if $l \leqslant 2I$. Thus nuclei of spin $I \geqslant 1$ will have quadrupole moments, nuclei with $I \geqslant 2$ moments of order 4, etc.... The term $l = 0$ of the electrostatic interaction between electron and nucleus clearly corresponds to the coupling with a point charge Ze. Since the nuclear radius R is much smaller than the electronic radius a, the various terms of W_E as given by equation (3), decrease rapidly, roughly as $(R/a)^l$. This explains why there is little experimental evidence of electrostatic interactions with $l > 2$. There is no reason, however, to doubt their existence. If diagonal elements $A_l^m = (\psi_n \,|\, \mathscr{A}_l^m \,|\, \psi_n)$ have not been clearly observed for $l > 2$, the existence of off-diagonal elements $(\psi_n \,|\, \mathscr{A}_4^m \,|\, \psi_{n'})$ between two nuclear states ψ_n and $\psi_{n'}$ has been demonstrated by the observation of multipolar electronic transitions such as E_4, between two such states.

Only quadrupole interactions will be considered from now on. The components $\mathscr{A}_2^m$ of the operator nuclear quadrupole moment can be rewritten as:

$$\mathscr{A}_2^{\pm 2} = \frac{\sqrt{6}}{4} \sum_i e_i (x_i \pm iy_i)^2,$$

$$\mathscr{A}_2^{\pm 1} = \frac{\sqrt{6}}{2} \sum_i e_i z_i (x_i \pm iy_i),$$

$$\mathscr{A}_2^0 = \tfrac{1}{2} \sum_i e_i (3z_i^2 - r_i^2). \tag{9}$$

According to the theorem on tensor operators the $\mathscr{A}_2^m$ have for a nuclear state of spin I, within the manifold of the $2I+1$ substates, $I_z = m$, the same matrix elements as the Hermitian tensor operator formed from the components of the vector $\mathbf{I}$ (we omit the subscript $l = 2$):

$$Q^{\pm 2} = \alpha(\sqrt{6}/4)(I_\pm)^2, \qquad Q^{\pm 1} = \alpha(\sqrt{6}/4)\{I_z I_\pm + I_\pm I_z\},$$

$$Q^0 = \tfrac{1}{2}\alpha(3I_z^2 - I(I+1)).$$

The constant α is determined, for instance, from the condition that Q^0 and $\mathscr{A}_2^0$ have the same expectation value in the substate $I_z = I$ noted as $|II\rangle$. The usual convention is to represent by the symbol eQ the quantity

$$eQ = \left(II \,\bigg|\, \sum_{i=1}^{A} e_i (3z_i^2 - r_i^2) \,\bigg|\, II \right).$$

Ch. VI ELECTRON-NUCLEUS INTERACTIONS

Writing
$$eQ = 2(II | \mathscr{A}_2^0 | II) = 2(II | Q^0 | II)$$
$$= \alpha(II | 3I_z^2 - I(I+1) | II), \qquad (10)$$

we get
$$\alpha = \frac{eQ}{I(2I-1)}$$

and
$$Q^{\pm 2} = \frac{eQ}{I(2I-1)} \frac{\sqrt{6}}{4} I_\pm^2,$$

$$Q^{\pm 1} = \frac{eQ}{I(2I-1)} \frac{\sqrt{6}}{4} \{I_z I_\pm + I_\pm I_z\},$$

$$Q^0 = \frac{eQ}{I(2I-1)} \tfrac{1}{2}\{3I_z^2 - I(I+1)\}. \qquad (11)$$

The components $\mathscr{B}_2^m$ of the electronic tensor (6) can be rewritten in the same way as the $\mathscr{A}_2^m$ in equation (9). It then becomes apparent that

$$\mathscr{B}_2^0 = \frac{1}{2} \sum_{i=1}^{N} \frac{e_i(3z_i^2 - r_i^2)}{r_i^5} = \frac{1}{2}\left(\frac{\partial^2 \mathscr{V}}{\partial z^2}\right)_{r=0} = \tfrac{1}{2}\mathscr{V}_{zz} \qquad (12)$$

where $\mathscr{V}(x,y,z)$ is the electrostatic potential produced by the electrons at the point x, y, z. $\mathscr{V}_{zz}$ in equation (12) is still an electron operator. Its expectation value V_{zz}, which is a *number*, is obtained by taking the expectation value $V_{zz} = \langle \psi_e | \mathscr{V}_{zz} | \psi_e \rangle$ over an electron wave function. It is easily found that

$$\mathscr{B}_2^{\pm 2} = \frac{1}{2\sqrt{6}}(\mathscr{V}_{xx} - \mathscr{V}_{yy} \pm 2i\mathscr{V}_{xy}),$$

$$\mathscr{B}_2^{\pm 1} = \frac{1}{\sqrt{6}}(\mathscr{V}_{xz} \pm i\mathscr{V}_{yz}). \qquad (13)$$

The Hamiltonian (7) describing the quadrupole coupling can be written

$$\mathscr{H}_2 = \sum_{m=-2}^{2} \mathscr{A}_2^m \mathscr{B}_2^{-m} = \sum_{m=-2}^{2} Q^m \mathscr{B}_2^{-m}, \qquad (14)$$

where Q^m and $\mathscr{B}_2^m$ are given by the formulae (11), (12), (13). It is easily verified from these equations that (14) can be rewritten:

$$\mathscr{H}_2 = \sum_{j,k} \left(\frac{\partial^2 \mathscr{V}}{\partial x_j \partial x_k}\right)_{r=0} Q_{jk}, \qquad (15)$$

where
$$Q_{jk} = \frac{eQ}{6I(2I-1)}\{\tfrac{3}{2}(I_j I_k + I_k I_j) - \delta_{jk} I(I+1)\}$$

is a traceless cartesian tensor. The traceless tensor operator $\left(\dfrac{\partial^2 \mathscr{V}}{\partial x_j \partial x_k}\right)_{r=0}$

is called the electric field gradient tensor at the origin. Actually (15) is derived much more directly by writing the classical interaction (1) as

$$W_E = \int \rho_n(r_n) V(r_n) \, dr_n, \qquad (16)$$

where $V(r_n)$ is the potential produced by the electron at the point r_n inside the nucleus.

Expanding

$$V(r) = V(0) + \sum_j x_j \left(\frac{\partial V}{\partial x_j}\right)_0 + \tfrac{1}{2} \sum_{j,k} x_j x_k \left(\frac{\partial^2 V}{\partial x_j \, \partial x_k}\right)_0, \qquad (17)$$

and, using the fundamental theorem to replace the traceless tensor $x_j x_k - (r^2/3)\delta_{jk}$ by a tensor proportional to it:

$$\tfrac{1}{2}(I_j I_k + I_k I_j) - \tfrac{1}{3} I(I+1)\delta_{jk},$$

the direct derivation of (15) is straightforward.

The expansion in spherical harmonics has the advantage over the Taylor expansion (17) of exhibiting clearly in the equations (11) the various non-vanishing matrix elements of the interaction. It is also the only unambiguous way of defining nuclear electric multipoles higher than the second (a rather academic advantage here). In the following we shall use indifferently the expressions (14) and (15) for the quadrupole interaction Hamiltonian. It has already been mentioned that the quadrupole interaction represented by the terms $l = 2$ of equation (7) is very small and is indeed detectable only because it splits an otherwise degenerate level. First-order perturbation theory is therefore applicable and only matrix elements of the perturbing Hamiltonian inside the degenerate manifold are to be considered. As far as the nuclear states are concerned this is precisely the meaning of the replacement of the operators $\mathscr{A}_2^m$ by the equivalent operators Q^m of equation (11), acting on the parameters describing the orientation of the nuclear spin.

As for the electron state it may or may not be degenerate. It will be degenerate in free atoms or molecules of non-zero angular momentum J, the degeneracy being connected with the existence of a rotational symmetry. In the absence of the quadrupole interaction there is then a degenerate manifold of $(2J+1)$ electron wave functions. Inside this manifold electron operators such as $\mathscr{B}_2^m$ or $\mathscr{V}_{jk}$ can be replaced by tensor operators constructed from the components J_x, J_y, J_z in the same way as the operators Q^m of (11) or Q_{jk} of equation (15) are built from the components of the angular spin I.

Thus for a single electron with an orbital momentum l describable by a set of $(2l+1)$ wave functions $f(r)Y_l^m(\theta,\phi)$, it will be possible to replace

$$\mathscr{V}_{jk} = -\frac{e(3x_j x_k - r^2\delta_{jk})}{r^5},$$

by the equivalent tensor

$$E_{jk} = \beta\{\tfrac{3}{2}(l_j l_k + l_k l_j) - \delta_{jk} l(l+1)\}, \tag{18}$$

where β is easily computed to be

$$\beta = e\left\langle\frac{1}{r^3}\right\rangle\frac{2}{(2l+3)(2l-1)}, \tag{19}$$

with

$$\left\langle\frac{1}{r^3}\right\rangle = \int_0^\infty f^2(r)\frac{1}{r^3}r^2\,dr.$$

The coupling Hamiltonian can be rewritten for that case as

$$\mathscr{H} = e^2 Q\left\langle\frac{1}{r^3}\right\rangle\frac{1}{(2l+3)(2l-1)I(2I-1)} \times$$
$$\times [3(\mathbf{l}.\mathbf{I})^2 + \tfrac{3}{2}(\mathbf{l}.\mathbf{I}) - l(l+1)I(I+1)]. \tag{20}$$

To write (20), use has been made of the commutation relations of the components of the vectors **l** and **I**. For an s-electron ($l=0$) the quadrupole interaction vanishes. Since these electrons are the only ones with a non-vanishing wave function at the nucleus, the neglect of the penetration of the electron inside the nucleus in the derivation of the multipole coupling is justified. More generally, for any atomic or molecular system with a total angular momentum J, the quadrupole coupling can be written as

$$\mathscr{H} = e^2 Q\xi[3(\mathbf{J}.\mathbf{I})^2 + \tfrac{3}{2}(\mathbf{J}.\mathbf{I}) - J(J+1)I(I+1)], \tag{21}$$

where the constant ξ depends on the details of the electronic structure of the atom or the molecule.

On the other hand, in bulk matter there is in general, with the exception of a few paramagnetic substances such as the rare earths, no orbital degeneracy left, and the operators $\mathscr{V}_{jk}$ (or $\mathscr{B}_2^m$) can be replaced by their expectation values V_{jk} taken over the single wave function representing the non-degenerate electronic state. As a consequence of this fact the electric field gradient at the nucleus is treated classically in bulk matter and quantum-mechanically in free atoms or molecules.

If we introduce the constants

$$V^0 = \tfrac{1}{2} V_{zz} = \tfrac{1}{2} \langle \mathscr{V}_{zz} \rangle$$

$$V^{\pm 1} = \frac{1}{\sqrt{6}} (V_{zx} \pm i V_{zy})$$

$$V^{\pm 2} = \frac{1}{2\sqrt{6}} (V_{xx} - V_{yy} \pm 2i V_{xy}), \tag{22}$$

the quadrupole interaction in bulk matter can be written

$$\mathscr{H} = \sum_m Q^m V^{-m}, \tag{23}$$

where the Q^m are given in equation (11). So far the orientation of the axes $Oxyz$ has been arbitrary. If we choose as axes of coordinates $OXYZ$ the principal axes of the symmetrical tensor V_{ij}, so that $V_{XZ} = V_{YZ} = V_{XY} = 0$, label these axes so that $|V_{ZZ}| \geqslant |V_{XX}| \geqslant |V_{YY}|$, and define $eq = V_{ZZ}$, $\eta = (V_{XX} - V_{YY})/V_{ZZ}$, the quadrupole Hamiltonian becomes

$$\mathscr{H} = \frac{e^2 q Q}{4I(2I-1)} \{3 I_Z^2 - I(I+1) + \tfrac{1}{2}\eta (I_+^2 + I_-^2)\}. \tag{24}$$

The case $\eta = 0$ corresponds to an axial symmetry of the nuclear surroundings. Since because of Laplace equation $V_{XX} + V_{YY} + V_{ZZ} = 0$, if the surroundings of the nucleus have cubic symmetry it follows that $V_{XX} = V_{YY} = V_{ZZ} = 0$ and the quadrupole coupling vanishes. As a result of the previous definitions, $0 \leqslant \eta \leqslant 1$.

The theoretical problems connected with the existence of quadrupole couplings represented by the Hamiltonian (24) are of two types.

On the one hand, there is the influence of the Hamiltonian (24), alone or combined with Zeeman and spin-spin couplings, on the energy levels, relaxation, and resonance line widths of nuclear spins. This will be considered at some length in later chapters. The other problem is the calculation of the constants q and η (required for an experimental determination of the nuclear quadrupole moment Q), either from first principles, or from other experimental evidence about the substance under consideration. This difficult problem belongs to solid-state physics and to theoretical chemistry rather than to nuclear magnetism proper, and only a few comments will be made here, referring the reader to the references (3) and (4) for further details.

B. Ionic crystals

In ionic crystals the individual ions are assumed to a first approximation to have spherical symmetry and their quadrupole couplings with their own nuclei vanish. The electric-field gradients at a nucleus would thus originate solely from charges external to the ion, neighbouring ions in crystals with symmetry lower than cubic, or imperfections in cubic crystals. These charges, supposed for simplicity to be fixed in space, produce at the nucleus of the ion an electric-field gradient described by the numerical tensor V_{jk}^e (the superscript e is for external) and the corresponding quadrupole coupling is, according to (15),

$$\mathscr{H}^e = \sum_{jk} Q_{jk} V_{jk}^e.$$

There is, however, an additional contribution to the quadrupole coupling that is due to the distortion of the spherical electronic shell of the ion by the external charges. Let $\mathscr{V}_{jk}$ be the *operator* describing the electric-field gradient produced by the electrons of the ion at the nucleus. The expectation values $V_{jk}^0 = (\psi_0 | \mathscr{V}_{jk} | \psi_0)$ vanish because of the spherical symmetry of the unperturbed wave function of the ion, ψ_0, calculated in the absence of the external charges. If the effect of the external charges on the ion can be treated as a perturbation, the new ionic wave function will be $\psi_0 + \psi_1$ to a first approximation, where ψ_1 is small and without loss of generality may be assumed orthogonal to ψ_0. The new values of the ionic-field gradient will be

$$V_{jk} = (\psi_0 + \psi_1 | \mathscr{V}_{jk} | \psi_0 + \psi_1) \cong 2(\psi_0 | \mathscr{V}_{jk} | \psi_1). \qquad (25)$$

Let $\mathscr{H}_0$ and E_0 be the unperturbed Hamiltonian and energy of the ion, $\mathscr{H}_1$ the perturbing electrostatic coupling between the external charges and the ion, E_1 the first-order change in the energy. From the equation

$$(\mathscr{H}_0 + \mathscr{H}_1)(\psi_0 + \psi_1 + \ldots) = (E_0 + E_1 + \ldots)(\psi_0 + \psi_1 + \ldots)$$

we get
$$(\mathscr{H}_0 - E_0)\psi_1 = (E_1 - \mathscr{H}_1)\psi_0,$$
$$\psi_1 = (\mathscr{H}_0 - E_0)^{-1}(E_1 - \mathscr{H}_1)\psi_0.$$

The value V_{jk} of the ionic field gradient is, by (25),

$$V_{jk} = 2(\psi_0 | \mathscr{V}_{jk} (\mathscr{H}_0 - E_0)^{-1} (E_1 - \mathscr{H}_1) | \psi_0). \qquad (26)$$

Since ψ_0 and $\mathscr{H}_0 - E_0$ are spherically symmetrical and $\mathscr{V}_{jk}$ is a tensor operator of order $l = 2$, only the part of $\mathscr{H}_1$ that transforms as a spherical harmonic of order 2 will give a contribution to (26). A multipole expansion of the electrostatic interaction $\mathscr{H}_1$ shows this part to be

$$\tfrac{1}{6} \sum_{jk} V_{jk}^e q_{jk},$$

where q_{jk} is a component of the operator electronic quadrupole moment of the ion, given by

$$q_{jk} = -e \sum (3 x_j^s x_k^s - (r^s)^2 \delta_{jk}).$$

Carrying this over into equation (26), we obtain a linear relation between the external field gradient V^e_{jk} and the gradient V_{jk} due to the distorted ion

$$V_{jk} = \sum_{j'k'} \gamma_{jk,j'k'} V^e_{j'k'},$$

where $\quad \gamma_{jk,j'k'} = 2(\psi_0 | \mathscr{V}_{jk}(\mathscr{H}_0 - E_0)^{-1} \tfrac{1}{6} q_{j'k'} | \psi_0).$

Again, because of the tensor character of $\mathscr{V}_{jk}$ and $q_{j'k'}$,

$$\gamma_{jk,j'k'} = \gamma \delta_{jj'} \delta_{kk'}.$$

The induced gradient $V_{jk} = \gamma V^e_{jk}$, and the total gradient is

$$(1+\gamma) V^e_{jk}.$$

γ is called the anti-shielding factor. Its values have been computed by various workers.† Small and negative for ions isoelectronic with the helium atom, it then becomes positive and very large for heavy ions ($\gamma \simeq 4$ for Na+, 70 for Rb+, 140 for Cs+).

As in most calculations of atomic structures, a variation method can be used as an alternative to perturbation theory (5). A modified function $\psi = \alpha(\psi_0 + \psi_1)$ is used to describe the ground state of the ion in the presence of external charges. The function ψ_0 is the unperturbed wave function of the ground state of the ion, α a normalization coefficient, and ψ_1 a trial function the form of which is chosen from considerations of symmetry and also simple expediency. The expectation value $(\psi | \mathscr{H} | \psi)$, where $\mathscr{H}$ now includes the interaction of the external charges with the ion, is minimized with respect to the parameters contained in ψ_1. When ψ_1 has been determined in this way, the expectation value of the interaction of the ion with the nuclear quadrupole moment is then calculated with the corrected function ψ.

It is sometimes argued‡ that the induced quadrupole coupling is the sum of two terms:

(α) The nuclear quadrupole moment polarizes the ionic shell which acquires a quadrupole moment proportional to the nuclear quadrupole moment Q. This electronic quadrupole moment then interacts with the external gradient V^e_{ij} to give a coupling bilinear in Q and V^e_{ij}.

(β) The external gradient also polarizes the ionic shell which, once distorted, produces at the nucleus a gradient proportional to V^e_{ij}. This gradient then interacts with the nuclear moment Q giving a second term bilinear in V^e_{ij} and Q, which is shown to be equal to the previous one and must be added to it. This argument, although illuminating, is somewhat misleading for it gives the impression that in the calculation

† See reference (3), p. 350. ‡ See reference (3), p. 347.

it is necessary to take into account the distortion of the electron wave function by the nuclear quadrupole moment. This is not so. The electron wave functions of the ion, distorted by the external charges and the expectation value $(\psi_0+\psi_1 | \mathscr{V}_{ij} | \psi_0+\psi_1)$ of the ionic field gradient, are calculated in the absence of the nuclear quadrupole moment. The two effects (α) and (β) of the previous picture correspond to the existence of two cross-product terms in

$$(\psi_0+\psi_1 | \mathscr{V}_{ij} | \psi_0+\psi_1).$$

C. Molecular crystals

A molecular solid may be considered in a first and crude approximation as an assembly of independent molecules. Therefore the strengths of the quadrupole couplings as measured in the solid state by magnetic resonance should not be very different from those obtained from the fine structure of rotational spectra in the gaseous molecules. This is indeed the case, the differences in the measured coupling constants e^2qQ between the solid and the gas being seldom larger than 10 per cent.

In a second step the quadrupole coupling of a nucleus of the free molecule is compared to that of the corresponding free atom. There the differences may be considerable since they are mostly affected by the nature of the bond between the atoms of the molecule. Molecules where ionic bonds are predominant and where the electronic surroundings of a given nucleus are almost spherical are likely to have much smaller quadrupole couplings than molecules with covalently bound atoms. A considerable amount of work, often of a crudely empirical nature, has been spent in attempts to relate the molecular quadrupole couplings to those of the free atoms and also to correlate the former with other molecular properties.[†]

Finally, the problem of evaluating the quadrupole couplings in free atoms themselves is by no means straightforward. Although in principle it reduces to the calculation of expectation values $\langle 1/r^3 \rangle$ for electrons outside closed shells as given by formula (19), it is greatly complicated by the polarization of the closed shells by the valence electrons. This is a problem similar to the one considered in connexion with ionic crystals, with the further complication that the polarizing charge, i.e. the valence electron, is not fixed but may penetrate inside the closed shells. This effect has been calculated by various authors (6) and results in a corrective factor $(1+R)$ on the field gradient, calculated

[†] See reference (4), p. 119.

disregarding the polarization. Positive as well as negative values of R may occur. In contrast with γ, $|R|$ is less than unity even for heavy atoms.† Unfortunately there is no way at present in which these calculated values of R can be checked. As a consequence there is a significant uncertainty over the values of most nuclear quadrupole moments.

II. Magnetic Interactions
A. The coupling Hamiltonian

A logical procedure would be to develop the theory of the magnetic interactions between the electron and the nucleus along the same lines as for the electrostatic interactions; that is, to assign to the electrons and to the nucleus electric-current densities (rather than charge densities as in the previous section) and to calculate their interactions according to the laws of classical electromagnetism. One would thus define for the nucleus magnetic multipole operators which like the electric ones would be tensor operators of integer order l.

If one recalls the opposite parity properties of the electric field (a polar vector) and of the magnetic field (an axial vector) it is understandable that even, rather than odd, values of l are forbidden for permanent magnetic multipoles, by the assumption of a well-defined parity of nuclear-energy states. The first non-vanishing nuclear multipole is thus a magnetic dipole, the next a magnetic octopole, etc.

Although the existence of magnetic octopoles has been established by atomic beam methods (7, 8), they have never been observed by means of magnetic resonance in bulk matter. Furthermore, the description of the magnetic properties of a nucleus as those of a system of currents is more complicated and at the same time, in our present state of knowledge, much less satisfying than the description of its electrostatic properties as those of a system of charges. We shall therefore be content to describe the magnetic properties of the nucleus as those of a magnetic dipole $\boldsymbol{\mu}_N = \gamma_N \hbar \mathbf{I}$. The reason why the magnetic dipole is collinear with the spin vector $\mathbf{I}$ is again that, within the manifold of the substates of a given nuclear state of spin I, all tensor operators of given l (vectors in the present case) have the same matrix elements. Magnetic fields of impossibly high values, of the order of 10^{16} gauss or more, would have to be applied to the nucleus before its magnetic energy $-\boldsymbol{\mu}_N \mathbf{H}$ became comparable to the interval between two different nuclear energy states, invalidating the approximation $\boldsymbol{\mu}_N = \gamma_N \hbar \mathbf{I}$.

† See a table of values of R in reference (3), p. 362.

Ch. VI   ELECTRON-NUCLEUS INTERACTIONS   171

The interaction of the nuclear dipole μ_N with the electronic shell is small even compared with atomic-energy splittings (let alone the nuclear ones) and will be computed by a perturbation method.

The behaviour of an electron in a magnetic field **H** is obtained by replacing the momentum **p** by $\mathbf{p}+(e/c)\mathbf{A}$ in its Hamiltonian, where **A** is the magnetic vector potential defined by

$$\operatorname{div}\mathbf{A} = 0, \qquad \operatorname{curl}\mathbf{A} = \mathbf{H}.$$

According to classical electromagnetic theory a magnetic dipole $\boldsymbol{\mu}$ produces at a point removed from it by a vector **r**, a magnetic field deriving from a vector potential

$$\mathbf{A} = \frac{\boldsymbol{\mu}\wedge\mathbf{r}}{r^3} = \operatorname{curl}\left(\frac{\boldsymbol{\mu}}{r}\right). \tag{27}$$

Near the dipole the vector potential **A** has a singularity of order r^{-2} and $\mathbf{H} = \operatorname{curl}\mathbf{A}$ a singularity of order r^{-3}, so some care must be exercised in the calculation of its interaction with an electron. In the non-relativistic Pauli description of the electron the Hamiltonian in the presence of **A** is

$$\mathscr{H} = \frac{1}{2m}\left(\mathbf{p}+\frac{e}{c}\mathbf{A}\right)^2 + 2\beta\mathbf{s}\cdot\operatorname{curl}\mathbf{A}, \tag{28}$$

where β is a Bohr magneton and **s** the electron spin. In a first-order perturbation calculation the only terms of (28) to be retained are those linear in **A**:

$$\mathscr{H}_1 = \frac{e}{2mc}(\mathbf{p}\cdot\mathbf{A}+\mathbf{A}\cdot\mathbf{p}) + 2\beta\mathbf{s}\cdot\operatorname{curl}\mathbf{A}.$$

This can be written by (27) as

$$\mathscr{H}_1 = 2\beta\frac{\mathbf{l}\cdot\boldsymbol{\mu}}{r^3} + 2\beta\mathbf{s}\cdot\operatorname{curl}\operatorname{curl}\left(\frac{\boldsymbol{\mu}}{r}\right), \tag{29}$$

where $\hbar\mathbf{l} = \mathbf{r}\wedge\mathbf{p}$ is the orbital momentum of the electron.

The spin-dependent part of (29) gives

$$\mathscr{H}_1^s = 2\beta\mathbf{s}\cdot\left[\boldsymbol{\nabla}\wedge\left(\boldsymbol{\nabla}\wedge\frac{\boldsymbol{\mu}}{r}\right)\right]$$
$$= 2\beta[(\mathbf{s}\cdot\boldsymbol{\nabla})(\boldsymbol{\mu}\cdot\boldsymbol{\nabla})-(\mathbf{s}\cdot\boldsymbol{\mu})\nabla^2]\frac{1}{r}, \tag{30}$$

which for reasons to appear presently we rewrite as

$$\mathscr{H}_1^s = 2\beta[(\mathbf{s}\cdot\boldsymbol{\nabla})(\boldsymbol{\mu}\cdot\boldsymbol{\nabla})-\tfrac{1}{3}(\mathbf{s}\cdot\boldsymbol{\mu})\nabla^2]\left(\frac{1}{r}\right) - \frac{4\beta}{3}(\mathbf{s}\cdot\boldsymbol{\mu})\nabla^2\left(\frac{1}{r}\right). \tag{31}$$

The magnetic interaction of the nuclear moment with the electron spin $W_m^s = (\psi_e\,|\,\mathscr{H}_1^s\,|\,\psi_e)$ is obtained by multiplying (31) by the electronic density $\rho = \psi_e^* \psi_e$ and integrating over the electron coordinates. For $r \neq 0$ $\mathscr{H}_1^s$, as given by (31), is a regular function where the first term is equal to $2\beta[3(\mathbf{s}.\mathbf{r})(\boldsymbol{\mu}.\mathbf{r})/r^5 - \mathbf{s}.\boldsymbol{\mu}/r^3]$, which is the usual dipole-dipole interaction, and the second term vanishes because of Laplace's equation. When r tends toward zero we may remark that the first term $\mathscr{H}_1^{s'}$ of (31) behaves under a rotation of the coordinate system as a spherical harmonic of order 2. Hence if ψ_e is expanded in a sum of spherical harmonics, $\psi_e = \sum_l a_l \psi^{(l)}$, the only non vanishing contributions to $(\psi_e | \mathscr{H}_1^{s'} | \psi_e)$ will come from terms $(\psi^{(l)} | \mathscr{H}_1^{s'} | \psi^{(l')})$ such that $l+l' \geqslant 2$. It is well known that a wave function $\psi^{(l)}$ is of order r^l near the origin so that in the matrix element

$$(\psi^{(l)} | \mathscr{H}_1^{s'} | \psi^{(l')}) = \int \psi^{(l)*} \mathscr{H}_1^{s'} \psi^{(l')} r^2 \, dr \, d\Omega$$

the integrand varies as $r^{(l+l'+2-3)}$ and the corresponding integral always remains finite since $l+l' \geqslant 2$. According to the theory of the Coulomb potential the second term of (31) is equal to $\tfrac{16}{3}\pi\beta(\mathbf{s}.\boldsymbol{\mu})\delta(\mathbf{r})$ and by integration gives

$$\tfrac{16}{3}\pi\beta(\mathbf{s}.\boldsymbol{\mu}) \, |\psi_e(0)|^2,$$

which is finite for s electrons and zero for the others. The Hamiltonian for the magnetic interaction of the electron with the nucleus can then be written without ambiguity as

$$\mathscr{H}_1 = 2\beta\gamma\hbar\mathbf{I}.\left[\frac{1}{r^3} - \frac{\mathbf{s}}{r^3} + 3\frac{\mathbf{r}(\mathbf{s}.\mathbf{r})}{r^5} + \tfrac{8}{3}\pi\,\mathbf{s}\delta(\mathbf{r})\right]. \tag{32}$$

If several electrons surround the nucleus, the interaction Hamiltonian is the sum of the contributions of the individual electrons. Although the expression (32) has been derived for the purpose of calculating its expectation value $(\psi_e | \mathscr{H}_1 | \psi_e)$, it is clear that it also gives unambiguous results for off-diagonal matrix elements $(\psi_e | \mathscr{H}_1 | \phi_e)$, between, say, the ground state and an excited state of the electronic system. Use will be made of this to calculate some effects of $\mathscr{H}_1$, using second-order perturbation theory.

The vector operator

$$\mathbf{H}_e = -2\beta\left[\frac{1}{r^3} - \frac{\mathbf{s}}{r^3} + \frac{3\mathbf{r}(\mathbf{s}.\mathbf{r})}{r^5} + \tfrac{8}{3}\pi\,\mathbf{s}\delta(\mathbf{r})\right] \tag{33}$$

can be called the field produced by the electron at the nucleus. For an atomic electron of orbital momentum l and total angular momentum $j = l \pm \tfrac{1}{2}$ it is possible inside the manifold j to replace in the Hamiltonian (32), $\mathscr{H}_1 = -\gamma_n \hbar \mathbf{H}_e.\mathbf{I}$, the vector $-\gamma\hbar\mathbf{H}_e$, by a vector $a_j\mathbf{j}$ proportional to $\mathbf{j}$. The value of a_j is obtained by writing that the expectation values $\langle a_j \mathbf{j}.\mathbf{j}\rangle$ and $\langle -(\gamma\mathbf{H}_e.\mathbf{j})\rangle$ are equal. Using (33) and remembering that $\mathbf{j} = \mathbf{l}+\mathbf{s}$, that $\mathbf{r}.\mathbf{l} = (1/\hbar)\mathbf{r}.(\mathbf{r}\wedge\mathbf{p}) = 0$, and that for a spin $s = \frac{1}{2}$ relations such as $s_x^2 = \frac{1}{4}$, $s_x s_y = -s_y s_x$, $s_x s_y = \frac{1}{2}i s_z$ are valid, it is easily found that

$$a_j = \tfrac{16}{3}\pi\,\beta\gamma\hbar|\psi(0)|^2 \quad \text{for an } s\text{-electron} \tag{34}$$

and
$$a_j = 2\beta\gamma\hbar\Big\langle\frac{1}{r^3}\Big\rangle\frac{l(l+1)}{j(j+1)} \quad \text{if } l \neq 0. \tag{35}$$

For a free atom or a paramagnetic molecule with many electrons in a state of total angular momentum J it is also possible to write $\mathscr{H}_1 = a_J \mathbf{I}.\mathbf{J}$ for the magnetic coupling of the electrons with the nucleus, where the value of a_J, to be determined by the same method as for a single electron, will depend on the electronic structure of the atom or the molecule.

In bulk matter the manifestations of the interaction (32) are manifold and greatly dependent on the nature of the substance. They can change the phenomenon of nuclear magnetic resonance in two ways, by introducing changes in the energy levels of the nuclear spin system and also by providing powerful relaxation mechanisms for the coupling of this system to the lattice. Only the first aspect of the electron-nucleus magnetic coupling will be considered in this chapter, the relaxation phenomenon being deferred till Chapters VIII and IX.

B. The effect of electron-nucleus coupling in diamagnetic substances

(a) General

The overwhelming majority of nuclear-resonance experiments are performed on diamagnetic samples, that is on substances without either spin or orbital electron paramagnetism. The lack of electron paramagnetism in, for instance, a molecule embedded in a molecular solid, or in an ion belonging to a crystal or a solution, corresponds to vanishing expectation values of $(\psi_0 | L_q | \psi_0)$ and $(\psi_0 | S_q | \psi_0)$ for all components of the total orbital and spin angular momentum of the molecule or the ion, of which ψ_0, a function of both orbital and spin coordinates, describes the ground state.

In spite of an apparent symmetry there is an important difference between the behaviour of the orbital and spin momentum in diamagnetic substances and consequently between the ways in which they affect the resonance of nuclear spins. The forces between electrons, atoms, and molecules are essentially electrostatic forces, and the magnetic spin-dependent forces are negligible compared to them. It follows that to a good approximation the total spin S of an electronic system is a good quantum number and the $(2S+1)$ substates of a state of given S have approximately the same energy. Hence if the expectation values of the three components of the vector $\mathbf{S}$ vanish in the non-degenerate ground state, necessarily $|\mathbf{S}|^2 = S(S+1) = 0$ for that state.

A similar conclusion cannot be made for the orbital momentum which for a molecule or an ion embedded in bulk matter is not in general a good quantum number. The vanishing of $\langle L_x \rangle$, $\langle L_y \rangle$, $\langle L_z \rangle$ in a non-degenerate electronic ground state does not necessarily mean that the total orbital momentum L has the value zero, nor indeed any well-defined value in that state. Conversely, it is easily shown that the lack of orbital degeneracy is a sufficient condition for the vanishing of the three expectation values $\langle L_x \rangle$, $\langle L_y \rangle$, $\langle L_z \rangle$: or, as it is called, the 'quenching' (9) of the orbital momentum. Let ψ be a non-degenerate wave function of the electronic system. The Hamiltonian, the sum of the kinetic and electrostatic energy of the electrons, is real and ψ can also be assumed real, for otherwise its real and imaginary parts would separately be eigenfunctions of the Hamiltonian with the same energy, and this would be contrary to the assumption of non-degeneracy. Since the function ψ is real, the expectation value

$$(\psi \mid L_z \mid \psi) = i\left(\psi \left| \sum_k x_k \frac{\partial}{\partial y_k} - y_k \frac{\partial}{\partial x_k} \right| \psi \right)$$

is necessarily imaginary, but it is also real since L is a hermitian operator, and hence $(\psi \mid L_z \mid \psi)$ must vanish.

Before discussing the various effects of the electron-nucleus coupling, we return to a problem discussed briefly in Chapter I, namely that nuclear orbital motion does not contribute to nuclear paramagnetism, in contrast to the situation in electron paramagnetism.

We should notice first that this contribution is not always negligible. When the nuclear orbital motion is not quenched, as for a hydrogen molecule in a beam where collisions are negligible, nuclear orbital and spin paramagnetism are of comparable magnitude.

In bulk matter the orbital nuclear paramagnetism is quenched in the same manner as the electron orbital paramagnetism. The main difference between nuclei and electrons is that for electrons, part of the orbital paramagnetism may be 'unquenched', that is, brought back by spin-orbit coupling. It is the relative smallness of spin-orbit coupling in the nuclear orbital motion, caused by the large nuclear mass, that makes such 'unquenching' negligible in nuclear magnetism.

In a diamagnetic substance the first-order effect of the electron-nucleus coupling represented by the Hamiltonian (32) (or by a sum of such Hamiltonians for the various electrons surrounding the nucleus) vanishes, as a consequence of the quenching of the orbital momentum and of the zero value of the total spin S. The electron-nucleus interaction does, however, have the following effects.

(i) The resonance frequency is different for a nucleus embedded in bulk matter and for a 'bare' nucleus. The origin of this frequency shift is twofold:

First, in an applied magnetic field H_0, the Larmor precession of the electronic charges around it is equivalent to an electric current producing at the nucleus a magnetic field H_d which adds to the applied field and is proportional to it.

Secondly, the applied field H_0 polarizes the electronic shells. The shells distorted in this way produce at the nucleus a magnetic field H_p also proportional to H_0.

The total field 'seen' by the nucleus is $H = H_0 + H_d + H_p = H_0(1-\sigma)$, where σ is a relative resonance frequency shift independent of the magnitude of H_0. This shift depends on the distribution of the electrons around the nucleus and naturally has different values in different chemical compounds, whence its name of chemical shift.

(ii) Spin-spin indirect interactions distinct from the usual dipolar interactions can originate via the electrons in the following way. A nuclear moment μ_1 produces a field which distorts the electronic shells. The shells distorted in this way produce a field H' proportional to the magnitude of μ_1 at the site of another nuclear moment μ_2. Its coupling with μ_2 thus results in a bilinear interaction between μ_1 and μ_2.

The principle of the calculation of the chemical shift and of the indirect spin-spin interactions will now be given.

(b) *Calculation of the chemical shift*

This calculation is largely based on reference (**10**) where a complete theory of the chemical shift was first given. Given a molecule containing N nuclear moments and n electrons, its Hamiltonian in the presence of an applied magnetic field H_0 can be written

$$\mathscr{H} - V = \frac{1}{2m} \sum_{k=1}^{n} \left(\mathbf{p}_k + \frac{e}{c}\mathbf{A}_k^0 + \frac{e}{c}\sum_{q=1}^{N} \mathbf{A}_k^q \right)^2 +$$

$$+ 2\beta \sum_{k=1}^{n} \mathbf{s}_k \cdot \operatorname{curl} \mathbf{A}_k^0 + 2\beta \sum_{k=1}^{n} \sum_{q=1}^{N} \mathbf{s}_k \cdot \operatorname{curl} \mathbf{A}_k^q. \quad (36)$$

In this formula $\mathbf{A}_k^0 = \frac{1}{2}(\mathbf{H}_0 \wedge \mathbf{r}_k)$ is the value, at the position $\mathbf{r}_k$ of the kth electron, of the vector potential of the external field $\mathbf{H}_0$,

$$\mathbf{A}_k^q = (\boldsymbol{\mu}_q \wedge \mathbf{r}_{qk})/r_{qk}^3$$

is the value, at the same point, of the vector potential produced by the nuclear moment $\boldsymbol{\mu}_q$ situated at $\mathbf{R}_q$. The origin of the vectors $\mathbf{r}_k$ and $\mathbf{R}_q$ is left unspecified. Clearly all physical results must be independent of the choice of origin, a consequence of the general principle of gauge invariance in electromagnetism. For an isolated atom it is natural to choose for origin the nucleus of that atom. In a molecule it may be convenient to select as origin the particular nucleus $\boldsymbol{\mu}_q = \boldsymbol{\mu}$ for which the frequency shift is being calculated. V is the electrostatic energy of the system. For brevity we have omitted the magnetic couplings between the electrons, since they are irrelevant to the problem, and we have also omitted the Zeeman couplings of the nuclei with the applied field H_0 and their dipolar couplings, since they are well known from the previous chapters.

An expansion of (36) gives the following terms:

$$\mathscr{H} - V - T - D = (Z_L + Z_S) + (O_1 + S_1 + S_2) + (O_2 + O_3), \qquad (37)$$

where
$$T = \frac{1}{2m} \sum_{k=1}^{n} p_k^2$$

is the kinetic energy of the electrons and

$$D = \frac{e^2}{mc^2} \sum_{k=1}^{n} (\mathbf{A}_k^0)^2$$

is their diamagnetic energy. These terms play no part in what follows. $Z_S = 2\beta \mathbf{H}_0 \cdot \mathbf{S}$ is the spin Zeeman energy of the electrons in the applied field;

$$Z_L = \frac{e}{2mc} \sum_{k=1}^{n} (\mathbf{p}_k \cdot \mathbf{A}_k^0 + \mathbf{A}_k^0 \cdot \mathbf{p}_k) = \beta \mathbf{H}_0 \cdot \mathbf{L}$$

is the orbital Zeeman energy of the electrons. $O_1 + S_1 + S_2$ is the sum of the magnetic couplings (32) between the n electrons and the N nuclei.

$$O_1 = 2\beta \sum_{k=1}^{n} \sum_{q=1}^{N} \frac{\boldsymbol{\mu}_k \cdot \mathbf{l}_{qk}}{r_{qk}^3} \quad \text{with} \quad \hbar \mathbf{l}_{qk} = \mathbf{r}_{qk} \wedge \mathbf{p}_k,$$

$$S_1 = 2\beta \sum_{k=1}^{n} \sum_{q=1}^{N} \frac{1}{r_{qk}^3} \left\{ \frac{3(\mathbf{s}_k \cdot \mathbf{r}_{qk})(\boldsymbol{\mu}_q \cdot \mathbf{r}_{qk})}{r_{qk}^2} - (\mathbf{s}_k \cdot \boldsymbol{\mu}_q) \right\},$$

$$S_2 = \frac{16\pi}{3} \beta \sum_{k=1}^{n} \sum_{q=1}^{N} (\mathbf{s}_k \cdot \boldsymbol{\mu}_q) \, \delta(\mathbf{r}_{qk}).$$

O_2 and O_3 are two terms not met with previously, the first is bilinear with respect to $\mathbf{H}_0$ and $\boldsymbol{\mu}_q$ and the second bilinear in $\boldsymbol{\mu}_q$ and $\boldsymbol{\mu}_{q'}$.

$$O_2 = \frac{e^2}{2mc^2} \sum_{k=1}^{n} \sum_{q=1}^{N} (\mathbf{H}_0 \wedge \mathbf{r}_{qk}) \cdot (\boldsymbol{\mu}_q \wedge \mathbf{r}_{qk})/r_{qk}^3.$$

The term O_2 represents the coupling (already mentioned) between the nuclear moments and the magnetic field of the currents induced by the Larmor precession of the electrons in the applied field H_0.

$$O_3 = \frac{e^2}{2mc^2} \sum_{k,q,q'} \frac{(\boldsymbol{\mu}_q \wedge \mathbf{r}_{qk}) \cdot (\boldsymbol{\mu}_{q'} \wedge \mathbf{r}_{q'k})}{r_{qk}^3 r_{q'k}^3}.$$

The chemical shift (and also the indirect interactions to be calculated in the next section) correspond to small modifications of the energy of the system and are calculated by perturbation theory. Furthermore the smallness of the chemical shift is such that it is usually unobservable except in liquid (or gaseous) samples where the molecules rotate rapidly. Without specifying how this rotation should be described (quantum-mechanically in gases or classically in most liquids) we shall represent the ground state of the molecule by the symbol $|0\lambda\rangle$, where λ refers to the orientation of the molecule and $|0\rangle$ to its other degrees of freedom (electronic and possibly vibrational states). In a diamagnetic substance the only terms on the right-hand side of (37) for which the expectation value $\langle 0\lambda| \ |0\lambda\rangle$ does not vanish are O_2 and O_3. The former, bilinear in $\boldsymbol{\mu}_q$ and $\mathbf{H}_0$, clearly provides a contribution to the chemical shift. If we select as origin the particular nuclear moment of the nucleus N_0 for which the shift is being calculated so that $\mathbf{r}_{0k}$ becomes just $\mathbf{r}_k$, the first-order change in energy is

$$\langle O_2 \rangle = \frac{e^2}{2mc^2} \left(0\lambda \left| \sum_k \frac{(\mathbf{H}_0 \wedge \mathbf{r}_k) \cdot (\boldsymbol{\mu} \wedge \mathbf{r}_k)}{r_k^3} \right| 0\lambda \right). \tag{38}$$

This expression can be rewritten as $\langle O_2 \rangle = \boldsymbol{\mu} \cdot \Sigma_d \cdot \mathbf{H}_0$. (The subscript d stands for diamagnetic.) In this expression the tensor Σ_d can be decomposed into a traceless part Σ'_d with

$$\Sigma_d'^{pq} = -\frac{e^2}{2mc^2} \left(0\lambda \left| \sum_k \frac{x_k^p x_k^q}{r_k^3} - \frac{1}{3} \frac{\delta^{pq}}{r_k} \right| 0\lambda \right) \tag{38'}$$

and a scalar part Σ'' given by $\Sigma''^{pq} = \sigma_d \delta_{pq}$,

$$\sigma_d = \frac{e^2}{3mc^2} \left(0\lambda \left| \sum_k \frac{1}{r_k} \right| 0\lambda \right). \tag{38''}$$

σ_d is independent of the rotational state of the molecule, which need not be specified, and can be written

$$\sigma_d = \frac{e^2}{3mc^2}\left(0\left|\sum_k \frac{1}{r_k}\right|0\right). \tag{39}$$

For the traceless part (38′), the average of its values over the various rotational states $|\lambda\rangle$ clearly vanishes, and in a liquid where the molecule passes very rapidly from one state $|\lambda\rangle$ to another, so that only the average over the states is observable, Σ'_d brings no contribution to the chemical shift, and the change in the Zeeman energy of the nucleus is $\sigma_d\boldsymbol{\mu}.\mathbf{H}_0$. In contrast to this in molecular-beam experiments where collisions are negligible and molecules are in well-defined rotational states $|\lambda\rangle$, the effect of the anisotropic part Σ'_d is observable. The constant σ_d is always positive *decreasing* the applied field H_0 by an amount $\sigma_d \mathbf{H}_0$, whence its name of shielding constant. A rough estimate of the order of magnitude of σ_d follows from the remark that $e^2/mc^2 = r_0$, the classical radius of the electron $= a_0/(137)^2$, where a_0 is the radius of the first Bohr orbit of the hydrogen atom. If one assumes that $(0\,|\,1/r_k\,|\,0)$ is of the order of $1/a_0$, values of σ_d between 10^{-4} and 10^{-5} must be expected.

It is possible to obtain another energy term Δ bilinear in $\boldsymbol{\mu}$ and $\mathbf{H}_0$, and thus contributing to the chemical shift, by combining, through second-order perturbation theory, a term A of (37) proportional to $\boldsymbol{\mu}$ with a term B proportional to $\mathbf{H}_0$. According to the formula

$$\Delta = \sum_n{}' \frac{(0\,|\,A\,|\,n)(n\,|\,B\,|\,0)}{E_0 - E_n} + \text{c.c.}, \tag{40}$$

A and B must both have non-vanishing matrix elements between the ground state $|0\rangle$ and an excited electronic state $|n\rangle$ (for brevity the orientation index λ is omitted). Since there are three terms proportional to $\boldsymbol{\mu}$, namely O_1, S_1, S_2, and two terms proportional to H_0, namely Z_L and Z_S, there are *a priori* six possible combinations to be used in (40). However, it is easy to see that (O_1, Z_L) is the only combination different from zero.

The lack of contribution from Z_S is due to the fact that since the ground state is an eigenstate of $S = 0$, $Z_S\,|\,0) = 0$. The operator B in (40) is thus necessarily Z_L. We define the operator

$$C = \sum{}' \frac{|n)(n|}{E_0 - E_n}. \tag{41}$$

If the forces and therefore the energies are independent of the spins,

C is a purely orbital operator. Furthermore, if one assumes that the energy levels of the excited states do not depend on the orientation of the molecule, it is clear that the operator C is invariant under rotation. The contribution to (40) from S_1 and S_2 can then be written:

$$\Delta = (0|(S_1+S_2)CZ_L|0)+\text{c.c.},\qquad(42)$$

and, since $|0)$ is an eigenstate of $S = 0$, and S_1 and S_2 are linear functions of the components of the electronic spins, (42) vanishes. If we reintroduce the orientation index λ for the ground state and choose as origin of coordinates the nuclear moment μ considered, so that $\hbar l_{0k} = \mathbf{r}_{0k} \wedge \mathbf{p}_k$ becomes just $\hbar l_k$, we obtain

$$\Delta = 2\beta^2 \sum_{n,k}{}' \frac{(0\lambda|\mathbf{L}.\mathbf{H}_0|n)(n|\boldsymbol{\mu}.\mathbf{l}_k/r_k^3|0\lambda)}{E_0-E_n}+\text{c.c.}$$

$$= 2\beta^2 \sum_k \left(0\lambda\left|(\mathbf{L}.\mathbf{H}_0)C\left(\frac{\boldsymbol{\mu}.\mathbf{l}_k}{r_k^3}\right)\right|0\lambda\right)+\text{c.c.}\qquad(43)$$

(43) can also be rewritten as a tensor coupling $\boldsymbol{\mu}.\Sigma_p.\mathbf{H}_0$ (the subscript p stands for paramagnetic), with a traceless part for Σ_p of

$$\Sigma_p'^{mn} = 2\beta^2 \sum_k \left(0\lambda\left|L^m C\frac{l_k^n}{r_k^3} - \frac{1}{3}\frac{LC.\mathbf{1}_k}{r_k^3}\right|0\lambda\right)+\text{c.c.}\qquad(43')$$

and a scalar part $\Sigma_p''^{mn} = \sigma_p \delta^{mn}$, where

$$\sigma_p = \frac{2\beta^2}{3}\sum_k \left(0\left|\frac{\mathbf{L}.C\mathbf{1}_k+\mathbf{1}_k.C\mathbf{L}}{r_k^3}\right|0\right).\qquad(43'')$$

The index λ is omitted in $(43'')$ because the scalar

$$[(\mathbf{L}.C\mathbf{1}_k)+(\mathbf{1}_k.C\mathbf{L})]/r_k^3$$

is independent of it.

In the same way as for the diamagnetic correction, the contribution of the anisotropic part Σ_p' to the frequency shift vanishes in a liquid.

The calculation of the paramagnetic shielding constant σ_p is far more difficult than that of σ_d for it requires, through the operator C given in (41), the knowledge of the excited states of the molecule. A rough estimate of the magnitude of σ_p can be made using the so-called closure approximation, where the operator $C = \sum'|n)(n|/(E_0-E_n)$ is replaced by a single constant $-1/\Delta E$, where ΔE has the meaning of an average excitation energy of the molecule. The closure approximation is correct if the contribution of one particular state to the sum (40) far exceeds that of all the others, for instance if this state is abnormally close to the ground state. It is little more than a shorthand notation in the general case. An order of magnitude estimate of

$$\sigma_p = -\frac{4\beta^2}{3\Delta E}\sum_k\left\langle 0\left|\frac{\mathbf{L}.\mathbf{l}_k}{r_k^3}\right|0\right\rangle \qquad (43''')$$

can be made by assuming that $1/r_k^3$ is of the order of $1/a_0^3$, where a_0 is the radius of the first Bohr orbit, and that the excitation energy is of the order of the ionization energy $\frac{1}{2}e^2/a_0$ of the hydrogen atom. It is found that $|\sigma_p| \sim \sigma_d \sim 1/(137)^2$. The real meaning to be attached to that estimate is that although σ_p is a second-order term, there is no reason to expect it to be of a smaller order of magnitude than the first-order correction.

If we look back at the expression (39) for σ_d, it seems surprising that the contributions to it from the various electrons fall off as slowly as $1/r_k$ so that even electrons localized on atoms of the molecule far removed from the nucleus considered affect the chemical shift appreciably. The explanation is that the paramagnetic term σ_p also contains contributions from distant atoms that partly cancel those contained in σ_d. Both σ_p and σ_d depend on the origin chosen for the vectors $\mathbf{r}_k$ in the expression $\frac{1}{2}(\mathbf{H}_0 \wedge \mathbf{r}_k)$ of the vector potential $\mathbf{A}_k^0$, whilst their sum $\sigma_p+\sigma_d$, which is a measurable quantity, is naturally independent of it. If instead of the nucleus N_0 another origin (the centre of mass of the molecule, for instance) is chosen, σ_d and σ_p become

$$\sigma_d = \frac{e^2}{3mc^2}\left\langle 0\left|\sum_k \frac{\mathbf{r}_k.\mathbf{r}_{0k}}{r_{0k}^3}\right|0\right\rangle,$$

$$\sigma_p = \frac{2\beta^2}{3}\sum_k\left\langle 0\left|\frac{(\mathbf{L}C.\mathbf{l}_{0k})+(\mathbf{l}_{0k}C.\mathbf{L})}{r_{0k}^3}\right|0\right\rangle. \qquad (44)$$

It would be desirable to modify the expressions for σ_d and σ_p given by (39) and (43'') or (43''') in such a way that their sum did not appear as a small difference of two large quantities. The key to such calculations of $\sigma_d+\sigma_p$ is the availability of a satisfactory description for the electronic states of the molecules, a problem which belongs to the domain of theoretical chemistry and is outside the scope of this book. The reader is referred to the considerable literature of which the references (11) to (14) are but a small selection, for a description of calculations of σ in the framework of the theory of molecular orbitals and their comparison with experiment. The contribution of σ_p to the frequency shift is particularly large when the molecule has excited electronic states in the neighbourhood of the ground state. If their distance to the ground state is comparable to kT, and if thermal transitions between those states occur sufficiently rapidly, an average, temperature-dependent chemical shift will be observed. The criterion for the required transition rate will be given in Chapter X. Such effects have been observed in diamagnetic compounds of cobalt (15, 16).

For linear molecules in $^1\Sigma$ states with angular momentum $\mathbf{J}$ it is possible to correlate the paramagnetic shielding constant σ_p with the magnetic field $H'\mathbf{J}$ produced by the rotation of the molecule, at the site of a nuclear moment $\mathbf{\mu}$ of the molecule, resulting in an interaction $-H'(\mathbf{\mu}.\mathbf{J})$ (13). The field $H'\mathbf{J}$ has a component $H'_n\mathbf{J}$ contributed by the rotation of the nuclear charges and another $H'_e\mathbf{J}$ contributed by the electrons.

$H'_n\mathbf{J}$ is easily computed by writing that it is equal to $\sum_i (\mathbf{v}_i/c) \wedge \mathbf{\mathcal{E}}_i$, where $\mathbf{v}_i$ are the rotational velocities of all the other nuclei N_i of the molecule with respect to the nucleus N^0 under consideration, and $\mathbf{\mathcal{E}}_i$ are the electrostatic fields that their charges $Z_i e$ produce at N_0. Let $\mathbf{n}a_i$ be the vectors $\mathbf{N}_0\mathbf{N}_i$ where $\mathbf{n}$ is the unit vector of the axis of the linear molecule

$$\mathbf{\mathcal{E}}_i = -\frac{eZ_i\mathbf{n}a_i}{|a_i|^3}, \qquad \mathbf{v}_i = \mathbf{\omega} \wedge \mathbf{n}a_i,$$

where $\mathbf{\omega} = \hbar\mathbf{J}/I$ is the vector representing the rotation of the molecule and I its moment of inertia. We obtain

$$H'_n\mathbf{J} = \sum_i \frac{\mathbf{v}_i}{c} \wedge \mathbf{\mathcal{E}}_i = \left(\frac{e\hbar}{Ic}\sum_i \frac{Z_i}{|a_i|}\right)\mathbf{J}. \tag{45}$$

The computation of the electronic part $H'_e\mathbf{J}$ is more involved. Let $\mathbf{L}$ be the operator electronic angular momentum of the molecule (the centre of mass being taken as the origin) and $\mathbf{K} = \mathbf{J}-\mathbf{L}$ the angular momentum arising from the rotation of the molecule as a rigid body. The energy of this rotation corresponds to a Hamiltonian

$$\frac{\hbar^2}{2I} \times |\mathbf{K}|^2 = \frac{\hbar^2}{2I}|\mathbf{J}-\mathbf{L}|^2.$$

The cross term $-(\hbar^2/I)\mathbf{J}.\mathbf{L}$ has zero expectation value in the ground $^1\Sigma$ state but it has matrix elements between this state and excited electronic states of the molecule, just as the Zeeman Hamiltonian $Z_L = \beta\mathbf{L}.\mathbf{H}$ does. It is thus possible by means of second-order perturbation theory to obtain contributions to the energy of the molecule that are bilinear in $\mathbf{J}$ and $\mathbf{\mu}$. In full analogy to the formula (43) this contribution can be written

$$\Delta' = -\frac{\hbar^2}{I}2\beta \sum_{n,k}{}' \frac{(0\lambda | \mathbf{L}.\mathbf{J} | n)(n | \mathbf{\mu}.\mathbf{l}_{0k}/r_{0k}^3 | 0\lambda)}{E_0-E_n} + \text{c.c.}$$

$$= -\frac{2\beta\hbar^2}{I} \sum_k \left(0\lambda \left| (\mathbf{L}.\mathbf{J})C\left(\frac{\mathbf{\mu}.\mathbf{l}_{0k}}{r_{0k}^3}\right)\right| 0\lambda\right) + \text{c.c.} \qquad (46)$$

The second equation (46) can be rewritten as

$$\Delta' = \mathbf{\mu}.(0\lambda | \mathbf{V} | 0\lambda),$$

where $\mathbf{V}$ is a vector operator. Inside the manifold of the ground state, the vector $\mathbf{V}$ can be replaced by a vector $-H'_e \mathbf{J}$ proportional to $\mathbf{J}$, where H'_e can be obtained by a method used several times previously:

$$H'_e = -\frac{1}{J(J+1)}(0\lambda | \mathbf{V}.\mathbf{J} | 0\lambda). \qquad (47)$$

Since the value of H'_e is independent of the molecular orientation λ, an average can be taken over all orientations of the molecule:

$$H'_e = \frac{\hbar^2 2\beta}{IJ(J+1)}\overline{\left(0\lambda\left|(\mathbf{L}.\mathbf{J})C\sum_k \frac{(\mathbf{l}_{0k}.\mathbf{J})}{r_{0k}^3}\right|0\lambda\right)} + \text{c.c.} \qquad (48)$$

$$= \frac{\hbar^2 2\beta}{IJ(J+1)} \sum_n{}' \left\{\overline{\frac{(0\lambda | \mathbf{L}.\mathbf{J} | n)\left(n \left| \sum_k (\mathbf{l}_{0k}.\mathbf{J}/r_{0k}^3) \right| 0\lambda\right)}{E_0-E_n}} + \text{c.c.}\right\}. \qquad (48')$$

Inside the symbol $(0\lambda | \quad | 0\lambda)$ of equation (48) terms such as J_X^2, J_Y^2, J_Z^2 and $J_X J_Y$, $J_Y J_Z$, $J_Z J_X$ appear. If the axes $OXYZ$ are chosen fixed with respect to the molecular frame (OZ along the axis of the linear molecule), it is well known (17) that J_X, J_Y, J_Z commute with the components along the same axes of all electronic operators such as $\mathbf{L}$ and $\mathbf{l}_{0k}/r_{0k}^3$. The average in (48) can then be taken over the components of $\mathbf{J}$ independently and cross products $J_X J_Y$, $J_Y J_Z$, $J_Z J_X$ disappear. Furthermore, since $\mathbf{J}$ is perpendicular to the axis of the molecule, $\bar{J}_Z^2 = 0$, $\bar{J}_X^2 = \bar{J}_Y^2 = \frac{1}{2}J(J+1)$. (48') can then be rewritten

$$H'_e = \frac{\beta\hbar^2}{I}\sum_k\sum_n{}' \frac{1}{E_0-E_n}\left\{\overline{(0\lambda | L_X | n)\left(n\left|\frac{l_{0kX}}{r_{0k}^3}\right|0\lambda\right)} + \right.$$
$$\left. + \overline{(0\lambda | L_Y | n)\left(n\left|\frac{l_{0kY}}{r_{0k}^3}\right|0\lambda\right)} + \text{c.c.}\right\}. \qquad (49)$$

Since L_Z commutes with the electronic Hamiltonian in a linear molecule, $(0\lambda | L_Z | n)$ vanishes and an extra term proportional to

$$(0\lambda | L_Z | n)\left(n\left|\frac{l_{0kZ}}{r_{0k}^3}\right|0\lambda\right)$$

may be introduced into (49) without changing the results. We obtain

$$H'_e = \frac{\beta\hbar^2}{I} \sum_k \left(0\left|\mathbf{L}\cdot C\frac{\mathbf{l}_{0k}}{r_{0k}^3} + \frac{\mathbf{l}_{0k}}{r_{0k}^3}\cdot C\mathbf{L}\right|0\right) \tag{50}$$

which when compared with (44) gives $\sigma_p = (2\beta I/3\hbar^2)H'_e$. Taking into account the contribution $H'_n = H' - H'_e$ given by (45), we have

$$\sigma_p = \frac{2\beta I}{3\hbar^2}\left(H' - \frac{e\hbar}{Ic}\sum_i \frac{Z_i}{|a_i|}\right). \tag{51}$$

It should be emphasized that in order for (51) to be valid σ_p and therefore correspondingly σ_d as given by (44), must be defined with respect to the centre of mass of the molecule rather than to the nucleus N_0.

For the hydrogen molecule using a wave function calculated by Nordsieck for the calculation of σ_d by (44) and the value of σ_p obtained from the value $H' = 27$ gauss measured by molecular beam methods, it is found (10) that

$$\sigma_d = 3\cdot24\times 10^{-5},$$
$$\sigma_p = -0\cdot56\times 10^{-5},$$
$$\sigma = 2\cdot68\times 10^{-5}.$$

Finally, it can be mentioned that just as has been done for induced quadrupole gradients, variational methods have been applied to the calculation of σ_p. A perturbed function $\psi = \psi_0 + \psi_1$ is used, where ψ_1 is a small perturbing trial function containing parameters determined so as to minimize the energy of the molecule in the presence of an applied field H. The expectation value of the coupling of the electrons described by the perturbed wave function ψ, with a nuclear spin, is then computed. A value of σ_p of the order of $-0\cdot5\times 10^{-5}$ in good agreement with the value (51) is found for a particular choice of the trial function (18) for the hydrogen molecule.

(c) *Indirect interaction between nuclear spins in diamagnetic substances*

The existence of bilinear couplings between nuclear spins as distinct from their dipolar couplings was first suggested by the discovery made around 1950 by several groups of workers that in liquid samples some nuclear resonance lines had a multiplet structure, with separations between the various components that were independent of the field. This independence of the value of the field demonstrated that the various components of a multiplet could not be assigned to nuclei with different chemical environments and therefore different chemical shifts, for then the intervals between the lines would have been directly proportional to the field.

The fact that such structures were observed in liquids where the molecules rotate very rapidly in a random fashion showed that in contradistinction to the case of solid samples the dipolar spin-spin interaction could not be responsible for the structure observed. Indeed, any bilinear interaction between two nuclear spins $\mathbf{I}_A$ and $\mathbf{I}_B$ can be written $\hbar \mathbf{I}_A \cdot \mathscr{I}_{AB} \cdot \mathbf{I}_B$, where $\mathscr{I}_{AB}$ is a tensor (it is convenient to introduce the constant $\hbar$ so that the components of the tensor $\mathscr{I}_{AB}$ have the dimensions of a frequency). The tensor $\mathscr{I}$ can be split up into a traceless part $\mathscr{I}'_{pq} = \mathscr{I}_{pq} - \delta_{pq} J$ and a diagonal part $\mathscr{I}''_{pq} = \delta_{pq} J$, where $J = \tfrac{1}{3}\mathrm{tr}(\mathscr{I})$. It has already been mentioned in the previous section that under the conditions of rapid molecular rotation that prevail in a liquid the traceless part of a tensor is not observed. Since a dipolar interaction is precisely described by a traceless tensor it has no effect on the energy levels of the nuclear spins and cannot be the cause of the structure observed.

Just as for the quadrupole interactions described in Section I, the study of the indirect spin-spin couplings can be divided into two parts: first, the calculation of the tensors $\mathscr{I}_{AB}$ from first principles starting from the electronic structure of the molecules; second, given these tensors, or, at least in liquids, their diagonal parts J, the calculation of the relative positions, intensities, and whenever possible line widths, of the components of the multiplet structure caused by these interactions, generally in the presence of chemical shifts. The first part only will be dealt with now, the multiplet structures will be examined in Chapter XI.

Since the first explanation of the indirect interactions by a distortion of the electronic wave function (19) (whence the name of indirect interactions) there have been many attempts to correlate the strengths of these interactions with the electronic structure of the molecules. Although straightforward in principle, as will appear shortly, these calculations are in practice beset by the same difficulties as those of the chemical shifts, namely lack of a satisfactory description for the excited states of molecules (and sometimes for their ground state as well). As in the previous section we will outline the principles of the theory, leaving out details that belong to the domain of theoretical chemistry and can be found in the references (20) to (23)—a far from exhaustive list.

(1) *The orbital coupling*

The Hamiltonian (37) contains a term bilinear with respect to the nuclear moments of the molecules, namely the term O_3, and the corresponding nuclear coupling is simply $\langle O_3 \rangle = (\psi_0 | O_3 | \psi_0)$, where ψ_0 is the wave function of the ground state of the molecule.

Although a first-order term, $\langle O_3 \rangle$ is very small. This is essentially due to the fact that O_3 is a sum of *one-electron* operators

$$(\mathbf{\mu}_A \wedge \mathbf{r}_{Ak}) \cdot (\mathbf{\mu}_B \wedge \mathbf{r}_{Bk})/r_{Ak}^3 r_{Bk}^3,$$

so that when the first factor $(\mathbf{\mu}_A \wedge \mathbf{r}_{Ak})/r_{Ak}^3$ is relatively large, the electron k being near the nuclear moment μ_A, the second one is necessarily small. Furthermore, the contribution to $\langle O_3 \rangle$ of the inner atomic shells where one of the factors is very large is considerably decreased by an angular factor. Let $|\alpha_A\rangle$ be an inner orbit of the atom A with linear dimensions much smaller than the internuclear distance AB. The distance of an electron of this orbit to the nucleus B is practically constant and equal to $AB = R$ so that

$$\left\langle \alpha_A \left| \frac{(\mathbf{\mu}_A \wedge \mathbf{r}_A)}{r_A^3} \cdot \frac{(\mathbf{\mu}_B \wedge \mathbf{r}_B)}{r_B^3} \right| \alpha_A \right\rangle \cong \left\langle \alpha_A \left| \frac{\mathbf{\mu}_A \wedge \mathbf{r}_A}{r_A^3} \right| \alpha_A \right\rangle \cdot \frac{\mathbf{\mu}_B \wedge \mathbf{R}}{R^3} = 0, \quad (52)$$

since the orbit $|\alpha_A\rangle$ has a well-defined parity with respect to the nucleus A. Besides the first-order term $\langle O_3 \rangle$ other contributions to the spin-spin interactions may be obtained by combining through a second-order perturbation calculation two of the following terms O_1, S_1, S_2. The terms thus obtained are of the form $(0 | PCQ | 0)$ where P and Q are one of the three operators O_1, S_1, S_2, and C is the operator already defined in the previous section,

$$C = \sum_n{}' \frac{|n)(n|}{E_0 - E_n}.$$

Of the six combinations that can be formed that way, those which mix orbital and spin terms, namely (O_1, S_1) and (O_1, S_2), give a vanishing result. The argument is similar to that already used for the chemical shift: if spin-dependent forces (such as the spin-orbit coupling) are negligible, C is a purely orbital operator and since the ground state $|0)$ is an eigenstate of the total spin with the eigenvalue zero, a term such as $(0 | O_1 CS_{1,2} | 0)$ vanishes. Furthermore, if the experiment is performed in a liquid so that only the scalar part of the coupling is observed, cross terms $(0 | S_1 CS_2 | 0)$ also vanish because of the different transformation properties of S_1 and S_2.

A rough estimate of the order of magnitude of the total contribution of the orbital terms $\langle O_3 \rangle + \langle O_2 C O_1 \rangle$ can be made as follows **(23)**. If we assume that the dimensions of the two atoms A and B containing the nuclear spins $\mathbf{\mu}_A$ and $\mathbf{\mu}_B$ are appreciably smaller than the nuclear distance $AB = R$, the dipolar field produced by the spin μ_A is approximately uniform over the atom B and has components

$$H_\beta = -\mathscr{E}_{\beta\gamma}\mu_A^\gamma,$$

where $\quad\quad\quad \mathscr{E}_{\beta\gamma} = R^{-5}\{R^2\delta_{\beta\gamma} - 3R_\beta R_\gamma\}.$ (53)

By definition of the shielding tensor $\Sigma_B^{\alpha\beta}$ for the atom B, the coupling of the nuclear spin μ_B with this field is

$$-\mu_B^\alpha\{\delta_{\alpha\beta} - \Sigma_B^{\alpha\beta}\}H_\beta, \quad\quad (54)$$

whence an indirect coupling:

$$-\mu_B^\alpha \Sigma_B^{\alpha\beta} \mathscr{E}_{\beta\gamma} \mu_A^\gamma.$$

A similar term containing the shielding tensor $\Sigma_A^{\alpha\beta}$ of the atom A must be added to it. This shows that the ratio of the orbital indirect coupling to the direct one $-\mu_A^\alpha \mu_B^\beta \mathscr{E}_{\alpha\beta}$ is of the order of the shielding constant σ which is always a very small number. Since in certain cases the indirect coupling may even exceed the direct one, clearly the orbital coupling is not its dominant part. Some of the approximate ways of estimating the indirect couplings will now be considered.

(2) *The Heitler–London approximation*

In this approximation, a chemical bond between two atoms A and B is described by a two-electron wave function

$$\psi(1,2) = \frac{1}{\sqrt{\{2(1+\Delta^2)\}}}\{\phi_A(1)\phi_B(2) + \phi_B(1)\phi_A(2)\}\chi(1,2). \quad (55)$$

In this formula ϕ_A and ϕ_B are normalized atomic orbitals belonging respectively to the atoms A and B, or linear combinations of such orbitals, and $\chi(1,2)$ is a function of the two electron spins $\mathbf{s}_1, \mathbf{s}_2$, describing a singlet spin state (note incidentally that (55) is not a Slater determinant but a sum of two such determinants). The constant

$$\Delta = \int \phi_A^*(1)\phi_B(1)\,d\tau$$

is a measure of the overlap of the two orbitals. With the closure approximation where the operator C is replaced by a constant, the indirect coupling between the two nuclear spins μ_A and μ_B is given by

$$-\frac{1}{\Delta E}(\psi(1,2)|\{\mathscr{H}_A(1) + \mathscr{H}_B(2) + \mathscr{H}_A(2) + \mathscr{H}_B(1)\}^2|\psi(1,2)). \quad (56)$$

(We omit the index λ for the orientation of the molecule.)

$\mathscr{H}_A$ and $\mathscr{H}_B$ represent terms containing respectively μ_A and μ_B in the sum $O_1 + S_1 + S_2$. Naturally only those terms that are bilinear in

μ_A and μ_B have to be selected from (56). Expanding (56) and taking into account (55), six types of terms are obtained:

$$(\phi_A | \mathcal{H}_A | \phi_A) \cdot (\phi_B | \mathcal{H}_B | \phi_B)$$
$$(\phi_A | \mathcal{H}_B | \phi_A) \cdot (\phi_B | \mathcal{H}_A | \phi_B)$$
$$(\phi_A | \mathcal{H}_A | \phi_B) \cdot (\phi_A | \mathcal{H}_B | \phi_B)$$
$$(\phi_A | \mathcal{H}_A \mathcal{H}_B | \phi_A)$$
$$(\phi_A | \mathcal{H}_A \mathcal{H}_B | \phi_B)$$
$$(\phi_B | \mathcal{H}_A \mathcal{H}_B | \phi_B)$$

where all the matrix elements are one-electron integrals. A Hamiltonian such as $\mathcal{H}_A$ is large if the electron is near the nucleus of the atom A and small if it is elsewhere. It is reasonable therefore to expect the first term to be much larger than the others, and to keep only this one in a first approximation. If the orbits ϕ_A and ϕ_B are represented by real wave functions so that the orbital momentum is quenched, as is usually the case in diamagnetic substances, $(\phi_A | O_1 | \phi_A) = 0$ and only spin terms remain. The expectation value

$$(\phi_A | \mathcal{H}_A | \phi_A) = (\phi_A | S_1^A + S_2^A | \phi_A)$$

is then necessarily of the form $\mathbf{I}_A \cdot \gamma \hbar \mathcal{T}_A \cdot \mathbf{s}$, where $\mathcal{T}_A$ is a certain tensor that would describe the hyperfine structure of the atom A if one could neglect the distortion of the orbit ϕ_A by the presence of the atom B. The interaction between the two spins $\mathbf{I}_A$ and $\mathbf{I}_B$ can then be written $\mathbf{I}_A \cdot \hbar \mathcal{I}_{AB} \cdot \mathbf{I}_B$, where the tensor $\mathcal{I}_{AB}$ is given by

$$\mathbf{I}_A \cdot \hbar \mathcal{I}_{AB} \cdot \mathbf{I}_B = \frac{-2\hbar^2 \gamma_A \gamma_B}{(1+\Delta^2)\Delta E} (\chi(1,2) | \{(\mathbf{I}_A \cdot \mathcal{T}_A \cdot \mathbf{s}_1)(\mathbf{I}_B \cdot \mathcal{T}_B \cdot \mathbf{s}_2)\} | \chi(1,2)) \tag{57}$$

Since $\chi(1,2)$ is a singlet spin function the following relations hold:

$$(\chi | s_{1x} s_{2x} | \chi) = \tfrac{1}{3}(\chi | \mathbf{s}_1 \cdot \mathbf{s}_2 | \chi) = -\tfrac{1}{4},$$
$$(\chi | s_{1x} s_{2y} | \chi) = 0,$$

whence
$$\hbar \mathcal{I}_{AB} = \frac{-\hbar^2 \gamma_A \gamma_B}{2(1+\Delta^2)\Delta E} \{\mathcal{T}_A \cdot \mathcal{T}_B\}, \tag{58}$$

where $\{\mathcal{T}_A \cdot \mathcal{T}_B\}$ is a symmetrized tensor product of the tensors $\mathcal{T}_A$ and $\mathcal{T}_B$.

It is apparent from (58) that the order of magnitude of $\mathcal{I}$ is $\nu_A \nu_B / \nu_0$, where ν_A and ν_B are the magnitudes of the hyperfine structures of either atom on the frequency scale, and $\nu_0 = \Delta E/\hbar$ an average distance (on the same scale) between the ground state and an excited state of the molecule. Important $\mathscr{I}$ couplings are therefore to be expected when the atoms A and B are heavy atoms with large hyperfine structures or when the molecule has accidentally low-lying excited states.

The above theory can be applied to the HD molecule where the isotropic part J of the coupling, the only part that was measurable in an experiment performed on a gaseous sample under pressure (24), was found equal to $2\pi \times 43$ c/s. (The reason why the experiment could not be performed on H_2 will appear in Chapter XI.) The best way to compare this result with the theory is to start from (58) where ΔE is considered as unknown and to see whether a reasonable value is obtained for it. If $1s$ orbitals of the atoms H and D are taken for ϕ_A and ϕ_B, $\mathscr{T}_A$, $\mathscr{T}_B$, and $\mathscr{I}_{AB}$ become scalars and (58) gives

$$J = \frac{\gamma_D}{\gamma_H} \frac{(\omega_H)^2}{2(1+\Delta^2)\Delta E/\hbar}, \qquad (59)$$

where $\omega_H/2\pi$ is the hyperfine structure interval for the hydrogen atom. From $J = 2\pi \times 43$ c/s,

$$\omega_H = 2\pi \times 1420 \text{ Mc/s}, \qquad \Delta^2 = 0.55, \qquad \gamma_D/\gamma_H = 0.156,$$

it is found that $\Delta E \sim 10$ eV.

Instead of the Heitler–London approximation the much better wave function of James and Coolidge can be used for a description of the ground state of the molecule HD and for the calculation of J from the equation (56). The value of ΔE obtained (20) in this way is 19 eV. For the distance of the excited triplet state, to the ground state, James and Coolidge find 9 eV. Although for the description of the ground state the wave function of James and Coolidge is undoubtedly much better than the Heitler–London approximation, the uncertainties introduced by the use of the closure approximation are such that any further discussion of this discrepancy seems fruitless.

It is interesting to speculate about the nature of the indirect spin-spin coupling in another substance, namely solid copper acetate, in spite of the fact that owing to experimental difficulties it has not been observed so far.

Electron resonance experiments (25) have demonstrated that in a crystal of copper acetate, the copper ions Cu^{++}, with electronic spins $\frac{1}{2}$, are assembled in pairs and that an electronic exchange coupling $A\mathbf{s}_1.\mathbf{s}_2$ exists between these spins. The sign of A is positive and its magnitude of the order of $\frac{1}{20}$ of electron volt. Each pair of ions thus has a diamagnetic ground state $S = 0$ and a paramagnetic excited state $S = 1$, where paramagnetic resonance can be observed. There is an obvious analogy with the molecule HD and the indirect interaction can be calculated immediately from formula (58) where one makes $\Delta E = A$. The comparison with HD calls for the following comments:

(i) The energy gap ΔE is $\frac{1}{20}$ of an electron volt rather than of the order of 10 electron volts, so that other things being equal an indirect coupling 200 times larger should be expected. From the known values of the hyperfine couplings for isolated Cu^{++} ions, measured in a different copper salt, values of $\mathscr{I}$ of the order of $2\pi \times 7500$ c/s, about 20 times larger than the strength of the ordinary dipolar coupling between the two nuclear spins of a pair, should be expected.

(ii) Since most of the contribution to the coupling $\mathscr{I}$ comes from a single excited state, the closure approximation is well justified.

(iii) The two tensors $\mathscr{T}_A$ and $\mathscr{T}_B$ describing the hyperfine structure of the individual copper ions are very anisotropic, the ratio between the largest principal value and the smallest one being of the order of 7. The axes of the two tensors $\mathscr{T}_A$ and $\mathscr{T}_B$ for the two ions of the pair are parallel and the resulting anisotropy for the tensor $\mathscr{I}$ given by (58) could be of the order of 50.

A third example where the Heitler–London approximation can be used is provided by the analysis (26) of the fine structure of the quadrupole resonance spectrum of solid iodine (27).

The orbitals ϕ_A and ϕ_B are assumed to be linear combinations of the $5s$ and $5p_\sigma$ atomic orbitals of the atoms of the diatomic iodine molecule:

$$\phi_A = \sqrt{s}\,\phi_{5sA} + \sqrt{(1-s)}\,\phi_{5p_\sigma A},$$
$$\phi_B = \sqrt{s}\,\phi_{5sB} + \sqrt{(1-s)}\,\phi_{5p_\sigma B}, \quad (60)$$

where the constant s expresses the s-character of the wave function. The tensor $\mathscr{I}_{AB}$ representing the indirect coupling can be computed from the formula (58). The hyperfine tensors $\mathscr{T}_A$ and $\mathscr{T}_B$ are easily obtained by taking the expectation values of the Hamiltonian (32) over the wave functions (60). Taking the z-axis along the axis of the iodine molecule it is found

$$\mathscr{T}_z/\beta = s\frac{16\pi}{3}|\psi_{5s}(0)|^2 + (1-s)\frac{8}{5}\left\langle\frac{1}{r^3}\right\rangle_{5p},$$

$$\mathscr{T}_x/\beta = \mathscr{T}_y/\beta = s\frac{16\pi}{3}|\psi_{5s}(0)|^2 - (1-s)\frac{4}{5}\left\langle\frac{1}{r^3}\right\rangle_{5p}, \quad (61)$$

and then from (58),

$$\mathscr{I}_z = \frac{\hbar\gamma^2}{2\Delta E}\mathscr{T}_z^2, \quad \mathscr{I}_x = \mathscr{I}_y = \frac{\hbar\gamma^2}{2\Delta E}\mathscr{T}_x^2,$$

where the overlap Δ between the two orbitals ϕ_A and ϕ_B is neglected.

The somewhat uncertain experimental results are consistent with

$$\mathscr{I}_z = \mathscr{I}_x = \mathscr{I}_y = 2\pi \times 3000 \text{ c/s}.$$

The quantity $\langle 1/r^3 \rangle_{5p}$ is known accurately from the spectrum of atomic iodine, $|\phi_{5s}(0)|^2$ can be computed by means of screened atomic wave functions, and the s-character of the bond determined either from the anisotropy or the absolute magnitude of the tensor $\mathscr{I}$. The isotropy $\mathscr{I}_z = \mathscr{I}_x$ would require $s = 1$, and taking $\Delta E = 5$ eV this would lead to the unacceptable value $\mathscr{I}_z = 2\pi \times 78$ kc/s.

Since the magnitude of the tensor $\mathscr{I}$ is a more sensitive function of s than its anisotropy, the authors (26) choose $s = 0{\cdot}22$, which from their assumptions should give

$$\mathscr{I}_z/2\pi = 6 \text{ kc/s}, \qquad \mathscr{I}_x/2\pi = \mathscr{I}_y/2\pi = 2{\cdot}4 \text{ kc/s},$$

a compromise between the observed anisotropy and absolute magnitude of the coupling. The agreement of $s = 0{\cdot}22$ with other chemical evidence is reasonable.

(3) *The method of molecular orbitals*

The method of molecular orbitals, which is an approach to the problem of chemical bonding different from the Heitler–London approximation, also provides an alternative approach to the problem of indirect spin-spin couplings (22).

In principle the method of molecular orbitals is an extension to molecules of the self-consistent field method for atoms. Instead of a central self-consistent field as in atoms, a self-consistent field is sought that has the symmetry of the molecule. In practice this is too difficult, and forgetting about the self-consistency one is often content to represent each molecular orbital as a linear combination of orbitals belonging to the different atoms of the molecule. In the ground state of a diamagnetic molecule we may assume that p orthogonal orbitals $\psi_1,..., \psi_p$ are available to house $2p$ electrons. If we take the spin into account a Slater determinant Ψ_0 can be constructed from the $2p$ one-electron states (two per orbit).

The second-order indirect interaction can then be computed by the relation

$$\sum_n{}' \sum_{NN'} \sum_{k,k'=1}^{2p} \frac{(\Psi_0 | \mathscr{H}_N(k) | n)(n | \mathscr{H}_{N'}(k') | \Psi_0)}{E_0 - E_N}, \qquad (62)$$

where $\mathscr{H}_N(k)$ is the Hamiltonian (32) of the electron k with respect to the nucleus N. If the closure approximation is made, (62) becomes

$$-\frac{1}{\Delta E} \sum_{N,N'} \sum_{k,k'} (\Psi_0 | \mathscr{H}_N(k) \mathscr{H}_{N'}(k') | \Psi_0). \qquad (63)$$

The expression (63) can be expanded by means of the usual rules for the expansion of Slater determinants. The terms obtained are of the following types:

one-particle matrix elements

$$(\psi_p \mid \mathcal{H}_N \mathcal{H}_{N'} \mid \psi_p),$$

direct, two-particle matrix elements

$$(\psi_p(1)\psi_q(2) \mid \mathcal{H}_N(1)\mathcal{H}_{N'}(2) \mid \psi_p(1)\psi_q(2)),$$

exchange, two-particle matrix elements

$$(\psi_p(1)\psi_q(2) \mid \mathcal{H}_N(1)\mathcal{H}_{N'}(2) \mid \psi_q(1)\psi_p(2)).$$

Since the orbitals ψ_p are not necessarily localized on any particular atom there is no simple criterion about the relative magnitude of those terms, as in the Heitler–London approximation. Because each molecular orbital is a linear combination

$$\psi_p = \sum_{N,q} C^p_{Nq} \phi_{Nq},$$

where the ϕ_{Nq} are atomic orbitals of the atom N, (63) can be computed if the C^p_{Nq} are known. (See reference (22) for further developments of this method.) An important extension of this method will be considered in the section on metals (II C (b)) where the orbitals of the conduction electrons extend over the whole sample.

C. The effect of electron-nucleus coupling in paramagnetic substances

(a) *Non-metals*

· (1) *Nature of the coupling*

Consider an atom (or an ion) with a single electron outside closed shells, embedded in bulk matter, and assume first that the spin-orbit coupling is negligible compared with the electrostatic forces that the atom (or ion) experiences from its surroundings. A general theorem (28) predicts that under fairly general conditions its ground state will have no orbital degeneracy. (There still remains a twofold degeneracy corresponding to the two orientations of the electronic spin.) According to an argument given in Section II B (a) the orbital wave function ϕ of this state is real and the expectation value $(\phi \mid \mathbf{l} \mid \phi)$ of the orbital momentum vanishes. The expectation value of the Hamiltonian (32) $(\phi \mid S_1 + S_2 \mid \phi)$ is clearly a tensor coupling $\hbar\gamma \mathbf{I} . \mathcal{T} . \mathbf{S}$ between the nuclear and the electronic spin (a fact already mentioned in Section II B (c).

If ϕ is expanded in spherical harmonics $\phi = \sum_l a_l \phi_l$ its isotropic or s-part $a_0 \phi_0$ will contribute a scalar term $\hbar A \mathbf{I}.\mathbf{S}$ to this coupling, where

$$A = \frac{16\pi}{3}\beta\gamma|a_0|^2|\phi_0(0)|^2, \qquad (64)$$

whereas the other components ϕ_l will bring a traceless contribution given by

$$\mathbf{I}.\mathscr{T}'.\mathbf{S} = 2\beta \sum_{l,l'} a_l a_{l'}^* \left(\phi_{l'} \left| 3\frac{(\mathbf{I}.\mathbf{r})(\mathbf{S}.\mathbf{r})}{r^5} - \frac{(\mathbf{I}.\mathbf{S})}{r^3} \right| \phi_l \right) \qquad (65)$$

with $\qquad |l-l'| = 0, 2.$

Thus, for instance, if ϕ is a wave function $p_\sigma = f(r) Y_1^0(\theta,\varphi)$,

$$\mathscr{T} = \mathscr{T}' = \begin{pmatrix} -\tfrac{1}{2}P \\ & -\tfrac{1}{2}P \\ & & P \end{pmatrix} \quad \text{with} \quad P = \frac{8\beta}{5}\left\langle \frac{1}{r^3} \right\rangle_p. \qquad (66)$$

For a wave function $d = f(r) Y_2^0(\theta,\varphi)$

$$P = \frac{8\beta}{7}\left\langle \frac{1}{r^3} \right\rangle_p, \quad \text{etc.}$$

s-electrons usually have much larger hyperfine couplings than electrons with $l \neq 0$ so that even a small admixture of ϕ_0 in the expansion of ϕ has a large influence on the tensor $\mathscr{T}$.

These results can be generalized in several ways.

First, if spin-dependent forces such as spin-orbit coupling are not negligible so that the orbital momentum is not completely quenched, a general theorem due to Kramers (29) states that in the absence of an applied magnetic field there is still a twofold degeneracy left in the ground state. Although neither of the two electronic wave functions spanning the twofold degenerate manifold of the ground energy level can then be written as a single product of an orbital wave function by a spin function, it is still possible, using the same argument as in Chapter II, to represent the Hamiltonian (32) inside this manifold as a bilinear form $\gamma\hbar \mathbf{S}.\mathscr{T}.\mathbf{I}$, $\mathbf{S}$ being now a fictitious or effective spin $\tfrac{1}{2}$ (30). Incidentally, in that case the Zeeman coupling of the electronic magnetic moment with an applied field also becomes an anisotropic tensor coupling $\beta \mathbf{H}.g.\mathbf{S}$. The two tensors g and $\mathscr{T}$ usually have the same principal directions, determined from the symmetry of the surroundings, but clearly have a quite different anisotropy. (For instance, if $\phi = p_\sigma$ the orbital momentum is quenched and the tensor g is isotropic, whereas $\mathscr{T}$ is given by (66).)

Secondly, for atoms with more than one electron outside closed shells, with spins coupling to a total spin S, it is still possible to define a hyperfine coupling $\gamma_n \hbar \mathbf{I} . \mathscr{T} . \mathbf{S}$. The best known example is that of doubly ionized manganese, Mn^{++}, which has an electronic spin $S = \frac{5}{2}$ and a large isotropic hyperfine structure, entirely determined by an admixture of electronic s-orbits from excited configurations (30).

Thirdly, the magnetic coupling of an unpaired electron with a nuclear spin I that does *not* belong to the same atom may be considered. If the distance R of the nuclear spin $\mathbf{I}$ from the atom is large compared with the dimensions of the latter it is permissible to treat the field $\mathbf{H}_n$ produced by the nuclear moment $\gamma \hbar \mathbf{I}$ as uniform over the atom. The electron nucleus coupling is then obtained by replacing $\mathbf{H}$ by $\mathbf{H}_n$ in the Zeeman tensor coupling $\beta \mathbf{H} . g . \mathbf{S}$. If, on the other hand, the electronic wave function ϕ has a non-vanishing value at the site of the nuclear spin $\mathbf{I}$, some care must be exercised in calculating the hyperfine interaction in order to avoid spurious infinities. For a single unpaired electron with quenched orbital momentum this coupling is given by

$$\frac{16\pi}{3}\gamma\hbar\beta(\mathbf{I}.\mathbf{S})|\phi(\mathbf{R})|^2 + 2\beta\gamma\hbar\mathbf{I}.\mathrm{grad}_R \int \frac{\mathrm{div}(\mathbf{S}\rho(r))}{|\mathbf{r}-\mathbf{R}|} d^3r. \qquad (67)$$

The first term is the isotropic contact coupling whilst the second, where $\rho = |\phi|^2$, is the coupling of the nuclear magnetic moment $\gamma\hbar\mathbf{I}$ with the magnetic field that classical magnetostatics defines inside a continuous distribution of magnetization of density $\mathbf{M} = -2\beta\rho\mathbf{S}$. This is the density of magnetization associated with the electron spin.

(2) *Observability of nuclear resonance*

(α) *Nuclear and electronic spins on the same atom.* If the nucleus of interest is that of a paramagnetic atom (or ion) the electronic magnetic field at the nucleus, as defined by formula (33), is usually orders of magnitude larger than the applied field H_0. The nuclear Zeeman coupling $-\gamma\hbar\mathbf{H}.\mathbf{I}$ is a small perturbation of what is now the main Hamiltonian of the combined system electron-plus-nucleus, namely the electronic Zeeman energy $\beta\mathbf{H}.g.\mathbf{S}$ plus the hyperfine coupling $\gamma\hbar\mathbf{I}.\mathscr{T}.\mathbf{S}$.

Assume for simplicity an isotropic magnetic tensor (quenched orbital momentum), an isotropic hyperfine coupling $\hbar A \mathbf{I}.\mathbf{S}$, nuclear and electronic spins $\frac{1}{2}$. In a strong field, i.e. a field H such that $\beta H \gg \hbar A$, the four energy levels of the system are as represented in Fig. VI, 1.

A transition such as $a \to b$ or $c \to d$ where the nuclear spin flips, the orientation of the electron spin remaining unchanged, is the analogue of an ordinary nuclear flip in an applied field H_0.

It should be remarked, however, that a 'nuclear' transition such as $a \to b$, induced by an r.f. field H_1, is actually an electron transition in the sense that its transition probability is proportional to

$$W_{ab} \propto |\langle a|-2\beta H_1 S_x + \gamma \hbar H_1 I_x | b\rangle|^2 \propto (4\beta^2 q^2 - 4\beta\gamma\hbar q + \gamma^2 \hbar^2)$$

$$\sim \frac{(\tfrac{1}{4}A^2 - \tfrac{1}{2}A\gamma H_0 + \gamma^2 H_0^2)}{H_0^2}. \tag{68}$$

$\tfrac{1}{2}A$, being the precession frequency of the nuclear spin in the electron field at the nucleus, is usually much larger than its Zeeman Larmor

——————— $|a\rangle = |+,+\rangle$

——————— $|b\rangle = p|+,-\rangle + q|-,+\rangle$

——————— $|d\rangle = |-,-\rangle$

——————— $|c\rangle = -q|+,-\rangle + p|-,+\rangle$

FIG. VI, 1. Energy levels of an electron spin S coupled to a nuclear spin $I = \tfrac{1}{2}$ by an isotropic coupling $\hbar A\mathbf{I}\cdot\mathbf{S}$ in a strong magnetic field. A symbol such as $|+-\rangle$ represents a state where $S_z = +\tfrac{1}{2}$, $I_z = -\tfrac{1}{2}$. The admixture coefficients are

$$q \simeq \frac{A\hbar}{4\beta H} \ll 1, \qquad p = (1-q^2)^{\tfrac{1}{2}} \simeq 1.$$

frequency γH_0, and the term $\tfrac{1}{4}A^2$, of electronic origin, is overwhelmingly predominant in the transition $a \to b$. This distinction is not purely academic for it has the important consequence that the transition probability is proportional to H_0^{-2}, rather than independent of H_0 as in an ordinary nuclear Zeeman transition.

A distinction must be made between the possibility of inducing such a transition by means of an applied r.f. field and the detection of the fact that it has occurred, by its reaction on the circuit, as explained in Chapter I.

The electromagnetic detection is considerably hampered by the line width of the transition. This originates in the shortness of the lifetime of either state caused by strong electron relaxation, and also in the local variations of the coupling constant A, to which the intervals $a \to b$ and $c \to d$ are proportional. No detection of a transition under those conditions between two hyperfine levels of a paramagnetic ion in the paramagnetic state has been reported so far.

On the other hand, it is perfectly feasible to drive such a transition by a sufficiently strong r.f. field and to detect its occurrence indirectly through its effect on the intensity of a strong electronic transition such as, say, $a \to c$. Considerable use has been made of this technique, which has received the name of ENDOR (electron-nuclear double resonance), to measure hyperfine splittings with great accuracy. We shall not describe it here, for it belongs to the domain of electron resonance and is outside the scope of this book.

If the lifetime τ of the electron spin in a given state becomes very short it is conceivable that eventually only the average value of the electron field at the nucleus becomes observable (an average value that is not zero because of the different populations of the states with $S_z = +\frac{1}{2}$ and $S_z = -\frac{1}{2}$). Very short lifetimes of electron spin states may occur if strong exchange couplings of the form $\hbar K \mathbf{S}_1 . \mathbf{S}_2$ exist between neighbouring electron spins. Such couplings, well known in electron magnetism, are electrostatic in nature and may correspond to values of K of the order of $2\pi 10^{12}$ c/s or even more. As explained in Chapter V such a coupling induces flip-flops between neighbouring electron spins at a frequency of the order of $K \sim 1/\tau$. A given nuclear spin thus 'sees' an electronic field fluctuating with that frequency. Anticipating the results of Chapter X it may be stated that the condition for such an averaging to occur is $A\tau \ll 1$, and that the width of the corresponding nuclear resonance line is of the order of $A^2\tau$, which should be much less than the Zeeman Larmor frequency γH_0 for the nuclear resonance to be observable.

An example of a situation where these conditions are fulfilled is provided by the resonance of Co^{59} in KCoF$_3$ and CoO, in the paramagnetic state (31).

The resonance observed in KCoF$_3$ has a Lorentz shape as shown in Fig. VI, 2, with an inverse half width at half intensity,

$$(\Delta\omega)^{-1} = T_2 = 1 \cdot 43 \times 10^{-6} \text{ sec.}$$

The nuclear Hamiltonian can be written

$$\mathcal{H} = -\gamma\hbar \mathbf{I} . (\mathbf{H} + \mathbf{H}_e),$$

where $\mathbf{H}_e$ is the operator electronic field at the nucleus given by eqn. (33) (summed over the seven unpaired electrons of the paramagnetic ion Co^{++}). Because of the strong exchange coupling between the electron spins the electronic field 'seen' by a nucleus of Co^{59} fluctuates at an estimated average rate of $\omega_e \sim 2\pi \times 10^{12}$ sec^{-1}, which is much faster than the instantaneous Larmor precession $A = \gamma H_e$ of the nucleus in the electronic field.

The nuclear Hamiltonian can then be replaced by

$$\langle \mathcal{H} \rangle = -\gamma\hbar \mathbf{I} \{\mathbf{H} + \langle \mathbf{H}_e \rangle\},$$

where $\langle \mathbf{H}_e \rangle$, the average value of $\mathbf{H}_e$ for a sample in thermal equilibrium, is much smaller than the instantaneous value $\mathbf{H}_e$ and even than the applied field H_0, displacing the resonance frequency by a few per cent from the value observed in diamagnetic compounds (itself strongly affected by the chemical shifts, see references (15) and (16) of this chapter).

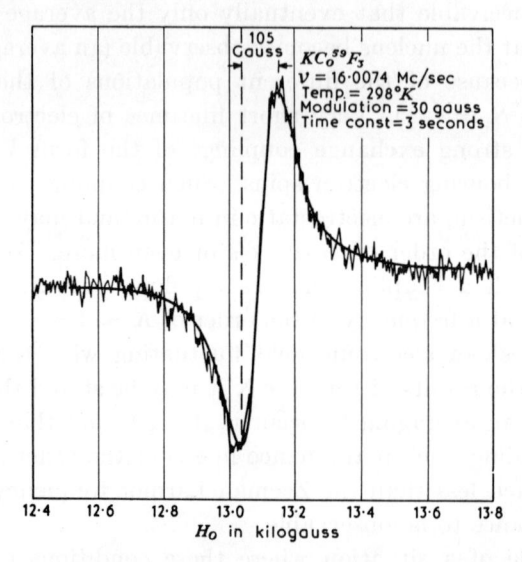

FIG. VI, 2. Copy of recorder trace of the derivative of $KCo^{59}F_3$ absorption. The full curve is calculated for a Lorentz line shape.

The fluctuating part $H_e - \langle H_e \rangle \simeq H_e$ causes a broadening of the line of the order of $\gamma \langle H_e^2 \rangle / \omega_e$ (in gauss) in qualitative agreement with the observed width.

Similar results are obtained in CoO.

(β) *Nuclear and electronic spins on different atoms.* If the nuclear and the electronic spins are on different atoms, the strength $\hbar A$ of their coupling is considerably decreased and the conditions $A\tau \ll 1$, $A^2\tau \ll \gamma H_0$, enabling a sharp average value of the electron field to be 'seen' by the nuclear spin, are much more easily met.

We may first dismiss the situation where the paramagnetic atoms (or ions) of the sample are impurities in such small concentrations that the *instantaneous* electronic field is smaller than the nuclear local field for all nuclei except a small minority of those abnormally close to a paramagnetic impurity. Although, as will appear in Chapter IX, these impurities play an important role in the spin-lattice relaxation of the nuclear spin system, their influence on its energy levels and on the non-saturated nuclear line shape is negligible.

On the other hand, in normal paramagnetic crystals where the relative positions of paramagnetic ions and 'resonant' nuclei are well defined, it is possible to observe a shift in the nuclear resonance line, due to the average electronic field superimposed upon the applied field. 'Resonant' nuclei occupying non-equivalent positions in the unit cell experience different shifts and the resonance line can have a complex structure.

A first example of such a structure is provided by the nuclear resonance of protons belonging to the water molecules in a crystal of $CuSO_4 \cdot 5H_2O$ (32).

The paramagnetic ion Cu^{++}, with one electron missing from the $3d$ shell, can be considered to a first approximation as a free spin with a magnetic moment $2\beta S$. (A 20 per cent anisotropy in the Zeeman tensor g is introduced by the incomplete quenching of the orbital momentum.) The average distance of a proton to a copper ion is of the order of 2·5 Å and the instantaneous electronic field H_e 'seen' by a proton is of the order of 600 gauss. There exists between neighbouring Cu^{++} ions an exchange coupling $\hbar K \mathbf{S}_1 \cdot \mathbf{S}_2$ where the constant K has been estimated to be of the order of 10^{11}. The quantity

$$A\tau \sim \gamma_p H_e/K$$

(where γ_p is that of the proton) is very small, of the order of 10^{-4}, and the electronic field H_e can thus be safely replaced by its average value $\langle H_e \rangle$. The contribution $(\gamma_p H_e/K) \cdot H_e$ of the copper ions to the proton line width is a fraction of a gauss, negligible in comparison to the dipolar line width due to the proton-proton couplings.

The average field $\langle \mathbf{H}_e \rangle$ is that produced by a classical dipole with a magnetic moment $\langle \mathbf{\mu} \rangle$. Neglecting the incompletely quenched orbital magnetism, except for the lowest temperatures, according to Curie's law,

$$\langle \mathbf{\mu} \rangle = \frac{\beta^2 \mathbf{H}}{kT}. \tag{69}$$

It is easy to show that to take into account the anisotropic character of the magnetic tensor of the Cu^{++} ion, $\langle \mathbf{\mu} \rangle$ must be replaced by $\tfrac{1}{4}\beta^2(g^2\mathbf{H}/kT)$, where g^2 is a tensor, the square of the tensor g (32).

There are two non-equivalent Cu^{++} ions in the unit cell and thus, *a priori*, twenty non-equivalent protons, and for an arbitrary orientation of the applied field $\mathbf{H}_0$ there should be twenty different values of the electronic field $\langle H_e \rangle$ and twenty components to the proton nuclear resonance line. This number is reduced to ten by the existence of a centre of symmetry in the unit cell. These ten lines, with separations proportional to H_0 and inversely proportional to the temperature according to (69) (except for the very low temperatures below 2° K), and with a complicated but entirely calculable dependence on the orientation of the applied field, have been resolved at the temperatures of liquid helium. Each line should actually be a doublet owing to the dipolar interactions between the two protons of a water molecule, as will be explained in detail in Chapter VII. The line width prevented a complete resolution of these doublets being made (33).

Finally, it should be mentioned that the electron dipolar field $\langle \mathbf{H}_e \rangle$ at a given nucleus includes contributions from all the Cu^{++} ions of the sample and thus, as is well known, depends on the shape of the sample. This dependence can be calculated by classical magnetostatics assuming a density of magnetization $\mathbf{M} = N\langle \mathbf{\mu} \rangle$, where N is the number of unit cells per unit volume and $\langle \mathbf{\mu} \rangle$ the average magnetic moment per unit cell.

A second example (34) is provided by the proton resonance in $CuCl_2 \cdot 2H_2O$. The pattern of the nuclear resonance spectrum is simpler than in copper sulphate for there are only four non-equivalent protons in the unit cell, and therefore four doublets (the doublet structure corresponding as before to the interaction between two protons of a water molecule). If we neglect the anisotropy (of the order of 10 per cent) of the magnetic tensor g of the copper ion in $CuCl_2 \cdot 2H_2O$, the angular dependence of the paramagnetic shift for each proton is $(1 - 3\cos^2\theta)$, where θ is the angle between the applied field and the vector $\mathbf{r}$, joining the proton to the ion, in agreement with experiment.

A third example of nuclear resonance in a paramagnetic crystal is that of F^{19} in MnF_2 (35). The structural data relevant for our purpose can be summarized as follows: there are two types of fluorines in the unit cell and two resonance lines are observed. Each unit cell contains four fluorines, two of each type, and two Mn^{++} ions. Each fluorine has a bond of one type with two Mn^{++} (type I) and a bond of another type (type II) with one Mn^{++}.

As in the two previous examples, it is the strong exchange coupling between the electronic spins of the Mn^{++} ions that reduces the electronic magnetic field at the fluorines to its average value, causing a well-defined shift of the fluorine nuclear resonance line. The novel feature is the considerable relative magnitude of the resonance shift which when the field rotates in the (001) plane includes an isotropic part of 7·34 per cent (at 77° K) and an anisotropic dipolar part about ten times smaller. This result is in at least qualitative agreement with the observed electron resonance spectrum of Mn++ diluted in ZnF$_2$ (36), where extra lines were observed which were caused by a coupling of the Mn++ electron spin with the nuclear spins of F^{19}, of a magnitude consistent with that of the shift observed in the nuclear resonance of F^{19} in MnF$_2$. The explanation for such a large coupling must be sought in the fact that the bond Mn++—F$^-$ is not purely ionic and that as a consequence there is an appreciable density of unpaired spins on the site of each fluorine nucleus, giving rise to an isotropic contact coupling between the electron spins and the nuclear spins of fluorine. For detailed discussion of this problem see reference (36).

CuCl$_2$.2H$_2$O, and MnF$_2$ become antiferromagnetic at 4·3° K and 68° K respectively. The nuclear resonance phenomenon in the antiferromagnetic state will be considered in Section II D of this chapter.

The spin-lattice relaxation of nuclear spins in magnetic substances will be examined in Chapter IX.

(b) *Metals*

(1) *The frequency shift in metals*

The coupling of the conduction electrons in metals with the nuclear spins, although described by the same Hamiltonian as in non-metals, has some special features owing to the following peculiarities of the conduction electrons.

(α) The conduction electrons are not localized. A given conduction electron described by a wave function $\phi_{\mathbf{k}}(\mathbf{r}) = e^{i\mathbf{k}\cdot\mathbf{r}} U_{\mathbf{k}}(\mathbf{r})$, where $U_{\mathbf{k}}(\mathbf{r})$ is a periodic function of the lattice normalized in the volume V of the sample, has the same probability of being found in the neighbourhood of any nuclear spin of the metal. Conversely, each nuclear spin 'sees' simultaneously the magnetic fields produced by all the conduction electrons of the metal.

(β) For the electronic densities N that exist in metals, the conduction electrons behave as a degenerate Fermi gas even at room temperatures. In the presence of a magnetic field $\mathbf{H}$, all orbits are filled up by two electrons with opposite spins, except those near the top of the Fermi distribution. The excess $n = V(N_- - N_+)$ of electrons with spins antiparallel to the applied field is $n = \beta H g(E_F)$, where $g(E)\,dE$ is the number of electronic states in the interval E, $E+dE$ and $E_F = kT_F$ is the Fermi energy. In the free electron model

$$g(E_F) = \frac{3NV}{2E_F}.$$

The electronic spin magnetization is

$$M = \frac{\beta n}{V} = \beta^2 H g(E_F)/V$$

and the paramagnetic susceptibility per unit volume

$$\chi_p = \frac{M}{H} = \beta^2 \frac{g(E_F)}{V}. \tag{70}$$

For free electrons $\chi_p = 3N\beta^2/2kT_F$.

This value of χ_p can be contrasted with $\chi = N\beta^2/4kT$ for bound electrons. The susceptibility of conduction electrons is practically independent of the temperature and, since T_F is of the order of 10^4 to 10^5 °K, this susceptibility is much smaller than that of non-metallic paramagnetic solids at all temperatures. It may be worth while pointing out that the two features, non-localization of the orbits, and small, temperature-independent susceptibility, due to the pairing off of the spins, are not necessarily present simultaneously. Thus in semiconductors the conduction electrons have their orbits spread over the whole sample but, because of their small number in the conduction band, their distribution in energy among those orbits is a Boltzmann one and their spins are not paired off to the same extent as in metals.

The coupling of a *given* nuclear spin I with the electrons is obtained by summing the expectation values of the hyperfine couplings (32) of this spin with all the conduction electrons. Assume for simplicity that the electronic orbital momentum is completely quenched, a restriction that may easily be lifted if necessary. In the one-electron description of the conduction electrons, orbits with two electrons bring no contribution to the hyperfine coupling and this coupling can then be written

$$\gamma \hbar \mathbf{I} \cdot \sum_{\substack{\text{unfilled}\\\text{orbits}}} \left(\phi_k \left| 2\beta \left\{ \frac{3(\mathbf{r}_k \cdot \mathbf{s}_k)\mathbf{r}_k}{r_k^5} - \frac{\mathbf{s}_k}{r_k^3} + \frac{8\pi}{3} \mathbf{s}_k \delta(\mathbf{r}_k) \right\} \right| \phi_k \right) \tag{71}$$

where $\mathbf{r}_k$ and $\mathbf{s}_k$ are the position and the spin of the unpaired electron in the orbit ϕ_k. As mentioned in Section II C (a), the expression $(\phi_k| \quad |\phi_k)$ can be written $\mathscr{T}_k \cdot \mathbf{s}_k$ where $\mathscr{T}_k$ is a tensor with components that depend on the nature of the orbital ϕ_k, and (71) becomes

$$\gamma \hbar \mathbf{I} \cdot \sum_{\substack{\text{unfilled}\\\text{orbits}}} \mathscr{T}_k \cdot \mathbf{s}_k. \tag{72}$$

Assume first that $\mathscr{T}_k$ has approximately the same value for all the unfilled orbitals near the top of the Fermi distribution; the expression (72) becomes

$$\gamma \hbar \mathbf{I} \cdot \mathscr{T} \cdot \sum_{\substack{\text{unfilled}\\\text{orbits}}} \mathbf{s}_k = \gamma \hbar \mathbf{I} \cdot \mathscr{T} \cdot \mathbf{S}, \tag{73}$$

where **S** is the total electronic spin of the metallic sample. In the presence of an applied field H_0 we have

$$2\beta \mathbf{S} = -V\mathbf{M} = -V\chi_p \mathbf{H}_0,$$

and the coupling (73) can be rewritten

$$-V\frac{\gamma\hbar}{2\beta}\mathbf{I}\cdot\chi_p\mathscr{T}\cdot\mathbf{H}_0. \tag{74}$$

The nuclear spin **I** 'sees' an internal field $V\chi_p\mathscr{T}\cdot\mathbf{H}_0/2\beta$ superimposed on the applied field H_0 and causing a paramagnetic shift of the nuclear resonance (known in the literature as the Knight shift, after W. D. Knight who was the first to observe it in copper in 1949) proportional to the magnitude of the applied field H_0.

If the tensors $\mathscr{T}_k$ corresponding to the various orbits ϕ_k are not identical, the tensor $\mathscr{T}$ is an average of the tensors $\mathscr{T}_k$ over all orbits near the Fermi surface.

There is a fundamental difference between the Knight shift and a paramagnetic shift of the nuclear resonance in a non-metallic substance such as copper sulphate (besides the very different temperature dependence). There, each nuclear spin 'sees' the magnetic field of only one electron spin and it is permissible to replace this field by its *time* average because of the very frequent flips of that spin, caused by exchange or very short spin-lattice relaxation time. In a metal, where the conduction electrons are not localized, each nuclear spin 'sees' comparable magnetic fields from all the electrons of the sample at the same time, and the average is an *ensemble* average.

The one-electron description is inaccurate in so far as it disregards the spatial correlations between the electrons, as evidenced for instance by the grossly incorrect results it yields for the magnetic susceptibility of the conduction electrons. It may therefore be worth while showing in a purely formal way that the existence of a tensor coupling of the type (74), between the nuclear spin and the applied field H_0, is in fact independent of that description. Let $\rho(1,2,...,N)$ be the statistical operator describing the N conduction electrons in the presence of a magnetic field H_0, where indices such as 1, 2, etc., represent the degrees of freedom of the orbit and the spin for the electrons 1, 2, etc.

The coupling energy between these electrons and the nuclear spin I can be written

$$\mathrm{tr}\{\rho(1,2,...,N)[\mathscr{H}_1(1)+\mathscr{H}_1(2)+...+\mathscr{H}_1(N)]\}, \tag{75}$$

where $\mathscr{H}_1$ is the hyperfine Hamiltonian (32). Since this Hamiltonian is a one-electron operator, (75) can be rewritten

$$N \underset{\text{over (1)}}{\text{trace}} \{\rho_1(1)\mathscr{H}_1(1)\} \quad \text{where } \rho_1(1) \text{ is } \underset{\text{over } 2,3...,N}{\text{trace}} \{\rho_1(1,2,...,N)\}.$$

The assumption of a quenched orbital momentum and a field-independent susceptibility leads necessarily for the spin-dependent part of $\rho_1(1)$ to an expression of the form $\Phi(\mathbf{r}_1)(\mathbf{H}_0 . \mathbf{S}_1)$, where $\Phi(\mathbf{r}_1)$ is an orbital operator independent of H_0. Since the spin-dependent part of $\mathscr{H}_1$ can be written $\mathbf{I}.a(\mathbf{r}_1).\mathbf{S}_1$, where $a(\mathbf{r})$ is an orbital tensor operator, the electron nucleus coupling can be written

$$\underset{\text{over 1}}{\text{trace}} \{\Phi(\mathbf{r}_1)(\mathbf{H}_0 . \mathbf{S}_1)(\mathbf{I}.a(r_1).\mathbf{S}_1)\} = \mathbf{I}.\mathscr{D}.\mathbf{H}_0,$$

where the tensor $\mathscr{D}$ is

$$\mathscr{D} = \tfrac{1}{4} \underset{\text{over 1}}{\text{trace}} \{\Phi(\mathbf{r}_1)a(\mathbf{r}_1)\},$$

which is a relation of the type (74). The generalization for the incompletely quenched orbital momentum where χ_p also becomes a tensor is straightforward.

If the symmetry of the electronic environment of the nuclear spin I is no lower than cubic only the scalar parts of the tensors $\mathscr{D}$ and $\mathscr{T}$ are different from zero. In the one-electron picture this corresponds to the fact that only the s-part of the periodic function $U_k(\mathbf{r}) = e^{-i\mathbf{k}.\mathbf{r}}\phi_k(\mathbf{r})$ contributes to the shift. In that case the coupling (74) becomes

$$-V\gamma\hbar(\mathbf{I}.\mathbf{H})\chi_p \frac{8\pi}{3}\langle|\phi_k(0)|^2\rangle_F, \tag{76}$$

where the symbol $\langle \ \rangle_F$ means that an average is made over all orbits at the top of the Fermi distribution.

Instead of the wave function ϕ_k it is more convenient to introduce $\psi_k(\mathbf{r}) = \sqrt{(V/\Omega)}\phi_k(\mathbf{r})$; $W_k(\mathbf{r}) = \sqrt{(V/\Omega)}U_k(\mathbf{r})$ which are normalized in the atomic volume Ω. If we introduce the susceptibility per unit mass $\tilde{\chi}_p$, and the atomic mass M, (76) becomes

$$-\frac{8\pi}{3}\langle|\psi_k(0)|^2\rangle_F \tilde{\chi}_p M\gamma\hbar(\mathbf{I}.\mathbf{H}_0),$$

which, since the Zeeman nuclear energy in the external field is $-\gamma\hbar(\mathbf{I}.\mathbf{H}_0)$, corresponds to a relative positive frequency shift K (37):

$$K = \frac{\Delta\nu}{\nu_0} = \frac{\Delta H}{H_0} = \frac{8\pi}{3}\langle|\psi_k(0)|^2\rangle_F \tilde{\chi}_p M. \tag{77}$$

Since s-electrons have a much larger hyperfine structure than the other electrons, even for metals with non-cubic structure, the isotropic part of the Knight shift given by (77) is in general much larger than the anisotropic part. The general trend of the Knight shift (with some exceptions) is to increase with the increasing atomic number, going from $2 \cdot 5 \times 10^{-4}$ for Li^7 to $2 \cdot 5 \times 10^{-2}$ for Hg^{199}. A table of the known values of the Knight shift can be found in the reference (38).

When quoting a value of this shift, the compound with respect to which the Knight shift is computed should be specified because of the existence of the chemical shifts. Chemical shifts naturally exist in the metals themselves, including a contribution from the ion cores and another from the diamagnetism of the conduction electrons, but they are usually very much smaller than the Knight shift.

The comparison of the theoretical value (77) with the measured value of the shift can be made in several ways. Crude qualitative estimates can be made by calculating the paramagnetic susceptibility from the formula (70) for free electrons, and assuming that $\langle |\psi_k(0)|^2 \rangle_F$ is equal to the square of the atomic wave function $|\psi_A(0)|^2$ which can often be obtained from the atomic hyperfine structure.

More refined procedures consist in:

(i) measuring the paramagnetic susceptibility χ_p directly or calculating it by an improved theory;

(ii) calculating $\langle |\psi_k(0)|^2 \rangle$ directly in the metal. In the literature the quantities $\langle |\psi_k(0)|^2 \rangle_F$ and $|\psi_A(0)|^2$ are usually represented by the symbols P_F and P_A and the ratio P_F/P_A by the letter ξ.

The paramagnetic susceptibility has been measured directly in lithium and sodium (39), relating it by the Kramers–Krönig formulae to the area under the absorption curve for the spin resonance of the conduction electrons. (An absolute calibration was made by comparing this area with that of the nuclear resonance curve in the same sample at the same frequency.) The measured values

$$(\chi_p)_{Li} = (2 \cdot 08 \pm 0 \cdot 1) 10^{-6} \text{ at } 300° \text{ K}$$
and
$$(\chi_p)_{Na} = (0 \cdot 95 \pm 0 \cdot 1) 10^{-6} \text{ at } 77° \text{ K}$$

compare favourably with the theoretical values of $1 \cdot 87 \times 10^{-6}$ and $0 \cdot 85 \times 10^{-6}$ respectively, obtained by a theory that takes into account the correlations between electrons (40). This gives confidence in the predictions of this theory for the metals where no experimental values of χ_p are available. From the measured value of the shift, and the measured or theoretical value of χ_p, a value can be obtained for P_F (or $\xi = P_F/P_A$, if P_A is known) giving interesting information on the electronic structure of the metal. If we call ξ_K the value of P_F/P_A, obtained from (77), where K is measured, χ_p taken from experiment (39), and P_A from spectroscopic data, a comparison can be made with ξ_{theor} obtained from first principles. The latter has been computed for Li and Na (41) and the agreement with ξ_K is quite good:

$$(\xi_K)_{\text{Li}} = 0{\cdot}44, \qquad (\xi_{\text{theor}})_{\text{Li}} = 0{\cdot}49 \pm 0{\cdot}05,$$
$$(\xi_K)_{\text{Na}} = 0{\cdot}70, \qquad (\xi_{\text{theor}})_{\text{Na}} = 0{\cdot}80 \pm 0{\cdot}03.$$

Further details on these points can be found in reference (38).

The volume dependence of the Knight shift has also been studied and compared with the theory (42). Pressures up to 10,000 kg/cm² have been applied to alkali metals, resulting in a change in volume of the order of 7 per cent for Li and 25 per cent for Cs at 10,000 kg/cm². The changes in volume as a function of the pressure could be deduced from earlier measurements of Bridgman. The detailed tabulated results of the corresponding changes in K can be found in reference (42). At the largest pressures $(\Delta K/K)/(\Delta V/V)$ is of the order of:

Li	Na	Rb	Cs
+0·15	+0·1	−0·3	−1·7

Using theoretical expressions for χ_P and P_F as indicated above, their variation with the volume can be computed and the variation that results for K can be compared with experiment. The general agreement is fair (42).

Finally, after an analysis of the data on the variation of K at constant pressure as a function of temperature, the conclusion has been made (42) that this variation was not entirely due to the change in volume accompanying the change in temperature, but that an explicit dependence of K on the temperature existed as well. There is no doubt that an experimental study of the Knight shift in metals and alloys does bring important information on their electronic structure, but any further consideration of these points is outside the scope of this book.

If the symmetry of the environment of a nuclear spin is lower than cubic, the tensor $\mathscr{T}$ describing the electron-nucleus coupling in (73) or (74) is not a scalar and the Knight shift will depend on the direction of the applied field with respect to the crystalline axes. To the isotropic term $-\gamma \hbar K (\mathbf{I} . \mathbf{H})$ we must add an anisotropic one:

$$-\gamma \hbar \{ K'_X I_X H_X + K'_Y I_Y H_Y + K'_Z I_Z H_Z \},$$

where OX, OY, OZ are the principal axes of the tensor $\mathscr{T}$ and $K'_Z = -(K'_X + K'_Y)$. For instance, if the electronic wave function at the top of the Fermi distribution behaves near the nucleus as $\sqrt{s}\phi_s + \sqrt{(1-s)}\phi_{p\sigma}$, we shall have $K'_X = K'_Y = -\tfrac{1}{2}K'_Z$ and

$$K'_Z/K = \frac{1-s}{s}\frac{3}{10\pi}\left\langle\frac{1}{r^3}\right\rangle_p \Big/ |\phi_s(0)|^2, \qquad (78)$$

usually a small number, as pointed out previously. Since the anisotropic internal field, a vector with components $\begin{pmatrix} K'_X H_X \\ K'_Y H_Y \\ K'_Z H_Z \end{pmatrix}$, is much smaller than the applied field H_0, only its component parallel to the applied field is effective in shifting the resonance frequency. This component,

$$\frac{1}{H_0}\{K'_X H_X^2 + K'_Y H_Y^2 + K'_Z H_Z^2\},$$

can be written

$$H_0\{K'_Z \cos^2\theta + K'_X \sin^2\theta \cos^2\varphi + K'_Y \sin^2\theta \sin^2\varphi\}, \qquad (79)$$

where θ and φ specify the direction of H_0 with respect to the axes $OXYZ$. In particular, if

$$K'_X = K'_Y = K'_\perp = -\tfrac{1}{2}K'_Z = -\tfrac{1}{2}K'_\parallel,$$

(79) can be written $\tfrac{1}{2}H_0 K'_\parallel(3\cos^2\theta - 1)$, and the relative frequency shift becomes

$$\Delta H/H = K + \tfrac{1}{2}K'_\parallel(3\cos^2\theta - 1).$$

Since the nuclear resonance experiments in metals are performed on small particles, there is a random orientation of the crystalline axes inside the sample and the anisotropic shift results in a broadening proportional to the applied field. The resonance frequency (counted from its average value $\nu_0(1+K)$) is

$$\nu = \tfrac{1}{2}\Delta\nu'_\parallel(3\cos^2\theta - 1), \qquad (80)$$

with $\Delta\nu'_\parallel = K'_\parallel \nu_0$. Naturally, $K'_\parallel$ and thus $\Delta\nu'_\parallel$ can have either sign. Since all the values of $u = \cos\theta$ are equally probable if the orientation of the small particles is random, the fraction $g(\nu)d\nu$ of particles such that the frequency has a value between ν and $\nu+d\nu$, is proportional to $|du| = |du/d\nu|d\nu$ and the line shape $g(\nu)$ is

$$g(\nu) \sim \frac{1}{|d\nu/du|} = \frac{1}{3|\Delta\nu'_\parallel u|} \alpha \left(1 + \frac{2\nu}{\Delta\nu'_\parallel}\right)^{-\tfrac{1}{2}}, \qquad (81)$$

where, from (80), ν may have values between $\Delta\nu'_\parallel$ and $-\tfrac{1}{2}\Delta\nu'_\parallel$ outside of which $g(\nu)$ vanishes. The corresponding theoretical line shape is represented in Fig. VI, 3 for $\Delta\nu'_\parallel > 0$ (broken curve). The real curve, which naturally is never infinite since there are other causes of broadening, is given by $G(\nu) = \int f(\nu')g(\nu-\nu')\,d\nu'$, where $f(\nu)$ describes the shape of the resonance curve in the absence of anisotropic shift (full curve in Fig. VI, 3). The asymmetrical shape and the linear increase of the width of the line with the field are characteristic of anisotropic magnetic broadening.

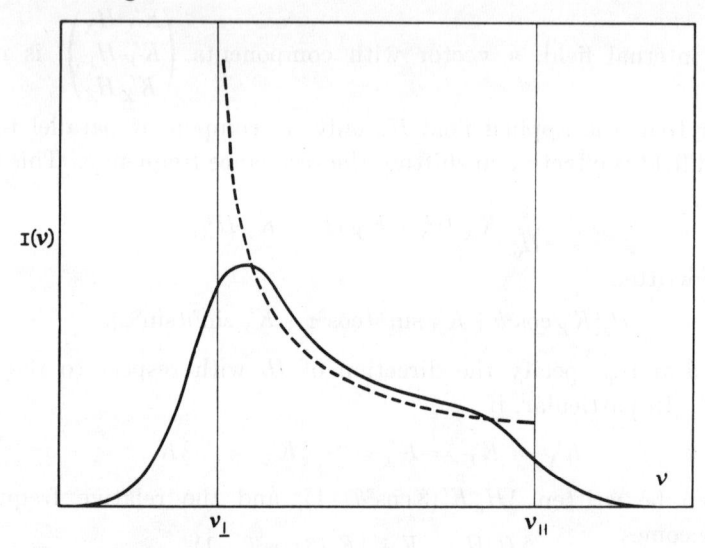

FIG. VI, 3. The theoretical line shape in a polycrystalline powder due to the anisotropy of the Knight shift in an axially symmetric conductor (broken curve). The full curve gives the experimental line shape, if dipolar interaction and other causes of symmetric line broadening are superimposed.

An analysis of the nuclear resonance data in white tin (43), which has a tetragonal structure, leads to $K = 75 \cdot 7 \times 10^{-4}$, $K' = 2 \cdot 3 \times 10^{-4}$. Similarly, in the nuclear resonance of thallium (44) it is found that in the oxide
$$K = 55 \times 10^{-4}, \qquad K' = 12 \times 10^{-4},$$
and in the metal
$$K = 156 \times 10^{-4}, \qquad K' = 16 \cdot 6 \times 10^{-4}.$$

The calculation of the line shape for a symmetry lower than tetragonal ($K'_X \neq K'_Y$) can be found in reference (43).

(2) *The indirect interactions in metals*

As in diamagnetic molecules, indirect interactions between nuclear spins operate through an admixture of the excited electronic states into the ground state but the actual calculation of their strength is sufficiently different to warrant a separate description (45). The method used is really that of molecular orbitals outlined previously, the molecular orbitals being the Bloch wave functions $\phi_{\mathbf{k}}(\mathbf{r}) = e^{i\mathbf{k}\cdot\mathbf{r}}U_{\mathbf{k}}(\mathbf{r})$, spread equally over all the atoms of the sample and normalized over its volume. The ground electronic state is, in our approximation, a Slater determinant constructed from these one-electron orbits each one filled with two electrons, up to the Fermi energy E_F.

We shall not use the closure approximation, for a simple description of an excited electronic state is provided by the promotion of an electron from an energy state $E_{\mathbf{k}} < E_F$ to a state $E_{\mathbf{k}'} > E_F$. The contribution from the electrons in the inner shells is neglected, since their excitation requires too much energy. Excited states where more than one electron is promoted need not be considered, for they are not coupled to the ground state by the one-electron operators O_1, S_1, S_2 that describe the coupling of the nuclear spins with the electrons. Of those, the contact term S_2 brings by far the largest contribution, and terms quadratic in S_2 will be considered first. Next come cross terms between S_2 on the one hand and S_1 and O_1 on the other, but it is easy to show by an argument used several times already that the cross terms (S, O) vanish if the orbital momentum is quenched.

The contribution of S_2 to the indirect interaction between two nuclear spins $\mathbf{I}_N$ and $\mathbf{I}_{N'}$ having positions $\mathbf{R}_N$ and $\mathbf{R}_{N'}$ can be written

$$\left(\frac{16\pi\beta\hbar}{3}\right)^2 \gamma_N \gamma_{N'} {\sum_{\mathbf{k}\mathbf{k}'}}' \frac{(\mathbf{k}|\delta(\mathbf{r}_N)(\mathbf{S}\cdot\mathbf{I}_N)|\mathbf{k}')(\mathbf{k}'|\delta(\mathbf{r}_{N'})(\mathbf{S}\cdot\mathbf{I}_{N'})|\mathbf{k})}{E_{\mathbf{k}}-E_{\mathbf{k}'}} + \text{c.c.}, \tag{82}$$

where $|\mathbf{k}\rangle$ is the wave function $e^{i\mathbf{k}\cdot\mathbf{r}}U_{\mathbf{k}}(\mathbf{r})\chi$, χ being a function of the spin, and $\mathbf{r}_N = \mathbf{r} - \mathbf{R}_N$, $\mathbf{r}_{N'} = \mathbf{r} - \mathbf{R}_{N'}$. The summation is made over the filled orbits for the index $\mathbf{k}$ and over the empty ones for $\mathbf{k}'$. If we define the operator

$$C_{\mathbf{k}} = {\sum_{\mathbf{k}'}}' \frac{|\mathbf{k}')(\mathbf{k}'|}{E_{\mathbf{k}}-E_{\mathbf{k}'}}, \tag{82'}$$

which is purely orbital if spin forces are neglected, the summation (82) can be written

$$\sum_{\mathbf{k}} (\mathbf{k}|\delta(\mathbf{r}_N)(\mathbf{S}\cdot\mathbf{I}_N)C_{\mathbf{k}}(\mathbf{S}\cdot\mathbf{I}_{N'})\delta(\mathbf{r}_{N'})|\mathbf{k}). \tag{83}$$

If $C_{\mathbf{k}}$ is purely orbital the summation over the electron spin states can be made separately, giving

$$\underset{\substack{\text{over electron}\\ \text{spins}}}{\text{trace}} \{(\mathbf{S}\cdot\mathbf{I}_N)(\mathbf{S}\cdot\mathbf{I}_{N'})\} = \tfrac{1}{2}(\mathbf{I}_N\cdot\mathbf{I}_{N'})$$

and demonstrating that the coupling (82) between $\mathbf{I}_N$ and $\mathbf{I}_{N'}$ is purely scalar. To proceed further the following approximations are made: the energy E_k is given by $E_k = \hbar^2 k^2/2m^*$ (m^* is the effective mass) and the number of orbital states $Z(\mathbf{k})\,d^3k$ in the space of the wave vectors is $V d^3k/(2\pi)^3$ as for the free electrons. The excited states $E_{k'}$ are supposed to extend from $E_F = E_{k_0} = \hbar^2 k_0^2/2m^*$ to infinity. Finally, it is remarked that the main contribution to (82) comes from values of $\mathbf{k}$ and $\mathbf{k}'$ such that $|k^2-k'^2|$ is very small, $|\mathbf{k}| \simeq |\mathbf{k}'| \simeq |\mathbf{k}_0|$. The slowly varying function

$$|\Delta_{kk'}|^2 = (U_\mathbf{k}\,|\,\delta(\mathbf{r}_N)\,|\,U_{\mathbf{k}'})(U_{\mathbf{k}'}\,|\,\delta(\mathbf{r}_{N'})\,|\,U_\mathbf{k}) = |U_\mathbf{k}(0)|^2 |U_{\mathbf{k}'}(0)|^2 \quad (84)$$

is replaced by its average value on the Fermi surface

$$|\Delta_{kk'}|^2 \simeq \frac{\xi^2 \Omega^2}{V^2} |\psi_A(0)|^4,$$

where Ω is the atomic volume, $\psi_A(\mathbf{r})$ the wave function for the free atom, and ξ has been defined as P_F/P_A in the previous section. Introducing

$$a_N = \frac{16\pi}{3}\beta\,\hbar\gamma_N |\psi_A(0)|^2,$$

the hyperfine coupling constant for the free atom, the scalar coupling coefficient $\hbar J_{NN'}$ becomes

$$\hbar J_{NN'} = \tfrac{1}{2} a_N\, a_{N'} \frac{\xi^2 \Omega^2}{(2\pi)^6} \int_0^{k_0} d^3k \int_{k_0}^{\infty} \frac{d^3k'\, e^{i(\mathbf{k}'-\mathbf{k}).\mathbf{R}_{NN'}}}{(\hbar^2/2m^*)(k^2-k'^2)} + \text{c.c.}, \quad (85)$$

which with a little algebra gives

$$\hbar J_{NN'} = \frac{m^* a_N\, a_{N'} \xi^2 \Omega^2}{4(2\pi)^3 \hbar^2 R_{NN'}^4}\{2k_0\,R_{NN'}\cos(2k_0\,R_{NN'}) - \sin(2k_0\,R_{NN'})\}. \quad (86)$$

To appreciate the order of magnitude of (86) it should be noted that for nearest neighbours $R_{NN'}^3$ is of the order of Ω. Because the factor ξ is of the order of unity we obtain

$$\hbar J_{NN'} \sim \frac{a_N a_{N'}}{4E_F}\left(\frac{k_0\, R_{NN'}}{2\pi}\right)^3.$$

This is of the same order of magnitude as was found previously in molecules with the Fermi energy E_F replacing the excitation energy ΔE, apart from a dimensionless factor $(k_0 R_{NN'}/2\pi)^3 = (R_{NN'}/\lambda_0)^3$, where λ_0 is the wavelength of an electron with the Fermi energy. This factor is of order unity in most metals, a fact which expresses the possibility of an electronic wave packet with an energy no larger than the Fermi energy being localized in the neighbourhood of a given atom.

Another difference from indirect coupling in molecules is the long range and the variable sign of the interaction (86).

The calculation of the cross term between S_2 and $S_1 = \mathbf{S}.a'(\mathbf{r}).\mathbf{I}$, where $a'(\mathbf{r})$ is a traceless tensor, is far more difficult and less reliable than that of the isotropic coupling given above (44) and we shall only discuss it briefly.

This term is proportional to $\mathscr{A}_{NN'} + \mathscr{A}_{N'N} + $ c.c. where

$$\mathscr{A}_{NN'} = \sum_{\mathbf{k},\mathbf{k'}}{}' \frac{(\mathbf{k}|\delta(\mathbf{r}_N)(\mathbf{S}.\mathbf{I}_N)|\mathbf{k'})(\mathbf{k'}|\mathbf{S}.a'(\mathbf{r}_{N'}).\mathbf{I}_{N'}|\mathbf{k})}{E_\mathbf{k}-E_\mathbf{k'}}.$$

Summing over the electronic spins as previously, we obtain a traceless tensor coupling $\hbar \mathbf{I}_N . \mathscr{I}'_{NN'} . \mathbf{I}_{N'}$ with the tensor $\mathscr{I}'_{NN'}$ proportional to $\mathscr{T}_{NN'} + \mathscr{T}_{N'N} + $ c.c., where

$$\mathscr{T}_{NN'} = \sum_{\mathbf{k},\mathbf{k'}}{}' \frac{(\phi_\mathbf{k}|\delta(\mathbf{r}_N)|\phi_\mathbf{k'})(\phi_\mathbf{k'}|a'(\mathbf{r}_{N'})|\phi_\mathbf{k})}{E_\mathbf{k}-E_\mathbf{k'}}.$$

There is no general group theoretical justification to the assertion sometimes made that $\mathscr{I}'_{NN'}$ is necessarily a pseudo-dipolar coupling of the form

$$\hbar \mathbf{I}_N . \mathscr{I}'_{NN'} . \mathbf{I}_{N'} = B_{NN'}\left\{\frac{3(\mathbf{I}_N.\mathbf{R}_{NN'})(\mathbf{I}_{N'}.\mathbf{R}_{NN'})}{R^2_{NN'}} - (\mathbf{I}_N.\mathbf{I}_{N'})\right\} \quad (87)$$

although it may reduce to such a form in special cases, as for instance, as shown in reference (44), in the so-called spheric approximation. This approximation assumes that the function $U_\mathbf{k}(\mathbf{r})$ is invariant under rotation around an axis parallel to the vector $\mathbf{k}$ and that the energy $E_\mathbf{k}$ is independent of the direction of the vector $\mathbf{k}$.

The ratio $\mathscr{I}'_{NN'}/J_{NN'}$ can be expected to be of the order of

$$(U_{k_0}|S_1|U_{k_0})/(U_{k_0}|S_2|U_{k_0}),$$

a number usually much smaller than unity, which justifies the neglect of the term quadratic in S_1.

Finally, it might be mentioned that the method used to calculate the indirect interactions in metals can be extended to insulators. The model used for the insulator is that of a full valence band of width $\hbar^2 k_0^2/2m^*$, separated from an empty conduction band by a gap of width $\hbar^2 K^2/2m^*$. For the scalar interaction a calculation similar at every stage to that performed for metals can be made and, if it is assumed for simplicity that $K^2 \gg k_0^2$, it is found (44) that

$$\hbar J_{NN'} = \frac{1}{4\pi^3} \frac{m^*}{\hbar^2} a_N a_{N'} \frac{\xi^2 \Omega^2}{R^4_{NN'}} \{(k_0 R_{NN'})\cos(k_0 R_{NN'}) - \sin(k_0 R_{NN'})\} e^{-KR_{NN'}}. \quad (88)$$

The main difference between the interaction (88) and that given in metals by the equation (86) is the short range of the coupling in the insulator caused by the exponential factor $\exp\{-KR_{NN'}\}$. This is physically plausible since in an insulator the electrons are practically localized on a given atom and thus have little probability of coupling two nuclear spins belonging to remote atoms.

A comparison of some of the results of this section with experiment will be made in Chapter X.

D. Nuclear resonance in antiferromagnetic and ferromagnetic substances

Nuclear magnetic resonance has been observed in substances where the electron spins are in an antiferromagnetic or a ferromagnetic state. A thorough discussion of this phenomenon would entail a discussion of electron ferromagnetism and antiferromagnetism which is outside the scope of this book, and only a greatly simplified description of the principle and the results of the experiments will be given.

An unsophisticated picture of an antiferromagnetic, adequate for a qualitative understanding of the nuclear resonance experiments to be outlined in this section, and valid approximately in low magnetic fields and well below the antiferromagnetic transition point (Néel temperature), is that of two spin sublattices R_1 and R_2, the electron spins of each sublattice being aligned parallel to each other and antiparallel to those of the other sublattice. Their common direction is that of a preferred direction in the crystal (crystalline axis). Under these conditions a nuclear spin in the crystal will 'see' besides the applied field, the large electronic fields H_e existing inside the sample.

The first difference from the paramagnetic case is that in the antiferromagnetic state the electron spins do not reorient themselves rapidly any more. It is then the instantaneous value of the electron field H_e, which is constant in time, rather than its time average $\langle H_e \rangle$, which is much smaller than H_e in the paramagnetic state, that determines the shift of the nuclear Larmor frequency.

The second difference is that two nuclear spins which have equivalent positions in the crystal do not necessarily have identical magnetic properties in the antiferromagnetic state since, owing to the existence of two electron spin sublattices, the translational symmetry of the system is lower than that of the crystal, which exists in the paramagnetic state.

While the first of the differences listed would occur between a paramagnetic and a ferromagnetic state as well, the second is characteristic of an antiferromagnetic arrangement.

A nuclear spin will in general be appreciably nearer to the electron spins of one sublattice, say R_1, than to the spins of R_2, and its frequency shift will be essentially proportional to the electron magnetization M_1 of R_1. This has to be contrasted with conventional electronic susceptibility measurements where only the algebraic sum M_1+M_2 of the magnetizations of the two sublattices, much smaller than either M_1 or M_2, can be obtained.

Nuclear resonance in the antiferromagnetic state was first observed for the water protons of $CuCl_2.2H_2O$ at the temperature of liquid helium (34).

In a weak magnetic field H_0 the electronic spins of either sublattice are approximately oriented along the a-axis of the crystal, regardless of the direction of H_0. The component H'_e of the local electronic field along the applied field, the only one that matters if $H_0 \gg H'_e$, produced by a given magnetic ion Cu at a given proton P will be

$$H'_e = \pm\mu_e(3\cos\theta\cos\alpha - \cos\psi)r^{-3}, \qquad (89)$$

where θ and ψ are the angles of the applied field with the Cu–P vector and the crystalline a-axis along which the electron spins are aligned respectively, and α is the angle between the Cu–P vector and the a-axis. The formula (89) has to be contrasted with the angular dependence of the electronic field in the paramagnetic case:

$$(H'_e)_{\text{param}} \propto (1 - 3\cos^2\theta)r^{-3}.$$

As already stated, two protons that occupy equivalent positions in the unit cell and thus 'see' the same local electronic field in the paramagnetic state, may 'see' opposite local fields in the antiferromagnetic state and have opposite frequency shifts. There should thus be twice as many resonance frequencies as in the paramagnetic state, forming a pattern that is symmetrical with respect to the Larmor frequency of the free proton.

When the magnetic field is rotated with respect to the crystal in the ab plane, the variation of the Larmor frequency shift H'_e/γ_P with the angle through which the field is rotated, should have, according to (89), a period of 360°, rather than of 180° as in the paramagnetic state.

It is these two very striking features of the proton resonance that led to the discovery that $CuCl_2.2H_2O$ became antiferromagnetic at 4·3° K.

Later work performed on this crystal by the nuclear-resonance method gave a considerable amount of information on the dependence of its antiferromagnetic properties on the temperature, the magnitude, and the orientation of the applied field, etc. (46).

A second example of nuclear resonance in an antiferromagnetic substance is that of F^{19} in MnF_2, which is antiferromagnetic below 68° K (47).

It was found that the nuclear resonance signal observed in paramagnetic MnF_2 above the Néel temperature of 68° K, disappeared at that temperature without broadening or shifting its position. This disappearance was explained by a rapid increase at the transition point of the expectation value $\langle S \rangle$ of the electron spin and therefore also of the electronic local field at the fluorine nucleus. This displaced the nuclear resonance frequency outside the range of the nuclear spectrometer. As has already been mentioned in Section II C, in MnF_2 the coupling between the electron spins and the spin of F^{19} is strong because of a finite density of unpaired electrons at the fluorine nucleus.

In contrast to $CuCl_2 \cdot 2H_2O$, in the antiferromagnetic state, the local field is much larger, rather than weaker, than the applied field.

Each nuclear spin of F^{19} is coupled to the electronic spins of two Mn^{++} ions by a tensor coupling $\mathbf{I} \cdot A^I \cdot \mathbf{S}$ and to the electronic spin of another Mn^{++} ion by a different coupling $\mathbf{I} \cdot A^{II} \cdot \mathbf{S}$. The components of the tensors A^I and A^{II} can be determined from an analysis of the nuclear frequency shifts in the paramagnetic state and it is possible to predict approximately the frequency of the nuclear transition in the antiferromagnetic state.

For an applied field H_0 parallel to the [001] direction of the crystal, chosen as z-axis, the paramagnetic nuclear frequency shift is

$$\delta\omega = (2A_z^I + A_z^{II})\langle S_z \rangle, \quad \text{where} \quad \langle S_z \rangle = \frac{\gamma_e \hbar S(S+1)}{3kT} H_0, \quad (90)$$

from which A_z^I and A_z^{II} can be determined.

In the antiferromagnetic state, $\langle S_z \rangle$ becomes approximately $(\pm \tfrac{5}{2})$ depending on the sublattice to which the Mn^{++} ion belongs. For each fluorine the two Mn^{++} ions with which it has the coupling A^I belong to the same sublattice and the ion Mn^{++} coupled to it by A^{II} belongs to the opposite sublattice.

The privileged direction for the alignment of the electron spins in the antiferromagnetic state is the axis [001] and the nuclear resonance frequency, for an applied field parallel to [001], becomes

$$\omega_{\pm} \simeq \pm[(2A_z^I - A_z^{II})\tfrac{5}{2} \pm \gamma_n H_0]. \quad (91)$$

Two resonance lines are, according to (91), observed in the antiferromagnetic state at the frequency $|\omega_\pm|$. The field-independent part of (91) is, from the values of A_z^I and A_z^II extracted from (90), of the order of $2\pi \times 177$ Mc/s. The experimental value at the lowest temperature of $1\cdot3°$ K is of the order of $2\pi \times 160$ Mc/s. In view of the crudeness of the model used for the antiferromagnetic state, the agreement is gratifying.

Nuclear resonance has also been observed in antiferromagnetic CoF_2 for the nuclear spin of Co^{59} (**48**).

This experiment differs from those performed in $CuCl_2$ and MnF_2 in so far as the nuclear resonance is observed for *the* nucleus of the magnetic ion itself. The spin of Co^{59} is $\frac{7}{2}$ and the field it 'sees' is the sum of a large electronic internal field and of the applied H_0.

As for F^{19} in MnF_2, one would expect two resonance lines, but for the fact that there is a quadrupole coupling between the spin of Co^{59} and the electronic shells of the ion, and each of these lines is split into $2I = 7$ components. Quantitatively, there are appreciable discrepancies between the observed positions of the 14 lines and the predictions of the simplified model of an antiferromagnetic, which is not too surprising in view of the crudeness of the latter.

Nuclear resonance has also been observed in the ferromagnetic state (**49, 50, 51**).

A very strong resonance of Co^{59} (spin $I = \frac{7}{2}$) in face-centred cubic metallic cobalt occurs at 213 Mc/s in zero applied field (**49**) at $300°$ K. The electronic field at the site of the cobalt nucleus corresponding to that frequency is 213·400 gauss, in good agreement with measurements of specific heat in bulk hexagonal cobalt between $0\cdot3°$ K and $1°$ K.

The resonance of Fe^{57} (spin $I = \frac{1}{2}$) has been observed in metallic iron in an isotopically enriched sample (78 per cent of Fe^{57}) (**50**) and in natural iron (2·25 per cent of Fe^{57}) (**51**).

The search for the resonance of Fe^{57} has been greatly facilitated by the approximate knowledge of the resonance frequency obtained from a study of the recoilless emission and absorption of γ-rays linking the ground state and the first excited state of Fe^{57} (Mössbauer effect) (**52**).

At $295°$ K the resonance frequency in iron is 45·49 Mc/s and the local field 330·000 gauss.

A very remarkable feature of the resonances observed in cobalt and iron is the magnitude of the signal. Considering the width of the line (~ 400 kc/s in cobalt, ~ 50 kc/s in iron) and the relative smallness of the r.f. fields applied ($\ll 1$ gauss) the size of the signal can only be explained by assuming that the r.f. field actually 'seen' by the nuclear spins is several orders of magnitude larger than the one applied.

The discussion of this remarkable phenomenon as well as of other features of the resonance such as the origin of the line width, the dependence of the frequency, the line width and the relaxation time, on the temperature and on the value of the applied field, delve too deeply into the nature and properties of electronic ferromagnetism to be undertaken here.

There seems to be little doubt that nuclear resonance can be a powerful tool for the study of electronic ferromagnetism and antiferromagnetism.

REFERENCES

1. J. H. SMITH, E. M. PURCELL, and N. F. RAMSEY, *Phys. Rev.* **108**, 120, 1957.
2. A. R. EDMONDS, *Angular Momentum in Quantum Mechanics*, Princeton University Press, 1957.
3. M. H. COHEN and F. REIF, *Solid State Physics*, **5**, Academic Press, New York, 1957.
4. T. P. DAS and E. L. HAHN, *Solid State Physics*, suppl. I, Academic Press, New York, 1958.
5. —— and R. BERSOHN, *Phys. Rev.* **102**, 733, 1956.
6. R. M. STERNHEIMER, ibid. **105**, 158, 1957.
7. V. JACCARINO, J. G. KING, R. A. SATTEN, and H. H. STROKE, ibid. **94**, 1798, 1954.
8. P. KUSCH and T. G. ECK, ibid., p. 1799, 1954.
9. J. H. VAN VLECK, *Theory of Electric and Magnetic Susceptibilities*, Clarendon Press, Oxford, 1932.
10. N. F. RAMSEY, *Phys. Rev.* **78**, 699, 1950.
11. A. SAIKA and C. P. SLICHTER, *J. Chem. Phys.* **22**, 26, 1954.
12. J. A. POPLE, *Proc. Roy. Soc.* A, **239**, 541, 550, 1957.
13. H. M. MCCONNELL, *J. Chem. Phys.* **27**, 226, 1957.
14. W. G. SCHNEIDER, H. J. BERNSTEIN, and J. A. POPLE, ibid. **28**, 601, 1958.
15. W. G. PROCTOR and F. C. YU, *Phys. Rev.* **81**, 20, 1951.
16. R. FREEMAN, G. R. MURRAY, and R. E. RICHARDS, *Proc. Roy. Soc.* A, **242**, 455, 1957.
17. J. H. VAN VLECK, *Rev. Mod. Phys.* **23**, 213, 1951.
18. T. P. DAS and R. BERSOHN, *Phys. Rev.* **104**, 849, 1956.
19. N. F. RAMSEY and E. M. PURCELL, ibid. **85**, 143, 1952.
20. —— ibid. **91**, 303, 1953.
21. H. S. GUTOWSKY, D. W. MCCALL, and C. P. SLICHTER, *J. Chem. Phys.* **21**, 279, 1953.
22. H. M. MCCONNELL, ibid. **24**, 460, 1956.
23. J. A. POPLE, *Molecular Physics*, **1**, 216, 1958.
24. H. Y. CARR and E. M. PURCELL, *Phys. Rev.* **88**, 415, 1952.
25. B. BLEANEY and K. D. BOWERS, *Proc. Roy. Soc.* A, **214**, 451, 1952.
26. T. ITOH and K. KAMBE, *J. Phys. Soc. Japan*, **12**, 763, 1957.

27. S. Kojima, S. Ogawa, S. Hagiwara, Y. Abe, and B. Minematsu, *J. Phys. Soc. Japan*, **11**, 964, 1956.
28. H. A. Jahn, *Proc. Roy. Soc.* A, **164**, 117, 1937.
29. H. A. Kramers, *Proc. Acad. Sci. Amst.* **33**, 959, 1930.
30. A. Abragam and M. H. L. Pryce, *Proc. Roy. Soc.* A, **205**, 135, 1951.
31. R. G. Shulman, *Phys. Rev. Letters*, **2**, 459, 1959.
32. N. Bloembergen, *Physica*, **16**, 95, 1950.
33. N. J. Poulis, ibid. **17**, 392, 1951.
34. —— and G. E. C. Hardeman, ibid. **18**, 201, 1952.
35. R. G. Shulman and V. Jaccarino, *Phys. Rev.* **108**, 1219, 1957.
36. M. Tinkham, *Proc. Roy. Soc.* A, **236**, 535, 1956.
37. C. H. Townes, C. Herring, and W. D. Knight, *Phys. Rev.* **77**, 852, 1950.
38. W. D. Knight, *Solid State Physics*, **2**, 93, 1956.
39. R. T. Schumacher and C. P. Slichter, *Phys. Rev.* **101**, 58, 1956.
40. D. Pines, ibid. **95**, 1090, 1954.
41. W. Kohn, ibid. **96**, 590, 1954.
    T. Kjeldaas Jr. and W. Kohn, ibid. **101**, 66, 1956.
42. G. B. Benedek and T. Kushida, *J. Phys. Chem. Solids*, **5**, 241, 1958.
43. N. Bloembergen and T. J. Rowland, *Acta Metallurgica*, **1**, 731, 1953.
44. —— *Phys. Rev.* **97**, 1679, 1955.
45. M. A. Ruderman and C. Kittel, ibid. **96**, 99, 1954.
46. N. J. Poulis and G. E. C. Hardeman, *Physica*, **19**, 391, 1953.
47. V. Jaccarino and R. G. Shulman, *Phys. Rev.* **107**, 1196, 1957.
48. —— *Phys. Rev. Letters*, **2**, 163, 1959.
49. A. C. Gossard and A. M. Portis, ibid. **3**, 164, 1959.
50. A. C. Gossard, A. M. Portis, and W. J. Sandle, *J. Phys. Chem. Solids*, 1961.
51. C. Robert and J. M. Winter, *C.R. Acad. Sci.* **250**, 383, 1960.
52. S. S. Hanna, J. Heberle, C. Littlejohn, G. J. Perlow, R. S. Preston, and D. H. Vincent, *Phys. Rev. Letters*, **4**, 177, 1960.

VII

FINE STRUCTURE OF RESONANCE LINES—QUADRUPOLE EFFECTS

Le homard demande à être coupé vivant
BRILLAT-SAVARIN

In Chapter IV we gave a description of the shape of a single Zeeman resonance line in a solid, broadened by dipolar couplings between the spins. In the present chapter we consider spectra that, because of particularly strong dipolar couplings or because of quadrupole couplings, exhibit a structure. Under certain conditions this structure may disappear and a single line be observed, as in imperfect cubic crystals or in so-called 'pure quadrupole' resonance for spins smaller than $I = 2$.

We shall suppose in this chapter that atomic and molecular motions caused by thermal excitations can be neglected. The effects of such motions on the form of the spectra will be discussed later.

I. Fine Structure caused by Dipolar Coupling

It sometimes happens that nuclear spins in solids form small groups within which the spin separations are distinctly smaller than those between two neighbouring groups. Since the dipole-dipole interaction decreases rapidly with distance, to a first approximation one may consider such a group as an isolated system and calculate its energy levels in the presence of an applied field H_0. The resonance line of such a system exhibits a fine structure which becomes increasingly complicated as the number of spins making up the group is increased. This spectrum of discrete lines is broadened by the influence of the spins belonging to neighbouring groups, but nevertheless a fine structure may remain and its study may provide more complete information than does the simple consideration of the second moment of the resonance line.

We shall examine some examples in order to illustrate the way in which this information is obtained.

A. Rigid lattice

(a) *Two identical spins (two protons)* $\mathbf{I}^1$ *and* $\mathbf{I}^2$

The Hamiltonian of such a pair is written as

$$\mathcal{H} = -\gamma\hbar H_0(I_z^1+I_z^2)+\frac{\gamma^2\hbar^2}{r^3}[\mathbf{I}^1.\mathbf{I}^2-3(\mathbf{I}^1.\mathbf{n})(\mathbf{I}^2.\mathbf{n})]. \tag{1}$$

In (1) $r = |\mathbf{r}_{12}|$ and $\mathbf{n} = \mathbf{r}_{12}/r$.

We consider only the phenomena observed in strong fields where the first, or Zeeman, term of (1) is much more important than the dipole interaction. We can then treat the last term as a perturbation and keep only that part $A+B$ defined by the equation (18) in Chapter IV. We can then replace (1) by the approximate Hamiltonian

$$\mathscr{H} = -\gamma\hbar H_0(I_z^1+I_z^2)+\frac{\gamma^2\hbar^2}{r^3}[\tfrac{3}{2}(1-3\cos^2\theta)I_z^1 I_z^2-\tfrac{1}{2}(1-3\cos^2\theta)\mathbf{I}^1.\mathbf{I}^2]. \tag{2}$$

The eigenstates of the Zeeman Hamiltonian $-\gamma\hbar H_0(I_z^1+I_z^2)$ are the four states $|++\rangle$, $|+-\rangle$, $|-+\rangle$, $|--\rangle$, where the symbols $|..\rangle$ have an obvious meaning. It is more convenient to take instead their linear combinations: the three triplet states

$$|1\rangle = |++\rangle, \qquad |0\rangle = \frac{1}{\sqrt{2}}[|+-\rangle+|-+\rangle], \qquad |-1\rangle = |--\rangle$$

that have the Zeeman energies $-\gamma\hbar H_0$, 0, $+\gamma\hbar H_0$, and the singlet state $(1/\sqrt{2})[|+-\rangle-|-+\rangle]$ with Zeeman energy zero.

As may readily be verified, the last state, being coupled to the triplet states by neither the field nor the dipole interaction, may be completely ignored.

The three triplet states then have the energies

$$E_{-1} = \langle -1|\mathscr{H}|-1\rangle = \quad \gamma\hbar H_0+\frac{\gamma^2\hbar^2}{4r^3}(1-3\cos^2\theta),$$

$$E_0 = \langle 0|\mathscr{H}|0\rangle = \quad -\frac{\gamma^2\hbar^2}{2r^3}(1-3\cos^2\theta),$$

$$E_1 = \langle 1|\mathscr{H}|1\rangle = -\gamma\hbar H_0+\frac{\gamma^2\hbar^2}{4r^3}(1-3\cos^2\theta). \tag{3}$$

There are two resonant frequencies ω' and ω'', given by

$$\hbar\omega' = E_{-1}-E_0 = \gamma\hbar H_0+\frac{3}{4}\frac{\gamma^2\hbar^2}{r^3}(1-3\cos^2\theta),$$

$$\hbar\omega'' = E_0-E_1 = \gamma\hbar H_0-\frac{3}{4}\frac{\gamma^2\hbar^2}{r^3}(1-3\cos^2\theta). \tag{4}$$

If we operate at a fixed frequency ω and vary the applied field H_0, resonances appear at the following field values:

$$H_0 = H^*\pm\alpha(3\cos^2\theta-1), \tag{5}$$

where
$$H^* = \frac{\omega}{\gamma} = \frac{2\pi\nu}{\gamma}, \qquad \alpha = \frac{3}{4}\frac{\gamma\hbar}{r^3}.$$

If the sample under study is a single crystal, the angle which the magnetic field makes with the vector $\mathbf{r}_{12}$ has a well-defined value and the preceding results can be applied directly to the determination of the length and orientation of the vector which joins the two protons.

This has been done (1) for the protons in gypsum. The crystalline structure of gypsum has been determined from X-ray data, but these data do not give the position of the protons.

In the experiment described a single crystal was cut in the shape of a cylinder whose axis was perpendicular to the (001) plane of the crystal; the cylinder could be rotated about this axis. The magnetic field H_0, perpendicular to the axis of rotation, could thus assume all orientations in the plane (001). The angle θ that the line PP makes with H_0 is given by

$$\cos\theta = \cos\delta \cos(\Phi - \Phi_0), \qquad (6)$$

where Φ_0 and δ are the angles of PP with the axis [100] and the plane (001). Observation of the separation ΔH of the two resonance lines given by (5) as a function of the angle θ,

$$\Delta H = 2\alpha(3\cos^2\theta - 1) = \frac{3\gamma\hbar}{2r^3}(3\cos^2\theta - 1), \qquad (7)$$

permits the determination of r, δ, and Φ_0 by (6) and (7). For gypsum there are two possible directions for the vector PP in the unit cell. They are symmetric with respect to the plane (010). The formula equivalent to (6) for the second pair is

$$\cos\theta = \cos\delta \cos(\Phi + \Phi_0). \qquad (6')$$

Fig. VII, 1 shows the dependence of ΔH on Φ for the two pairs of lines observed. The theoretical points were obtained by substituting $2\alpha = 10\cdot 8$ gauss, that is $r = 1\cdot 58$ Å, in (7) and (6') or (6).

Polycrystalline sample (powdered). Each crystal in the powder gives rise to two lines whose separation $h = H_0 - H^*$ with respect to the central value is given by

$$h = \epsilon\alpha(3u^2 - 1) \quad \text{with } \epsilon = \pm 1,\ u = \cos\theta. \qquad (8)$$

(In fact each crystal gives rise to two doublets corresponding to two kinds of water molecules in the unit cell, but the two kinds obviously give the same spectrum in a powder and it is enough to consider only one.)

As the crystals are oriented at random, the different values of $u = \cos\theta$ are equally probable, and the superposition of the lines arising from the individual crystals gives a continuous spectrum of density defined by $f(h) \sim |du/dh|$ or, by (8),

$$f(h) \sim \frac{1}{|u|} \sim \left(\frac{\epsilon h}{\alpha}+1\right)^{-\frac{1}{2}}.$$

The line $|-1\rangle \to |0\rangle$ which corresponds to $\epsilon = +1$, contributes to $f(h)$, in the interval $-\alpha < h < 2\alpha$: $f_+(h) = (h/\alpha+1)^{-\frac{1}{2}}$.

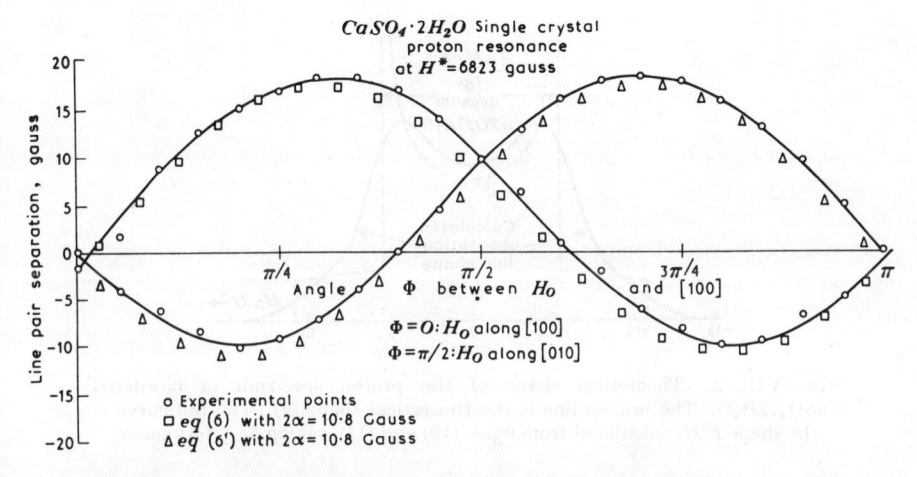

FIG. VII, 1. Line pair separation as a function of the angle Φ between H_0 and [100]. Because two directions exist for proton-proton lines in a gypsum single crystal, there are two similar curves differing in phase.

Similarly, the line $|0\rangle \to |+1\rangle$ gives the spectrum

$$f_-(h) = \left(-\frac{h}{\alpha}+1\right)^{-\frac{1}{2}}$$

in the interval $-2\alpha < h < \alpha$.

The total spectrum, defined by

$$\begin{aligned}
f(h) &= \left(-\frac{h}{\alpha}+1\right)^{-\frac{1}{2}} && (-2\alpha < h < -\alpha) \\
&= \left(-\frac{h}{\alpha}+1\right)^{-\frac{1}{2}} + \left(\frac{h}{\alpha}+1\right)^{-\frac{1}{2}} && (-\alpha < h < \alpha) \\
&= \left(\frac{h}{\alpha}+1\right)^{-\frac{1}{2}} && (\alpha < h < 2\alpha), \quad (9)
\end{aligned}$$

is shown by the broken line in Fig. VII, 2, where α has been put equal to 5·4 gauss as is appropriate for gypsum.

More precisely, this would be the shape of the spectrum if the elementary spectrum of each crystal were composed of two discrete lines. In fact each of these lines is broadened by the interaction of the protons with the neighbouring water molecules. The resulting spectrum will be given by

$$F(H) = \int_{-\infty}^{\infty} f(H_0-H^*)S(H-H_0)\,dH_0, \qquad (10)$$

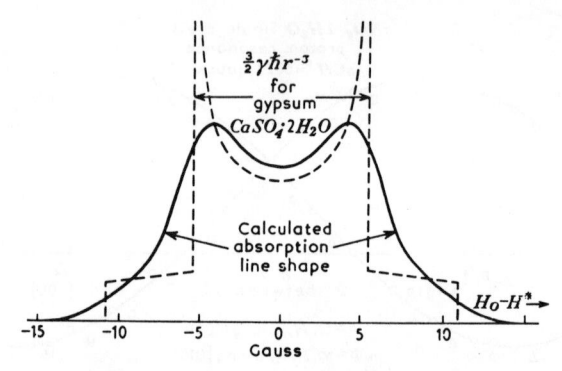

Fig. VII, 2. Theoretical shape of the proton spectrum in powdered $CaSO_4.2H_2O$. The broken line is the theoretical shape (9). The full curve is the shape $F(H)$ calculated from eqns. (10) and (11), taking $\beta = 1.54$ gauss.

where $S(H-H_0)$ represents the shape of an individual line of a crystal. One may take for $S(H-H_0)$ the Gaussian function

$$S(H-H_0) = \frac{1}{\beta\sqrt{2\pi}} \exp\left[-\frac{(H-H_0)^2}{2\beta^2}\right], \qquad (11)$$

which, as seen in Chapter IV, is not a bad approximation to represent dipole broadening. The full curve in Fig. VII, 2 was obtained by taking $\beta = 1.54$ gauss, which gave the best fit with the data.

To be strictly correct the fact that the root mean square deviation β for each crystal depends on the orientation of this crystal with respect to the magnetic field, and is therefore a function $\beta(\mathbf{H}_0/H_0)$, should have been taken into account. The experimental precision does not warrant this additional complication.

Dichlorethane. As another example of groups of two protons, consider 1,2 dichlorethane CH_2Cl-CH_2Cl (2). In this compound the magnetic interaction of the two protons bound to a single carbon dominates all the others. The shape of the resonance curve in Fig. VII, 3 shows clearly that a doublet is involved, and the separation

$3\gamma\hbar/2r^3 = 8\cdot 8$ gauss leads to a proton distance of approximately $1\cdot 69$ Å. To make this result more precise, one tries to represent this curve by a formula of the type (10) where f is given by (9) and S by (11). The constant β of (11) is determined as follows: the second moment of the resonance line is $\Delta H^2 = \frac{4}{5}\alpha^2 + \beta^2$, where $\alpha = \frac{3}{4}\gamma\hbar/r^3$ (the factor $\frac{4}{5}$ represents the mean value of $(1-3\cos^2\theta)^2$). This moment ΔH^2, determined

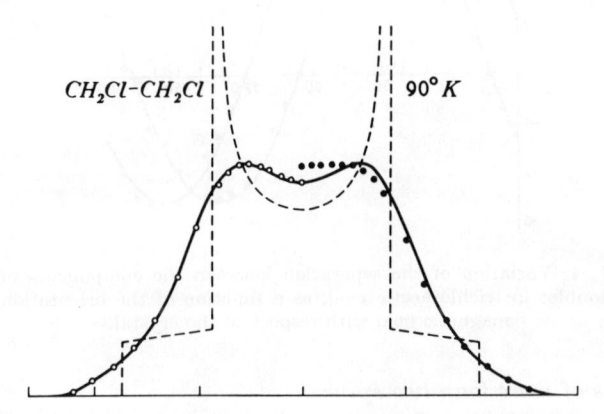

Fig. VII, 3. A two-spin system with $I = \frac{1}{2}$. The open circles in the left half are theoretical values computed for a proton separation of $1\cdot 70$ Å and the points in the right half are for $1\cdot 72$ Å.

by graphical integration of the experimental curve, is found to be equal to $18\cdot 2$ gauss². The theoretical curve is calculated using (9), (10), and (11) for $r = 1\cdot 70$ Å and $r = 1\cdot 72$ Å, taking each time the value $\Delta H^2 - \frac{4}{5}\alpha^2$ for β^2. From Fig. VII, 3 we see that the points on the left part of the curve, represented by open circles and corresponding to $r = 1\cdot 70$ Å, fit the experimental curve much better than the black points on the right for which $r = 1\cdot 72$ Å. $r = 1\cdot 70$ Å seems a reasonable value for the proton-proton distance.

As a last example we consider the trichloracetic acid where in a single crystal a spectrum similar to that of gypsum has been obtained (3). Fig. VII, 4 shows the distance in gauss between the two components of a doublet (there are two such doublets, corresponding to two proton pairs) as a function of the angle through which the crystal is rotated around a certain axis. The existence of proton pairs clearly demonstrates that trichloracetic acid is a dimer in the solid state. The distance r between the protons obtained from $2\alpha = \frac{3}{2}\gamma\hbar/r^3 = 2\cdot 56 \pm 0\cdot 06$ gauss is found to be $r = 2\cdot 56 \pm 0\cdot 02$ Å. The width of the lines $\simeq 0\cdot 3$ gauss is much smaller than in gypsum, the various pairs being farther removed from each other.

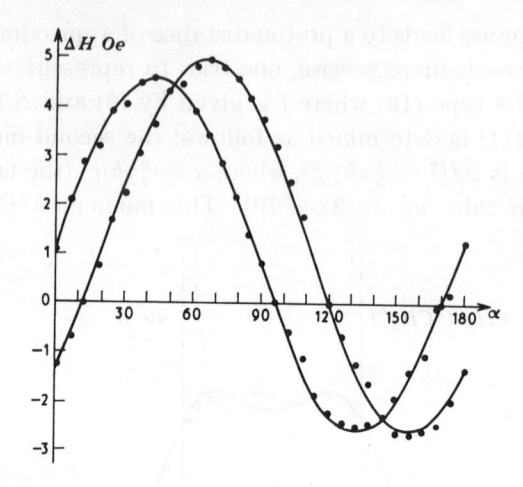

FIG. VII, 4. Variation of the separation between the components of either proton doublet in trichloracetic acid, as a function of the orientation of the magnetic field with respect to the crystal.

(b) Systems of more than two spins

The problem of three spins can be treated in the same fashion, but it is distinctly more complicated and we shall not carry out the calculation here. Fig. VII, 5 shows the form of the spectrum for a powdered crystal when the three spin $\frac{1}{2}$ nuclei are located at the three apices of an equilateral triangle (4). The top of the figure shows the spectrum of an isolated triangle. At the bottom, the full curve is a theoretical plot where we have taken into account the broadening due to neighbours by use of a formula of the type (10). The broken curve is an experimental curve corresponding to trichlorethane, CCl_3—CH_3, which is a good example of a system of three spins placed at the apices of an equilateral triangle. The difference in the shape of the line as compared with that observed for two spins (Fig. VII, 3) is clear. This qualitative difference has been put to advantage in the determination of the structure of diketene (5). The chemical and spectroscopic information about diketene is compatible with either one of the formulae below (or a combination of the two):

(a) CH_2=C—CH_2
        | |
        O—C=O

(b) CH_3—C=CH
           | |
           O—C=O.

The experimental curve which shows a central minimum proves without ambiguity that it has structure (a) where the protons are paired.

Even a few per cent of structure (b) would completely mask the central minimum.

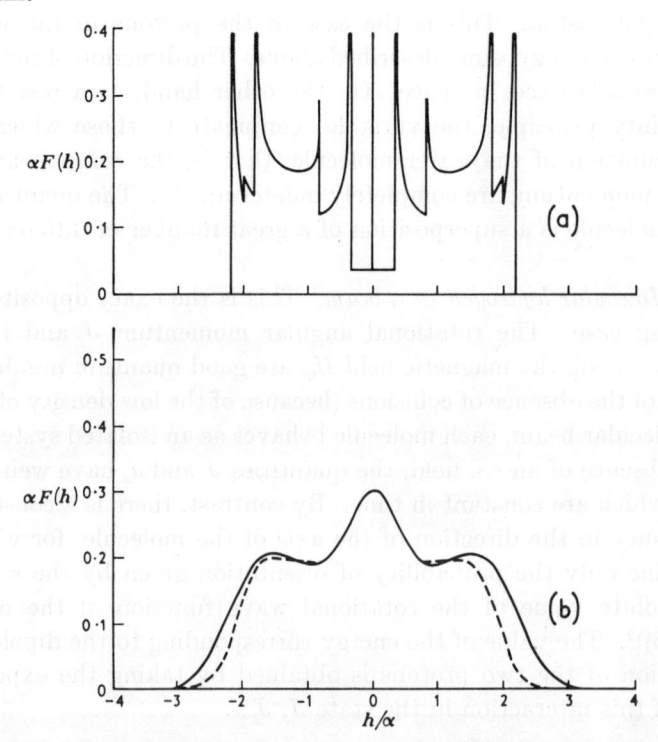

FIG. VII, 5. Line shape for three nuclei of spin $\frac{1}{2}$ at the corners of an equilateral triangle. (a) Isolated rigid triangle. (b) Rigid triangle broadened by neighbours. In (b) the broken line is the integrated experimental shape and the full line is calculated.

The system of four spins has been studied in detail in a similar way, especially in the case of the ammonia halides X—NH$_4$ where the four protons are located at the apices of a regular tetrahedron. Further details and references to earlier work can be found in reference (6).

B. Nuclear resonance in solid hydrogen (7)

We shall describe this experiment in some detail for it provides an interesting example of a situation where quantum effects play an important role in the description of the relative positions of two interacting protons.

(a) Introduction. System of two interacting protons

When magnetic resonance techniques are used to study a system of two interacting nuclear spins (two protons) in an applied field H_0, the pattern observed depends in a large measure on the nature of the environment of this system.

(i) *Rigid system.* This is the case of the protons in the water of crystallization in gypsum, described above. The direction of the proton-proton axis is fixed in space. On the other hand, as a result of the uncertainty principle, the variables conjugate to those which define the orientation of the water molecule, that is, the components of its angular momentum, are completely indeterminate. The quantum state of the molecule is a superposition of a great number of different states (J, J_z).

(ii) *Molecular hydrogen in a beam.* This is the exact opposite of the preceding case. The rotational angular momentum J and its component J_z along the magnetic field H_0, are good quantum numbers. As a result of the absence of collisions (because of the low density of matter in a molecular beam, each molecule behaves as an isolated system), and in the absence of an r.f. field, the quantities J and J_z have well-defined values which are constant in time. By contrast, there is a considerable uncertainty in the direction of the axis of the molecule, for which we can define only the probability of orientation given by the square of the absolute value of the rotational wave function of the molecule $|\psi_J^{m_J}(\theta, \varphi)|^2$. The value of the energy corresponding to the dipole-dipole interaction of the two protons is obtained by taking the expectation value of this interaction in the state J, J_z:

$$W_{m_J}^J = \langle \psi_J^{m_J} | \mathscr{H}_{\mathrm{dip}} | \psi_J^{m_J} \rangle. \tag{12}$$

Just as for the gypsum protons the resonance line will show a fine structure, although of a different nature. We recall that in beam experiments the small number of particles in the beam prevents detection of the resonance by electromagnetic methods, and instead 'trigger' detection by matter as described in Chapter I is used.

(iii) *Molecule of hydrogen in a gas under pressure.* Each molecule undergoes a great number of collisions per second (10^{10} at atmospheric pressure). The quantities J and J_z are still good quantum numbers during the time between collisions but can change in the course of a collision: moreover, changes of J_z which do not change the rotational energy of the molecule may be more frequent than changes of J, which do change it.

To calculate the dipole energy we must take the mean value of (12) over all the values of m_J and possibly of J, because the transitions due to collisions follow one another very rapidly, and only this mean value is observable. It is easy to see from symmetry considerations that such a mean value leads to a null result for the dipole energy. In a strong field H_0 a single structureless resonance line is observed. We must not however, conclude that in this case the dipole-dipole interaction has no observable effect. We shall see in Chapter VIII that it is responsible for the thermal relaxation of the protons in the gas.

(b) *Solid hydrogen*

Now consider the situation in solid hydrogen. One might think that, since we are dealing with a solid, the direction of the axis PP is fixed and that the line has the same appearance as in a rigid lattice, such as gypsum, the only difference being due to different proton-proton distances ($r = 1.58$ Å in water, 0.75 Å in hydrogen).

To test this assumption consider (in an extremely schematic fashion) the gradual passage from case (ii), which is that of the free molecule having a well-defined angular momentum, to case (i), which is that of the molecule having a fixed orientation in the solid, introducing the intermolecular actions as a perturbation. The free molecule has a total angular momentum J to which there corresponds an energy of rotation $BJ(J+1)$. (The quantity B/k, where k is the Boltzmann constant, has the value 86° K in hydrogen.) In the absence of the magnetic field H_0 this state has a $(2J+1)$-fold degeneracy, corresponding to $(2J+1)$ possible values of J_z.

The effect of the perturbation by the intermolecular interactions (which we shall represent by a mean electric potential V) is to lift this degeneracy partially or completely and to modify the form of the rotational wave functions which describe the orientation of the molecule. In zero order, if the energy of perturbation V is much smaller than the differences between the energies of the different rotational states of the free molecule, the eigenfunctions $\psi_J^{m_J}$ will be replaced by $2J+1$ linear combinations φ_J^k:

$$\varphi_J^k = \sum c_{m_J}^k \psi_J^{m_J}. \tag{13}$$

If the importance of the perturbation increases, it is necessary to add to (13) contributions $\psi_{J'}^{m_{J'}}$, that arise from the other rotational states. The case of the molecule rigidly fixed in space is a limiting case, where the perturbation treatment breaks down since in this case the rotational wave function of the molecule oriented in a fixed direction $\Omega_0 = (\theta_0, \varphi_0)$ can be written

$$\delta(\Omega - \Omega_0) = \sum_{J, m_J} Y_J^{m_J*}(\Omega_0) \cdot Y_J^{m_J}(\Omega)$$

and contains contributions of comparable weight from all the rotational states.

There are several reasons for believing that the intermolecular forces are particularly weak in solid hydrogen. One of the arguments invoked is the smallness of the heat of fusion, suggesting that the molecule can rotate almost freely in the solid. Another indication is the existence in *liquid* hydrogen of Raman frequencies in good agreement with those calculated from the energies of rotation of the free molecule.

(c) Ortho- *and* para-*hydrogen*

Because protons are particles of spin $\frac{1}{2}$ which obey Fermi statistics, their total wave function, the product of a spin function and a rotational function, must be antisymmetric.

The spin function is symmetric in the triplet state $I = 1$ and antisymmetric in the singlet state $I = 0$. For molecules in the triplet state, the rotational functions are antisymmetric, i.e. J is odd; the opposite occurs if the spin state is a singlet.

The phenomenon which dominates the study of hydrogen at low temperatures is the extreme slowness of the transitions between the singlet and triplet states because of the weakness of the intermolecular magnetic forces, which alone are capable of producing this conversion. At low temperatures we can thus consider the molecules in the singlet and triplet states as two different species labelled *para*-hydrogen and *ortho*-hydrogen respectively.

It is the existence of these two distinct species which makes possible the observation of the magnetic resonance in solid hydrogen. Indeed, at temperatures of 1° to 2° K where the experiments are made, if all the hydrogen molecules could reach thermal equilibrium, they would practically all be in the lower state $J = 0$, $I = 0$, as a result of the smallness of the Boltzmann factors $\exp[-BJ(J+1)/kT]$ which correspond to the other states. Since the state $J = 0$, $I = 0$ has neither spin nor orbital magnetic moment, no resonance can be observed there. Actually, if hydrogen gas, initially at a high temperature where thermal equilibrium could be established, is brought down rapidly to a temperature of 1° or 2° K, only transitions where the parity of J does not change occur, *ortho*-molecules remaining *ortho* and the same for *para*-molecules. Consequently, the ratio of *ortho*- to *para*-states at low temperature will be the same as it was at higher temperatures, namely three, in accordance with the statistical weights of the triplet and singlet states.

In the solid, the *para*-molecules will all be in the state $J = 0$ and the *ortho*-molecules, the only ones with nuclear magnetism, will be in the lowest state compatible with the *ortho*-quality, that is, in the state $J = 1$. More precisely, this will be true if one can neglect the intermolecular actions. But if these are small with respect to the energy difference between the state of rotation $J = 1$ and the nearest *ortho*-state which is $J = 3$, equal to $10B$, we may use the zero-order approximation and describe the state of an *ortho*-molecule in the solid by a linear combination of the three functions $\psi_1^1, \psi_1^0, \psi_1^{-1}$.

(d) Crystalline potential

In order to determine the coefficients of these combinations it is sufficient to know the symmetry of the environment of a given *ortho*-molecule. This environment is composed of *para*-molecules which have spherical symmetry and of other *ortho*-molecules. At temperatures slightly below 14° K, the melting-point of hydrogen, each *ortho*-molecule goes rapidly from one state m_J to another and presents to its neighbours an average appearance which has spherical symmetry.

The crystalline symmetry of the system is then the same as that of pure *para*-hydrogen, that is of closely packed spheres (close-packed hexagonal) and the degeneracy in m_J is not lifted. At lower temperatures, when the rate of transitions between the various values of m_J is lower, the lack of sphericity of the *ortho*-molecules makes itself felt and the average potential existing at the site of an *ortho*-molecule does have a lower symmetry. This is a co-operative transition; the fact that the *ortho*-molecules rotate less freely gives rise to a non-cubic crystalline potential which in turn hinders the free rotation of the *ortho*-molecules.

As a result of the random distribution of the *ortho*- and *para*-molecules, the environments of two different *ortho*-molecules as well as the crystalline potentials in which they lie are different.

The potential which corresponds to a given *ortho*-molecule is a function $V(\theta, \varphi)$ of the orientation of the molecule. In the zero-order approximation, only those matrix elements of V inside the manifold $J = 1$ spanned by the three functions $\psi_1^{0,\pm 1}$ are of significance. If we expand $V(\theta, \varphi)$ in spherical harmonics, only the harmonics of order 2 make a contribution (the zero-order harmonic is a constant, the first-order harmonic gives zero by virtue of symmetry, and the higher order harmonics give zero on account of the properties of orthogonality of these spherical functions). The potential $V(\theta, \varphi)$ is therefore a quadratic form in the direction cosines $\lambda_\xi, \lambda_\eta, \lambda_\zeta$ of the axis of the molecule. This quadratic form can be diagonalized by a suitable choice of coordinate axes ξ, η, ζ and put in the form

$$V = A_\xi \lambda_\xi^2 + A_\eta \lambda_\eta^2 + A_\zeta \lambda_\zeta^2. \tag{14}$$

In this form it is seen immediately that the eigenfunctions or linear combinations (13) which diagonalize (14) are:

$$\phi_\xi = \sqrt{\frac{3}{4\pi}} \lambda_\xi, \qquad \phi_\eta = \sqrt{\frac{3}{4\pi}} \lambda_\eta, \qquad \phi_\zeta = \sqrt{\frac{3}{4\pi}} \lambda_\zeta. \tag{15}$$

If the three constants A_ξ, A_η, A_ζ are different, as will generally be the case, the degeneracy will be completely lifted. It should be made clear that the values of the constants A_ξ, A_η, A_ζ and the orientations of the frame $O\xi\eta\zeta$ change from one molecule to another.

(e) *Magnetic resonance in a strong field*

The magnetic interactions which exist in a hydrogen molecule placed in an applied field H_0 are very weak in comparison to the potential (14) (0·001° K instead of 1° K) and, therefore, can be replaced by their expectation values in the three states (15) according to first order perturbation theory.

These interactions can be written (in the notation used in studies of the free molecule H_2 by beam techniques):

$$\mathcal{H} = \mathcal{H}_{ab} + \mathcal{H}_{cd},$$
$$\mathcal{H}_{ab} = -h(aI_z + bJ_z),$$
$$\mathcal{H}_{cd} = -h\left\{c\mathbf{I}\cdot\mathbf{J} + 5d\left[3\frac{(\mathbf{I}^1\cdot\mathbf{r})(\mathbf{I}^2\cdot\mathbf{r})}{r^2} - \mathbf{I}^1\cdot\mathbf{I}^2\right]\right\}, \tag{16}$$

$$a = \frac{\gamma_P H_0}{2\pi} = 4\cdot258 H_0 \text{ kc/s}, \qquad b = \frac{\gamma_J H_0}{2\pi} = 0\cdot617 H_0 \text{ kc/s},$$

$$c = \frac{\gamma_P H'}{2\pi} = 113\cdot8 \text{ kc/s}, \qquad d = \frac{1}{5}\frac{\hbar\gamma_P^2}{2\pi}\langle r^{-3}\rangle = 57\cdot68 \text{ kc/s}.$$

In this formula the first and second terms of $\mathcal{H}_{ab}$ represent the Zeeman energies of the magnetic moment of the spin and of the orbital magnetic moment due to the rotation of the molecule respectively. The first and second terms of $\mathcal{H}_{cd}$ represent the spin orbit coupling (the interaction of the nuclear moments with the magnetic field produced by the rotation of the molecule) and the dipole interaction of the two nuclear moments, respectively.

We may easily show that the effect of the operators $-bJ_z$ and $-c\mathbf{I}\cdot\mathbf{J}$ is zero. Indeed, as stated at the beginning of this section, we must calculate the expectation values of these operators in the three states described by the wave functions (15). The expectation values of all the components of J in any one of these states is zero; all the components of J are pure imaginary operators in the representation (θ, φ). Their expectation values taken over the real functions (15) are then imaginary numbers, although, being expectation values of hermitian operators, they must also be real. They are, therefore, zero. This so-called quenching of the orbital momentum has already been discussed in Chapter VI.

Now that the magnetic rotational effects have been eliminated in this way the only difference between the theory of dipolar interaction in solid hydrogen, and that given above for a rigid lattice, resides in the fact that the orientation of the vector PP is not fixed in space any more. Instead, its orientation with respect to crystalline axes is described by a probability distribution equal to the square of one of the wave functions (15). The crystalline axes themselves have a random orientation with respect to the applied field H_0.

Formula (5) which gives the values at resonance of the applied field H_0 may be used provided that $3\cos^2\theta - 1$ is replaced by the mean value $\langle 3\cos^2\theta - 1 \rangle$ taken over one of the three functions (15).

Let Θ and Φ be the polar angles defining the orientation with respect to the applied field H_0 of one of the principal axes, for example ζ, of the crystalline field that exists in the neighbourhood of a given *ortho*-molecule. The angle θ that the direction PP makes with the magnetic field is given by

$$\cos\theta = \lambda_\zeta \cos\Theta + \lambda_\xi \sin\Theta\cos\Phi + \lambda_\eta \sin\Theta\sin\Phi$$

or, by substituting

$$\lambda_\zeta = \cos\delta, \quad \lambda_\xi = \sin\delta\cos\omega, \quad \lambda_\eta = \sin\delta\sin\omega,$$

$$\cos\theta = \cos\delta\cos\Theta + \sin\delta\sin\Theta\cos(\Phi-\omega). \tag{17}$$

The mean value of $3\cos^2\theta - 1$ in the state ψ_ζ is given by

$$\langle 3\cos^2\theta - 1 \rangle_\zeta = \frac{3}{4\pi} \int (3\cos^2\theta - 1)\cos^2\delta\sin\delta\, d\omega d\delta$$

where $\cos\theta$ is given by (17). This integration is elementary and gives

$$\langle 3\cos^2\theta - 1 \rangle_\zeta = \tfrac{2}{5}(3\cos^2\Theta - 1). \tag{18}$$

(18) shows that for a molecule of *ortho*-hydrogen in a given state ψ_ζ (or ψ_ξ or ψ_η) the resonance frequencies are the same as for a pair of protons whose axis, fixed in space, makes the same angle with the field H_0 as the axis ζ (or ξ or η), provided that the dipole-dipole interaction is reduced in the ratio 2/5 The shape of the resonance line is the same

as for rigid proton pairs in a powder, the constant $\alpha = \tfrac{3}{4}\gamma\hbar/r^3$ being replaced by $\tfrac{3}{10}\gamma_P \hbar \langle r^{-3}\rangle = 3\pi d/\gamma_P$.

The quantity d read from the resonance curve can be compared with the quantity d_f obtained from measurements made by applying the beam technique to a free molecule. The experiment gives for d_s (the parameter d in the solid state):

$$d_s = 54 \cdot 2 \pm 0 \cdot 5 \text{ kc/s}, \quad \text{whence } d_s/d_f = 0 \cdot 94.$$

This proves both that the model is essentially correct and that the separation of the two protons (or more precisely the expectation value of $\langle r^{-3}\rangle$ taken over the vibrational wave function of the molecule) is slightly different in the solid and in the gas.

No assumption has been made about the relative values of the thermodynamic energy kT and the energy differences $E_\xi - E_\zeta$ and $E_\eta - E_\zeta$. Each molecule may be found in any one of three states E_ξ, E_η, E_ζ, with the probabilities given by the Boltzmann factors. The shape of the resonance curve is independent of the values of these factors.

The central line. We said above that experiment confirms the prediction of the theory. In fact things are a little more complicated.

At 4° K one sees a single line. As the temperature is lowered the intensity of this line diminishes, and simultaneously the characteristic side peaks of the two-proton structure appear. At the lowest temperature attained, which is 1·16° K, some vestiges of the central line still remain. The explanation offered is the following: at 4° K an *ortho*-molecule makes rapid transitions between the three states E_ξ, E_η, E_ζ. If the frequency of these transitions is greater than the frequency interval $3d$ which gives the fine structure of the line, only the mean effect of the dipole interaction taken over the three states ψ_ξ, ψ_η, ψ_ζ is observable. This mean is zero and a single line is observed, exactly as in case (iii) of the introduction. It is not unreasonable to assume, as suggested by an approximate calculation (7), that the frequency of these transitions will be a rapidly-increasing function of the ratio $kT/(E_\xi - E_\zeta)$. The possibility of simultaneously observing the central line and side peaks at the lowest temperatures is explained by the fact that the energy differences $E_\xi - E_\zeta$, etc., which vary from one molecule to another, are small compared with kT for certain molecules which, therefore, are making rapid transitions among the three states ξ, η, ζ and give rise to the central line, while for other molecules the inverse is true and only the lateral peaks are observable.

(f) Magnetic resonance in zero field

Assume now that no external field is applied to the sample. Because of the quenching of the orbital momentum **J**, the Hamiltonian (16) for an *ortho*-molecule in a state say ϕ_ζ given by (15) reduces to

$$5dh\langle\phi_\zeta | \mathbf{I}^1 . \mathbf{I}^2 - 3(\mathbf{I}^1 . \mathbf{n})(\mathbf{I}^2 . \mathbf{n}) | \phi_\zeta\rangle. \qquad (19)$$

According to (19) an average of the dipolar coupling has to be taken over the orientations of the unit vector $\mathbf{n} = \mathbf{r}_{12}/|\mathbf{r}_{12}|$ with a statistical weight equal to $|\phi_\zeta|^2$. Because of the form of ϕ_ζ, the result of this average is invariant under rotation around the axis $O\zeta$, and therefore

$$M = I_\zeta^1 + I_\zeta^2$$

is a good quantum number for (19). Furthermore, (19) does not change with a change of M into $-M$, and inside the manifold $I = 1$ of the spin states of an *ortho*-molecule its trace vanishes. We may then write

$$5dh\langle\phi_\zeta | \mathbf{I}^1 . \mathbf{I}^2 - 3(\mathbf{I}^1 . \mathbf{n})(\mathbf{I}^2 . \mathbf{n}) | \phi_\zeta\rangle = p(3M^2 - 2), \qquad (20)$$

the constant p being obtained by taking the expectation values on both sides of (20) in, say, the spin state $M = 1$, that is, for

$$I_\zeta^1 = I_\zeta^2 = \tfrac{1}{2}.$$

We find

$$p = \frac{5dh}{4}\langle\phi_\zeta | 1 - 3\lambda_\zeta^2 | \phi_\zeta\rangle = \frac{15dh}{4}\int_0^1 (1 - 3\lambda_\zeta^2)\lambda_\zeta^2 \, d\lambda_\zeta = -hd. \qquad (21)$$

The frequency $(1/h)(E_0 - E_{\pm 1}) = -3p/h = 3d$, of the zero field transition between the states $M = \pm 1$ and the state $M = 0$, is independent of the orbital state ϕ_ξ, ϕ_η, ϕ_ζ of the *ortho*-molecule. This transition has indeed been observed at a frequency of 165.7 ± 1 kc/s, leading to $d = 55.2$ kc/s in good agreement with the value 54.7 ± 0.7 kc/s deduced from high field measurements.

(g) Magnetic resonance in HD and D_2

In solid HD all molecules are in the lowest state $J = 0$, where the expectation value of the dipolar coupling proton-deuteron vanishes, and a single resonance line should be and has been observed for the proton.

Solid D_2 has also been investigated. There, since the spin of deuterium is one, making it obey Bose statistics, the existing states should be: $J = 0$, $I = 0, 2$, and $J = 1$, $I = 1$. The state $J = 0$, $I = 2$ should give and did actually give a single resonance line. No fine structure that would correspond to the state $J = 1$, $I = 1$ was observed. The weakness of the signal may account for it. (It may be pointed out that the main part of the fine structure in D_2 is due, as is known from molecular beam experiments, to the quadrupole interaction of the spins with the molecular electric field gradient rather than to the weak dipolar interactions.)

II. Energy Levels of Nuclear Spins in the Presence of Quadrupole Interactions

In Chapter VI several equivalent expressions were given for the Hamiltonian describing the interaction of the nuclear quadrupole moment with the electric-field gradients existing at the nucleus. The purpose of the present section is to give a description of the energy levels and magnetic resonance spectra resulting from these interactions in the solid state. Since the pioneer work (8) this subject has received considerable attention and we shall be content to give a general outline referring the reader for further details to two important reviews of the subject of quadrupole interactions containing a wealth of information (9, 10). The influence of quadrupole interactions on relaxation phenomena will be discussed in Chapters VIII and IX.

In the presence of an applied magnetic field **H**, the Hamiltonian of a nuclear spin with a quadrupole moment can be written (formula (24) of Chapter VI):

$$\mathcal{H} = -\gamma\hbar\mathbf{H}.\mathbf{I} + \frac{e^2qQ}{4I(2I-1)}[3I_Z^2 - I(I+1) + \tfrac{1}{2}\eta(I_+^2 + I_-^2)]. \quad (22)$$

In (22) the axes X, Y, Z are the principal axes of the tensor V_{ij} that describes the electric field gradient with, as explained in Chapter VI,

$$|V_{ZZ}| \geqslant |V_{YY}| \geqslant |V_{XX}|, \quad eq = V_{ZZ}, \quad \eta = \frac{V_{XX} - V_{YY}}{V_{ZZ}}, \quad 0 \leqslant \eta \leqslant 1.$$

It may sometimes be convenient to use another frame of reference $Oxyz$, in particular one where Oz is parallel to **H**. Standard formulae for the transformation of components of tensor operators $3I_Z^2 - I(I+1)$, I_+^2, and I_-^2 will give the form of the Hamiltonian in such a frame.

To obtain the eigenstates and eigenvalues of (22) in the most general case when the Zeeman and quadrupole interactions are comparable in strength, it is necessary to solve a secular equation of order $2I+1$, that is already of third order for the smallest spin possessing a quadrupole moment: $I = 1$. Such a situation occurs very seldom, since most experiments deal with cases when the Zeeman coupling is either large or small in comparison with the quadrupole coupling. The first or 'high field' situation is treated in reference (9), the second in (10).

A. High magnetic fields

(a) *Energy levels in single crystals*

It is more convenient for this problem to choose the z-axis along the magnetic field. We shall assume for simplicity most of the time that the quadrupole gradient has cylindrical symmetry so that η vanishes. We may then without loss of generality choose the axis Oz parallel to **H** in the plane XOZ, and write

$$I_Z = I_z \cos\theta + I_x \sin\theta$$

which, introduced into (22), gives

$$\mathcal{H} = \mathcal{H}_Q + \mathcal{H}_M,$$

$$\mathcal{H}_M = -\gamma\hbar H I_z,$$

$$\mathcal{H}_Q = \frac{e^2qQ}{4I(2I-1)}\begin{bmatrix} \frac{1}{2}(3\cos^2\theta-1)(3I_z^2-I(I+1)) \\ +\frac{3}{2}\sin\theta\cos\theta[I_z(I_+ + I_-)+(I_+ + I_-)I_z] \\ +\frac{3}{4}\sin^2\theta(I_+^2+I_-^2) \end{bmatrix}. \quad (23)$$

The operators $I_\pm$ in (23) are defined as $I_x \pm iI_y$ with respect to the frame $Oxyz$ of the magnetic field, which is referred to in the following as Σ^H, while they are $I_X \pm iI_Y$ in (22), with respect to the crystal frame Σ^P.

The expression (23) has been written in a way that immediately shows up the diagonal ($\Delta m = 0$) and off-diagonal ($\Delta m = \pm 1, \pm 2$) matrix elements of the quadrupole Hamiltonian $\mathcal{H}_Q$ which will be treated as a perturbation.

The various energy levels of (23) can be written as

$$E_m = E_m^{(0)} + E_m^{(1)} + E_m^{(2)} + \dots,$$

where $E_m^{(p)}$ represents the contribution to the energy, of the perturbation of order p.

Introducing for brevity

$$\nu_Q = \frac{3e^2qQ}{h2I(2I-1)}, \qquad a = I(I+1), \qquad \mu = \cos\theta, \qquad \nu_L = \frac{\gamma H}{2\pi}, \quad (23')$$

we get from (23) by standard second-order perturbation theory:

$$E_m^{(0)} = -\gamma\hbar H m = -h\nu_L m,$$

$$E_m^{(1)} = \tfrac{1}{4}h\nu_Q(3\mu^2-1)(m^2-\tfrac{1}{3}a),$$

$$E_m^{(2)} = -h\left(\frac{\nu_Q^2}{12\nu_L}\right)m\begin{bmatrix}\tfrac{3}{2}\mu^2(1-\mu^2)(8m^2-4a+1) \\ +\tfrac{3}{8}(1-\mu^2)^2(-2m^2+2a-1)\end{bmatrix}. \quad (24)$$

(For higher-order terms see reference (11).) Because of the changes $E_m^{(1)}$ and $E_m^{(2)}$ in the energy levels, instead of a single resonance frequency

$$\nu_L = \frac{E_{m-1}^{(0)} - E_m^{(0)}}{h}$$

there are now several resonance frequencies:

$$\nu_m = \frac{E_{m-1} - E_m}{h} = \nu_L + \nu_m^{(1)} + \nu_m^{(2)} + \ldots$$

and as many resonance lines. The first-order change in the frequency, $\nu_m^{(1)}$, is given by

$$\nu_m^{(1)} = \frac{E_{m-1}^{(1)} - E_m^{(1)}}{h} = -\nu_Q(m-\tfrac{1}{2})\frac{3\mu^2 - 1}{2}. \tag{25}$$

The first-order shift $\nu_m^{(1)}$ vanishes for $m = \tfrac{1}{2}$. For half-integer spins the frequency of the central transition $-\tfrac{1}{2} \leftrightarrow \tfrac{1}{2}$ is not shifted in first order by the quadrupole interaction. The frequencies of the other lines are, however, shifted and satellite lines corresponding to transitions $m \leftrightarrow (m-1)$ with $m \neq \tfrac{1}{2}$ appear on each side of the central line. The intensities of the various lines proportional to $|\langle m | I_x | m-1 \rangle|^2$ are $3:4:3$ for spin $\tfrac{3}{2}$; $5:8:9:8:5$ for spin $\tfrac{5}{2}$, etc.

In a single crystal where $\mu = \cos\theta$ has a well-defined value, there are $2I$ lines with positions given in first order by (25). The value of the nuclear spin I is thus obtained most directly from the number of components in the spectrum.

The second-order shift $\nu_m^{(2)} = (E_{m-1}^{(2)} - E_m^{(2)})/h$ is immediately obtainable from (24). In particular, for the central line $-\tfrac{1}{2} \leftrightarrow \tfrac{1}{2}$ it is given by

$$\nu_{\frac{1}{2}}^{(2)} = \frac{-\nu_Q^2}{16\nu_L}(a-\tfrac{3}{4})(1-\mu^2)(9\mu^2-1). \tag{26}$$

It is worth pointing out that $E_m^{(2)}$ is an odd function of m. Consequently, the second-order frequency shifts:

$$\nu_m^{(2)} = \frac{1}{h}(E_{m-1}^{(2)} - E_m^{(2)}), \qquad \nu_{-m}^{(2)} = \frac{1}{h}(E_{-m}^{(2)} - E_{-(m-1)}^{(2)})$$

are equal and their difference

$$\Delta\nu^{(2)} = \nu_m^{(2)} - \nu_{-m}^{(2)}$$

vanishes.

Thus the first-order formula

$$\Delta\nu = \nu_Q(m-\tfrac{1}{2})(3\mu^2-1), \tag{27}$$

if used to express the distance between the two satellite lines $(m-1) \leftrightarrow m$ and $-m \leftrightarrow -(m-1)$, is correct in second order also. The comparison with experiment of the theoretical results given by the formulae (25) and (26) is illustrated in Figs. VII, 6 and VII, 7 (8).

Fig. VII, 6 shows the frequency interval in kc/s between the central line and the symmetrically placed satellites in the spectrum of Na^{23} with spin $I = \frac{3}{2}$, observed from a single crystal of NO$_3$Na. The smooth curve is $83 \cdot 5(3\cos^2\theta - 1)$ kc/s whence, by (25), $\nu_Q = 2 \times 83 \cdot 5$ kc/s;

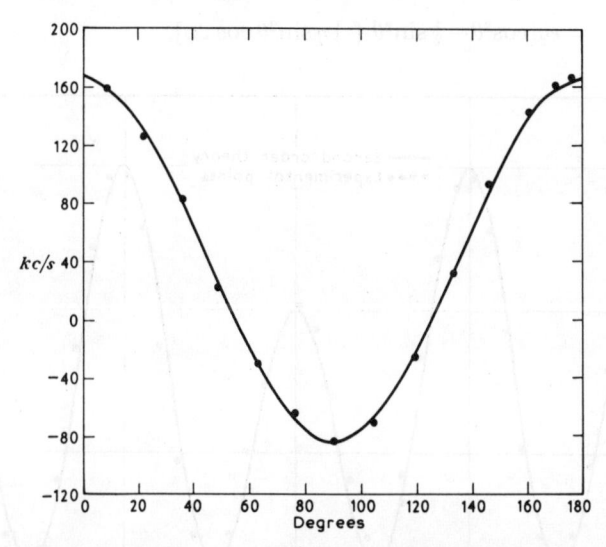

FIG. VII, 6. The frequency interval, in kc/s, between the central line and the symmetrically placed satellites in NO$_3$Na as a function of θ, the angle between H_0 and the crystal axis. The smooth curve is $83 \cdot 5(3\cos^2\theta - 1)$ kc/s.

$e^2qQ/h = 2\nu_Q = 334$ kc/s. In this experiment, $\nu_L \simeq 7 \cdot 17$ Mc/s, the ratio ν_Q/ν_L is a small number, and second-order effects are unimportant. Fig. VII, 7 shows the second-order frequency shift of the central line of Al^{27} ($I = \frac{5}{2}$) in Al$_2$O$_3$ as a function of the angle between the magnetic field and the crystalline axis. The experimental points represent the measured shift $\nu_{\frac{1}{2}}^{(2)}$ multiplied by $32\nu_L/\nu_Q^2(a - \frac{3}{4}) = 4\nu_L/\nu_Q^2$. This dimensionless quantity should, according to (26), be equal to $2(1-\mu^2)(1-9\mu^2)$ represented by the smooth curve. The agreement between theory and experiment is very good. The experimental value of ν_Q is 718 kc/s, a non-negligible fraction of the Zeeman frequency $\nu_L \simeq 2 \cdot 6$ Mc/s.

The generalization of the above results to the case of field gradients with less than cylindrical symmetry ($\eta \neq 0$) is straightforward but lengthy. Thus, the first-order change in energy is given by

$$E_m^{(1)} = \frac{eQ}{4I(2I-1)}[3m^2 - I(I+1)]V_{zz},$$

where Oz is the direction of the applied magnetic field **H**, defined with respect to the crystalline frame Σ^P by the polar angles θ and φ. Using the definition $\eta = (V_{XX}-V_{YY})/V_{ZZ}$, we get

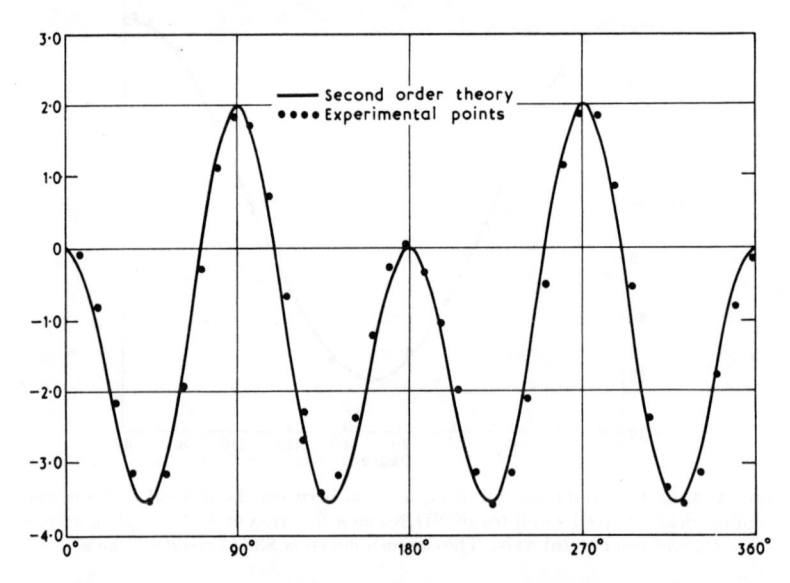

$$V_{zz} = \cos^2\theta\, V_{ZZ} + \sin^2\theta\cos^2\varphi\, V_{XX} + \sin^2\theta\sin^2\varphi\, V_{YY}$$
$$= V_{ZZ}[\cos^2\theta + \tfrac{1}{2}(\eta-1)\sin^2\theta\cos^2\varphi - \tfrac{1}{2}(\eta+1)\sin^2\theta\sin^2\varphi]$$
$$= eq[\cos^2\theta - \tfrac{1}{2}\sin^2\theta + \tfrac{1}{2}\eta\sin^2\theta\cos 2\varphi], \tag{28}$$

FIG. VII, 7. A plot of the frequency shift of the central line of Al^{27} in Al_2O_3, multiplied by $4\nu_L/\nu_a^2$, as a function of the angle between the magnetic field and the crystal axis. The full curve is the theoretical curve $2(1-\mu^2)(1-9\mu^2)$.

from which we obtain immediately

$$E_m^{(1)} = \tfrac{1}{4}h\nu_Q[3\cos^2\theta - 1 + \eta\sin^2\theta\cos 2\varphi](m^2 - \tfrac{1}{3}a). \tag{28'}$$

Similar formulae are easily obtained for higher-order terms but we shall not write them out explicitly.

If we use single crystals that can be rotated with respect to the applied field, it is possible from the spectral shapes observed for different orientations, to determine the position of the principal axes of the field gradient in the crystal and the values of the parameters η and eqQ.

Different ways of analysing the data in order to obtain that information have been devised (9).

(b) Imperfect cubic crystals

(1) Powder pattern

The formulae (25), (26), and (28'), which give the nuclear resonance frequencies in the presence of quadrupole interactions, are valid for single crystals where the angle θ, or the angles θ and φ for non-symmetrical field gradients, between the applied field **H** and the crystalline frame, have well-defined values. In a powder made of many small crystals oriented at random a broad pattern is obtained. Its shape can be calculated by adding the contributions to (25), (26), (28') from all orientations.

The first-order pattern for a symmetrical gradient is obtained from (25) by a method very similar to that already used in Chapter VI for anisotropic Knight shifts or in Section I of this chapter for dipolar fine structures.

If we set $(\nu_m - \nu_L)/|m - \tfrac{1}{2}| = x$ ($m \neq \tfrac{1}{2}$), the line shape $p(x)$ is given by
$$p(x)\,dx = \tfrac{1}{2}d\mu = \tfrac{1}{2}dx/|dx/d\mu|.$$
From (25) we get

$$p(x) = \tfrac{1}{3}[\tfrac{1}{3} + \tfrac{2}{3}x]^{-\tfrac{1}{2}} \quad (-\tfrac{1}{2} \leqslant x \leqslant 1) \text{ if } m > \tfrac{1}{2},$$
$$= \tfrac{1}{3}[\tfrac{1}{3} - \tfrac{2}{3}x]^{-\tfrac{1}{2}} \quad (-1 \leqslant x \leqslant \tfrac{1}{2}) \text{ if } m \leqslant 0,$$
$$p(x) = 0 \quad \text{otherwise.} \tag{29}$$

The broken line in Fig. VII, 8 (**12**) shows the powder pattern for $I = \tfrac{3}{2}$, and the full curve, the same pattern modified by dipolar broadening. Similar curves for the second-order broadening and also for asymmetrical gradients are found in reference (**9**). It is clear that most of the information obtainable from single crystals is lost in a powder.

(2) First-order broadening in imperfect crystals

A situation where, of necessity, only powder patterns are obtainable is that of imperfect cubic crystals. In a perfect cubic crystal the quadrupole interaction vanishes and a single Zeeman line is observed. Imperfections of the crystal, created by dislocations, strains, vacancies, interstitials, foreign atoms, etc., create, at the position of a nucleus, quadrupole gradients which vary not only in orientation but also in magnitude from site to site and have a considerable influence on the shape of the resonance line. An abundant literature exists on the subject; most of it is reviewed in the references (**9**) and (**12**), and only the most salient aspects of the problem will be discussed here. In particular, no attempt will be made to correlate in any detail the shape

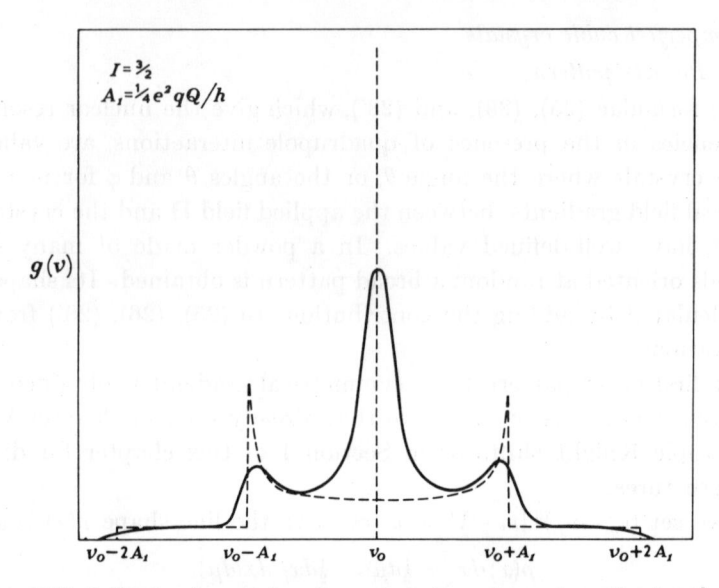

FIG. VII, 8. First-order quadrupole perturbation. Line shape for $I = \frac{3}{2}$ in powdered samples of axially symmetric crystals (broken line). With dipolar broadening superimposed the full curve results. Frequently the satellites are spread out over such a large frequency range that the wings become unobservable.

and intensity of the observed resonance with the nature and concentration of defects, density of dislocations, etc., this problem being outside the range of nuclear magnetism proper. For half-integer nuclear spins, which are much more numerous than the integer spins $I \neq 0$, the study of quadrupole effects in imperfect cubic crystals is dominated by the fact that, as seen previously, the frequency of the central transition $\frac{1}{2} \leftrightarrow -\frac{1}{2}$ is unaffected in first order by the imperfections of the crystal.

For a given number of nuclear spins in the crystal and a given detection device, let Δ be the line width in c/s above which the nuclear signal falls below the noise level, δ the dipolar width of the line in a perfect crystal, ν_L the Larmor frequency, and $h\overline{|\nu_Q|}$ the average magnitude of the quadrupole couplings due to imperfections. If the conditions

$$\Delta < \overline{|\nu_Q|}, \qquad \delta \gg \nu_Q^2/\nu_L \tag{30}$$

are realized simultaneously, the satellites will be completely wiped out by the first order quadrupole broadening due to imperfections, whereas the central line $-\frac{1}{2} \leftrightarrow \frac{1}{2}$ will be practically unaffected.

A single narrow resonance line will be observed in that case, and the quadrupole broadening will manifest itself only through a loss in intensity, the relative intensity of the central line being

$$0{\cdot}4 \text{ for spins } \tfrac{3}{2}, \qquad \frac{9}{9+2\times(8+5)} = \frac{9}{35} \text{ for spin } \tfrac{5}{2}, \text{ etc.}$$

Thus the line intensities of Br^{79} (spin $\tfrac{3}{2}$), Br^{81} (spin $\tfrac{3}{2}$), and I^{127} (spin $\tfrac{5}{2}$) in single crystals of KBr and KI were found to be respectively about 0·4 times and 0·3 times the expected intensities (13). Similarly, it was found (14) that in filings of metallic copper ($I = \tfrac{3}{2}$), annealing the sample increased the signal by a factor 2·5. This was interpreted by assuming that in the annealed metal the full intensity was observed, whereas in the original sample the existence of dislocations (eliminated by annealing) led to the disappearance of satellite lines through first-order quadrupole broadening. Incidentally, it may be pointed out that if the conditions (30) are satisfied, the dipolar width of the central components is actually smaller than that of the full line in a perfect crystal. The reason, as explained in Chapter IV, is that simultaneous flips: $-\tfrac{1}{2} \leftrightarrow \tfrac{1}{2}$, $m \to m-1$ with $m \neq \tfrac{1}{2}$, will not occur between two neighbouring spins since because of quadrupole broadening they correspond to different frequencies. The dipolar second moment of the central line is given by formulae (65) of Chapter IV.

If the first-order broadening is not so large as to prevent the satellite lines from being observed, their width and shape can be computed if definite assumptions are made about the nature of the crystal defects causing the broadening.

Assuming point defects distributed at random in the crystal, the shape and the width of a line broadened by first order quadrupole interactions due to these defects can be easily predicted in the two limiting cases of large and small defect concentrations (9).

Let $\overline{\nu^2}$ be the mean square frequency displacement for a given transition $m \to m-1$ ($m \neq \tfrac{1}{2}$) caused by the defects surrounding a nuclear spin. Since the crystal is on the average cubic, the average frequency shift $\bar{\nu}$ vanishes. If the concentration of defects is so large that each nuclear spin is influenced appreciably by, say, as many as ten different defects, it is possible to use the central limit theorem and to state that the line shape is gaussian with an r.m.s. width $(\overline{\nu^2})^{\tfrac{1}{2}}$.

We may ask why in a crystal lattice, where a spin has many nearest and next nearest neighbours, the dipolar line shape does depart appreciably from a gaussian one. The answer is that because of the mutual flip term in the dipolar interaction the various broadening agents, namely the spins surrounding a given spin, are not independent of each other and the central limit theorem is not valid. On the other hand, in electron spin resonance, if the line is broadened by unresolved hyperfine structure owing to the coupling of the electron spin with several identical nuclear spins, the shape is very nearly gaussian. There the interaction among nuclear spins is negligible in comparison with their coupling with the electron spin, and and they may be considered as independent broadening agents.

For small concentrations, assume for definiteness that the defects are point charges of either sign in equal proportions. For a nuclear spin separated from a charged defect by a vector $\mathbf{r}$ making an angle θ with the applied field H, the frequency shift of a satellite will be proportional to the component of the field gradient parallel to the field produced by the charged defect, that is, to $(1-3\cos^2\theta)/r^3$. We can write such a frequency displacement as

$$\nu = \pm \tfrac{1}{2}\nu_c a^3 \frac{1-3\cos^2\theta}{r^3}, \tag{31}$$

where ν_c has the dimension of a frequency and a is the distance between two nearest neighbours in the lattice.

The formula (31) is formally identical with formula (58) of Chapter IV, which gives the frequency displacement of a spin coupled by dipolar interaction to another spin in the theory of dipolar broadening under great dilution. We can take over the result of the calculation performed there and conclude that the line has a Lorentz shape with a width Γ that results immediately from formula (63) of Chapter IV, if $\tfrac{1}{2}\nu_c a^3$ replaces $\tfrac{3}{4}\gamma^2\hbar$.

$$\Gamma = \frac{4\pi^2}{9\sqrt{3}}\nu_c a^3 n, \tag{32}$$

where n is the volume concentration of the defects. If polarization effects could be neglected, the frequency ν_c in (31) would be given by

$$h\nu_c = (2m-1)\frac{3e^2Q}{2I(2I-1)}a^{-3}.$$

Actually, as explained in Section I of Chapter VI, ν_c should be multiplied by the so-called anti-shielding factor $(1+\gamma)$ that corresponds to the distortion of the electronic closed shells of the ion that contains the nuclear spin, by the external gradient of the charged defect. Values of $|\gamma|$ larger than a hundred have been obtained from theory for the heavier ions. Values of comparable order of magnitude are required for a given density of defects to explain the disappearance of satellite lines in imperfect cubic crystals, and also, as will appear in Chapter IX, to account for observed values of relaxation times. To take into account the macroscopic polarization of the crystal by a point-charge defect, ν_c should also be multiplied by an extra factor $(2\epsilon+3)/5\epsilon$ (9), where ϵ is the dielectric constant of the crystal. As ϵ varies from unity to infinity, this factor only changes from unity to $\frac{2}{5}$ and is thus much less important than $(1+\gamma)$.

(3) *Transient methods, multiple echoes*

The C.W. nuclear resonance method applied to the study of quadrupole interactions in imperfect cubic crystals has a serious limitation: if the conditions (30) are satisfied, that is, if the satellites are wiped out and the central line unaffected, only an upper and a lower limit can be set for the strength of the random quadrupole interactions.

Since the quadrupole broadenings are essentially static in nature it is possible to apply to their study a transient spin echo method (15) which overcomes that limitation to a certain extent. Since, apart from the extra information on imperfect crystals that it is capable of providing, it also has some very interesting features of its own, this method will be described in some detail. Because the symmetry of the quadrupole broadening is different from that of the magnetic broadening caused by an inhomogeneous magnetic field, the usual unsophisticated explanation of the damping of the free precession and of the formation of the echo in purely classical terms of precessing moments cannot be used and a quantum mechanical description is necessary. The quadrupole coupling $\mathcal{H}_Q$ seen by a given spin can in first order be replaced by its part $\tilde{\mathcal{H}}_Q$ that commutes with the Zeeman Hamiltonian $-\gamma\hbar H_0 I_z$:

$$\tilde{\mathcal{H}}_Q = \frac{eQ}{4I(2I-1)}[3I_z^2-I(I+1)]V_{zz} = a\hbar I_z^2 + \text{c.c.} \tag{33}$$

where, according to (28),

$$\hbar a = \frac{3e^2qQ}{4I(2I-1)}[\cos^2\theta - \tfrac{1}{2}\sin^2\theta + \tfrac{1}{2}\eta\sin^2\theta\cos 2\varphi],$$

q, η, θ, and φ being random variables describing the orientation and the strength of the field gradient at the various nuclear sites. From their distribution laws there is a certain distribution $f(a)$ normalized to unity for the random variable a itself. From the shape of the echo some information about $f(a)$ should be obtainable. In accordance with the conditions (30), we may write

$$\overline{a^2} = \int a^2 f(a)\,da \gg \delta^2, \qquad \overline{a^2}/\nu_L \ll \delta.$$

In the following we neglect the dipolar width δ, thus assuming an infinitely sharp central line. In the frame rotating at the Larmor frequency $\omega_L = -\gamma H_0$, the effective field vanishes and the Hamiltonian of the spin system reduces to $\mathcal{H}_Q$ as given by (33).

The signal proportional to the amplitude of the voltage induced in a coil by the precessing magnetization is proportional to

$$S(t) = \mathrm{tr}[\rho(t) I_+], \tag{34}$$

where $\rho(t)$ is the density matrix describing the motion of the spin system in the rotating frame. When this system is in thermal equilibrium ρ is a function of I_z only and (34) vanishes as it should. After an r.f. pulse, occurring at time $t = 0$, the density matrix $\rho(t)$ at time t is related to its value $\rho(0)$ immediately after the pulse, through

$$\rho(t) = \exp\left[-\frac{i}{\hbar}\mathcal{H}_Q t\right]\rho(0) \exp\left[\frac{i}{\hbar}\mathcal{H}_Q t\right]$$

$$= \exp[-ia I_z^2 t]\rho(0) \exp[ia I_z^2 t], \tag{35}$$

and the signal of free precession $S(t)$ following the pulse is

$$S(t) = \mathrm{tr}[e^{-ia I_z^2 t}\rho(0) e^{ia I_z^2 t} I_+]$$

$$= \sum_m \langle m | \rho(0) | m+1 \rangle [I(I+1) - m(m+1)]^{\frac{1}{2}} e^{ia(2m+1)t}. \tag{36}$$

The term $m = -\frac{1}{2}$ of the sum (36) is time-independent and in contrast with the magnetic broadening case $S(t)$ does not decay to zero but tends instead towards a finite limit

$$S(\infty) = \langle -\tfrac{1}{2} | \rho(0) | \tfrac{1}{2} \rangle (I+\tfrac{1}{2}). \tag{37}$$

This time-independent signal is the Fourier transform of the central line $\frac{1}{2} \to -\frac{1}{2}$, assumed to be infinitely sharp. In practice, however, it will decay in a time of the order of $(2\pi\delta)^{-1}$ and our neglect of the dipolar width δ is only valid for times $t \ll (2\pi\delta)^{-1}$. The positions and shapes of the echoes are obtained as follows. We assume that a second pulse of negligible duration Δt is applied at a time τ after the first pulse. The effect of such a pulse can be described by means of a unitary operator R that transforms the density matrix $\rho(\tau)$ existing immediately before the pulse into

$$\rho'(\tau) = R\rho(\tau) R^{-1} = R e^{-ia I_z^2 \tau}\rho(0) e^{ia I_z^2 \tau} R^{-1}. \tag{38}$$

At time $t > \tau$ we shall have

$$\rho(t) = e^{-ia I_z^2 (t-\tau)}\rho'(\tau) e^{ia I_z^2 (t-\tau)}$$

Ch. VII   QUADRUPOLE EFFECTS   243

and the signal $S(t) = \text{tr}[\rho(t)I_+]$ will be

$$S(t) = \text{tr}[e^{-iaI_z^2(t-\tau)}Re^{-iaI_z^2\tau}\rho(0)e^{iaI_z^2\tau}R^{-1}e^{iaI_z^2(t-\tau)}I_+]$$

$$= \sum_{mm'm''} \langle m | R | m'' \rangle \langle m'' | \rho(0) | m' \rangle \langle m' | R^{-1} | m+1 \rangle \times$$

$$\times [I(I+1)-m(m+1)]^{\frac{1}{2}}\exp\{ia[(2m+1)(t-\tau)-(m''^2-m'^2)\tau]\}. \quad (39)$$

The term of the sum, (39), that corresponds to $2m+1 = m''^2-m'^2 = 0$ gives a time-independent signal E_0. If for particular values of t, $S(t)$ is independent of a, signals from all the nuclei are 'in phase' and an echo appears. According to (39) this occurs for

$$\frac{t-\tau}{\tau} = \frac{m''^2-m'^2}{2m+1} = k. \quad (40)$$

We consider specifically the case of a spin $I = \frac{5}{2}$, which is that of I^{127} on which the experiment was performed in a crystal of KI (15). The possible values of k are easily found to be $k = \frac{1}{2}, 1, \frac{3}{2}, 2, 3$. Thus in principle five echoes at most can be expected, occurring at various intervals after the second pulse given by (40).

Allowed echoes. To estimate their relative intensities assumptions must be made about the nature of the r.f. pulses. We suppose first that the amplitude H_1 of the r.f. field is so large that during a pulse the existence of quadrupole couplings may be disregarded. This can be expressed through the condition

$$|\omega_1| = |\gamma H_1| \gg (\overline{a^2})^{\frac{1}{2}}.$$

The effect of a pulse of duration Δt is then to rotate all the spins through the same angle $\varphi = -\gamma H_1 \Delta t$ around an axis Oy of the rotating frame. In particular, if a first pulse of 90° is applied to a spin system in thermal equilibrium described by a density matrix

$$\rho_{\text{equil}} \sim e^{\gamma\hbar I_z H/kT} \simeq 1 + \frac{\gamma\hbar I_z H}{kT},$$

immediately after the pulse

$$\rho(0) \sim 1 + \frac{\gamma\hbar I_x H}{kT},$$

which we may write as $\rho(0) \sim I_x$ since the unity term in the expansion of $\rho(0)$ clearly gives no contribution to the signal. The operator $\rho(0) \sim I_x$ has non-vanishing matrix elements

$$\langle m'' | \rho(0) | m' \rangle \quad \text{for} \quad |m''-m'| = 1$$

only, which, as can be seen from (40), forbids the values $k = \frac{3}{2}$ and 3. To get the shapes and amplitudes $E_k(t)$ of the remaining three echoes we must integrate the signal (39) over the distribution $f(a)$ of the quadrupole frequencies:

$$E_k(t) = \sideset{}{'}\sum_{mm'm''} \langle m \mid R \mid m''\rangle\langle m'' \mid \rho(0) \mid m'\rangle\langle m' \mid R^{-1} \mid m+1\rangle \times$$

$$\times [I(I+1)-m(m+1)]^{\frac{1}{2}} \int_{-\infty}^{\infty} f(a)\exp\{i(2m+1)a[t-(k+1)\tau]\}\,da. \quad (41)$$

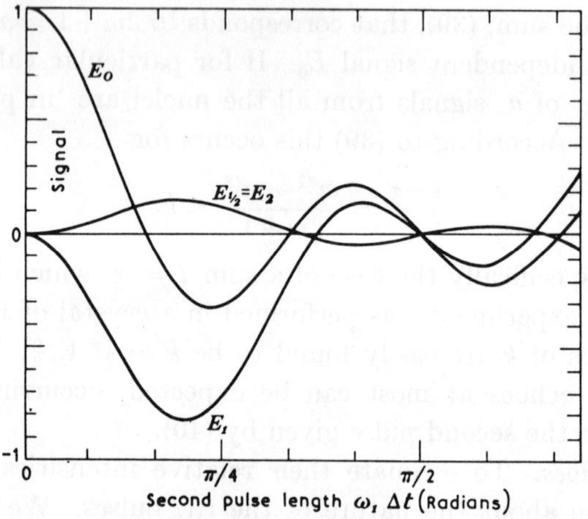

FIG. VII, 9. Calculated values of allowed echoes and time-independent tail after a sequence of a 90° pulse followed by a φ pulse, as a function of $\varphi = \gamma H_1 \Delta t$. The value of the time-independent tail before the second pulse is the signal unit.

The summation $\sum'$ in (41) must be restricted to values such that

$$m''^2 - m'^2 = k(2m+1).$$

With our assumption of large r.f. fields the matrix elements such as $\langle m \mid R \mid m''\rangle$ are simply those of an operator, rotation of a spin $I = \frac{5}{2}$ through an angle φ around an axis Oy, and are given by

$$\langle m \mid R \mid m'\rangle = \langle m \mid \mathscr{D}^{\frac{5}{2}}(0,\varphi,0) \mid m'\rangle. \quad (42)$$

See ref. (16). The relative values of $E_k(t)$ computed as a function of the angle φ of the second pulse, the first being 90°, are represented in Fig. VII, 9. They are symmetrical with respect to the value $\varphi = \frac{1}{2}\pi$ and, with the exception of E_0, vanish for $\varphi = \pi$, in contrast with the magnetic broadening case where the amplitude of the echo is a maximum when the second pulse has the amplitude π. The shapes of the two echoes $E_{\frac{1}{2}}$ and E_2 are given by

$$E_{\frac{1}{2}} = 2\sqrt{10}\langle \tfrac{5}{2} \mid R \mid \tfrac{1}{2}\rangle\langle \tfrac{3}{2} \mid R \mid \tfrac{3}{2}\rangle \int_{-\infty}^{\infty} f(a)\cos[4a(t-\tfrac{3}{2}\tau)]\,da,$$

$$E_2 = 2\sqrt{10}\langle \tfrac{5}{2} \mid R \mid \tfrac{1}{2}\rangle\langle \tfrac{3}{2} \mid R \mid \tfrac{3}{2}\rangle \int_{-\infty}^{\infty} f(a)\cos[2a(t-3\tau)]\,da. \quad (43)$$

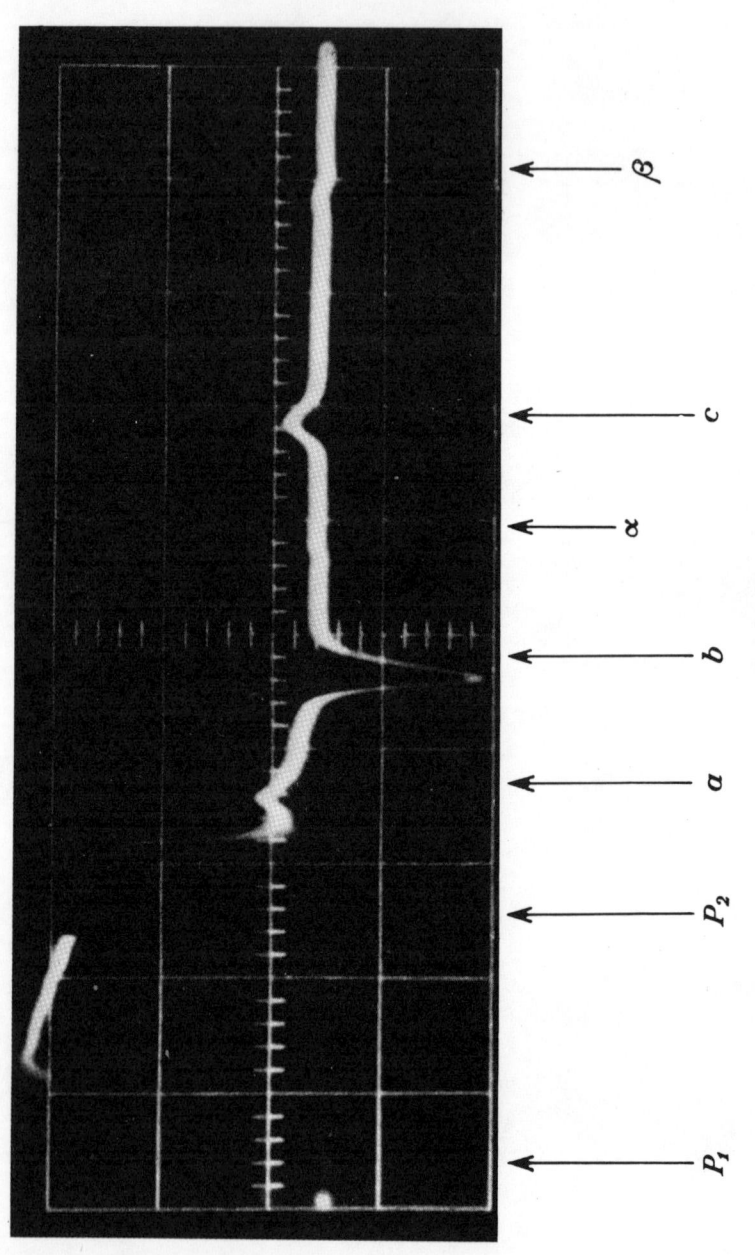

FIG. VII, 10. The five echoes of I^{127} in a single crystal of potassium iodide:

$P_1 = 90°$ pulse
$P_2 = 35°$ pulse
a, b, c = allowed echoes
α, β = forbidden echoes

The total trace is 500 μ long.

It appears from (43) that these two echoes have the same amplitude but that E_2 is twice as broad as $E_{\frac{3}{2}}$. We omit a similar but slightly more complex formula for $E_1(t)$ which is a sum of several terms analogous to (43).

Forbidden echoes. Actually the ratio $(\overline{a^2})^{\frac{1}{2}}/|\gamma H_1|$ is not entirely negligible, the effect of the first pulse is not a pure rotation, and $\rho(0)$ is not strictly equal to I_x. To first order we may write

$$\rho(0) = I_x + \frac{a}{\gamma H_1}\rho_2,$$

where ρ_2 has non-vanishing matrix elements for $\Delta m = \pm 2$, which carried into (39) give 'forbidden echoes' at the times $t-\tau = \frac{3}{2}\tau$ and 3τ. As functions of time their amplitudes are proportional to

$$E_{\frac{3}{2}}(t) \sim \int_{-\infty}^{\infty} \frac{a}{\gamma H_1} f(a)\sin[4a(t-\tfrac{5}{2}\tau)]\,da,$$

$$E_3(t) \sim \int_{-\infty}^{\infty} \frac{a}{\gamma H_1} f(a)\sin[2a(t-4\tau)]\,da. \qquad (44)$$

It is seen from (44) that since $f(a)$ is an even function of a these echoes are not bell-shaped curves like the allowed echoes but rather derivatives of such curves.

Fig. VII, 10 (15) shows the five echoes of I^{127} for a first pulse of $90°$ and a second pulse of $35°$. The number, positions, shapes, and relative amplitudes of the five echoes are in satisfactory agreement with the above theory. Since the decay time $(2\pi\delta)^{-1}$ of the signals, caused by the dipolar broadening, was of the order of 1 millisec, an interval τ between pulses of the order of 125 μs was used.

Fig. VII, 11 shows the inverse width of the echo E_1, as a function of the amplitude H_1 of the r.f. field for three different samples of KI. Since, as appears from (41), the shape of an echo is the Fourier transform of the distribution of quadrupole couplings $f(a)$, the inverse width of the echo is a measure of the strength of quadrupole couplings.

It is interesting to notice on Fig. VII, 11 that for the smaller values of H_1, the inverse width is proportional to H_1 and is independent of the sample. This is understandable, for as long as $|\gamma H_1|$ is small compared with the width $\overline{(a^2)}^{\frac{1}{2}}$ of the distribution of frequencies, only the fraction $|\gamma H_1|/(\overline{a^2})^{\frac{1}{2}}$ of the spins, for which the spread in frequencies is of the order of $|\gamma H_1|$, is affected by the pulse and contributes to the signal. On the other hand, for $|\gamma H_1| \gg (\overline{a^2})^{\frac{1}{2}}$ the intrinsic features of the distribution $f(a)$ will show up, and the echo method appears as a powerful tool for the study of strains and defects in cubic crystals. From the instrumental point of view it is imperative that the pulses be coherent since the amplitudes of the various echoes combine coherently with the ever present time-independent signal $E_0(t)$.

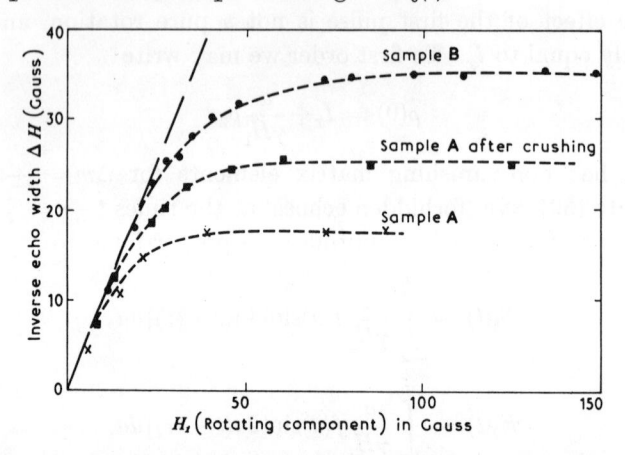

FIG. VII, 11. Inverse echo width ΔH as a function of the r.f. field H_1. In the limit of large H_1, ΔH is a measure of the average interaction between the quadrupole moment of iodine and the random gradients due to the defects in the KI crystals.

If, instead of large r.f. fields H_1, small r.f. fields are used, they should flip only those spins that are in the states $\pm\frac{1}{2}$. This problem can be treated by the technique of the fictitious spin $\frac{1}{2}$ as explained in Chapter II. As stated there for spin $I = \frac{5}{2}$ the maximum decay tail should be observed for a 30° pulse $|\gamma H_1 \Delta t| = \frac{1}{6}\pi$ rather than for a 90° pulse.

(4) *Second-order quadrupole broadening in imperfect crystals*

As the strength of quadrupole interactions caused by crystal defects increases, the central line $-\frac{1}{2} \to \frac{1}{2}$ also begins to get affected. From the formula (26) that gives the displacement of the frequency of a given nucleus experiencing a quadrupole gradient (assumed symmetrical for simplicity) one expects the line to become asymmetrical with its centre of gravity shifted below the unperturbed frequency ν_L. Quantitative predictions about the shape of the line as a function of the nature and the concentration of crystal defects are difficult and uncertain (**9**).

Experimental evidence of second-order broadening was obtained first in ionic crystals (**13**). In KBr the resonance lines of Br[79] and Br[81] became broader and asymmetrical with more intensity on the low-frequency side of the line when the crystals were subjected to linear compression resulting in about 22 per cent decrease in length because of plastic flow. The resonance line of I^{127} in KI was also made somewhat asymmetrical through compression and its intensity was reduced by a few per cent. These effects have been explained by an increase in the density of dislocations due to the compression.

Large quadrupole gradients are created in the neighbourhood of foreign atoms introduced into ionic crystals or metals. We saw earlier that in metallic copper cold work could reduce the signal to a fraction 0·4 of that observed in well-annealed metal, in agreement with the assumption that such mechanical treatment was unable to affect the central transition $-\frac{1}{2} \rightarrow \frac{1}{2}$.

To reduce the signal below the 0·4 value, which corresponds to the disappearance of satellites, zinc or silver has been added to copper (14). Curiously enough, as the concentration of foreign atoms increases, the copper line neither broadens nor becomes asymmetrical but simply decreases in intensity. Fig. VII, 12 shows the signal of Cu^{63} in the alloy as a function of the concentration of the foreign atom Zn or Ag. The explanation proposed is that of an all-or-nothing effect of a foreign atom on the resonance of a neighbouring copper. If we assume that the electric field gradient produced by a foreign atom at a distance r is proportional to r^{-3}, the second-order frequency shift of the central line of a nuclear spin of copper caused by an impurity at that distance, will be proportional to r^{-6}. In a face-centred cubic lattice the ratios of the frequency displacements caused by impurities in positions of first, second, and third nearest neighbours of a copper atom will be 27:3·37:1. It is thus conceivable that this displacement will be much larger than the dipolar width for say, the nearest neighbour and appreciably smaller for the others. If this is so, all copper atoms which have no impurity atoms among their first nearest neighbours will give an essentially undisplaced signal, whereas the other atoms will move outside the observable range. For small concentrations c of impurities, the probability of a copper atom having no impurities among its z nearest neighbours will be $(1-c)^z$ and the intensity of the signal as a function of $1-c$, should have the slope z on a doubly logarithmic plot. Experimental points on such a plot do indeed give a straight line but its slope turns out to be 18, the sum of the numbers of first and second nearest neighbours in an f.c.c. lattice, showing that the critical radius separating impurities that throw the resonance frequency out of observation from those that leave it inside the dipolar width, passes between the second and the third nearest neighbours.

On the other hand, in alloys of aluminium and zinc (spin of Al^{27} $\frac{5}{2}$) it is found that as the concentration in zinc is increased the intensity of the aluminium signal tends towards the limiting value of 9/35 of the signal in pure aluminium, showing that even the deformation caused by a foreign atom nearest neighbour of an aluminium atom is unable

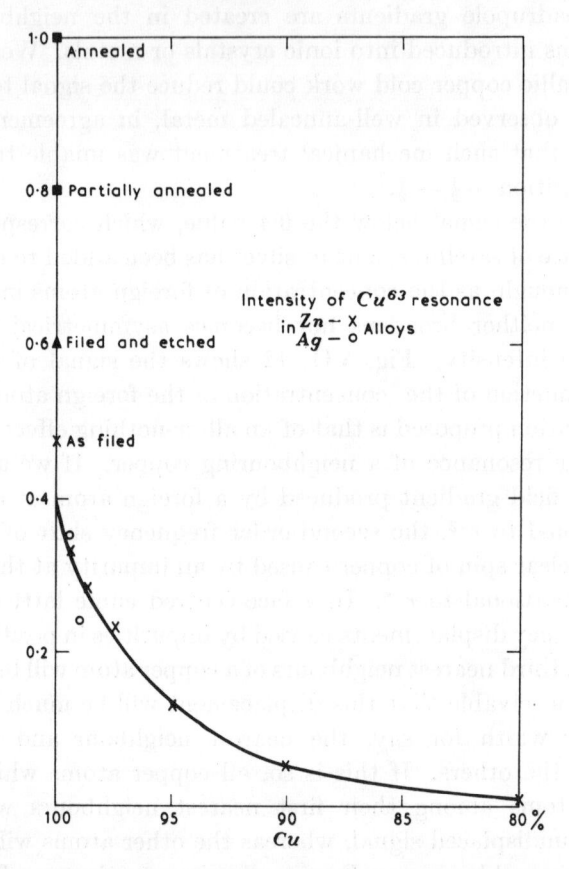

Fig. VII, 12. The intensity of the Cu^{63} magnetic resonance in annealed and cold-worked copper and in alloys with zinc and silver. By cold work the contribution of the transitions $m_I = \frac{3}{2} \to \frac{1}{2}$, $m_I = -\frac{1}{2} \to -\frac{3}{2}$ is washed out. Annealing restores the total intensity of the nuclear resonance.

to shift appreciably the frequency of the central line. There seems to be evidence in that alloy of an 'all-or-nothing' effect for the satellites broadened in first order only. This seems difficult to understand since the frequency displacement as a function of the distance r from the impurity varies as r^{-3} and has a much slower variation than for the central line. Any detailed discussion of these problems would necessarily

B. Low magnetic fields

We consider now the situation where the Zeeman energy of the nuclear spin is a small fraction of its quadrupole energy. The important limiting case, when the applied magnetic field vanishes altogether, described in the literature (somewhat improperly as pointed out in Chapter I) as pure quadrupole resonance, will be considered first.

(a) Zero field spectra

A magnetic r.f. field of the right frequency may induce magnetic dipole transitions between the energy levels of the quadrupole Hamiltonian $\mathcal{H}_Q$ given by the second term of (22):

$$\mathcal{H}_Q = A[3I_Z^2 - I(I+1) + \tfrac{1}{2}\eta(I_+^2 + I_-^2)], \tag{45}$$

where
$$A = \frac{e^2qQ}{4I(2I-1)}.$$

The components I_Z, I_+, I_- of the nuclear spin are referred to the molecular frame Σ^P and the energy levels and resonance frequencies are independent of the orientation of the crystal in space. It is therefore clear that the zero field resonance can be (and has been) observed in powders as well as in single crystals. It has, in particular, been observed in metals where skin depth requirements forbid the use of single crystals. The energy levels of (45) are most easily found for a symmetrical gradient when η vanishes and are given by

$$E_m = \frac{e^2qQ}{4I(2I-1)}[3m^2 - I(I+1)]. \tag{46}$$

There is a twofold degeneracy, the levels $I_Z = \pm m$ having the same energy. The selection rule for the transitions is $|\Delta m| = 1$. According to (46) there is a single resonance line for spins $I = 1$ and $\tfrac{3}{2}$ with frequencies given respectively by

$$\nu_1 = \frac{3}{4}\frac{e^2qQ}{h}, \qquad \nu_{\frac{3}{2}} = \frac{1}{2}\frac{e^2qQ}{h}.$$

For $I = 2$ there are two lines corresponding to the transitions $|1| \to 0$ and $|2| \to |1|$ with frequencies

$$\nu_2' = \frac{1}{8}\frac{e^2qQ}{h}, \qquad \nu_2'' = 3\nu_2'.$$

For $I = \frac{5}{2}$ there are two lines corresponding to the transitions $|\frac{3}{2}| \to |\frac{1}{2}|$ and $|\frac{5}{2}| \to |\frac{3}{2}|$ with frequencies

$$\nu'_{\frac{1}{2}} = \frac{3}{20}\frac{e^2qQ}{h}, \qquad \nu''_{\frac{1}{2}} = 2\nu'_{\frac{1}{2}}.$$

More generally, for two lines corresponding to transitions $|m| \to |m|-1$ and $|m'| \to |m'|-1$, the ratio of their frequencies is

$$\frac{2|m|-1}{2|m'|-1}.$$

If the asymmetry parameter η is different from zero, the behaviour of the energy levels is quite different for integer and half-integer spins.

(1) *Integer spins*

For integer spins, the $\pm m$ degeneracy of the axial symmetry case is lifted completely. In order to solve the secular equation for (45) it is convenient to notice that the asymmetry term $\frac{1}{2}\eta(I_+^2+I_-^2)$ does not couple together even states $(|m\rangle+|-m\rangle)/\sqrt{2}$ and odd states $(|m\rangle-|-m\rangle)/\sqrt{2}$. Furthermore, states of the same parity are only coupled if Δm is an even integer. It follows from these remarks that the secular determinant of order $2I+1$ may be factorized into equations of lower order, the highest being clearly $(\frac{1}{2}I)+1$ if I is an even integer, and $(\frac{1}{2}I)+\frac{1}{2}$ if I is odd.

Thus for $I = 1$ it follows from (45) that the eigenstates are the states $|0\rangle$ and $|\xi_\pm\rangle = (|1\rangle\pm|-1\rangle)/\sqrt{2}$ with energies

$$E_0 = -\tfrac{1}{2}e^2qQ, \qquad E_{\xi_\pm} = \tfrac{1}{4}e^2qQ(1\pm\eta).$$

It is worth noting that the three states $|\xi_\pm\rangle$ and $|0\rangle$ are eigenstates of I_Y, I_X, and I_Z with the same eigenvalue zero, of these operators. For $I = 2$ and 3 there is no equation in the secular determinant of order higher than second and the energies and eigenstates can be calculated exactly. For higher integer spins a numerical solution is necessary.

If η is small and the anisotropy term is treated as a perturbation, it follows from (45) that for $I > 1$ the level $m = \pm 1$, which is degenerate in a symmetrical gradient, is the only one to be split in first order.

Finally, it may be useful to remark that when $\eta \neq 0$ the quadrupole Hamiltonian of an integer spin I is formally identical with the rotational Hamiltonian of an asymmetric top molecule, a problem treated in detail in the literature (**17**). The existence of an asymmetry of the gradient is observed through an increase in the number of lines, caused by the lifting of the $\pm m$ degeneracy. For instance, for a spin $I = 1$

three lines are observable in principle, corresponding to the three transitions: $|\xi_+\rangle \to |0\rangle$, $|\xi_-\rangle \to |0\rangle$, $|\xi_+\rangle \to |\xi_-\rangle$. Their frequencies are:

$$\omega_{YZ} = \frac{3}{4}\frac{e^2qQ}{\hbar}(1+\tfrac{1}{3}\eta), \qquad \omega_{XZ} = \frac{3}{4}\frac{e^2qQ}{\hbar}(1-\tfrac{1}{3}\eta), \qquad \omega_{XY} = \frac{e^2qQ}{\hbar}\tfrac{1}{2}\eta. \tag{47}$$

The lack of symmetry between the expressions (47) of these three frequencies is due to our singling out of the Z-axis in the definition of q and η. If η is small, the frequency of the last line is much lower than the other two and may be difficult to observe. From a measurement of two of the three frequencies (47), e^2qQ and η can both be obtained.

(2) *Half-integer spins*

For half-integer spins the asymmetry term $\tfrac{1}{2}\eta(I_+^2+I_-^2)$ is unable to lift the twofold degeneracy of the eigenstates (although $m = I_Z$ is clearly no longer a good quantum number). This result, which is a special case of a general theorem due to Kramers, can be understood in an elementary fashion as follows:

(i) all matrix elements $(m' | \mathcal{H}_Q | m'')$ vanish unless
$$|m'-m''| = 0, 2;$$

(ii) for every matrix element
$$(m' | \mathcal{H}_Q | m'') = (-m'' | \mathcal{H}_Q | -m');$$

(iii) because of (i), if $|\xi\rangle = \sum_m c_m |m\rangle$ is an eigenstate of $\mathcal{H}_Q$, the difference between any two values of m in the sum is an *even* integer;

(iv) because of (ii), the state vector $|\bar{\xi}\rangle = \sum_m c_m |-m\rangle$ is an eigenstate of $\mathcal{H}_Q$ with the same energy as $|\xi\rangle$. This energy level is thus twofold degenerate unless $|\xi\rangle$ and $|\bar{\xi}\rangle$ are the same vector, which cannot be for then $|\xi\rangle$ would contain in its expansion both $|m\rangle$ and $|-m\rangle$. This is impossible for $m-(-m) = 2m$ is an odd integer. $|\bar{\xi}\rangle$ is called the Kramers conjugate of $|\xi\rangle$.

Thus for $I = \tfrac{3}{2}$ there are still only two levels and a single line when $\eta \neq 0$ and it is impossible to find out from the resonance frequency whether the gradient is symmetrical or not.

The secular equations for the energy levels of a half-integer spin in an asymmetrical gradient of order $I+\tfrac{1}{2}$ are easily found from (45). For $I = \tfrac{3}{2}$, and writing
$$x = \frac{E}{A}, \qquad A = \frac{e^2qQ}{4I(2I-1)},$$

the secular equation is $\quad x^2-3\eta^2-9=0,\quad$ (48)

and the single resonance frequency

$$\nu = \frac{e^2qQ}{2h}(1+\tfrac{1}{3}\eta^2)^{\frac{1}{2}}.$$

For $I = \tfrac{5}{2}$, and writing $x = E/2A$, the secular equation is

$$x^3-7(3+\eta^2)x-20(1-\eta^2) = 0. \quad (48')$$

An expansion of the energy levels limited to the lowest power in η^2, valid for $\eta < 0.1$, gives

$$E_{\pm\frac{5}{2}} \simeq A(10+\tfrac{5}{9}\eta^2), \qquad E_{\pm\frac{3}{2}} \simeq A(-2+3\eta^2), \qquad E_{\pm\frac{1}{2}} \simeq A(-8-\tfrac{32}{9}\eta^2).$$

A symbol such as $E_{\pm\frac{5}{2}}$ represents the energy of the states that reduce to $m = \pm\tfrac{5}{2}$ when $\eta = 0$. The two transition frequencies become

$$\nu'_{\frac{3}{2}} \simeq \frac{3e^2qQ}{20h}(1+\tfrac{59}{54}\eta^2), \qquad \nu''_{\frac{3}{2}} \simeq \frac{3e^2qQ}{10h}(1-\tfrac{11}{54}\eta^2),$$

whence $\qquad\qquad\qquad \dfrac{\nu''_{\frac{3}{2}}}{\nu'_{\frac{3}{2}}} \simeq 2(1-\tfrac{35}{27}\eta^2).$

For a spin $I = \tfrac{5}{2}$, the asymmetry of the gradient is thus immediately apparent from the ratio of the two frequencies. Secular equations for higher spins, more accurate expressions for the energy levels, and considerations on the relative intensities of the various lines can be found in reference (10).

We shall represent by a notation such as $|\widetilde{\pm m}\rangle$ the degenerate manifold which reduces to $|\pm m\rangle$ when η vanishes. If η is not small m is not even approximately a good quantum number and extra transitions, forbidden in a symmetrical gradient, such as $|\widetilde{\tfrac{5}{2}}\rangle \to |\widetilde{\tfrac{1}{2}}\rangle$, can be observed in principle. We shall disregard them in what follows.

Quadrupole frequencies vary within very large limits that depend on the magnitude of the quadrupole nuclear moment and the nature of the compound studied. For a survey of the very considerable experimental evidence available, the reader is again referred to reference (10). Small quadrupole splittings below, say, 1 Mc/s are more conveniently studied in high magnetic fields if single crystals are available. The largest frequencies observed so far are those of I^{127} in covalent compounds. Thus, in solid ICl, values of e^2qQ/h of the order of 3000 Mc/s and resonance frequencies of the order of 450 Mc/s and 900 Mc/s have been observed. Even larger quadrupole splittings exist in uranyl compounds, as appears from the fact that α particles emitted by nuclei of U^{233} at 1° K show appreciable anisotropy.

(b) *Zeeman splittings of quadrupole levels*

In an experiment performed in zero field, only the constants e^2qQ and η can be accurately measured (the latter only if $I \neq \tfrac{3}{2}$). In order to determine the orientation of the principal axes of the electric-field gradient tensor, with respect to the laboratory frame, it is necessary to create a privileged direction in this frame by applying to the sample a magnetic field H. The coupling of H with the nuclear magnetic moment splits or shifts the quadrupole energy levels by an amount that depends on the relative orientation of the magnetic field and the crystal frame and therefore results in a broadening of the lines unless single crystals are used. The common practice in these experiments is to use small magnetic fields $\leqslant 100$ gauss. The Zeeman energy is then much smaller than the quadrupole energy and a perturbation treatment is sufficient.

(1) *Integer spins*

Let θ and φ define the orientation of the magnetic field H with respect to the crystalline frame. The Zeeman Hamiltonian can be written

$$\mathscr{H}_M = -\gamma\hbar H(I_Z \cos\theta + I_X \sin\theta \cos\varphi + I_Y \sin\theta \sin\varphi). \qquad (49)$$

If the gradient is symmetrical so that I_Z is a good quantum number the degeneracy of the levels $I_Z = \pm m$ is lifted by the magnetic field and the splitting is given by first-order perturbation theory as

$$E_{-m} - E_m = 2\gamma\hbar H m \cos\theta.$$

The simplicity of this formula is due to the fact that for an integer spin $|m-(-m)|$ is at least 2 so that inside the degenerate manifold $|\pm m\rangle$, $\mathscr{H}_M$ has no off-diagonal matrix elements $(m\,|\,\mathscr{H}_M\,|-m)$.

If $\eta \neq 0$ there is no degeneracy left in zero field and the effect of the magnetic field is simply to displace the levels without introducing extra lines. The calculation of the displacements requires a knowledge of the eigenstates ξ_i of the quadrupole Hamiltonian $\mathscr{H}_Q$. The displacements are then given in first order by the expectation values $\langle \xi_i\,|\,\mathscr{H}_M\,|\,\xi_i\rangle$. If η is so small that the anisotropy splitting is comparable to the Zeeman splitting this treatment breaks down and $\mathscr{H}_M$ must be diagonalized inside the almost degenerate manifold of the two states $|\xi_+\rangle$ and $|\xi_-\rangle$, which reduces to the manifold $|\pm m\rangle$ for zero anisotropy. If $m = 1$, it is possible and simpler to start from the degenerate manifold $|\pm 1\rangle$ and inside this manifold diagonalize the sum

$$-\gamma\hbar\mathbf{H}\cdot\mathbf{I} + \tfrac{1}{2}A\eta(I_+^2 + I_-^2).$$

Thus for spin $I = 1$ the secular equation

$$\begin{vmatrix} -\gamma\hbar H\cos\theta - x & A\eta \\ A\eta & \gamma\hbar H\cos\theta - x \end{vmatrix} = 0,$$

where x is the distance from the degenerate level $|\pm 1\rangle$, gives

$$x = \pm(\gamma^2\hbar^2 H^2 \cos^2\theta + A^2\eta^2)^{\frac{1}{2}}$$

and a splitting (in cycles)

$$\Delta\nu = \frac{2}{h}(\gamma^2\hbar^2 H^2 \cos^2\theta + A^2\eta^2)^{\frac{1}{2}}.$$

A remarkable feature of this result is that it does not depend on the azimuthal angle φ of the magnetic field.

(2) *Half-integer spins*

As already mentioned, half-integer spins are more numerous and have been studied much more thoroughly than integer spins for quadrupole effects.

Assuming first that $\eta = 0$, for degenerate states $|\pm m\rangle$ the same argument as for integer spins shows that if $m \neq \pm\frac{1}{2}$ the changes in energy are given by the first-order formula

$$E_m = -\gamma\hbar H m \cos\theta.$$

The case $m = \pm\frac{1}{2}$ requires a special treatment since there the Zeeman Hamiltonian has off-diagonal matrix elements between the states $|\pm\frac{1}{2}\rangle$. The secular equation for the change in energy caused by the magnetic field, can be written

$$\begin{vmatrix} -\frac{1}{2}\gamma\hbar H\cos\theta - x & \frac{1}{2}\gamma\hbar H\sin\theta(I+\frac{1}{2}) \\ \frac{1}{2}\gamma\hbar H\sin\theta(I+\frac{1}{2}) & \frac{1}{2}\gamma\hbar H\cos\theta - x \end{vmatrix} = 0, \tag{50}$$

whence $\quad x = \mp\frac{1}{2}f\gamma\hbar H\cos\theta \quad$ with $\quad f = [1+(I+\frac{1}{2})^2\tan^2\theta]^{\frac{1}{2}}. \tag{50'}$

The eigenstates $|+\rangle$ and $|-\rangle$ which reduce to $|+\frac{1}{2}\rangle$ and $|-\frac{1}{2}\rangle$, respectively, when the applied field is parallel to the crystalline axis, are given by

$$|+\rangle = |\tfrac{1}{2}\rangle\cos\delta + |-\tfrac{1}{2}\rangle\sin\delta,$$

$$|-\rangle = -|\tfrac{1}{2}\rangle\sin\delta + |-\tfrac{1}{2}\rangle\cos\delta,$$

$$\tan\delta = \left(\frac{f-1}{f+1}\right)^{\frac{1}{2}}. \tag{50''}$$

The resonance spectrum observed with an r.f. field perpendicular to the crystalline axis, which in zero field contained the $I-\frac{1}{2}$ transitions $|m| \to |m|-1$, in the presence of a magnetic field is modified as follows:

for $|m| > \frac{3}{2}$, the single frequency observed in zero field is split into two,

$$\omega_{\pm|m|} = \frac{3A}{\hbar}(2|m|-1) \pm \gamma H \cos\theta.$$

On the other hand, the single line $\pm\frac{3}{2} \to \pm\frac{1}{2}$ is replaced now by four transitions:

$$|\tfrac{3}{2}\rangle \to |+\rangle, \quad |-\tfrac{3}{2}\rangle \to |-\rangle, \quad |\tfrac{3}{2}\rangle \to |-\rangle, \quad |-\tfrac{3}{2}\rangle \to |+\rangle.$$

The first two transitions are called in the literature the α lines, the last two the β lines, and their frequencies computed from (50') are given by

$$\omega_\alpha = \frac{6A}{\hbar} - \frac{3-f}{2}\gamma H \cos\theta,$$

$$\omega_\beta = \frac{6A}{\hbar} - \frac{3+f}{2}\gamma H \cos\theta,$$

$$\omega_{\alpha'} = \frac{6A}{\hbar} + \frac{3-f}{2}\gamma H \cos\theta,$$

$$\omega_{\beta'} = \frac{6A}{\hbar} + \frac{3+f}{2}\gamma H \cos\theta. \tag{51}$$

The intensities of the four lines are easily calculated from the matrix elements $|\langle\pm\tfrac{3}{2}|I_X|\pm\rangle|^2$ and depend on the orientation of the magnetic field. In particular, for $\theta = 0$, $f = 1$, $\delta = 0$, $|\pm\rangle = |\pm\tfrac{1}{2}\rangle$, the β lines vanish and a single pair of lines is observed. For $\theta = \tfrac{1}{2}\pi$, the splitting of the $\pm\tfrac{3}{2}$ levels vanishes in first order and again only a single pair is observed but for a different reason. The coalescence of the α and β lines is complete for very small fields only and as the field increases they separate. Finally the α lines collapse together for $f = 3$, that is, for $\tan\theta = 4\sqrt{2}/(2I+1)$. In particular, for $I = \tfrac{3}{2}$ this happens for $\tan\theta = \sqrt{2}$ or $1 - 3\cos^2\theta = 0$. From a study of the frequencies (51) as a function of the relative orientation of the magnetic field and the crystal, the axis of the gradient can be located. Fig. VII, 13 (**18**) shows the spectrum observed from Cl^{35} ($I = \tfrac{3}{2}$) in a single crystal of sodium chlorate. The unit cell contains four chlorines and the directions of the symmetrical electric-field gradients that they 'see', are those of the four body diagonals of a cube. The figure displays (*a*) the unsplit line in zero field, (*b*) the Zeeman pattern in a field of 50 gauss applied along a face diagonal of the cube. The angle θ is 90° for two of the four chlorines in the unit cell and 35° 16' for the other two. For the first two chlorines the α and β lines coincide as expected, but are separated for the other two, so we have a spectrum of six lines. They reduce to three if the field is applied along an edge of the cube, for then the condition

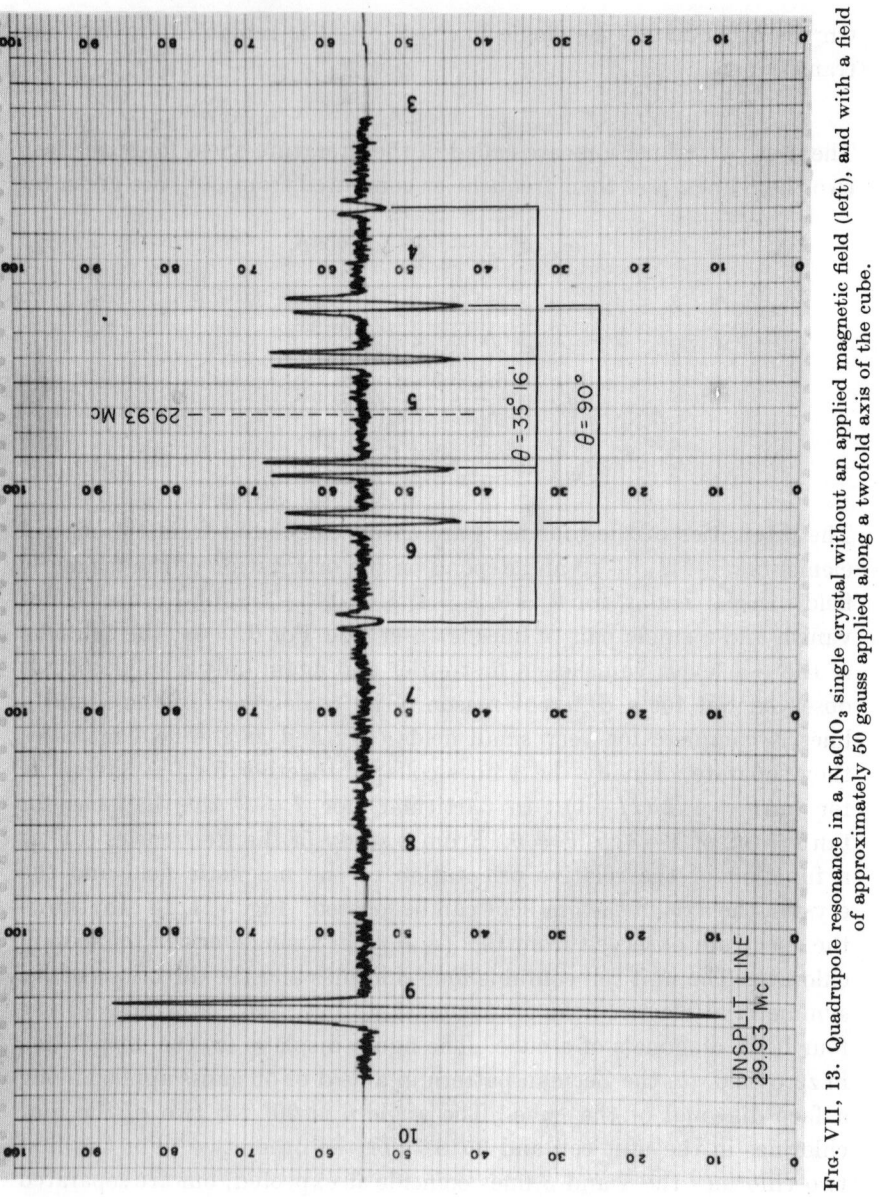

FIG. VII, 13. Quadrupole resonance in a NaClO₃ single crystal without an applied magnetic field (left), and with a field of approximately 50 gauss applied along a twofold axis of the cube.

$1-3\cos^2\theta = 0$ is fulfilled for the four chlorines of the unit cell and the lines come together.

Finally, there is the low-frequency transition $|+\rangle \to |-\rangle$ with a frequency
$$\omega = f\gamma H \cos\theta = \gamma H[\cos^2\theta + (I+\tfrac{1}{2})^2 \sin^2\theta]^{\frac{1}{2}}.$$
The Zeeman splitting of this line can be described by an apparent gyromagnetic factor $\gamma'(\theta)$ which is equal to $\gamma(I+\tfrac{1}{2})$ for $\theta = \tfrac{1}{2}\pi$.

The case of an asymmetrical gradient is fairly complicated and we shall be content to explain how the spectrum can be calculated, referring again to (10) for results.

Let $|\xi\rangle = \sum_m c_m |m\rangle$ and its Kramers conjugate $|\bar{\xi}\rangle = \sum_m c_m |-m\rangle$ be two eigenstates spanning a twofold degenerate energy level of $\mathcal{H}_Q$. The calculation of the amplitudes c_m requires the solution of a secular equation such as (48) or (48'), and can be performed exactly for $I = \tfrac{3}{2}$ only. For small values of η perturbation methods can be used. The Zeeman splitting is obtained by diagonalizing the Zeeman Hamiltonian given by (49) inside this manifold:

$$\langle \xi | \mathcal{H}_M | \xi \rangle = -\langle \bar{\xi} | \mathcal{H}_M | \bar{\xi} \rangle = -\tfrac{1}{2}\gamma \hbar H a \cos\theta,$$
$$\langle \xi | \mathcal{H}_M | \bar{\xi} \rangle = \langle \bar{\xi} | \mathcal{H}_M | \xi \rangle^* = \tfrac{1}{2}\gamma \hbar H \sin\theta (b e^{i\varphi} + c e^{-i\varphi}), \qquad (52)$$

where the constants a, b, c, calculated from the amplitudes c_m, are functions of the asymmetry parameter η. From (52) we find for the Zeeman energy displacement

$$x = \pm\tfrac{1}{2}\gamma \hbar H[a^2 \cos^2\theta + \sin^2\theta(b^2 + c^2 + 2bc \cos 2\varphi)]^{\frac{1}{2}}. \qquad (52')$$

If η is not very small, two pairs of lines will be observed for transitions between manifolds such as $|\pm\tfrac{5}{2}\rangle$ and $|\pm\tfrac{3}{2}\rangle$, etc., instead of one as in the axial case.

(c) *Transient methods*

Pulse experiments, analogous to free precession and spin echo studies in Zeeman resonance, have been performed on quadrupole levels in small or zero magnetic fields (19).

A full treatment of these problems is fairly complicated and the principle rather than the detail of the calculations will be given. A symmetrical field gradient and half-integer spins will be assumed.

(1) *Transient magnetization in zero field*

The Hamiltonian describing a situation where a magnetic r.f. field is applied at right angles to the crystalline axis, is given by

$$\mathcal{H} = A[3I_z^2 - I(I+1)] - 2\gamma H_1 I_x \cos\omega t. \qquad (53)$$

The frequency of the r.f. field is equal to the frequency

$$\omega_0 = \frac{1}{\hbar}[E_{\pm m} - E_{\pm(m-1)}] = \frac{3A}{\hbar}(2m-1) \qquad (54)$$

of a quadrupole transition.

We saw in Chapter II that when two states $|a\rangle$ and $|b\rangle$ of a system were linked by a resonant r.f. perturbation of frequency $\omega = (E_a - E_b)/\hbar$, it was possible to disregard all the other energy levels of the system and describe its behaviour (steady state as well as transient) as that of a fictitious spin $\frac{1}{2}$ placed in an appropriate fictitious magnetic field $\mathbf{H}'$.

In the present problem, in spite of the twofold quadrupole degeneracy, we may assign a fictitious spin $\mathbf{s}'$ to the couple of states $|m\rangle$, $|m-1\rangle$, excited only by one rotating component of the linearly polarized r.f. field, and a second fictitious spin $\mathbf{s}''$ to the couple $|-m\rangle$, $|-(m-1)\rangle$. This may be done because the r.f. field does not mix the two couples together.

Consider the first fictitious spin $\mathbf{s}'$. In accordance with the treatment outlined in Chapter II, we identify the states $|m\rangle$ and $|m-1\rangle$ with the eigenstates $|+\frac{1}{2}\rangle$ and $|-\frac{1}{2}\rangle$ of the component s'_ζ of $\mathbf{s}'$. The fictitious magnetic field $\mathbf{H}'$ is the sum of a d.c. component H'_ζ and a transverse rotating field H'_1. If we assign to the fictitious spin $\mathbf{s}'$ the gyromagnetic factor γ of the real spin $\mathbf{I}$, the components of $\mathbf{H}'$ are defined by

$$H'_\zeta = -\frac{E_m - E_{m-1}}{\hbar\gamma} = -\frac{\omega_0}{\gamma},$$

$$\langle \tfrac{1}{2} | H'_1 s'_\xi | -\tfrac{1}{2}\rangle = \langle m | H_1 I_x | m-1 \rangle,$$

or $\qquad H'_1 = \alpha H_1 \quad \text{with} \quad \alpha = [I(I+1) - m(m-1)]^{\frac{1}{2}}$

$$(\sqrt{3} \text{ for } I = \tfrac{3}{2}). \qquad (55)$$

The density matrix for the manifold $|m\rangle$, $|m-1\rangle$ can be written as

$$\rho = \mathbf{m}' \cdot \mathbf{s}' + \text{const.}, \qquad (56)$$

where $\mathbf{m}'$ is a c-vector which has been shown in Chapter II to satisfy the well known equation of motion of a classical magnet in the fictitious field $\mathbf{H}'$:

$$\frac{d\mathbf{m}'}{dt} = \gamma \mathbf{m}' \wedge \mathbf{H}'. \qquad (57)$$

The components I_x, I_y, I_z of the spin $\mathbf{I}$ are related to those of the fictitious spin $\mathbf{s}'$ inside the manifold $|m\rangle$, $|m-1\rangle$ through

$$I_x = \alpha s'_\xi, \qquad I_y = \alpha s'_\eta, \qquad I_z = s'_\zeta, \qquad (58)$$

and their expectation values, according to (56) and (58), are given by

$$\langle I_x \rangle = \mathrm{tr}\,\rho I_x = \tfrac{1}{2}\alpha m'_\xi, \qquad \langle I_y \rangle = \tfrac{1}{2}\alpha m'_\eta, \qquad \langle I_z \rangle = \tfrac{1}{2}m'_\zeta. \qquad (59)$$

The problem of calculating the magnetization resulting from any combination of pulses of the r.f. field is now virtually solved.

Pulses of the r.f. field H_1 are translated into pulses of the fictitious r.f. field H'_1, the classical motion of the vector $\mathbf{m}'$ is immediately obtained from (57), and then the magnetization $\mathbf{M} = N\gamma\hbar \langle \mathbf{I} \rangle$ obtained from (59). Actually, in this way one obtains only the part $\mathbf{M}'$ that corresponds to the couple $|m)$, $|m-1)$, to which has to be added the part $\mathbf{M}''$ calculated from the motion of the c-vector $\mathbf{m}''$ associated with the second spin $\mathbf{s}''$.

Thus, t seconds after a pulse of very short duration t_w we obtain from (57)

$$m'_\xi = m'_0 \sin(\gamma H'_1 t_w) \sin \omega_0 t, \qquad m'_\eta = m'_0 \sin(\gamma H'_1 t_w) \cos \omega_0 t,$$

$$m'_\zeta = m'_0 \cos(\gamma H'_1 t_w), \qquad (60)$$

where m'_0, the value of m'_ζ for $t = 0$, is, by (56), the difference between the thermal equilibrium populations of the states $|m)$ and $|m-1)$ at $t = 0$, and is thus equal to $\hbar\omega_0/(2I+1)kT$. From (55) and (59) we obtain the components of the magnetization:

$$M'_x = \frac{N\gamma\hbar^2\omega_0}{2(2I+1)kT}\alpha \sin(\alpha\gamma H_1 t_w)\sin \omega_0 t,$$

$$M'_y = \frac{N\gamma\hbar^2\omega_0}{2(2I+1)kT}\alpha \sin(\alpha\gamma H_1 t_w)\cos \omega_0 t,$$

$$M'_z = \frac{N\gamma\hbar^2\omega_0}{2(2I+1)kT}\cos(\alpha\gamma H_1 t_w). \qquad (61)$$

To the values (61) must be added those resulting from the r.f. excitation of the couple $|-m)$, $|-(m-1)\rangle$ calculated by means of the spin $\mathbf{s}''$ and the vector $\mathbf{m}''$. We can assign to $\mathbf{s}''$ the same gyromagnetic factor γ as to $\mathbf{s}'$ (and $\mathbf{I}$) and keep inside the manifold $|-m)$, $|-(m-1)\rangle$ the same relations as (59) (with $\mathbf{m}'$ replaced by $\mathbf{m}''$) if we identify the state $|+\tfrac{1}{2}\rangle$ of $\mathbf{s}''_\zeta$ with $|-(m-1)\rangle$ and $|-\tfrac{1}{2}\rangle$ with $|-m)$. It follows that $H''_\zeta = -H'_\zeta$ and the fictitious Larmor frequency of the spin $\mathbf{s}''$ is opposite to that of $\mathbf{s}'$. The contribution $\mathbf{M}''$ of the couple $|-m)$, $|-(m-1))$ to the magnetization, comes immediately from (61) by changing ω_0 into $-\omega_0$, whence

$$M''_x = M'_x, \qquad M''_y = -M'_y, \qquad M''_z = -M'_z,$$

and a total magnetization

$$M_x(t) = \frac{N\gamma\hbar^2\omega_0}{(2I+1)kT}\alpha\sin(\alpha\gamma H_1 t_w)\sin\omega_0 t,$$

$$M_y(t) = M_z(t) = 0. \tag{62}$$

The transient magnetization following a pulse produced by a linearly polarized r.f. field is also linearly polarized along the same direction in contrast to the Zeeman resonance case. It should be added that the magnetizations **M'** and **M"** that add to give **M** are not simple mathematical fictions for they can be excited and observed separately if a true rotating field (produced, for instance, by means of two orthogonal coils out of phase by 90°) is used instead of a linear r.f. field.

All the features of the transient magnetization can be predicted from (61) and (62). The equivalent of a 90° pulse that makes the transverse magnetization maximum has a duration t_w given by

$$\alpha\gamma H_1 t_w = \tfrac{1}{2}\pi \quad \text{or, for } I = \tfrac{3}{2}, \quad \gamma H_1 t_w = \pi/2\sqrt{3}.$$

The 180° pulse that reverses the transverse magnetization has twice that duration.

The damping by a distribution of pseudo-Larmor frequencies, due for instance to the irregularities in the quadrupole interaction A, the refocusing by a '180° pulse' at time τ, and the appearance of an echo at time 2τ, are immediately understood in terms of the precessing vectors **m'** and **m"**. The production of such echoes in sodium chlorate and sodium bromate is described in (**19**). It is worth pointing out that the above treatment can be applied to asymmetrical gradients as well. For instance, let $|\widetilde{\pm\tfrac{3}{2}}\rangle$ and $|\widetilde{\pm\tfrac{1}{2}}\rangle$ be the eigenstates spanning the two degenerate levels of a spin $I = \tfrac{3}{2}$ in an asymmetrical gradient. If an r.f. field $2H_1(pI_x+qI_y+rI_z)\times\cos\omega t$ is applied to the system, it is always possible to replace the states $|\widetilde{\pm\tfrac{3}{2}}\rangle$ by a linear combination of these states, $|\widetilde{\widetilde{\pm\tfrac{3}{2}}}\rangle$, and similarly the states $|\widetilde{\pm\tfrac{1}{2}}\rangle$ by two states $|\widetilde{\widetilde{\pm\tfrac{1}{2}}}\rangle$, such that the r.f. Hamiltonian has no matrix elements between $|\widetilde{\widetilde{\tfrac{3}{2}}}\rangle$ and $|\widetilde{\widetilde{-\tfrac{1}{2}}}\rangle$ nor between $|\widetilde{\widetilde{-\tfrac{3}{2}}}\rangle$ and $|\widetilde{\widetilde{\tfrac{1}{2}}}\rangle$. From then on the analysis proceeds as in the symmetrical case, associating a fictitious spin **s'** with the couple $|\widetilde{\widetilde{\tfrac{3}{2}}}\rangle$, $|\widetilde{\widetilde{\tfrac{1}{2}}}\rangle$ and **s"** with the couple $|\widetilde{\widetilde{-\tfrac{3}{2}}}\rangle$, $|\widetilde{\widetilde{-\tfrac{1}{2}}}\rangle$.

(2) *Transient magnetization in a small magnetic field* H_0

The previous analysis must be modified when a magnetic field lifts the degeneracy of the quadrupole levels. We consider briefly a few cases of increasing complexity.

(α) If $m > \frac{3}{2}$, the eigenstates of the Hamiltonian $\mathcal{H}_Q - \gamma\hbar\mathbf{H}.\mathbf{I}$ are the same as in the absence of a magnetic field and the analysis by means of the fictitious spins $\mathbf{s}'$ and $\mathbf{s}''$ is still valid, the only difference from (1) being that the two magnetizations $\mathbf{M}'$ and $\mathbf{M}''$ precess now in opposite directions with frequencies that differ by $2\gamma H_0 \cos\theta$, where θ is the angle of the applied field with the crystalline axis. If the amplitude H_1 of the r.f. field is much smaller than H_0, only $\mathbf{M}'$ or $\mathbf{M}''$ can be excited and the motion of the magnetization is a simple precession. If $H_1 \gg H_0$, both $\mathbf{M}'$ and $\mathbf{M}''$ can be excited at the same time and beats at the frequency $2\gamma H_0 \cos\theta$ between $\mathbf{M}'$ and $\mathbf{M}''$ will be observed in $M_x(t)$ (or $M_y(t)$ which does not vanish any more).

(β) If $m = \frac{3}{2}$, a complication arises from the fact that the eigenstates $|\pm\frac{1}{2}\rangle$ of I_z are replaced by the states $|\pm)$ of equation (50''). If $H_1 \ll H_0$, only one of the four transitions α, α', β, β' with frequencies given by (51) is excited and the description by a fictitious spin $\frac{1}{2}$ is still possible.

Using (50'') it is easily found that after a pulse of duration t_w, the amplitude of the precessing magnetization is for, say, the transition α,

$$M_\perp = \frac{N\gamma\hbar^2\omega_\alpha}{2(2I+1)kT}[I(I+1)-\tfrac{3}{4}]^{\frac{1}{2}}\cos\delta\sin\{\cos\delta[I(I+1)-\tfrac{3}{4}]^{\frac{1}{2}}\gamma H_1 t_w\},$$

where δ and ω_α are given by (50'') and (51).

If, on the other hand, the r.f. amplitude $H_1 \gg H_0$, and the four transitions α, α', β, β' are excited simultaneously, the analysis by fictitious spins $\frac{1}{2}$ breaks down and a general formalism similar to that of Section II A (b 3) must be used instead. Although straightforward in principle, the analysis of this problem is too lengthy to be presented here.

APPENDIX

Sign of the quadrupole coupling

We saw in Chapter III that in a crossed-coils experiment it was possible to determine the sign of a nuclear gyromagnetic ratio γ. The question naturally arises, whether it is possible to measure the sign eqQ of a quadrupole interaction in a nuclear magnetic resonance experiment. It is generally possible to be reasonably certain of the sign, if not the magnitude of $eq = \partial^2 V/\partial z^2$, so that the knowledge of the sign of qQ would lead to that of Q itself, a result of interest to the nuclear physicist.

Such a determination is clearly possible at very low temperatures, when the departure of the Boltzmann exponential $e^{-h\nu/kT}$ from its first-order expansion $1 - h\nu/kT$ is appreciable. For instance, if $h\nu \gg kT$, for a spin $I = \frac{5}{2}$ the transition $|\pm\frac{5}{2}\rangle \to |\pm\frac{3}{2}\rangle$ will be much less or much more intense than the transition at half the frequency $|\pm\tfrac{3}{2}\rangle \to |\pm\tfrac{1}{2}\rangle$, depending on whether qQ is positive or negative. On the other hand, at room temperature, by which we mean a temperature where $\exp(-h\nu/kT)$ is indistinguishable from $1-(h\nu/kT)$, the sign of a quadrupole interaction cannot be determined by a nuclear magnetic resonance experiment. This impossibility holds whatever type of magnetic field, d.c. or r.f., transient or permanent, linearly or circularly polarized, is applied to the sample. A general proof of this statement is given below.

Let $\mathscr{H} = \mathscr{H}_0 + \mathscr{H}_1(t)$ be the Hamiltonian of the spin system, where

$$\mathscr{H}_0 = \mathscr{H}_Q - \gamma\hbar\mathbf{H}_0 \cdot \mathbf{I},$$

and

$$\mathscr{H}_1(t) = -\gamma\hbar\mathbf{I} \cdot \mathbf{H}_1(t)$$

represents the coupling of the spin with time-dependent magnetic fields applied starting at $t = 0$ when the spin system is in thermal equilibrium. The nuclear signal is proportional to $d\mathbf{M}/dt$ with

$$\mathbf{M} = N\gamma\hbar\langle\mathbf{I}\rangle = N\gamma\hbar\,\mathrm{tr}\,\rho\mathbf{I}, \quad \text{where} \quad \rho = \frac{1+\sigma}{2I+1}$$

is the density matrix of the spin system. The assumption of room temperature is expressed by the expansion of the equilibrium density matrix as

$$\rho_0 = \rho(0) = \frac{1+\sigma_0}{2I+1} = \frac{1-(\mathscr{H}_0/kT)}{2I+1}. \tag{63}$$

Let $\mathscr{H}^+$, $\mathscr{H}_0^+$, σ^+, σ_0^+, etc., be quantities that correspond to a given sign of qQ and $\mathscr{H}^-$, etc., those obtained by reversing this sign. We are going to show that $\mathbf{M}$ is independent of this sign. From the form of $\mathscr{H}$ and the assumption (63) we get

$$\mathscr{H}^+(\mathbf{I},t) = -\mathscr{H}^-(-\mathbf{I},t), \qquad \sigma_0^+(\mathbf{I}) = -\sigma_0^-(-\mathbf{I}). \tag{64}$$

$\sigma^+(t)$ is a solution of

$$i\frac{d\sigma^+}{dt} = -[\mathscr{H}^+(\mathbf{I},t), \sigma^+] \quad \text{with} \quad \sigma^+(0) = -\frac{1}{2I+1}\frac{\mathscr{H}_0^+(\mathbf{I})}{kT}. \tag{65}$$

Similarly, σ^- obeys the equation

$$i\frac{d\sigma^-}{dt} = -[\mathscr{H}^-(\mathbf{I},t), \sigma^-] \quad \text{with} \quad \sigma^-(0) = -\frac{1}{2I+1}\frac{\mathscr{H}_0^-(\mathbf{I})}{kT}, \tag{66}$$

which, according to (65), can be rewritten as

$$i\frac{d\sigma^-}{dt} = [\mathscr{H}^+(-\mathbf{I},t), \sigma^-] \quad \text{with} \quad \sigma^-(0) = \frac{1}{2I+1}\frac{\mathscr{H}_0^+(-\mathbf{I})}{kT}. \tag{66'}$$

Let $\sigma^+(t) = f(\mathbf{I},t)$ be the solution of (64). Introduce a new set of variables I'_x, I'_y, I'_z, with $\mathbf{I}' = -\mathbf{I}$. They differ from I_x, I_y, I_z by the sign of their commutation relations

$$[I'_x, I'_y] = -iI'_z \quad \text{instead of} \quad [I_x, I_y] = iI_z.$$

If in (66') we write $i' = -\sqrt{(-1)} = -i$ we get the equation

$$i'\frac{d\sigma^-}{dt} = -[\mathscr{H}^+(\mathbf{I}',t), \sigma^-], \qquad \sigma^-(0) = -\frac{1}{2I+1}\frac{\mathscr{H}_0^+(\mathbf{I}')}{kT}. \tag{67}$$

for σ^-. The solution of (67) as compared with (65) is clearly

$$\sigma^- = -f(\mathbf{I}',t),$$

whence the magnetization

$$\mathbf{M}^-(t) = \mathrm{tr}\,\sigma^-\mathbf{I} = \mathrm{tr}f(\mathbf{I}',t).\mathbf{I}'$$
$$= \mathrm{tr}f(\mathbf{I},t).\mathbf{I} = \mathbf{M}^+(t). \tag{68}$$

REFERENCES

1. G. E. Pake, *J. Chem. Phys.* **16**, 327, 1948.
2. H. S. Gutowsky, G. B. Kistiakowsky, G. E. Pake, and E. M. Purcell, ibid. **17**, 972, 1949.
3. M. Goldman, *J. Phys. Chem. Solids*, **7**, 165, 1958.
4. E. R. Andrew and R. Bersohn, *J. Chem. Phys.* **18**, 159, 1950.
5. P. T. Ford and R. E. Richards, *Disc. Faraday Soc.* **19**, 193, 1955.
6. R. Bersohn and H. S. Gutowsky, *J. Chem. Phys.* **22**, 651, 1954.
7. F. Reif and E. M. Purcell, *Phys. Rev.* **91**, 631, 1953.
8. R. V. Pound, ibid. **79**, 685, 1950.
9. M. H. Cohen and F. Reif, *Solid State Physics*, **5**, 321, 1957.
10. T. P. Das and E. L. Hahn, ibid. Suppl. I, 1958.
11. G. M. Volkoff, *Canad. J. Phys.* **31**, 820, 1953.
12. N. Bloembergen, *Report of the Conference on Defects in Crystalline Solids*, Physical Society, 1954.
13. G. D. Watkins and R. V. Pound, *Phys. Rev.* **89**, 658, 1953.
14. N. Bloembergen and T. J. Rowland, *Acta Metallurgica*, **1**, 731, 1953.
15. I. Solomon, *Phys. Rev.* **110**, 61, 1958.
16. E. Wigner, *Gruppentheorie*, Chap. XV, Friedrich Vieweg und Sohn, Braunschweig, 1931.
17. C. H. Townes and A. L. Schawlow, *Microwave Spectroscopy*, McGraw-Hill, N.Y., 1955.
18. R. Livingston, *Science*, **118**, 61, 1953.
19. M. Bloom, E. L. Hahn, and B. Herzog, *Phys. Rev.* **97**, 1699, 1955.

VIII

THERMAL RELAXATION IN LIQUIDS AND GASES

Fugit irreparabile tempus

I. Introduction

A. Coupling of the nuclear spins with the radiation field

In this chapter, the elementary mechanisms which ensure the coupling of a system of nuclear spins with the surrounding medium or 'lattice' will be studied and the principles of the calculation of the spin-lattice relaxation times T_1 will be given.

A noticeable feature of this problem is the weakness of the interactions to which nuclear spins are subjected from the surrounding medium, and, as a consequence, the sometimes considerable lengths of the relaxation times. The nuclear relaxation times observed experimentally, depending on the magnitude of the nuclear moments, the nature of the sample, and its temperature, vary approximately between 10^{-5} and 10^5 seconds. These are considerable durations on the time scale of the phenomena usually encountered in atomic and nuclear physics. It will be seen, however, that these times do not exceed the theoretical expectations and that most of the conceivable relaxation mechanisms lead to even longer relaxation times.

As a starting-point consider the probability W per unit time of the transition from the upper into the lower energy state, through spontaneous emission of a photon of energy $\hbar\omega = \gamma\hbar H_0$, for a nuclear spin $\tfrac{1}{2}$ with a magnetic moment $\boldsymbol{\mu} = \gamma\hbar\mathbf{I}$ placed in a d.c. field H_0. It is given by the well-known formula for a magnetic dipole radiating in free space:

$$W = \frac{4}{3}\frac{1}{\hbar}\frac{1}{\lambdabar^3}\{|(+|\mu_x|-)|^2 + |(+|\mu_y|-)|^2\} = \frac{2}{3}\frac{\gamma^2\hbar}{\lambdabar^3}, \tag{1}$$

where $\lambdabar = c/\omega = \lambda/2\pi$ and λ is the wavelength of the emitted photon. For a proton in a magnetic field of 7500 gauss, i.e. for a Larmor frequency $\omega/2\pi$ of 30 Mc/s, equation (1) gives

$$W \simeq 10^{-25} \text{ sec}^{-1}.$$

The coupling of the nuclear spins with the radiation field would thus appear to be a negligible phenomenon. It should be remembered, however, that spontaneous emission determines the transition probability of a system coupled with electromagnetic radiation, only under the assumption that no photons are present in the space surrounding it, and that an extra transition probability nW caused by induced emission (or absorption if the system is in the lower state) also exists. The number n is the number of photons present at the frequency considered, per mode of oscillation of the radiation field. In free space the relation between n and the energy $E_F^\perp(\nu)$ of the part of the radiation field polarized at right angles to the applied d.c. field per unit volume and unit frequency range is

$$E_F^\perp(\nu) = n(\nu)\frac{8\pi h}{\lambda^3} \times \frac{2}{3}, \qquad (2)$$

where the subscript F in $E_F^\perp(\nu)$ stands for free space. In a coil or a resonator of volume V and quality factor Q tuned at a frequency ν_0, the corresponding energy density E_R is a Lorentz function of ν, of half width at half intensity $\nu_0/2Q$, so normalized that

$$V \int E_R(\nu)\,d\nu = n(\nu_0)h\nu_0,$$

and therefore given by the formula

$$E_R(\nu) = \frac{1}{\pi}\frac{\nu_0}{2Q}\frac{n(\nu_0)h\nu_0}{(\nu-\nu_0)^2+(\nu_0/2Q)^2}\frac{1}{V}; \qquad (3)$$

whence

$$E_R(\nu_0) = \frac{1}{\pi}\frac{2Qn(\nu_0)h}{V} \qquad (4)$$

and

$$E_R(\nu_0)/E_F^\perp(\nu_0) = \frac{1}{4\pi^2}\frac{Q\lambda^3}{V}\frac{3}{2}. \qquad (5)$$

The enhancement factor (5) can be a very large number; thus for $Q = 100$, $V = 1$ c.c., $\lambda = 10^3$ cm ($\nu = 30$ Mc/s), it is equal to 4.10^9.

If the radiation field surrounding the sample is in thermal equilibrium at a temperature T, the number of photons n is given by Planck's formula $n = [\exp(h\nu/kT)-1]^{-1}$ which, since $h\nu/kT$ is always a very small number in nuclear magnetism (10^{-6} for $\nu = 30$ Mc/s at room temperature), can be safely replaced by its first-order expansion $(kT/h\nu)$. The transition probability for induced emission or absorption under those conditions becomes

$$W_1 = \frac{kT}{h\nu}\frac{3}{2}\frac{1}{4\pi^2}\frac{Q\lambda^3}{V}\frac{2\gamma^2\hbar}{3\lambda^3} = 2\pi Q\frac{kT}{h\nu}\frac{1}{\hbar}\frac{\gamma^2\hbar^2}{V}, \qquad (6)$$

which for the previous example (protons in a field of 7500 gauss, $Q = 100$, $V = 1$ c.c., $T = 300°$ K) gives $W_1 \sim 10^{-10}$ sec^{-1}.

It is easy to show that an *incoherent* coupling of nuclear spins with a radiation field in thermal equilibrium would be a mechanism of thermal relaxation for the former, although it would be exceedingly weak. For a spin coupled with such a field, the probability of losing energy by going from the higher state $|a)$ into the lower state $|b)$ through emission of a photon, is larger than the probability for the opposite transition by the factor

$$\frac{W_{a\to b}}{W_{b\to a}} = \frac{\text{emission}}{\text{absorption}} = \frac{n+1}{n} = \exp\left(\frac{h\nu}{kT}\right). \tag{7}$$

The steady-state condition for an assembly of spins with populations N_a and N_b will be $N_a W_{a\to b} = N_b W_{b\to a}$, which by (7) leads to the Boltzmann equilibrium distribution

$$N_a/N_b = \exp(-h\nu/kT).$$

The foregoing discussion has an academic character not only because of the smallness of W_1 given by formula (6), but also because the assumption of *incoherent* coupling between *non-interacting* spins and the radiation fields is invalid when they are separated by distances that are small compared with the wavelength.

For a macroscopic sample, the radiation field of the wavelength required is always, whatever its origin, thermal or not, uniform over a large number of spins and, according to the fundamental equation $d\mathbf{M}/dt = \gamma \mathbf{M} \wedge \mathbf{H}$, cannot change the magnitude of the magnetization $\mathbf{M}$ and bring it to a finite value if it starts from an unpolarized state.

This discussion should not be construed as meaning that in the absence of an applied r.f. field the coupling of the nuclear spins of a macroscopic sample with the radiation field is a negligible phenomenon. The very possibility of detecting the free precession of nuclear spins, the corresponding radiation damping, the existence of auto-oscillators or masers using nuclear precession are spectacular manifestations of the opposite. All these phenomena, however, are connected with coherent radiation when there is a definite relationship between the phases of the wave functions describing the individual spins, and, as was shown in Chapter III, these phenomena change the orientation rather than the magnitude of the nuclear magnetization of the sample and therefore are not mechanisms of relaxation capable of bringing a spin system into equilibrium at a finite temperature. Because the coupling with the radiation field is hopelessly inadequate as a relaxation mechanism we turn towards the coupling of the spin system with another material system, the 'lattice'.

B. Coupling of the spin system with the lattice

This coupling can be described in two different ways. In the first of these descriptions, the lattice is considered as a quantum mechanical system possessing energy levels which, however, because of the large number of degrees of freedom of the lattice, form a quasi-continuous spectrum. The interaction Hamiltonian $\hbar\mathcal{H}_1$ between the lattice and the spins has matrix elements $(f, s \mid \hbar\mathcal{H}_1 \mid f', s')$ corresponding to transitions which take the lattice from a state $|f\rangle$ to a state $|f'\rangle$, and the system of spins from a state $|s\rangle$ to a state $|s'\rangle$. Conservation of energy requires

$$E_f + E_s = E_{f'} + E_{s'}.$$

This description of the lattice entails two fundamental hypotheses:

(1) The lattice is assumed to be in thermal equilibrium with itself. This is equivalent to saying:

(a) the state of the lattice is not a state $|f\rangle$ of well-defined energy nor even a superposition $\sum_f C_f |f\rangle$ of such states, but a statistical mixture defined by the probabilities P_f of finding the lattice in the states $|f\rangle$;

(b) the probabilities P_f are proportional to the Boltzmann factors

$$\exp(-E_f/kT).$$

(2) The specific heat of the lattice is assumed to be infinite. This is equivalent to saying that the probabilities P_f are not modified by the exchange of energy which takes place between the lattice and the spins (this assumption is not always justified at very low temperatures). From these hypotheses one can deduce the relation

$$\frac{W_{s \to s'}}{W_{s' \to s}} = \exp\{(E_s - E_{s'})/kT\}, \tag{8}$$

where $W_{s \to s'}$ represents the transition probability between two well-defined states $|s\rangle$ and $|s'\rangle$ of the spin system, induced by the coupling with the lattice. Indeed, a quantity such as $W_{s \to s'}$ is the sum

$$W_{s \to s'} = \sum_{f,f'} P_f W_{f, s \to f', s'},$$

where $|f\rangle$ and $|f'\rangle$ are all the states of the lattice such that

$$E_f + E_s = E_{f'} + E_{s'}.$$

P_f is the probability $\exp(-E_f/kT)$ of finding the lattice in the state $|f\rangle$, and $W_{f, s \to f', s'}$ the probability of going from a well-defined (discrete) state $|f\rangle|s\rangle$ of the entire spin-lattice system to another state $|f'\rangle|s'\rangle$. According to the general principles of quantum mechanics, these individual probabilities are equal to the inverse probabilities. One has therefore

$$\frac{W_{s\to s'}}{W_{s'\to s}} = \frac{\sum_{f,f'} \exp(-E_f/kT) W_{f,s\to f',s'}}{\sum_{f,f'} \exp(-E_{f'}/kT) W_{f',s'\to f,s}} = \exp\left(\frac{E_s - E_{s'}}{kT}\right) \quad (9)$$

which, from the steady-state condition $N_s W_{s\to s'} = N_{s'} W_{s'\to s}$, leads to a Boltzmann distribution for the populations of the spin system.

In the second method, the lattice is described classically. Lattice parameters which occur in the interaction Hamiltonian are then given functions of time. To the statistical character of the description of the lattice through probabilities P_f in the first method, here there corresponds the assumption that these functions are random functions described by their probability distributions, as will be explained shortly. In this method, the relation (8) does not appear automatically but must be introduced as an *ad hoc* hypothesis.

The first, more rigorous, method can be applied whenever one has a lattice model susceptible of a simple quantum mechanical description. We shall apply it to the relaxation produced by lattice vibrations in crystalline solids and to nuclear relaxation in metals, caused by the interaction of nuclear spins with conduction electrons. The quantum mechanical description becomes necessary at very low temperatures when only a small number of degrees of freedom of the lattice are excited.

The second method is particularly well adapted to the study of relaxation in liquids, where the Brownian motion of molecules is responsible for the random variation of the lattice parameters.

It will be shown later that it is possible to cast these two approaches into a single general formalism.

II. Relaxation in Liquids and Gases
A. General

Various properties of nuclear magnetism undergo deep changes when liquid (or gaseous) samples are used. These changes are brought about by the existence of rapid molecular motions of large amplitude and random character, inside these samples. These motions include rotational tumbling of individual molecules, relative translational motion of molecules and even, because of chemical exchange, migrations of atoms or groups of atoms from one molecule to another. The main consequences of this state of affairs for nuclear magnetism are the following.

The couplings between the nuclear spins, described classically by local magnetic fields, are considerably reduced by these motions and to a first approximation disappear completely. From a practical point of view this results in very narrow resonance lines (infinitely narrow in a first approximation). The theoretical approach is also greatly simplified by this circumstance. It was shown in Chapter IV on dipolar broadening in rigid solids that because of the tight coupling between nuclear spins exemplified by frequent flip-flops between neighbours, the correct approach to nuclear magnetism in solids was a collective one, where a single large spin-system with many degrees of freedom was considered, rather than a collection of individual spins. A spin temperature, distinct from the lattice temperature, could be defined for such a system. On the other hand, in liquids where the spin-spin coupling is weak and comparable to the coupling of the spins with the lattice (or sometimes much weaker when quadrupole nuclear moments are present), it is legitimate to consider individual spins, or at most groups of spins inside a molecule, as separate systems coupled independently to a thermal bath, the lattice.

Electrostatic couplings of nuclear quadrupole moments in non-cubic environments with electric-field gradients, that in solids change the energy spectrum of the nuclear spins to the extent of making a resonance observable only in zero magnetic field or in single crystals, are also greatly reduced in liquids by the molecular motion, sometimes sufficiently so as to leave a simple Zeeman spectrum to be observed.

When the first-order effects of local magnetic fields or electrostatic field gradients have thus been removed, the main problem of the theory of nuclear magnetism in liquids (or gases) is to find out how in the second approximation these rapidly fluctuating fields or gradients induce transitions between the energy levels of the individual spins (or spin systems) and broaden the resonance lines, which are infinitely sharp in the first approximation.

The physical picture underlying this description and first proposed by Bloembergen, Purcell, and Pound (BPP) (1), is best understood in the case of a fluctuating electric gradient acting on the quadrupole moment of a nucleus. The Fourier spectrum of the time-dependent components of this gradient may contain with a non-vanishing intensity the frequency ω_0 corresponding to the transition between two levels of the nuclear spin system, a transition which thus has a finite probability of being induced by the fluctuating field gradient. Since the energy required for this transition is provided by a system that is in thermal equilibrium, these transitions are weighted by Boltzmann factors, as explained in the introduction, and can bring the nuclear spins into thermal equilibrium with the liquid. At the same time a broadening of the resonance lines occurs which for very rapid molecular motion is of the order of $(1/T_1)$ on the frequency scale but with a numerical factor that only an elaborate theory can provide. The shape of the spectrum of the random molecular motion can seldom be predicted quantitatively, but fortunately the results are not very sensitive functions of that shape and rather crude models give reasonable agreement with experiment.

A general feature of these spectra is that the more disorderly the molecular motion and the shorter the correlation time which expresses the duration of a correlation between two configurations of a nuclear environment at two different times, the broader the frequency spectrum.

Similar considerations can be made concerning the magnetic coupling between nuclear spins. A simple image is to consider each spin as 'seeing' a fluctuating magnetic field produced by a neighbour which induces transitions among its levels. It is more correct to consider a system of two or more such spins and the time-dependent coupling among them as inducing transitions between the energy levels of the combined system. A consistent quantitative theory of these phenomena will be given in the next sections.

B. Definitions

Let $y(t)$ be a function of a parameter t (which we shall identify with the time).

We shall say that it is a random function if the value y which it takes at each instant t is a random variable subject to a law of probability $p(y, t)$ depending on the parameter t.

The average value of the random function at an instant t, represented by $\overline{y(t)}$, is defined by

$$\overline{y(t)} = \int y\, p(y, t)\, dy. \tag{10}$$

If $f(y)$ is a given function of y, f will also be a random function of t and we shall have

$$\overline{f(t)} = \int p(y, t) f(y)\, dy. \tag{11}$$

The values $y(t)$ of the random function corresponding to the various different times t are not in general independent random variables, but show a correlation. The only correlation which we shall use is that which corresponds to two different times t_1 and t_2. We define the function $p(y_1, t_1; y_2, t_2)$ as the probability of y taking the value y_1 at the instant t_1 and y_2 at the instant t_2.

A function which has a slightly different meaning and is represented by $P(y_1, t_1; y_2, t_2)$ is the probability that y takes the value y_2 at time t_2 when we know that it takes the value y_1 at time t_1. There is the obvious relation
$$p(y_1, t_1; y_2, t_2) = P(y_1, t_1; y_2, t_2) p(y_1, t_1). \tag{12}$$

We shall call the expression $G(t_1, t_2)$ defined by

$$\begin{aligned} G(t_1, t_2) &= \overline{f(t_1) f^*(t_2)} \\ &= \iint p(y_1, t_1; y_2, t_2) f(y_1) f^*(y_2)\, dy_1\, dy_2 \\ &= \iint p(y_1, t_1) P(y_1, t_1; y_2, t_2) f(y_1) f^*(y_2)\, dy_1\, dy_2 \end{aligned} \tag{13}$$

the auto-correlation function of the random function $f(y)$ relative to the times t_1 and t_2.

An important class of random functions, the only ones we shall consider, are the stationary random functions which are invariant under a change of the origin of time.

For this class of functions $p(y, t)$ is in fact a time-independent function $p(y)$ and the functions $p(y_1, t_1; y_2, t_2)$, $P(y_1, t_1; y_2, t_2)$, and $G(t_1, t_2)$ depend on t_1 and t_2 only through the difference $t_2 - t_1 = \tau$. We may then write

$$\begin{aligned} G(\tau) &= \iint p(y_1, y_2, \tau) f(y_1) f^*(y_2)\, dy_1\, dy_2 \\ &= \iint p(y_1) P(y_1, y_2, \tau) f(y_1) f^*(y_2)\, dy_1\, dy_2. \end{aligned} \tag{14}$$

A correlation time τ_c is defined somewhat loosely by the condition that $G(\tau)$ is very small for $|\tau| \gg \tau_c$.

It is clear from the definition of $p(y_1, y_2, \tau)$ that

$$p(y_1, y_2, \tau) = p(y_2, y_1, -\tau)$$

and thus, according to (14),

$$G(-\tau) = G^*(\tau). \tag{14'}$$

If it is further supposed that there is a symmetry between the past and future, so that

$$p(y_1, y_2, -\tau) = p(y_1, y_2, \tau),$$

we deduce immediately from (14) that

$$G(-\tau) = G^*(\tau) = G(\tau). \tag{15}$$

The auto-correlation function is then both an even function and a real function of τ. We shall introduce the following Fourier transforms of G or spectral densities:

$$j(\omega) = \int_0^\infty G(\tau)e^{-i\omega\tau}\,d\tau,$$

$$J(\omega) = 2\int_0^\infty G(\tau)\cos(\omega\tau)\,d\tau = \int_{-\infty}^\infty G(\tau)e^{-i\omega\tau}\,d\tau,$$

$$k(\omega) = \int_0^\infty G(\tau)\sin(\omega\tau)\,d\tau, \tag{15'}$$

whence $j(\omega) = \tfrac{1}{2}J(\omega) - ik(\omega)$. It is clear that J and k are real and that $J(-\omega) = J(\omega)$. From the second of the relations (15') we get

$$G(\tau) = \frac{1}{2\pi}\int_{-\infty}^\infty J(\omega)e^{i\omega\tau}\,d\omega. \tag{16}$$

If in (16) $\tau = 0$, it follows that

$$G(0) = \overline{|f|^2} = \frac{1}{2\pi}\int_{-\infty}^\infty J(\omega)\,d\omega. \tag{17}$$

Frequently $f(t)$ will represent a component of a fluctuating magnetic field. The quadratic quantity $G(0) = \overline{|f|^2}$ will then be the magnetic energy corresponding to that component and (17) may be interpreted as an expansion of this energy into the spectrum of frequencies.

Similarly, for two different random functions f_a and f_b, a cross-correlation function

$$G_{ab}(\tau) = \overline{f_a(t)f_b^*(t+\tau)} \tag{18}$$

can be introduced with spectral densities defined similarly.

Quantum mechanical operators with matrix elements that are random functions of time will be called random operators.

C. Motion of a system subject to a perturbation which is a random function of time

(a) Transition probability

Consider a system S (for instance the spin system) having levels $|\alpha\rangle,\ldots,|\beta\rangle$, etc., with energies $\hbar\alpha$, $\hbar\beta$, etc.

If we apply to this system a perturbation represented by a time-dependent Hamiltonian $\hbar\mathscr{H}_1(t)$ (for instance the coupling with a lattice), the state of the system can be represented by $|\xi\rangle = \sum_\alpha C_\alpha(t)e^{-i\alpha t}|\alpha\rangle$, where the coefficients C_α obey the Schrödinger equation

$$i\frac{dC_\alpha}{dt} = \sum_\beta (\alpha\,|\,\mathscr{H}_1(t)\,|\,\beta)e^{i\omega_{\alpha\beta}t}C_\beta \tag{19}$$

in which $\omega_{\alpha\beta} = \alpha - \beta$. We seek the transition probability $P_{\alpha\beta}(t)$ which is the value of $|C_\alpha(t)|^2$ at time t, knowing that at time zero all the coefficients $C_{\beta'}$ are zero except one $C_\beta(0) = 1$, or, more exactly, the probability of transition per unit time

$$W_{\alpha\beta} = \frac{dP_{\alpha\beta}}{dt}.$$

In a perturbation treatment (19) is integrated, putting

$$C_{\beta'}(t) = C_{\beta'}(0) = \delta_{\beta\beta'}$$

on the right-hand side of (19). We get

$$C_\alpha(t) = \frac{1}{i} \int_0^t (\alpha | \mathcal{H}_1(t') | \beta) e^{i\omega_{\alpha\beta}t'} dt', \tag{20}$$

$$W_{\alpha\beta} = dP_{\alpha\beta}/dt = C_\alpha \frac{dC_\alpha^*}{dt} + \text{c.c.}$$

$$= \int_0^t (\beta | \mathcal{H}_1(t) | \alpha)(\alpha | \mathcal{H}_1(t') | \beta) e^{i\omega_{\alpha\beta}(t'-t)} dt' + \text{c.c.} \tag{21}$$

Now suppose that $\mathcal{H}_1(t)$ is a random operator, as for example the dipolar coupling between two spins in relative Brownian motion. The quantity $W_{\alpha\beta}$ defined by (21) will also be a random function. The observable quantity will be the average $\overline{W_{\alpha\beta}(t)}$ taken over a statistical ensemble. We get

$$\overline{W_{\alpha\beta}} = \int_0^t \overline{(\alpha | \mathcal{H}_1(t') | \beta)(\beta | \mathcal{H}_1(t) | \alpha)} e^{i\omega_{\alpha\beta}(t'-t)} dt' + \text{c.c.} \tag{22}$$

The quantity under the bar is the correlation function $G_{\alpha\beta}$ of the random function $(\alpha | \mathcal{H}_1(t) | \beta)$ and if the latter is stationary, the quantity under the integral sign depends only on the difference

$$t - t' = \tau.$$

It follows that

$$\overline{W_{\alpha\beta}} = \int_0^t G_{\alpha\beta}(\tau) e^{-i\omega_{\alpha\beta}\tau} d\tau + \text{c.c.} = \int_{-t}^t G_{\alpha\beta}(\tau) e^{-i\omega_{\alpha\beta}\tau} d\tau. \tag{23}$$

As always in time-dependent perturbation theory, we consider the value of $\overline{W_{\alpha\beta}}$ at the end of a time t much larger than $1/\omega_{\alpha\beta}$. The limits of the integration in (23) are then replaced by $\pm\infty$ and we get $W_{\alpha\beta}$ for the transition probability (where we henceforth omit the bar)

$$W_{\alpha\beta} = J_{\alpha\beta}(\omega_{\alpha\beta}), \tag{24}$$

where $J_{\alpha\beta}(\omega)$ is the Fourier transform of $G_{\alpha\beta}(\tau)$. In the simple case where $\mathcal{H}_1(t)$ can be written as a single product: $\mathcal{H}_1(t) = AF(t)$, where A is an operator acting on the variables of the system S and $F(t)$ a random function, (24) becomes

$$W_{\alpha\beta} = |(\alpha|A|\beta)|^2 J(\omega_{\alpha\beta}), \qquad (24')$$

where
$$J(\omega) = \int_{-\infty}^{\infty} g(\tau) e^{-i\omega\tau} \, d\tau$$

and
$$g(\tau) = \overline{F(t)F(t+\tau)}.$$

(b) The master equation for populations

The rate of change of the populations $P_\alpha = \overline{|C_\alpha|^2}$ of the energy levels $|\alpha)$ of the system S is then described by the usual differential system or master equation

$$\frac{dP_\alpha}{dt} = \sum_\beta W_{\alpha\beta} P_\beta - P_\alpha \sum_\beta W_{\beta\alpha}. \qquad (25)$$

The derivation of the master equation (25) which describes an irreversible dissipative behaviour, starting from the Schrödinger equation (19), which is invariant under time reversal, is a difficult and not completely solved problem of the general theory of irreversible processes, which goes far beyond the framework of nuclear magnetism, and the proof sketched above lays no claim to completeness or rigour. Because of the symmetry relation $W_{\alpha\beta} = W_{\beta\alpha}$, (25) can be rewritten

$$\frac{dP_\alpha}{dt} = \sum_\beta W_{\alpha\beta}(P_\beta - P_\alpha), \qquad (25')$$

which makes it clear that its steady state solution is $P_\alpha = P_\beta = 1/A$, where A is the number of degrees of freedom of the system S ($2I+1$ for a spin I), whereas for a system coupled with a lattice in thermal equilibrium, it was shown in the introduction (eqn. (9)) that the steady state solution should be $P_\alpha = C \exp(-\hbar\alpha/kT)$. The system (25') will still give the correct description of the rate of change in populations provided the variables P_α in it stand for $p_\alpha \exp(\hbar\alpha/kT)$, where the variables p_α are the true populations. In the high temperature approximation, which is almost always valid in nuclear magnetism,

$$P_\alpha = p_\alpha \exp\left(\frac{\hbar\alpha}{kT}\right) \simeq p_\alpha\left(1 + \frac{\hbar\alpha}{kT}\right) \simeq p_\alpha + \frac{1}{A}\frac{\hbar\alpha}{kT}. \qquad (26)$$

To the same approximation, the difference $p_\alpha^0 - p_\beta^0$ of two equilibrium populations can be written

$$p_\alpha^0 - p_\beta^0 = \frac{\exp(-\hbar\alpha/kT) - \exp(-\hbar\beta/kT)}{\sum_\alpha \exp(-\hbar\alpha/kT)} \simeq -\frac{\hbar}{A}\left(\frac{\alpha-\beta}{kT}\right). \qquad (26')$$

(25′) can then be written

$$\frac{dp_\alpha}{dt} = \sum_\beta W_{\alpha\beta}[(p_\beta - p_\beta^0) - (p_\alpha - p_\alpha^0)], \qquad (26'')$$

which shows that the equation (25′) may be used to describe the trend of the system S toward thermal equilibrium if the variables P_α represent the departures of the populations from the thermal equilibrium values rather than the populations themselves. Once the system (25) is solved and the populations known, it is possible for any physical quantity represented by an operator Q that commutes with the main Hamiltonian $\hbar\mathscr{H}_0$ and thus has a well-defined eigenvalue Q_α for each eigenstate $|\alpha\rangle$ of $\mathscr{H}_0$, to calculate its expectation value $\langle Q(t)\rangle$, which represents the observable macroscopic value of the physical quantity. It is given by

$$\langle Q(t)\rangle = \sum_\alpha Q_\alpha p_\alpha(t).$$

Thus, for instance, the magnetization $M_z(t)$ of a system of nuclear spins I in a large magnetic field H_0 will be

$$M_z(t) = \langle \mathscr{M}_z(t)\rangle = N\gamma\hbar \sum_{m=-I}^{I} m p_m(t), \qquad (27)$$

where m is an eigenvalue of I_z along H_0.

From an inspection of the system (25) it is clear that for a system S with more than two degrees of freedom the time-dependent solutions $p_\alpha(t)$ will be in general expressed through a superposition of several decreasing exponentials and it is not at all obvious that their linear combination (27) will be expressed by a single exponential as is usually implied in the definition of the spin-lattice relaxation time T_1, by the relation

$$\frac{dM_z}{dt} = -\frac{1}{T_1}(M_z - M_0).$$

Furthermore, the use of the system (25) necessarily involves the assumption that the behaviour of the statistical ensemble of systems S is describable through the populations of their energy states. This precludes the description of a situation where a certain coherence exists between the phases of the amplitudes of these states, that is to say, where the density matrix has off-diagonal elements. Such a situation exists, for instance, for a system of nuclear spins after a 90° pulse when a non-vanishing transverse magnetization has been established, which the previous formalism is unable to predict. A more general formalism containing the system (25) as a special case must then be introduced.

(c) *The master equation for the density matrix*

We start from the equation of motion of the density matrix σ for the system S:

$$\frac{1}{i}\frac{d\sigma}{dt} = -[\mathcal{H}_0 + \mathcal{H}_1(t), \sigma] \tag{28}$$

where the perturbing Hamiltonian $\hbar\mathcal{H}_1(t)$ is a stationary random operator. In the interaction representation with

$$\sigma^* = e^{i\mathcal{H}_0 t}\sigma e^{-i\mathcal{H}_0 t}, \qquad \mathcal{H}_1^*(t) = e^{i\mathcal{H}_0 t}\mathcal{H}_1(t)e^{-i\mathcal{H}_0 t}$$

the equation (28) becomes

$$\frac{1}{i}\frac{d\sigma^*}{dt} = -[\mathcal{H}_1^*(t), \sigma^*]. \tag{29}$$

(29), integrated by successive approximations up to the second order, gives

$$\sigma^*(t) = \sigma^*(0) - i\int_0^t [\mathcal{H}_1^*(t'), \sigma^*(0)]dt' - \int_0^t dt' \int_0^{t'} dt''[\mathcal{H}_1^*(t'), [\mathcal{H}_1^*(t''), \sigma^*(0)]], \tag{30}$$

whence, taking the time derivative of (30),

$$\frac{d\sigma^*}{dt} = -i[\mathcal{H}_1^*(t), \sigma^*(0)] - \int_0^t dt'[\mathcal{H}_1^*(t), [\mathcal{H}_1^*(t'), \sigma^*(0)]] \tag{31}$$

or, introducing in the integral a new variable $\tau = t - t'$,

$$\frac{d\sigma^*}{dt} = -i[\mathcal{H}_1^*(t), \sigma^*(0)] - \int_0^t d\tau[\mathcal{H}_1^*(t), [\mathcal{H}_1^*(t-\tau), \sigma^*(0)]]. \tag{32}$$

Since $\mathcal{H}_1(t)$ is a random operator, so is σ^* according to the equation (32), and the observable behaviour of a statistical ensemble of systems S will be described by an average density operator $\overline{\sigma^*}$ which obeys an equation obtained by taking an ensemble average on both sides of equation (32) over all the random Hamiltonians $\hbar\mathcal{H}_1$. It can always be assumed that $\overline{\mathcal{H}_1(t)} = 0$, for all average matrix elements $\overline{(\alpha\,|\,\mathcal{H}_1(t)\,|\,\beta)}$, if different from zero, can be included into a redefined unperturbed Hamiltonian $\hbar\mathcal{H}_0$. It will be assumed further that

(a) it is permissible to neglect the correlation between $\mathcal{H}_1^*(t)$ and $\sigma^*(0)$ in the averaging of (32) and average them separately;

(b) it is then permissible to replace $\sigma^*(0)$ by $\sigma^*(t)$ on the right-hand side of (32);

(c) it is permissible to extend the upper limit of the integral (32) to $+\infty$;

(d) it is permissible to neglect all unwritten higher-order terms on the right-hand side of (32).

A justification of these assumptions for a limited class of systems, based on the shortness of the correlation time of $\hbar\mathcal{H}_1$, will be discussed presently but we may remark that the four previous assumptions are also used, although seldom formulated explicitly, in the usual derivation of the system (25) which corresponds to the special case of a density matrix that commutes with the unperturbed Hamiltonian $\hbar\mathcal{H}_0$. With these assumptions and omitting the bar on σ^* which will henceforth stand for the average density matrix, equation (32) becomes

$$\frac{d\sigma^*}{dt} = -\int_0^\infty d\tau \overline{[\mathcal{H}_1^*(t), [\mathcal{H}_1^*(t-\tau), \sigma^*(t)]]}. \tag{33}$$

If we transcribe (33) in a matrix notation in a representation where the basic states are the eigenstates $|\alpha\rangle$, $|\beta\rangle$ of the unperturbed Hamiltonian $\hbar\mathcal{H}_0$ with eigenvalues $\hbar\alpha$, $\hbar\beta$, a straightforward calculation gives

$$\frac{d}{dt}\sigma^*_{\alpha\alpha'} = \sum_{\beta\beta'} e^{i(\alpha-\alpha'-\beta+\beta')t} R_{\alpha\alpha',\beta\beta'} \sigma^*_{\beta\beta'}, \tag{34}$$

where, because of the stationary character of the random Hamiltonian $\hbar\mathcal{H}_1(t)$, the coefficients $R_{\alpha\alpha',\beta\beta'}$ are independent of time.

According to the equation (33) the density matrix σ^* would be time-independent in the absence of the perturbing Hamiltonian $\hbar\mathcal{H}_1$ and, since the latter is a small perturbation, the variation of σ^* with time is slow.

According to a previous discussion (eqn. (23) of Chapter IV), the effect on σ^* of terms with rapidly varying exponentials $e^{i(\alpha-\alpha'-\beta+\beta')t}$ should be negligible compared to that of secular terms for which $\alpha-\alpha'-\beta+\beta' = 0$, and the summation in (34) may be restricted to those. The differential system (34) then becomes a system with constant coefficients which is the generalized master equation

$$\frac{d}{dt}\sigma^*_{\alpha\alpha'} = \sum_{\beta\beta'}{}' R_{\alpha\alpha',\beta\beta'} \sigma^*_{\beta\beta'} \tag{35}$$

where the summation $\sum'_{\beta\beta'}$ is restricted to states of energies $\hbar\beta$ and $\hbar\beta'$ such that
$$\beta-\beta' = \alpha-\alpha'. \tag{36}$$

As for the more special case of the equations (25), the semi-classical treatment of the coupling with the lattice as a random perturbation should be corrected by the replacement of $\sigma^*(t)$ by $\sigma^*(t)-\sigma_0^*$ where

$$\sigma_0^* = \sigma_0 = \exp[-\hbar\mathcal{H}_0/kT]/\mathrm{tr}\{\exp[-\hbar\mathcal{H}_0/kT]\}.$$

It will be understood from now on that in all linear equations for σ^* or σ, it is really $\sigma^*-\sigma_0$ or $\sigma-\sigma_0$ that is calculated. It will be seen later how a quantum mechanical description of the lattice permits us to do away with this *ad hoc* assumption.

It is easily seen that, because of the condition (36), if all off-diagonal elements $\sigma^*_{\alpha\alpha'}$ such that the energies $\hbar\alpha$ and $\hbar\alpha'$ are different, vanish at a time t, that is, if the density operator σ^* commutes at that time with the main Hamiltonian $\hbar\mathcal{H}_0$, it will commute with $\hbar\mathcal{H}_0$ at any later time. This is a satisfactory feature since the non-commutation of σ^* (or σ) with $\hbar\mathcal{H}_0$ corresponds to the existence of a phase coherence inside the system, and one does not expect such coherence to be introduced into a system through a coupling described by a random Hamiltonian.

In the presence of degeneracy of the main Hamiltonian $\hbar\mathcal{H}_0$, two different eigenstates $|\alpha)$ and $|\alpha')$ can have equal energies $\hbar\alpha = \hbar\alpha'$ and matrix elements $\sigma^*_{\alpha\alpha'}$ and coefficients such as $R_{\alpha\alpha',\beta\beta'}$ are not uniquely defined. It can be verified that the previous statement that all off-diagonal elements $\sigma^*_{\alpha\alpha'}$ vanish for $t' > t$ if they vanish at time t, can be extended to this case by a suitable choice of the basic states $|\alpha)$.

If all off-diagonal elements vanish, the generalized master equation (35) reduces to the master equation (25) for the populations.

The passage from the equations (35) for σ^* to those for σ is straightforward, it is sufficient to add a term $-i[\mathcal{H}_0, \sigma]_{\alpha\alpha'}$ on the right-hand side of (35) and replace σ^* everywhere by σ.

We shall not write the coefficients $R_{\alpha\alpha',\beta\beta'}$ explicitly. They can be easily obtained from equation (33) or from an operator transcription of the master equation to be given now.

(d) *The master equation in operator form*

The random Hamiltonian $\hbar\mathcal{H}_1(t)$ can be expanded as

$$\mathcal{H}_1(t) = \sum_q F^{(q)}(t) A^{(q)}, \tag{37}$$

where the $F^{(q)}$ are random functions of time and the $A^{(q)}$ are operators acting on the variables of the system S.

We introduce the correlation functions

$$g_{qq'}(\tau) = \overline{F^{(q)}(t) F^{(q')*}(t+\tau)}$$

and the spectral densities

$$J_{qq'}(\omega) = \int_{-\infty}^{\infty} g_{qq'}(\tau) e^{-i\omega\tau}\, d\tau, \qquad j_{qq'}(\omega) = \int_0^{\infty} g_{qq'}(\tau) e^{-i\omega\tau}\, d\tau.$$

If the $F^{(q)}$ are complex functions and the $A^{(q)}$ non-hermitian operators, the hermiticity of $\mathcal{H}_1$ requires that to each term $F^{(q)} A^{(q)}$ there be associated a term $F^{(q)*} A^{(q)\dagger}$. We make the convention $F^{(-q)} = F^{(q)*}$,

$A^{(-q)} = A^{(q)\dagger}$. The expansion (37) of $\mathcal{H}_1(t)$ can be supplemented by introducing the following definitions:

$$\left.\begin{aligned} e^{i\mathcal{H}_0 t}A^{(q)}e^{-i\mathcal{H}_0 t} &= A^{(q)}(t) = \sum_p A_p^{(q)}e^{i\omega_p^{(q)}t} \\ e^{i\mathcal{H}_0 t}A^{(-q)}e^{-i\mathcal{H}_0 t} &= A^{(-q)}(t) = \sum_p A_p^{(-q)}e^{i\omega_p^{(-q)}t} \\ \text{with} \quad \omega_p^{(-q)} &= -\omega_p^{(q)} \end{aligned}\right\}, \tag{38}$$

$$\mathcal{H}_1^*(t) = e^{i\mathcal{H}_0 t}\mathcal{H}_1(t)e^{-i\mathcal{H}_0 t} = \sum_{p,q} F^{(q)}A_p^{(q)}e^{i\omega_p^{(q)}t}, \tag{39}$$

where the $A_p^{(q)}$ are operators acting on the variables of the system S.

Replacing $\mathcal{H}_1^*$ in (33) by its expression (39), we obtain

$$\frac{d\sigma^*}{dt} = -\sum_{q,q',p,p'} e^{i(\omega_p^{(q)}+\omega_{p'}^{(q')})t}[A_{p'}^{(q')}, [A_p^{(q)}, \sigma^*(t)]] \int_0^\infty g_{q,-q'}(\tau)e^{-i\omega_p^{(q)}\tau}\,d\tau. \tag{40}$$

Neglecting the non-secular terms and assuming for simplicity

$$g_{qq'}(\tau) = \delta_{qq'}g_q(\tau),$$

which will be realized in all the examples to be considered later, we get

$$\frac{d\sigma^*}{dt} = -\sum_{q,p}[A_p^{(-q)}, [A_p^{(q)}, \sigma^*(t)]]\int_0^\infty g_q(\tau)e^{-i\omega_p^{(q)}\tau}\,d\tau, \tag{41}$$

where $g_q(\tau) = \overline{F^{(q)}(t)F^{(q)*}(t+\tau)}$ is a real and even function of τ.

$$\int_0^\infty g_q(\tau)e^{-i\omega_p^{(q)}\tau}\,d\tau = \tfrac{1}{2}\int_{-\infty}^\infty g_q(\tau)e^{-i\omega_p^{(q)}\tau}\,d\tau - i\int_0^\infty g_q(\tau)\sin(\omega_p^{(q)}\tau)\,d\tau$$

$$= \tfrac{1}{2}J_q(\omega_p^{(q)}) - ik_q(\omega_p^{(q)}).$$

It can be shown that the imaginary term $ik_q(\omega_p^{(q)})$ results in a very small shift in the energy of the system S which can be included in a redefined unperturbed Hamiltonian $\mathcal{H}_0$ and thus dropped from the relaxation equation (41), which now reads

$$\frac{d\sigma^*}{dt} = -\tfrac{1}{2}\sum_{q,p} J_q(\omega_p^{(q)})[A_p^{(-q)}, [A_p^{(q)}, \sigma^*]]. \tag{42}$$

This is the master equation in the operator form.

A situation which has received the name of 'extreme narrowing' is that where the correlation time τ_c is so short that all the products $\omega_p^{(q)}\tau_c$ are very small numbers and all spectral densities $J_q(\omega_p^{(q)})$ are practically independent of the frequency and equal to $J_q(0)$ (white spectrum approximation). In that case, reverting from the general equation (40) to the corresponding equation for

$$\sigma = e^{-i\mathcal{H}_0 t}\sigma^* e^{i\mathcal{H}_0 t},$$

we obtain the very compact relation

$$\frac{d\sigma}{dt} = -i[\mathcal{H}_0, \sigma] - \sum_{q,q'} j_{q,-q'}(0)[A^{(q')}, [A^{(q)}, \sigma]] \qquad (42')$$

and, with the simplifying assumption $j_{q,-q'} = \delta_{q,-q'} j_q$,

$$\frac{d\sigma}{dt} = -i[\mathcal{H}_0, \sigma] - \sum_{q} j_q(0)[A^{(-q)}, [A^{(q)}, \sigma]]$$

$$\cong -i[\mathcal{H}_0, \sigma] - \tfrac{1}{2} \sum_q J_q(0)[A^{(-q)}, [A^{(q)}, \sigma]]. \qquad (42'')$$

The master equation for extreme narrowing can be rewritten in a still different form, starting from the equation (33):

$$\frac{d\sigma^*}{dt} = -\int_0^\infty \overline{[\mathcal{H}_1^*(t), [\mathcal{H}_1^*(t-\tau), \sigma^* - \sigma_0]]}\, d\tau \cong -\tfrac{1}{2} \int_{-\infty}^\infty \overline{[\ , [\ \ ]]}\, d\tau.$$

From the relation

$$\frac{d\sigma}{dt} = -i[\mathcal{H}_0, \sigma] + e^{-i\mathcal{H}_0 t}\frac{d\sigma^*}{dt} e^{i\mathcal{H}_0 t},$$

we get

$$\frac{d\sigma}{dt} = -i[\mathcal{H}_0, \sigma] - \tfrac{1}{2} \int_{-\infty}^\infty \overline{[\mathcal{H}_1(t), [e^{-i\mathcal{H}_0\tau}\mathcal{H}_1(t-\tau)e^{i\mathcal{H}_0\tau}, \sigma - \sigma_0]]}\, d\tau \qquad (42''')$$

and, for very short correlation times $e^{-i\mathcal{H}_0\tau} \cong 1$,

$$\overline{\mathcal{H}_1(t)\mathcal{H}_1(t-\tau)} \cong 2\tau_c \overline{\mathcal{H}_1(t)\mathcal{H}_1(t)}\, \delta(\tau),$$

where τ_c is the correlation time; whence

$$\frac{d\sigma}{dt} = -i[\mathcal{H}_0, \sigma] - \tau_c \overline{[\mathcal{H}_1(t), [\mathcal{H}_1(t), \sigma - \sigma_0]]}. \qquad (42^{iv})$$

(e) Macroscopic differential equations

If a given operator Q acts on the variables of the system S, the quantity observed in an experiment performed on a macroscopic sample containing a collection of systems S is $q(t) = \langle Q \rangle = \text{tr}\{\sigma(t)Q\}$. In most cases, it is more convenient to calculate

$$q^*(t) = \langle Q \rangle^* = \text{tr}\{\sigma^*(t)Q\}$$

in order to exhibit the slow variation of Q due to the coupling of the system S with the lattice, rather than the fast motion due to the main Hamiltonian $\hbar\mathcal{H}_0$. In the special case when Q commutes with $\hbar\mathcal{H}_0$ and is a constant of the unperturbed motion the two results coincide.

It is sometimes possible to obtain an equation of motion for $\langle Q \rangle^*$ directly without solving the master equation.

From the operator equation (33), multiplying both sides by Q and taking the trace, we get

$$\frac{dq^*}{dt} = -\{a^* - a_0\}, \tag{43}$$

where
$$a^* = \langle \mathscr{A} \rangle^* = \text{tr}\{\mathscr{A}\sigma^*\},$$
$$a_0 = \text{tr}\{\mathscr{A}\sigma_0\},$$

and the operator $\mathscr{A}$ is defined as

$$\mathscr{A} = \int_0^\infty d\tau \overline{[\mathscr{H}_1^*(t-\tau), [\mathscr{H}_1^*(t), Q]]}. \tag{44}$$

As remarked previously, we have replaced

$$\sigma^* \quad \text{by} \quad \sigma^* - \sigma_0.$$

If we use the expansion (39) for $\mathscr{H}_1^*(t)$ with the simplifying assumption $j_{q,q'} = \delta_{qq'} j_q$, we get

$$\mathscr{A} = \tfrac{1}{2} \sum_{q,p} J_q(\omega_p^{(q)})[A_p^{(q)}, [A_p^{(-q)}, Q]]. \tag{45}$$

In the extreme narrowing case we get a very simple equation for

$$q = \langle Q \rangle = \text{tr}\{Q\sigma\},$$
$$\frac{dq}{dt} = \text{tr}\{i[\mathscr{H}_0, Q]\sigma\} - (b - a_0), \tag{45'}$$

with $b = \langle B \rangle = \text{tr}\{B\sigma\}$ and

$$B = \tfrac{1}{2} \sum_q J_q(0)[A^{(q)}, [A^{(-q)}, Q]]. \tag{46}$$

According to (42^{iv}), B can also be written

$$B = \tau_c \overline{[\mathscr{H}_1(t), [\mathscr{H}_1(t), Q]]}.$$

(f) Summary of the notation introduced in this section

$\hbar \mathscr{H}_0$ Hamiltonian of the system

$\hbar \mathscr{H}_1(t)$ random operator representing the coupling with the lattice

$\mathscr{H}_1^*(t) = e^{i\mathscr{H}_0 t} \mathscr{H}_1(t) e^{-i\mathscr{H}_0 t}$

$\sigma(t)$ average density operator for the system S

$\sigma^*(t) = e^{i\mathscr{H}_0 t} \sigma(t) e^{-i\mathscr{H}_0 t}$

σ_0 equilibrium density operator
$$= \exp(-\hbar \mathscr{H}_0/kT)/\text{tr}\{\exp(-\hbar \mathscr{H}_0/kT)\}$$

$|\alpha\rangle, |\beta\rangle;\ \hbar\alpha, \hbar\beta$ eigenstates and eigenvalues of $\hbar\mathscr{H}_0$

$\mathscr{H}_1(t) = \sum_q F^{(q)}(t) A^{(q)}$: $F^{(q)}$ = random function; $A^{(q)}$ = spin operator

$g_{qq'}(\tau) = \overline{F^{(q)}(t) F^{(q')*}(t+\tau)} = \overline{F^{(q)}(t) F^{(-q')}(t+\tau)}$

$$A^{(-q)} = A^{(q)\dagger}, \qquad F^{(-q)} = F^{(q)*}$$

$$j_{qq'}(\omega) = \int_0^\infty e^{-i\omega\tau} g_{qq'}(\tau)\, d\tau$$

$$j_q(\omega) = j_{qq}(\omega)$$

$$J_q(\omega) = \int_{-\infty}^\infty g_{qq}(\tau) e^{-i\omega\tau}\, d\tau$$

$$k_q(\omega) = \int_0^\infty g_{qq}(\tau)\sin(\omega\tau)\, d\tau$$

$$A^{(q)}(t) = e^{i\mathscr{H}_0 t} A^{(q)} e^{-i\mathscr{H}_0 t} = \sum_p A_p^{(q)} e^{i\omega_p^{(q)} t}$$

$$\mathscr{H}_1^*(t) = \sum_q F^{(q)}(t) A^{(q)}(t) = \sum_{q,p} F^{(q)}(t) A_p^{(q)} e^{i\omega_p^{(q)} t}$$

(g) *Justification of the four assumptions leading to the generalized master equation*

We start from the equation (32) where the value of t has not yet been specified.

(i) The random function $\sigma^*(0)$ depends on the behaviour of $\mathscr{H}_1$ before the time $t=0$, and, since there is a correlation between $\mathscr{H}_1$ taken at different times, there is in principle a correlation between $\mathscr{H}_1^*(t)$ and $\mathscr{H}_1^*(t-\tau)$ on the one hand and $\sigma^*(0)$ on the other. It is clear that the correlation between $\mathscr{H}_1^*(t)$ and $\mathscr{H}_1^*(t-\tau)$ is negligible unless τ is of the order of τ_c at most, but then the correlation between the two of them and $\sigma^*(0)$ becomes negligible if $t \gg \tau_c$. It is thus permissible to average over $\sigma^*(0)$ and $\mathscr{H}_1^*(t)\mathscr{H}_1^*(t-\tau)$ independently in the integral (32) if $t \gg \tau_c$.

(ii) The increase $\{\sigma(t)-\sigma(0)\}/\sigma(0)$ is of the order of

$$\frac{t}{|\sigma(0)|}\left|\frac{d\sigma}{dt}\right| \simeq \left|t \int_0^t \overline{\mathscr{H}_1^*(t)\mathscr{H}_1^*(t-\tau)}\, d\tau\right| \simeq t\overline{|\mathscr{H}_1|^2}\tau_c.$$

If the conditions $t \gg \tau_c$, $t\overline{|\mathscr{H}_1|^2}\tau_c \ll 1$, are compatible it is permissible to replace $\overline{\sigma^*(0)}$ in the integral by $\overline{\sigma^*(t)}$.

(iii) The contribution to the integral $\int_0^t \overline{\mathscr{H}_1^*(t)\mathscr{H}_1^*(t-\tau)}\, d\tau$ of values of $\tau \gg \tau_c$ is negligible. It is thus permissible to extend the upper limit to infinity.

(iv) The contribution from higher-order terms can be neglected, for the nth term A_n is deduced from the term A_{n-1} through

$$A_n(t) = -i\int_0^t \left[\mathscr{H}_1^*(t), A_{n-1}(t')\right] dt' \qquad (46')$$

and it is apparent that $|A_n/A_{n-1}| \cong [\overline{|\mathcal{H}_1|^2 \tau_c^2}]^{\frac{1}{2}}$, which is very small if the correlation time is very short.

To conclude, the equation (33) is valid if $[\overline{|\mathcal{H}_1|^2 \tau_c^2}]^{\frac{1}{2}}$ is a very small number.

For the validity of the master equation (35) with constant coefficients a further assumption is required: all differences $\alpha-\beta$ or even combined differences $(\alpha-\alpha')-(\beta-\beta')$ between the energies (on the frequency scale) of the unperturbed Hamiltonian $\hbar\mathcal{H}_0$, unless identically zero, must be large compared with the constant $1/T \cong \overline{|\mathcal{H}_1|^2 \tau_c}$ which gives the relative rate of change of σ^*.

D. Quantum mechanical formulation of the problem

The semi-classical treatment where the coupling with the lattice is represented by random functions suffers from several defects, the main one being that, for the spin system, it always leads to a steady state described by an infinite temperature.

It will be shown that a quantum mechanical description of the lattice can be cast in a form very similar to that of the semi-classical description, but will lead for a spin system to a finite temperature equal to that of the lattice.

We start with a time-independent Hamiltonian

$$\hbar\mathcal{H} = \hbar(\mathcal{H}_0 + \mathcal{F} + \mathcal{H}_1), \tag{47}$$

where $\hbar\mathcal{H}_0$ and $\hbar\mathcal{F}$ are the unperturbed Hamiltonians of the spin system S and the lattice, respectively, with eigenstates $|\alpha\rangle$ and $|f\rangle$, and $\hbar\mathcal{H}_1$ describes the perturbing coupling between them and contains parameters of both the spin system and the lattice.

$\mathcal{H}_1$ can be expanded as

$$\mathcal{H}_1 = \sum_q F^{(q)} A^{(q)}, \tag{48}$$

where the $F^{(q)}$ and the $A^{(q)}$ are respectively lattice and spin operators.

To pursue the parallel with the previous semi-classical formalism we define
$$\mathcal{H}_1(t) = e^{i\mathcal{F}t}\mathcal{H}_1 e^{-i\mathcal{F}t} = \sum_q F^{(q)}(t) A^{(q)}, \tag{49}$$

with $$F^{(q)}(t) = e^{i\mathcal{F}t} F^{(q)} e^{-i\mathcal{F}t}$$

and
$$\mathcal{H}_1^*(t) = e^{i\mathcal{H}_0 t}\mathcal{H}_1(t) e^{-i\mathcal{H}_0 t} = \sum_q F^{(q)}(t) A^{(q)}(t) = \sum_{q,p} F^{(q)}(t) A_p^{(q)} e^{i\omega_p^{(q)} t}. \tag{49'}$$

The similarity in form with the notation of the previous sections is complete.

To understand how the description of the spin-lattice coupling leads to a finite temperature for the spin system, consider for simplicity the case where the expansion (48) contains a single term $\mathscr{H}_1 = FA$ which induces in the spin system S a probability per unit time $W_{\alpha\beta}$ of passing from a state $|\beta\rangle$ to a state $|\alpha\rangle$, which differ in energy by $\omega_{\alpha\beta} = \alpha - \beta$. We consider first the more detailed transition $|\beta,f\rangle \to |\alpha,f'\rangle$ of the combined system spins plus lattice:

$$W_{\alpha f',\beta f} = \int_0^t (\beta,f|\mathscr{H}_1|\alpha,f')(\alpha,f'|\mathscr{H}_1|\beta,f)e^{-i[\alpha-\beta+f'-f](t-t')}\,dt' + \text{c.c.}, \quad (50)$$

which can be made very similar in appearance to formula (2) by using

$$\mathscr{H}_1(t) = e^{i\mathscr{F}t}\mathscr{H}_1 e^{-i\mathscr{F}t} = Ae^{i\mathscr{F}t}Fe^{-i\mathscr{F}t} = AF(t),$$

$$W_{\alpha f',\beta f} = \int_0^t (\beta,f|\mathscr{H}_1(t)|\alpha,f')(\alpha,f'|\mathscr{H}_1(t-\tau)|\beta,f)e^{-i\omega_{\alpha\beta}\tau}\,d\tau + \text{c.c.}$$

$$= |(\alpha|A|\beta)|^2 \int_0^t (f|F(t)|f')(f'|F(t-\tau)|f)e^{-i\omega_{\alpha\beta}\tau}\,d\tau + \text{c.c.} \quad (51)$$

The total probability $W_{\alpha\beta} = \sum_{f,f'} P(f)W_{\alpha f',\beta f}$, where $P(f) = ae^{-\hbar f/kT}$ is the probability of finding a lattice at a temperature T, in any initial state $|f\rangle$, is given by

$$W_{\alpha\beta} = |(\alpha|A|\beta)|^2 \int_{-\infty}^{\infty} e^{-i\omega_{\alpha\beta}\tau} \sum_{f,f'} P(f)(f|F(t)|f')(f'|F(t-\tau)|f)\,d\tau. \quad (52)$$

The discrete summation over the index f should actually, because of the continuous spectrum of the lattice, be replaced by an appropriate integration $\int \eta(f)\,df$, where $\eta(f)$ is the density of lattice states. We will continue to write symbolically $\sum_f$ for simplicity. The expression

$$\sum_{f,f'} P(f)(f|F(t)|f')(f'|F(t-\tau)|f)$$

which from the definition of $F(t)$ is clearly independent of t, can be written
$$g(\tau) = \text{tr}_f\{F(t)\mathscr{P}(\mathscr{F})F(t+\tau)\}, \quad (53)$$

where $\mathscr{P}(\mathscr{F})$ is the statistical operator

$$\mathscr{P}(\mathscr{F}) = ae^{-\hbar\mathscr{F}/kT} = \exp\left(\frac{-\hbar\mathscr{F}}{kT}\right)\Big/\text{tr}_f\left\{\exp\left(\frac{-\hbar\mathscr{F}}{kT}\right)\right\}. \quad (54)$$

$g(\tau)$ is the quantum mechanical analogue of the classical correlation function $g(\tau)$ of a classical random function $F(t)$, defined previously as $g(\tau) = \overline{F(t).F(t+\tau)}$, where the bar represented an ensemble average over the probability distribution of the random function. Defining

$$J(\omega) = \int_{-\infty}^{\infty} g(\tau)e^{-i\omega\tau}\,d\tau, \quad (55)$$

we obtain
$$W_{\alpha\beta} = |(\alpha\,|\,A\,|\,\beta)|^2 J(\omega_{\alpha\beta}), \qquad (56)$$
which is formally identical with formula (24′). There is, however, an important difference because now
$$J(-\omega) = \exp(\hbar\omega/kT) J(\omega),$$
and according to (56) a lattice induced transition where the lattice gains the energy $\hbar\omega$ is more probable than the opposite one by a factor $\exp(\hbar\omega/kT)$. This is seen from the definitions (53) and (55) of $J(\omega)$:
$$\begin{aligned} J(\omega) &= a \int_{-\infty}^{\infty} \sum_{f,f'} |(f\,|\,F\,|\,f')|^2 e^{-\hbar f/kT} e^{i(f-f'-\omega)\tau}\, d\tau \\ &= 2\pi a \sum_f |(f\,|\,F\,|\,f-\omega)|^2 e^{-\hbar f/kT}, \\ J(-\omega) &= 2\pi a \sum_f |(f\,|\,F\,|\,f+\omega)|^2 e^{-\hbar f/kT}, \end{aligned} \qquad (57)$$
or, since the summation over f is actually a continuous integration from $-\infty$ to $+\infty$, replacing $f+\omega$ by f we get
$$J(-\omega) = 2\pi a \sum_f |(f-\omega\,|\,F\,|\,f)|^2 e^{-\hbar(f-\omega)/kT} = e^{\hbar\omega/kT} J(\omega). \qquad (58)$$

We now pass on to the more general problem of deriving a master equation describing the motion of the spin system S, analogous to the equations (34), (35), or (42) of this chapter. A density matrix ρ now describes the behaviour of the combined quantum mechanical system: spins+lattice. Its transform in the interaction representation
$$\rho^* = e^{i(\mathcal{H}_0+\mathcal{F})t} \rho\, e^{-i(\mathcal{H}_0+\mathcal{F})t}$$
obeys the equation
$$\frac{i\, d\rho^*}{dt} = -[\mathcal{H}_1^*(t), \rho^*], \qquad (59)$$
where $\mathcal{H}_1^*(t)$ is defined by equation (49′). A forward integration of (59) leads to an equation similar in form to equation (32):
$$\frac{d\rho^*}{dt} = -i[\mathcal{H}_1^*(t), \rho^*(0)] - \int_0^t d\tau\, [\mathcal{H}_1^*(t), [\mathcal{H}_1^*(t-\tau), \rho^*(0)]]$$
$$+ \text{ higher-order terms.} \qquad (60)$$

Since all the observations are performed on the spin system, all the relevant information is contained in the reduced density matrix
$$\sigma^* = \text{tr}_f\{\rho^*\}$$
with matrix elements $(\alpha\,|\,\sigma^*\,|\,\alpha') = \sum_f (f\alpha\,|\,\rho^*\,|\,f\alpha')$. We make the fundamental assumption that the lattice, because of its very large heat capacity, remains in thermal equilibrium so that $\rho^*(t) = \mathcal{P}(\mathcal{F})\sigma^*(t)$, where $\mathcal{P}(\mathcal{F})$ is the statistical operator (54).

In order to obtain an equation for the rate of change of the spin density matrix σ^*, we perform on both sides of (60) the operation trace with respect to the lattice parameters f.

Assume first that the temperature of the lattice is infinite so that the statistical operator $\mathscr{P}(\mathscr{F})$ is proportional to the unit operator and $\rho^*(0) = a\sigma^*(0)$. $\quad a = [\text{tr}_f\{\exp(-\hbar\mathscr{F}/kT)\}]^{-1}$

becomes in that case $1/L$, where L is the number (astronomically large) of degrees of freedom of the lattice. We shall represent by a bar the operation $a\,\text{tr}_f\{\ \}$. In that case we get

$$\frac{d\sigma^*}{dt} = -i[\overline{\mathscr{H}_1^*(t), \sigma^*(0)}] - \int_0^t d\tau\,[\overline{\mathscr{H}_1^*(t), [\mathscr{H}_1^*(t-\tau), \sigma^*(0)]}]. \quad (61)$$

This equation is formally identical to the equation (32), and using the expansion (49) for $\mathscr{H}_1(t)$ a master equation of exactly the same form as equations (40) and (42) can be obtained for σ^*.

The only change is that correlation functions of classical random functions $F^{(q)}(t)$ are replaced by correlation functions of operators $F^{(q)}$, defined by

$$g_{qq'}(\tau) = \overline{F^{(q)}(t)F^{(-q')}(t+\tau)} = \frac{1}{L}\sum_{f,f'}(f|F^{(q)}|f')(f'|F^{(-q')}|f)e^{i(f'-f)\tau}, \quad (62)$$

a special case of the definition (53) given for a finite temperature of the lattice. The conditions of validity of the master equation, relative to the shortness of the correlation time, are formulated in the same way as in Section II C (g).

The semi-classical treatment of relaxation is thus formally equivalent to the quantum mechanical one for the limiting case of infinite lattice temperatures.

The case of a finite lattice temperature is more complex, for then the lattice operators $F^{(q)}$ and $\mathscr{P}(\mathscr{F})$ do not commute and it is necessary to expand the double commutator on the right-hand side of (60) into four different terms and consider each of them separately. This situation has been studied in great detail (2, 3), and has been shown to lead again to a linear master equation for σ^* which is, however, more complex than (33) or (42). Furthermore, generalized correlation functions of the forms (53) occur in it, with spectral densities $J(\omega)$ having the property

$$J(-\omega) = \exp(\hbar\omega/kT)J(\omega), \quad (63)$$

with, as a consequence, a steady state solution of the form

$$\sigma_0^* = \sigma_0 = \exp(-\hbar\mathscr{H}_0/kT)/\text{tr}\{\exp(-\hbar\mathscr{H}_0/kT)\}.$$

For simplicity we shall first demonstrate this on the assumption (actually seldom realized in practice) that the lattice temperature is sufficiently high to allow a linear expansion of $\exp(-\hbar\mathscr{F}/kT)$ into $1-(\hbar\mathscr{F}/kT)$ and that the state of the spin system described by the density matrix $\sigma^*(t)$ is never very remote from one of equal populations of all spin energy levels. Then

$$\rho^*(t) = \sigma^*(t)\mathscr{P}(\mathscr{F}) \simeq a\left\{\sigma^*(t) - \frac{1}{A}\frac{\hbar\mathscr{F}}{kT}\right\}, \tag{64}$$

where A is the number of degrees of freedom of the spin system.

As a consequence, on the right-hand side of the master equation for σ^* there appears an extra term

$$\int_0^\infty \overline{\left[\mathscr{H}_1^*(t), \left[\mathscr{H}_1^*(t-\tau), \frac{\hbar\mathscr{F}}{kT}\frac{1}{A}\right]\right]} \, d\tau \tag{65}$$

or, neglecting small imaginary terms,

$$\frac{1}{2}\int_{-\infty}^\infty \overline{\left[\mathscr{H}_1^*(t), \left[\mathscr{H}_1^*(t-\tau), \frac{\hbar\mathscr{F}}{kT}\frac{1}{A}\right]\right]} \, d\tau.$$

It is easily verified that

$$\int_{-\infty}^\infty [\mathscr{H}_1^*(t-\tau), \mathscr{H}_0+\mathscr{F}]\,d\tau = i\int_{-\infty}^\infty \frac{d}{d\tau}[\mathscr{H}_1^*(t-\tau)]\,d\tau = 0,$$

a consequence of the conservation of total energy $\mathscr{F}+\mathscr{H}_0$. It is permissible to replace $\mathscr{F}$ in (65) by $-\mathscr{H}_0$ and, since a unit operator commutes with everything, to rewrite it as

$$\int_0^\infty \overline{\left[\mathscr{H}_1^*(t), \left[\mathscr{H}_1^*(t-\tau), \left(1-\frac{\hbar\mathscr{H}_0}{kT}\frac{1}{A}\right)\right]\right]} \, d\tau,$$

thus obtaining for the master equation

$$\frac{d\sigma^*}{dt} = -\int_0^\infty \overline{[\mathscr{H}_1^*(t), [\mathscr{H}_1^*(t-\tau), \sigma^*-\sigma_0]]} \, d\tau. \tag{66}$$

The empirical rule whereby in the relaxation equation obtained by the semi-classical method σ^* should be replaced by $\sigma^*-\sigma_0$ is thus justified.

This proof is clearly inadequate in most situations since the very broad and unnecessary requirement $\exp(-\hbar\mathscr{F}/kT) \simeq 1-(\hbar\mathscr{F}/kT)$ leads through (62) to expressions for the correlation function, and thus to values for the relaxation times, that are temperature-independent. Actually the much less stringent assumption

$$\left|\frac{\hbar \mathcal{H}_0}{kT}\right| \ll 1, \quad \left|\sigma^* - \frac{1}{A}\right| \ll 1$$

need only be made. Starting from

$$\frac{d\sigma^*}{dt} = -\mathrm{tr}_f\left\{\int_0^t [\mathcal{H}_1^*(t), [\mathcal{H}_1^*(t-\tau), \rho^*]]\, d\tau\right\}$$

$$\cong -\mathrm{tr}_f\left\{\tfrac{1}{2} \int_{-\infty}^{\infty} [\mathcal{H}_1^*(t), [\mathcal{H}_1^*(t'), \sigma^* \mathcal{P}(\mathcal{F})]]\, dt'\right\}, \qquad (66')$$

where $\mathcal{P}(\mathcal{F}) = ae^{-\beta \mathcal{F}}$ and $\beta = \hbar/kT$, (66′) can be rewritten as

$$-\mathrm{tr}_f\Big\{\tfrac{1}{2}\int_{-\infty}^{\infty}[\mathcal{H}_1^*(t),[\mathcal{H}_1^*(t'),\sigma^*]]\,dt'\cdot\mathcal{P}(\mathcal{F})-$$

$$-\tfrac{1}{2}\int_{-\infty}^{\infty}[\mathcal{H}_1^*(t'),\mathcal{P}(\mathcal{F})]\,dt'\cdot\sigma^*\mathcal{H}_1^*(t)+\tfrac{1}{2}\mathcal{H}_1^*(t)\sigma^*\int_{-\infty}^{\infty}[\mathcal{H}_1^*(t'),\mathcal{P}(\mathcal{F})]\,dt'\Big\}.$$

$$(66'')$$

Consider the matrix element

$$\left(\alpha f \left| \int_{-\infty}^{\infty} [\mathcal{H}_1^*(t'), \mathcal{P}(\mathcal{F})]\, dt' \right| \alpha' f'\right)$$

$$= \int_{-\infty}^{\infty} e^{i(f+\alpha-f'-\alpha')t'}\,dt'(\alpha f|\mathcal{H}_1|\alpha'f')a(e^{-\beta f'} - e^{-\beta f})$$

$$= 2\pi a \delta(f+\alpha-f'-\alpha')(\alpha f|\mathcal{H}_1|\alpha'f')(e^{-\beta f'} - e^{-\beta f}). \quad (66''')$$

Since
$$\beta(f-f') = \beta(\alpha'-\alpha) \ll 1,$$

(66‴) can be rewritten as

$$\int_{-\infty}^{\infty} e^{i(f+\alpha-f'-\alpha')t'}\,dt'(\alpha f|\mathcal{H}_1|\alpha'f')ae^{-\beta f'}(1 - e^{-\beta(\alpha'-\alpha)})$$

$$\cong \int_{-\infty}^{\infty} e^{i(f+\alpha-f'-\alpha')t'}\,dt'(\alpha f|\mathcal{H}_1|\alpha'f')ae^{-\beta f'}\beta(\alpha'-\alpha)$$

$$\cong \int_{-\infty}^{\infty} (\alpha f|[\mathcal{H}_1^*(t'), \beta\mathcal{H}_0]\mathcal{P}(\mathcal{F})|\alpha'f')\,dt'$$

$$\cong \int_{-\infty}^{\infty} (\alpha f|\mathcal{P}(\mathcal{F})[\mathcal{H}_1^*(t'), \beta\mathcal{H}_0]|\alpha'f')\,dt'.$$

We can thus replace $\int_{-\infty}^{\infty} [\mathcal{H}_1^*(t'), \mathcal{P}(\mathcal{F})] dt'$ in (66″) by means of

$$\int_{-\infty}^{\infty} [\mathcal{H}_1^*(t'), \beta\mathcal{H}_0]\mathcal{P}(\mathcal{F}) dt' \cong -A \int_{-\infty}^{\infty} [\mathcal{H}_1^*(t'), \sigma_0]\mathcal{P}(\mathcal{F}) dt',$$

where $\quad\sigma_0 \cong \frac{1}{A}\{1-\beta\mathcal{H}_0\} \cong e^{-\beta\mathcal{H}_0}/\mathrm{tr}\{e^{-\beta\mathcal{H}_0}\}.$

(66″) can thus be rewritten as

$$-\mathrm{tr}_f\Big\{\tfrac{1}{2} \int_{-\infty}^{\infty} [\mathcal{H}_1^*(t), [\mathcal{H}_1^*(t'), \sigma^*]]\, dt' \mathcal{P}(\mathcal{F}) +$$
$$+ \tfrac{1}{2} \int_{-\infty}^{\infty} [\mathcal{H}_1^*(t'), \sigma_0] A\sigma^* \mathcal{H}_1^*(t) \mathcal{P}(\mathcal{F})\, dt' -$$
$$- \tfrac{1}{2} \int_{-\infty}^{\infty} \mathcal{H}_1^*(t) \sigma^* A [\mathcal{H}_1^*(t'), \sigma_0]\, dt'\, \mathcal{P}(\mathcal{F})\Big\}.$$

In the last two terms we replace $A\sigma^*$ by unity within the approximation $|\sigma^* - 1/A| \ll 1$, whence (66″) becomes

$$-\mathrm{tr}_f\Big\{\tfrac{1}{2} \int_{-\infty}^{\infty} [\mathcal{H}_1^*(t), [\mathcal{H}_1^*(t'), \sigma^* - \sigma_0]] \mathcal{P}(\mathcal{F})\, dt'\Big\}.$$

The definition (62) of the correlation function should be replaced by the following:

$$g_{qq'}(\tau) = \overline{F^{(q)}(t) F^{(-q')}(t+\tau)}$$
$$= \mathrm{tr}_f\{e^{i\mathcal{F}t} F^{(q)} e^{-i\mathcal{F}t} e^{i\mathcal{F}(t+\tau)} F^{(-q')} e^{-i\mathcal{F}(t+\tau)} \mathcal{P}(\mathcal{F})\}$$
$$= \frac{1}{\mathrm{tr}_f\{e^{-\hbar\mathcal{F}/kT}\}} \sum_{f,f'} (f|F^{(q)}|f')(f'|F^{(-q')}|f) e^{-i(f-f')\tau} e^{-\hbar f/kT},$$

where the dependence on the lattice temperature is apparent.

E. Relaxation by dipolar coupling

The dipole-dipole interaction between two spins I and S can be written

$$\hbar\mathcal{H}_1 = \sum_q F^{(q)} A^{(q)}, \tag{67}$$

where the $F^{(q)}$ are random functions of the relative positions of two spins and the $A^{(q)}$ are operators acting on the spin variables with the convention $F^{(q)} = F^{(-q)*}$; $A^{(q)} = A^{(-q)\dagger}$.

$$F^{(1)} = \frac{\sin\theta\cos\theta\, e^{-i\varphi}}{r^3}, \quad F^{(2)} = \frac{\sin^2\theta\, e^{-2i\varphi}}{r^3}, \quad F^{(0)} = \frac{1-3\cos^2\theta}{r^3}, \tag{68}$$

$$A^{(0)} = \alpha\{-\tfrac{2}{3} I_z S_z + \tfrac{1}{6}(I_+ S_- + I_- S_+)\},$$
$$A^{(1)} = \alpha\{I_z S_+ + I_+ S_z\}, \tag{69}$$
$$A^{(2)} = \tfrac{1}{2}\alpha I_+ S_+, \quad \alpha = -\tfrac{3}{2}\gamma_I \gamma_S \hbar.$$

We assume an isotropic random motion for the orientation of the vector **r**, so that

$$\overline{F^{(q)}(t)F^{(q')*}(t+\tau)} = \delta_{qq'}\,G^{(q)}(\tau),$$
$$J^{(q)}(\omega) = \int_{-\infty}^{\infty} G^{(q)}(\tau)e^{-i\omega\tau}\,d\tau. \tag{70}$$

The main Hamiltonian $\hbar\mathcal{H}_0$ is given by

$$\mathcal{H}_0 = \omega_I I_z + \omega_S S_z. \tag{71}$$

We propose to derive the equations of motion for the quantities $\langle \mathbf{I}\rangle$ and $\langle \mathbf{S}\rangle$ proportional to the macroscopic magnetizations. It should be noted that if the spins I and S are like spins, only the sum $\langle \mathbf{I}\rangle + \langle \mathbf{S}\rangle$ is observed, whereas $\langle \mathbf{I}\rangle$ and $\langle \mathbf{S}\rangle$ are observed separately if the spins are unlike.

(a) Like spins

We derive first for the motion of the longitudinal magnetization proportional to $\langle I_z + S_z \rangle$, an equation of the type

$$\frac{d}{dt}\langle I_z + S_z\rangle = -(a_z - a_0),$$

where $a_z = \text{tr}\{\mathcal{A}_z \sigma^*\}$ and the operator $\mathcal{A}_z$ is given by the formula (45) with $Q = I_z + S_z$. For like spins,

$$\mathcal{H}_0 = \omega_I(I_z + S_z),$$
$$e^{i\mathcal{H}_0 t}A^{(0)}e^{-i\mathcal{H}_0 t} = A^{(0)},$$
$$e^{i\mathcal{H}_0 t}A^{(1)}e^{-i\mathcal{H}_0 t} = A^{(1)}e^{i\omega_I t},$$
$$e^{i\mathcal{H}_0 t}A^{(2)}e^{-i\mathcal{H}_0 t} = A^{(2)}e^{2i\omega_I t}.$$

According to the formula (45) (writing I' instead of S to emphasize the fact that the spins are like), $\mathcal{A}_z$ is given by

$$\mathcal{A}_z = \tfrac{1}{2}J^{(1)}(\omega_I)\{[A^{(-1)},[A^{(1)},I_z+I'_z]]\}+\text{h.c.}+$$
$$+\tfrac{1}{2}J^{(2)}(2\omega_I)\{[A^{(-2)},[A^{(2)},I_z+I'_z]]\}+\text{h.c.}, \tag{72}$$

where h.c. stands for hermitian conjugate. From the standard commutation relations between the components of angular momentum, we get

$$[A^{(-1)},[A^{(1)},I_z+I'_z]]$$
$$= 2\alpha^2 I_z I'^2_z + 2\alpha^2 I'_z I^2_z - \alpha^2(I_+ I'_- + I_- I'_+)(I_z+I'_z), \tag{73}$$
$$[A^{(-2)},[A^{(2)},I_z+I'_z]]$$
$$= \alpha^2 I_z(I'^2_x + I'^2_y + I'_z) + \alpha^2 I'_z(I^2_x + I^2_y + I_z). \tag{73'}$$

In the approximation for high temperatures where $\sigma^* - \sigma_0$ is an infinitely small quantity of the first order, quantities such as $\langle I_x \rangle$, $\langle I_y \rangle$, $\langle I_z \rangle$ are also small quantities of the first order and to the same approximation

$$\langle I_z \cdot I'^2_x \rangle \cong \langle I_z \rangle \frac{I(I+1)}{3},$$

$$\langle I_z \cdot I'_z \rangle \cong \langle I_z \cdot I'_x \rangle \cong 0.$$

We then obtain from (69), (73), and (73'),

$$\langle \mathscr{A}_z \rangle \cong \frac{2\alpha^2}{3} I(I+1) \langle I_z + I'_z \rangle \{ J^{(1)}(\omega_I) + J^{(2)}(2\omega_I) \}. \tag{74}$$

It is easily verified that for the special case of spins $\frac{1}{2}$ equation (74) is valid rigorously.

The macroscopic equation for spin-lattice relaxation has the following form:

$$\frac{d}{dt} \langle I_z + I'_z \rangle = -\frac{1}{T_1} \{ \langle I_z + I'_z \rangle - \langle I_z + I'_z \rangle_0 \}, \tag{75}$$

with

$$\frac{1}{T_1} = \tfrac{3}{2} \gamma^4 \hbar^2 I(I+1) \{ J^{(1)}(\omega_I) + J^{(2)}(2\omega_I) \}. \tag{76}$$

This result is immediately generalized to the case when each spin I interacts with several identical spins, provided their motions are not correlated, and the equation

$$\frac{dM_z}{dt} = -\frac{1}{T_1} (M_z - M_0)$$

is still valid with T_1 given by

$$\frac{1}{T_1} = \tfrac{3}{2} \gamma^4 \hbar^2 I(I+1) \sum_k{}' \{ J^{(1)}_{ik}(\omega_I) + J^{(2)}_{ik}(2\omega_I) \}. \tag{77}$$

Equation (75) demonstrates the non-trivial fact that the time variation of $M_z - M_0$ caused by the relaxation is indeed expressed by a single exponential. It should also be remarked that this result is independent of the assumption, usually made in the derivation, and unnecessary, that the state of the spin systems is describable by populations.

The time-dependence of the amplitude of the precessing magnetization in the plane perpendicular to H_0 can be investigated by the same method. Assuming that for $t = 0$ the magnetization is along the x-axis of the laboratory frame, its amplitude is represented at time t by the operator

$$I_x \cos(\omega_I t) + I_y \sin(\omega_I t) = e^{-i\mathscr{H}_0 t} I_x e^{i\mathscr{H}_0 t}.$$

Its expectation value is

$$\mathrm{tr}\{ e^{-i\mathscr{H}_0 t} I_x e^{i\mathscr{H}_0 t} \sigma \} = \mathrm{tr}\{ I_x \sigma^* \} = \langle I_x \rangle^*,$$

and its motion is described by the equation

$$\frac{d}{dt}\langle I_x+I'_x\rangle^* = -\text{tr}\{\mathscr{A}_x\sigma^*\} = -\langle\mathscr{A}_x\rangle^*,$$

where $\mathscr{A}_x$ is given by

$$\mathscr{A}_x = \tfrac{1}{2}J^{(1)}(\omega_I)[A^{(-1)},[A^{(1)},I_x+I'_x]]+\text{h.c.}+$$
$$+\tfrac{1}{2}J^{(2)}(2\omega_I)[A^{(-2)},[A^{(2)},I_x+I'_x]]+\text{h.c.}+$$
$$+\tfrac{1}{2}J^{(0)}(0)[A^{(0)},[A^{(0)},I_x+I'_x]]. \qquad (78)$$

The principle of the calculation is exactly the same as for the relaxation along the z-axis and for that reason only the result will be given:

$$\frac{d}{dt}\langle I_x+I'_x\rangle^* = -\frac{1}{T_2}\langle I_x+I'_x\rangle^*,$$

where $\qquad \dfrac{1}{T_2} = \gamma^4\hbar^2 I(I+1)\{\tfrac{3}{8}J^{(2)}(2\omega_I)+\tfrac{15}{4}J^{(1)}(\omega_I)+\tfrac{3}{8}J^{(0)}(0)\}. \qquad (79)$

For more than two spins in uncorrelated relative motion, a generalization of (79) similar to (77) can be made.

If the correlation time for the random change of $\mathscr{H}_1(t)$ is very short, much shorter than the Larmor period, all spectral densities $J^{(k)}(\omega)$ become independent of ω, in the frequency range of interest, and equal to $J^{(k)}(0)$.

If, furthermore, the same very short correlation time is assumed for all the random quantities concerned,

$$J^{(0)}:J^{(1)}:J^{(2)} = \overline{|F^{(0)}|^2}:\overline{|F^{(1)}|^2}:\overline{|F^{(2)}|^2} = 6:1:4.$$

A comparison of (76) and (79) then shows immediately that $T_1 = T_2$. A more fundamental proof of this equality can be obtained from the relations (44) and (45). For infinitely short correlation times the integral (44) which determines the rate of change of $\langle Q\rangle^* = \text{tr}\{Q\sigma^*\}$ becomes proportional to the value of the integrand for $\tau = 0$, $[\mathscr{H}_1^*(t),[\mathscr{H}_1^*(t),Q]]$, where, since $\mathscr{H}_1(t)$ is a stationary random operator, it is permissible to replace $\mathscr{H}_1(t)$ by $\mathscr{H}_1(0)$, thus obtaining

$$\overline{[e^{i\mathscr{H}_0 t}\mathscr{H}_1(0)e^{-i\mathscr{H}_0 t},[e^{i\mathscr{H}_0 t}\mathscr{H}_1(0)e^{-i\mathscr{H}_0 t}Q]]}. \qquad (80)$$

It turns out, however, that (80) is actually independent of t and thus equal to
$$\overline{[\mathscr{H}_1(0),[\mathscr{H}_1(0),Q]]}. \qquad (81)$$

This is a consequence of the independence of $\mathscr{H}_1$ of the choice of the axis of quantization for the spins and of the fact that the operator $e^{i\mathscr{H}_0 t} = e^{i\omega_I(I_z+I'_z)t}$ represents a rotation of the interacting spins around the z-axis.

It is then clear that, since there is no preferred direction for $\mathcal{H}_1$, the results obtained by replacing Q in (81) by I_x, I_y, I_z are deduced from each other by the permutation of the indices x, y, z and that therefore $T_1 = T_2$. One should beware of the following fallacious statement which is sometimes encountered. In the extreme narrowing case, when all the products $\omega_{\alpha\beta}\tau_c$ are negligible, the motion of the system submitted to the random perturbation is in the interaction representation the same as if $\mathcal{H}_0$ did not exist. This amounts to claiming that (80) = (81), which is not true in general. An example to illustrate this point will be seen shortly in connexion with the coupling between unlike spins. Even for like spins, the widespread belief that under very rapid relative random motion, the relaxation caused by dipolar coupling is describable by a single relaxation time, is unwarranted. As an example, of little practical importance for reasons that will appear shortly, but of some theoretical interest, consider the relaxation caused by the mutual dipolar coupling of three or four identical spins $\frac{1}{2}$ placed at the corners of a rigid equilateral triangle or regular tetrahedron rotating in a random fashion. The dipolar Hamiltonian can be written

$$\hbar \mathcal{H}_1 = \hbar \sum_{i<k,q} F_{ik}^{(q)} A_{ik}^{(q)}, \tag{82}$$

where the F and A are defined by the relations (68) and (69) and the indices i and k label the various spins of the molecule. The complication here as compared with the assumption of uncorrelated relative motions of the interacting spins, that leads to the simple formula (77), arises from the fact that because the molecule moves as a rigid body there is a correlation between the relative motions of various pairs of spins, expressed by relations

$$\overline{F_{ij}^{(q)}(t) F_{i'j'}^{(q)*}(t+\tau)} \neq 0.$$

Skipping a calculation that is lengthy (4) but, once a correct description of the random motion of a solid body has been obtained, follows as a straightforward development of the general formalism, we give the result, in the limit when $\omega_I \tau_c \ll 1$, for the time variation of the z component of the magnetization M_z under the assumption of an initial Boltzmann distribution of spin populations. For three spins it is found that

$$M_z(t) - M_0 = \{M_z(0) - M_0\}(ae^{-\alpha t/T_1} + be^{-\beta t/T_1}), \tag{83}$$

where T_1 is the relaxation time in the absence of correlation between relative motions of the various pairs of spins as given by the formula

(77), and the coefficients a, α, b, β are given by

$$a = \frac{183 - 23\sqrt{61}}{366} = 0.008,$$
$$\alpha = 3(19 - \sqrt{61})/80 = 0.42,$$
$$b = \frac{183 + 23\sqrt{61}}{366} = 0.992,$$
$$\beta = 3(19 + \sqrt{61})/80 = 1.005.$$
(84)

It appears from these values that the single exponential obtained by making $a = 0$, $b = \beta = 1$ in (83) is an excellent approximation to the exact formula and that the departure from a single exponential, proportional to a, would be very difficult to observe experimentally. The equation (83) also describes, under the same assumptions, the variation of the longitudinal magnetization for a system of four spins $\frac{1}{2}$, with

$$a = (92 - 13\sqrt{46})/184 = 0.02,$$
$$\alpha = (73 + 2\sqrt{46})/60 = 1.45,$$
$$b = (92 + 13\sqrt{46})/184 = 0.979,$$
$$\beta = (73 - 2\sqrt{46})/60 = 0.99.$$
(85)

Although the coefficient a in the corrective term is somewhat larger than in the case of three spins, the experimental observation of this term is unlikely to be any easier because of its faster decay caused by a larger value of α.

(b) *Unlike spins*

The main Hamiltonian is $\hbar \mathcal{H}_0 = \hbar\{\omega_I I_z + \omega_S S_z\}$ with $\omega_I \neq \omega_S$, and the quantities $A_p^{(q)}$ of equation (38) are given by the following relations:

$$\left.\begin{aligned}
e^{i\mathcal{H}_0 t} A^{(0)} e^{-i\mathcal{H}_0 t} &= -\frac{2\alpha}{3} I_z S_z + \tfrac{1}{6}\alpha I_+ S_- \, e^{i(\omega_I - \omega_S)t} + \tfrac{1}{6}\alpha I_- S_+ \, e^{-i(\omega_I - \omega_S)t} \\
\text{whence} & \\
A_1^{(0)} &= -\frac{2\alpha}{3} I_z S_z, \quad A_2^{(0)} = \tfrac{1}{6}\alpha I_+ S_-, \quad A_3^{(0)} = \tfrac{1}{6}\alpha I_- S_+ \\
\omega_1^{(0)} &= 0, \quad \omega_2^{(0)} = \omega_I - \omega_S, \quad \omega_3^{(0)} = \omega_S - \omega_I \\
e^{i\mathcal{H}_0 t} A^{(1)} e^{-i\mathcal{H}_0 t} &= \alpha I_+ S_z \, e^{i\omega_I t} + \alpha I_z S_+ \, e^{i\omega_S t} \\
A_1^{(1)} &= \alpha I_+ S_z, \quad A_2^{(1)} = \alpha I_z S_+ \\
\omega_1^{(1)} &= \omega_I, \quad \omega_2^{(1)} = \omega_S \\
e^{i\mathcal{H}_0 t} A^{(2)} e^{-i\mathcal{H}_0 t} &= \tfrac{1}{2}\alpha I_+ S_+ \, e^{i(\omega_I + \omega_S)t} \\
\omega_1^{(2)} &= \omega_I + \omega_S
\end{aligned}\right\}. \quad (86)$$

In the equation
$$\frac{d\langle I_z\rangle}{dt} = -\{a_z^I - a_0^I\},$$
where
$$a_z^I = \langle \mathscr{A}_z^I\rangle^* = \mathrm{tr}\{\mathscr{A}_z^I \sigma^*\},$$
$\mathscr{A}_z^I$ is given by the formulae (45) and (86):

$$2\mathscr{A}_z^I = \frac{\alpha^2}{36} J^{(0)}(\omega_I - \omega_S)[I_- S_+, [I_+ S_-, I_z]] + \mathrm{h.c.} +$$
$$+ \alpha^2 J^{(1)}(\omega_I)[I_- S_z, [I_+ S_z, I_z]] + \mathrm{h.c.} +$$
$$+ \frac{\alpha^2}{4} J^{(2)}(\omega_I + \omega_S)[I_- S_-, [I_+ S_+, I_z]] + \mathrm{h.c.}$$

Computing the various commutators and neglecting second order terms as explained previously, we find

$$\langle \mathscr{A}_z^I\rangle^*$$
$$= \gamma_I^2 \gamma_S^2 \hbar^2 \Big[\langle I_z\rangle S(S+1)\Big\{ \frac{J^{(0)}(\omega_I - \omega_S)}{12} + \tfrac{3}{2} J^{(1)}(\omega_I) + \tfrac{3}{4} J^{(2)}(\omega_I + \omega_S) \Big\} +$$
$$+ \langle S_z\rangle I(I+1)\Big\{ \frac{-J^{(0)}(\omega_I - \omega_S)}{12} + \tfrac{3}{4} J^{(2)}(\omega_I + \omega_S) \Big\} \Big].$$

An identical equation is obtained for $\langle \mathscr{A}_z^S\rangle^*$ by interchanging the letters I and S. We thus obtain the coupled equations

$$\frac{d\langle I_z\rangle}{dt} = -\frac{1}{T_1^{II}}(\langle I_z\rangle - I_0) - \frac{1}{T_1^{IS}}(\langle S_z\rangle - S_0),$$
$$\frac{d\langle S_z\rangle}{dt} = -\frac{1}{T_1^{SI}}(\langle I_z\rangle - I_0) - \frac{1}{T_1^{SS}}(\langle S_z\rangle - S_0),$$
(87)

with

$$\frac{1}{T_1^{II}} = \gamma_I^2 \gamma_S^2 \hbar^2 S(S+1)\{\tfrac{1}{12} J^{(0)}(\omega_I - \omega_S) + \tfrac{3}{2} J^{(1)}(\omega_I) + \tfrac{3}{4} J^{(2)}(\omega_I + \omega_S)\},$$
$$\frac{1}{T_1^{IS}} = \gamma_I^2 \gamma_S^2 \hbar^2 I(I+1)\{-\tfrac{1}{12} J^{(0)}(\omega_I - \omega_S) + \tfrac{3}{4} J^{(2)}(\omega_I + \omega_S)\},$$
(88)

and similar equations for T_1^{SS} and T_1^{SI} by interchanging the indices I and S. A very remarkable feature appears in equation (87). The polarizations of the spins S and I are coupled so that an r.f. field at a frequency say ω_I will affect $\langle S_z\rangle$ while acting upon $\langle I_z\rangle$. The important physical consequences of this circumstance will be discussed later. Under conditions of extreme narrowing it is easily verified that the ratio
$$\left(\frac{1}{T_1^{II}}\right) \Big/ \left(\frac{1}{T_1^{IS}}\right) = \frac{2S(S+1)}{I(I+1)},$$
and if we define
$$i_z = \langle I_z\rangle / I(I+1),$$
$$s_z = \langle S_z\rangle / S(S+1),$$

(87) can be rewritten

$$\frac{di_z}{dt} = -\frac{1}{T_1^{II}}\{(i_z-i_0)+\tfrac{1}{2}(s_z-s_0)\},$$
$$\frac{ds_z}{dt} = -\frac{1}{T_1^{SS}}\{(s_z-s_0)+\tfrac{1}{2}(i_z-i_0)\}. \tag{88'}$$

An interesting situation arises when the spin S is an electronic spin belonging to an ion or molecule with a hyperfine structure due to a static coupling of the spin **S** with another nuclear spin **K** represented by a Hamiltonian $\hbar A\mathbf{S}\cdot\mathbf{K}$.

Under conditions of extreme narrowing the first equation (87) describing the rate of change of $\langle I_z\rangle$ is still valid and thus has the same form as in the absence of the coupling $\hbar A\mathbf{S}\cdot\mathbf{K}$. This appears most clearly in the equation (45′) where we make $Q = I_z$, for I_z commutes with the extra term $A(\mathbf{S}\cdot\mathbf{K})$ introduced in the Hamiltonian $\mathcal{H}_0$, and B as defined by (46) does not depend on **K** if the direct coupling of **I** with **K** is neglected. Examples of this situation will be discussed later.

Analogous equations can be established for the amplitudes of the precessing magnetization of the spins I and S, which are proportional to $\langle I_x\rangle^* = \mathrm{tr}\{I_x\sigma^*\}$ and $\langle S_x\rangle^* = \mathrm{tr}\{S_x\sigma^*\}$, respectively. We find from (45) and (86) that

$$\frac{d}{dt}\langle I_x\rangle^* = -\langle \mathscr{A}_x^I\rangle^*,$$

where

$$2\mathscr{A}_x^I = \frac{4\alpha^2}{9}J^{(0)}(0)[I_zS_z,[I_zS_z,I_x]]+$$
$$+\frac{\alpha^2}{36}J^{(0)}(\omega_I-\omega_S)[I_-S_+,[I_+S_-,I_x]]+\text{h.c.}+$$
$$+\alpha^2 J^{(1)}(\omega_I)[I_-S_z,[I_+S_z,I_x]]+\text{h.c.}+$$
$$+\alpha^2 J^{(1)}(\omega_S)[I_zS_-,[I_zS_+,I_x]]+\text{h.c.}+$$
$$+\frac{\alpha^2}{4}J^{(2)}(\omega_I+\omega_S)[I_-S_-,[I_+S_+,I_x]]+\text{h.c.}$$

Computing the various commutators we find

$$\frac{d\langle I_x\rangle^*}{dt} = -\frac{\langle I_x\rangle^*}{T_2^I},$$

with

$$\frac{1}{T_2^I} = \gamma_I^2\gamma_S^2\hbar^2 S(S+1)\{\tfrac{1}{6}J^{(0)}(0)+\tfrac{1}{24}J^{(0)}(\omega_I-\omega_S)+$$
$$+\tfrac{3}{4}J^{(1)}(\omega_I)+\tfrac{3}{2}J^{(1)}(\omega_S)+\tfrac{3}{8}J^{(2)}(\omega_I+\omega_S)\}, \tag{89}$$

and also a similar formula for $\langle S_x\rangle^*$ by interchanging the indices I and S. It is easily checked that, as for like spins, if the correlation time is very short, $1/T_2^I = 1/T_1^{II}$. Still, the isotropy of the phenomenon is not complete since while the amplitudes of the longitudinal magnetizations of the two spin species proportional to $\langle I_z \rangle$ and $\langle S_z \rangle$ are coupled together, those of the transverse magnetizations which precess at very different frequencies ω_I and ω_S are not.

The $\frac{3}{2}$ effect. In the 'extreme narrowing' case, comparing the values of T_2 given by (79) for coupling between like spins and by (89) for unlike spins it is easily found that, all things being equal (in particular spins),

$$(1/T_2)_{\text{like}}/(1/T_2)_{\text{unlike}} = \tfrac{3}{2}. \tag{90}$$

Coupling between like spins is more efficient for damping transverse magnetization than coupling between unlike spins, an effect which resembles very much a similar one for broadening in a rigid lattice.

(c) Correlation functions resulting from random molecular rotation or translation

It remains for us to choose a model for representing the disordered motion of the spin-carrying nuclei and to calculate the correlation functions of the three random functions $F^{(0)}$, $F^{(1)}$, and $F^{(2)}$, as well as the Fourier transforms of these correlation functions.

It is convenient to introduce the reduced correlation functions

$$\tilde{g}^{(i)}(\tau) = G^{(i)}(\tau)/G^{(i)}(0)$$

and their Fourier transforms $\tilde{\mathscr{J}}^{(i)}(\omega)$.

A crude but convenient assumption that we will often make and in certain cases be able to justify, is that the reduced correlation function $\tilde{g}(\tau)$ is the same for $G^{(0)}$, $G^{(1)}$, $G^{(2)}$ and can be represented by $\exp(-|\tau|/\tau_c)$, where the constant τ_c, called the correlation time, is a characteristic of the medium. The reduced Fourier transform then becomes

$$\tilde{\mathscr{J}}(\omega) = \frac{2\tau_c}{1+\omega^2\tau_c^2}. \tag{91}$$

It is apparent from this formula that for a given frequency ω, $\tilde{\mathscr{J}}(\omega)$ is maximum for $\tau_c = 1/\omega$. Therefore for a constant power

$$G(0) = \frac{1}{2\pi}\int_{-\infty}^{\infty} J(\omega)\,d\omega,$$

available in the relaxation spectrum, the rate of relaxation transitions induced is a maximum when the correlation time τ_c is of the order of $1/\omega$, i.e. of the Larmor period.

It falls down and reduces to zero for both τ_c very short or very long, compared with $1/\omega$. The existence of such an optimum is physically understandable and indeed more general than the special assumption of the form (91) for the spectral density. Roughly speaking the power spectrum has a cut-off for a frequency $\omega^* \sim 1/\tau_c$ and is more or less uniform for $\omega < \omega^*$. If ω^* is much smaller than the Larmor frequency ω, there is no power available for the transition and $1/T_1$ is very small. If $\omega^* \gg \omega$ the power spectrum is spread uniformly over a very large band and since its total intensity is constant, $J(\omega)$ and $1/T_1$ will accordingly be small. An optimum should thus exist for an intermediate value of $\omega^* \sim 1/\tau_c$.

Among the interactions responsible for spin-lattice relaxation it is useful to distinguish between intramolecular interactions inside a molecule and intermolecular interactions among spins of different molecules.

In the interior of a molecule, the variation of dipolar coupling of spins arises almost uniquely from the rotation of the molecule, the variation of the distance between the spins due to vibrations being negligible. In the interactions between molecules, their relative translation as well as their characteristic rotations must be considered. In order to avoid excessive complication we shall neglect the latter with respect to the former.

As an example we consider a molecule, such as water, containing two identical spins $\frac{1}{2}$.

We shall assume that the various Brownian motions of the molecules are correctly described by a diffusion equation.

(1) *Rotation*

Let θ and φ be the polar angles defining the direction of the proton-proton axis and $\Psi(\theta, \varphi, t) = \Psi(\Omega, t)$ be the probability of finding this axis in the direction Ω at time t.

We suppose, following Debye, that the rotation of the molecule can be compared to that of a rigid sphere of radius a in a medium of viscosity η and that it can be described by a diffusion equation

$$\frac{\partial \Psi}{\partial t} = \frac{D_s}{a^2} \Delta_s \Psi, \qquad (92)$$

where the operator Δ_s is the Laplacian operator on the surface of a sphere. The diffusion constant for rotation D_s is given by Stokes's formula

$$D_s = \frac{kT}{8\pi a \eta}. \qquad (93)$$

To calculate the correlation functions of the random functions $F^{(0)}$, $F^{(1)}$, $F^{(2)}$ by formula (14), we need the function $P(\Omega, \Omega_0, t)$ which is the probability that the axis of the two spins has the orientation Ω at time t, when we know that it has the orientation Ω_0 at time zero.

$P(\Omega, \Omega_0, t)$ is the solution of (92) which satisfies the initial condition

$$\Psi(\Omega, 0) = \delta(\Omega - \Omega_0). \tag{94}$$

We seek the solution of (92) in the form of an expansion in spherical harmonics:
$$\Psi(\Omega, t) = \sum c_l^m(t) Y_l^m(\Omega). \tag{95}$$

By substituting (95) into (92) and taking into account the relation $\Delta_s Y_l^m(\Omega) = -l(l+1) Y_l^m(\Omega)$ as well as the orthogonality of the spherical harmonics, it follows that

$$\frac{dc_l^m}{dt} = -\frac{D_s}{a^2} l(l+1) c_l^m \tag{96}$$

or, introducing τ_l through

$$\frac{D_s}{a^2} l(l+1) = 1/\tau_l, \tag{97}$$

it follows that
$$c_l^m(t) = c_l^m(0) e^{-t/\tau_l}. \tag{98}$$

The expansion of the function $\delta(\Omega - \Omega_0)$ in spherical harmonics is well known (and, moreover, can be obtained immediately from the definition of the δ function):

$$\delta(\Omega - \Omega_0) = \sum_{l,m} Y_l^{m*}(\Omega_0) Y_l^m(\Omega), \tag{99}$$

whence $\quad c_l^m(0) = Y_l^{m*}(\Omega_0)$

and $\quad P(\Omega, \Omega_0, t) = \sum_{l,m} Y_l^{m*}(\Omega_0) Y_l^m(\Omega) e^{-t/\tau_l}. \tag{100}$

By applying (13), in which one should note that the *a priori* probability $p(\Omega_0)$ is constant and equal to $1/4\pi$, it follows for the correlation function of a random function that

$$G(t) = \frac{1}{4\pi} \int\int F^*(\Omega) F(\Omega_0) \sum_{m,l} Y_l^{m*}(\Omega_0) Y_l^m(\Omega) e^{-t/\tau_l} d\Omega d\Omega_0. \tag{101}$$

The random functions $F^{(1)}$ and $F^{(2)}$ are related to the normalized spherical harmonics $Y_l^m(\Omega)$ by

$$F^{(1)}(\Omega) = \frac{1}{b^3} \sqrt{\frac{8\pi}{15}} Y_2^{(1)}(\Omega),$$

$$F^{(2)}(\Omega) = \frac{1}{b^3} \sqrt{\frac{32\pi}{15}} Y_2^{(2)}(\Omega), \tag{102}$$

where b is the distance between the two spins in the molecule.

By substituting these expressions in (101), it follows immediately that

$$G^{(1)}(t) = \frac{1}{b^6} \frac{2}{15} e^{-|t|/\tau_2}; \qquad G^{(2)}(t) = \frac{1}{b^6} \frac{8}{15} e^{-|t|/\tau_2}, \tag{103}$$

from which

$$J^{(1)}(\omega) = \frac{1}{b^6}\frac{4}{15}\frac{\tau_2}{1+\omega^2\tau_2^2}; \qquad J^{(2)}(\omega) = \frac{1}{b^6}\frac{16}{15}\frac{\tau_2}{1+\omega^2\tau_2^2}. \qquad (104)$$

It is seen that, in the case of rotation, the form $e^{-|t|/\tau_2}$ of the correlation function is a consequence of the diffusion equation.

According to (76) the spin-lattice relaxation time due to rotation is given by

$$\left(\frac{1}{T_1}\right)_{\text{rot}} = \frac{2}{5}\frac{\gamma^4\hbar^2}{b^6}I(I+1)\left[\frac{\tau_2}{1+\omega^2\tau_2^2}+\frac{4\tau_2}{1+4\omega^2\tau_2^2}\right]. \qquad (105)$$

If the time τ_2 is very much shorter than the Larmor period $2\pi/\omega$, it follows that

$$\left(\frac{1}{T_1}\right)_{\text{rot}} = \frac{2\gamma^4\hbar^2}{b^6}I(I+1)\tau_2,$$

or, for spins $\frac{1}{2}$,

$$= \frac{3}{2}\frac{\gamma^4\hbar^2}{b^6}\tau_2 = 2\pi\frac{\gamma^4\hbar^2}{b^6}\frac{a^3\eta}{kT}. \qquad (106)$$

The correlation time $\tau_2 = 4\pi\eta a^3/3kT$ is related to the correlation time τ_1 of Debye by the formula $\tau_1 = 3\tau_2$. Indeed, Debye was interested in the orientation of molecules with a permanent dipole moment and the random function whose correlation he studied was spherical harmonic of order 1. The theory is the same except that in (97), it is necessary to set $l = 1$, from which $\tau_1 = 3\tau_2$.

(2) Translation

We again suppose that the diffusion equation is valid for describing the motion of molecules. To describe the spin-spin interaction of spins on different molecules we may either consider the interactions between individual spins, or add up the spins on each molecule and consider the interactions between the resultant molecular spins. To the approximation which we shall make, it is easy to see that the two methods give identical results. Indeed, $1/T_1$ is proportional to the product $NI(I+1)$ where N is the density of spins per cm³.

If we consider the individual spins $\frac{1}{2}$ this product is equal to $3N/4$. If, on the other hand, we consider the molecular spins, one-quarter of the molecules have spin zero and contribute nothing to the relaxation, while three-quarters have spin 1 and, as the density of the molecules is half that of the spins $\frac{1}{2}$, they contribute

$$\tfrac{3}{4}\tfrac{1}{2}N\,I(I+1) = \frac{3N}{4}$$

as before.

The solution of the diffusion equation

$$\frac{\partial \Psi(\mathbf{r},t)}{\partial t} = D \Delta \Psi,$$

where $\Psi(\mathbf{r},0) = \delta(\mathbf{r}-\mathbf{r}_0)$, is given by the classical formula of diffusion theory,

$$\Psi(\mathbf{r},\mathbf{r}_0,t) = (4\pi Dt)^{-3/2} \exp\left\{-\frac{(\mathbf{r}-\mathbf{r}_0)^2}{4Dt}\right\}. \tag{107}$$

If $\mathbf{r}$ represents, not the radius vector of the molecule which diffuses relative to a fixed point, but the distance $\mathbf{r}_1-\mathbf{r}_2$ between two identical molecules which diffuse relative to each other, it is clear that the only change is the replacement of Dt by $2Dt$ in (107), leading to the expression

$$P(\mathbf{r},\mathbf{r}_0,t) = (8\pi Dt)^{-3/2} \exp\left\{-\frac{(\mathbf{r}-\mathbf{r}_0)^2}{8Dt}\right\}. \tag{108}$$

The correlation function of any one of the three functions $F^{(0)}$, $F^{(1)}$, $F^{(2)}$ will be given by

$$G^{(m)}(t) = \alpha^{(m)} N (8\pi Dt)^{-3/2} \int\int \frac{Y_2^{m*}(\Omega_0)}{r_0^3} \frac{Y_2^m(\Omega)}{r^3} e^{-|\mathbf{r}-\mathbf{r}_0|^2/8Dt} d^3r_0\, d^3r, \tag{109}$$

where $\quad \alpha^{(1)} = \dfrac{8\pi}{15}, \quad \alpha^{(2)} = \dfrac{32\pi}{15}, \quad \alpha^{(0)} = \dfrac{48\pi}{15}.$

In the integration (109), r and r_0 cannot go below a lower limit d which is the minimum distance of approach of molecules. If we treat the molecules as spheres of radius a, $d = 2a$.

To integrate (109) we replace $\exp(-|\mathbf{r}-\mathbf{r}_0|^2/8Dt)$ by its Fourier expansion

$$\exp\left(-\frac{|\mathbf{r}-\mathbf{r}_0|^2}{8Dt}\right) = (2\pi)^{-3/2}(4Dt)^{3/2}\int \exp(-2Dt\rho^2)e^{i\boldsymbol{\rho}\cdot(\mathbf{r}-\mathbf{r}_0)}\, d^3\rho; \tag{110}$$

then we replace $e^{i\boldsymbol{\rho}\cdot\mathbf{r}}$ and $e^{-i\boldsymbol{\rho}\cdot\mathbf{r}_0}$ by their classical expansion

$$e^{i\boldsymbol{\rho}\cdot\mathbf{r}} = 4\pi\left(\frac{\pi}{2\rho r}\right)^{1/2} \sum_{m,l} i^l Y_l^{m*}(\Omega) Y_l^m(\Omega') J_{l+\frac{1}{2}}(\rho r), \tag{111}$$

where Ω' specifies the orientation of the vector $\boldsymbol{\rho}$ and the $J_{l+\frac{1}{2}}$ are Bessel functions. There is an analogous expansion for $e^{-i\boldsymbol{\rho}\cdot\mathbf{r}_0}$. In substituting (110) and (111) into (109) and taking account of orthogonality properties of the spherical harmonics, it follows that

$$G(t) = N \int_0^\infty \rho \exp(-2Dt\rho^2) \left[\int_d^\infty \frac{J_{\frac{5}{2}}(\rho r)}{r^{\frac{5}{2}}} dr\right]^2 d\rho. \tag{112}$$

The integral in the brackets is classical and equal to $\rho^{\frac{1}{2}}(\rho d)^{-\frac{1}{2}}J_{\frac{3}{2}}(\rho d)$, from which, by writing $\rho d = u$,

$$G(t) = \frac{N}{d^3} \int_0^\infty [J_{\frac{3}{2}}(u)]^2 \exp\left(\frac{-2D}{d^2} u^2 t\right) \frac{du}{u}. \tag{113}$$

Its Fourier transform $J(\omega)$ is immediately calculable:

$$J(\omega) = \frac{N}{dD} \int_0^\infty \frac{[J_{\frac{3}{2}}(u)]^2}{u^3} \frac{du}{1+(\omega^2\tau^2/u^4)} = \frac{N}{dD} \int_0^\infty [J_{\frac{3}{2}}(u)]^2 \frac{u\, du}{u^4+\omega^2\tau^2}, \tag{114}$$

where we define $\tau = d^2/2D$.

If we take $d = 2a$ and for D the Stokes formula for translational diffusion $D = kT/6\pi a\eta$ valid for a rigid sphere, we find

$$\tau = 12\pi a^3 \eta / kT = 9\tau_2.$$

For simplicity we make the assumption, often realized in practice, that $\omega\tau \ll 1$. For a discussion of the general case see Chapter X, Section V B. It is then permissible to replace in the integral (114) $1+\omega^2\tau^2/u^4$ by unity except for very small values of u which make a negligible contribution to the integral.

The integral $\int_0^\infty [J_{\frac{3}{2}}(u)]^2 u^{-3}\, du$ is classical and equal to $\frac{2}{15}$, from which

$$J(\omega) = J(0) = \frac{2}{15} \frac{N}{dD} = \frac{2\pi}{5} \frac{N\eta}{kT}. \tag{114'}$$

According to (109),

$$J^{(1)} = \frac{8\pi}{15} J, \qquad J^{(2)} = \frac{32\pi}{15} J,$$

and

$$\left(\frac{1}{T_1}\right)_{\text{transl}} = \tfrac{3}{2}\gamma^4\hbar^2 I(I+1)\{J^{(1)}+J^{(2)}\}$$

or, for spins $\frac{1}{2}$,

$$\left(\frac{1}{T_1}\right)_{\text{transl}} = \frac{6\pi^2}{5} \gamma^4\hbar^2 \frac{N\eta}{kT} = \frac{\pi}{5} \frac{N\gamma^4\hbar^2}{aD}. \tag{115}$$

If we compare (106) with (115), it follows that,

$$\left(\frac{1}{T_1}\right)_{\text{transl}} \Big/ \left(\frac{1}{T_1}\right)_{\text{rot}} = \frac{3\pi}{5} \frac{Nb^6}{a^3}. \tag{116}$$

This treatment of the relaxation due to the relative translational motion of molecules is easily made to apply to the important problem of nuclear relaxation through dissolved paramagnetic impurities. It has been shown previously (eqns. (87) and (89)) that while coupling with a spin of a different nature did lead for a nuclear spin to a single transverse relaxation time the situation was more complex for the longitudinal relaxation where a system of coupled equations (87) was necessary.

However, if the spin S is an electronic spin, it has relaxation mechanisms of its own, much stronger than those due to its coupling with the nuclear spin I and it is permissible to assume that $\langle S_z \rangle = S_0$ is a permanent relationship and neglect the term proportional to $\{\langle S_z \rangle - S_0\}$ in the first equation (87). The longitudinal relaxation of the spins I can then be described by a single relaxation time as well: $T_1 = T_1^{II}$ of formula (88).

In formula (88) the quantity $\gamma_S^2 \hbar^2 S(S+1)$ represents the square of the magnetic moment of the paramagnetic ion and this formula is valid without modifications only for ions in an S state, like Mn^{++} or Fe^{+++}, which have no orbital magnetism. For other ions $\gamma_S^2 \hbar^2 S(S+1)$ must be replaced by the quantity $\langle \mu^2 \rangle$, the mean square of the magnetic moment of the ion. To use the results of paramagnetic electronic resonance experiments or susceptibility measurements to calculate $\langle \mu^2 \rangle$, it is necessary to recall that in principle the three types of experiment, nuclear relaxation produced by ions, electronic paramagnetic resonance, and susceptibility, measure different quantities. For a paramagnetic ion with energy levels $E_1, E_2, ..., E_n$, the quantity $\langle \mu^2 \rangle$ which occurs in nuclear relaxation is defined by

$$\langle \mu^2 \rangle = \sum_i p_i \langle \mu^2 \rangle_i, \tag{117}$$

where $\langle \mu^2 \rangle_i$ is the expectation value of the operator M^2 in the state E_i and $p_i = \exp(-E_i/kT)/\sum_i \exp(-E_i/kT)$ is the population of that state.

The electronic resonance must often be observed at low temperature and thus gives information on the ground state only, such as the gyromagnetic factor g_0 of this state, from which one can only deduce the first term $\langle \mu^2 \rangle_0$ of (117).

Finally, the measurements of susceptibility yield a more complex quantity which is the mean value of $\langle M_z \rangle$ in the presence of an applied field $H_z = H_0$. Van Vleck has shown that if the separation from the first excited state $E_1 - E_0 \gg kT$, the susceptibility is composed of two terms of which the first or Curie term is given by $N\langle \mu^2 \rangle_0 / 3kT$ and the second or Van Vleck term is independent of the temperature and given by the expression

$$\frac{2N}{3} \sum_{n \neq 0} \frac{|(0|M_z|n)|^2}{E_n - E_0}.$$

If $E_n - E_0$ is not large compared with kT, the variation of susceptibility with temperature is much more complex.

The inverse of the nuclear relaxation time will be proportional to the susceptibility of the ion only at temperatures where the Van Vleck term is negligible.

To calculate T_1^{II} we make the assumptions, for which simplicity is the chief justification, that each ion may be treated as a sphere of the same radius as a water molecule, that the diffusion equation describes their motion correctly, and that the constant $\tau = d^2/2D$ is small compared with the Larmor period of the paramagnetic ions (a debatable assumption in high fields).

The formula (88),
$$\frac{1}{T_1} = \gamma_I^2 \langle \mu^2 \rangle \left\{ \frac{J^{(0)}}{12} + \tfrac{3}{2} J^{(1)} + \tfrac{3}{4} J^{(2)} \right\},$$
where
$$J^{(0)} : J^{(1)} : J^{(2)} = 6 : 1 : 4,$$
then gives
$$\frac{1}{T_1} = \frac{16\pi^2}{15} N_{\text{ion}} \langle \mu^2 \rangle \frac{\gamma_I^2 \eta}{kT} = \frac{16\pi^2}{15} N_{\text{ion}} \gamma_S^2 \hbar^2 S'(S'+1) \frac{\gamma_I^2 \eta}{kT}, \quad (118)$$
where S' is a dimensionless number defined by
$$\gamma_S^2 \hbar^2 S'(S'+1) = \langle \mu^2 \rangle.$$

(118) can be modified to take account of the difference in radius and mobility between the paramagnetic ions and the water molecules. If we keep the hypothesis that their motion is correctly described by the diffusion of rigid spheres of radii a_1 and a_2, respectively, in a medium of viscosity η, it is sufficient to replace d in (114) by $a_1 + a_2$ and D by $\tfrac{1}{2}(D_1 + D_2)$, and as a consequence multiply (118) by
$$4a_1 a_2 / (a_1 + a_2)^2.$$

In this calculation we have supposed that the only cause of variation of the magnetic field produced by the ion at the position of the nucleus was their relative displacement. We must add to this the variation due to the electronic relaxation of the ionic spin characterized by the time θ. We can try to take account of this by multiplying the characteristic function $G(t)$, given by (113), by a supplementary factor $e^{-t/\theta}$ representing the correlation between two values of the spin of the ion taken at two times separated by t.

This amounts to replacing the exponent $2Du^2 t/d^2$ in (113) by
$$\left(\frac{2Du^2}{d^2} + \frac{1}{\theta} \right) t.$$

As the principal contribution to the integral (113) comes from values of u of the order of unity, one sees that the effect of θ will be observable for $\theta \leqslant d^2/2D \simeq 10^{-10}$ sec, and can become important for solutions with a viscosity much greater than that of water.

Section II E contains the essentials of the theory required for a description of the relaxation phenomena caused by dipolar couplings in liquids. In view of the crudeness of the physical models utilized (molecules represented by solid spheres moving according to Stokes's macroscopic law, for instance), the utility of making elaborate calculations, tracking down coefficients of order unity, while the order of magnitude of the expected results could in fact be obtained much more simply, may well be questioned. The justification for the more painstaking approach is the following. Certain essential features of the phenomena, such as ratios of relaxation times or the nature of the coupling between polarizations of different species of interacting spins, are rather insensitive to the details of the physical model and can therefore be submitted to a much more accurate experimental test than would seem possible at first. Furthermore, an unambiguous definition of the quantity to be measured itself sometimes requires a rather careful theory. Finally, in order to investigate the validity of a model, however crude, it is preferable whenever possible to avoid the extra uncertainty introduced through an incorrect mathematical treatment.

The comparison of the theory of relaxation in liquids with experiment will be deferred until after some other mechanisms of relaxation in liquids have been examined.

F. Other mechanisms of relaxation in liquids

(a) General

The mechanism of relaxation through dipolar coupling is an example of a class of processes which can be described as follows.

A given spin I can be subject to several types of couplings with its surroundings or with other spins, describable by means of tensors, which are functions of the lattice parameters. Thus a bilinear coupling between two spins $\mathbf{I}$ and $\mathbf{S}$ can be written in tensor form $\mathbf{I} \cdot \hbar \mathscr{A}_m \cdot \mathbf{S}$. The dipolar interaction is the best known example of such a coupling.

A bilinear coupling between the applied field H and the spin I (distinct from the Zeeman coupling $-\gamma \hbar \mathbf{I} \cdot \mathbf{H}$) would be described by a coupling $\mathbf{H} \cdot \hbar \mathscr{A}_s \cdot \mathbf{I}$. Such a coupling corresponds to the existence of an anisotropic shift of the Larmor frequency of the nuclear spin.

Finally, for spins larger than $\frac{1}{2}$ a very important mechanism is the coupling of electric field gradients with the quadrupole moment of the nucleus, which can be written in tensor form $\mathbf{I} \cdot \hbar \mathscr{A}_q \cdot \mathbf{I}$.

Depending on whether the semi-classical description of Section C or the quantum mechanical treatment of Section D is used, the components of the tensors $\mathscr{A}$ are random functions of time or operators acting on the lattice variables.

To understand how relaxation occurs, assume, for example, that the tensor $\mathscr{A}$ has well-defined, constant components with respect to axes rigidly bound to the molecule containing the nuclear spin I under consideration. Since in a nuclear resonance experiment we are concerned with the orientation of the nuclear spin with respect to axes fixed in the laboratory frame, the components $\mathscr{A}_{ij}$ of the tensor $\mathscr{A}$ in the laboratory frame are obtained as linear combinations of the components $\mathscr{A}_{ij}^0$ in the frame of the molecule through the usual formulae, where the coefficients are functions of the relative orientation of the two frames of reference.

If the molecule is subject to rapid random rotation, the $\mathscr{A}_{ij}(t)$ become random functions of time and a relaxation mechanism is thus provided as explained in Section II C.

Parallel to this relaxation, which is a second-order effect, there appears a first-order effect causing a frequency shift. For example, the coupling
$$\mathbf{H}.\mathscr{A}.\mathbf{I} = \sum_{i,k} H_i \mathscr{A}_{ik}(t) I_k$$
will have an average $\sum_{i,k} H_i \overline{\mathscr{A}_{ik}(t)}.I_k$ which for an isotropic rotation reduces to $A \sum_i H_i I_i = A\mathbf{H}.\mathbf{I}$, where $3A$ is the trace, which is invariant under rotation, $\overline{\mathscr{A}_{11}+\mathscr{A}_{22}+\mathscr{A}_{33}} = \mathscr{A}_{11}^0+\mathscr{A}_{22}^0+\mathscr{A}_{33}^0$ of the tensor $\mathscr{A}$.

(b) (1) *Scalar spin-spin coupling*

Consider first the effects of the bilinear coupling $\mathbf{I}.\hbar\mathscr{A}.\mathbf{S}$. Such a coupling, different from the pure dipolar coupling, can exist between an electronic and a nuclear spin if, as was shown in Chapter VI, the electron wave function has a non-vanishing value at the position of the nucleus. The best known example is that of a nucleus belonging to a paramagnetic ion, molecule, or free radical which bears the unpaired electron spin. Then, however, the coupling of the nuclear moment with the electronic spin is usually so strong as to overshadow the Zeeman coupling of the nuclear moment with the applied field. The resonant transitions observed are essentially electron transitions between states of the combined system electron+nucleus, and their study belongs to electronic magnetism and is outside the scope of this book.

There are, however, instances when a molecule bearing the nuclear spin under study gets attached to the outside of a paramagnetic ion, molecule, or free radical, under conditions such that this coupling is a small perturbation of the Zeeman energy for the nuclear spin and resonant transitions between nuclear Zeeman levels can be induced and detected.

A bilinear coupling between two nuclear spins inside a molecule, distinct from their dipolar coupling and originating as a second-order effect from their hyperfine coupling with the electrons, can also exist, as was mentioned in Chapter VI and as will be considered in greater detail in Chapter XI.

Whatever the nature of the second spin S and the origin of the bilinear coupling, the tènsor $\mathscr{A}$ can always be written as a sum of a traceless tensor $\mathscr{A}'$ and a diagonal tensor $\mathscr{A}^0$ with $\mathscr{A}^0_{ik} = A\delta_{ik}$. Under random rotation of the molecule, the first-order effect of $\mathscr{A}'$ vanishes on the average and its second-order effect is to induce relaxation. This was discussed in great detail for the case where $\mathscr{A}'$ had the form of a dipolar interaction. We shall not pursue this study any farther. In principle the most general coupling described by a traceless symmetrical tensor $\mathscr{A}'$ depends on five independent constants rather than three as for the pure dipolar case and thus does not really deserve the name of pseudo-dipolar coupling that it is sometimes given in the literature. However, as far as the problem of relaxation by random isotropic motion is concerned, this distinction is somewhat academic; it is easily shown using general arguments of rotational invariance that the various aspects of the relaxation phenomenon are the same for the most general traceless tensor $\mathscr{A}'$ as for the more restricted case of a dipolar coupling; in particular, the coupled equations (87) are still valid and the ratio T_1^{II}/T_1^{IS} is the same as in the dipolar case.

We are thus left with a scalar coupling $\hbar A \mathbf{I}.\mathbf{S}$. The case when I and S are two identical spins $\mathbf{I}_1$ and $\mathbf{I}_2$ can be dismissed straight away, for then the coupling $\hbar A \mathbf{I}_1.\mathbf{I}_2$ commutes with the unperturbed Hamiltonian

$$\hbar \mathscr{H}_0 = \hbar \omega_0 (I_z^1 + I_z^2),$$

and with the r.f. Hamiltonian

$$\hbar \mathscr{H}_{\rm rf} = -\hbar \gamma H_1 (I_x^1 + I_x^2)\cos \omega t,$$

and is unobservable. (This statement requires some qualification if $\mathscr{H}_0$ contains extra coupling terms between I_1 and I_2 on the one hand and on the other, unlike spins S, a point to be discussed in Chapter XI.)

Suppose then that I and S are unlike spins, and assume that:
  (i) the relaxation time of either spin is long compared with the inverse $1/A$ of the frequency A;
  (ii) there are no causes, such as chemical exchange between molecules, which would lead to a time-dependence of the coupling constant A between two given spins I and S: at least any exchange time constant τ_e would be much longer than $1/A$.

The resonance line of each spin then acquires a multiplet structure due to its coupling with the other. A detailed study of these couplings will be given in Chapter XI.

(b) (2) *Scalar relaxation of the first kind*

If either $1/T_1$ or $1/\tau_e$ is much larger than A, the multiplet structure disappears, each spin giving rise to a single resonance line, and the scalar coupling $\hbar A\mathbf{I}\cdot\mathbf{S}$ can become a mechanism of relaxation. We assume first that the chemical exchange dominates so that τ_e is much shorter than the T_1 of either spin. The scalar coupling constant of a specified spin I with a given spin S_i becomes a random function of time $A_i(t)$ which has only two values: A when I and S_i are on the same molecule and zero when they are not. It is clear that $\overline{|A_i(t)|^2} = P_i A^2$, where P_i is the probability that the spin I will be found on the same molecule as the spin S_i, and that the reduced correlation function of $A_i(t)$ is $\overline{A_i(t)\cdot A_i(t+\tau)}/[\overline{A_i(t)}]^2 = e^{-\tau/\tau_e}$, which is the probability that I and S_i are still on the same molecule at the time $t+\tau$ when we know that they were there at time t. The calculation of the relaxation times T_1 produced by this time-dependent mechanism can be obtained immediately from the analogous calculation performed previously for the part of the dipolar interaction proportional to $(I_+S_- + I_-S_+)$, which, from the formulae (68) and (69), could be written as

$$-\tfrac{1}{4}\gamma_I\gamma_S\hbar F_0(t)(I_+S_- + I_-S_+).$$

This has to be replaced in the case of scalar coupling and chemical exchange by
$$\tfrac{1}{2}\hbar A_i(t)(I_+S_{i-} + I_-S_{i+}). \tag{119}$$

Equation (119) leads to the relaxation equations, derived from (87) by adding the contributions from the various spins S_i,

$$\frac{d\langle I_z\rangle}{dt} = -\frac{1}{T_1^{II}}\left\{\langle I_z\rangle - I_0 - \frac{I(I+1)}{S(S+1)}(\langle S_z\rangle - S_0)\right\},$$

$$\frac{1}{T_1^{II}} = \frac{S(S+1)}{3} J_{\text{exch}}(\omega_I - \omega_S),$$

$$J_{\text{exch}}(\omega) = \sum_i P_i \int_{-\infty}^{\infty} e^{-i\omega\tau}\, d\tau\, \overline{A_i(t)A_i(t+\tau)}. \tag{120}$$

Assuming for simplicity that $\sum_i P_i = 1$, i.e. that the time spent by a spin I being uncoupled to any spin S is negligible,

$$\frac{1}{T_1^{II}} = \frac{2A^2}{3} \frac{\tau_e}{1+(\omega_I-\omega_S)^2\tau_e^2} S(S+1). \tag{121}$$

The transverse relaxation time T_2^I can be calculated in the same way by a slight modification of the formula (89):

$$\frac{1}{T_2^I} = \frac{A^2}{3} S(S+1)\left\{\tau_e + \frac{\tau_e}{1+(\omega_I-\omega_S)^2\tau_e^2}\right\}. \tag{122}$$

If both dipolar and scalar couplings are present, their contributions to the relaxation times can be added independently; there are no interference terms, any average product of matrix elements such as

$$\overline{(\alpha\mid\mathcal{H}_1^{\text{dip}}(t)\mid\beta)(\beta'\mid\mathcal{H}_1^{\text{scal}}(t')\mid\alpha')}$$

vanishes because of the different transformation laws of $\mathcal{H}_1^{\text{dip}}$ and $\mathcal{H}_1^{\text{scal}}$ under rotation. It was mentioned in Chapter VI that for two nuclear spins I and S the indirect scalar coupling $\hbar A\mathbf{I}\cdot\mathbf{S}$ is usually (with the exception of very heavy atoms) much smaller than their ordinary dipolar coupling. This does not mean that the corresponding relaxation mechanism is negligible, for the correlation time τ_e may be much longer than the rotational correlation time and this fact, as has been explained in Section II E (c), may enhance considerably the relative relaxing effectiveness of the scalar coupling as compared with that of the dipolar coupling. A consequence of this fact, apparent in the formulae (121) and (122), is that, since $(\omega_I-\omega_S)\tau_e$ is not vanishingly small, T_1 and T_2 may be field-dependent and T_1/T_2 appreciably larger than unity. A specific example will be discussed in Section III B (a).

(b) (3) *Scalar relaxation of the second kind*

We now consider the opposite situation when the spin S has a relaxation time due to some mechanism other than the coupling $\hbar A\mathbf{I}\cdot\mathbf{S}$, much shorter than both the chemical exchange time constant τ_e and the inverse $1/A$ of the frequency A. The second assumption implies that the local magnetic field $A\mathbf{S}(t)/\gamma_I$, produced by the spin S and 'seen' by the spin I, fluctuates at a rate that is fast compared with the frequency A of the splitting it would produce among the energy levels of I, if it did not fluctuate. As a consequence, physically evident (and to be derived quantitatively in a later chapter), only the average value of this coupling is observed and the spin I has a single Zeeman frequency and a single resonance line.

The problem of the relaxation of the spin I can be approached as follows: the 'spin' system is the spin I, the system of the spin S being lumped with the 'lattice' with which it is assumed to be in equilibrium because of its short relaxation time. The treatment of Section II D can be applied.

The perturbing Hamiltonian $\hbar \mathcal{H}_1$ responsible for the relaxation of the spin I is
$$\hbar \mathcal{H}_1 = \hbar A \mathbf{I}.\mathbf{S} = \hbar \sum_q F^{(q)} A^{(q)},$$
where the 'spin' operators $A^{(q)}$ are
$$A^{(0)} = I_z, \qquad A^{(1)} = I_+, \qquad A^{(-1)} = I_-,$$
and the 'lattice' operators $F^{(q)}$
$$F^{(0)} = AS_z, \qquad F^{(1)} = \tfrac{1}{2}AS_-, \qquad F^{(-1)} = \tfrac{1}{2}AS_+.$$

The unperturbed Hamiltonian $\hbar \mathcal{H}_0 = \hbar \omega_I I_z$, whence
$$e^{i\mathcal{H}_0 t} A^{(0)} e^{-i\mathcal{H}_0 t} = A_1^{(0)} = I_z,$$
$$e^{i\mathcal{H}_0 t} A^{(1)} e^{-i\mathcal{H}_0 t} = I_+ e^{i\omega_I t} = A_1^{(1)} e^{i\omega_I t},$$
$$e^{i\mathcal{H}_0 t} A^{(-1)} e^{-i\mathcal{H}_0 t} = I_- e^{-i\omega_I t} = A_1^{(-1)} e^{-i\omega_I t},$$

and the macroscopic equation for the average magnetization $\langle \mathbf{I} \rangle$ in the frame rotating at the frequency ω_I can be written
$$\frac{d\langle \mathbf{I} \rangle}{dt} = -\tfrac{1}{2} \sum_{q,p} J_q(\omega_p^{(q)}) \{\langle [A_p^{(q)}, [A_p^{(-q)}, \mathbf{I}]] \rangle - \langle \;\; \rangle_0\}, \qquad (123)$$

where $\langle \; \rangle_0$ means the average when thermal equilibrium is established. Equation (123) gives
$$\frac{d\langle I_z \rangle}{dt} = -\frac{1}{T_1} \{\langle I_z \rangle - I_0\},$$
$$\frac{d\langle I_x \rangle}{dt} = -\frac{1}{T_2} \langle I_x \rangle. \qquad (124)$$

From (123) it is easily found that
$$\frac{1}{T_1} = 2J^{(1)}(\omega_I),$$
where
$$J^{(1)}(\omega) = \int_{-\infty}^{\infty} G^{(1)}(\tau) e^{-i\omega\tau} \, d\tau, \qquad (125)$$
$$G^{(1)}(\tau) = \tfrac{1}{4} A^2 \, \overline{S_+(0).S_-(\tau)},$$

where the average represented by the bar is a 'lattice' average since the spins S are assumed to be a part of the lattice, given by the formula (62) of Section II D.

Similarly,
$$\frac{1}{T_2} = J^{(1)}(\omega_I) + \tfrac{1}{2} J^{(0)}(0),$$
where
$$J^{(0)}(\omega) = \int_{-\infty}^{\infty} G^{(0)}(\tau) e^{-i\omega\tau}\, d\tau, \qquad (126)$$
$$G^{(0)}(\tau) = A^2 \overline{S_z(0) S_z(\tau)}.$$

If we assume that the motion of the spin S, for which the coupling with the spin I is a negligible perturbation, can be described by Bloch's equations, with relaxation times τ_1 and τ_2, we may assume

$$G^{(1)}(\tau) = \tfrac{1}{4} A^2 \overline{S^+(0) S^-(0)} e^{i\omega_S \tau} e^{-\tau/\tau_2} \simeq \frac{A^2 S(S+1)}{6} e^{i\omega_S \tau} e^{-\tau/\tau_2}$$

and similarly
$$G^{(0)}(\tau) = \frac{A^2 S(S+1)}{3} e^{-\tau/\tau_1},$$

whence
$$\frac{1}{T_1} = \frac{2A^2}{3} S(S+1) \frac{\tau_2}{1+(\omega_I-\omega_S)^2 \tau_2^2},$$
$$\frac{1}{T_2} = \frac{A^2}{3} S(S+1) \left\{ \frac{\tau_2}{1+(\omega_I-\omega_S)^2 \tau_2^2} + \tau_1 \right\}. \qquad (127)$$

It will be seen that these formulae have exactly the same form as the equations (121) and (122) for relaxation caused by chemical exchange if we make $\tau_1 = \tau_2 = \tau_e$.

This indeed is not very surprising. One can argue that all the nuclear spin I 'sees' is a fluctuating field $\mathbf{H}_S(t) = A\mathbf{S}/\gamma_I$ with a correlation time $\tau_e = \tau_1 = \tau_2$ and it neither 'knows' nor 'cares' whether this variation is contained in A or in S. The fact that a term proportional to $\langle S_z \rangle - S_0$ is present in equation (120) and absent in the first equation (124) is explained by the fact that in the latter case the spin S was considered as part of the lattice, which in our treatment of the spin-lattice coupling is assumed to be permanently in thermal equilibrium, so that $\langle S_z \rangle - S_0$ vanishes.

There is, however, a rather subtle difference between the two mechanisms of relaxation through scalar coupling, which has not always been fully appreciated. It has already been mentioned that in the case of coupled equations of the type

$$\frac{d\langle I_z \rangle}{dt} = -\frac{1}{T_1}\left\{ \langle I_z \rangle - I_0 + \frac{\xi I(I+1)}{S(S+1)}[\langle S_z \rangle - S_0] \right\} \qquad (128)$$

or
$$\frac{d\langle i_z \rangle}{dt} = -\frac{1}{T_1}\{\langle i_z \rangle - i_0 + \xi(\langle s_z \rangle - s_0)\} \qquad (128')$$

with
$$\langle i_z \rangle = \frac{\langle I_z \rangle}{I(I+1)}, \qquad \langle s_z \rangle = \frac{\langle S_z \rangle}{S(S+1)},$$

it was possible to modify $\langle I_z \rangle$ by applying an r.f. field at the frequency ω_S of the spin S. Thus a very strong r.f. field, by making $\langle S_z \rangle = 0$ would lead, for a steady-state solution of (128), to

$$\langle I_z \rangle = I_0 + \xi S_0 I(I+1)/S(S+1),$$

which may be much larger than I_0 if $|S_0| \gg |I_0|$. For instance, we would have from the equations (88), $\xi = \frac{1}{2}$ for a dipolar coupling in the extreme narrowing case and, from the equations (120), $\xi = -1$ for a time-dependent scalar coupling $\hbar A(t) \mathbf{I}.\mathbf{S}$.

The situation is different when the relaxation of I is produced by a coupling $A\mathbf{I}.\mathbf{S}(t)$ where the time-dependence of S is due to a strong relaxation mechanism of this spin. To understand this point, consider the simple case of two spins $S = I = \frac{1}{2}$. The levels of the combined system represented below are not pure states for either spin since the coupling $A\mathbf{I}.\mathbf{S}$ mixes the states $|+-\rangle$ and $|-+\rangle$, where the first sign refers to S_z and the second to I_z.

$\|1\rangle$ ——	$\|+,+\rangle$	the small coefficient
$\|2\rangle$ ——	$\beta\|+,-\rangle + \alpha\|-,+\rangle$	$\alpha \simeq \frac{1}{2} A/(\omega_S - \omega_I)$
$\|3\rangle$ ——	$\beta\|-,+\rangle - \alpha\|+,-\rangle$	$\beta = (1-\alpha^2)^{\frac{1}{2}}$
$\|4\rangle$ ——	$\|-,-\rangle$	

The mechanism of the relaxation of I is the following. Since the spin S is strongly coupled to the lattice, transitions occur at a fast rate between the states $|1\rangle \leftrightarrow |3\rangle$ and $|2\rangle \leftrightarrow |4\rangle$ (flip of a spin S). However, because the states $|2\rangle$ and $|3\rangle$ are not pure, transitions also occur at a rate slower by an order α^2 between states $|2\rangle \leftrightarrow |3\rangle$ (simultaneous flip of I and S) and also between $|1\rangle \leftrightarrow |2\rangle$ and $|3\rangle \leftrightarrow |4\rangle$ (flip of a spin I alone). If only the simultaneous flips $|2\rangle \leftrightarrow |3\rangle$ occurred one would indeed have $\xi = -1$ in the equation (128) but because of the transitions $|1\rangle \leftrightarrow |2\rangle$ and $|3\rangle \leftrightarrow |4\rangle$, $|\xi|$ is actually smaller.

It can be shown, using the general formalism of this chapter, that for any relaxation mechanism of the spin S (not necessarily $\frac{1}{2}$) that has a white spectrum and is such that the relaxation of S alone is describable by Bloch's equations with $1/\tau_1 = 1/\tau_2 \gg \omega_I, \omega_S$, equation (128) is still valid with $\xi = -\frac{1}{2}$. The proof although straightforward is too long to be reproduced here.

The resonance frequency of the spin I undergoes a shift $(A/\gamma_I)S_0$ (in gauss) because of the coupling $\hbar A \mathbf{I}.\mathbf{S}$. This shift may be appreciable if S is an electronic spin. Examples will be cited later.

(c) *Quadrupole relaxation in liquids through molecular reorientation*

The tensor coupling $\mathbf{I} \cdot \mathscr{A} \cdot \mathbf{I}$ between the nuclear spin and the electric field gradient at the nucleus can be rewritten

$$\mathscr{H}_1 = \sum_m F^{(m)}(\Omega) A^{(m)}(\mathbf{I}),$$

where the lattice functions $F^{(m)}$ and the spin operators $A^{(m)}$ transform under rotation as the spherical harmonics $Y_2^{(m)}$ of order two. Ω stands for the three Euler angles α, β, γ defining the orientation of the molecule with respect to the laboratory frame.

We can write
$$A^{(0)}(\mathbf{I}) = 3I_z^2 - I(I+1),$$
$$A^{(\pm 1)}(\mathbf{I}) = \tfrac{1}{2}\sqrt{6}(I_z I_\pm + I_\pm I_z), \qquad (129)$$
$$A^{(\pm 2)}(\mathbf{I}) = \tfrac{1}{2}\sqrt{6}\, I_\pm^2.$$

Since the unperturbed Hamiltonian $\hbar\mathscr{H}_0$ is $\hbar\omega_0 I_z$ it is clear that

$$e^{i\mathscr{H}_0 t} A^{(m)}(\mathbf{I}) e^{-i\mathscr{H}_0 t} = e^{im\omega_0 t} A^{(m)}(\mathbf{I})$$

and therefore according to equation (45) the expectation value

$$p^* = \mathrm{tr}(P\sigma^*)$$

of any spin operator P obeys the differential equation

$$\frac{dp^*}{dt} = -(b^* - b_0), \qquad (130)$$

where
$$b^* = \mathrm{tr}\{B\sigma^*\}$$

and
$$B = \tfrac{1}{2}\sum_m \overline{|F^{(m)}(\Omega)|^2}\,\tilde{\mathscr{J}}(m\omega_0)[A^{(-m)},[A^{(m)},P]]. \qquad (131)$$

$\tilde{\mathscr{J}}(\omega)$ is the Fourier transform of the reduced correlation function

$$\tilde{G}(t) = \frac{\overline{F^{(m)}(t)F^{(m)*}(t+\tau)}}{\overline{|F^{(m)}(t)|^2}},$$

supposed to be the same for all $F^{(m)}$. If $\tilde{G}(\tau) = e^{-|\tau|/\tau_c}$,

$$\tilde{\mathscr{J}}(\omega) = \frac{2\tau_c}{1+\omega^2\tau_c^2}.$$

The functions $F^{(m)}(\Omega)$ are related to the function $F^{(m)}(0)$ in the frame of the molecule, through

$$F^{(m)}(\Omega) = \sum_{m'} a_{mm'}(\Omega) F^{(m')}(0), \qquad (132)$$

where the random character of $F^{(m)}(\Omega)$ appears in the coefficients $a_{mm'}(\Omega)$.

It is clear that
$$\overline{a_{mm'}(\Omega) a_{mm''}^*(\Omega)} = \frac{1}{2l+1}\delta_{m'm''},$$

where $l = 2$, and thus

$$B = \tfrac{1}{10}\Big\{\sum_{m'}|F^{(m')}(0)|^2\Big\} \times \sum_{m} \tilde{\mathscr{J}}(m\omega_0)[A^{(-m)},[A^{(m)},P]]. \tag{133}$$

With the notation of Chapter VI, in the frame $Ox'y'z'$ of the molecule, the quadrupole Hamiltonian $\hbar\mathscr{H}_1$ can be written

$$\hbar\mathscr{H}_1 = \frac{eQ}{4I(2I-1)}\frac{\partial^2 V}{\partial z'^2}\Big\{A^{(0)} + \frac{\eta}{\sqrt{6}}(A^{(2)}+A^{(-2)})\Big\}, \tag{134}$$

where η is the asymmetry parameter defined previously.

$$B = \frac{1}{160}\Big(\frac{eQ}{I(2I-1)}\frac{\partial^2 V}{\partial z'^2}\Big)^2 \Big(1+\frac{\eta^2}{3}\Big)\frac{1}{\hbar^2}\sum_m \tilde{\mathscr{J}}(m\omega_0)[A^{(-m)},[A^{(m)},P]]. \tag{135}$$

By making $P = I_z$, it is easily found from (129) and the usual commutation relations that

$$[A^{(-1)},[A^{(1)},I_z]] = \tfrac{3}{2}\{16I_z^3 - I_z[8I(I+1)-2]\},$$
$$[A^{(-2)},[A^{(2)},I_z]] = \tfrac{3}{2}\{-16I_z^3 + I_z[16I(I+1)-8]\}.$$

Unless $\tilde{\mathscr{J}}(2\omega) = \tilde{\mathscr{J}}(\omega)$ (extreme narrowing), B contains terms proportional to I_z^3 and a relaxation equation of the type

$$\frac{d\langle I_z\rangle}{dt} = -\frac{1}{T_1}\{\langle I_z\rangle - I_0\} \tag{136}$$

does not exist. On the other hand, in the extreme narrowing case the terms in I_z^3 do cancel and (136) is valid with

$$\frac{1}{T_1} = \frac{3}{40}\frac{2I+3}{I^2(2I-1)}\Big(1+\frac{\eta^2}{3}\Big)\Big(\frac{eQ}{\hbar}\frac{\partial^2 V}{\partial z'^2}\Big)^2 \tau_c, \tag{137}$$

where $\tilde{\mathscr{J}}(2\omega_0) = \tilde{\mathscr{J}}(\omega_0) = \tilde{\mathscr{J}}(0) = 2\tau_c$.

Another special case when (136) is valid is that of a spin $I = 1$ for then $I_z^3 = I_z$ and it is found that

$$\frac{1}{T_1} = \frac{3}{80}\Big(1+\frac{\eta^2}{3}\Big)\Big(\frac{eQ}{\hbar}\frac{\partial^2 V}{\partial z'^2}\Big)^2 \{\tilde{\mathscr{J}}(\omega_0) + 4\tilde{\mathscr{J}}(2\omega_0)\}, \tag{138}$$

which for extreme narrowing reduces to a special case of (137). The same method can be applied to the calculation of T_2, replacing P by I_x (or I_+) in (135). In the extreme narrowing case it is found that the equation

$$\frac{d\langle I_x\rangle}{dt} = -\frac{1}{T_2}\langle I_x\rangle$$

is valid with $T_2 = T_1$ and is given by (137). We do not give the derivation, since this result is predictable *a priori* by an argument similar to that of Section E (eqn. (81)). The case $I = 1$ requires a special treatment.

It is easy to show that for $I = 1$

$$[I_z^2, [I_z^2, I_+]] = I_+,$$
$$[I_+^2, [I_-^2, I_+]] = 4I_+,$$
$$[I_z I_+ + I_+ I_z, [I_z I_- + I_- I_z, I_+]] = 6I_+,$$
$$[I_z I_- + I_- I_z, [I_z I_+ + I_+ I_z, I_+]] = 4I_+,$$

and therefore

$$\frac{1}{T_2} = \frac{1}{160}\left(\frac{eQ}{\hbar}\frac{\partial^2 V}{\partial z'^2}\right)^2\left(1+\frac{\eta^2}{3}\right)\{9\tilde{\mathscr{J}}(0)+15\tilde{\mathscr{J}}(\omega_0)+6\tilde{\mathscr{J}}(2\omega_0)\}, \quad (139)$$

which for

$$\tilde{\mathscr{J}}(0) = \tilde{\mathscr{J}}(\omega_0) = \tilde{\mathscr{J}}(2\omega_0)$$

is equal to $1/T_1$ as given by (138).

(d) Relaxation through anisotropic chemical shift combined with molecular reorientation

It was shown in Chapter VI that the Zeeman coupling $-\gamma\hbar\mathbf{H}.\mathbf{I}$ of a d.c. magnetic field H with a nuclear spin in a molecule had to be supplemented by a small corrective term $-\gamma\hbar\mathbf{H}.\mathscr{A}.\mathbf{I}$, where $\mathscr{A}$ was the chemical shift tensor with components having definite values in a frame bound to the molecule. Under rotational tumbling of the molecule, the trace $3\delta_0 = \mathscr{A}_{11}+\mathscr{A}_{22}+\mathscr{A}_{33}$ of this tensor is observed as a small frequency shift: $\Delta\omega = \delta_0\omega_0$ of the Larmor frequency.

The study of these shifts as a means of investigation of molecular structure has been extensively developed and will be considered in more detail in Chapter XI. The traceless part $\mathscr{A}'$ of this tensor, with components

$$\delta_{z'}, \quad \delta_{x'} = -\tfrac{1}{2}(1-\eta)\delta_{z'}, \quad \delta_{y'} = -\tfrac{1}{2}(1+\eta)\delta_{z'}$$

in the frame of the molecule is, because of the molecular tumbling, a relaxation mechanism of the nuclear resonance. The corresponding Hamiltonian $\hbar\mathscr{H}_1$ can be written in any frame as

$$\mathscr{H}_1 = \sum_m F^{(m)} A^{(m)},$$

with

$$A^{(0)} = 3H_z I_z - (\mathbf{I}.\mathbf{H}), \quad (140)$$
$$A^{(\pm 1)} = \tfrac{1}{2}\sqrt{6}\{H_z I_\pm + I_z H_\pm\},$$
$$A^{(\pm 2)} = \tfrac{1}{2}\sqrt{6}\, H_\pm I_\pm,$$

and

$$H_\pm = H_x \pm i H_y.$$

In particular, in the laboratory frame $Oxyz$, where
$$H_x = H_y = 0, \qquad H_z = H_0.$$
$$A^{(0)} = 2H_0 I_z, \qquad A^{(\pm 1)} = \tfrac{1}{2}\sqrt{6}\,H_0 I_\pm, \qquad A^{(\pm 2)} = 0.$$

On the other hand, in the molecular frame $Ox'y'z'$, $\mathscr{H}_1 = \sum_m F^{(m)}(0) A^{(m)}$ or
$$\mathscr{H}_1 = \gamma\{H_{z'} I_{z'} \delta_{z'} + H_{x'} I_{x'} \delta_{x'} + H_{y'} I_{y'} \delta_{y'}\}$$
$$= \tfrac{1}{2}\gamma \delta_{z'}\{3 H_{z'} I_{z'} - (\mathbf{I}.\mathbf{H}) + \tfrac{1}{2}\eta(H_+ I_+ + H_- I_-)\},$$
whence
$$F^{(0)}(0) = \tfrac{1}{2}\gamma \delta_{z'}, \qquad F^{(\pm 1)}(0) = 0, \qquad F^{(\pm 2)}(0) = \frac{\gamma \delta_{z'}}{2\sqrt{6}}\eta.$$

The equation of motion of $\langle \mathbf{I} \rangle$ comes immediately from eqn. (133):
$$\frac{d\langle I_{z'}\rangle}{dt} = -\frac{\gamma^2 H_0^2}{40}\delta_{z'}^2\left(1+\frac{\eta^2}{3}\right)\left(\frac{\sqrt{6}}{2}\right)^2 \tilde{\mathscr{J}}(\omega_0) \times$$
$$\times \{\langle [I_+,[I_-,I_z]] + [I_-,[I_+,I_z]]\rangle - \langle\ \rangle_0\},$$
whence
$$\frac{1}{T_1} = \tfrac{6}{40}\gamma^2 H_0^2 \delta_{z'}^2\left(1+\frac{\eta^2}{3}\right)\tilde{\mathscr{J}}(\omega_0). \tag{141}$$

Similarly,
$$\frac{d\langle I_+\rangle}{dt} = -\frac{\gamma^2 H_0^2 \delta_{z'}^2}{40}\left(1+\frac{\eta^2}{3}\right)\langle(\tfrac{1}{2}\sqrt{6})^2 \tilde{\mathscr{J}}(\omega_0)[I_+,[I_-,I_+]] + 4\tilde{\mathscr{J}}(0)[I_z,[I_z,I_+]]\rangle,$$
whence
$$\frac{1}{T_2} = \tfrac{1}{40}\gamma^2 H_0^2 \delta_{z'}^2\left(1+\frac{\eta^2}{3}\right)\{3\tilde{\mathscr{J}}(\omega_0) + 4\tilde{\mathscr{J}}(0)\}. \tag{142}$$

A remarkable feature of these formulae is that even in the extreme narrowing case when $\tilde{\mathscr{J}}(0) = \tilde{\mathscr{J}}(\omega_0)$,
$$\left(\frac{1}{T_2}\right)\bigg/\left(\frac{1}{T_1}\right) = \tfrac{7}{6} \neq 1.$$

G. Nuclear relaxation in gases

Only the principles will be given in this section and the comparison with experiment deferred till Section III D.

(a) The H_2 molecule—diatomic molecules

The behaviour of an isolated H_2 molecule in a magnetic field can be described by the Hamiltonian $\hbar \mathscr{H}$, where
$$\mathscr{H} = \omega_I(I_z^1 + I_z^2) + \omega_J J_z + \omega'(\mathbf{I}^1 + \mathbf{I}^2).\mathbf{J} + \omega''\{\mathbf{I}^1.\mathbf{I}^2 - 3(\mathbf{I}^1.\mathbf{n})(\mathbf{I}^2.\mathbf{n})\}. \tag{143}$$
$\mathbf{I}^1$ and $\mathbf{I}^2$ are the spins of the two protons with $\mathbf{I}^1 + \mathbf{I}^2 = \mathbf{I}$. $\mathbf{J}$ is the rotational angular momentum of the molecule, $\omega_I = -\gamma_I H_0$ is the proton Larmor frequency in the applied field H_0, $\omega_J = -\gamma_J H_0$ is the Larmor frequency of the rotational magnetic moment of the molecule,

$\omega' = \gamma_I H'$ is the strength (in frequency units) of the coupling between the magnetic moments of the protons and the magnetic field produced at their positions by the rotation of the molecule, $\omega'' = 2\gamma_I H'' = \gamma_I^2 \hbar/b^3$ is the strength (in frequency units) of the dipolar coupling between the protons, b is their distance, and $\mathbf{n}$ is the unit vector $\mathbf{b}/b$.

The constants ω' and ω'' (or H' and H'') have been measured with great precision in atomic beam experiments where transitions between various levels of the Hamiltonian (143) were observed. The experimental values are

$$H' = 27 \text{ gauss},$$
$$H'' = 34 \text{ gauss}.$$

Under the pressures customarily used for the observation of nuclear resonance signals in hydrogen gas (of the order of an atmosphere or more), each molecule undergoes a large number of collisions per second. The effect of these collisions is much stronger on the rotational angular momentum $\mathbf{J}$, which is sensitive to strong electric forces acting during a collision, than on the spin vectors $\mathbf{I}^1$ and $\mathbf{I}^2$, which, at least for spins $\frac{1}{2}$, are sensitive only to the much weaker magnetic forces. As a consequence of the very rapid reorientation of the molecule because of collisions, all structure disappears from the spectrum that can be observed in atomic beam experiments, where there are no collisions. There remains a single Zeeman resonance line corresponding to the Hamiltonian $\omega_I(I_z^1 + I_z^2)$ which is the only part of (143) unaffected by the collisions.

The ground state of the hydrogen molecule is the non-magnetic state $I = 0$, $J = 0$, which is unobservable by magnetic resonance methods.

The first excited state is $J = 1$, $I = 1$. The condition $I = 1$ follows from the Fermi statistics obeyed by the protons; the state $J = 1$ is odd and $I = 1$ even, with respect to the interchange of the two protons, and thus the total wave function is odd, as required for fermions. The next excited state is $J = 2$, $I = 0$ where no resonance can be observed under pressure (although it is observable in the atomic beams) and then follows the state $J = 3$, $I = 1$, etc.

The molecules in the states $I = 1$, J odd, are called *ortho*-hydrogen molecules, those in the states $I = 0$, J even, *para*-hydrogen molecules. The probability of a collision inducing a transition between an *ortho*- and a *para*-state is very small because it corresponds to a highly forbidden transition from a singlet to a triplet state of the nuclear spins. We shall neglect them in what follows. The rotational energy of the hydrogen molecule is given by the formula $BJ(J+1)$ where the constant B when expressed in degrees Kelvin is 86. The difference in energy between the *ortho*-states $J = 1$ and $J = 3$ is thus $10B$ or $860°$ K. At room temperature the ratio of molecules in those two states is

$$\tfrac{7}{3}\exp(-\tfrac{860}{300}) \simeq 0{\cdot}13$$

and the ratio decreases rapidly with T. The assumption that all *ortho*-molecules are in the state $J = 1$ and thus that the collisions which change the magnitude of the vector $\mathbf{J}$ can be neglected in comparison with those that only reorient $\mathbf{J}$, leaving its magnitude unchanged, which is qualitatively correct at room temperature, becomes an excellent approximation below liquid-air temperatures. We shall make it for simplicity in what follows.

Under those conditions it is reasonable to assume, following an argument used in Chapter II, that the three components of the vector $\langle \mathbf{J} \rangle$, the expectation value of $\mathbf{J}$, follow Bloch equations of the usual type

$$\frac{d\langle \mathbf{J}\rangle}{dt} = \gamma_J \mathbf{H}_0 \wedge \langle \mathbf{J}\rangle - \frac{1}{\tau_c}\{\langle \mathbf{J}\rangle - J_0\}, \tag{144}$$

where
$$\mathbf{J}_0 = \frac{\gamma_J \hbar J(J+1)}{3kT}\mathbf{H}_0.$$

If the collisions are 'strong' collisions for the vector $\mathbf{J}$ in the sense of Chapter II, that is, such that an average taken over molecules immediately after a collision gives $\bar{J}_x = \bar{J}_y = 0$; $\bar{J}_z = J_0$, τ_c is the average time τ between two collisions. There are reasons, to be discussed later, for believing that τ_c is actually a good deal longer, the cross-section for disorientation of $\mathbf{J}$ being smaller than the cross-section for collision as defined in the kinetic theory of gases.

Now consider the relaxation mechanisms for the spins $\mathbf{I}^1$ and $\mathbf{I}^2$.

There is a first mechanism due to the scalar coupling

$$\omega_J(\mathbf{I}^1 + \mathbf{I}^2).\mathbf{J} = \omega_J \mathbf{I}.\mathbf{J}.$$

This problem is formally identical to the one treated in Section F (*b*) (3), of relaxation of a spin $\mathbf{I}$ coupled by a scalar interaction $\hbar A \mathbf{I}.\mathbf{S}$ to a spin $\mathbf{S}$, itself very strongly coupled to the lattice. The relaxation times T_1^{sc} and T_2^{sc} (sc for scalar) of the spin $\mathbf{I}$ are given by the formula (127) of this chapter where $\tau_1 = \tau_2 = \tau_c$, S is replaced by J, ω_S by ω_J, and A by ω'. Thus

$$\frac{1}{T_1^{\text{sc}}} = \frac{2\omega'^2 J(J+1)}{3}\frac{\tau_c}{1+(\omega_J - \omega_I)^2 \tau_c^2},$$

$$\frac{1}{T_2^{\text{sc}}} = \frac{\omega'^2 J(J+1)}{3}\left\{\tau_c + \frac{\tau_c}{1+(\omega_J - \omega_I)^2 \tau_c^2}\right\} \tag{145}$$

or, since in practice $|(\omega_J-\omega_I)\tau_c| \ll 1$,

$$\frac{1}{T_1^{sc}} = \frac{1}{T_2^{sc}} = \tfrac{2}{3}\omega'^2 J(J+1)\tau_c. \tag{145'}$$

As in Section F(b)(3), we assume in this calculation that the vector **J** is a 'lattice' operator, as distinguished from **I** which is a spin operator.

The second relaxation mechanism is the dipolar coupling between protons. As we saw in the example of water this coupling exists between protons on different molecules as well as between protons in the same molecule. It is easily seen that the effect of the former is negligible in hydrogen gas. Even for water it is smaller than that due to the dipolar coupling inside a molecule and, in addition, it is proportional to the density of the molecules, i.e. orders of magnitude smaller in hydrogen gas than in water. It should be remarked, however, that the magnetic coupling between different molecules is the only one capable of inducing an *ortho-para*-transition.

The remaining relaxation mechanism due to the modulation by a random rotation of the dipolar coupling between the two proton spins of a hydrogen molecule, is thus very similar to the one studied for water and we may use the formula (76)

$$\frac{1}{T_1^d} = \frac{3}{2}\frac{\gamma^4\hbar^2 I^1(I^1+1)}{b^6}\{J^{(1)}(\omega_I)+J^{(2)}(2\omega_I)\}, \tag{146}$$

where $J^{(1)}$ and $J^{(2)}$ are the spectral densities of the random functions

$$F^{(1)}(\theta,\varphi) = \sin\theta\cos\theta\, e^{-i\varphi}, \qquad F^{(2)}(\theta,\varphi) = \sin^2\theta\, e^{-2i\varphi}, \tag{146'}$$

where the angles θ and φ describe the orientation of the vector $\mathbf{n} = \mathbf{b}/b$. The spectral densities $J^{(i)}(\omega)$ in (146) are Fourier transforms of correlation functions $\overline{F^{(i)}(t).F^{(i)*}(t+\tau)}$ where the $F_i(t)$ given by (146') are 'lattice' operators. In the analogous calculation for water we had assumed $\overline{F^{(i)}(t).F^{(i)*}(t+\tau)} = \overline{|F^{(i)}(t)|^2}g(\tau)$ where $g(\tau)$ is a reduced correlation function, and taken for, say, $\overline{|F^{(1)}(t)|^2} = \overline{\sin^2\theta\cos^2\theta}$ the classical average value $\tfrac{2}{15}$. Since we assume that for all states of the molecule the magnitude of the rotational angular momentum has a well-defined value J, one might be tempted to replace this average by

$$\frac{1}{2J+1}\mathrm{tr}\{\sin^2\theta\cos^2\theta\} = \frac{1}{2J+1}\sum_{J_z=-J}^{J}\langle JJ_z|\sin^2\theta\cos^2\theta|JJ_z\rangle. \tag{147}$$

Such a procedure would be incorrect. Since (147) contains matrix elements such as $\langle J|\sin\theta\cos\theta\, e^{i\varphi}|J'\rangle\langle J'|\sin\theta\cos\theta\, e^{-i\varphi}|J\rangle$ with $J \neq J'$, it would amount to giving the same weight to collisions that change J as to those that leave J unchanged, an assumption we rejected on physical grounds. We can get around this difficulty if we replace the tensor operators $F^{(1)}$ and $F^{(2)}$ by different tensor operators $\mathscr{F}^{(1)}$ and $\mathscr{F}^{(2)}$, that are equal to $F^{(1)}$ and $F^{(2)}$ inside the manifold $J = $ const. but have no off-diagonal matrix elements linking this manifold with manifolds $J' \neq J$. According to a general theorem on tensor operators, quoted in Chapter VI, such operators are uniquely defined within the manifold $J = $ const. and are given by

$$\mathscr{F}^{(0)} = 3J_z^2 - J(J+1) \quad (=)_J a_J F^{(0)} = a_J(3\cos^2\theta - 1),$$
$$\mathscr{F}^{(1)} = \tfrac{1}{2}(J_z J_- + J_- J_z) \quad (=)_J a_J \sin\theta \cos\theta \, e^{-i\varphi}, \qquad (148)$$
$$\mathscr{F}^{(2)} = J_-^2 \qquad\qquad (=)_J a_J \sin^2\theta \, e^{-2i\varphi}.$$

The symbol $(=)_J$ means that the right- and left-hand sides of (148) have the same matrix elements inside the manifold J. If we take the expectation value of, say, both $\mathscr{F}^{(0)}$ and $a_J F^{(0)}$ in the state $J_z = J$, we find

$$a_J = \frac{\langle 3J_z^2 - J(J+1)\rangle_{J_z=J}}{\langle 3\cos^2\theta - 1\rangle_{J_z=J}} = \frac{J(2J-1)}{-2J/(2J+3)} = -\frac{(2J-1)(2J+3)}{2}. \quad (149)$$

We can now safely replace $\overline{|F_i(t)|^2}$ by $[1/(2J+1)]\,\mathrm{tr}\{\mathscr{F}\mathscr{F}^+\}$ since all matrix elements such as $\langle J|\mathscr{F}|J'\rangle\langle J'|\mathscr{F}^+|J\rangle$ with $J' \neq J$ are automatically eliminated by the nature of $\mathscr{F}$. It should be clearly understood that the real reason for the substitution (148) is a physical one, rather than just mathematical convenience.

We then find that

$$\overline{|\mathscr{F}^{(2)}|^2} = \frac{1}{(2J+1)}\,\mathrm{tr}\{J_+^2 J_-^2\} = \tfrac{2}{15}J(J+1)(2J-1)(2J+3),$$
$$\overline{|\mathscr{F}^{(1)}|^2} = \frac{1}{4(2J+1)}\,\mathrm{tr}\{(J_z J_+ + J_+ J_z)(J_- J_z + J_z J_-)\} \qquad (150)$$
$$= \tfrac{1}{30}J(J+1)(2J-1)(2J+3).$$

From (146), (147), and (150) we get, assuming extreme narrowing,

$$\frac{1}{T_1^d} = \frac{3}{2}\frac{I^1(I^1+1)}{a_J^2}\,\omega''^2 \cdot 2\tau_c\{\overline{|\mathscr{F}^{(1)}|^2} + \overline{|\mathscr{F}^{(2)}|^2}\} \qquad (151)$$

(the index d stands for dipolar) or, for $I^1 = \tfrac{1}{2}$,

$$\frac{1}{T_1^d} = \tfrac{3}{2}\omega''^2 \tau_c \frac{J(J+1)}{(2J-1)(2J+3)} = \frac{3}{2}\frac{\gamma_I^4 \hbar^2}{b^6}\tau_c\frac{J(J+1)}{(2J-1)(2J+3)}$$

and, for $J = 1$, $\qquad\qquad\qquad\qquad\qquad\qquad\qquad\qquad\qquad\qquad$ (151')

$$\frac{1}{T_1^d} = \tfrac{3}{5}\omega''^2 \tau_c = \tfrac{12}{5}\gamma_I^2 H''^2 \tau_c.$$

It is interesting to notice that if in the first formula (151′) we make J very large we find that $1/T_1^d = \frac{3}{8}\gamma_I^4(\hbar^2\tau_c/b^6)$ which does not reduce to the value $\frac{3}{2}(\gamma^4\hbar^2/b^6)\tau_c$ given for water in (106) where classical Brownian rotation was assumed. This is not surprising since the replacement of the operators $F^{(i)}$ by the $\mathscr{F}^{(i)}$ forbade collisions where J could change, thus making the Brownian motion of the molecule very different from the classical one, even for very large J.

The total relaxation time T_1 is given for $J = 1$ by

$$\frac{1}{T_1} = \frac{1}{T_1^{sc}} + \frac{1}{T_1^d} = \tau_c\left\{\frac{4\omega'^2}{3} + \frac{3\omega''^2}{5}\right\} \tag{152}$$

or, substituting the known values of ω' and ω'',

$$\frac{1}{T_1} = 2\cdot74 \times 10^{12}\tau_c. \tag{152'}$$

As explained earlier, there are no interference terms between the scalar coupling $\omega'\mathbf{I}\cdot\mathbf{J}$ and the dipolar coupling of the protons. An interesting situation arises when one observes the proton resonance in the HD molecule. There, since the two nuclei are different, in the lower state $J = 0$ the nuclear resonance of the proton can be observed, but the two relaxation mechanisms, just described, are lacking and the proton relaxation time should be accordingly long. We postpone the discussion of the experimental results on H_2 and HD till Section III D (a).

Finally, it should be remarked that in writing the Hamiltonian (143) we have omitted the indirect scalar coupling $\hbar J \mathbf{I}^1 \cdot \mathbf{I}^2$. It is unobservable in H_2, but, since it is unaffected by collisions because it is a scalar, it has been observed in HD. Details will be given in Chapter XI.

We have treated specifically the case of the molecule H_2 because of its simplicity and of the abundance of experimental evidence available. It is clear that relaxation in other diatomic molecules, both homonuclear and heteronuclear, could be studied by the same methods.

A special situation occurs when one or both nuclei have a quadrupole moment, for then the random variation of the electric quadrupole gradient due to collisions generally becomes the main mechanism of relaxation. The description of that mechanism is very similar to that outlined for liquids in Section II F (c).

If the temperature is such that the molecules in the sample are present in states with many different values of J, as will be the case for the heavier molecules, a classical treatment of the random rotation should in general be a good approximation. If, on the other hand, a single value of J exists and collisions towards states with different values of

J are infrequent, the second order spherical harmonics $F^{(m)}(\theta, \varphi)$ must be replaced by the tensors $\mathscr{F}^{(m)}(\mathbf{J})$ as explained in connexion with the H_2 molecules. The lack at the present time of experimental evidence does not warrant a detailed discussion.

(b) Relaxation in monatomic gases

For a monatomic gas, with atoms in an electronic $^1S_0$ state (to exclude the existence of electronic magnetism), the relaxation of a nuclear spin $\frac{1}{2}$ is due to the coupling that exists during a collision between two nuclear magnetic moments. The order of magnitude of the relaxation time due to that mechanism can be estimated as follows. If τ is the mean collision time, each atom undergoes an average of $1/\tau$ collisions per second. In the interval between collisions the magnetic interactions are so small as to be negligible. If d is the distance of closest approach of two atoms and v their relative velocity the duration t of a collision is of the order of d/v. Since this time is very short it is reasonable to assume that the probability amplitude q for a transition of the nuclear spin is of the order of t times the strength (in frequency units) of the Hamiltonian representing the magnetic coupling between the nuclear spins of the colliding atoms:

$$q \sim \frac{t\gamma\gamma'\hbar}{d^3} \sim \frac{\gamma\gamma'\hbar}{d^2 v}. \tag{153}$$

(In (153), by writing $\gamma\gamma'$ rather than γ^2 we have allowed for the possible existence of two different spin species in the gas.) Since the effects of successive collisions are incoherent, the transition probability W per unit time, and also $1/T_1$, will be of the order of

$$\frac{1}{T_1} \cong q^2 \frac{1}{\tau} \sim \frac{\gamma^2 \gamma'^2 \hbar^2}{v^2 d^4} \frac{1}{\tau}. \tag{154}$$

A feature of (154) that contrasts strikingly with, say, the formula (152) established for a diatomic gas, is the inverse dependence on the collision time. In the case of hydrogen the magnetic interactions exist even in the absence of collisions and the collisions, by making these interactions time-dependent, make relaxation possible. The variation $1/T_1 \propto \tau_c$ is due to the shortness of the correlation time as explained at the beginning of Section II E (c). On the other hand, in monatomic gases the interactions only exist during a collision and it is natural for the transition probability to be proportional to their frequency $1/\tau$. This dependence $1/T_1 \propto 1/\tau$ should not be confused with a similar dependence existing according to (91) for $\omega\tau_c \gg 1$ in very viscous liquids or in certain solids. There, the limiting situation $\tau \to \infty$ corresponds to a time-independent magnetic interaction (frozen lattice) whereas in a monatomic gas $\tau \to \infty$ would mean absence of interactions (free atoms).

It has been suggested that an additional magnetic interaction may possibly occur during a collision owing to the polarization of the electronic shells.

This effect can be described in simple terms as follows. During the time $t \simeq d/v$ of a collision the two atoms may be considered as forming a diatomic molecule. The distortion of the electronic shells corresponding to that situation may, to a good approximation, be assumed instantaneous. The extra coupling induced between nuclear spins via the distortion of the electronic shells is nothing but the indirect coupling described in some detail in Chapter VI. Since the two atoms are in $^1S_0$ states when at some distance from each other, this coupling would probably be mostly scalar, and so could only be a relaxation mechanism for unlike nuclear spins. It might conceivably be a stronger relaxation mechanism than the normal dipolar coupling, for heavy atoms.

For spins larger than $\frac{1}{2}$, the existence of a nuclear quadrupole moment should be the main relaxation mechanism. We shall be content to remark again that the concept of a diatomic molecule existing during a time $t \simeq d/v$ is helpful and that the ratio of magnetic relaxation to quadrupole relaxation in monatomic gases should be comparable to that existing in stable diatomic molecules.

III. COMPARISON BETWEEN THEORY AND EXPERIMENT

Since the pioneering work of Bloembergen, Purcell, and Pound (1), referred to in what follows as BPP, where the essentials of the microscopic theory of relaxation in liquids were first outlined and compared with experiment, a considerable body of experimental evidence has been gathered by many workers in the field. No attempt will be made here even to summarize the results thus obtained. Instead, a few experiments will be singled out as most representative examples of situations where quantitative theoretical predictions could be compared with experimental results. A general remark can be made in this connexion. The predictions of the theory rest on hypotheses made on the nature of both systems, the spins and the lattice. Since the spin system is considerably simpler than the lattice (at least in liquids), predictions that depend specifically on the nature of the spin system rest on much firmer ground than those that depend on the structure of the lattice.

A. Dipolar coupling between like spins

(a) Short correlation times, relative values of T_1

For correlation times τ_c much shorter than the Larmor period of the spins, it was found that the spectral densities, and thus the inverse relaxation times also, were proportional to τ_c. Furthermore, if it is assumed that the diffusion equation describes the random motions of the molecules correctly, τ_c and thus also $1/T_1$ are proportional to the inverse $1/D$ of the self-diffusion coefficient or, by the Stokes formula, to the ratio η/T, where η is the macroscopic viscosity and T the temperature.

The proton relaxation time has been measured in a large number of organic liquids (1) and although for a given temperature the product ηT_1 is by no means the same in all the compounds studied, it certainly varies much less than either factor, η or T_1. Table I shows the values found for several hydrocarbons.

TABLE I

Liquid	η (centipoises)	T_1 (sec)	ηT_1
Petroleum ether	0·48	3·5	1·68
Ligroin	0·79	1·7	1·34
Kerosene	1·55	0·7	1·09
Light machine oil	42	0·075	0·31
Heavy machine oil	260	0·013	3·38
Mineral oil	240	0·007	1·68

A systematic error which was particularly important for the longer relaxation times, was introduced into the early measurements (1) by the failure to remove adequately dissolved gaseous oxygen, which being paramagnetic shortened T_1 appreciably. In fact no organic liquid with a proton relaxation time longer than 4 sec was found at that time although, in later measurements, values of T_1 as long as, say, 20 sec were found for benzene.

In view of the considerable differences in molecular structure among the various substances listed, a better test of the dependence of T_1 on η/T is obtained by producing a variation of η/T through a change in the temperature of a given substance. Fig. VIII, 1 (1) shows a logarithmic plot of T_1 as a function of η/T for ethyl alcohol, measured at two different frequencies, 29 Mc/s and 4·8 Mc/s. Both the slope −1 of the curve, and the independence of T_1 from the Larmor frequency are in agreement with the theory.

In a careful experiment (5), T_1 was measured in water as a function of temperature between 0° C and 100° C with T_1 (0°) = 1·59 sec and T_1(100°) = 11·55 sec. In the same range of temperatures the self-diffusion coefficient D was measured accurately by the spin-echo method

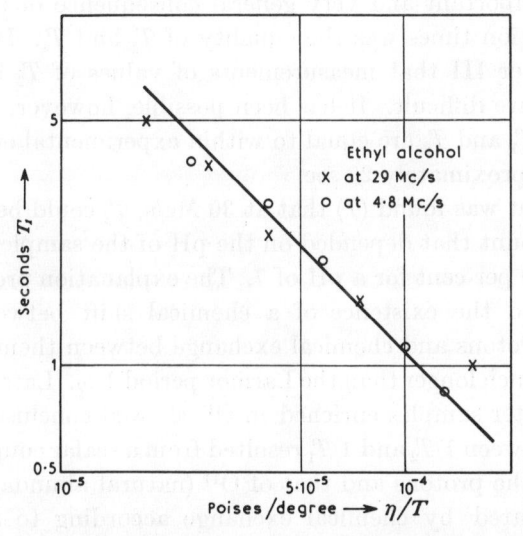

FIG. VIII, 1. Relaxation time for protons in ethyl alcohol, measured at 29 Mc/sec and at 4·8 Mc/sec.

A, as explained in Chapter III. The ratio $(D\eta/T)/(D\eta/T)_{25}$ was found to be constant to within 10 per cent and the ratio $(T_1\eta/T)/(T_1\eta/T)_{25}$ constant to within 15 per cent over that range of temperatures. Since the estimated experimental error over the former ratio is 7 per cent, and only 2 per cent over the latter, it would appear that the law $D \propto (\eta/T)^{-1}$ is verified more accurately than the law $T_1 \propto (\eta/T)^{-1}$. This effect, if real, is not too surprising for η and D are both physical quantities that are essentially related to the translational motion of the molecules, whereas the rotation of the molecules is an important relaxation mechanism.

Similar conclusions were reached in a study of T_1 as a function of pressure (6) up to 10,000 atmospheres, the diffusion coefficient D being also measured. For instance, for toluene it was found that in that range of pressures the product $D\eta$ remained constant within experimental error in spite of an increase of η by a factor 100. At the same time the corresponding decrease of T_1 was only by a factor 14. This was interpreted by assuming that an increase in pressure was less effective in hindering the Brownian rotational motion of molecules which contributes appreciably to the relaxation, than the translational motion to which the quantities D and η are directly related. Unfortunately the conclusions of this interesting experiment are somewhat marred by the unknown contribution to $1/T_1$ from dissolved oxygen.

Another important and very general consequence of the theory for short correlation times was the equality of T_1 and T_2. It was pointed out in Chapter III that measurements of values of T_2 longer than a few seconds are difficult. It has been possible, however, to check that for benzene T_1 and T_2 are equal to within experimental error and have a value of approximately 20 sec.

For water it was found (7) that at 30 Mc/s, T_2 could be shorter than T_1 by an amount that depended on the pH of the sample and could be as large as 30 per cent for a pH of 7. The explanation proposed at first (7) postulated the existence of a chemical shift between two non-equivalent protons and chemical exchange between them with a time constant τ_e much longer than the Larmor period $1/\omega$. Later experiments (8), using water samples enriched in O^{17} showed conclusively that the difference between $1/T_2$ and $1/T_1$ resulted from a scalar coupling between the spins of the protons and that of O^{17} (natural abundance 0·037 per cent), modulated by chemical exchange according to a mechanism discussed in II F (b).

(b) *Absolute values of T_1*

The two contributions to $1/T_1$ in water, resulting respectively from the rotational and translational motions of the water molecules and given respectively by (106) and (115), can be added together and rewritten as

$$\frac{1}{T_1} = \left(\frac{a}{b}\right)^2 \left(\frac{\gamma^4 \hbar^2}{3Db^4}\right) \left\{1 + \frac{3\pi}{5} \frac{Nb^6}{a^3}\right\}, \tag{155}$$

where a is the radius of the hard sphere to which the molecule is approximated in the Stokes formula, b is the distance between protons, D the self-diffusion coefficient of water, and N the number of protons per c.c.

Considerable uncertainty exists over the value to be given to a, which clearly is not a well-defined molecular parameter.

If we assume for argument's sake that the structure of water is one of hard spheres in hexagonal close packing, a value of $a = 1·74 \times 10^{-8}$ cm is obtained. The other parameters in (155) are well known:

$b = 1{\cdot}58 \times 10^{-8}$, $N = 6{\cdot}75 \times 10^{22}$, and $D = 1{\cdot}85 \times 10^{-5}$ sec/cm^2 (5), and (155) yields $T_1 = 3{\cdot}7$ sec, the rotational contribution to $1/T_1$ being about three times the translational one. Since the experimental value is 3·6 sec, such an agreement, in view of the crudeness of the model and of the arbitrariness in the choice of the value of a, is clearly accidental. It is gratifying, however, to obtain the correct order of magnitude for T_1.

A study of the relaxation time T_1 of the protons in a solution of methane CH_4 in liquid CS_2 has been reported (9). The sample was under pressure to obtain sufficient concentration, about 5 per cent by volume. This low concentration and the magnetically inert solvent made the intramolecular four spin interaction dominant. It was shown in Section II E (a) that in spite of the correlation in the interactions of the several spin pairs in such a case, the relaxation should follow a simple exponential almost exactly, with a time constant T_1 just 1 per cent larger than if there were no correlation in the motion of the several pairs of spins in the rigid molecule. The experimental value $T_1 = 6{\cdot}5$ sec, together with the measured viscosity, was used to compute an effective molecular radius, entering the theory through the use of the Stokes relation. The value found was 1·71 Å. This compares with a mean value of 1·91 Å for several methods of specifying molecular diameters from gas kinetics. The agreement is as good as could be expected.

(c) Long correlation times

Formula (105) predicts that on a logarithmic scale T_1 as a function of τ_c should have a slope equal to -1 for $\tau_c \ll 1/\omega$, $+1$ for $\tau_c \gg 1/\omega$, and pass through a minimum proportional to ω when $\omega \tau_c \cong 0{\cdot}6$. The same behaviour would be predicted, if the Stokes relation were believed, for a logarithmic plot of T_1 against η/T. All these features were verified qualitatively with glycerine, where it was possible to vary the parameter η/T in a ratio of the order of 10^4 by changing the temperature (1).

The measurements were made at two frequencies, 29 Mc/s and 4·8 Mc/s. For both, a minimum in the value of T_1 was observed, larger and occurring for a smaller value of η/T for the larger frequency, thus demonstrating that the physical assumptions of the theory were essentially correct. An exact agreement with formula (105) based on the assumption of a single correlation time in a substance as complex as glycerine is hardly to be expected.

B. Coupling between unlike spins

(a) Single irradiation methods

A conclusive proof of the fact that the relaxation of protons in water is caused by their mutual dipolar coupling was provided by a measurement of the T_1 of protons in mixtures of H_2O—D_2O of various concentrations (**10**). As the percentage of heavy water is increased, the relaxation time of protons also increases, because of the smaller magnetic moment of the deuterons, to which they are associated by dipolar coupling. Since in the mixture the protons exchange rapidly, a single relaxation time is observed for the protons. (It is clear that in the absence of such exchange there would be protons with fast relaxation times in HOH molecules, and protons with slow relaxation in HOD molecules.) If it is assumed that the rotational and translational motions are correctly described by correlation times proportional to (η/T), the relaxation time of the protons in the mixture should be given by

$$\frac{1}{T_1} = \eta \left(\frac{1}{T_1}\right)_W [\alpha + (1-\alpha)R]. \tag{156}$$

In (156), $(T_1)_W$ is the relaxation time in pure water, η the ratio of the viscosity of the mixture to that of ordinary water, α the volume concentration of ordinary water in heavy water, and R a coefficient equal to

$$R = \frac{2}{3} \frac{\gamma_D^2}{\gamma_P^2} \frac{I_D(I_D+1)}{I_P(I_P+1)} = \frac{16}{9} \frac{\gamma_D^2}{\gamma_P^2} = 0 \cdot 042. \tag{157}$$

The factor $\frac{2}{3}$ expresses the lesser efficiency of unlike spins in inducing relaxation, described earlier as the $\frac{3}{2}$ effect. From (156) it appears that a plot of $f(\alpha) = 1/\eta T_1$ against α should be a straight line intersecting the ordinate axis at the point $f(0) = R(1/T_1)_W$. Fig. VIII, 2 shows an excellent fit of $1/\eta T_1$ to a straight line that can be extrapolated to $\alpha = 0$ to give $R = 0 \cdot 056 \pm 0 \cdot 010$. The experimental errors and the uncertainty in the assumptions made in writing (156) are unfortunately too large to provide conclusive evidence of the $\frac{3}{2}$ effect.

The contribution to the reciprocal of the relaxation time, due to the presence of paramagnetic ions in a liquid sample, is given by formula (118). The proportionality of that contribution to the concentration N of the ions is well established (except for some ions such as Cr^{++} at very low concentrations). Fig. VIII, 3 of reference (**11**) shows T_1 and T_2 for protons in water as a function of the concentrations of ferric ions $FeNH_4(SO_4)_2 + 12H_2O$. If we express T_1 in seconds and N in number of ions per c.c., the product NT_1 is approximately constant and equal

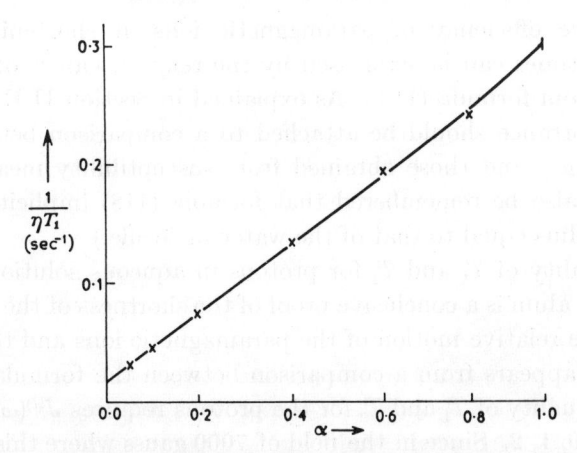

FIG. VIII, 2. Reciprocal of the product of the proton relaxation time, T_1, and the relative viscosity, η, plotted against the relative volume concentration, α, of ordinary water in H_2O—D_2O mixtures.

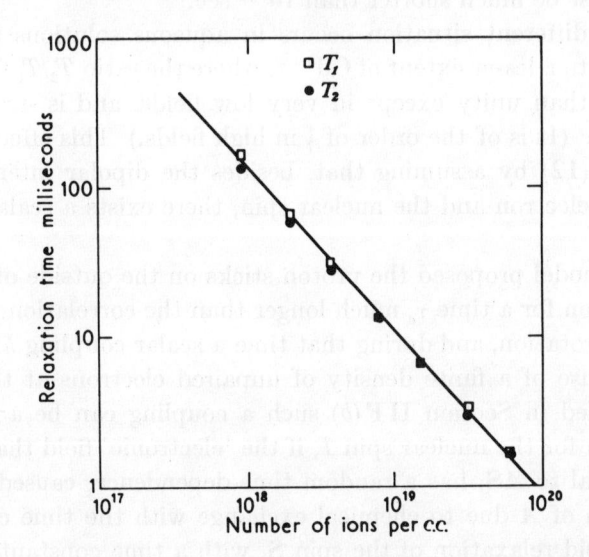

FIG. VIII, 3. Longitudinal and transverse relaxation times of protons at room temperature in paramagnetic solutions of ferric ions $(FeNH_4(SO_4)_2 + 12H_2O)$ of different concentrations.

to 10^{17}. If from (118) we attempt to calculate $\langle \mu^2 \rangle = \gamma_S^2 \hbar^2 S'(S'+1)$ taking $N \simeq 10^{17}$, $\eta = 0\cdot 01$, $T = 300°$, we find $S' \simeq \frac{7}{2}$.

Since the spin S of the ferric ion is $\frac{5}{2}$, such a close agreement is no doubt accidental in view of the crudeness of the model. Still the correctness of the absolute order of magnitude of the result is gratifying.

The relative efficiency of paramagnetic ions in shortening nuclear relaxation times can be expressed by the relative values of their $\langle\mu^2\rangle$ obtained from formula (118). As explained in Section II E (c), not too much importance should be attached to a comparison between these values of $\langle\mu^2\rangle$ and those obtained from susceptibility measurements. (It should also be remembered that formula (118) implicitly assumes an ionic radius equal to that of the water molecule.)

The equality of T_1 and T_2 for protons in aqueous solutions of ferric ammonium alum is a conclusive proof of the shortness of the correlation time for the relative motion of the paramagnetic ions and the protons. Indeed, as appears from a comparison between the formulae (88) and (89), the equality of T_1 and T_2 for the protons requires $J^{(i)}(\omega_S) = J^{(i)}(0)$ where $i = 0, 1, 2$. Since in the field of 7000 gauss where this particular experiment was performed, $\omega_S \gg \omega_I$ is of the order of $1 \cdot 25 \times 10^{11}$, this requires that the correlation time for the relative motion electron-proton must be much shorter than 10^{-11} sec.

A very different situation occurs in aqueous solutions of salts of Mn^{++}, and to a lesser extent of Gd^{+++}, where the ratio T_2/T_1 for protons is smaller than unity except in very low fields, and is strongly field-dependent. (It is of the order of $\frac{1}{7}$ in high fields.) This effect has been explained (12) by assuming that, besides the dipolar interaction between the electron and the nuclear spin, there exists a scalar coupling $\hbar A \mathbf{I}.\mathbf{S}$.

In the model proposed the proton sticks on the outside of the paramagnetic ion for a time τ_e much longer than the correlation time τ_c for molecular rotation, and during that time a scalar coupling $\hbar A \mathbf{I}.\mathbf{S}$ may exist because of a finite density of unpaired electrons at the proton. As explained in Section II F (b) such a coupling can be a relaxation mechanism for the nuclear spin I, if the 'electronic' field that it 'sees', proportional to $A\mathbf{S}$, has a random time-dependence, caused either by a variation of A due to chemical exchange with the time constant τ_e or to a rapid relaxation of the spin $\mathbf{S}$, with a time constant τ_S, where the faster of the two processes takes over. If either time constant is appreciably longer than the electron Larmor period it is possible for the scalar interaction to shorten the proton T_2 appreciably, while leaving its T_1 practically unchanged in high fields. On the other hand, in fields sufficiently low for the electronic Larmor period to be appreciably longer than the time constant for the scalar interaction, the contribution of the latter to nuclear T_1 and T_2, and thus also T_1 and T_2 themselves, would become equal.

The electron relaxation time T_S, of the order of 10^{-9} sec for Mn^{++}, is probably faster than τ_e and thus is the time constant for this process. In other paramagnetic salts such as the ferric alum mentioned above, T_S is probably shorter by a factor of a hundred or more than in Mn^{++}, and in that case the contribution of the scalar coupling (if it exists) to the proton T_2 is probably negligible in comparison with that of the dipolar coupling, thus explaining the observed equality of nuclear T_1 and T_2 even in high fields.

The assumption of a scalar interaction $\hbar A\mathbf{I}\cdot\mathbf{S}$ also has the advantage of providing, as already mentioned at the end of Section II F (b), a natural explanation for the shift of the nuclear resonance frequency observed in concentrated solutions of paramagnetic ions (13).

An even more convincing proof of the existence of an important scalar electron-proton interaction in solutions of Mn^{++} will be described in the next section.

Another example of relaxation of a spin I caused by a scalar coupling $\hbar A\mathbf{I}\cdot\mathbf{S}$ with a different (nuclear) spin S that possesses a quadrupole moment and thus has a very short relaxation time $(T_1)_s = (T_2)_s = \tau$, is provided by the analysis of experiments performed on $CHCl_3$, PCl_3, and PBr_3 (14).

The relaxation times of the spin I are given by

$$\left(\frac{1}{T_1}\right)_I = DD + \frac{2A^2}{3} S(S+1) \frac{\tau}{1+(\omega_I-\omega_S)^2\tau^2},$$
$$\left(\frac{1}{T_2}\right)_I = DD + \frac{A^2}{3} S(S+1)\left\{\frac{\tau}{1+(\omega_I-\omega_S)^2\tau^2} + \tau\right\}, \tag{158}$$

where the first term DD represents the contribution to $1/T_1$ and $1/T_2$ from the dipolar coupling with the spins S. This contribution is the same for T_1 and T_2 and independent of the nuclear frequency because of the very short correlation time of the dipolar coupling (extreme narrowing). The second term in each equation (158) represents the effects of the scalar coupling as predicted by equation (127).

The equations (158) contain three unknown quantities DD, τ, A, which could be determined for each compound listed, in the following manner.

$CHCl_3$. *The spin I is the proton, the spin S the chlorine*

Experimentally $T_1 = 40$ sec, $T_2 = 10$ sec, both quantities being independent of the frequency. The inference is that $(\omega_I-\omega_S)\tau \gg 1$ and $1/[1+(\omega_I-\omega_S)^2\tau^2] \ll 1$.

We then get

$$\left(\frac{1}{T_1}\right)_I = DD, \qquad \left(\frac{1}{T_2}\right)_I - \left(\frac{1}{T_1}\right)_I = \frac{A^2}{3}S(S+1)\tau.$$

Fortunately, τ, the relaxation time of chlorine, can be obtained from the measurement of its line width, which for Cl^{35} leads to $\tau = 17\cdot 5$ μsec, so we get the three parameters:

$$\frac{1}{DD} = 40 \text{ sec}, \qquad \left(\frac{A}{2\pi}\right)_{Cl^{35}} = 5\cdot 5 \text{ sec}^{-1}, \qquad \tau = 17\cdot 5 \text{ }\mu\text{sec}.$$

Actually there are two chlorine isotopes Cl^{35} and Cl^{37} both with spins $\tfrac{3}{2}$ and an isotopic ratio of 3 to 1. They have different magnetic and quadrupole moments and thus correspond to different values of A and τ.

However, the ratio

$$(A^2\tau)_{35}/(A^2\tau)_{37} = \left(\frac{\gamma_{35}}{\gamma_{37}}\right)^2 \left(\frac{Q_{37}}{Q_{35}}\right)^2 \simeq 0\cdot 9$$

and the contributions to $(1/T_2)_I - (1/T_1)_I$ from either spin differ little.

Within experimental accuracy the complications due to the presence of two isotopes are disregarded.

The coupling with Cl^{37} should be of the order of

$$\left(\frac{A}{2\pi}\right)_{Cl^{37}} \simeq 5\cdot 5 \times \left(\frac{\gamma_{35}}{\gamma_{37}}\right) \simeq 9\cdot 3 \text{ c/s}.$$

PCl$_3$. *The spin I is P^{31}*

It is found in a similar way that

$$(T_1)_I = 4 \text{ sec}, \quad (T_2)_I = 9 \text{ millisec}, \quad (\tau)_{35} = 34 \text{ }\mu\text{sec},$$

whence

$$\left(\frac{A}{2\pi}\right)_{35} \simeq 260 \text{ c/s}.$$

PBr$_3$. *The spin S is that of bromine*

The bromine signal could not be observed, which pointed to a very short value of its relaxation time τ. This was confirmed by the existence of a frequency-dependence of T_1 given by the following set of values:

Frequency (Mc/s)	16	8	4	3	2
T_1	3·8 sec	2 sec	0·8 sec	0·5 sec	0·3 sec

Using the first formula (158) we can deduce

$$\frac{1}{DD} = 3\cdot 8 \text{ sec}, \qquad \tau = 0\cdot 32 \text{ }\mu\text{sec}, \qquad \frac{A}{2\pi} = \simeq 700 \text{ sec}^{-1}.$$

The second equation (158) then gives $T_2 = 30$ millisec for T_2 at 16 Mc/s, which checked with a direct measurement of that quantity.

The small differences between the contributions of two bromine isotopes Br[79] and Br[81] are again disregarded here.

(b) Double irradiation methods

Such methods, where the 'driving' by an r.f. field of a transition of the spin system is accompanied by the detection of another transition of this system at a different frequency, have in recent years become an important tool in studies of nuclear magnetism. The first experiment of this type (**15**), that demonstrated the reality of quadrupole relaxation in solids, will be described in Chapter IX. Often the two frequencies will be the Larmor frequencies of spins I and S belonging to two different species. The driving of the spins S will affect the spins I because of the existence of a coupling between them, and the behaviour of the spins I while the spins S are being excited (or after, in transient experiments) may provide interesting information on the nature of that coupling. In particular, if the spin S is an electronic spin, considerable enhancements of the polarization of the spins I or dynamic polarization, as we shall call it, may result. Such methods, first proposed (**16**) for, and applied (**17**) to metals (see Chapter IX) will be considered here in connexion with a study of relaxation mechanisms in liquids.

(1) *The* HF *molecule* (**11, 18**)

It was shown in Sections II E (*b*) and II F (*b*) that if the relaxation of spins I and S of two different species was caused by their mutual coupling, either dipolar or scalar, made time-dependent by the Brownian motion of the molecules, or by the chemical exchange between molecules respectively, the free motion of their magnetizations, caused by relaxation, was described by a set of coupled equations (at least for the longitudinal components). In order to facilitate the comparison with the experimental results of (**11**) and (**18**) (where a derivation of the theoretical results using a different approach can also be found), we shall modify our notation and rewrite the equations of motion for the magnetization as follows:

$$\frac{d\langle I_z\rangle}{dt} = -\rho\{\langle I_z\rangle - I_0\} - \sigma\{\langle S_z\rangle - S_0\},$$
$$\frac{d\langle S_z\rangle}{dt} = -\rho'\{\langle S_z\rangle - S_0\} - \sigma'\{\langle I_z\rangle - I_0\}, \qquad (159)$$

where ρ, σ, ρ', σ' are what we have defined as $(1/T_1^{II})$, $(1/T_1^{IS})$, $(1/T_1^{SS})$,

$(1/T_1^{SI})$ respectively, given by (88) for the dipolar coupling and by (121) for the scalar spin-spin coupling.

We make the following simplifying assumptions, well verified in liquid HF to which the theory will be applied first:

(i) The dipolar interaction between spins belonging to different molecules can be neglected.

(ii) During the chemical exchange, the time each spin S spends being unattached to a spin I is negligible.

Under these assumptions the spectral densities $J^{(0)}(\omega)$, $J^{(1)}(\omega)$, $J^{(2)}(\omega)$ that appear in the formulae giving the dipolar relaxation times can be written as follows:

$$J^{(0)}(\omega) = \frac{24\delta}{15} \frac{1}{1+\omega^2\tau_c^2},$$

$$J^{(1)}(\omega) = \frac{4\delta}{15} \frac{1}{1+\omega^2\tau_c^2}, \quad (159')$$

$$J^{(2)}(\omega) = \frac{16\delta}{15} \frac{1}{1+\omega^2\tau_c^2},$$

$\delta = \gamma_I^2 \gamma_S^2 \hbar^2 \tau_c / b^6$, τ_c being the correlation time for the random rotation of the molecule and b the distance HF in the molecule.

For the scalar coupling $\hbar A \mathbf{I} \cdot \mathbf{S}$ with a time constant τ_e for chemical exchange, (121) and (122) apply. Under those conditions we have

$$\rho = S(S+1)\left\{\frac{\delta}{15}\left[\frac{2}{1+(\omega_I-\omega_S)^2\tau_c^2}+\frac{6}{1+\omega_I^2\tau_c^2}+\frac{12}{1+(\omega_I+\omega_S)^2\tau_c^2}\right]+\right.$$
$$\left.+\frac{2A^2}{3}\frac{\tau_e}{1+(\omega_I-\omega_S)^2\tau_e^2}\right\},$$

$$\sigma = I(I+1)\left\{\frac{\delta}{15}\left[\frac{-2}{1+(\omega_I-\omega_S)^2\tau_c^2}+\frac{12}{1+(\omega_I+\omega_S)^2\tau_c^2}\right]-\right.$$
$$\left.-\frac{2A^2}{3}\frac{\tau_e}{1+(\omega_I-\omega_S)^2\tau_e^2}\right\},$$
(160)

$$\frac{1}{T_2^I} = S(S+1)\times$$
$$\times\left\{\frac{\delta}{15}\left[4+\frac{1}{1+(\omega_I-\omega_S)^2\tau_c^2}+\frac{3}{1+\omega_I^2\tau_c^2}+\frac{6}{1+\omega_S^2\tau_c^2}+\frac{6}{1+(\omega_I+\omega_S)^2\tau_c^2}\right]+\right.$$
$$\left.+\frac{A^2}{3}\left[\tau_e+\frac{\tau_e}{1+(\omega_I-\omega_S)^2\tau_e^2}\right]\right\}. \quad (161)$$

Similar formulae may be found for ρ', σ', and $1/T_2^S$ simply by interchanging the indices I and S. If either spin has an extra relaxation mechanism besides their mutual coupling, an extra contribution has to be added to ρ and $1/T_2^I$ (or ρ' and $1/T_2^S$). In most liquids the dipolar correlation time τ_c is sufficiently short for products such as $|\omega_{I,S}\tau_c|$ to be negligible (extreme narrowing) and (161) can be rewritten as

$$\rho = S(S+1)\left\{\frac{4\delta}{3} + \frac{2A^2}{3}\frac{\tau_e}{1+(\omega_I-\omega_S)^2\tau_e^2}\right\},$$

$$\sigma = I(I+1)\left\{\frac{2\delta}{3} - \frac{2A^2}{3}\frac{\tau_e}{1+(\omega_I-\omega_S)^2\tau_e^2}\right\}, \quad (162)$$

$$\frac{1}{T_2^I} = S(S+1)\left\{\frac{4\delta}{3} + \frac{A^2}{3}\tau_e\left[1 + \frac{1}{1+(\omega_I-\omega_S)^2\tau_e^2}\right]\right\}.$$

In HF, where $I = S = \frac{1}{2}$,

$$\rho = \rho' = \delta + \frac{A^2}{2}\frac{\tau_e}{1+(\omega_I-\omega_S)^2\tau_e^2},$$

$$\sigma = \sigma' = \tfrac{1}{2}\delta - \frac{A^2}{2}\frac{\tau_e}{1+(\omega_I-\omega_S)^2\tau_e^2}, \quad (163)$$

$$\frac{1}{T_2^I} = \frac{1}{T_2^S} = \frac{1}{T_2} = \delta + \frac{A^2}{4}\left\{\tau_e + \frac{1}{1+(\omega_I-\omega_S)^2\tau_e^2}\right\}.$$

There are therefore in the equations (163) three independent parameters ρ, σ, T_2 and from their measurement the three physical quantities δ, A, τ_e can be obtained. One of these quantities, namely τ_e, can be varied at will within large limits simply by changing by a very little the amount of water contained in HF, which acts as a catalyst for the exchange of protons between the molecules. The following experiments have been performed.

(α) After applying a 180° pulse to one of the spins, say S, the decay of the longitudinal magnetization of either spin was studied. The experimental technique has been described in Chapter III: 90° pulses applied at variable times t, after the initial 180° pulses, give signals proportional to $I_z(t)$ or $S_z(t)$. When the system (159) is solved with $\rho = \rho'$, $\sigma = \sigma'$ under the initial conditions $\langle S_z \rangle = -S_0$, $\langle I_z \rangle = I_0$, existing immediately after a 180° pulse, it is found that

$$\langle I_z \rangle - I_0 = -\tfrac{1}{2}S_i[e^{-t/D_1} - e^{-t/T_1}],$$
$$\langle S_z \rangle - S_0 = \tfrac{1}{2}S_i[e^{-t/D_1} + e^{-t/T_1}], \quad (164)$$

with $\quad T_1 = \dfrac{1}{\rho+\sigma}, \quad D_1 = \dfrac{1}{\rho-\sigma}, \quad S_i = \langle S_z \rangle_i - S_0 = -2S_0.$

Whereas $\langle S_z \rangle - S_0$, the sum of two exponentials, has the usual decay shape, $\langle I_z \rangle - I_0$ has a very remarkable behaviour, since starting from zero it passes through a maximum (or a minimum if $\sigma < 0$) before decaying to zero. Figs. VIII, 4 and VIII, 5 show a plot of $[\langle S_z \rangle - S_0]/S_i$ and $-[\langle I_z \rangle - I_0]/S_i$ as a function of time, after a 180° pulse applied to the spins S, for a certain sample of HF.

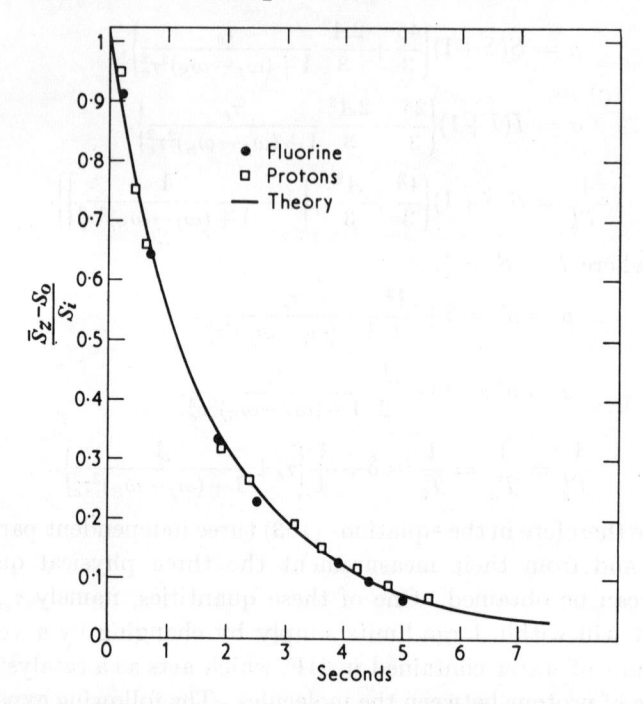

FIG. VIII, 4. Motion of the longitudinal magnetization of the nuclear spins S, after they have been excited by a 180° pulse. The full curve represents the theoretical expression

$$\frac{\langle S_z \rangle - S_0}{S_i} = \frac{1}{2}\left[\exp\left(\frac{-t}{D_1}\right) + \exp\left(\frac{-t}{T_1}\right)\right],$$

with $T_1 = 1 \cdot 27$ sec, $D_1 = 2 \cdot 25$ sec.

The full curves represent $[\langle S_z \rangle - S_0]/S_i$ and $-[\langle I_z \rangle - I_0]/S_i$ as given by (164) with $T_1 = 1 \cdot 27$ sec, $D_1 = 2 \cdot 25$ sec, adjusted for the best fit with the experimental points, shown on the same figures. The agreement between the theoretical equations (164) and experiment is as good as could be expected.

(β) In the same sample, a 90° pulse applied to either spin was followed by a decay of its transverse magnetization, observed by the spin echo method, and in agreement with theory, describable by a single exponential with the same time constant for either spin

$$T_2' = 0 \cdot 43 \pm 0 \cdot 15 \text{ sec.}$$

(γ) A strong r.f. field was applied continuously to one of the spins, say S, making $\langle S_z \rangle = 0$, and the steady-state value of $\langle I_z \rangle$ given by

$$\langle I_z \rangle_{\text{steady}} = I_0 + \frac{\sigma}{\rho} S_0 \qquad (165)$$

was observed. A value of $\langle I_z \rangle$ larger than I_0 by about 30 per cent was found.

Had the spin-spin interaction been purely dipolar, as was believed before the experiment was performed, the expected results should have been, according to (163),

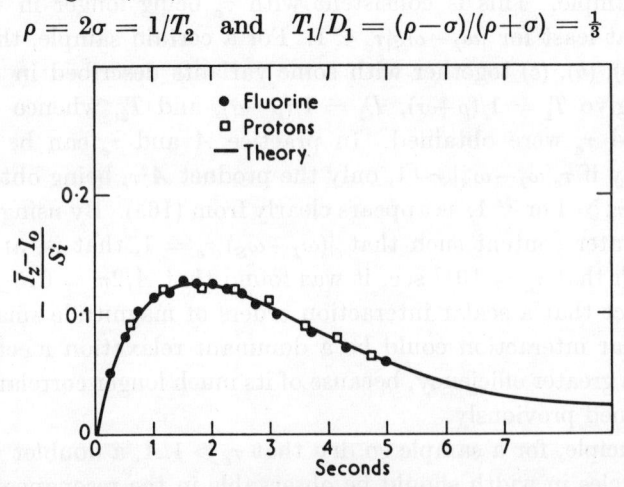

$$\rho = 2\sigma = 1/T_2 \quad \text{and} \quad T_1/D_1 = (\rho-\sigma)/(\rho+\sigma) = \tfrac{1}{3}$$

FIG. VIII, 5. Motion of the longitudinal magnetization of the nuclear spins I, after the spins S of the other species have been excited by a 180° pulse. The full curve represents the theoretical expression

$$-\frac{\langle I_z \rangle - I_0}{S_i} = \tfrac{1}{2}\left[\exp\left(\frac{-t}{D_1}\right) - \exp\left(\frac{-t}{T_1}\right)\right],$$

where T_1 and D_1 are the same as in Fig. VIII, 4.

instead of the experimental ratio $1{\cdot}27/2{\cdot}25 = 0{\cdot}56$. The disagreement for T_2 was even worse. The theory predicts $T_2/T_1 = \tfrac{2}{3}$ instead of the experimental value $0{\cdot}43/1{\cdot}27$. Finally, there should have been an increase

$$\frac{\langle I_z \rangle_{\text{steady}} - I_0}{S_0} = \frac{\sigma}{\rho} = \tfrac{1}{2},$$

instead of the observed $\tfrac{1}{3}$.

The fact that these discrepancies are due to the existence of a scalar coupling $\hbar A \mathbf{I}.\mathbf{S}$ modulated by chemical exchange, itself catalysed by small amounts of water, was dramatically demonstrated by the following experiment. After the variation $I_z(t)$ following a 180° pulse on S had been observed on a very pure sample of HF, the latter was unsealed for a few seconds to absorb some atmospheric moisture, sealed again, and then the experiment was repeated. Figs. VIII, 6 and VIII, 7 show the amplitudes of signals due to 90° pulses on I at various intervals for both experiments. In the purer sample $\langle I_z(t) \rangle - I_0$ is negative and passes through a minimum, whilst it is positive and passes through a maximum in the sample with a larger water content. According to (164) the situation in the pure sample corresponds to a negative value of σ or, by (163), to relaxation being more strongly influenced by the scalar coupling than by the dipolar one; it is the opposite for the more humid sample. This is consistent with τ_e being longer in the purer sample, at least for $|\omega_I - \omega_S|\tau_e \ll 1$. For a certain sample, the experiments (a), (b), (c) together with some variants described in (18) were able to give $T_1 = 1/(\rho+\sigma)$, $D_1 = 1/(\rho-\sigma)$, and T_2, whence by (163), δ, A, and τ_e were obtained. In practice A and τ_e can be obtained separately if $\tau_e|\omega_I - \omega_S| \sim 1$, only the product $A^2\tau_e$ being obtainable if $|\omega_I - \omega_S|\tau_e \gg 1$ or $\ll 1$, as appears clearly from (163). By using a sample with a water content such that $|(\omega_I - \omega_S)|\tau_e = 1$, that is, at the field used such that $\tau_e \sim 10^{-7}$ sec, it was found that $A/2\pi = 615 \pm 50$ c/s.

The fact that a scalar interaction orders of magnitude smaller than the dipolar interaction could be a dominant relaxation mechanism is due to its greater efficiency, because of its much longer correlation time, as explained previously.

In principle, for a sample so dry that $\tau_e \gg 1/A$, a doublet structure of 615 cycles in width should be observable in the resonance of either spin. This, however, would require an impossibly small water content in the sample.

The reader is referred to (18) for further information on the very interesting relaxation processes in HF and their interpretation in terms of chemical physics. It is sufficient to say that they yield further confirmation of the general correctness of the BPP theory of relaxation in liquids.

(2) *Coupling between a nuclear spin and an electronic spin*

Experiments of dynamic nuclear polarization very similar in principle to those described in Section (b) (1) but more spectacular in some respects because of considerable difference in the Larmor frequencies of the two interacting spins, have been performed on solutions of free radicals or ions with unpaired electronic spins.

(α) *Dipolar coupling.* In a solution of naphthalene in 1, 2 dimethoxy-ethane (19), partial saturation of the electron resonance of the free

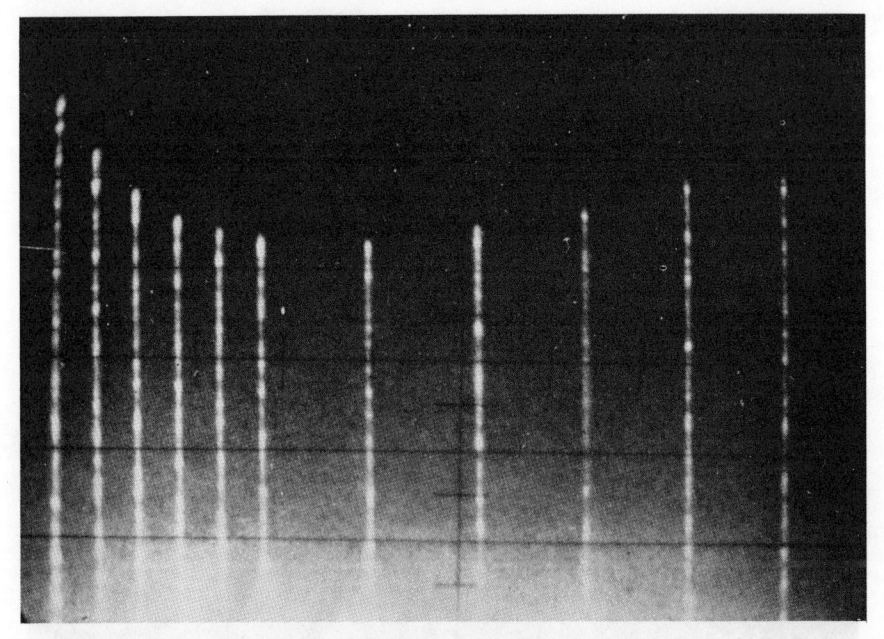

FIG. VIII, 6. Amplitude of tails of 90° pulses at fluorine Larmor frequency following a 180° pulse at hydrogen frequency, obtained with a very pure sample of hydrofluoric acid.

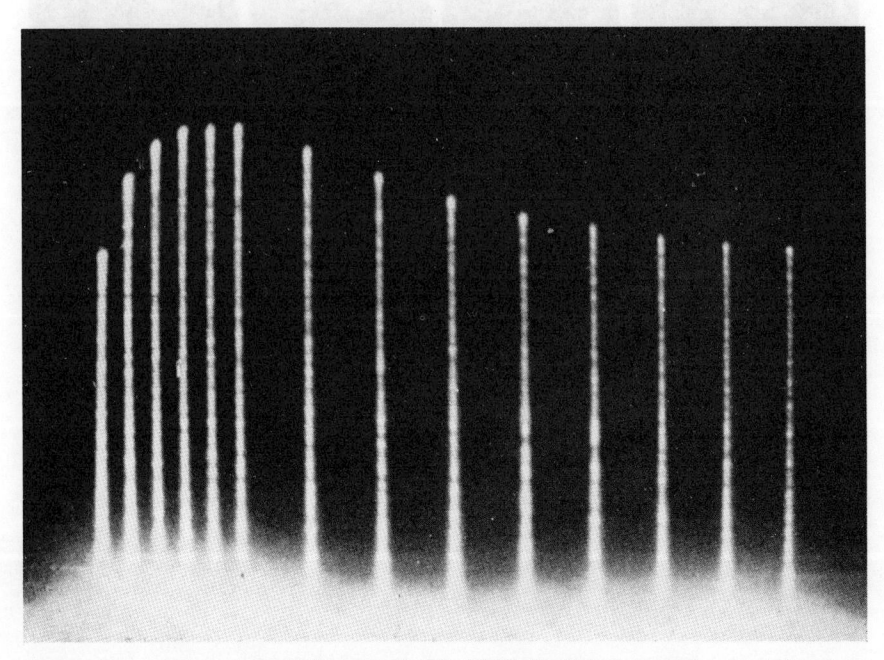

FIG. VIII, 7. The same experiment as in Fig. VIII, 6, after the sample of acid has been unsealed for a few seconds to absorb atmospheric moisture.

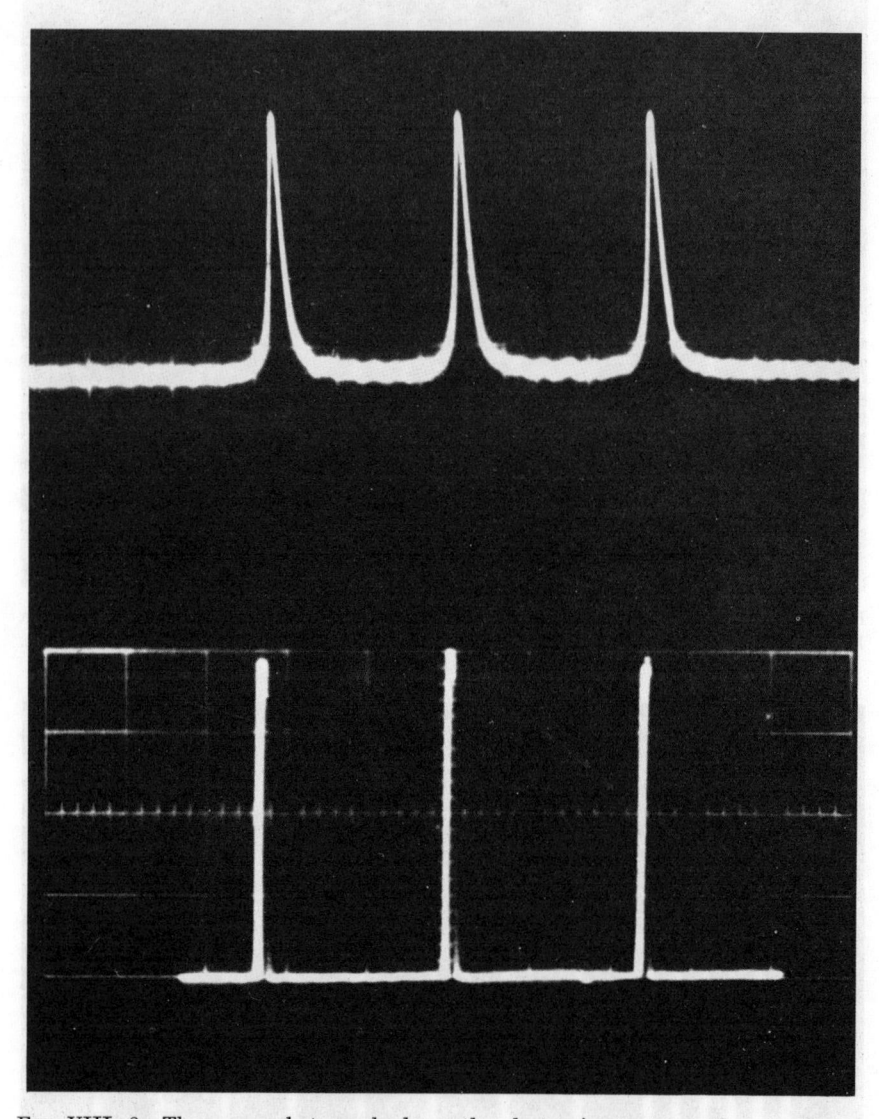

FIG. VIII, 8. The upper photograph shows the electronic spectrum of peroxylamine disulphonate in an aqueous solution, observed in a field of 3000 gauss. The lower photograph shows the proton maser signal, observed when the corresponding electronic line of the upper photograph is saturated.

radical formed by negative ionization of naphthalene, performed in a field of 17·8 gauss, led to a change in the polarization of the protons of the solvent by a factor of the order of −60.

From the power available to saturate the electron resonance at 50 Mc/s, assuming the validity of Bloch equations with $T_1 = T_2$ for the electron resonance, it was possible since its width was known to compute a steady-state value for $\langle S_z \rangle - S_0 \cong -S_0/5$. This, according to the first equation (159), yielded

$$\langle I_z \rangle = I_0 - \frac{\sigma}{\rho}\{\langle S_z \rangle - S_0\} \cong I_0 + \frac{\sigma}{\rho}\frac{S_0}{5}.$$

Since $S_0/I_0 \cong -660$ for protons, the observed value $\langle I_z \rangle / I_0 \cong -60$ is consistent with $\sigma/\rho = \frac{1}{2}$, that is, with a purely dipolar interaction between the electron and the nuclear spin, in the absence of 'leakage', that is, of any significant contribution to nuclear relaxation from a mechanism other than the coupling with the electrons. The relative narrowness of the electron resonance line of the naphthalene ion that permits its saturation with reasonable r.f. power is due to strong exchange couplings between electron spins and thus requires the use of concentrated solutions. The proton relaxation time is accordingly short and the proton line is broad (a fraction of a gauss), an inconvenient feature for some applications such as masers.

This drawback does not exist for another free radical ion $ON(SO_3)_2$ (peroxylamine disulphonate) where exchange narrowing is absent and narrow electron resonance lines are observed at great dilution (20). The electronic spectrum, shown in Fig. VIII 8, top, has three lines owing to the hyperfine coupling of the electron spin **S** with the nuclear spin **K** of N^{14} in the ion.

The intervals between the lines are 13 gauss or 36·4 Mc/s wide. Saturating one of these lines will lead to an increase in the absolute value and to a change in the sign of the nuclear polarization of the spins I of the protons in water where the disulphonate can be dissolved.

It has been remarked earlier (after the equation (88′)), that if the relative motion of the electronic and the nuclear spin is very rapid, the first of the equations (87) or (159) is still valid even in the presence of a hyperfine coupling $\hbar A \mathbf{K} \cdot \mathbf{S}$ between the electron spin and a nuclear spin **K** (N^{14} here) belonging to the same ion. In order to compute the steady state nuclear polarization

$$\langle I_z \rangle = I_0 - \frac{\sigma}{\rho}\{\langle S_z \rangle - S_0\}, \tag{165′}$$

it is necessary to calculate the value of $\langle S_z \rangle$ when one of the lines of the electronic spectrum of the ion is saturated. In a high field, S_z and K_z, the components along the field of the spins of the electron and of N^{14}, are good quantum numbers. If one assumes that the saturation of the transition $|\tfrac{1}{2}, K_z \rangle \to |-\tfrac{1}{2}, K_z \rangle$ leaves unaffected the populations of the other levels of the ion (purely electronic relaxation of the ion), such saturation clearly leads to

$$\langle S_z \rangle = \tfrac{2}{3} S_0 \quad \text{and} \quad \langle S_z \rangle - S_0 = -\tfrac{1}{3} S_0,$$

whence, by (165') with $\sigma/\rho = \tfrac{1}{2}$,

$$\frac{\langle I_z \rangle}{I_0} \simeq \frac{1}{6} \frac{S_0}{I_0} \simeq -110. \tag{166}$$

Enhancements of nuclear proton polarization of approximately half that value were observed in aqueous solutions of peroxylamine disulphonate in a field of 3000 gauss (**21**). Fig. III, 18 showed the normal and the enhanced proton signals. The sample was placed in a coil, itself contained in a microwave cavity. The microwave power fed into the cavity from a continuous wave magnetron was not sufficient for a complete saturation of the electron line, and so the enhancement of the nuclear polarization was smaller than the value $\tfrac{1}{6}(S_0/I_0)$ predicted by (166).

The applied field could be made sufficiently homogeneous to permit auto-oscillation of the maser type of the proton spins as explained in Chapter III. Fig. VIII, 8, bottom, represents proton signals (**21**) obtained in the following manner. With the magnetron oscillating at a fixed frequency ν_S, the nuclear generator was disconnected and the terminals of the r.f. coil surrounding the sample were simply connected to an amplifier. The field H_0 was swept very slowly so that the frequency of either one of the three electronic lines of the ion could be successively made equal to ν_S. When the resonance condition was realized for one of these lines, the proton nuclear magnetization along the field reached a large negative value $\simeq -50 \chi_P H_0$ and an auto-oscillation started that lasted until the sweep of the field took the electron-line out of resonance. The voltage induced during the maser oscillation by the precessing protons in the r.f. coil, amplified and detected, appears on the photograph as a signal. There are three such signals, one for each electron line.

Dynamic polarization experiments were also performed in low fields, between 20 and 80 gauss (**22, 23**). The electronic energy scheme shown in Fig. VIII, 9 and described by the well-known Breit–Rabi formulae,

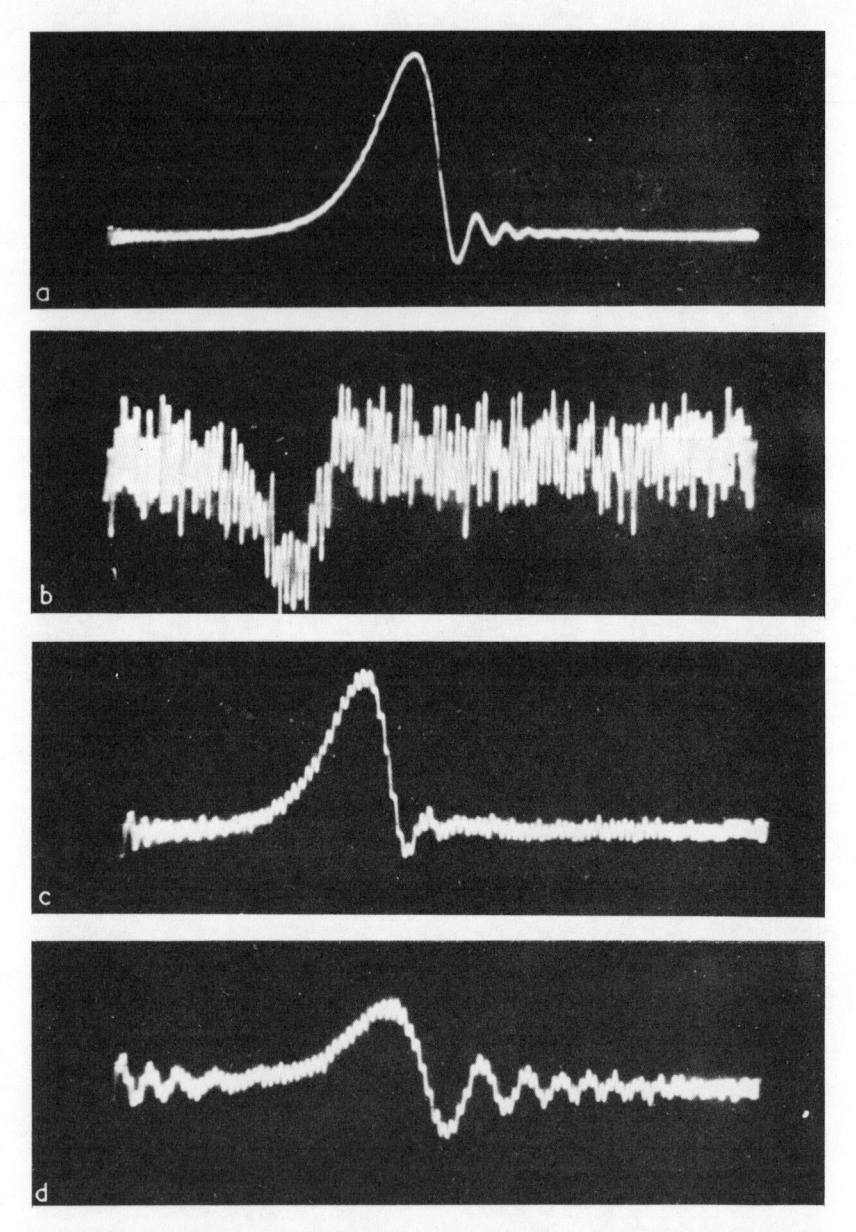

FIG. VIII, 10. Dynamic polarization in an aqueous solution of peroxylamine disulphonate at 72 gauss. (a) Enhanced proton signal. (b) Normal proton signal. The gain is 50 times larger than in (a). Note the reversal of the signal. (c) Enhanced signal of F^{19} added to the solution. (d) Enhanced signal of Li^7. The normal signals of Li^7 and F^{19} are lost in the noise.

becomes more complex and many more electronic transitions can be driven to induce dynamic nuclear polarization. The corresponding electronic frequencies are of the order of 150 Mc/s and complete saturation of an electronic transition could be achieved more easily than in the microwave range. It was thus possible to obtain enhancements of nuclear polarization of the order of 100 for the protons of the solvent

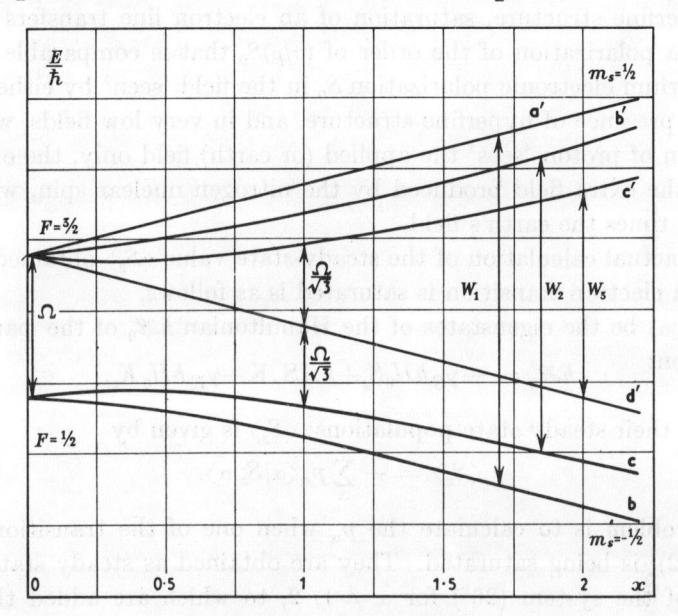

FIG. VIII, 9. Energy levels of peroxylamine disulphonate in a low magnetic field. The zero field splitting $\frac{1}{2}\Omega/\pi$ is equal to 56 Mc/s. The scale on the x-axis is
$$x = \frac{[\gamma_e - \gamma(N^{14})]H_0}{\Omega}.$$

(22). Enhancements of that order or larger could also be obtained for Li^7 and F^{19} by dissolving compounds containing these nuclei in an aqueous solution of peroxylamine disulphonate. Fig. VIII, 10 shows the normal and enhanced signals of protons, and the enhanced signals of Li^7 and F^{19} (the normal signals are below the noise level), in a field of 72 gauss.

It has been remarked earlier that because of the existence of hyperfine structure in the ion, complete saturation of even one electron line led to a value of $\langle S_z \rangle \neq 0$ and thus in high fields to a value of
$$|\langle S_z \rangle - S_0| \cong |S_0/3|,$$
corresponding to a smaller nuclear dynamic polarization than in the absence of hyperfine structure.

On the other hand, in very low fields such as the earth's field, where S_0 is accordingly small, saturation of a transition such as $a' \to b$ (Fig. VIII, 9) leads to a value of $|\langle S_z \rangle| \gg S_0$ and thus to a dynamic nuclear polarization much larger than would obtain in the absence of hyperfine structure (24). A physical explanation of this fact, which is illuminating provided it is not taken too literally, is as follows. In the absence of hyperfine structure, saturation of an electron line transfers to the nuclei a polarization of the order of $(\sigma/\rho)S_0$ that is comparable to the equilibrium electronic polarization S_0 in the field 'seen' by either spin. In the presence of hyperfine structure, and in very low fields, whereas the spin of proton 'sees' the applied (or earth) field only, the electron 'sees' the extra field produced by the nitrogen nuclear spin, which is several times the earth's field.

The actual calculation of the steady-state value $\langle S_z \rangle$ obtained when a given electron transition is saturated is as follows.

Let $|\alpha\rangle$ be the eigenstates of the Hamiltonian $\hbar \mathcal{H}_0$ of the paramagnetic ion:

$$\hbar \mathcal{H}_0 = -\gamma_S \hbar H_0 S_z + \hbar A \mathbf{S} \cdot \mathbf{K} - \gamma_K \hbar H_0 K_z, \qquad (167)$$

and p_α their steady state populations; $\langle S_z \rangle$ is given by

$$\langle S_z \rangle = -\sum_\alpha p_\alpha \langle \alpha | S_z | \alpha \rangle. \qquad (168)$$

The problem is to calculate the p_α when one of the transitions, say $|1\rangle \to |2\rangle$, is being saturated. They are obtained as steady state solutions of the system (26") for $\alpha \neq 1, 2$, to which are added the two equations

$$\frac{dp_1}{dt} = \sum_\beta W_{1\beta}[(p_\beta - p_\beta^0) - (p_1 - p_1^0)] - V(p_1 - p_2),$$
$$\frac{dp_2}{dt} = \sum_\beta W_{2\beta}[(p_\beta - p_\beta^0) - (p_2 - p_2^0)] - V(p_2 - p_1). \qquad (169)$$

The W are the transition probabilities due to relaxation, whereas V is that caused by the applied saturating r.f. field. In the limit $V \to \infty$, $p_1 - p_2 \to 0$, the system (169) must be replaced by the two equations

$$p_1 = p_2 = p,$$
$$0 = \sum_\beta (W_{1\beta} + W_{2\beta})(p_\beta - p_\beta^0) + \sum_\beta W_{1\beta}(p_1^0 - p) + \sum_\beta W_{2\beta}(p_2^0 - p), \qquad (169')$$

of which the second is the sum of the two equations (169). To calculate the p_α a knowledge of the $W_{\alpha\beta}$, that is, of the relaxation mechanism for the paramagnetic ion, is required. Conversely, from a measurement of $\langle I_z \rangle$ it is possible to obtain $\langle S_z \rangle$ by (165), whence, by (168), some information on the p_α and thus also on the $W_{\alpha\beta}$ and the electronic relaxation mechanism may result. In this manner it is possible in principle, and sometimes in practice, to determine not only the electron resonance frequencies but also the electronic relaxation mechanism, without ever detecting an electronic resonance.

Such a study was made (23) in order to discriminate between two possible relaxation mechanisms for the ion of disulphonate peroxylamine:

(i) a purely electronic mechanism that leaves the nitrogen spin unaffected;

(ii) relaxation by an anisotropic traceless hyperfine coupling. As explained in Section II F (a), such a coupling averaged out to first order by the random Brownian rotation of the ion in solution does not affect the energy levels of the ion, but provides a relaxation mechanism. By measuring ratios such as $\langle I_z \rangle_{a'b}/\langle I_z \rangle_{b'c}$ (where $\langle I_z \rangle_{a'b}$ means the steady-state value of $\langle I_z \rangle$ when the transition $a' \to b$ of the electronic spectrum is saturated), for which the theory predicts different values for different relaxation mechanisms assumed, it was possible to conclude that the contribution of the anisotropic hyperfine structure to the relaxation of the ion was small (23).

(β) *Scalar coupling.* An early observation of dynamical polarization in a liquid where the electron nucleus coupling was of the scalar type (25) was made in sodium ammonia solutions, where a very narrow electron resonance line can be observed and easily saturated (26).

An enhancement of the proton polarization of the order of

$$\langle I_z \rangle / I_0 \cong -\tfrac{2}{3} S_0 / I_0 \cong 400$$

was obtained in fairly concentrated solutions in a field of 11·7 gauss, demonstrating that the electron-proton coupling was predominantly scalar. In a certain model used to explain the behaviour of these solutions (27) it is assumed that unpaired electrons are trapped in large 'cavities' on the outside of which the protons attach themselves and are coupled to the electrons inside the cavity by a scalar s-electron coupling. The correlation time of this interaction would thus be the time that a given proton remains attached to a given cavity (interaction modulated by chemical exchange). The fact that the polarization $|\langle I_z \rangle|$ is larger than $|\tfrac{1}{2} S_0|$ is an argument in favour of that interpretation, as discussed in Section II F (b) (3).

A somewhat different situation prevails in solutions of the paramagnetic ion Mn++, where the random modulation of the electron nucleus interaction $\hbar A\mathbf{I}.\mathbf{S}$ is caused by the short relaxation time T_S of the electron spin (12). In fields of the order of a few gauss, where a dynamical polarization experiment was performed (28), the electronic spins $S = \frac{5}{2}$ and the nuclear spin $K = \frac{5}{2}$ of Mn++ couple to a total angular momentum F which takes integer values from 0 to 5 and is a good quantum number as well as its component F_z along the applied field. The transitions saturated are $\Delta F = 0$, $\Delta F_z = \pm 1$, which correspond to a frequency half of that of a free electron in the same field ($g_F = 1$). If one assumes that $1/T_S$, the inverse electron relaxation time, is small compared with the frequencies of the transitions $|\Delta F| = 1$, the smallest of which is 265 Mc/s and the others are multiples of that value, flips of the proton spin I accompanied by 'off-diagonal' transitions $|\Delta F| = 1$ of the paramagnetic ion are relatively rare and can be neglected. The problem is then formally identical to that of a fictitious electronic spin S' of gyromagnetic ratio $\gamma'_e = \frac{1}{2}\gamma_e$ coupled by a scalar interaction $\hbar A\mathbf{I}.\mathbf{S}'$ to a proton spin $\mathbf{I}$, the relaxation of which is caused by frequent flips of the spin S'. Under those conditions the maximum enhancement of the nuclear polarization should be

$$\frac{\langle I_z \rangle}{I_0} = -(\tfrac{1}{2})\frac{\gamma'_e}{\gamma_I} = -\tfrac{1}{4}\frac{\gamma_e}{\gamma_I} \simeq -160. \qquad (170)$$

The origin of the extra factor $\frac{1}{2}$ in (170) was explained in Section II F (b) (3). The experimental accuracy does not seem to permit a definite statement about the absolute magnitude of the large enhancements observed for the proton polarization. Furthermore, the electronic line width $1/T_S$ is larger than the Larmor frequency $|\gamma'_e H_0|$ of the transition saturated, and the r.f. field H_1 is comparable to the d.c. field H_0, a condition under which the simple theory of Section II F(b) may not apply.

In order to obtain a clear situation it would be interesting to perform an experiment similar to that described in reference (18) for HF, on a molecule with strong scalar coupling, between two nuclear spins I and S. If S has a relaxation time, due for instance to a quadrupole moment, faster than the chemical exchange time τ_e, and if the dipolar relaxation is negligible this would parallel exactly the situation described in Section II F (b) (3).

In any case the experiment on Mn++, by the *sign* observed for the dynamic proton polarization, demonstrates clearly that the electron proton coupling is indeed predominantly scalar, confirming the conclusions of reference (12).

(γ) *General remarks on dynamic polarization.* It should be noticed that the phenomenon of dynamical polarization is more general and rests on fewer hypotheses than the coupled equations such as (87) that we used previously to demonstrate its existence. In particular, it is independent of the high-temperature approximation where the Boltzmann exponential is replaced by its first order expansion.

Let us assume first that the coupling between a nuclear spin **I** and an electronic spin **S** is scalar, and is the only relaxation mechanism for the spin I (although naturally not for the spin S). Let $N_\pm$ and $n_\pm$ be the populations of the states $\pm\frac{1}{2}$ for either spin (assumed to be $\frac{1}{2}$, for simplicity). A nuclear flip can only occur if an electronic flip occurs in the opposite direction. The steady state condition for the spins I, expressing that there are equal numbers of nuclear spin flips in opposite directions, can be written

$$N_+ n_- W_{(+-)\to(-+)} = N_- n_+ W_{(-+)\to(+-)}. \tag{171}$$

Quantities such as $W_{(+-)\to(-+)}$ are probabilities for transitions where the energy is exchanged with a lattice in thermal equilibrium and, as explained in Section I B,

$$W_{(+-)\to(-+)}/W_{(-+)\to(+-)} = \exp\{(E_{+-}-E_{-+})/kT\} = \exp\{\hbar(\omega_S-\omega_I)/kT\}, \tag{172}$$

where $\omega_S = -\gamma_S H_0$ and $\omega_I = -\gamma_I H_0$ are the Larmor frequencies of the spins S and I in the applied field H_0. If an external r.f. field at the frequency ω_S saturates the electron resonance, making $N_+ = N_-$, it appears from (171) and (172) that

$$n_+/n_- = \exp\{\hbar(\omega_S-\omega_I)/kT\}, \tag{173}$$

rather than $\exp\{-\hbar\omega_I/kT\}$ as when the spins I are in thermal equilibrium. The departure of (n_+/n_-) from unity is considerably increased, and in the high-temperature approximation the nuclear polarization is augmented by a factor

$$-\frac{\omega_S-\omega_I}{\omega_I} \simeq -\frac{\omega_S}{\omega_I} = -\frac{\gamma_S}{\gamma_I}.$$

Since γ_S is negative this corresponds to a positive dynamic polarization for nuclei with positive magnetic moments.

If the interaction (**S**, **I**) is not purely scalar, there may exist transition probabilities such as $W_{(\pm,+)\to(\pm,-)}$ or $W_{(++)\to(--)}$ where the electron spin does not flip at all, or where both spins flip in the same direction.

The equation (171) must then be generalized by addition to either side of the equation of contributions from these processes.

The fact that the maximum enhancement by dynamic polarization is $+\frac{1}{2}\gamma_S/\gamma_I$ for a dipolar coupling, under extreme narrowing conditions, rather than $-\gamma_S/\gamma_I$ as for the scalar coupling, results from a competition between the various transition probabilities W present in a dipolar interaction.

A slightly different description (29) of the physical nature of dynamic polarization can be given that has the advantage of being more directly applicable to the situation where, as in metals, the electrons obey Fermi rather than Boltzmann statistics (16).

The origin of the Boltzmann ratio $P_a/P_b = \exp\{-(E_a-E_b)/kT\}$ between the populations of two levels E_a and E_b of a system coupled to a 'lattice' at temperature T, lies in the fact that the lattice receives or surrenders the energy $|E_a-E_b|$ when the system makes the transition $E_a \leftrightarrow E_b$.

Now consider an electronic and a nuclear spin coupled by a scalar coupling. In a simultaneous flip $(+-) \to (-+)$ the lattice receives the energy $\hbar(\omega_S-\omega_I)$. Then the electron which has a fast relaxation mechanism of its own flips back, receiving back from the lattice the energy $\hbar\omega_S$. Thus the net energy provided by the lattice for a nuclear flip $|-) \to |+)$ is $\hbar\omega_I$ and the ratio of the nuclear populations

$$\frac{P_+}{P_-} = \exp\left(-\frac{\hbar\omega_I}{kT}\right)$$

as it should. Suppose now that the lattice never gives back the fraction of the energy which it received during the simultaneous electron-nucleus flip, because the back flip $|-) \to |+)$ of the electron is caused by an r.f. field that short-circuits the relaxation process of the electron (this is what saturation of electron resonance really means). Then a net flip of the nuclear spin $|-) \to |+)$ represents a net gain in energy $\hbar(\omega_S-\omega_I)$ for the lattice and so there is a steady-state ratio

$$P_+/P_- = \exp\{\hbar(\omega_S-\omega_I)/kT\}$$

for the nuclear spin populations.

The generalization of this argument to dipolar coupling is straightforward.

C. Electric quadrupole relaxation in liquids

For spins $I > \frac{1}{2}$ the coupling of the nuclear quadrupole moment with the fluctuating electric fields that exist in the liquid state is almost always the main relaxation mechanism. Thus it was possible in an early measurement to associate each of the two resonances of Br[79] and Br[81] with the correct isotope in spite of their equal abundance (**30**). The two lines had unequal widths and it was possible to attribute the broader line to the isotope with the larger quadrupole moment, known from microwave studies of rotational spectra to be Br[79].

Even for the deuteron, in spite of the smallness of its quadrupole moment ($2 \cdot 8 \times 10^{-27}$ cm^2), the relaxation time turns out to be much shorter than would be expected from a magnetic dipolar interaction with the neighbouring spins. In pure D_2O the magnetic relaxation time would be longer than the relaxation time for protons in pure H_2O, $T_1 \cong 3 \cdot 6$ sec, by a factor

$$(\gamma_H/\gamma_D)^4 \frac{I_H(I_H+1)}{I_D(I_D+1)} = 675$$

(neglecting small changes in viscosity and density). In a mixture of 50 per cent of D_2O and H_2O where an experimental value $T_1 = 0 \cdot 5$ sec was found (**1**) for deuterons, the magnetic relaxation time T_1^H of the deuteron due to the HD interaction would still be longer than T_1 for protons in pure water by a factor

$$2 \times \tfrac{3}{2} \times \left(\frac{\gamma_H}{\gamma_D}\right)^2 \simeq 135.$$

Finally, it may be remarked that the number of ferric ions to dissolve per c.c. in heavy water in order to shorten the relaxation time of deuterons by a factor two, should be (according to results of Section III B (a)) $N = 2 \times 10^{17} \times (\gamma_H/\gamma_D)^2 \simeq 10^{19}$, a concentration for which the proton relaxation time in H_2O is shortened by a factor $\cong 400$.

To calculate the absolute value of $1/T_1$ for deuterium by the formula (137):

$$\frac{1}{T_1} = \frac{3}{8}\left(1+\frac{\eta^2}{3}\right)\left(\frac{eQ}{\hbar}\frac{\partial^2 V}{\partial z'^2}\right)^2 \tau_c,$$

where η is the asymmetry parameter and $\partial^2 V/\partial z'^2$ the largest value of the field gradient in the molecular frame, we may tentatively use for these parameters values obtained from solid state experiments on water of crystallization in $LiSO_4 \cdot D_2O$, namely

$$\frac{1}{\hbar}eQ\frac{\partial^2 V}{\partial z'^2} = 246 \text{ kc/s}, \qquad \eta = 0 \cdot 1.$$

If we again use the value $\tau_c = 2a^2/9D$ for the rotational correlation time τ_c, where D is the translational self-diffusion coefficient and a the radius of the sphere to which the D_2O molecule is assimilated by use of the Stokes formula, and take for $D = 1\cdot 85 \times 10^{-5}$ cm²/sec the value measured for H_2O and for $a = 1\cdot 74 \times 10^{-8}$ the value estimated for H_2O from volumetric considerations, we find

$$\frac{1}{T_1} \cong \frac{1}{12} \frac{a^2}{D}\left(\frac{eQ}{\hbar}\frac{\partial^2 V}{\partial z'^2}\right)^2 \cong 4, \qquad T_1 = T_2 \simeq 0\cdot 25 \text{ sec},$$

of the same order of magnitude as the experimental value.

A series of measurements of T_1 and T_2 in the liquid state on molecular chlorine (31) compounds, where the values of the electric-field gradients were known from solid-state measurements, can also be used to test the validity of (137). The value of T_1 was obtained from the saturation behaviour of the resonance line and, T_2 being the main broadening mechanism, from the measurement of the line width. The following compounds were investigated: $TiCl_4$, $VOCl_3$, CrO_2Cl_2, $SiCl_4$. Table II gives for each compound T_1, T_2, $(eQ/\hbar)(\partial^2 V/\partial z'^2)$, and the value of τ_c computed using eqn. (137).

TABLE II

	T_1 millisec	T_2 millisec	$\frac{eQ}{\hbar}\frac{\partial^2 V}{\partial z'^2}$ Mc/s	τ_c sec
$TiCl_4$	0·31	0·35	12	$0\cdot 51 \times 10^{-11}$
$VOCl_3$	0·12	0·14	22	$0\cdot 32 \times 10^{-11}$
CrO_2Cl_2	0·08	0·092	32	$0\cdot 26 \times 10^{-11}$
$SiCl_4$	0·046	0·052	40	$0\cdot 29 \times 10^{-11}$

Within experimental error $T_1 = T_2$ and the values of τ_c deduced from (137) are within the range of correlation times usually found for non-viscous liquids.

There are few other examples in the literature, where careful measurements of T_1 and T_2 caused by quadrupole relaxation under controlled conditions have been made. Such observations are obscured by complex physico-chemical effects that small concentrations of foreign ions can exert on the relaxation time and line width of a given ion in solution.

To terminate this review of experimental evidence on relaxation phenomena in liquids we can mention an example of relaxation believed to be due to anisotropic chemical shift (32). The magnetization of C^{13} in CS_2 has a relaxation time of one minute, which is relatively short considering the small concentration of nuclei with non-zero spin (1 per cent of C^{13}). This must be contrasted with the relatively long relaxation time of C^{13} in CCl_4 in spite of the presence of the nuclear magnetic moments of Cl^{35} and Cl^{37}. This contrast is thought to be due to the presence in the molecule CS_2, and to the absence in the much more symmetrical molecule CCl_4, of anisotropic chemical shift acting as a relaxation mechanism. The obvious check of changing the resonance frequency ω and investigating whether $1/T_1 \propto \omega^2$, does not seem to have been made.

D. Nuclear relaxation in gases

(a) *Nuclear relaxation in hydrogen gas*

An early observation of the proton resonance (33) in hydrogen gas under pressure had led to a value of $T_1 \simeq 0.015$ sec at 10 atm and room

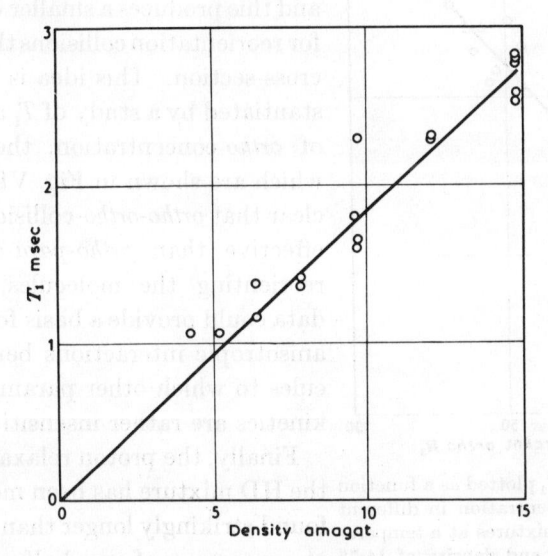

Fig. VIII, 11. T_1 plotted as a function of the density of normal H_2 gas (75 per cent *ortho*, 25 per cent *para*) at 20·4° K.

temperature which, from (152'), gives $\tau_c = 2.45 \times 10^{-11}$. This, contrasted with the collision time τ as defined in the kinetic theory of gases which under the same conditions of temperature and pressure is $\tau = 10^{-11}$, would suggest that as far as relaxation of the vector $\langle \mathbf{J} \rangle$ is concerned the collisions are not completely 'strong' in the sense of Section II G (a).

A study of the proton relaxation in gaseous H_2 as a function of density and *ortho-para*-concentration, was made at low temperature (20° K) (34) and the following results obtained. Fig. VIII, 11 shows a plot of T_1 as a function of density for a mixture of 75 per cent *ortho-* and 25 per cent *para*-hydrogen, where the density unit, the amagat, corresponds to 2.7×10^{19} molecules per c.c. and is the mass per c.c. at $0°$ C and 1 atm. Within experimental error T_1 is proportional to the density, in accordance with (152), since it is reasonable to assume the proportionality of the collision frequency to the density. However, the absolute value of τ_c is longer than the collision time τ, as defined from kinetic theory, by a factor of the order of six. This leads us to assume (34) that only the anisotropic part of the forces acting between colliding molecules is effective in reorientating the vector $\langle \mathbf{J} \rangle$, and this produces a smaller cross-section for reorientation collisions than the total cross-section. This idea is further substantiated by a study of T_1 as a function of *ortho*-concentration, the results of which are shown in Fig. VIII, 12. It is clear that *ortho-ortho*-collisions are more effective than *ortho-para*-collisions in reorienting the molecules, and these data could provide a basis for a study of anisotropic interactions between molecules to which other parameters of gas kinetics are rather insensitive.

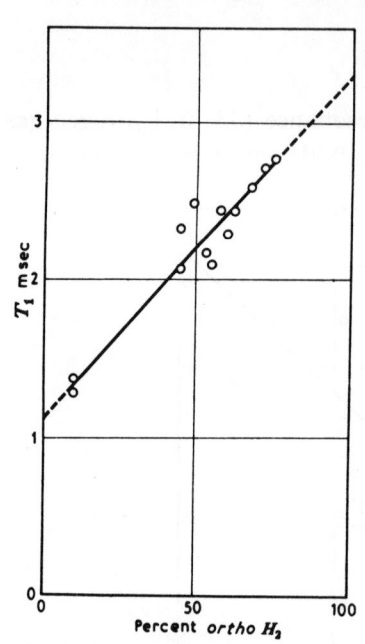

FIG. VIII, 12. T_1 plotted as a function of *ortho*-H$_2$ concentration in different *ortho-para* gas mixtures at a temperature of $20.4°$ K and density of 14.75 amagat.

Finally, the proton relaxation time in the HD mixture has been measured and found strikingly longer than in H$_2$, since at a pressure of one-half of an atmosphere (density 7 amagat) it was 2 sec. The reason for such a long relaxation time, namely the absence of rotational angular momentum in the lowest state $J = 0$ of the HD molecule, has already been given in Section II G (a). At $20°$ K the proportion of HD molecules in the state $J = 1$ is of the order of 1/200. If we assume that relaxation by a mechanism similar to that in H$_2$ occurs only while the molecule is in this state, a relaxation time of the same order of magnitude as the experimental value can be arrived at.

(b) *Nuclear relaxation in liquid hydrogen*

The reason why this topic is considered in this section rather than in the one devoted to liquids is that the model of Brownian motion used for liquids based on the Stokes formula turns out to be completely inadequate for liquid hydrogen, which has a relaxation behaviour somewhat similar to that of the gas (35). Fig. VIII, 13 shows experimental values of T_1 in liquid hydrogen as a function of temperature for various *ortho-para*-proportions. Applying tentatively the formula (152'), we find from a value $T_1 = 0.18$ sec, obtained in liquid hydrogen at $20.4°$ K

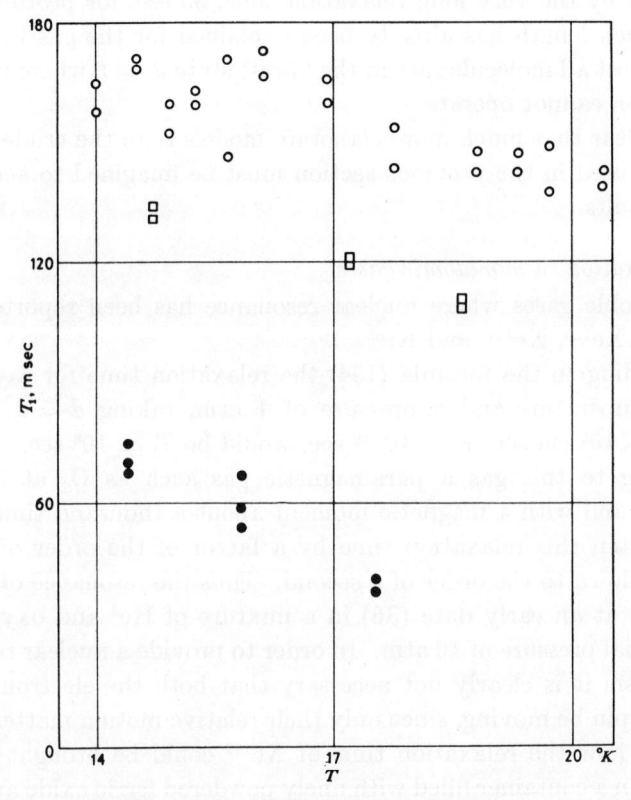

FIG. VIII, 13. T_1 plotted as a function of temperature for protons in liquid H_2 samples having different *ortho*-concentrations.

○ 55 per cent *ortho*-H_2 ● 5 per cent *ortho*-H_2 □ 41 per cent *ortho*-H_2

for a concentration of 75 per cent of *ortho*-hydrogen (not shown in Fig. VIII, 12), $\tau_c \cong 2 \times 10^{-12}$ sec. This compares favourably with the value 1.6×10^{-12} obtained by using the measured value $\eta = 1.4 \times 10^{-4}$ in the Stokes formula and taking $a \cong 2 \times 10^{-8}$.

There, however, the agreement ends since the approximation of the small sphere moving in a viscous medium cannot account for the dependence of T_1 on the *ortho-para*-concentration. The temperature-dependence where η/T, and thus also in the Stokes model τ_{rot}, decrease with increasing temperature, is also wrong. In using (152′) we have neglected the so-called translational contribution to $1/T_1$ from spins in different molecules, which is easily seen to be very small.

Even more striking evidence of the fact that the total orbital momentum J remains a good quantum number in liquid hydrogen is provided by the very long relaxation time, 35 sec, for protons in HD, where such length has already been explained for the gas by the fact that almost all molecules are in the lowest state $J = 0$ where rotational relaxation cannot operate.

It is clear that much more elaborate models than the crude approximations used in the previous section must be imagined to account for these results.

(c) *Relaxation in monatomic gases*

The noble gases where nuclear resonance has been reported so far are He^3, Xe^{129}, Xe^{131}, and Kr^{83}.

According to the formula (154) the relaxation time for say He^3, at room temperature and a pressure of 1 atm, taking $d = 2.10^{-8}$ cm, $v = 1\cdot 4 \times 10^5$ cm/sec, $\tau_c \cong 10^{-10}$ sec, would be $T_1 \cong 10^6$ sec.

Adding to this gas a paramagnetic gas such as O_2 at the same pressure and with a magnetic moment about a thousand times larger, will shorten this relaxation time by a factor of the order of 10^6 and bring it down to the order of a second. Thus the resonance of He^3 was observed at an early date (36) in a mixture of He^3 and oxygen each at a partial pressure of 10 atm. In order to provide a nuclear relaxation mechanism it is clearly not necessary that both the electron and the nuclear spin be moving, since only their relative motion matters, which explains how the relaxation time of Xe^{129} could be brought down to 10^{-2} sec in a container filled with finely powdered ferric oxide and xenon gas at 12 atm (37).

In pure xenon at a pressure of 50 atm, surprisingly short relaxation times of the order of 10^{-2} sec for Xe^{131}, which has a spin $I = \frac{3}{2}$ and a quadrupole moment, and about 400 sec for Xe^{129}, which has spin $I = \frac{1}{2}$ (38), were observed.

In Xe^{131} an order of magnitude calculation of the electric field gradient produced at the nucleus during a collision through a polarization of the electron shells, leads to a theoretical value for the quadrupole relaxation time of

$$T_1 = 7\cdot 7 \times 10^{-2} \text{ sec} \quad \text{for } p = 58 \text{ atm},$$
$$T_1 = 2\cdot 9 \times 10^{-2} \text{ sec} \quad \text{for } p = 76 \text{ atm}.$$

The fast variation of T_1 with pressure is due to the departure of xenon at these pressures from the perfect gas behaviour.

Whether the relaxation time of Xe^{129} can be explained by an indirect coupling between nuclear spins during a collision remains an open question.

REFERENCES

1. N. BLOEMBERGEN, E. M. PURCELL, and R. V. POUND, *Phys. Rev.* **73**, 679, 1948.
2. R. K. WANGSNESS and F. BLOCH, ibid. **89**, 728, 1953.
3. F. BLOCH, ibid. **102**, 104, 1956.
4. P. S. HUBBARD, ibid. **109**, 1153, 1958.
5. J. H. SIMPSON and H. Y. CARR, ibid. **111**, 1201, 1958.
6. G. B. BENEDEK and E. M. PURCELL, *J. Chem. Phys.* **22**, 2003, 1954.
7. S. MEIBOOM, Z. LUZ, and D. GILL, ibid. **27**, 1411, 1957.
8. S. MEIBOOM, *Bull. Am. Phys. Soc.* **5**, 176, 1960.
9. P. S. HUBBARD, Ph.D. Thesis, Harvard University.
10. W. A. ANDERSON and J. T. ARNOLD, *Phys. Rev.* **101**, 511, 1956.
11. I. SOLOMON, ibid. **99**, 559, 1955.
12. N. BLOEMBERGEN, *J. Chem. Phys.* **27**, 572, 1957.
13. W. C. DICKINSON, *Phys. Rev.* **81**, 717, 1951.
14. J. WINTER, *C.R. Acad. Sci.* **249**, 1346, 1959.
15. R. V. POUND, *Phys. Rev.* **79**, 685, 1950.
16. A. W. OVERHAUSER, ibid. **92**, 411, 1953.
17. T. R. CARVER and C. P. SLICHTER, ibid., p. 212, 1953.
18. I. SOLOMON and N. BLOEMBERGEN, *J. Chem. Phys.* **25**, 261, 1956.
19. L. H. BENNETT and H. C. TORREY, *Phys. Rev.* **108**, 499, 1957.
20. G. E. PAKE, J. TOWNSEND, and S. I. WEISSMAN, ibid. **85**, 682, 1952.
21. E. ALLAIS, *C.R. Acad. Sci.* **246**, 2123, 1958.
22. A. LANDESMAN, ibid., p. 1538, 1958.
23. —— *J. Phys. Rad.* **20**, 937, 1959.
24. A. ABRAGAM, J. COMBRISSON, and I. SOLOMON, *C.R. Acad. Sci.* **245**, 157, 1958.
25. T. R. CARVER and C. P. SLICHTER, *Phys. Rev.* **102**, 975, 1956.
26. C. A. HUTCHISON and R. C. PASTOR, *Rev. Mod. Phys.* **25**, 285, 1953.
27. J. KAPLAN and C. KITTEL, *J. Chem. Phys.* **21**, 1429, 1953.
28. R. S. CODRINGTON and N. BLOEMBERGEN, ibid. **29**, 600, 1958.
29. C. KITTEL, *Phys. Rev.* **95**, 589, 1954.
30. R. V. POUND, ibid. **72**, 1273, 1947.
31. Y. MASUDA, *J. Phys. Soc. Japan*, **11**, 670, 1956.
32. H. M. MCCONNELL and C. H. HOLM, *J. Chem. Phys.* **25**, 1289, 1956.
33. E. M. PURCELL, R. V. POUND, and N. BLOEMBERGEN, *Phys. Rev.* **70**, 986, 1946.
34. M. BLOOM, *Physica*, **23**, 237, 1957.
35. —— ibid., p. 378, 1957.
36. H. L. ANDERSON and A. NOVICK, *Phys. Rev.* **73**, 919, 1948.
37. W. G. PROCTOR and F. C. YU, ibid. **81**, 20, 1951.
38. E. BRUN, J. OESER, H. H. STAUB, and C. G. TELSCHOW, *Helv. phys. Acta*, **27**, 173, 1954.

IX

THERMAL RELAXATION AND DYNAMIC POLARIZATION IN SOLIDS

Hence you long-legg'd spinners, hence !
A MIDSUMMER-NIGHT'S DREAM.

IN this chapter we extend the study of nuclear relaxation mechanisms to solids. The problem here is essentially the same as that for liquids and gases, namely to calculate the probability of a flip of a nuclear spin caused by its coupling with the thermal motion of a 'lattice'. In the same way as for liquid samples, this flip can always be visualized as resulting from a fluctuating magnetic field or a fluctuating electric-field gradient 'seen' by the nuclear spin under consideration. For some internal motions such as translational diffusion of atoms or hindered rotation of molecules, that take place in solids, the description used for liquids can be taken over with very little change.

There are, however, some significant differences. The internal motions in solids will often have much smaller amplitudes and/or much longer correlation times than in liquids. This has important consequences for the values of the relaxation times.

It is sometimes possible in solids to obtain relatively simple quantum mechanical models of the 'lattice' and to perform a realistic calculation of the relaxation times, using the quantum mechanical approach, which in liquids had a rather formal character. This approach becomes a necessity at very low temperatures when few degrees of freedom of the 'lattice' are excited.

For nuclei with spins larger than $\frac{1}{2}$ the existence of quadrupole interactions in nuclear environments with lower than cubic symmetry (due to crystal structure or to crystal imperfections) modifies the spacings between the spin energy levels and creates new situations, not met with in liquids.

Finally, in solids the tight coupling that exists between nuclear spins has important consequences. In liquids we were able to give a similar treatment both to the calculation of T_1, which measures the time required for the diagonal matrix elements of the density matrix of the spin system (populations) to reach their thermal equilibrium values, and also to that of T_2, which is the decay time for the off-diagonal matrix elements.

In solids, the establishment of a thermal equilibrium between the spin system and the lattice, under certain conditions outlined in Chapter V, can be broken into two steps; first, the spin system reaches an internal thermal equilibrium with a spin temperature T_S in a time T_2, for which only a qualitative definition can be given since the approach of the spin system to equilibrium has no reason to be and in fact is not exponential; secondly, the spin temperature T_S tends towards the lattice temperature T with a time constant $T_1 \gg T_2$ which can be defined much less ambiguously than T_2, since the decay of a single parameter, the spin temperature (or rather its reciprocal, as will appear shortly), is involved.

A general expression for T_1 under the assumption of the existence of a spin temperature will be derived in the next section in connexion with the problem of nuclear relaxation in metals, which we shall consider first, as possibly the best example of a situation where the quantum mechanical features of the 'lattice' play an important role.

I. Conduction Electrons and Spin-lattice Relaxation in Metals

We saw in Chapter VI that in metals the hyperfine coupling between electronic and nuclear spins produced a modification of the energy levels of the nuclear spin system, expressed by a change in the nuclear Larmor frequency (Knight shift) and by the appearance of the so-called indirect couplings between the nuclear spins. We consider now a dynamical effect of this hyperfine coupling which is a powerful mechanism for nuclear spin-lattice relaxation. We shall assume in its evaluation that the hyperfine coupling is the scalar contact interaction

$$\hbar \mathcal{H} = -\frac{8\pi}{3} \gamma_e \gamma_n \hbar^2 \delta(r_I)(\mathbf{I}\cdot\mathbf{S}), \tag{1}$$

disregarding the usually much smaller dipolar coupling of the nuclear spins with the spins of the electrons, as well as their coupling with the orbital moments of the electrons.

In connexion with the contributions of the neglected hyperfine interactions the following point should, however, be made. As explained in Chapter VI, the existence of a dipolar hyperfine coupling could manifest itself through an anisotropy of the Knight shift, and the incomplete quenching of the orbital electronic momentum, responsible for the orbital hyperfine coupling, through a departure of the electronic g-factor from the spin-only value. The absence of these effects in the Knight shift and the g-factor does *not* permit us to conclude that the dipolar and orbital hyperfine couplings are ineffective for nuclear relaxation. The Knight shift has a tensor dependence on the orientation of the applied d.c. field H_0, and any nuclear environment with at least cubic symmetry will necessarily lead to an isotropic Knight shift whatever the character of the individual electronic wave functions.

For the probabilities of relaxation transitions it is the squares of the off-diagonal matrix elements of the dipolar hyperfine interaction that count and those may well be different from zero even if the anisotropic Knight shift vanishes. An analogous argument holds for the orbital coupling.

In that connexion, we saw many examples in Chapter VI of situations where, in spite of the absence of first-order effects, the dipolar and orbital hyperfine couplings were, through the squares of their off-diagonal elements, responsible for effects such as indirect spin-spin couplings and chemical shifts. A value of T_1 calculated on the basis of a purely scalar coupling, the magnitude of which is deduced from the isotropic Knight shift, could then be longer than the real one.

Nuclear relaxation by conduction electrons is not restricted to metals and exists in semiconductors also. However, in the latter it competes with another type of relaxation, that by fixed paramagnetic impurities to be described in Section II of this chapter, and we shall postpone the discussion of semiconductors till then.

A. An elementary calculation of the relaxation time

The relaxation mechanism originating in the scalar interaction (1) works as follows: this interaction can induce a simultaneous flip of the electron and nuclear spins in opposite directions, the energy $\hbar(\omega_e-\omega_n)$ (where $\omega_e = -\gamma_e H_0$ and $\omega_n = -\gamma_n H_0$ are the electronic and nuclear Larmor frequencies) required for such a flip being provided by an equal change in the kinetic energy of the electron. Two consequences, both important for the nuclear relaxation mechanism, follow from the Fermi statistics obeyed by the conduction electrons in a metal. First, the average kinetic energy of the electrons is much larger than the thermal energy kT and is of the same order of magnitude as the Fermi energy E_F; secondly, because of the Pauli principle most conduction electrons cannot take or give up the small energy $\hbar(\omega_e-\omega_n)$, and only the fraction (kT/E_F) on top of the Fermi distribution contributes to the nuclear relaxation process. The order of magnitude of the probability of a nuclear spin flip can be evaluated as follows. The electronic field produced by the conduction electron at a nucleus can be considered as a fluctuating local field with a correlation time τ_c. If we assume on the average one conduction electron per atomic volume, the order of magnitude of τ_c, which is roughly the duration for which a conduction electron can be localized on a given atom, is by a well-known quantum mechanical argument $\sim \hbar/E_F$, where E_F is the Fermi energy.

Since for a random perturbation $\hbar \mathcal{H}_1(t)$, with a very short correlation time τ_c, the transition probability is of the order of $|\overline{\mathcal{H}_1^2}|\tau_c$, we find

$$\frac{1}{T_1} \sim |\overline{\mathcal{H}_1^2}| \frac{\hbar}{E_F} \frac{kT}{E_F} \sim \left(\frac{8\pi}{3}\right)^2 \gamma_e^2 \gamma_n^2 \hbar^3 |\psi(0)|^4 \frac{kT}{E_F^2}, \qquad (2)$$

which is the correct formula to within a dimensionless numerical factor of order unity. In (2), $\psi(r)$ is the electronic wave function normalized to unity in an atomic volume, and the factor kT/E_F takes account of the reduction through the Pauli principle in the number of conduction electrons that participate in the relaxation process. A more accurate calculation will be presented now.

We assume that nuclear spins are $I = \frac{1}{2}$ and that the applied field is sufficiently high for the nuclear spin-spin energy to be negligible in comparison with the nuclear Zeeman energy. Then the decay or growth of the nuclear magnetization, proportional to the difference $p_+ - p_-$ of the populations of the states $I_z = \pm\frac{1}{2}$, is clearly describable by a single exponential, and a single relaxation time can be defined for the nuclear spin system. For the electrons we assume that their own relaxation time is sufficiently short for us to consider their spins as being constantly in equilibrium with the lattice, and that the temperature is sufficiently high for the electronic Zeeman energy $-\gamma_e \hbar H_0$ to be much smaller than kT. Under those assumptions, electrons with spins up or down have approximately the same Fermi distribution function

$$f(E) = \frac{1}{1+\exp\{(E-E_F)/kT\}}. \qquad (2')$$

The probability of an electron making a transition from a state of kinetic energy E to a state of energy E' must be weighted by the factor $f(E)[1-f(E')]$, which is the simultaneous probability for the initial state to be occupied and for the final state to be empty prior to the transition. If the transition is that which involves a simultaneous electron-nuclear spin flip, the change in kinetic energy being very small, the assumption $E' \simeq E$ is legitimate and $f(E)[1-f(E')]$ may be safely replaced by

$$f(E)[1-f(E')] = \frac{\exp\{(E-E_F)/kT\}}{[1+\exp\{(E-E_F)/kT\}]^2} = -kT\frac{df}{dE} \simeq kT\,\delta(E-E_F). \tag{3}$$

The last approximate equality in (3) results from the fact that, since E_F is much larger than kT, $f(E)$ is practically the Heaviside unit-step function (with a minus sign) and its derivative is thus a δ function.

The probability $w_{(+-)\to(-+)}$ of a simultaneous electron-nuclear spin flip can be written

$$\frac{2\pi}{\hbar}|(i\,|\,\hbar\mathcal{H}\,|f)|^2\delta(E_i-E_f), \tag{4}$$

with
$$|i) = U_\mathbf{k}(\mathbf{r})e^{i\mathbf{k}\mathbf{r}}|+,-),$$
$$|f) = U_{\mathbf{k}'}(\mathbf{r})e^{i\mathbf{k}'\mathbf{r}}|-,+),$$
$$E_i-E_f = E_\mathbf{k}-E_{\mathbf{k}'}+\hbar(\omega_e-\omega_n) \simeq E_\mathbf{k}-E_{\mathbf{k}'}, \tag{5}$$
$$\hbar\mathcal{H}_1 = -\frac{8\pi}{3}\gamma_e\gamma_n\hbar^2\delta(\mathbf{r}_I)\{I_z s_z+\tfrac{1}{2}(I_+ s_-+I_- s_+)\}.$$

$\mathbf{k}$ and $\mathbf{k}'$ are the wave vectors of the Bloch wave functions

$$\phi_\mathbf{k} = U_k(\mathbf{r})e^{i\mathbf{k}\mathbf{r}}$$

describing the initial and the final electronic orbits and are normalized to unity in the volume V of the sample. The symbol $|+,-)$ describes the state $s_z = +\tfrac{1}{2}$, $I_z = -\tfrac{1}{2}$. We get

$$w_{(+-)\to(-+)} \simeq \frac{2\pi}{\hbar}\left\{\frac{8\pi}{3}\gamma_e\gamma_n\hbar^2\right\}^2|(\mathbf{k}\,|\,\delta(\mathbf{r}_I)\,|\,\mathbf{k}')|^2\tfrac{1}{4}\delta(E_\mathbf{k}-E_{\mathbf{k}'}). \tag{6}$$

To get the total probability $W_{(+-)\to(-+)}$ of a simultaneous flip, we must multiply (6) by
$$Z(\mathbf{k})Z(\mathbf{k}')f(E)[1-f(E')],$$

where $Z(\mathbf{k})$ is the density of states in the $\mathbf{k}$ space, and integrate over $d^3k\,.\,d^3k'$. We assume for simplicity that the Fermi surface has spherical symmetry in the $\mathbf{k}$ space, $\rho(E)\,dE$ being the number of states (of a given spin) in the interval dE, in the neighbourhood of $E = E_F$. Taking into account (3) we get

$$\left(\frac{1}{T_1}\right)_0 = 2W_{(+-)\to(-+)} = 2\times\frac{2\pi}{\hbar}\left(\frac{8\pi}{3}\gamma_e\gamma_n\hbar^2\right)^2\tfrac{1}{4}|\phi_F(0)|^4 kT\{\rho(E_F)\}^2$$

$$= \frac{64\pi^3}{9}\gamma_e^2\gamma_n^2\hbar^3|\phi_F(0)|^4\{\rho(E_F)\}^2 kT. \tag{7}$$

We write $(1/T_1)_0$ rather than $1/T_1$ to recall the restrictive assumptions used to derive (7) (nuclear spins $\frac{1}{2}$ and high applied field). If the assumption of spherical symmetry of the Fermi surface is not valid, $|\phi_F(0)|^4$ must be replaced by an average of $|\phi_{\mathbf{k}}(0)|^2 \cdot |\phi_{\mathbf{k}'}(0)|^2$ over the Fermi surface.

The resemblance between the crude estimate (2) and the more careful one (7) can be seen if one remembers that, if $\psi_F(\mathbf{r})$ are normalized in the atomic volume $\Omega = V/N$, $|\psi_F(0)|^2 = N|\phi_F(0)|^2$, and that $\rho(E_F) \simeq N/E_F$. For free electrons, where $\rho(E_F) = 3N/4E_F$, (7) can be rewritten

$$\left(\frac{1}{T_1}\right)_0 = 4\pi^3 \gamma_e^2 \gamma_n^2 \hbar^3 |\psi_F(0)|^4 \frac{T}{T_F} \frac{1}{kT_F},$$

where $T_F = E_F/k$ is called the Fermi temperature.

The most remarkable feature of (7) is the proportionality of $1/T_1$ to the temperature T, which is to be contrasted with the much faster increase at low temperatures of relaxation times connected with lattice motions, such as thermal vibrations, diffusion, molecular rotation, etc.

Before comparing the theory with experiment we shall free ourselves from the two assumptions of nuclear spins $\frac{1}{2}$ and high applied fields.

B. Nuclear relaxation time and spin temperature

As recalled in the introduction to this chapter, a statistical description of a system of interacting nuclear spins by a spin temperature T_S, possibly different from that of the lattice, that is, by a density matrix $\rho \propto \exp(-\beta \mathcal{H}_0)$ with $\beta = 1/kT_S$ is, under certain conditions, a good approximation. Since the state of the spin system is then described by a single constant β it is reasonable to assume that the spin-lattice relaxation, that is, the trend of the spin temperature towards the lattice temperature, should be described by a single constant T_1 according to the relation

$$\frac{d\beta}{dt} = -\frac{1}{T_1}(\beta - \beta_0), \tag{8}$$

where $\beta_0 = 1/kT$ and T is the temperature of the lattice. It is clear that because of the Curie law, (8) coincides in high fields with the usual definition of T_1 through

$$\frac{dM_z}{dt} = -\frac{1}{T_1}(M_z - M_0).$$

With the assumption of spin temperature it will be possible to calculate the spin-lattice relaxation times for arbitrary nuclear spins and in arbitrary applied fields. This is done by calculating in two different ways the rate of change $d\bar{E}/dt$ of the average energy $\bar{E} = \text{tr}\{\rho \mathcal{H}_0\}$ of the nuclear spin system.

With the approximation, valid in practically all experimental situations, of high spin and lattice temperature

$$\bar{E} = \text{tr}\{\rho\mathcal{H}_0\} = \frac{\text{tr}\{\mathcal{H}_0 e^{-\beta\mathcal{H}_0}\}}{\text{tr}\{e^{-\beta\mathcal{H}_0}\}} \cong -\frac{\beta\langle\mathcal{H}_0^2\rangle}{\langle 1\rangle}, \qquad (9)$$

where $\langle\mathcal{H}_0^2\rangle = \text{tr}\{\mathcal{H}_0^2\}$, $\langle 1\rangle$ is the trace of the unit operator, and use has been made of $\langle\mathcal{H}_0\rangle = 0$. Thus

$$\frac{d\bar{E}}{dt} = -\frac{d\beta}{dt}\frac{\langle\mathcal{H}_0^2\rangle}{\langle 1\rangle}. \qquad (10)$$

The second evaluation of $d\bar{E}/dt$ is as follows (1). Let p_m be the populations (and p_m^0 the equilibrium populations) of the eigenstates $|m\rangle$ of the nuclear spin system (unknown in low fields). With the assumption of a high spin temperature

$$p_m = \frac{1 - \beta E_m}{\langle 1\rangle}.$$

The diagonal part of the master equation, which gives the rate of change of the populations, can be written

$$\frac{dp_m}{dt} = \sum_n W_{mn}\{(p_n - p_n^0) - (p_m - p_m^0)\}, \qquad (11)$$

where $W_{mn} = W_{nm}$ is the transition probability from the state $|m\rangle$ to the state $|n\rangle$, induced by the hyperfine coupling (1) summed over all electron and nuclear spins. Equation (11) can be rewritten as

$$\frac{dp_m}{dt} = \frac{1}{\langle 1\rangle}(\beta_0 - \beta)\sum_n W_{mn}(E_n - E_m). \qquad (11')$$

Multiplying both sides of (11') by E_m and summing over m, we get

$$\frac{d\bar{E}}{dt} = \frac{1}{\langle 1\rangle}(\beta_0 - \beta)\sum_{n,m} W_{mn} E_m(E_n - E_m)$$

or, from the relation $W_{mn} = W_{nm}$,

$$\frac{d\bar{E}}{dt} = -\frac{1}{2}\frac{1}{\langle 1\rangle}(\beta_0 - \beta)\sum_{n,m} W_{mn}(E_n - E_m)^2. \qquad (12)$$

A comparison of (8), (10), and (12) gives

$$\frac{1}{T_1} = \frac{1}{2}\frac{\sum_{n,m} W_{mn}(E_n - E_m)^2}{\langle\mathcal{H}_0^2\rangle}. \qquad (13)$$

Equation (13) is quite general and its validity is by no means restricted to nuclear spins relaxed by conduction electrons.

It will be shown now that although each individual transition probability W_{mn} cannot be computed in general since the states $|m\rangle$ of the nuclear spin system are not known, the expression (13) can be made to appear as a trace and calculated explicitly.

The elementary probability w_{mn} of a conduction electron passing from a state $|\mathbf{k}\rangle|s\rangle$ to a state $|\mathbf{k}'\rangle|s'\rangle$, $|s\rangle$ and $|s'\rangle$ being electron spin states, is given by

$$w_{mn} = \frac{2\pi}{\hbar}\left(\frac{8\pi}{3}\gamma_e\gamma_n\hbar^2\right)^2 \sum_{p,q}(\mathbf{k}|\delta(\mathbf{r}_p)|\mathbf{k}')(\mathbf{k}'|\delta(\mathbf{r}_q)|\mathbf{k}) \times$$
$$\times\{(m|\mathbf{I}_p|n).(s|\mathbf{s}|s')\}\{(s'|\mathbf{s}|s).(n|\mathbf{I}_q|m)\}\delta(E_i-E_f), \quad (14)$$

where $\mathbf{I}_p$ and $\mathbf{I}_q$ are two nuclear spins separated from the electron by $\mathbf{r}_p$ and $\mathbf{r}_q$. Multiplying (14) by $Z(\mathbf{k}).Z(\mathbf{k}')f(E_i)\{1-f(E_f)\}$, integrating over d^3k and d^3k', and summing over the spin states $|s\rangle$ and $|s'\rangle$, we get with the same simplifying assumptions as in Section A, namely high lattice temperature and spherical Fermi surface, the total probability W_{mn}:

$$W_{mn} = \sum_{p,q} a_{pq}(m|\mathbf{I}_p|n).(n|\mathbf{I}_q|m) \quad (15)$$

with
$$a_{pq} = \frac{64}{9}\pi^3\hbar^3\gamma_e^2\gamma_n^2|\phi(0)|^4\frac{\sin^2(k_F R_{pq})}{(k_F R_{pq})^2}kT[\rho(E_F)]^2, \quad (15')$$

where k_F, the wave number at the Fermi surface, is defined by

$$\hbar^2 k_F^2/2m = E_F.$$

To prove (15) use has been made of

$$\sum_{s,s'}(s|s_\alpha|s')(s'|s_\beta|s) = \mathrm{tr}\{s_\alpha s_\beta\} = \tfrac{1}{2}\delta_{\alpha\beta},$$

where s_α and s_β are components of the spin $\mathbf{s}$. From the expression (13) for $1/T_1$ and (15) for W_{mn} we get immediately

$$\frac{1}{T_1} = -\frac{1}{2}\sum_{p,q} a_{pq}\frac{\langle[\mathscr{H}_0,\mathbf{I}_p].[\mathscr{H}_0,\mathbf{I}_q]\rangle}{\langle\mathscr{H}_0^2\rangle}. \quad (16)$$

The relation (16) could be obtained immediately from the general master equation (66) of Chapter VIII:

$$\frac{d\sigma^*}{dt} = -\frac{1}{2}\int_{-\infty}^{\infty}\overline{[\mathscr{H}_1^*(t),[\mathscr{H}_1^*(t-\tau),\sigma^*-\sigma_0]]}\,d\tau$$

which, if there is a spin temperature, reads

$$\mathscr{H}_0\frac{d\beta}{dt} = \tfrac{1}{2}(\beta_0-\beta)\int_{-\infty}^{\infty}\overline{[\mathscr{H}_1^*(t),[\mathscr{H}_1^*(t-\tau),\mathscr{H}_0]]}\,d\tau.$$

Multiplying both sides by $\mathscr{H}_0$ and taking the trace with respect to the nuclear spin variables, we get

$$\frac{d\beta}{dt} = -\frac{1}{2}\frac{(\beta_0-\beta)}{\langle\mathscr{H}_0^2\rangle}\int_{-\infty}^{\infty}\overline{\langle[\mathscr{H}_1^*(t),\mathscr{H}_0][\mathscr{H}_1^*(t-\tau),\mathscr{H}_0]\rangle}\,d\tau, \qquad (17)$$

which from the assumed form of the scalar hyperfine Hamiltonian $\hbar\mathscr{H}_1$, shows immediately that $1/T_1$ should be given by a relation of the form of (16), if the correlation time of $\mathscr{H}_1(t)$ is very short.

Among the coefficients a_{pq} given by (15'), which can be rewritten as $a_k = a_{(p-q)}$ since they depend only on the distance R_{pq}, the coefficient a_0 is the largest and, unless the wavelength $\lambda_F = 2\pi/k_F$ is abnormally long, all the other coefficients a_k can be neglected for a first approximation. In the description which uses the concept of the local field produced at the nuclei by the electrons, the neglect of the a_k for $k \neq 0$ corresponds to the assumption that the local electronic fields at two different nuclei are incoherent. From (15') we see that $a_0 = (1/T_1)_0$, as given by (7), whence from (16)

$$\frac{1}{T_1} = \left(\frac{1}{T_1}\right)_0 \frac{\langle -\sum_p [\mathscr{H}_0, \mathbf{I}_p]^2 \rangle}{2\langle\mathscr{H}_0^2\rangle}. \qquad (18)$$

If we write $\mathscr{H}_0 = Z + \mathscr{H}_{SS}$, where Z is the Zeeman energy and

$$\mathscr{H}_{SS} = \mathscr{H}_d + \mathscr{H}_{ex}$$

is the sum of the dipolar spin-spin coupling and of an indirect scalar spin-spin coupling (if it exists), it is easily found from (18) that

$$\frac{1}{T_1} = \left(\frac{1}{T_1}\right)_0 \frac{\langle Z^2 + 2\mathscr{H}_{SS}^2\rangle}{\langle Z^2 + \mathscr{H}_{SS}^2\rangle}. \qquad (19)$$

This leads to a rather interesting conclusion (based solely on the assumption of incoherence between local fields 'seen' by two different nuclear spins): in high fields where $\langle Z^2 \rangle \gg \langle \mathscr{H}_{SS}^2 \rangle$ we have

$$(1/T_1) = (1/T_1)_0,$$

whilst in fields much smaller than the local field

$$(1/T_1) = 2(1/T_1)_0.$$

The opposite and less usual extreme of a complete correlation between the local fields 'seen' by neighbouring spins, that is, of a Fermi wavelength much larger than a lattice spacing, is expressed by the equality of all coefficients a_{pq} in (16) and yields

$$\frac{1}{T_1} \propto -\sum_{p,q}\frac{\langle[\mathscr{H}_0,\mathbf{I}_p]\cdot[\mathscr{H}_0,\mathbf{I}_q]\rangle}{\langle\mathscr{H}_0^2\rangle} \propto \frac{\langle Z^2+3\mathscr{H}_d^2\rangle}{\langle Z^2+\mathscr{H}_d^2+\mathscr{H}_{ex}^2\rangle}. \qquad (20)$$

It is easy to verify that for a dipolar coupling $\mathcal{H}_d$ between like spins, the expectation value $\langle \mathcal{H}_d^2 \rangle$ is $5 \langle \tilde{\mathcal{H}}_d^2 \rangle$, where $\tilde{\mathcal{H}}_d$ is the truncated dipolar Hamiltonian used for the calculation of the second moment of the nuclear resonance line in Chapter IV. It follows that if scalar spin-spin coupling is absent and a single spin species is present in the sample, the equations (19) and (20) can be replaced by

$$TT_1 \propto [H^2 + \tfrac{5}{3}\Delta H^2]/[H^2 + \tfrac{5}{3}\beta \Delta H^2], \qquad (20')$$

where β is 2 for uncorrelated and 3 for strongly correlated electronic fields at the nuclei, and ΔH^2 is the second moment of the resonance line.

An essential feature of the above calculations is the neglect of correlations between the individual conduction electrons. This approach is inadequate for an evaluation of the relaxation time in the superconducting state of the metal, where according to the present status of the theory, strong correlations exist between electrons.

It is not possible to give an account of the calculation of T_1 in the superconducting state without going into details of the theory of superconductivity that are outside the scope of this book. This calculation, which can be found in reference (1), predicts for the relaxation rate $1/T_1$ in zero field, a steep rise below the critical temperature T_c, followed by a decrease as the temperature goes down.

Finally, an important relation can be established between the relaxation time T_1 as given by (7) and the Knight shift given by eqn. (77) of Chapter VI:

$$T_1 \left(\frac{\Delta H}{H_0}\right)^2 = \frac{1}{\pi(kT)} \frac{\chi_p'^2 N^2}{\gamma_e^2 \gamma_n^2 \hbar^3 [\rho(E_F)]^2}, \qquad (21)$$

where χ_p' is the paramagnetic susceptibility per conduction electron. If the model of independent electrons is used then, as is well known (and will be shown in the next section), we have

$$\chi_p' = \frac{(\gamma_e \hbar)^2}{2N} \rho(E_F), \qquad (21')$$

whence the so-called Korringa relation:

$$T_1 \left(\frac{\Delta H}{H_0}\right)^2 = \frac{\hbar}{4\pi kT}\left(\frac{\gamma_e}{\gamma_n}\right)^2. \qquad (22)$$

The independent electron approximation is known to give incorrect results for the electronic paramagnetic susceptibility in contrast with the so-called collective theory (2). It would therefore seem advisable to use in (21) those values χ_s and ρ_s for the susceptibility and the density of states, taken from this theory. Eqn. (22) must then be replaced by

$$T_1\left(\frac{\Delta H}{H_0}\right)^2 = \frac{\hbar}{4\pi kT}\left(\frac{\gamma_e}{\gamma_n}\right)^2\left(\frac{\chi_s}{\chi_0}\right)^2\left[\frac{\rho_0(E_F)}{\rho_s(E_F)}\right]^2 \qquad (22')$$

where ρ_0 and χ_0 refer to the independent electron approximation.

C. Dynamic nuclear polarization in metals (the Overhauser effect)

(a) Fermi statistics and non-equilibrium electron spin distribution

In the calculations above it was assumed that the electronic spins were in thermal equilibrium with the lattice and that the fluctuating fields they produced at the nuclei could be considered as a part of the 'lattice' (as explained in Section II F (b) (3) of Chapter VIII). There are, however, situations where the electronic spins are not in equilibrium with the lattice, as for instance in dynamic polarization experiments when they are being driven by an r.f. field at the electronic Larmor frequency. Such situations have already been examined in Chapter VIII where, using the semi-classical model of random functions, it was shown that, in a liquid, for a nuclear spin **I** coupled to an electronic spin **S** by a bilinear coupling $\mathbf{I}.\mathscr{A}(t).\mathbf{S}$, the rate of change of

$$\langle i_z \rangle = \langle I_z \rangle / I(I+1)$$

is given by the formula

$$\frac{d\langle i_z \rangle}{dt} = -\frac{1}{T_1}\{\langle i_z \rangle - i_0 + \xi(\langle s_z \rangle - s_0)\}, \qquad (23)$$

where $\langle s_z \rangle = \langle S_z \rangle / S(S+1).$

For a scalar coupling and a very short correlation time, it was shown that $\xi = -1$.

At first sight we should expect equation (23) to hold, with the same value of ξ, for nuclear spins in a metal, where these conditions are realized for their coupling with conduction electrons. That this is actually not so, as will appear shortly, is due to the fact that, because of the exclusion principle, the conduction electrons in metals follow Fermi statistics. It is sometimes argued that this complication can be removed if instead of considering statistics of individual electrons the Gibbs statistical approach is taken, as explained in Chapter V. The macroscopic system made of all the electrons of a sample, when in thermal equilibrium, obeys Boltzmann statistics and is described by a statistical operator $\rho \propto \exp\{-\hbar\mathscr{F}/kT\}$, where $\hbar\mathscr{F}$ is the total Hamiltonian of the electrons including their interactions.

Although this statement is undoubtedly correct, it should be used with some circumspection as exemplified by the following erroneous calculation.

If in order to calculate the electron spin susceptibility we repeat a reasoning used in Chapter III, and write the electron Hamiltonian in the presence of a magnetic field H_0 as $\hbar\mathcal{H} = \hbar\mathcal{F} - H_0 M_z$, we get

$$\langle M_z \rangle = \frac{\text{tr}\{\exp(-\hbar\mathcal{H}/kT) \cdot M_z\}}{\text{tr}\{\exp(-\hbar\mathcal{H}/kT)\}}. \tag{24}$$

If we assume that $\mathcal{F}$ is independent of the spins and thus commutes with M_z and that $|M_z H_0| \ll kT$, both legitimate assumptions, we get

$$\langle M_z \rangle \simeq \frac{\text{tr}\{M_z^2\}}{\text{tr}\{1\}} \cdot \frac{H_0}{kT}, \tag{24'}$$

that is, the Curie law, well known to be invalid for the paramagnetism of conduction electrons. The error is due to the fact that the sum over states expressed by the traces (24) or (24') should be restricted according to the Pauli principle to eigenstates of $\mathcal{H}$ that are completely antisymmetrical with respect to the orbital and spin coordinates of all electrons, whereas the trace technique takes in all eigenstates of $\mathcal{H}$, antisymmetrical or not. If the trace technique is to be retained the operation $\text{tr}\{\rho M_z\}$ should be replaced by $\text{tr}\{P\rho M_z P\}$, where P is the projection operator over antisymmetrical states only. This operator clearly contains spin variables, thus making it possible for the result (24') to be wrong. For that reason, whilst keeping the general Gibbs approach in mind, and using it with caution for general arguments, we shall return for practical calculations to the one-electron description.

In the presence of a magnetic field $H_z = H_0$, the Fermi distribution function $f = [1 + \exp\{(E - E_F)/kT\}]^{-1}$ describing electrons in thermal equilibrium has to be replaced by two functions $f_\pm$, one for each sign of S_z,

$$f_\pm = \left[1 + \exp\left(\frac{E \pm \tfrac{1}{2}\hbar\omega_e - E_F}{kT}\right)\right]^{-1}, \tag{25}$$

where E represents the kinetic energy of the electron and $\omega_e = -\gamma_e H_0$. If $|\hbar\omega_e| \ll kT$, $f_\pm \simeq f \pm \tfrac{1}{2}\hbar\omega_e(df/dE)$, where f is given by (2'), and df/dE by (3),
$$f_\pm = f \mp (\tfrac{1}{2}\hbar\omega_e)\delta(E - E_F).$$

The total magnetization of the N electrons of the sample will be

$$M = \tfrac{1}{2}\gamma_e \hbar \int \rho(E)[f_+ - f_-]\,dE = H_0(\tfrac{1}{2}\gamma_e^2 \hbar^2)\rho(E_F) \tag{25'}$$

in accordance with (21').

The distribution (25) describes electronic spins in thermal equilibrium. The electronic spin relaxation times, being of the order of 10^{-10} second or longer, are much longer than the relaxation times τ relative to the electronic kinetic energy, which are of the order of 10^{-13} sec. It is therefore a reasonable approximation to assume that electrons of either spin are in equilibrium among themselves and are described by two distributions:

$$f_\pm(E) = \left\{1+\exp\left(\frac{E\pm\tfrac{1}{2}\hbar\omega_e - E_F^\pm}{kT}\right)\right\}^{-1} \tag{26}$$

with values $E_F^\pm$ for their Fermi energies, which are usually different and become equal when the spins are in equilibrium with the other degrees of freedom.

The disappearance of electron paramagnetism caused by the saturation of the electron resonance corresponds to $f_+ = f_-$, or

$$\tfrac{1}{2}\hbar\omega_e - E_F^+ = -\tfrac{1}{2}\hbar\omega_e - E_F^-, \qquad E_F^+ - E_F^- = \hbar\omega_e. \tag{26'}$$

If the departure of $E_F^\pm$ from the equilibrium value E_F is small compared with kT, the conservation of the total number of the electrons leads to $E_F^+ + E_F^- = 2E_F$, whence, writing $E_F^\pm = E_F \pm \tfrac{1}{2}\epsilon$, we get

$$f_\pm(E) = f\{E\pm\tfrac{1}{2}(\hbar\omega_e-\epsilon)\} \simeq f(E) \pm (\tfrac{1}{2}\hbar\omega_e - \tfrac{1}{2}\epsilon)\frac{df}{dE}$$

$$\simeq f \mp \tfrac{1}{2}(\hbar\omega_e - \epsilon)\delta(E-E_F). \tag{27}$$

The expectation value $\langle S_z \rangle$ for one electron is obtained from (27):

$$\langle S_z \rangle = \frac{1}{2N}\int (f_+ - f_-)\rho(E)\,dE = \frac{1}{2N}(\epsilon - \hbar\omega_e)\rho(E_F), \tag{28}$$

whence

$$\epsilon = E_F^+ - E_F^- = \frac{2N}{\rho(E_F)}\langle S_z \rangle + \hbar\omega_e = \frac{2N}{\rho(E_F)}\{\langle S_z \rangle - S_0\}, \tag{28'}$$

where the equilibrium electron polarization S_0 is given by

$$S_0 = \frac{-\hbar\omega_e \rho(E_F)}{2N} = \frac{\gamma_e \hbar H_0 \rho(E_F)}{2N}. \tag{28''}$$

Equation (28') can then be rewritten as

$$E_F^+ - E_F^- = \hbar\omega_e \frac{[S_0 - \langle S_z \rangle]}{S_0}. \tag{28'''}$$

We shall define the saturation parameter s by the relation

$$E_F^+ - E_F^- = s\hbar\omega_e. \tag{28^{iv}}$$

If the temperature is sufficiently high for the expansion (27) to be valid, according to (28'''),
$$s = \frac{S_0 - \langle S_z \rangle}{S_0}.$$

On the other hand, whatever the temperature, we can still define s by the relation (28^{iv}). Thermal equilibrium corresponds to $E_F^+ = E_F^-$, or $s = 0$, equality of the populations of the two spins levels to $s = 1$, intermediate situations to $0 < s < 1$.

(b) *Dynamic polarization*

To demonstrate the possibility of dynamic polarization in metals (Overhauser effect) the assumption of high temperature is not necessary, as will be shown below.

Suppose for simplicity that the nuclear spins are $\frac{1}{2}$. The extension to nuclear spins $I > \frac{1}{2}$ is straightforward if spin-spin interactions maintain among them a spin-temperature as explained previously. The equation for the rate of change of n_+, the population of nuclei with spins up, normalized to $n_+ + n_- = 1$, can be written

$$\frac{dn_+}{dt} = \frac{2\pi}{\hbar} \int dE \{|(i \mid \hbar \mathcal{H}_1 \mid f)|^2\} \rho(E) \rho(E - \hbar(\omega_e - \omega_n)) \times$$
$$\times \{n_- f_+(E - \hbar(\omega_e - \omega_n))[1 - f_-(E)] - n_+ f_-(E)[1 - f_+(E - \hbar(\omega_e - \omega_n))]\}. \tag{29}$$

From the definitions (26) of f_+ and f_-, writing that the curly bracket in (29) vanishes, we obtain for the steady-state value of n_+/n_-

$$\left(\frac{n_+}{n_-}\right)_{\text{st}} = \exp\left\{\frac{E_F^+ - E_F^- - \hbar\omega_n}{kT}\right\}. \tag{30}$$

If the electron spins are in thermal equilibrium $E_F^+ = E_F^-$, and n_+/n_- is given by the nuclear Boltzmann factor, $\exp(-\hbar\omega_n/kT)$, as it should be. If, on the other hand, the electron spin polarization is made to vanish by a saturating r.f. field at the electron frequency ω_e, according to (26'), $E_F^+ - E_F^- = \hbar\omega_e$ and the nuclear polarization is greatly enhanced since $|\omega_e| \gg |\omega_n|$:

$$\left(\frac{n_+}{n_-}\right) = \exp\left\{\frac{\hbar(\omega_e - \omega_n)}{kT}\right\}. \tag{31}$$

For incomplete saturation represented by the saturation parameter s $(0 < s < 1)$

$$\left(\frac{n_+}{n_-}\right) = \exp\left\{\frac{\hbar(s\omega_e - \omega_n)}{kT}\right\}. \tag{31'}$$

This is the Overhauser effect (3).

If ω_e and ω_n have opposite signs (positive nuclear moments) the dynamic polarization has the same sign as the thermal equilibrium one; it has the reverse sign if $\omega_n < 0$. For nuclear spins $I > \tfrac{1}{2}$ the ratio (31) or (31') is that existing between the populations of two adjacent states: (n_m/n_{m-1}). It is seen that the dynamic polarization obtained is independent of the fact that the electron spins obey Fermi statistics rather than Boltzmann statistics as in paramagnetic solutions. The reader will easily convince himself that the general argument of Chapter VIII, Section III B (b) (2), predicting such dynamic polarization, is still valid.

(c) Coupled equations for nuclear and electron spin polarization

We again make the assumption, abandoned in Section C (b), of high lattice temperature.

To obtain a rate equation similar to (23) for the polarization of the nuclear spin, we make a first-order expansion of (29), assuming that $\langle I_z \rangle = \tfrac{1}{2}(n_+ - n_-)$, $\langle S_z \rangle - S_0 = \epsilon \rho(E_F)/2N$ (given by (28')), $\hbar \omega_e/kT$, and $\hbar \omega_n/kT = -4 I_0$ are small quantities.

With a little algebra we obtain from (29)

$$\frac{d\langle I_z \rangle}{dt} = \frac{2\pi}{\hbar} \int dE |(i|\hbar \mathcal{H}_1|f)|^2 \rho^2(E) \times$$
$$\times \left\{ -2 \langle I_z \rangle f(E)[1-f(E)] + \frac{df}{dE} \tfrac{1}{2}(-\epsilon + \hbar \omega_n) \right\}. \quad (32)$$

From equation (3), $\qquad f(1-f) = -kT \dfrac{df}{dE},$

and from equation (28'),

$$\epsilon = \frac{2N}{\rho(E_F)} \{\langle S_z \rangle - S_0\}.$$

(32) can thus be rewritten

$$\frac{d\langle I_z \rangle}{dt} = \frac{2\pi}{\hbar} \int dE |(i|\hbar \mathcal{H}_1|f)|^2 \rho^2(E) \times$$
$$\times f(E)(1-f(E)) \left\{ -2\{\langle I_z \rangle - I_0\} + \frac{N}{kT\rho(E_F)} \{\langle S_z \rangle - S_0\} \right\} \quad (33)$$

or $\qquad \dfrac{d\langle I_z \rangle}{dt} = -\dfrac{1}{T_1}\{\langle I_z \rangle - I_0 + \xi\{\langle S_z \rangle - S_0\}\}, \qquad (34)$

where $\qquad \xi = -\dfrac{N}{2kT\rho(E_F)}$

or, for free electrons, $\qquad \xi = \dfrac{-2T_F}{3T}.$

If we use the value (28″) for S_0 and, for the nuclear thermal equilibrium polarization, the Boltzmann value

$$I_0 = \frac{\gamma \hbar I(I+1)H_0}{3kT} \quad \text{or, for } I = \tfrac{1}{2}, \quad I_0 = \frac{\gamma \hbar H_0}{4kT},$$

(34) can be rewritten as

$$\frac{d\langle I_z\rangle}{dt} = -\frac{1}{T_1}\left\{\langle I_z\rangle - I_0 - \frac{\gamma_e}{\gamma_n}I_0\frac{\langle S_z\rangle - S_0}{S_0}\right\}$$

$$= -\frac{1}{T_1}\left\{\langle I_z\rangle - I_0\left(1 - s\frac{\gamma_e}{\gamma_n}\right)\right\}. \tag{34'}$$

In contradistinction with the equation (23), the equation (34′) is independent of the type of statistics followed by the electron spins and is equally valid for paramagnetic ions or free radicals in liquids and conduction electrons in metals. It is easy to see that if $I > \tfrac{1}{2}$, (34′) is still valid. In particular, complete saturation of the electron resonance which makes $\langle S_z\rangle = 0$, $s = 1$, leads to the steady-state value for $\langle I_z\rangle$,

$$\langle I_z\rangle = I_0\left(1 - \frac{\gamma_e}{\gamma_n}\right),$$

that is, to the same value as for a scalar coupling between an electronic and a nuclear spin in a liquid.

In equation (34) the fact that S_0 is smaller in the metal by a factor $3T/2T_F$ because of Fermi statistics is exactly compensated by the fact that $|\xi|$ is larger in the metal by the same amount. This compensation is naturally not fortuitous, and was to be expected from the general arguments on dynamic polarization.

As has already been discussed in Chapter VIII, if the nuclear spins have other relaxation mechanisms besides their coupling with the electrons, with a relaxation time T_1', an extra term or leakage term

$$-\frac{1}{T_1'}\{\langle I_z\rangle - I_0\}$$

must be added on the right-hand side of (34) or (34′), reducing the maximum dynamical polarization by the factor

$$f = \left(\frac{1}{T_1}\right)_e \bigg/ \left(\frac{1}{T_1}\right)_{\text{tot}}, \tag{34″}$$

where $(1/T_1)_e$ is the inverse relaxation time due to the coupling with the conduction electrons and $(1/T_1)_{\text{tot}} = (1/T_1)_e + 1/T_1'$ the total inverse relaxation time.

The steady-state value of $\langle I_z \rangle$ can thus be written

$$\langle I_z \rangle = I_0 \left\{ 1 - sf \frac{\gamma_e}{\gamma_n} \right\}. \tag{35}$$

D. Comparison with experiment

(a) Measurements of T_1

The first feature of the above theory to be checked experimentally is the proportionality of T_1 to $1/T$ over a wide range of temperatures.

Table I (4) shows a set of data obtained at liquid helium temperatures, using the fast passage technique described in Chapter III.

TABLE I

Metals	H (gauss)	$T_1 T$ (sec. deg.)	T (°K)
Li^7	800	43 ± 2	$4 \cdot 2 - 1 \cdot 3$
Na^{23}	1030	$5 \cdot 1 \pm 0 \cdot 3$	$4 \cdot 2 - 1 \cdot 15$
Al^{27}	1020	$1 \cdot 77 \pm 0 \cdot 1$	$4 \cdot 2 - 1 \cdot 2$
Cu^{63}	240	$1 \cdot 27 \pm 0 \cdot 1$	$4 \cdot 2 - 1 \cdot 35$
Cu^{65}	240	$1 \cdot 04 \pm 0 \cdot 1$	$2 \cdot 04 - 1 \cdot 35$

It is seen that the constancy of $T_1 T$ is very well verified in this range of temperatures where all other relaxation mechanisms such as quadrupole interactions or nuclear spin-spin coupling modulated by translational diffusion are probably quite inefficient.

On the other hand, if one attempts to verify the constancy of $T_1 T$ in the high-temperature range, care must be taken to subtract the contributions to $(1/T_1)$ due to these mechanisms. The procedure used for isolating in $(1/T_1)$ the contribution $(1/T_1)_e$ from conduction electrons will be described later. In measurements of T_1, using the method of a pulse of 180° followed by a pulse of 90°, described in Chapter III, the corrected product $T(T_1)_e$ turns out to be 4·8 for Na^{23} (measured at 9 Mc/s) and 44·6 for Li^7 (measured at 15 Mc/s) between roughly 0° C and 260° C (5). The agreement with the low-temperature values of Table I is excellent. For Li^6 in the same range of the temperatures it is found (5) that
$$(T_1)_e T = 290.$$

The ratio $(T_1 T)_{\text{Li}^7}/(T_1 T)_{\text{Li}^6} = 6 \cdot 5$ is in good agreement with the ratio $[\gamma(\text{Li}^7)/\gamma(\text{Li}^6)]^2 = 7$, to be expected if nuclear relaxation is caused by conduction electrons since the modulus $|\psi(0)|$ of the electronic wave function should be practically the same near a nucleus of Li^6 and Li^7.

A second point of the theory that can be compared with experiment is the dependence of T_1 on the applied field, as given by (20′) where β is the neighbourhood of 2.

This comparison should be made at temperatures that are sufficiently low for the other relaxation mechanisms such as diffusion that may be frequency-dependent to be negligible.

For Li^7 at 1·3° K and Na^{23} at 1·1° K an excellent agreement with experiment is reached as shown in Fig. IX, 1A by making $\beta = 2·2$ for Li^7 and 2·28 for Na^{23} in (20′) (4).

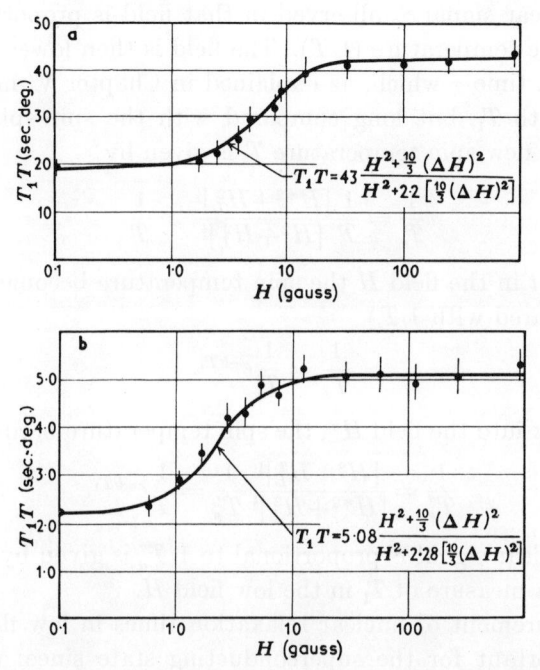

FIG. IX, 1A. (a) Magnetic field dependence of the nuclear spin relaxation time in Li^7 at 1·3° K. The experimental points are indicated by the dots and bars. The $(\Delta H)^2$ used is 4·3 gauss^2. (b) Magnetic field dependence of the nuclear spin relaxation time in Na^{23} at 1·1° K. The $(\Delta H)^2$ used is 0·63 gauss^2.
Note added at second impression. The theoretical curves were plotted using an incorrect formula. When the correct formula (20′) is used, the agreement with experiment is less good.

On the other hand, a serious disagreement between theory and experiment occurs in aluminium, where the experimental points can be made to fit the theory only if the second moment ΔH^2 is taken to be about three times the value predicted by theory or determined experimentally, and in copper where the experimental data do not appear to match the theoretical curve even in form. For either metal $(T T_1)$ decreases steadily by a factor of the order of 3, from high to very low field.

The agreement with theory for lithium and sodium gives reasonable confidence in the validity of the theory and leads us to believe that the discrepancies in copper and aluminium may be explained by the presence of quadrupole effects or impurities.

The principle of the measurement of the nuclear relaxation time in low fields is as follows.

Since the nuclear spins are in equilibrium with the lattice in a high field H^*, much larger than the local field H_L, the Curie law is valid and the nuclear signal S_0 observed in that field is proportional to the inverse lattice temperature $(1/T)$. The field is then lowered to a value $H \ll H^*$ in a time τ which, as explained in Chapter V, must be short compared with T_1, but long compared with the spin-spin relaxation time T_2. The new spin temperature T_s is given by

$$\frac{1}{T_s} = \frac{1}{T} \frac{[H^{*2}+H_L^2]^{\frac{1}{2}}}{[H^2+H_L^2]^{\frac{1}{2}}} \gg \frac{1}{T}.$$

After a time t in the field H the spin temperature becomes (neglecting $1/T$ as compared with $1/T_s$)

$$\frac{1}{T'_s} = \frac{1}{T_s} e^{-t/T_1}.$$

If we go back into the field H^*, the spin temperature becomes

$$\frac{1}{T''_s} = \frac{[H^2+H_L^2]^{\frac{1}{2}}}{[H^{*2}+H_L^2]^{\frac{1}{2}}} \frac{1}{T'_s} = \frac{1}{T} e^{-t/T_1}.$$

The new nuclear signal S proportional to $1/T''_s$ is given by $S = S_0 e^{-t/T_1}$ and affords a measure of T_1 in the low field H.

The measurement of nuclear relaxation times in low fields is particularly important for the superconducting state since: (i) superconductivity is destroyed in most superconductors by a magnetic field larger than a few hundred gauss; (ii) the magnetic field does not penetrate inside a superconductor.

In the measurements performed on aluminium (1), the metal is in the normal state in the field H^* where the nuclear signal is observed but it is superconducting in the low field where relaxation occurs. The results exhibit the dependence on temperature shown in Table II. In

TABLE II

Temperature (°K)	T/T_c	$(1/T_1)_s/(1/T_1)_n$
1·16	0·99	1·1±0·2
1·127	0·963	1·6±0·2
1·06	0·906	1·55±0·2
0·945	0·81	2·2±0·3
0·58	0·5	1

this table T_c is the critical temperature for the superconducting transition and the indices n and s refer to normal and superconducting states. This behaviour of $(1/T_1)_s/(1/T_1)_n$ is in agreement with the theory of superconductivity of Bardeen, Cooper, and Schrieffer.

For a check of the absolute value of T_1, given by the theory, we compare experimental values of T_1 with those extracted using experimental values of the Knight shift $\Delta H/H$ from the Korringa relation (22) or its improved version (22'). The results, reduced to a temperature of 300° K, are shown in Table III (2). The experimental values are taken

TABLE III

	Exp. (millisec)	Eqn. (22)	Eqn. (22')
Li[7]	150±5	88	232
Na[23]	15·9±0·3	10·3	18·1
Rb[85]	2·75±0·2	2·1	2·94
Cu[63]	4·1±0·6	2·3	4·0
Al[27]	6·0±0·1	5·1	6·5

from low-temperature measurements of reference (4), except for Rb[85] taken from (5).

In all cases the free electron theory gives values shorter than the experimental values. This discrepancy is in the wrong direction. As explained in Section A the theoretical values should be, if anything, longer than the experimental values since the theory neglects non-scalar couplings between electronic and nuclear spins. In that respect the theoretical values (22') are more satisfactory since they are nearer to and systematically longer than the experimental values.

(b) *Dynamic polarization experiments*

Finally, the dynamic nuclear polarization resulting from the saturation of the electron spin resonance was found to be in good agreement with theoretical predictions (6, 7). In these experiments the nuclear resonance was observed while the electron resonance was simultaneously saturated. A low nuclear frequency (50 kc/s) and accordingly low applied fields H_0 were used in order to bring the electron frequency into the hundred Mc/s range.

Although the low nuclear resonance frequency resulted in poor signal-to-noise ratio, this drawback was offset by the possibility of obtaining much greater electronic saturation in the 100 Mc/s range than in the microwave range, thanks to the larger r.f. power available, and also, as will appear shortly, because of the larger skin depth, at the lower frequency.

In Li⁷ the various parameters of the experiment were:

Applied field $H_e = 30\cdot 3$ gauss
Electron frequency $= 84$ Mc/s
Electron half width at half intensity $\Delta H = 2\cdot 5$ gauss
Maximum amplitude of the rotating component of the saturating r.f. field $H_1 = 3\cdot 3$ gauss
Temperature of the sample (increased by r.f. heating) $T = 70°$ C

The skin depth δ for the electronic frequency was of the order of $10\,\mu$ under these conditions and metallic particles of dimensions smaller than δ could be made without difficulty.

The normal signal of Li⁷ was below the noise level and was estimated by comparison with a proton signal in a sample of glycerine used as a calibrator. From the formula (35) (where we neglect unity as compared with $|\gamma_e/\gamma_n|$) we find the enhancement of the nuclear signal to be

$$A = \frac{\langle I_z \rangle}{I_0} = fs\left|\frac{\gamma_e}{\gamma_n}\right|.$$

If we assume that the electron magnetization obeys the Bloch equations with relaxation times τ_1 and τ_2 (we use the symbols τ_1, τ_2 for the electronic resonance, and T_1, T_2 for the nuclear resonance), we get

$$s = \frac{S_0 - \langle S_z\rangle}{S_0} = \frac{\gamma_e^2 H_1^2 \tau_1 \tau_2}{1 + \gamma_e^2 H_1^2 \tau_1 \tau_2}.$$

Therefore, plotting against $1/H_1^2$ the inverse $1/A$ of the observed enhancement A, a straight line should be and was obtained. The intersection of this line with the $(1/A)$ axis gives the inverse maximum enhancement reached for $H_1^2 \to \infty$, $s = 1$. It is equal to $|f(\gamma_e/\gamma_n)|^{-1}$. The enhancement $A_{\frac{1}{2}} = \frac{1}{2}(A_{\max})$ is reached for

$$\gamma^2 H_1^2 \tau_1 \tau_2 = \frac{\gamma H_1^2 \tau_1}{\Delta H}.$$

The fact that $A_{\frac{1}{2}}$ was actually reached approximately for $(H_1)_{\frac{1}{2}} = 2\cdot 3$ gauss $\simeq \Delta H$, demonstrated that to within experimental error, for the electron resonance,

$$\tau_1 = \tau_2 \quad \text{and} \quad \frac{S_0 - \langle S_z\rangle}{S_0} = \frac{1}{1 + [(H_1)_{\frac{1}{2}}/H_1]^2} \simeq \frac{1}{1 + (\Delta H/H_1)^2}.$$

The largest enhancement actually observed for an r.f. field, $H_1 = 3\cdot 3$ gauss, was approximately 110 with

$$s = \frac{1}{1 + (2\cdot 3/3\cdot 3)^2} \simeq 0\cdot 7.$$

The maximum enhancement to be expected for complete saturation is

$$A_{\max} = 110/0{\cdot}7 \simeq 157$$

and the leakage coefficient

$$f = \left|\frac{\gamma_e}{\gamma_n}\right| \bigg/ A_{\max} = \frac{157}{1690} \simeq 0{\cdot}09.$$

The extra relaxation mechanism responsible for this leakage is the translational diffusion of the lithium atoms. At 70° C the correlation time τ_c for this process is of the order of 10^{-7} sec, or less (5), and the product

$$\omega_n \tau_c = 2\pi \times 5 \,.\, 10^4 \times 10^{-7}$$

is a very small number. It follows that the contributions $1/T_1'$ and $1/T_2'$ of the diffusion process to longitudinal and transverse nuclear relaxation are equal (extreme narrowing). Since the same is true for the nuclear relaxation by conduction electrons, the total relaxation time $(T_1)_{\text{tot}} = (T_2)_{\text{tot}} = 6$ millisec (from the observed line width of Li^7). On the other hand, from Table I we find at 70° C $(T_1)_e \simeq 130$ millisec. This gives $f \simeq 6/130 \sim 0{\cdot}05$ as compared with a value of $0{\cdot}09$ found from the maximum enhancement $A_{\max}$. In view of the uncertainties involved in the evaluation of the various parameters the agreement can be considered as satisfactory.

In sodium the situation is somewhat different. The relaxation by conduction electrons is dominant and little leakage is to be expected. On the other hand, at 70° C the electronic line width was of the order of 12 gauss and the maximum value of H_1 at the electronic frequency of 124 Mc/s of the order of 1 gauss. The maximum enhancement expected $\simeq |\gamma_e/\gamma_n|(H_1/\Delta H)^2$ was thus of the order of 20. An enhancement of 10 was actually observed (7).

(c) *Dynamic nuclear polarization in metals at the temperatures of liquid helium*

In order to obtain nuclear polarizations of the order of a few per cent, dynamic polarization experiments have to be performed in fields of several thousands gauss and at liquid helium temperatures. Such experiments have a number of special features worth mentioning.

Each conduction electron 'sees' a nuclear field H_n produced by the nuclear spins. The value H_n of this field is given by

$$H_n = \frac{8\pi}{3} |\psi^2(0)| \gamma_n \hbar \langle I_z \rangle. \tag{36}$$

It is related to the Knight shift K, given by formula (77) of Chapter VI, by

$$H_n = \frac{KN}{\chi_p}\gamma_n \hbar \langle I_z \rangle, \qquad (37)$$

where χ_p is the electron spin susceptibility per unit volume and N the number of nuclei (and conduction electrons) per unit volume.

For instance, if one assumes a complete nuclear polarization $|\langle I_z \rangle| = I$ (which would be reached approximately by a complete Overhauser effect in an applied field of say 25,000 gauss, at 1° K), the following values are obtained for $|H_n|$.

For Li^7, where $K \simeq 2 \cdot 5 \times 10^{-4}$ and $\chi_p \simeq 2 \cdot 1 \times 10^{-6}$,

$$|H_n| \simeq 80 \text{ gauss}.$$

For Na^{23}, where $K \simeq 1 \cdot 1 \times 10^{-3}$ and $\chi_p \simeq 10^{-6}$,

$$|H_n| \simeq 310 \text{ gauss}.$$

H_n has the same sign as γ_n and would thus result in a positive shift of the electron frequency for positive nuclear moments.

Such an important shift can be used to detect the Overhauser effect and measure the amount of nuclear polarization (**3**, **10**). On the other hand, if $\langle I_z \rangle$ has its thermal equilibrium value,

$$\langle I_z \rangle = \frac{\gamma_n \hbar I(I+1)}{3kT} H_0,$$

the relative change in the electron frequency is then

$$D = \frac{H_n}{H_0} = \frac{KN}{\chi_p}\frac{\gamma_n^2 \hbar^2 I(I+1)}{3kT} = K\frac{\chi_{\text{nucl}}}{\chi_p}.$$

It is always positive and has the following values:

$$\text{for Li}^7, \qquad D \simeq \frac{5 \cdot 6}{T} \times 10^{-6},$$

$$\text{for Na}^{23}, \qquad D \simeq \frac{1 \cdot 45}{T} \times 10^{-5}. \qquad (37')$$

In an incomplete Overhauser effect with a saturation parameter s the relative electron frequency shift will be

$$D = \left(1 - s\frac{\gamma_e}{\gamma_n}\right) K \frac{\chi_{\text{nucl}}}{\chi_p}. \qquad (38)$$

It should be noted that (38) is valid as long as the enhanced nuclear polarization is still sufficiently small to have

$$\left|1 - s\frac{\gamma_e}{\gamma_n}\right| \frac{\hbar \omega_n}{kT} \ll 1.$$

Operating at very low temperatures has another advantage; the increase of the electron relaxation time makes the electron resonance easier to saturate.

Thus in sodium the electronic relaxation time $\tau_1 = \tau_2 = \tau$ is inversely proportional to the absolute temperature and at 4° K has the value 6×10^{-7} sec which corresponds to an electric line width of the order of 0·1 gauss (8). (There is reasonable evidence that the lack of dependence of τ on temperature, observed in lithium (8), is an effect of impurities.)

On the other hand, at low temperatures and for high applied fields there are difficulties due to the lack of penetration of the saturating microwave field inside the metal, because of the very small skin depth. The classical formula for the skin depth, $\delta = c/\sqrt{(2\pi\sigma\omega)}$ where c is the velocity of light and σ the conductivity of the metal, gives $\delta = 0\cdot1\,\mu$ for sodium at 10,000 Mc/s and 4° K.

Besides the difficulty of making such small particles, one must then consider the phenomenon of surface relaxation, due to collisions of the electron against the boundaries of the grains, which is more frequent as their size decreases. This introduces an extra cause of flip for the electron spin and may shorten τ considerably and broaden the electron line (8). The problem is further complicated by the diffusion of the conduction electrons in and out of the skin depth and also by the fact that the classical formula giving its value δ is no longer valid when the mean free path Λ of the electrons becomes much larger than δ.

A detailed discussion of these problems, which have been studied in connexion with the shape of the unsaturated electron resonance line (9), is outside the scope of this book. We shall be content to summarize briefly some results that can be obtained very simply if the Bloch equations describing the motion of the electron magnetization are supplemented by a diffusion term (Chapter III, eqn. 44), where the diffusion constant D is that of the conduction electrons, and are used together with the Maxwell equations to calculate the electron magnetization inside a metallic sample of dimensions large compared with the skin depth.

The most important fact is that while the microwave magnetic field is damped inside the metal over a distance of the order of the skin depth, the precessing transverse electron magnetization, and also the saturation $s = (M_0 - M_z)/M_0$ of its longitudinal component, penetrate much more deeply.

This penetration is measured by the diffusion length $L = (D\tau)^{\frac{1}{2}}$, where τ is the electron spin relaxation time and $D = \frac{1}{3}\Lambda v$, v being the average electron velocity. For sodium at $4°$ K, $\Lambda \sim 100\,\mu$, $v \sim 2 \times 10^8$ cm/sec, whence $D \simeq 2 \times 10^6$ cm^2/sec and $L = (D\tau)^{\frac{1}{2}} \simeq 1$ cm.

In other words, for a particle smaller than say $10\,\mu$, all components of the magnetization are uniform inside the grain within a tenth of 1 per cent. Since the r.f. field H_1 is localized in a depth of the order of δ whilst the r.f. magnetization is uniform over the dimensions d of the particle (with $d \gg \delta$), it is natural to expect, and a detailed calculation can show, that achieving a given degree of saturation s, which in an insulator would be obtained with an r.f. amplitude H'_1 such that $(\gamma H'_1 \tau)^2/(1+\gamma^2 H'^2_1 \tau^2) = s$, will require at the surface of the metal, under the conditions described, a field H''_1, d/δ times larger.

Finally, the theory of the anomalous skin depth shows that the value $\delta = c/\sqrt{(2\pi\sigma\omega)}$ should be replaced in the previous considerations by the anomalous skin depth $\Delta \simeq \delta(\Lambda/\delta)^{\frac{1}{3}}$.

It has been possible to obtain inside single crystals of LiF strongly irradiated with neutrons very small and very pure particles of metallic lithium of average size $1\,\mu$ and of average ESR line width 0·1 gauss.

Under these conditions appreciable saturation of the electron line could be obtained (10).

This line, displaced by the nuclear field by an amount given by (36), comes back to its original position at the rate of the nuclear relaxation $1/T_1$ (Fig. IX, 1 B).

The nuclear relaxation time of 9·8 sec found in this manner at $4·2°$ K is in good agreement with the results of Table I. From the displacement of the ESR line extrapolated to complete saturation a value of

$$\xi = \frac{\langle |\psi^2(0)| \rangle_F}{|\psi^2(0)|_{\text{atomic}}} = 0·44 \pm 0·015$$

could be obtained, in good agreement with the values quoted in Chapter VI (p. 204).

II. Nuclear Relaxation caused by Fixed Paramagnetic Impurities

The importance of the couplings between nuclear spins and unpaired electron spins (even in small concentrations), as nuclear relaxation mechanisms in liquids and gases, was discussed at some length in Chapter VIII. The effect of such couplings between nuclear spins and conduction electrons in metals was described in Section I of this chapter.

There is strong evidence, to be discussed later in this section, that in many non-metallic solids, such as ionic crystals, the nuclear

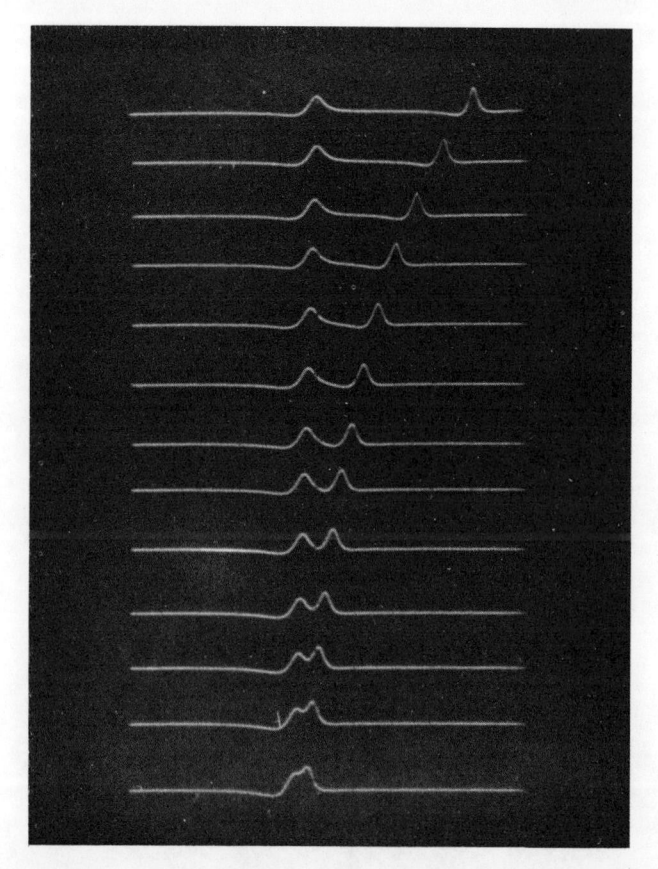

Fig. IX, I_B. Exponential decay of the dynamic nuclear polarization in Li metal at 4·2° K carrying the shifted E.S.R. line back to its normal position. Overall length of the trace 8·7 gauss, duration of the sweeps 0·02 sec, repetition time 2 sec. The undisplaced line on the left originates in metal grains too large for the E.S.R. line to be saturated and therefore to be displaced by the nuclear field.

relaxation is caused mainly by the presence of paramagnetic impurities in proportions sometimes as small as one part in a million.

The underlying theory is, however, different from that describing, say, the nuclear relaxation caused by dissolved paramagnetic ions in liquids. There, the rapid relative Brownian motion of nuclear and electron spins has two effects: first, the kinetic energy of thermal origin associated with this motion acts as a thermal bath for the spins, and its continuous spectrum ensures the conservation of energy when one or both of the interacting spins flip; secondly, the mobility of the electron spins inside the sample enables them, even under very small concentrations, to come at some time or other sufficiently near to the nuclear spins to relax them.

Both effects are lacking in crystals where electron and nuclear spins occupy fixed positions in space. The two mechanisms suggested to replace these effects are respectively electron relaxation and spin diffusion (11).

A. Theory

We have already discussed in Section II F of Chapter VIII the possibility of a bilinear coupling $\mathbf{S} . \hbar \mathscr{A} . \mathbf{I}$ acting as a nuclear relaxation mechanism in a liquid where the fluctuating character of the electronic field $\mathbf{H}_e(t) = -1/\gamma_I(\mathbf{S}.\hbar\mathscr{A})$ 'seen' by the nuclear spin I, is contained in the variation $\mathbf{S}(t)$ of the electronic spin, caused by electronic relaxation, rather than in the tensor $\hbar\mathscr{A}$ which remains constant in time. Since in a liquid the orientation, if not necessarily the length, of the vector $\mathbf{SI}$ joining the electronic and the nuclear spin, varies with a correlation time τ_c (of the order of 10^{-11} sec or shorter), the tensor coupling $\mathscr{A}$ can only remain time-independent if it is invariant under rotation, that is, if it is a scalar. The only process allowed by such a coupling is a simultaneous flip of the electronic and nuclear spins in opposite directions, a process that requires an energy $\hbar(\omega_S-\omega_I)$ to be taken from the spectrum of the fluctuating operator $\mathbf{S}(t)$, considered as a 'lattice' operator. As a consequence, the inverse of the nuclear relaxation time T_1 given by the first relation (127) of Chapter VIII, contains in the denominator the quantity $1+(\omega_S-\omega_I)^2\tau_2^2$, where τ_2 is the transverse electronic relaxation time.

A typical example of that situation was the relaxation of protons in water containing dissolved manganese ions Mn++ (reference (12) of Chapter VIII). For applied fields so weak that $|(\omega_S-\omega_I)\tau_2|$ is not a large number, the scalar coupling may be a more important nuclear relaxation mechanism in a liquid than the dipolar coupling, even if the latter has a larger absolute value, because the dipolar coupling is affected by the Brownian rotation which has a correlation time $\tau_c \ll \tau_2$.

On the other hand, in solids where the vector **SI** has a fixed orientation in space the dipolar coupling will be much more important than the scalar one, not only because it is usually much larger but also for the following reason. Among the various operators contained in a dipolar coupling there is the operator

$$C = -\tfrac{3}{2} \sin\theta \cos\theta \, e^{-i\phi} S_z I_+, \qquad (39)$$

where θ and ϕ define the orientation of the **SI** vector with respect to the applied field.

This operator may induce a flip of the nuclear spin unaccompanied by an electron flip, a process that requires an energy $\hbar\omega_I$ much smaller than the energy $\hbar(\omega_S \pm \omega_I)$ involved in the other processes that can occur through a bilinear coupling.

The nuclear relaxation time T'_1, due to this process is, by a calculation very similar to that outlined in Section II F (b) (3) of Chapter VIII (formula (125)), given by

$$\frac{1}{T'_1} = \frac{9}{2} \frac{\gamma_S^2 \gamma_I^2 \hbar^2 \sin^2\theta \cos^2\theta}{r^6} \int_{-\infty}^{\infty} S_z(0) S_z(t) e^{-i\omega_I t} \, dt$$

$$= \frac{3}{2} \frac{\gamma_S^2 \gamma_I^2 \hbar^2}{r^6} \sin^2\theta \cos^2\theta \, S(S+1) \frac{2\tau}{1+\omega_I^2 \tau^2}, \qquad (40)$$

where τ is the longitudinal electron relaxation time.

If $\omega_S \tau_2 \gg 1$, that is, if the electron resonance frequency is much larger than electron line width (which will be the case except in very low fields or for abnormally short electron relaxation times), it is clear that the process induced by the operator (39) will, in solids, by-pass all the other processes permitted by a bilinear coupling $\mathbf{S} \cdot \hbar \mathscr{A} \cdot \mathbf{I}$.

If in equation (40) we neglect the angular dependence, it can be rewritten as

$$\frac{1}{T'_1} = C r^{-6} \quad \text{with} \quad C \simeq \tfrac{2}{5} \gamma_S^2 \gamma_I^2 \hbar^2 S(S+1) \frac{\tau}{1+\omega_I^2 \tau^2}, \qquad (41)$$

or, for $|\omega_I \tau| = |\gamma_I H_0 \tau| \gg 1$,

$$\frac{1}{T'_1} = \frac{2}{5} \frac{S(S+1)}{\tau} \left(\frac{H_e}{H_0}\right)^2, \qquad (41')$$

where the electronic field H_e is defined as $H_e = \gamma_S \hbar / r^3$.

Consider as an example **(11)** potassium alum $KAl(SO_4)_2 \cdot 12H_2O$ where a small fraction of aluminium has been replaced by paramagnetic chromium Cr^{+++} (spin $S = \frac{3}{2}$). In a certain sample where the ratio $N(Al)/N(Cr)$ was 28,000, the proton resonance was observed in a field H_0 of 7500 gauss and the proton relaxation time T_1 was found by a saturation method to be 4 sec at 77° K. The electron relaxation time τ was estimated to be 5×10^{-7} sec. From (41′) we find that the electron field H_e for which T'_1 is equal to 4 sec is $H_e \sim 2$ gauss and the corresponding distance r_0 between the electronic and the nuclear spin is $r_0 \cong 20$ Å.

The value $H_e \sim 2$ gauss is of the same order of magnitude as the local field produced at a nucleus by the other nuclei, that is, of the order of the nuclear line width. Nuclei inside the critical radius r_0 where $H_e = H_{\text{local}}$ will have relaxation times shorter than 4 sec, but the nearer they are to the paramagnetic impurity, the larger is the local electronic field and consequently the larger the nuclear frequency shift (and broadening). Roughly speaking one may say that nuclei well inside the critical radius have short relaxation times but are not observed because their resonance frequency is shifted too much. On the other hand, if the concentration of paramagnetic impurities is small, the average distance R between an impurity and a nucleus, of the order of $N^{-\frac{1}{3}}$ (where N is the number of impurities per unit volume), will be much larger than r_0, and for most nuclei the relaxation time T'_1 given by (41′) will be much longer than the observed value.

Thus for the sample considered, R is of the order of

$$150 \text{ Å} \gg r_0 \cong 20 \text{ Å},$$

and for 90 per cent of the nuclei, the relaxation time as given by (41′) is longer than 10^4 sec.

There is a glaring discrepancy with the experimental value of 4 sec and it is clear that some mechanism must be imagined to carry to the remote nuclear spins the information on the lattice temperature dispensed by the electronic spins. This mechanism is the spin diffusion that occurs through the mutual flips between neighbouring spins and was shown in Chapter V to be described by a diffusion equation of the type

$$\frac{\partial p}{\partial t} = D \Delta p. \qquad (42)$$

The quantity p describes the nuclear polarization, or the spin temperature, at a point (actually in a small region having dimensions of the order of several lattice spacings) and the diffusion constant D is of the order of Wa^2, where a is the distance and W the probability of a flip-flop between nearest neighbours. Typical values of D and W are of the order of 10^{-13} cm²/sec and 10^3 sec⁻¹ respectively.

In the presence of paramagnetic impurities and of a saturating r.f. field H_1 which induces spin flips at a rate $A \simeq \gamma_I^2 H_1^2/\delta H$, where δH is the nuclear line width, the following transport equation can be written for $p(\mathbf{r}, t)$:

$$\frac{\partial p}{\partial t} = D\Delta p - C \sum_n \frac{1}{|\mathbf{r}-\mathbf{r}_n|^6}(p-p_0) - 2Ap, \qquad (43)$$

where the $\mathbf{r}_n$ are the positions of the impurities assumed to be distributed at random, and p_0 is the nuclear polarization for thermal equilibrium.

Since, according to (43), the magnetization is a function of time as well as space, we must first of all give an unambiguous definition of the relaxation time T_1.

An operational definition, convenient for the interpretation of saturation experiments, is $1/T_1 = 2A_{\frac{1}{2}}$, where $A_{\frac{1}{2}}$ is the value of the r.f. transition probability in equation (43) for which the steady-state value of the total nuclear magnetic moment of the sample is reduced by a factor 2 from its thermal equilibrium value (**11**). The steady-state solution of (43), where $A = A_{\frac{1}{2}}$, satisfies the relation

$$\int p(\mathbf{r})\,d^3r = \tfrac{1}{2}p_0 \int d^3r.$$

This definition of T_1 coincides with the usual definition derived from the elementary equation for r.f. saturation:

$$\frac{dn}{dt} = -\frac{1}{T_1}(n-n_0) - 2An.$$

We shall show that it is possible under fairly general conditions to obtain for T_1, thus defined, the very simple expression (**12, 27**)

$$2A_{\frac{1}{2}} = \frac{1}{T_1} = 4\pi NbD, \qquad (44)$$

where b is a certain length, which in many cases turns out to be of the order of the internuclear spacing a.

To solve the mathematical problem involved, consider first the steady state solution of the equation (43) in the absence of an r.f. field and with a single paramagnetic impurity, that is the solution of the equation

$$D\Delta p - \frac{C}{r^6}(p-p_0) = 0. \qquad (45)$$

For $r = 0$ one must have $p = p_0$, to express the fact that in the immediate neighbourhood of the impurity the nuclear spins are in thermal equilibrium. The solution of (45) that vanishes at infinity has the asymptotic form
$$p = p_0(b/r). \qquad (45')$$
For dimensional reasons b is necessarily equal to $(C/D)^{\frac{1}{4}}$, which has the dimension of a length, times a dimensionless factor, determined from the exact solution of (45), that matches (45′) at large distances and is equal to p_0 for $r = 0$. From the asymptotic expansion of this solution, which is a combination of Bessel functions, a straightforward but somewhat lengthy calculation yields
$$b = 0 \cdot 7(C/D)^{\frac{1}{4}}. \qquad (46)$$
From the expression (41) for C and the value $D \simeq Wa^2$, where a is the distance between two neighbouring nuclei, we get
$$b \sim a(H_e^0/H)^{\frac{1}{2}}(W\tau)^{-\frac{1}{4}}, \qquad (47)$$
where $H_e^0 = \gamma_e \hbar/a^3$ is the electronic local field at a distance of the order of a lattice spacing a.

For normal nuclear densities, a is of the order of a few Ångströms and H_e^0 at most a thousand gauss. $(H_e^0/H)^{\frac{1}{2}}$ will thus be rather smaller than unity and $(W\tau)^{-\frac{1}{4}}$, for $10^{-3} > \tau > 10^{-6}$ sec, will be of the order of a few units. In many cases the length b will thus be of the order of the average nuclear spacing a and much smaller than the average spacing between impurities: $R \sim N^{-\frac{1}{3}}$.

We consider now the diffusion equation in the presence of the r.f. field:
$$D \Delta p - \frac{C}{r^6}(p-p_0) - 2Ap = 0. \qquad (48)$$
With the same boundary condition as (45) the asymptotic form of its solution can be written
$$p = p_0 \frac{b}{r} e^{-k_0 r}, \quad \text{where} \quad k_0 = \left(\frac{2A}{D}\right)^{\frac{1}{2}}. \qquad (48')$$
The asymptotic length b in (48′) has the same value as in (46) because the product $k_0 b$ is very small, even for values of A sufficiently strong to induce appreciable nuclear saturation. Indeed, anticipating the result (44), the product $(k_0 b)$ where we make $A = A_{\frac{1}{2}}$ is equal to
$$(4\pi b^3 N)^{\frac{1}{2}} \sim \left(\frac{b}{R}\right)^{\frac{3}{2}} \ll 1.$$

The influence of the exponential $e^{-k_0 r}$ on the asymptotic solution (48'), or that of the term $2Ap$ in the equation (48), becomes appreciable only for $r \sim 1/k_0 \gg b$ and does not affect the determination of b.

It follows from the foregoing that except in small regions surrounding the impurities, which do not contribute to the nuclear signal, we can replace the exact equation (43) by a simpler equation, which has the same asymptotic features for distances far from the impurities, namely,

$$\frac{\partial p}{\partial t} = D\Delta p - 2Ap \qquad (|\mathbf{r}-\mathbf{r}_n| \geqslant b),$$

$$p = p_0 \qquad (|\mathbf{r}-\mathbf{r}_n| \leqslant b). \qquad (49)$$

We attempt to construct explicitly the steady-state solution of (49) by writing

$$p_s(\mathbf{r}) = p_0 \xi \sum_n g(\mathbf{r}-\mathbf{r}_n), \qquad (50)$$

where $g(\mathbf{r})$ is defined as

$$g(\mathbf{r}) = \frac{b}{r} e^{-k_0 r} \quad \text{for } r \geqslant b, \qquad g(\mathbf{r}) = e^{-k_0 b} \quad \text{for } r \leqslant b,$$

and ξ is a constant to be determined. The expression (50) does satisfy the approximate equation (49) for $|\mathbf{r}-\mathbf{r}_n| \geqslant b$. We now put it equal to p_0 for $|\mathbf{r}-\mathbf{r}_n| \leqslant b$. Take for origin one of the paramagnetic centres, say $\mathbf{r}_1$. We require that

$$p_0 \xi \left[e^{-k_0 b} + \sum_{n \neq 1} \frac{b}{|\mathbf{r}-\mathbf{r}_n|} e^{-k_0 |\mathbf{r}-\mathbf{r}_n|} \right] = p_0, \qquad (51)$$

for $r \leqslant b$, or, since $|r_n| \gg b$ and $k_0 b \ll 1$,

$$p_0 \xi \left[1 + \sum_{n \neq 1} \frac{b}{r_n} e^{-k_0 r_n} \right] = p_0. \qquad (52)$$

The equation (52) can be greatly simplified by noticing that the number n of impurities that contribute appreciably to the sum (52) is very large. It is of the order of the number of centres inside a sphere of a radius $1/k_0$, that is, $n \sim \frac{4}{3}\pi(N/k_0^3)$ or, with $k_0 = (2A/D)^{\frac{1}{2}}$ and

$$2A \simeq 2A_{\frac{1}{2}} = 4\pi NbD,$$

$$n \sim (Nb^3)^{-\frac{1}{2}} \sim \left(\frac{R}{b}\right)^{\frac{3}{2}} \gg 1.$$

Since very many centres contribute to (52) it is permissible to replace the sum

$$\sum_{n \neq 1} \frac{b}{r_n} e^{-k_0 r_n}$$

by the integral $\quad Nb \int \dfrac{e^{-k_0 r}}{r} d^3r = \dfrac{4\pi Nb}{k_0^2} = \dfrac{4\pi NbD}{2A}$

and to rewrite (52) as

$$p_0 \xi \left[1 + \dfrac{4\pi NbD}{2A}\right] = p_0, \quad \xi = \dfrac{2A}{2A+4\pi NbD}. \qquad (53)$$

The reduction in the total magnetization in the presence of the r.f. field is then obtained from (50) and (53):

$$\dfrac{\int p_s(r)\, d^3r}{p_0 \int d^3r} \cong \dfrac{N\xi p_0 \int g(r)\, d^3r}{p_0} = \dfrac{4\pi N\xi b}{k_0^2} = \dfrac{4\pi NbD}{2A+4\pi NbD}; \qquad (54)$$

whence $\quad 2A_{\frac{1}{2}} = \dfrac{1}{T_1} = 4\pi NbD.$

The approximations made in anticipating this result are thus justified *a posteriori*.

The three lengths that play an essential part in the process described by the equation (43) are

$$b \sim \left(\dfrac{C}{D}\right)^{\frac{1}{4}}, \quad R \sim N^{-\frac{1}{3}}, \quad \text{and} \quad \dfrac{1}{k_0} = L \sim R\left(\dfrac{R}{b}\right)^{\frac{1}{2}}.$$

The first might be called the scattering amplitude of a single impurity, the second the average distance between impurities, and the third the diffusion length during a relaxation time. The approximations made in the calculation rest on $b \ll R \ll 1/k_0$, which express respectively that outside a small troubled region, the nuclear magnetization diffuses freely and that a large number of impurities contribute to the nuclear magnetization in each point.

Instead of defining the relaxation time T_1 by means of the saturation behaviour of the nuclear resonance, it is possible to investigate the growth of the total nuclear magnetization starting say from an initial state at time $t = 0$, where it is zero. Starting from the equation (43), where we make $A = 0$, a solution is sought in the form

$$p(\mathbf{r}, t) = p_0[1+\theta(\mathbf{r})e^{-\mu t}], \qquad (55)$$

where $\theta(\mathbf{r})$ is a solution of the equation

$$D\Delta\theta - \sum_n \dfrac{C}{|\mathbf{r}-\mathbf{r}_n|^6}\theta + \mu\theta = 0 \qquad (56)$$

which must satisfy the condition

$$\int \theta(\mathbf{r})\, d^3r = -1 \quad \text{corresponding to} \quad \int p(\mathbf{r}, 0)\, d^3r = 0. \qquad (57)$$

The value of $\mu = 1/T_1$ is determined as the condition for the compatibility of (56) and (57). The calculation of μ is very similar to that of $2A_{\frac{1}{2}}$ performed previously, which is not surprising since (43), where we make $\partial p/\partial t = 0$, differs from (56) only by the change of $-2A$ into μ. Details of the calculation can be found in (12), together with the important result that it is indeed possible to find a solution of the form (55) and that the value $\mu = 1/T_1$ is the same as $2A_{\frac{1}{2}}$ given by (44).

The recovery of the total nuclear magnetization after, say, complete saturation should according to (55) be a single exponential with the same time constant T_1 as that measured by saturation.

B. Comparison with experiment

In an early attempt to test the foregoing theory, crystals were grown with known dilute magnetic content (11) (potassium alums with a known percentage of aluminium replaced by paramagnetic chromium). The equation (43) was solved numerically assuming for simplicity a *uniform* distribution of paramagnetic impurities, and the calculated values of $A_{\frac{1}{2}} = 1/T_1$ for the protons were plotted against τ. Two important features of the numerical values thus obtained, namely proportionality of $1/T_1$ to the concentration of impurities N and to $\tau^{-\frac{1}{2}}$, are in agreement with the explicit formulae (44) and (46) given for T_1 in Section II A.

The agreement with the experimental curves $1/T_1 = f(\tau)$ is qualitatively correct, the theory giving the right order of magnitude for T_1, although the experimental variation would be better described by $1/T_1 \propto \tau^{-m}$ with $m \simeq \frac{1}{2}$.

As another example of nuclear relaxation by dilute magnetic impurities some experiments on single crystals of lithium fluoride (13) may be quoted.

In a very pure single crystal at room temperature and in a field of 6400 gauss, the relaxation time of Li^7 was found to be 5 min and that of F^{19} 2 min. An irradiation of the crystal by X-rays, creating paramagnetic impurities such as F-centres, reduced these relaxation times to 30 sec for Li^7 and 10 sec for F^{19}.

Convincing evidence of the connexion between spin diffusion and relaxation was provided by a study of the relaxation time as a function of the orientation of this irradiated crystal with respect to the applied field.

In the simplified theory given in Section II A we neglected the anisotropy of the coefficient $D = Wa^2$ for spin diffusion. Actually the probability W of the mutual flip of two spins of, say, Li⁷ is the product of two factors: first, the square (B^2) of the matrix element of the Li⁷—Li⁷ dipolar interaction responsible for that flip; second, the shape function $g(\nu)$, inversely proportional to the line width of Li⁷, which in a crystal such as LiF with two magnetic ingredients is caused mainly by the interaction of Li⁷ with its nearest neighbours F¹⁹. When the angle of the field with the (100) plane of the crystal varies from 0° to 180°, it is found experimentally that T_1 is roughly proportional to the line width, the ratio between the maximum and the minimum value of either quantity being approximately 2.

The important point is that the squared matrix element B^2, originating in a coupling between like nuclei, and the line width caused mainly by a coupling between unlike nuclei, have a different angular dependence. However, in a crystal like CaF_2, where calcium nuclei have spin zero, these two quantities have the same origin and a similar angular dependence, and the angular variation of T_1 should be a good deal slower. The single exponential character of the growth of the signal towards equilibrium has also been well established in LiF.

Further corroboration of the theory is provided by the study of the relaxation time of F¹⁹ as a function of the nuclear frequency in a single crystal of LiF (for a fixed orientation of the crystal). According to eqn. (44) this variation should be

$$T_1 \propto [1+\omega_I^2 \tau^2]^{\frac{1}{4}}. \tag{57'}$$

T_1 was measured at 300° K and 77° K between 4 Mc/s and 42 Mc/s (**14**). In Fig. IX, 2, T_1^4, T_1^2, and T_1 have been plotted as functions of the square of the nuclear frequency at 300° K (curves A, B, C) and 77° K (curves A', B', C').

The fact that the curves A and A' are straight lines to within the experimental error is in excellent agreement with the relation (57').

From these measurements it is also possible to extract values for the electronic spin-lattice relaxation time τ of the paramagnetic impurities. In the crystal studied it is found, using (57'), that

$$\tau \cong 2\times 10^{-8} \text{ sec at } 300° \text{ K}; \quad \tau \cong 2\times 10^{-6} \text{ sec at } 77° \text{ K}.$$

These values should not be taken too seriously for there are several species of paramagnetic impurities in the crystal, F-centres, transition elements (iron and cobalt), etc.

As qualitative evidence of the spin diffusion mechanism it is interesting to compare the relaxation time measured in zero field with that extrapolated by (57') from high field measurements.

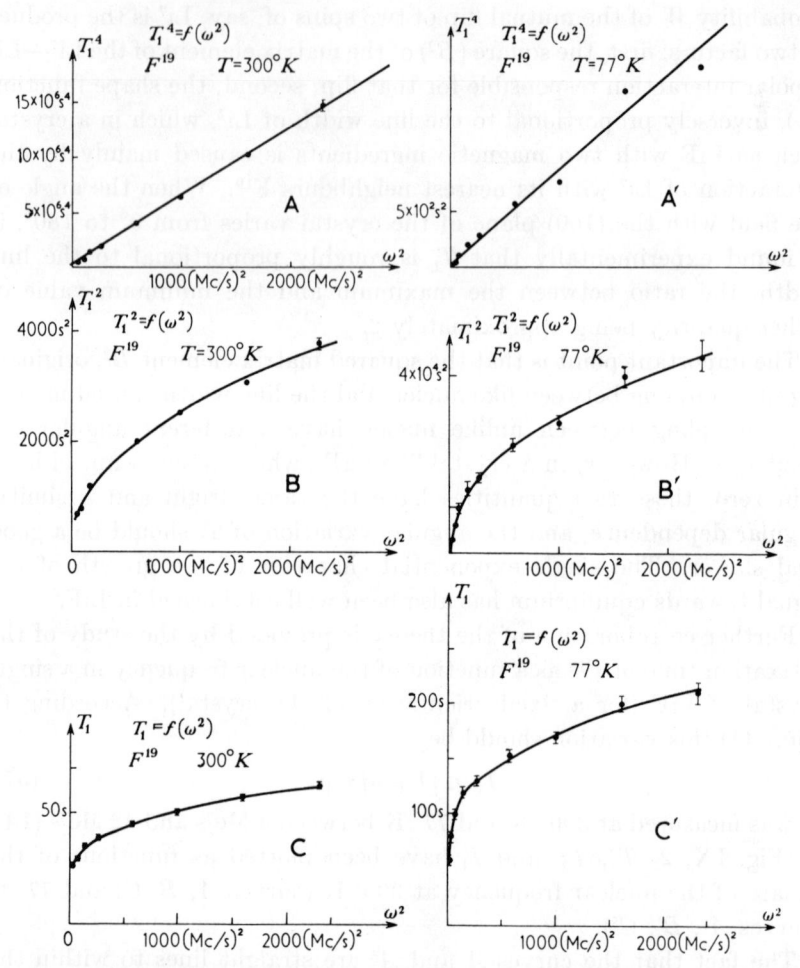

FIG. IX, 2. Spin-lattice relaxation of F^{19} in a single crystal of LiF, caused by paramagnetic impurities. The fourth, second, and first powers of T_1 have been plotted against the square of the frequency at 300° K (A, B, C) and at 77° K (A', B', C'). The fact that the plots A and A' are straight lines is in agreement with the theoretical formula (57′).

As has already been pointed out in Section I B there is no reason to expect the two values to be identical because of the role played by the spin-spin interaction in zero field. Still it is interesting to note that for F^{19} the ratio

$$\frac{T_1 \text{ (extrapolated)}}{T_1 \text{ (zero field)}}$$

was 5·6 in FLi and 2·3 only in F_2Ca at room temperature (14).

For FLi the fact that in zero field spin diffusion may occur between unlike spins (F and Li) as well as between like spins, in contradistinction to the high field case, is an extra mechanism for the shortening of the relaxation time that is lacking in F_2Ca where Ca^{40} has spin zero.

III. Magnetic Relaxation and Dynamic Polarization in Semiconductors and Insulators

A. Relaxation by conduction electrons in semiconductors

Two characteristic features of the nuclear relaxation by conduction electrons in metals, that make this mechanism very different from, say, relaxation by fixed paramagnetic impurities, are the Fermi statistics and the rapid motion of the electrons. The main consequences of these features are, respectively, the proportionality of the nuclear relaxation rate $1/T_1$ to the absolute temperature T, and the possibility of producing dynamic nuclear polarizations (Overhauser effect), as explained in Section I. Another, less essential, feature of this relaxation mechanism is the scalar character of the coupling $A(\mathbf{I}.\mathbf{S})$ assumed to exist between electronic and nuclear spins. The most spectacular change that would occur if this coupling were mainly dipolar as for p-electrons, rather than scalar as for s-electrons, would be a change in the sign of the dynamic enhancement of nuclear polarization, as explained in Chapter VIII.

A good example of a situation where conduction electrons obey Boltzmann rather than Fermi statistics is provided by semiconductors such as n-type silicon. The Boltzmann character of the statistics is determined by the small density of electrons in the conduction band making the electron gas non-degenerate.

We first calculate the nuclear relaxation time assuming spherical symmetry for the conduction energy band. We start from the formula (6) of Section I A where the assumptions on the statistics of the electrons have not yet been introduced. The elementary probability $w_{+,-}$ given by (6) must be summed over all the initial states $|\mathbf{k}\rangle$ weighted by the Boltzmann factors $e^{-\beta E}$, where $\beta = 1/kT$, and over all the final states $|\mathbf{k}'\rangle$, which, if we neglect the small change $\hbar(\omega_e - \omega_n)$ in the kinetic energy of the electron, must have the same energy as the initial state $|\mathbf{k}\rangle$.

We get

$$\frac{1}{2T_1} = W_{+-} = \frac{2\pi}{\hbar}\left\{\frac{8\pi}{3}\gamma_e\gamma_n\hbar^2\right\}^2 \frac{1}{4}\int |\phi_E(0)|^4 P(E)\rho^2(E)\,dE, \qquad (58)$$

where $\rho(E)$ is the density of states for electrons of a given spin orientation, $|\phi_E(0)|^2$ the electronic density at the nucleus for an electron of energy E (normalized to unity in a unit volume and thus a dimensionless quantity), and, according to Boltzmann statistics, $P(E) = ae^{-\beta E}$. The constant a is so chosen that

$$\int \rho(E) P(E)\, dE = a \int \rho(E) e^{-\beta E}\, dE = \tfrac{1}{2} N. \tag{59}$$

In (59), N is the total number of conduction electrons per unit volume and the normalization of (59) to $\tfrac{1}{2}N$, rather than N, results from the fact that only one-half of the electrons, those with spin $+$, may contribute to a nuclear flip from $-$ to $+$. In the free electron approximation the density of states $\rho(E)$ is given by

$$\rho(E) = \frac{4\pi p^2\, dp}{(2\pi\hbar)^3\, dE}, \quad \text{where} \quad E = \frac{p^2}{2m},$$

or $\qquad \rho(E) = \dfrac{1}{2\pi^2} \dfrac{1}{\hbar^3} (2m^3)^{\frac{1}{2}} E^{\frac{1}{2}}.$ (60)

If we assume that $|\phi_E(0)|^2$ has a value η independent of energy, and use (59), (58) becomes

$$\frac{1}{2T_1} = \frac{2\pi}{\hbar} \left\{ \frac{8\pi}{3} \gamma_e \gamma_n \hbar^2 \right\}^2 \frac{1}{4} \frac{\eta^2}{2\pi^2} \frac{(2m^3)^{\frac{1}{2}}}{\hbar^3} \tfrac{1}{2} N \frac{\int_0^\infty E e^{-\beta E}\, dE}{\int_0^\infty E^{\frac{1}{2}} e^{-\beta E}\, dE}$$

$$= \frac{32}{9} \pi N \eta^2 \gamma_e^2 \gamma_n^2 \left(\frac{m^3 kT}{2\pi} \right)^{\frac{1}{2}}. \tag{61}$$

In silicon a natural modification of (61) is the replacement of m^3 by the product $m_1 m_2 m_3$ of the anisotropic effective masses. Furthermore, since there are $l = 6$ equivalent minima in the conduction band of silicon, if interband transitions are included, the density of final states $\rho(E)$ and thus also the relaxation rate $1/T_1$ become l times larger.

The modified expression for $1/T_1$ reads

$$\frac{1}{T_1} = \frac{32}{9} \gamma_e^2 \gamma_n^2 N l \eta^2 (2\pi m_1 m_2 m_3 kT)^{\frac{1}{2}}. \tag{62}$$

The results of measurements of relaxation times of Si[29] at room temperature in n-type and p-type silicon are shown in Fig. IX, 3 (**15**). The relaxation rate is proportional to carrier concentration for high concentrations but approaches an asymptotic value in the purer samples. For comparable carrier concentrations the relaxation rate is smaller in p-type samples, which is to be expected because of the p nature, and smaller hyperfine interaction of the hole wave functions. From formula (62), where m_1, m_2, m_3 are known from cyclotron resonance experiments and $1/T_1$ is measured, the value of η can be extracted. The value $\eta = 186 \pm 18$, in reasonable agreement with theoretical predictions, is quoted in (15). (It should be noted, however, that $1/T_1$ as given by (62) is larger by a factor 2 than the corresponding formula given in (15) and smaller by a factor 4 than the expression given in formula (7) of reference (16).)

No experimental results on the variation of T_1 with temperature have been published so far. For phosphorus-doped, otherwise pure, silicon, all donors are already ionized at the temperature of liquid nitrogen and between 77° K and 300° K the number of carriers should not change and T_1 should be proportional to $T^{-\frac{1}{2}}$ if the foregoing theory is correct. A feature which is not completely understood is the considerable shortening of the relaxation time in zero field; thus for a certain sample of n-type silicon with a nuclear relaxation time of 5 min in high field, this time was 30 sec only in zero field (17).

The Overhauser effect is much easier to observe in n-type silicon (17) than in metals. Because of its low conductivity the skin depth problem is much less severe even at microwave frequencies, and because of the long nuclear relaxation time the measurement of the dynamic nuclear polarization need not be performed at the same time or even in the same magnet and at the same temperature as the saturation of the electron resonance. In a certain sample at room temperature the

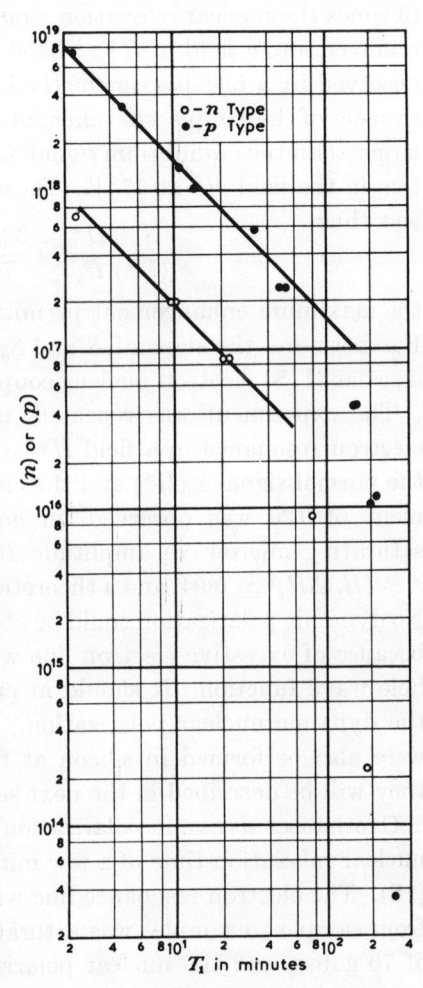

FIG. IX, 3. Si^{29} spin-lattice relaxation times plotted against the mobile carrier concentration.

conduction electron resonance line had a width of approximately 30 gauss, but at 77° K the electronic half width ΔH was only 4 gauss, which made it possible to achieve appreciable electronic saturation. The electron resonance was first saturated at 77° K in a field H_0 of 70 gauss for 15 min (3 times the nuclear relaxation time T_1). The sample was then quickly removed into a field of $H^* = 3250$ gauss and the signal S immediately observed by a fast passage method. When the loss of signal during the transfer of the sample was taken into account, S was found to be 25 times larger than the equilibrium signal $S_0(H^*)$ obtained after a long polarization in the field H^* at 77° K. The enhancement of nuclear polarization was thus

$$\frac{S}{S_0(H^*)} \frac{H^*}{H_0} = \frac{3250}{70} \times 25 \simeq 1200,$$

the maximum enhancement permitted by theory being $|\gamma_e/\gamma_n| \simeq 3300$. Furthermore, the signs of S and S_0 were opposite, in accordance with a scalar $(\mathbf{I}.\mathbf{S})$ electron nucleus coupling since $\gamma(\text{Si}^{29})$ is negative.

The experiment was repeated, using a magnetron to saturate the electron resonance in a field $H^* = 3250$ gauss. Fig. IX, 4 (**17**) shows the normal signal $S_0(H^*)$ and the enhanced signal $S(H^*)$. An enhancement of 120 was observed, in good agreement with an estimated saturating microwave amplitude $H_1 \sim 0.8$ gauss, a saturation factor $s \sim (H_1/\Delta H)^2 \simeq 0.04$, and a theoretical enhancement $3250 \times 0.04 \simeq 130$. No dynamic polarization could be observed in p-type silicon, presumably because of excessive electron line width, due to the p character of the hole wave function. It should in principle lead to an opposite sign in the dynamic nuclear polarization. Dynamic polarization experiments were also performed in silicon at the temperatures of liquid helium; they will be described in the next section.

Overhauser dynamic polarization of the nuclear spins of C^{13} with a nuclear relaxation time of a few minutes was also observed in graphite (**18**). The electron resonance line with a width of a few gauss (varying from sample to sample) was saturated at room temperature in a field of 70 gauss, and the nuclear polarization enhancement was measured by a technique similar to that described for silicon. Enhancements between 100 and 300 (maximum theoretical enhancement 2600) were observed.

B. Dynamic polarization by fixed paramagnetic impurities—solid state effect

The possibility of creating a nuclear dynamic polarization through saturation of the electron resonance has been demonstrated both theoretically and experimentally in situations where a rapid relative electron-nucleus motion exists in the sample (paramagnetic impurities in liquids, conduction electrons in metals and semiconductors), for different types of electron statistics (Fermi or Boltzmann), and for different types of electron-nucleus couplings (scalar or dipolar).

We examine now the possibility of producing a dynamic nuclear polarization in samples where the electronic spins have fixed positions in space.

This is an important problem, for, at the temperatures of liquid helium, necessary for the existence of the large electronic polarization required as a starting-point if appreciable nuclear polarization is to be achieved, there are few substances, excepting metals, where the electronic spins are not fixed in space.

It has been shown in Section II that the main nuclear relaxation mechanism resulted then from a process where only the nuclear spin flipped, the orientation of the electronic spin remaining unchanged. It is almost obvious (and can be shown by a detailed argument (19)) that under those conditions, the saturation of the electron resonance by a large r.f. field at the frequency ω_S, will *not* create a dynamic nuclear polarization since, as explained in Chapter VIII, the relaxation process does *not* involve the exchange with the lattice of large 'electronic' quanta $\hbar\omega_S$, but only of small 'nuclear' quanta $\hbar\omega_I$.

To understand how a dynamic nuclear polarization, different from the Overhauser effect, can be produced under these circumstances we go back to the principle of dynamic polarization outlined in Chapter VIII. There we first assumed a purely scalar (**I.S**) coupling. Then the steady-state condition for the nuclear spins (assumed $\frac{1}{2}$ for simplicity) was given by formula (171) of Chapter VIII:

$$N_+ n_- W_{(+-)\to(-+)} = N_- n_+ W_{(-+)\to(+-)}. \qquad (63)$$

The gist of the argument for nuclear dynamic polarization by the Overhauser effect was that since the transition probabilities W were induced by a coupling with a lattice in thermal equilibrium, the ratio $W_{(+-)\to(-+)}/W_{(-+)\to(+-)}$ was equal to $\exp\{\hbar(\omega_S-\omega_I)/kT\}$ and the saturation of the electron resonance, by making $N_+ = N_-$, led to a greatly enhanced nuclear polarization as expressed by

$$n_+/n_- = \exp\{\hbar(\omega_S-\omega_I)/kT\}.$$

If the interaction (**S, I**) is not purely scalar, extra transition probabilities such as $W_{(\pm,-)\rightleftarrows(\pm,+)}$ and $W_{(+,+)\rightleftarrows(-,-)}$ also occur, and their contribution must be added on both sides of (63). It is precisely the fact that $W_{(\pm,+)\rightleftarrows(\pm,-)}$ short-circuits all the other W if the relaxation occurs through fixed impurities, as explained in Section II of this chapter, that prevents the existence of Overhauser dynamic polarization.

Let us now assume that the transition probabilities $W_{(+-)\leftrightarrow(-+)}$ or $W_{(++)\leftrightarrow(--)}$ instead of originating in the spin-lattice coupling, are induced by an external r.f. source at the frequency

$$\Omega = \omega_S + \omega_I \quad \text{or} \quad \Omega = \omega_S - \omega_I$$

which supplies the energy required by such a simultaneous flip. The inverse transition probabilities $W_{a\to b}$ and $W_{b\to a}$ induced in such a manner are equal. We also suppose that the intensity of the source at frequency Ω is such that the corresponding rate of transitions is much faster than the nuclear relaxation rate $(1/T_1)_I$ but much slower than the electronic rate $(1/T_1)_S$. It is clear that the electronic populations $N_\pm$ will be practically unaffected by the radio-frequency and will keep their Boltzmann equilibrium values $(N_\pm)_0$, whereas the nuclear populations will be given by the relations

$$\begin{aligned}
n_+/n_- &= (N_+/N_-)_0 = \exp\left(-\frac{\hbar\omega_S}{kT}\right) \quad \text{if } \Omega = \omega_S - \omega_I, \\
n_+/n_- &= (N_-/N_+)_0 = \exp\left(+\frac{\hbar\omega_S}{kT}\right) \quad \text{if } \Omega = \omega_S + \omega_I,
\end{aligned} \quad (64)$$

and the nuclear polarization will be greatly enhanced.

The reality of this effect was first demonstrated in a crystal of LiF where the part of the electronic spin S was played by the spin of F^{19} and that of the nuclear spin I by the spin of Li^6, which has a very long relaxation time (more than a day) at room temperature. Fig. IX, 5 (20) shows the nuclear signal obtained by fast passage under the following circumstances:

Top: normal signal of Li^6 after polarization in a field of 12,000 gauss. (In view of the length of $T_1(Li^6)$, this polarization could only be achieved by thermal spin-spin mixing as explained in Chapter V.)

Middle: signal of Li^6 obtained after irradiation by a strong r.f. field H_1 at a frequency $\Omega = 2\pi \times 9.4$ Mc/s in an applied field $H_0 = 2800$ gauss such that $\Omega = \omega(F^{19}) - \omega(Li^6)$.

Bottom: signal of Li^6 obtained after irradiation by the same frequency but in a field $H_0 = 2000$ gauss, such that $\Omega = \omega(F^{19}) + \omega(Li^6)$.

The sizes and the signs of the signals observed are in agreement with the relations (64) which predict an enhancement of the signal of Li^6 by a factor $\pm\gamma(F^{19})/\gamma(Li^6) = \pm 6.5$.

The mechanism whereby a forbidden transition of a system of two spins, such as $W_{(+-)\to(-+)}$, is induced by an applied r.f. field at a frequency

$\Omega = \omega_S - \omega_I$ can be understood as follows, assuming $|\gamma_S| \gg |\gamma_I|$ for simplicity. A state such as $|+,-)$ is actually not pure since the term $S_z I_+$ of the dipolar spin-spin coupling mixes into it a contribution $\alpha|+,+)$, the admixture coefficient α being given by

$$\alpha = \frac{H_S}{2H_0}, \quad \text{where} \quad H_S = -\frac{3}{2}\frac{\gamma_S \hbar}{r^3}\sin\theta\cos\theta\, e^{-i\phi} \tag{65}$$

is the field produced at the spin I by the spin S. The state $|+,-)$ is thus actually a state $|\xi) = |+,-)+\alpha|+,+)$. Similarly, $|-,+)$ must be replaced by $|\eta) = |-,+)+\alpha|-,-)$. The transition probability W between the two states $|\xi)$ and $|\eta)$ induced by an r.f. field at a frequency Ω is thus smaller than the probability W_0 for an allowed transition of the spin S only (induced by an r.f. field of the same amplitude but at a frequency ω_S), by a factor $4\alpha^2 = (H_S/H_0)^2$. We get the same result for the forbidden transition $|+,+) \to |-,-)$. The condition

$$W \gg (1/T_1)_I,$$

which can be rewritten as $W_0(T_1)_I(H_S/H_0)^2 \gg 1$, was quite easy to satisfy in LiF, where $T_1(\mathrm{Li}^6)$ is very long and where at 2000 gauss and for an r.f. field H_1 of the order of one gauss, W was of the order of 2×10^{-2} sec$^{-1} \gg (1/T_1)_I$, and the full dynamic polarization of Li^6 could be reached in a minute or so.

When this method, which we shall call the 'solid-state effect', is applied to paramagnetic impurities in solids, we must take into account two further aspects of the problem, the first favourable, the second unfavourable for the achievement of dynamic polarization.

When the concentration of paramagnetic impurities becomes very small, the field H_S defined by (65) and the probability for a simultaneous electron-nucleus flip $W = W_0(H_S/H_0)^2$, are negligibly small except for nuclear spins in the neighbourhood of the impurity. Fortunately, just as in the problem of relaxation by paramagnetic impurities, treated in Section II, dynamic nuclear polarization can be transported away from the electronic spins to all the nuclear spins of the sample by spin diffusion. In fact, the two problems are formally identical, the r^{-6} dependence of W on the distance r between the two spins being the same as that of $1/T_1'$ in formula (40), and all the conclusions reached there can be taken over with very little change.

The combined effects of electron relaxation and spin diffusion were shown to result in a contribution to the rate of change of total nuclear polarization, equal to $-(1/T_1)\{\langle I_z\rangle - I_0\}$. The constant T_1 was given by the approximate formulae (44) and (46): $1/T_1 = 0{\cdot}7 \times 4\pi N C^{\frac{1}{4}} D^{\frac{3}{4}}$, where

$1/T_1' = Cr^{-6}$ was twice the transition probability induced by the direct coupling between an electron and a nucleus separated by a distance r. Similarly, the 'solid-state effect' should contribute to $d\langle I_z\rangle/dt$ a term

where
$$-V\left\{\langle I_z\rangle \pm \frac{\gamma_S}{\gamma_I} I_0\right\}$$
$$V = 0{\cdot}7 \times 4\pi N\Gamma^{\frac{1}{4}} D^{\frac{3}{4}}$$
$$\Gamma = W_0 \frac{9}{4} \frac{\gamma_S^2 \hbar^2 \sin^2\theta \cos^2\theta}{H_0^2}$$
(66)

Besides allowing the coefficient V to have appreciable values even when the 'direct' transition probability W is negligible for most nuclear spins, spin diffusion has another important consequence.

The direct transition probability $W = W_0(H_S/H_0)^2$ is proportional to the applied r.f. power, through W_0, and inversely proportional to the square of the applied field. In order to achieve large dynamic nuclear polarization it is natural to use high applied fields and accordingly high r.f. or microwave frequencies $\Omega = \omega_S \pm \omega_I$. Since the power available from microwave sources often goes down as the frequency goes up, it would seem to be very difficult to achieve high 'solid-state' dynamic polarizations on that account, but for spin diffusion. The relation (66) makes the dependence of the dynamic polarization rate V on the applied r.f. power P and the applied H_0 field much slower than that of W since it varies as $P^{\frac{1}{4}} H_0^{-\frac{1}{2}}$, a very favourable feature.

On the other hand, a difficulty that often occurs in 'solid-state' polarization is caused by the electronic line width $\Delta\omega_S$ being comparable to, or even larger than, the nuclear frequency $|\omega_I|$ so that the two frequencies $\Omega = \omega_S \pm \omega_I$ fall inside the electron resonance line itself. The answer to that difficulty is whenever possible to increase the applied field H_0, which increases ω_I but does not in general affect $\Delta\omega_S$.

If the condition $\Delta\omega_S > |\omega_I|$ cannot be fulfilled, the results may be expected to depend on the character of the homogeneous or inhomogeneous broadening of the electron line.

If the broadening is homogeneous, as it is in a system of like interacting spins, the effect of a strong r.f. field applied at a frequency Ω *within* the electronic line width cannot be predicted without a specific theory of saturation in solids, to be given in Chapter XII. The experimental evidence is too scarce at present to warrant a discussion of this case.

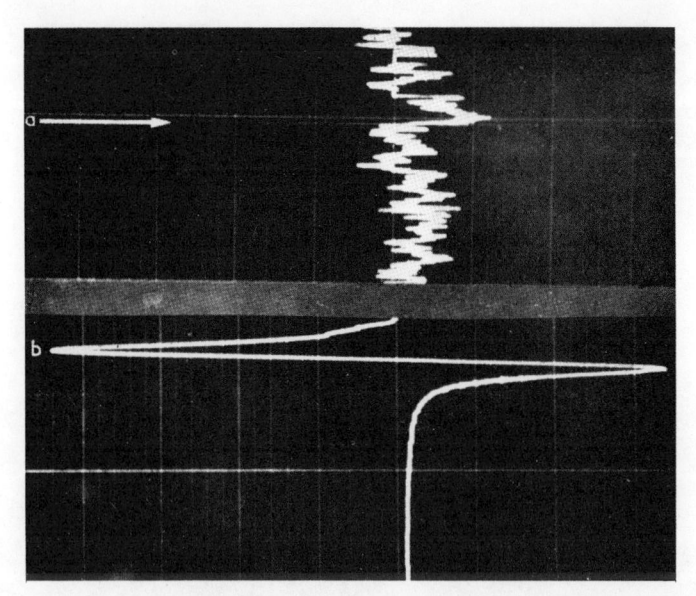

FIG. IX, 4. Dynamic polarization of Si^{92} in n-type silicon at 77° K in a field of 3300 gauss. (a) Normal signal. (b) Signal enhanced by partial saturation of the electron resonance. (Gain reduced by a factor 15.)

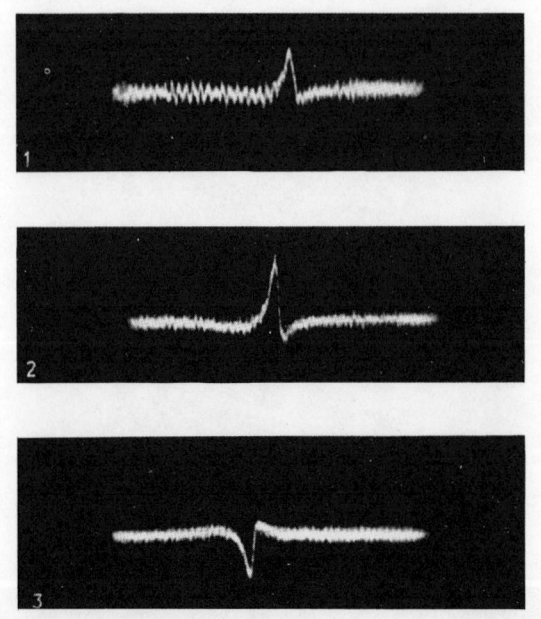

FIG. IX, 5. (1) Normal signal of Li^6 after polarization in a field of 12,000 gauss. (2) Signal of Li^6 after dynamic polarization in a field of 2800 gauss irradiating at a frequency $\Omega = \omega(F^{19}) - \omega(Li^6) \simeq 9\cdot4$ Mc/s. (3) Signal of Li^6 after dynamic polarization in a field of 2000 gauss while irradiating at a frequency
$$\Omega = \omega(F^{19}) + \omega(Li^6) \simeq 9\cdot4 \text{ Mc/s.}$$

If the broadening of the electron line is inhomogeneous, that is, it is due to the spread in the Larmor frequencies of individual narrow spin packets that can be saturated independently of each other, the situation is different.

Since the frequency Ω is applied inside the electron line, there will be spin packets with Larmor frequencies $\omega_S = \Omega - \omega_I$ and others with $\omega_S = \Omega + \omega_I$, and the net nuclear polarization will be proportional to $h(\Omega - \omega_I) - h(\Omega + \omega_I)$, where $h(\omega)$ is the shape function of the inhomogeneously broadened line that gives the relative weights of the individual spin packets. If $\Delta\omega_S \gg |\omega_I|$, the variation in the nuclear polarization observed by sweeping the applied frequency Ω (or the applied field H_0) will be proportional to the derivative $dh/d\omega$ of the electronic shape function.

Actually there seem to be few examples of electron lines that are strictly inhomogeneously broadened, since a certain amount of cross-saturation occurs by means of spin diffusion among the individual spin packets. We shall not discuss this complicated situation.

Dynamical polarization by the 'solid-state' effect has been observed in various samples in several laboratories. We describe a few typical examples, in a field which is rapidly expanding.

A good example of an inhomogeneously broadened electron line is provided by F-centres in ionic crystals such as, say, LiF where the line width of the order of a hundred gauss is well known to result from unresolved hyperfine couplings between the electronic spin and neighbouring nuclear spins. Enhancements of the nuclear polarization of either Li[7] or F[19] of the order of 20 have been observed between 1·6° K and 4° K in a field of 3300 gauss. The maximum enhancements, of opposite sign, occur on either side of the frequency $\Omega = \omega_S$ at approximately

$$\Omega = \omega_S \pm \tfrac{1}{2}\Delta\omega_S,$$

where the slope of the shape function is a maximum, rather than at the values $\omega_S \pm \omega_n$, in agreement with the inhomogeneous character of the broadening. Furthermore, the enhancement is the same for Li[7] and F[19] in spite of their very different γ. This is understandable, for the enhancement is proportional to

$$\frac{\gamma_S}{\gamma_I}\{h(\Omega+\omega_I)-h(\Omega-\omega_I)\} \propto \frac{\gamma_S}{\gamma_I}\gamma_I H_0 \frac{dh}{d\omega}, \qquad (67)$$

and thus approximately independent of γ_I (21) (Fig. IX, 6).

Enhancements of the order of 50 have been obtained at 4·2° K in a field of 12,000 gauss for the nuclear polarization of protons in polystyrene, where a paramagnetic free radical (diphenyl picryl hydrazil) had been dissolved in a proportion of one electron spin for 300 protons (**22**). Fig. IX, 7 (**22**) shows the normal and enhanced nuclear proton signal.

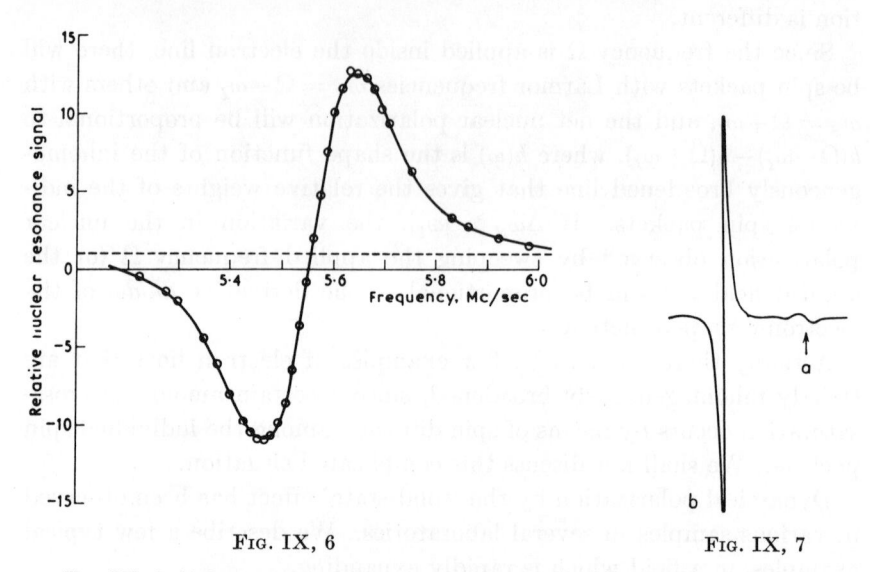

FIG. IX, 6.

FIG. IX, 7.

FIG. IX, 6. Enhancement of the nuclear resonance signal of Li⁷ at 1·6° K for a fixed microwave frequency as a function of the applied field. The curve is the derivative of the inhomogeneously broadened resonance line of the paramagnetic F-centres.

FIG. IX, 7. Signal from protons in polystyrene doped with diphenyl picryl hydrazil at 4·2° K and 12,000 gauss: (*a*) normal signal; (*b*) signal enhanced dynamically by the 'solid-state effect'.

The absolute value of the dynamic proton polarization in the experiment was $\langle I_z \rangle / I_0 \simeq 1 \cdot 5$ per cent. Although the electron spin concentration was too large for the assumptions on which the relation (66) is based to apply, the dependence of the nuclear polarization on the microwave power P was between $P^{\frac{1}{4}}$ and $P^{\frac{1}{2}}$, rather than proportional to P, as would have been expected in the absence of spin diffusion. The nuclear frequency $\omega_I = 2\pi \times 50$ Mc/s was much larger than the electron width $\Delta \omega_S = 2\pi \times 20$ Mc/s and the frequencies for maximum and opposite enhancements were separated by $2\omega_I$ rather than $\Delta \omega_S$ as in the previous example.

At the temperature of liquid helium, the behaviour of nuclear dynamic polarization of Si^{29} in phosphorus-doped silicon depends critically on the concentration of phosphorus atoms. This variation is closely related to the nature of the unpaired electrons introduced by the doping, which is revealed by the electron spin resonance spectrum (**23**).

For small concentrations ($< 5 \times 10^{16}$ per c.c.) of phosphorus atoms, the electronic spectrum exhibits two hyperfine structure lines with a separation of 42 gauss, which correspond to the two orientations of the spin $I = \frac{1}{2}$ of the phosphorus nucleus to which the unpaired electron

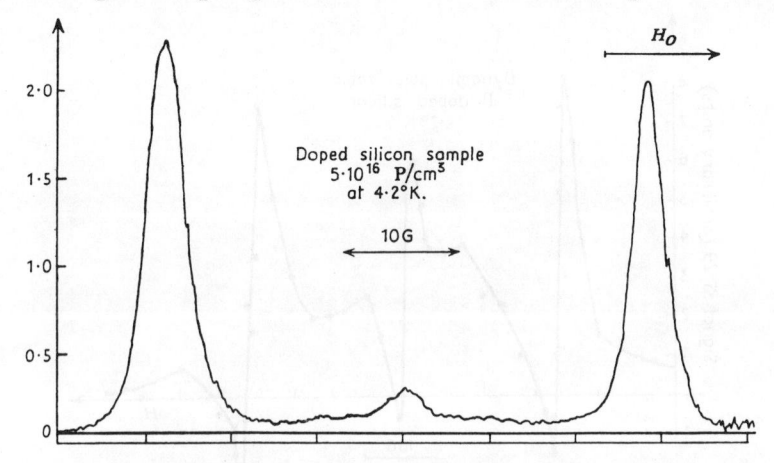

Fig. IX, 8. Electron resonance spectrum of phosphorus-doped silicon at 4·2° K.

is bound. As the concentration increases, a third line appears half-way between the other two, originating from clusters of two impurities. For still larger concentrations further lines appear, representative of clusters of more than two atoms, which grow at the expense of the two initial lines, until, for concentrations of 10^{18} impurities or more, a single electronic line appears very similar in appearance, width, and relaxation to the conduction electron line observable at 77° K. For resonance experiments at least, these electronic spins behave in the same way as do those of conduction electrons.

It is not unreasonable to expect a 'solid-state effect' in samples with low concentrations and a well-resolved hyperfine structure, and a straight Overhauser effect in heavily doped samples with a single electron resonance line.

Fig. IX, 8 (**24**) shows the electronic spectrum of a silicon sample containing approximately 5×10^{16} phosphorus atoms per c.c., observed at 4·2° K and 9200 Mc/s. Fig. IX, 9 represents the results of dynamic polarization experiments performed on the same sample. The strength of the nuclear resonance signal of Si^{29}, observed after a 5 minutes irradiation at 9200 Mc/s, is plotted in arbitrary units against the magnetic field in which the irradiation was performed. The + sign corresponds to dynamic polarizations opposite to the normal polarization. This pattern is interpreted as due to a 'solid-state effect', with the central peak corresponding to the central 'cluster' line of Fig. IX, 8, on which is superposed a broad Overhauser effect extending over the whole spectrum.

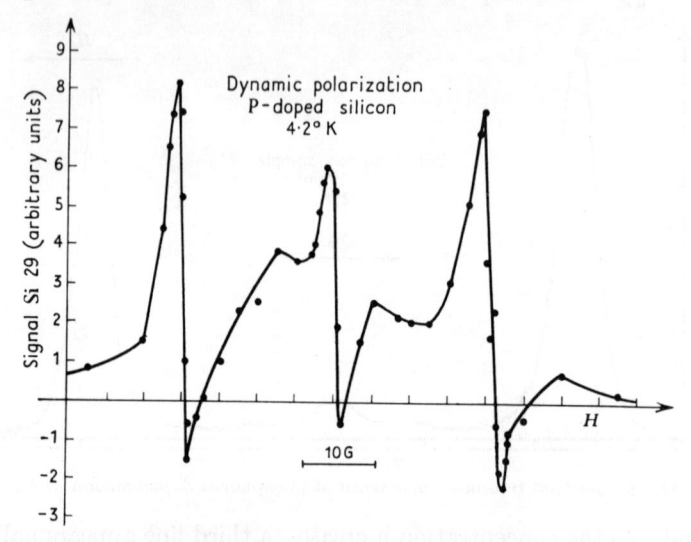

FIG. IX, 9. Dynamic polarization of Si^{29} in the sample with the electron spectrum of Fig. IX, 8.

The maximum enhancement observed, of the order of 20, was reached almost completely after two hours, that is, in a time much shorter than the nuclear relaxation time $T_1 \simeq 20$ hours, a further evidence of a 'solid state' rather than Overhauser effect. The failure to achieve a larger enhancement may be due to the fact that the nuclear frequency $\omega_n/2\pi \simeq 2\cdot 8$ Mc/s is smaller than the electronic line width $\Delta\omega_S/2\pi \simeq 7$ Mc/s, and, since the broadening is not perfectly inhomogeneous, the electron resonance is saturated, as explained previously.

In a strongly doped sample (concentration $\sim 2\times 10^{18}$) with a single electron resonance line, no 'solid-state' effect but only a straight Overhauser effect was observed as expected, with a maximum enhancement of the order of 200 (25).

Finally, it should be pointed out that an Overhauser effect rather than a 'solid-state' effect can be expected even for fixed paramagnetic impurities if their electron resonance line is considerably narrowed by strong exchange coupling between the electronic spins. The qualitative argument for such a surmise is that the reservoir of energy with a continuous spectrum, required for the Overhauser effect, is provided by the exchange energy.

Another way of putting it is to say that because of the frequent spin flips between neighbouring paramagnetic centres there is a rapid motion of the *orientation* of the electron spin, if not of the electron itself.

Such an exchange exists between the electronic spins of paramagnetic centres in some charcoals which have a very narrow electron line with a half width of 0·25 gauss. By saturating the electron line at room temperature an Overhauser enhancement by a factor 150 of the polarization of the protons present in the sample was observed (**18**).

IV. Relaxation by Thermal Vibrations in a Crystalline Lattice

In this section we consider the coupling of the nuclear spins with the thermal vibrations of a crystal lattice.

The effect of these vibrations is to create at the nuclei time-dependent magnetic fields or electric-field gradients which, according to a mechanism already considered many times in this book, induce transitions among the energy levels of the spin system that enable the populations of these levels to reach their thermal equilibrium values.

A detailed quantum mechanical description of lattice vibrations by means of lattice quanta or phonons, which has been extensively used in the theory of physical properties of crystals such as specific heat, thermal conductivity, etc., is available for the calculation of spin-lattice relaxation times in crystals. In this calculation, to be given below, we shall be content to outline the general principles, and to estimate the theoretical order of magnitude of the relaxation times.

The justification for such a procedure is different for magnetic and electric spin-lattice couplings. The effects of the former, which could in principle be estimated quite accurately, if desired, turn out to be hopelessly inadequate to account for the nuclear relaxation times actually observed in most crystals and no single instance is known where the observed relaxation could be assigned to this mechanism with certainty.

On the other hand, the electric coupling of the lattice vibrations with the nuclear quadrupole moments is much more important and is known to be responsible for nuclear relaxation in many crystals. Unfortunately, the actual calculation of relaxation times is made very uncertain by factors that are difficult to estimate because of effects such as polarization of the electronic shells, covalent bonding, etc.

The simplified procedure also has the advantage that the underlying physical principle is not obscured by the multitude of indices that appear inevitably in an accurate calculation.

A. Lattice vibrations and phonons

We begin by summarizing very briefly the description of acoustical lattice vibrations by phonons (high-frequency or so-called optical vibrations are of no interest to us).†

The position of a nucleus in, say, a cubic lattice with one nucleus per unit cell is defined by a vector $\mathbf{R} = \mathbf{r}+\mathbf{u}(\mathbf{r}) = \mathbf{p}a+\mathbf{u_p}$, where a is the lattice spacing and the components p_1, p_2, p_3 of the dimensionless vector $\mathbf{p}$ are integer numbers. $\mathbf{r} = \mathbf{p}a$ is the equilibrium position and $\mathbf{u}(\mathbf{r}) = \mathbf{u_p}$ the departure from equilibrium, of each atom (or ion).

The motion of the lattice is described by the time-dependence of the $3N$ components of the N vectors $\mathbf{u_p}(t)$ (N being the total number of atoms).

The theory of solids shows that whereas the time-dependence of the $3N$ components of the vectors $\mathbf{u_p}(t)$ is very complicated, it is possible to construct $3N$ linear combinations of these components, called normal coordinates, which behave as harmonic oscillators and are defined through the wave-like expansion

$$\mathbf{u}(\mathbf{r}) = \mathbf{u_p} = \sum_{\mathbf{f}} \sum_{j=1}^{3} q(\mathbf{f},j)\exp\left[i\frac{\mathbf{r}.\mathbf{f}}{a}\right]\mathbf{e}(\mathbf{f},j). \tag{68}$$

In this expansion the components f_1, f_2, f_3 of the phase vector $\mathbf{f}$ are dimensionless quantities, which can have discrete equidistant values in the interval $-\pi$, $+\pi$. The spacing between these values is determined by the condition that the total number of vectors $\mathbf{f}$ is equal to the number of atoms N. The density of points in the phase space $\mathbf{f}$ is therefore equal to $N/(2\pi)^3$ and, since the number N is very large, the variation of this vector in the $\mathbf{f}$ space may be considered as continuous. To each vector $\mathbf{f}$ there are attached three vectors $\mathbf{e}(\mathbf{f},j)$ describing three possible directions of polarization of the atomic vibration, and three normal coordinates $q(\mathbf{f},j)$. Each normal coordinate $q(\mathbf{f},j)$ obeys the equation of the harmonic oscillator:

$$\ddot{q}(\mathbf{f},j)+\omega^2(\mathbf{f},j)q(\mathbf{f},j) = 0. \tag{69}$$

† For a clear and rigorous description see reference (**26**).

A basic problem of the dynamics of crystalline lattices is the determination of their vibrational spectra, that is, of the dependence of the frequency $\omega(\mathbf{f},j)$ on the phase vector $\mathbf{f}$ and the polarization index j.

An approximation, which while admittedly crude is sufficient for our purpose, is Debye's assumption of a propagation velocity v that is independent of the direction of propagation and of the polarization of the wave and is expressed by the relation

$$\omega(\mathbf{f},j) = \frac{v}{a}|\mathbf{f}|. \tag{70}$$

The number $\sigma(\omega)\,d\omega$ of modes of oscillation having frequencies between ω and $\omega+d\omega$ is given by

$$\sigma(\omega)\,d\omega = \frac{3N}{8\pi^3}4\pi f^2\,df = \frac{3N}{8\pi^3}4\pi\frac{a^3}{v^3}\omega^2\,d\omega$$

or, since $Na^3 = V$ is the volume of the sample,

$$\sigma(\omega) = \frac{3V}{2\pi^2}\frac{\omega^2}{v^3}. \tag{71}$$

In order to keep the total number of modes equal to the number $3N$ of degrees of freedom of the lattice, the spectrum (71) must have an upper cut-off at a frequency Ω such that

$$\int_0^\Omega \sigma(\omega)\,d\omega = 3N, \quad \text{whence} \quad \sigma(\omega) = \frac{9N\omega^2}{\Omega^3} \quad \text{and} \quad \Omega = \frac{v}{a}(6\pi^2)^{\frac{1}{3}}. \tag{72}$$

The parameter Θ defined by the relation

$$k\Theta = \hbar\Omega \tag{73}$$

is called the Debye temperature of the crystal. To give an order of magnitude, Θ is of the order of 200° to 300° K for many substances. The Debye frequency Ω is thus of the order of $2\pi \times 10^{13}$ sec^{-1}.

The last step in the description of lattice vibrations is the quantization of the coordinates q of the lattice oscillators, which are considered as quantum mechanical operators. For that purpose the expansion (68) is replaced by the following:

$$\mathbf{u}_\mathbf{p} = \sum_{\mathbf{f},j} \mathbf{e}(\mathbf{f},j)\{q(\mathbf{f},j)e^{i(\mathbf{r}\cdot\mathbf{f}/a)} + q^\dagger(\mathbf{f},j)e^{-i(\mathbf{r}\cdot\mathbf{f}/a)}\}, \tag{74}$$

where q and $q^\dagger$ are quantum mechanical operators, hermitian conjugate to each other. For every mode $(\mathbf{f},j)$ there is a lattice oscillator with a frequency $\omega(\mathbf{f},j)$ and a set of equally spaced energy levels $(n+\tfrac{1}{2})\hbar\omega(\mathbf{f},j)$ where n is an integer. The only non-vanishing matrix elements of the operators $q(\mathbf{f},j)$ and $q^\dagger(\mathbf{f},j)$ are

$$(n\,|\,q\,|\,n+1) = (n+1\,|\,q^\dagger\,|\,n)^* = \left(\frac{\hbar}{2M}\right)^{\frac{1}{2}} \omega^{-\frac{1}{2}} e^{-i\omega t}\sqrt{(n+1)}, \quad (75)$$

where $M = Nm$ is the mass of the crystal and m the mass of a single atom.

It is convenient to describe a situation where a lattice oscillator $(\mathbf{f},j)$ is in a state of energy $|n\rangle$ by saying that there are n lattice quanta or phonons $(\mathbf{f},j)$ present in the crystal. With this convention q is an operator for absorption and $q^\dagger$ for creation, of phonons. Phonons obey Bose statistics so that the number $n(\mathbf{f},j)$ of phonons $(\mathbf{f},j)$ present in a crystal at temperature T will be given by Planck's law

$$n(\mathbf{f},j) = \left[\exp\!\left(\frac{\hbar\omega(\mathbf{f},j)}{kT}\right) - 1\right]^{-1}. \quad (76)$$

B. Transition probabilities induced by the spin-phonon coupling

The coupling between the lattice vibrations and the spins can be represented quite generally by a Hamiltonian

$$\hbar\mathscr{H}_1 = \hbar \sum_q F^{(q)} A^{(q)}, \quad (77)$$

where $F^{(q)}$ and $A^{(q)}$ are respectively lattice and spin operators. We can always assume that the operators $A^{(q)}$ are dimensionless, with matrix elements of order unity, and that consequently the $F^{(q)}$ have the dimensions of a frequency. An operator such as $F^{(q)}$, where we drop the index q for brevity, will in general be a function of the relative positions

$$\mathbf{R}_{12} = \mathbf{R}_1 - \mathbf{R}_2 = \mathbf{r}_1 - \mathbf{r}_2 + \mathbf{u}_1(\mathbf{r}_1) - \mathbf{u}_2(\mathbf{r}_2)$$

of neighbouring atoms (or ions) in the lattice. In the approximate treatment presented here we replace a component such as $(\mathbf{u}_1 - \mathbf{u}_2)_x$ by the first-order expansion $(\mathbf{r}_1 - \mathbf{r}_2) \cdot \partial \mathbf{u}_1/\partial x$. This approximation, which rests on the assumption that $\mathbf{u}$ does not vary appreciably over an interatomic distance a, is only correct for vibrations of wavelength $\lambda = 2\pi v/\omega$, appreciably larger than the interatomic spacing, but although it breaks down, according to (72), at the upper end of the spectrum when $\omega \simeq \Omega$, this will not affect appreciably the order of magnitude of the results.

We can then expand F as a function of the stresses

$$W_{ik} = \frac{1}{2}\!\left(\frac{\partial u_i}{\partial x_k} + \frac{\partial u_k}{\partial x_i}\right)$$

of the lattice, an expansion we write symbolically as

$$F = F_0 + F_1 W + F_2 W^2 + F_3 W^3 + \ldots . \tag{78}$$

It is clear that F_1 is a two-index tensor, $F_1 W$ standing for $\sum_{i,k} F_{1,ik} W_{ik}$, F_2 a four-index tensor, etc. Since the W are dimensionless quantities the coefficients F_1, F_2, F_3, etc., have the dimensions of a frequency. They are all of comparable magnitude, and the series (78) converges because of the smallness of the W. For instance, for a magnetic coupling between two spins,

$$\hbar \mathcal{H}_1 \sim \hbar F A \sim \frac{\gamma_1 \gamma_2 \hbar^2}{R^3} \left\{ \mathbf{I}_1 \cdot \mathbf{I}_2 - \frac{3(\mathbf{I}_1 \cdot \mathbf{R})(\mathbf{I}_2 \cdot \mathbf{R})}{R^2} \right\},$$

$$F_0 \sim \frac{\gamma_1 \gamma_2 \hbar}{r^3}; \quad F_1 \sim r \frac{\partial F_0}{\partial r} \sim F_0; \quad F_2 \sim r^2 \frac{\partial^2 F_0}{\partial r^2} \sim F_1, \quad \text{etc.} \tag{79}$$

In a quantized theory of lattice vibrations, F_1, F_2, etc., are c-numbers whereas the $W \sim \partial u/\partial x$ are, by the expansion (74), operators for emission or absorption of phonons. The matrix element of W for the emission of a phonon $(\mathbf{f}, j)$ of frequency ω, there being n such phonons present in the crystal, is according to (74) and (75) of the order of

$$\left[\frac{(n+1)\hbar}{2M\omega} \right]^{\frac{1}{2}} \frac{f}{a} \cong \left[\frac{(n+1)\hbar}{2M\omega} \right]^{\frac{1}{2}} \frac{\omega}{v}. \tag{80}$$

The first term of the expansion (78), $F_1 W$, permits the absorption or the emission of a single phonon (direct process). The second term, $F_2 W^2$, permits the emission or absorption of two phonons, or the absorption of one phonon and emission of another (Raman process).

Direct process. Consider the probability of a transition induced by the term $F_1 W$ between two spin states $|m\rangle$ and $|m'\rangle$ with an energy difference $\hbar \omega_0$ which will be equal to the energy of the phonon emitted or absorbed. The order of magnitude of the matrix element Y between the two states $|m, n\rangle$ and $|m', n+1\rangle$ of the combined spin-lattice system will be

$$Y \sim (m \mid A \mid m')(n \mid \hbar F_1 W \mid n+1)$$

or, with our assumption that A is of the order of unity, and using (80),

$$Y \sim \hbar F_1 \frac{\omega}{v} \left[\frac{(n+1)\hbar}{2M\omega} \right]^{\frac{1}{2}}. \tag{81}$$

The transition probability can be written $P_1 \sim (2\pi/\hbar)|Y|^2 \rho(E)$, where $E = \hbar \omega_0$ is the energy of the phonon emitted or absorbed, and $\rho(E) = (1/\hbar)\sigma(\omega)$ the density of final phonon states.

Using $\rho(E) = (1/\hbar)\sigma(\omega)$, where σ is given by (72),

$$P_1 \sim \frac{2\pi}{\hbar}|Y|^2\frac{\sigma(\omega_0)}{\hbar} \sim 2\pi F_1^2 \frac{\omega_0^2}{v^2}\frac{\hbar}{2M\omega_0}\frac{9N\omega_0^2}{\Omega^3}(n+1). \qquad (82)$$

The number n of phonons must be replaced by its thermal equilibrium value (76), which, since the nuclear spin energies $\hbar\omega_0$ are much smaller than kT even for the lowest temperatures obtainable, is approximately $kT/\hbar\omega_0 \gg 1$. The equation (82) can be rewritten

$$P_1 \sim 9\pi F_1 \frac{F_1 \omega_0^2}{\Omega^3}\frac{kT}{mv^2} \sim 9\pi\Omega\left(\frac{F_1}{\Omega}\right)^2\left(\frac{\omega_0}{\Omega}\right)^2\left(\frac{k\Theta}{mv^2}\right)\left(\frac{T}{\Theta}\right) \sim \frac{1}{T_1}, \qquad (83)$$

where T_1 is the relaxation time due to this process.

A characteristic feature of the direct process, apparent in (83), is the proportionality of $1/T_1$ to the square of the frequency ω_0 (or for a Zeeman resonance to the square of the applied field H_0) and to the absolute temperature T. We may thus write

$$\frac{1}{T_1} = CH_0^2 T.$$

We defer a discussion of the magnitude of T_1 until the next section, but it is apparent from formula (83) that it is exceedingly small whatever the nature of the coupling $\hbar FA$ between the nuclear spins and the lattice. Both the strength, in cycles, of the coupling F_1 and the resonance frequency ω_0 are smaller than Ω by factors of the order of 10^5–10^7 or more, and the last factor kT/mv^2 is also easily shown to be small. It can be written $kT/mv^2 = N_0 kT/N_0 mv^2 = RT/Av^2$, where N_0 is the Avogadro number 6.06×10^{23}, R is the Joule constant 8.4×10^7, and A the atomic mass in grammes. If, for instance, we take $A = 20$, $v = 2\times 10^5$ (a current value for the velocity of sound in crystals), $T = 300°$ K, we find $kT/mv^2 \sim 0.03$. The physical reason for the inefficiency of the direct process is that only phonons in the neighbourhood of the frequency ω_0 contribute to it, and that the corresponding spectral density of thermal energy, proportional to $\sigma(\omega_0)$, is very small.

Raman process. Among the two-phonon processes induced by the term $F_2 W^2$, we consider only the Raman process, the absorption of one phonon and the emission of another, which is overwhelmingly more important than the emission or absorption of two phonons. In the Raman process the frequencies ω and ω' of the phonons involved satisfy the relation $\omega-\omega' = \omega_0$, and ω can take all the values inside the phonon spectrum, from ω_0 to Ω. On the other hand, for emission or absorption of two phonons the relation $\omega+\omega' = \omega_0$ restricts the process to a very small fraction of the phonon spectrum.

The matrix element for a transition where there are n photons identical to the one emitted and n' like the one absorbed, is of the order of

$$Y \sim \hbar F_2 \frac{\omega}{v} \frac{\omega'}{v} \left[\frac{(n+1)\hbar}{2M\omega} \frac{n'\hbar}{2M\omega'} \right]^{\frac{1}{2}}. \tag{84}$$

For a given frequency ω of the first phonon, the transition probability is

$$\frac{2\pi}{\hbar} |Y|^2 \rho(E) = \frac{2\pi}{\hbar} |Y|^2 \frac{\sigma(\omega')}{\hbar}, \tag{85}$$

where the frequency ω' of the second phonon is given by $\omega' = \omega - \omega_0$. To obtain the total transition probability, (85) must be summed over all the possible frequencies ω of the first phonon, that is, it must be multiplied by $\sigma(\omega)\,d\omega$ and integrated over ω.

In this integration, since the difference $\omega - \omega' = \omega_0$ is much smaller than either ω or ω' except for a negligible fraction of the phonon spectrum, it is permissible to make $\omega' = \omega$. When the occupation numbers n and $n' \simeq n$ are replaced by their thermal equilibrium values (76), we get

$$P_2 \simeq \frac{2\pi}{\hbar} \int \sigma(\omega)\hbar^2 F_2^2 \frac{\omega^4}{v^4} \frac{n(n+1)\hbar^2}{4M^2\omega^2} \frac{\sigma(\omega)}{\hbar} d\omega$$

$$\simeq \frac{81\pi}{2} \left(\frac{F_2 \hbar}{mv^2} \right)^2 \int_0^\Omega \frac{e^{\hbar\omega/kT}}{(e^{\hbar\omega/kT}-1)^2} \frac{\omega^6}{\Omega^6} d\omega. \tag{86}$$

In the high-temperature limit, when $kT \gg \hbar\Omega$, we expand $e^{\hbar\omega/kT}$ into $1+\hbar\omega/kT$ and obtain

$$P_2 \simeq \frac{81\pi}{10} \left(\frac{kT}{mv^2} \right)^2 \frac{F_2^2}{\Omega} \simeq \frac{81\pi}{10} \left(\frac{\hbar F_2}{mv^2} \right)^2 \Omega \left(\frac{T}{\Theta} \right)^2$$

$$\simeq \frac{81\pi}{10} \left(\frac{F_2}{\Omega} \right)^2 \left(\frac{k\Theta}{mv^2} \right)^2 \left(\frac{T}{\Theta} \right)^2 \Omega. \tag{87}$$

Actually (87) turns out to be already a fairly good approximation for $T \approx \Theta$.

In the low-temperature limit, if we introduce the variable $\hbar\omega/kT = x$, (86) can be rewritten as

$$P_2 \simeq \frac{81\pi}{2} \left(\frac{F_2 \hbar}{mv^2} \right)^2 \left(\frac{T}{\Theta} \right)^7 \Omega \int_0^X \frac{x^6 e^x \, dx}{(e^x-1)^2}, \tag{88}$$

where

$$X = \frac{\hbar\Omega}{kT} = \frac{\Theta}{T}.$$

For $T \ll \Theta$ the integral tends toward the limit

$$\int_0^\infty \frac{x^6 e^x \, dx}{(e^x - 1)^2}$$

and the relaxation rate varies as T^7. It should be noted that the integral $\int_0^X$ converges rather slowly, and that the T^7 law is only valid when $T/\Theta \leqslant 0.02$ or less.

The characteristic features of the Raman process are thus independence of the relaxation rate on the spin resonance frequency ω_0, its variation as T^2 for $T \geqslant \Theta$ and as T^7 for $T \ll \Theta$.

If we compare the relaxation rate P_2 as given by (87) with that resulting from the direct process and given by (83), we get, assuming $F_1 \approx F_2$,

$$\frac{P_2}{P_1} \sim \frac{kT}{mv^2}\left(\frac{\Omega}{\omega_0}\right)^2, \tag{89}$$

that is, according to our estimate of kT/mv^2 and ω_0, a very large number, The reason for the relative importance of the Raman process, as explained previously, is that all the phonons of the spectrum do take part in it.

The question naturally arises as to whether higher-order terms such as $F_3 W^3$ in the expansion (78) will not give comparable or even larger contributions to the relaxation rate. Actually the whole procedure of expansion into powers of W turns out to be meaningful. The introduction of an extra phonon in the process results in an extra squared phonon matrix element, multiplied by $\sigma(\omega)\,d\omega$ and integrated over ω. This yields an extra factor in the transition probability of the order of

$$Z \sim \int \frac{9N\omega^2}{\Omega^3} d\omega \left[\frac{\omega}{v}\left(\frac{n\hbar}{M\omega}\right)^{\frac{1}{2}}\right]^2 \tag{90}$$

or, taking $n \sim kT/\hbar\omega$, $Z \sim (kT/mv^2)$, which has been shown previously to be appreciably smaller than unity.

It should be clearly understood that the direct process and the Raman process are both *first*-order processes in the sense of perturbation theory, the perturbing Hamiltonian being $\hbar A F_1 W$ and $\hbar A F_2 W^2$, respectively.

It has been suggested (27) that a *second*-order contribution to the relaxation rate may result from an interference between the spin-lattice term $\hbar A F_1 W$ and the anharmonic term $G_3 W^3$ in the lattice energy, responsible for thermal conductivity.

A typical matrix element for such a process would be of the form

$$Y = \frac{(m\phi \mid \hbar A F_1 W \mid m'\phi')(m'\phi' \mid G W^3 \mid m'\phi'')}{E_{m\phi} - E_{m'\phi'}}, \quad (91)$$

where $|m\rangle$ stands for a spin state and $|\phi\rangle$ for a lattice state. The state $|\phi'\rangle$ would differ from $|\phi\rangle$ by one phonon and $|\phi''\rangle$ would differ from $|\phi'\rangle$ by three phonons, either emitted or absorbed.

For instance, $|\phi\rangle$ could be defined by the numbers $|n_1, n_2, n_3, n_4\rangle$ of four phonons (among others), $|\phi'\rangle$ would be $|n_1+1, n_2, n_3, n_4\rangle$, and $|\phi''\rangle$ would be $|n_1+1, n_2-1, n_3+1, n_4-1\rangle$. The energy difference $E_{m\phi} - E_{m'\phi'}$ would be $\hbar(\omega_0 + \omega_1)$. Conservation of energy requires

$$\omega_2 + \omega_4 = \omega_1 + \omega_3 + \omega_0.$$

The transition probability is of the order of

$$P \sim \frac{2\pi}{\hbar} \int \sigma(\omega_1)\sigma(\omega_2)\sigma(\omega_3) \frac{\sigma(\omega_4)}{\hbar} |Y|^2 d\omega_1 d\omega_2 d\omega_3.$$

In the high-temperature range where $\hbar\omega/kT \ll 1$, an order of magnitude calculation gives

$$P \sim \Omega \left(\frac{G_3 \hbar F_1}{m^2 v^4}\right)^2 \left(\frac{T}{\Theta}\right)^4. \quad (92)$$

On the other hand, a crude estimate leads to $G_3 \sim G_2 \sim mv^2$, so that (92) can be rewritten

$$P \sim \Omega \left(\frac{\hbar F_1}{mv^2}\right)^2 \left(\frac{T}{\Theta}\right)^4,$$

which is possibly of the same order as (87) or even higher, above the Debye temperature.

Interference between terms of higher order either in the spin-lattice coupling $\hbar A F_m W^m$ or in the lattice energy $G_m W^m$ can be seen to lead to smaller contributions.

In view of the extreme crudeness of the estimate (92), it is interesting to investigate the temperature-dependence of the experimental relaxation rates to see whether the interference term, which does not seem to have been considered by other authors, is of any significance (see Section IV C (b) ref. (31)).

C. Magnetic and quadrupole relaxation by spin-phonon coupling

(a) *Magnetic relaxation*

A magnetic spin-spin coupling such as that written in (79) does contain terms proportional to $I_z^1 I_\pm^2$ or $I_\pm^1 I_\pm^2$ that have matrix elements, and may induce transitions, between different spin states. An expansion of (79), similar to (78), results in coupling constants $F_1, F_2 \sim \gamma_1 \gamma_2 \hbar r^{-3}$.

Taking $r \sim 2$ Å and $\gamma_1 = \gamma_2 = 2\pi \times 4 \cdot 2 \times 10^3$ (protons), we find $F_1 \sim F_2 \sim 2\pi \times 10^4$. For the direct process P_1, as given by (83), where we take $\Omega = 2\pi \times 10^{13}$, $\omega_0 = 2\pi \times 10^7$, $k\Theta/mv^2 = 0.03$, (83) yields

$$P_1 \sim 5 \times 10^{-17}\left(\frac{T}{\Theta}\right), \qquad (93)$$

which is of academic interest only.

However, if the spins are electronic rather than nuclear, as was assumed in the first detailed investigation of this problem (28), it follows from (83) that P_1 is increased by a factor $(\gamma_e/\gamma_n)^6$, which is of the order of 10^{18} or more, and, since it is of the order of $10^2(T/\Theta)$, P_1 becomes a measurable quantity.

For the Raman process we find by (87), with the same values of the parameters,

$$P_2 \simeq 1 \cdot 5 \times 10^{-6}\left(\frac{T}{\Theta}\right)^2, \qquad (94)$$

which is still much too small to account for any observed relaxation times.

In view of the smallness of the effects due to nuclear spin-spin couplings it may be worth while pointing to the existence of another nuclear magnetic relaxation mechanism, the spin-orbit coupling, suggested by several authors (27).

A nucleus vibrating in an electric field $\mathcal{E}$ with a velocity $\dot{\mathbf{u}}$ 'sees' a magnetic field

$$\mathbf{H} = \frac{\dot{\mathbf{u}}}{c} \wedge \mathcal{E} = \frac{m}{ec}[\dot{\mathbf{u}} \wedge \ddot{\mathbf{u}}],$$

to which it is coupled by a Hamiltonian $\mathcal{H} = -\gamma\hbar\mathbf{H}.\mathbf{I}$. $\mathcal{H}$ is an expression that is bilinear in the phonon operators q, and can cause relaxation by the Raman effect. Its matrix elements are of the order of

$$Y \sim (n, n' \mid \mathcal{H} \mid n+1, n'-1) \sim \gamma\hbar\omega^3 \frac{m}{ec}\frac{\hbar}{2M\omega}\sqrt{\{(n+1)n'\}}$$

and the corresponding transition probability, by a calculation identical to that leading to (86), is

$$P \sim \frac{2\pi}{\hbar}\int \sigma(\omega)|Y|^2\frac{\sigma(\omega)}{\hbar}\,d\omega \sim \frac{81\pi}{14}\Omega\left(\frac{\gamma kT}{ec}\right)^2.$$

Taking $\gamma \sim 2\pi \times 4 \cdot 2 \times 10^4$, $\Omega \simeq 2\pi \times 10^{13}$, we find

$$P \sim 10^{-11}T^2,$$

which is of an order of magnitude comparable to the contribution of the spin-spin coupling to the Raman process.

DYNAMIC POLARIZATION IN SOLIDS

To conclude, the magnetic nuclear spin-phonon coupling is a negligible relaxation mechanism.

(b) Quadrupole relaxation

Although there is conclusive evidence that in sufficiently pure ionic crystals the coupling of lattice vibrations with the quadrupole nuclear moment can be the main relaxation mechanism for nuclear spins $I > \frac{1}{2}$, there exist at present few data on nuclear relaxation times and fewer careful investigations of the temperature-dependence of T_1 (31).

The interpretation of data available from saturation experiments in cubic crystals is further complicated by the existence of crystal imperfections which, as explained in Chapter VII, prevent parts of the resonance line from being observed.

A crucial experiment demonstrating the existence of quadrupole relaxation was performed at an early date on a single crystal of $NaNO_3$, where the non-cubic environment of the Na^{23} nucleus (spin $\frac{3}{2}$) results in slightly different frequencies for the three transitions $\frac{3}{2} \to \frac{1}{2}$, $\frac{1}{2} \to -\frac{1}{2}$, $-\frac{1}{2} \to -\frac{3}{2}$, and a spectrum with three lines (29). In this experiment one of the lines could be saturated while another was observed simultaneously by means of a very weak r.f. field that did not perturb the populations.

Suppose first that the relaxation is magnetic and consequently that there are no relaxation transitions with $|\Delta m| = 2$. It is easy to show that the saturation of any one line cannot affect the intensities of the others.

We assume first that the transition $-\frac{3}{2} \to -\frac{1}{2}$ is being saturated by a strong r.f. field and introduce the populations $n_{\frac{3}{2}},..., n_{-\frac{3}{2}}$ of the four levels, and also the quantities $n'_{\frac{3}{2}} = n_{\frac{3}{2}} - (n_{\frac{3}{2}})_0$, which are departures of the populations from their thermal equilibrium value. These quantities satisfy the obvious relation

$$n'_{\frac{3}{2}} + n'_{\frac{1}{2}} + n'_{-\frac{1}{2}} + n'_{-\frac{3}{2}} = 0. \tag{95}$$

The equations for the rates of change,

$$\frac{dn'_{\frac{3}{2}}}{dt} = \frac{dn'_{\frac{3}{2}}}{dt} \quad \text{and} \quad \frac{dn'_{\frac{1}{2}}}{dt} = \frac{dn'_{\frac{1}{2}}}{dt}$$

where the r.f. transition probability does not appear, are

$$\frac{dn'_{\frac{3}{2}}}{dt} = -W_1(n'_{\frac{3}{2}} - n'_{\frac{1}{2}}),$$

$$\frac{dn'_{\frac{1}{2}}}{dt} = -W_1(n'_{\frac{1}{2}} - n'_{\frac{3}{2}}) - W_2(n'_{\frac{1}{2}} - n'_{-\frac{1}{2}}). \tag{96}$$

Under steady-state conditions the first equation (96) gives $n'_{3/2} = n'_{1/2}$, and then the second gives $n'_{1/2} = n'_{-1/2}$. It follows that the intensities of the central line $\frac{1}{2} \to -\frac{1}{2}$ and of the second satellite $\frac{3}{2} \to \frac{1}{2}$, proportional to $n_{1/2}-n_{-1/2}$ and $n_{3/2}-n_{1/2}$ respectively, have the same values as in the absence of the saturating field. If the central line $\frac{1}{2} \to -\frac{1}{2}$ were saturated, the first equation (96) would still be valid, giving $n'_{3/2} = n'_{1/2}$ and so an intensity of the satellites independent of the saturation of the central line.

We now assume quadrupole relaxation and saturation of the transition $-\frac{3}{2} \to -\frac{1}{2}$. The rates of change of populations are given by

$$\frac{dn'_{3/2}}{dt} = -W_1(n'_{3/2}-n'_{1/2})-W_2(n'_{3/2}-n'_{-1/2}),$$

$$\frac{dn'_{1/2}}{dt} = -W_1(n'_{1/2}-n'_{3/2})-W_2(n'_{1/2}-n'_{-3/2}),$$

$$\frac{dn'_{-1/2}}{dt} = -W_1(n'_{-1/2}-n'_{-3/2})-W_2(n'_{-1/2}-n'_{3/2})-V(n_{-1/2}-n_{-3/2}),$$

$$\frac{dn'_{-3/2}}{dt} = -W_1(n'_{-3/2}-n'_{-1/2})-W_2(n'_{-3/2}-n'_{1/2})-V(n_{-3/2}-n_{-1/2}). \qquad (97)$$

W_1 and W_2 are relaxation induced transitions with $|\Delta m| = 1$ and $|\Delta m| = 2$, respectively, and V is the transition probability induced by the r.f. field. We call x the ratio W_2/W_1. The equilibrium populations $(n_{3/2})_0,..., (n_{-3/2})_0$ are equidistant (if we neglect the effect of small quadrupole splittings on the populations). We define λ as the difference

$$\lambda = (n_{3/2})_0-(n_{1/2})_0 = (n_{1/2})_0-(n_{-1/2})_0 = (n_{-1/2})_0-(n_{-3/2})_0.$$

We note that there is no quadrupole induced transition $\frac{1}{2} \to -\frac{1}{2}$ because the corresponding matrix element $(+\frac{1}{2}|I_zI_++I_+I_z|-\frac{1}{2})$ vanishes.

Adding the first two relations (97) (under steady-state conditions), we get $n'_{3/2}+n'_{1/2} = n'_{-1/2}+n'_{-3/2}$ or, because of (95), $n'_{3/2}+n'_{1/2} = 0$. Complete saturation of the $-\frac{3}{2} \to -\frac{1}{2}$ transition ($V \to \infty$) gives

$$n_{-3/2}-n_{-1/2} = 0 \quad \text{or} \quad n'_{-3/2}-n'_{-1/2} = \lambda. \qquad (98)$$

From (98) and the first equation (97) we get immediately

$$n_{3/2}-n_{1/2} = \frac{2\lambda}{2+x} = \frac{2}{2+x}[(n_{3/2})_0-(n_{1/2})_0],$$

$$n_{1/2}-n_{-1/2} = \lambda\left(\frac{3}{2}+\frac{x}{2(2+x)}\right) = \left(\frac{3}{2}+\frac{x}{2(2+x)}\right)[(n_{1/2})_0-(n_{-1/2})_0]. \qquad (99)$$

In the presence of quadrupole relaxation, the saturation of a satellite weakens the other satellite and enhances the central line. In particular, if we assume $W_1 = W_2$, i.e. $x = 1$, the enhancement of the central line is $\frac{5}{3}$ and the decrease of the second satellite $\frac{2}{3}$.

Both the enhancement of the central line and the reduction of the second satellite were observed, demonstrating in a striking manner the reality of quadrupole relaxation. The corresponding experimental factors were approximately 2 and 0·6, in qualitative agreement with the assumption $W_1 = W_2$.

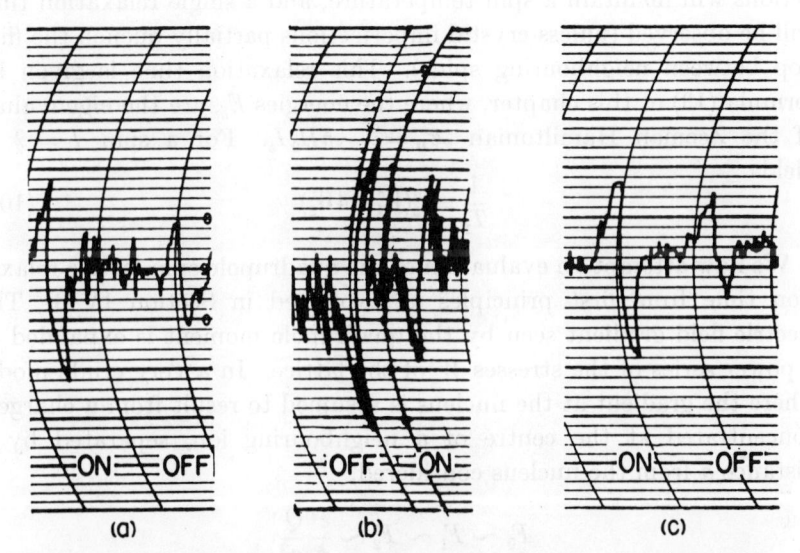

FIG. IX, 10. Demonstration of the existence of quadrupole relaxation. (a) Increase in the intensity of the central line of Na^{23} in $NaNO_3$ when a satellite line is saturated. This increase should be $\frac{5}{3}$ if $W_1 = W_2$. (b) Decrease of the lower satellite when the upper satellite is saturated. This decrease should be $\frac{2}{3}$ if $W_1 = W_2$. (c) Increase of a satellite when the central line is saturated. This increase should be $\frac{3}{2}$ if $W_1 = W_2$.

If the central line is saturated, a calculation similar to the previous one predicts for the satellites an enhancement by a factor

$$\frac{1+2x}{1+x}, \quad \text{or} \quad \tfrac{3}{2} \quad \text{for } x = 1.$$

The observed enhancement was of the order of 1·75. Although the experimental accuracy as exhibited by the signals shown in Fig. IX, 10 is not sufficient to permit a precise determination of x, all three experimental results point to a value of $x = W_2/W_1$ rather larger than unity, of the order of 1·5.

It can be shown, using the expression for the quadrupole coupling given in Chapter VI, that the assumption $W_1 = W_2$ would be valid only if the environment of a sodium nucleus had spherical symmetry, which is certainly not the case in a single crystal of NaNO$_3$.

For a cubic crystal where the static quadrupole interaction vanishes, the condition for the decay of $\langle I_z \rangle = \frac{3}{2}n_{\frac{3}{2}} + \ldots + (-\frac{3}{2})n_{-\frac{3}{2}}$ to be described by a single relaxation time is also $W_1 = W_2$, as can be seen from the system (97), where one makes $V = 0$. However, for a cubic crystal and pure Zeeman resonance, whatever the ratio W_2/W_1, spin-spin interactions will maintain a spin temperature, and a single relaxation time will be observed (unless crystal imperfections partially quench the flip-flop between neighbouring spins). This relaxation time is given by formula (13) of this chapter, where the energies E_m are the eigenvalues of the Zeeman Hamiltonian $\mathcal{H}_0 = -\gamma \hbar H_0 I_z$. For a spin $I = \frac{3}{2}$ it yields

$$\frac{1}{T_1} = \tfrac{2}{5}(W_1 + 4W_2). \tag{100}$$

We now attempt an evaluation of the quadrupole spin-lattice relaxation time from first principles as explained in Section IV B. The electric-field gradient seen by the quadrupole moment is expanded in a power series of the stresses W of the lattice. In a very crude model where the gradient at the nucleus is assumed to result from a charge e concentrated at the centre of a neighbouring ion, separated by a distance a from the nucleus considered,

$$F_0 \sim F_1 \sim F_2 \sim \frac{e^2}{\hbar}\frac{Q}{a^3}.$$

For a cubic crystal F_0 vanishes because the contributions from all the neighbours cancel each other, but the cubic symmetry is violated during the lattice vibrations and neither F_1 nor F_2 vanish.

Consider first the direct process. Taking

$$\Omega = 2\pi \times 10^{13}, \quad k\Theta/mv^2 \sim 0\cdot 03, \quad \omega_0 = 2\pi \times 10^7,$$

$$Q = 10^{-25} \text{ cm}^2, \quad a = 2 \text{ Å} = 2\times 10^{-8} \text{ cm},$$

we find $\quad F_1 \sim F_2 \sim e^2 Q/\hbar a^3 \sim 2\pi \times 4 \times 10^5,$

whence from (83), $\quad P_1 \sim \dfrac{1}{T_1} \sim 10^{-13}\left(\dfrac{T}{\Theta}\right).$ \hfill (101)

With the same parameters as for the direct process we find by (87)

$$P_2 \sim \frac{1}{T_1} \sim 2\times 10^{-3}\left(\frac{T}{\Theta}\right)^2. \tag{102}$$

Experimentally, relaxation rates of the order of 10^2 sec^{-1} have been observed at room temperature for bromine and iodine in alkali halides, and of the order of 10^{-1} for Na^{23}.

The estimate (102) would indicate that theoretical relaxation times are far too long, but its crudeness is such that it could easily be out by a factor of a hundred and a more careful calculation is necessary.

Such a calculation was performed for a cubic lattice using a simplified model where the crystalline field at a nucleus was assumed to arise from a number of equal point charges placed on the nearest neighbouring lattice sites (30). The principle of the calculation is identical to that outlined in Section IV B but attention is paid to numerical coefficients, angular factors, to the fact that the phonon wavelength may be comparable to the lattice spacing, etc. As an example, the value $T_1 \sim 4 \times 10^4$ sec is found for I^{127} in KI as compared with an experimental value 0·04, a discrepancy by a factor of the order of 10^6. The ratio $x = W_2/W_1$ is also calculated and found to vary between 1·43 and 0·82 as the applied field is rotated from the [001] to the [111] direction. It would be interesting to estimate this ratio for a hexagonal structure such as that of NO_3Na where the double irradiation method used in reference (29) could provide an accurate measurement of W_2/W_1.

On the other hand in a cubic environment the relaxation time T_1 calculated by formula (13) from the values of W_1 and W_2 given in reference (30) turns out to be isotropic, and to depend on the nuclear spin I through the same factor

$$f(I) = \frac{2I+3}{I^2(2I-1)}$$

as in a liquid (31) (formula (137) of Chapter VIII).

This dependence has been checked by careful measurements of the ratios of the relaxation times of Sb^{123} ($I = \frac{7}{2}$) and Sb^{121} ($I = \frac{5}{2}$) in AlSb, and of Rb^{85} ($I = \frac{5}{2}$) and Rb^{87} ($I = \frac{3}{2}$) in RbCl (31).

For Sb the theory gives

$$\frac{T_1(Sb^{121})}{T_1(Sb^{123})} = \left[\frac{Q(Sb^{123})}{Q(Sb^{121})}\right]^2 \frac{f(\frac{7}{2})}{f(\frac{5}{2})} = 1 \cdot 63 \times 0 \cdot 425 = 0 \cdot 69.$$

The experimental value of this ratio in AlSb is $0 \cdot 75 \pm 0 \cdot 10$. For Rb the theory gives

$$\frac{T_1(Rb^{87})}{T_1(Rb^{85})} = \left[\frac{Q(Rb^{85})}{Q(Rb^{87})}\right]^2 \frac{f(\frac{5}{2})}{f(\frac{3}{2})} = 4 \cdot 28 \times 0 \cdot 24 = 1 \cdot 027.$$

The experimental value of this ratio in RbCl is $1 \cdot 23 \pm 0 \cdot 40$.

The temperature dependence of the relaxation time T_1 of I^{127} in KI has been measured between 77° K and 800° K and found in very good agreement with the law $T_1 \propto T^{-2}$ expected for a Raman process. The contribution of a higher order process (27) that would lead to $T_1 \propto T^{-4}$ seems excluded by that experiment.

The fact that the theoretical relaxation times calculated on the basis of the point-charge model turn out to be much too long is not surprising. We saw in Chapter VI that when a charge e is placed outside a spherical ion at a distance a from the nucleus of this ion, the actual field gradient 'seen' by the nucleus is larger than the external gradient $3e/a^3$ by a factor which for the heavier ions was estimated to be of the order of a hundred. This enhancement of the gradient which results from a polarization effect or distortion of the electronic shells of the ion by the external charge was written as $(1+\gamma)$ in Chapter VI. The calculated relaxation times should thus be divided by factors of the order of $(1+\gamma)^2$. Since the discrepancy is in some cases still of the order of 10^6 it will not be removed entirely by the factor $(1+\gamma)^2$.

The ionic model also fails to account for results in sodium chloride where
$$Q(Cl^{35}) \sim 0.8 \times 10^{-25}, \quad Q(Na^{23}) \sim 1 \times 10^{-25},$$
and
$$T_1(Cl) \simeq T_1(Na) \simeq 10 \text{ sec.}$$
This contrasts with the calculated values $(1+\gamma) \sim 55$ for chlorine and $(1+\gamma) \sim 5$ for sodium, which, according to the previous argument, should lead to a ratio of their relaxation rates of the order of a hundred.

As an alternative explanation it has been suggested (32) that at least in some ionic crystals covalent effects might be important.

Consider a negative ion, say Br⁻, and a neighbouring positive ion, say K⁺. In a purely ionic structure both ions have closed electronic shells. In a hypothetical purely covalent structure one electron is missing from the p shell of the bromine atom and is localized on a neighbouring potassium atom.

The remaining hole in the p shell produces a very large gradient at the nucleus, proportional to $\langle r^{-3} \rangle$, which can be several orders of magnitude larger than a^{-3}. Even if in the actual structure the relative weight λ of the covalent structure is small, the gradient proportional to $\lambda \langle r^{-3} \rangle$ is still fairly large. Furthermore, what matters in a relaxation process is not the equilibrium value of the gradient (which incidentally if summed over all neighbours must vanish because of cubic symmetry) but its dependence on a small relative displacement of the neighbouring atoms. It is not unreasonable to think that the degree of covalency λ is a very sensitive function of the interatomic distance, which further enhances the efficiency of a covalent structure for relaxation. In favour of the hypothesis that a partly covalent structure is of importance in relaxation processes one may quote an interesting correlation between the measured relaxation time and the measured chemical shift. Fig. IX, 11 from reference (33) shows the correlation between the bromine relaxation time measured in different halides and the chemical shift (taken with respect to an aqueous solution of NaBr). It will be remembered that the chemical shift σ, defined as a relative change $(1-\sigma)$ of the nuclear resonance frequency, is the sum of a diamagnetic contribution σ_d that is always positive and of a paramagnetic contribution σ_p which vanishes for electronic closed shells such as exist for an ionic structure. The association of the shorter relaxation times with the larger negative shifts would point to an increasing influence of a covalent structure for either quantity.

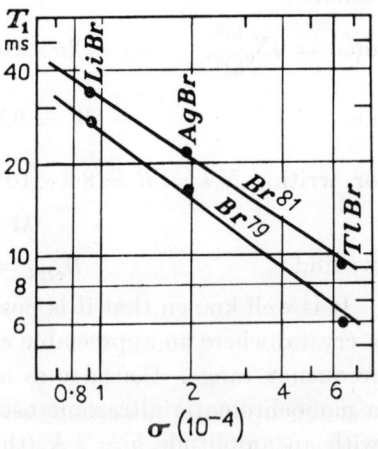

FIG. IX, 11. Relation between T_1 and the chemical shift of bromine in LiBr, AgBr, and TlBr.

In the calculation of relaxation times of reference (32) the degree of covalency λ was extracted from chemical shift data and its dependence on the interatomic distance R was assumed to be of the form $\exp(-R/\rho)$, which, according to the theory of ionic crystals of Born and Mayer, describes the variation of the repulsive potential between ions. The relaxation times, due to the Raman process, calculated from these data by a procedure otherwise similar to that of reference (30), are longer than the experimental value by factors of the order of three to ten.

D. Ultrasonic experiments

The extraordinary smallness of the relaxation transition rate caused by the direct process was explained earlier by the negligible density of energy in the phonon spectrum at the spin resonance frequency ω_0. Consider, for instance, a crystalline sample with a mass equal to the atomic mass M_0 containing $N_0 = 6 \cdot 06 \times 10^{23}$ atoms, and calculate the thermal energy W contained in an interval of 10 kc/s (a typical nuclear

spin line width) around a frequency of 10 Mc/s:
$$W = \sigma(\omega_0)\Delta\omega\,\bar{n}\hbar\omega_0,$$
where
$$\sigma(\omega) = 9N_0\frac{\omega^2}{\Omega^3}, \qquad \omega_0 = 2\pi\times 10^7, \qquad \Delta\omega = 2\pi\times 10^4, \qquad \bar{n} = kT/\hbar\omega_0;$$
i.e.
$$W = 9N_0 kT\left(\frac{\omega_0}{\Omega}\right)^2\left(\frac{\Delta\omega}{\Omega}\right)$$
or, writing $N_0 k = R = 8\cdot 4\times 10^7$ and taking
$$\Omega = 2\pi\times 10^{13},$$
we find
$$W_{\text{ergs}} = 7\cdot 5\times 10^{-13}T.$$

It is well known that it is possible to excite ultrasonic oscillations in a crystal where an appreciable energy is concentrated in a very narrow frequency range. Consider as an example the energy U contained in a monochromatic ultrasonic oscillation at a frequency of 10 Mc/s and with an amplitude $b = 2$ Å (this is easily realized in practice). This energy is of the order of $M\omega^2 b^2$ or, assuming $M = 20$ g (the same sample as before), $U \sim 30$ ergs. To estimate the rate of the transitions that this energy may induce between two levels of a system of nuclear spins having a width of 10 kc/s, we may multiply the relaxation rates, for the direct process, (93) or (101), by the ratio U/W of the ultrasonic energy to the thermal energy, that exists in the required band width. We find from (93) a magnetic transition probability of the order of
$$5\times 10^{-17}\left(\frac{T}{\Theta}\right)\times\frac{30}{7\cdot 5\times 10^{-13}T} \simeq \frac{2\times 10^{-3}}{\Theta} \sim 10^{-5}\,\text{sec}^{-1}.$$

Similarly, (101) yields a quadrupole transition probability $\sim 0\cdot 02\,\text{sec}^{-1}$.

We have calculated the ultrasonic transition probabilities in this roundabout fashion to demonstrate that their mechanism is essentially the same as that of a thermal direct process, the only difference being in the spectral density at the spin frequency ω_0.

The order of magnitude of the magnetic or electric ultrasonic probabilities, can also be very easily estimated directly.

Consider two nuclear spins (protons) separated by a distance $a \sim 2$ Å. A vibration of the nuclear spins of amplitude b at a frequency ω and with a wavelength $\lambda = 2\pi v/\omega$, causes a change in the relative distance of the nuclei $d \sim 2\pi(a/\lambda)b$. The local field $\sim \gamma\hbar/a^3$, produced by one spin at the other, then has an oscillating component
$$H_1 \sim \frac{\gamma\hbar}{a^3}\frac{d}{a} \sim \frac{\gamma\hbar}{a^3}\frac{2\pi b}{\lambda}.$$

The transition probability induced by that field will be $P_1 \sim (\gamma H_1)^2/\Delta\omega$, where $\Delta\omega$ is the line width. With

$$\gamma \sim 2\pi \times 4{\cdot}2 \times 10^3 \text{ (protons)}, \qquad a = b = 2 \times 10^{-8},$$
$$v = 2 \times 10^5, \qquad \omega = 2\pi \times 10^7,$$

we find directly $\qquad P_1 \sim 5 \times 10^{-6} \text{ sec}^{-1};$

that is of the same order of magnitude as before. A similar calculation is easily performed for an electric quadrupole transition.

(a) Quadrupole transitions

We saw in Section C that a calculation of quadrupole transition probabilities due to the Raman process, using the point-charge model, resulted in a gross underestimate of these probabilities.

It is reasonable to think that the same applies to direct process transitions and that the actual rates of quadrupole ultrasonic transitions could be a good deal larger than the order of magnitude estimate of 0·02 sec^{-1} that we made, and so could compete successfully with the quadrupole thermal relaxation rates caused by the Raman process. This has actually been confirmed experimentally.

The detection of the ultrasonic transitions can in principle be based on direct absorption of ultrasonic energy by the sample, resulting in an additional load on the ultrasonic generator when the resonance condition is satisfied.

Only a single experiment of that kind has been reported so far (**34**). In a single crystal of InSb, where the static quadrupole splitting of In^{115} (spin $\frac{9}{2}$) vanishes because of its cubic environment, ultrasonic vibrations were produced at the Larmor frequency of In^{115} and also at twice its Larmor frequency (transitions $|\Delta m| = 2$). The corresponding change in the mechanical Q (quality factor, not quadrupole moment!) of the sample resulted in an observable change in the electric impedance of a quartz transducer. This type of detection may prove of interest in the future for the detection of nuclear resonance in metals since the penetration of acoustical waves into the sample is not hampered by skin depth problems.

All other experiments in the field, reported so far, are based on ultrasonic saturation of the nuclear resonance, resulting in a change of the nuclear magnetization examined subsequently by transient magnetic resonance methods.

In a pioneer experiment (**35**), the existence of nuclear electric quadrupole transitions induced by acoustical vibrations was first demonstrated by a saturation of the zero magnetic field resonance of Cl^{35} in $NaClO_3$ at a frequency of 30·57 Mc/s.

In later experiments (**36, 37, 38**) acoustical transitions were induced between nuclear Zeeman levels in ionic cubic crystals where the static quadrupole interaction vanished (except for crystal imperfections). Although electric quadrupole transitions $|\Delta m| = 1$ and $|\Delta m| = 2$ have comparable intensities, the acoustical frequency chosen was systematically twice the nuclear Larmor frequency to make sure that the transition induced $|\Delta m| = 2$ was unambiguously electric quadrupole rather than magnetic dipole.

It has been shown in Chapter V for spins $I = \tfrac{3}{2}$ (but this result is easily generalized to higher spins) that if the nuclear spin levels are equidistant, spin-spin interactions maintain a distribution of their populations describable by a spin temperature, and that under these conditions, externally induced transitions $|\Delta m| = 2$, with a probability per unit time W, reduce the nuclear polarization to a steady-state value

$$\langle I_z \rangle = \frac{I_0}{1+8WT_1/5}. \tag{103}$$

A measurement of $\langle I_z \rangle$, immediately after the ultrasonic saturation, yields the value of W. The transition probability W can be computed exactly if the amplitudes of the changes δV_{ij} in the electric-field gradient at the nucleus, caused by the ultrasonic vibration, are known in addition to the nuclear quadrupole moment. The theoretical difficulty is that the actual relationship between the δV_{ij} and the stresses W_{ik} caused by the ultrasonic vibrations, as stated several times already, is *not* the one obtainable by using the point charge model.

The problem to which ultrasonic experiments may provide a solution is thus the determination of the four-index tensor S, relating the changes δV_{ij} to the stresses W_{kl}:

$$\delta V_{ij} = \sum_{k,l} S_{ij,kl} W_{kl}. \tag{104}$$

The introduction of the polarization factor $(1+\gamma)$ corresponds to the assumption that if $S'_{ij,kl}$ is the tensor calculated using the point charge model, the real tensor S is related to S' by a single constant:

$$S_{ij,kl} = (1+\gamma) S'_{ij,kl}. \tag{105}$$

The experimental determination of the tensor S. or under the simplified assumption (105), of the constant γ, requires the knowledge of the stresses W, that is, of the amplitude, the wavelength, and the polarization of the ultrasonic vibrations. Several methods have been used to estimate at least the amplitude of the vibrations.

In the experiment (36) performed on Na^{23} in NaCl, this amplitude was deduced from the acoustical energy U stored in the sample, the latter being correlated to the power P dissipated in the sample, by the relation
$$P = U/T_{\text{ph}},$$
where the constant T_{ph} is the phonon relaxation time, or decay time of an ultrasonic excitation at the frequency ω, and is determined independently. The value $1+\gamma \sim 1$, arrived at for Na^{23}, should not be taken too seriously in view of the considerable uncertainties involved in this evaluation.

In the experiment (37) performed on INa, ultrasonic saturation of the resonance of either spin was performed. No attempt was made to measure the absolute values of $\gamma(\text{I}^{127})$ (spin $\frac{5}{2}$) and of $\gamma(\text{Na}^{23})$ (spin $\frac{3}{2}$), which would require the knowledge of the amplitude of acoustical vibrations; instead, the ratio $\gamma(\text{I}^{127})/\gamma(\text{Na}^{23})$ was estimated from a relative comparison of their saturation behaviour and found to be 10·9 in qualitative agreement with the value 28 expected from a theoretical estimate of polarization effects. Unfortunately, the interpretation of the experimental results on which the estimate of $\gamma(\text{I}^{127})/\gamma(\text{Na}^{23})$ is based, is made very uncertain by the considerable broadening of the transitions $|\Delta m| = 2$ for I^{127}, due to crystal imperfections. The spin diffusion is partially quenched by the inequalities between the distances of Zeeman energy levels of I^{127}, created by these imperfections, and the effect of an ultrasonic saturation under these circumstances is difficult to predict.

In the experiment (38) an attempt was made to produce a well-defined mode of oscillation in a cylindrical rod cut from a single crystal of NaCl, and the amplitude of acoustical vibrations was obtained from a careful measurement of the displacements of the end of the rod. Ultrasonic saturation of Na^{23} and Cl^{35} was observed. The main conclusions reached in that study are that a relation such as (105) based on a purely ionic model of the crystal is not valid, and that covalent effects play an important part in the determination of the oscillating electric field gradients near the nucleus.

(b) *Magnetic transitions*

The ultrasonic modulation of the relative position of two nuclear spins may in principle induce transitions between their energy levels.

Assuming the spins to be different, with Larmor frequencies ω and ω', this modulation can be produced at either one of the frequencies ω, ω', $\omega+\omega'$, $\omega-\omega'$. The first two frequencies correspond to a flip of a single spin, the third and the fourth to simultaneous flips of either spin in the same or in opposite directions. The transition rates are expected to be small, as exhibited by our order of magnitude calculation.

However, experimental situations can be found where this effect could be observed and provide information of some interest. It would probably be possible, by creating in a crystal of LiF an acoustical oscillation at a frequency $\omega(F^{19}) \pm \omega(Li^6)$ and by taking advantage of the exceedingly long relaxation time of Li^6, to reproduce the 'solid-state' polarization described in Section III in this chapter and in reference (20).

More generally, large 'solid-state' dynamic nuclear polarization resulting from dipolar coupling between electron and nuclear spins could be produced if ultrasonic vibrations of sufficient intensity were available at microwave frequencies. This effect is different from the one described in Section III B, where a simultaneous flip of two nuclear spins was produced by a large magnetic r.f. field driving a forbidden transition. There, for a constant amplitude of the r.f. field, that is, for a constant stored electromagnetic energy, the transition probability went down as ω^{-2}, with increased resonance frequency. In the ultrasonic experiment, for a constant stored energy $Mb^2\omega^2$, where b is the amplitude of oscillation, the relative displacement of two nuclei with an equilibrium separation a is $d \approx 2\pi(a/\lambda)b \propto b\omega$, and the ultrasonic transition probability for a constant stored energy is independent of the frequency.

Another point in a hypothetical magnetic ultrasonic experiment is the following: the corresponding transition probabilities can, in contradistinction to quadrupole transitions, be easily calculated exactly from first principles if the ultrasonic amplitudes are known. Conversely, a measure of the ultrasonic magnetic transition rates would provide information on the vibrational intensity, that could then be used to interpret the result of quadrupole ultrasonic transitions observed in the same crystal. Lithium fluoride is an example in point. These considerations would not apply if indirect electron-coupled interactions existed between the nuclear spins, for their dependence on the relative positions of the nuclear spins is conditioned by the overlap of electronic wave functions and cannot be predicted in a simple way.

REFERENCES

1. L. C. HEBEL and C. P. SLICHTER, *Phys. Rev.* **113**, 1504, 1959.
2. D. PINES, *Solid State Physics*, **1**, 366, 1955.
3. A. W. OVERHAUSER, *Phys. Rev.* **92**, 411, 1953.
4. A. ANDERSON and A. C. REDFIELD, *Proc. Int. Conf. on Low Temperatures, Madison Wisconsin*, Aug. 1957.
5. D. F. HOLCOMB and R. E. NORBERG, *Phys. Rev.* **98**, 1074, 1955.
6. T. R. CARVER and C. P. SLICHTER, ibid. **92**, 212, 1953.
7. —— —— ibid. **102**, 975, 1956.
8. G. FEHER and A. F. KIP, ibid. **98**, 337, 1955.
9. F. J. DYSON, ibid., p. 349, 1955.
10. CH. RYTER, *Phys. Rev. Letters*, **5**, 10, 1960.
11. N. BLOEMBERGEN, *Physica*, **15**, 386, 1949.
12. P. G. DE GENNES, *Phys. Chem. Solids*, **3**, 345, 1958.
13. R. V. POUND, *J. Phys. Chem.* **57**, 743, 1953.
14. J. WINTER, *C.R. Acad. Sci.* **249**, 2192, 1959.
15. R. G. SHULMAN and B. J. WYLUDA, *Phys. Rev.* **103**, 1127, 1956.
16. N. BLOEMBERGEN, *Physica*, **20**, 1130, 1954.
17. A. ABRAGAM, J. COMBRISSON, and I. SOLOMON, *C.R. Acad. Sci.* **246**, 1035, 1958.
18. ——, A. LANDESMAN, and J. M. WINTER, ibid. **247**, 1849, 1958.
19. —— *Phys. Rev.* **98**, 1729, 1955.
20. —— and W. G. PROCTOR, *C.R. Acad. Sci.* **246**, 1258, 1958.
21. M. ABRAHAM, M. A. H. MCCAUSLAND, and F. N. H. ROBINSON, *Phys. Rev. Letters*, **2**, 449, 1959.
22. M. BORGHINI and A. ABRAGAM, *C.R. Acad. Sci.* **248**, 1803, 1959.
23. G. FEHER, R. C. FLETCHER, and E. A. GERE, *Phys. Rev.* **100**, 1784, 1955.
24. A. ABRAGAM, J. COMBRISSON, and I. SOLOMON, *C.R. Acad. Sci.* **247**, 2337, 1958.
25. J. COMBRISSON, to be published.
26. R. E. PEIERLS, *Quantum Theory of Solids*, Clarendon Press, Oxford, 1955.
27. G. R. KUTSISHVILI, *Publications of the Georgian Institute of Sciences IV*, **1**, 1956 (in Russian).
28. I. WALLER, *Z. Phys.* **79**, 370, 1932.
29. R. V. POUND, *Phys. Rev.* **79**, 685, 1950.
30. J. VAN KRANENDONK, *Physica*, **20**, 781, 1954.
31. R. L. MIEHER, *Phys. Rev. Letters*, **4**, 57, 1960.
32. K. YOSIDA and T. MORIYA, *J. Phys. Soc. Japan*, **11**, 33, 1956.
33. T. KANDA, ibid. **10**, 85, 1955.
34. M. MENES and D. I. BOLEF, *Phys. Rev.* **109**, 218, 1958.
35. W. G. PROCTOR and W. H. TANTTILA, ibid. **98**, 1854, 1955; **101**, 1757, 1956.
36. —— and W. ROBINSON, ibid. **104**, 1344, 1956.
37. D. A. JENNINGS, W. H. TANTTILA, and O. KRAUS, ibid. **109**, 1059, 1958.
38. E. F. TAYLOR and N. BLOEMBERGEN, ibid. **113**, 431, 1959.

X

THEORY OF LINE WIDTH IN THE PRESENCE OF MOTION OF THE SPINS

Je hais le mouvement qui déplace les lignes.
BAUDELAIRE (*Les Fleurs du mal*)

IN the last two chapters we have given a survey of the various relaxation processes that enable a nuclear spin system to come into thermal equilibrium with the other degrees of freedom of the sample or 'lattice'. These processes originate in various types of motions such as molecular rotation or translation, motion of conduction electrons, relaxation flips of the spins of paramagnetic impurities, considered as part of the 'lattice', and so on. In the present chapter we consider the influence of these motions on another aspect of nuclear magnetism, the width of the resonance lines. With a few exceptions these motions result in a narrowing of the resonance lines, whence the name of 'motion narrowing' generally used for a description of such phenomena.

We also include in this chapter a short survey of some relaxation processes in solids, namely those caused by internal motions such as hindered rotation of molecules, torsion oscillations, and translational diffusion. The reason for including this survey in this chapter rather than in Chapter IX, which was specifically devoted to relaxation in solids, is a practical one: in solids, the effects of such motions on line width are more striking and have been studied more extensively than their influence on the spin-lattice relaxation times, and investigations of both phenomena are usually presented together in the literature. It therefore seemed advisable to present theoretical and experimental results on relaxation times in solids where these types of motion prevail, after rather than before a theory of motion narrowing.

I. INTRODUCTION

A theory of dipolar line width in a rigid lattice, that is, in a sample where the lengths and orientations of the vectors describing the relative positions of the spins do not change in time, has been given in Chapter IV. The conclusion reached was that the assumption of a gaussian shape for the resonance curve was a fair, although by no means perfect, approximation, and that consequently its r.m.s. width available from the theory was not qualitatively different from the real width. This width, expressed in gauss, was of the order of the local field produced at the site of a spin by its neighbours, that is, in most practical cases, of the order of a gauss, to within a factor 10.

Neither the assumptions nor the results of the previous theory are valid for liquid or gaseous samples where fast relative motions exist among the spins and where the observed line widths are orders of magnitude smaller than in a rigid lattice. Furthermore, the shape of the resonance curve is much more nearly that of a Lorentz curve than that of a gaussian. Those are the facts to be explained.

The physical explanation of the motion narrowing which was given at an early date (1) is the following. If the spins are in rapid relative motion, the local field 'seen' by a given spin will fluctuate rapidly in time. Only its average value taken over a time long compared with the duration of a fluctuation will be observed and this average is much smaller than the instantaneous value of the local field, whence the narrowing. The rate of the fluctuations of the local field can be described by a correlation time τ_c already introduced in Chapter VIII. The question immediately arises: what is the criterion for stating that the fluctuation is fast, or in other words, with what frequency should the frequency of fluctuation, $1/\tau_c$, of the local field, be compared? There is a strong temptation to compare it with the Larmor frequency and to say that only an average, greatly reduced value of the local field would be observed if it had fluctuated many times during a Larmor period. The fallacy of this argument is made apparent by the use of a frame rotating at the Larmor frequency. In that frame the applied field is absent and the *only* field 'seen' by each spin is the local field which cannot be made to cancel in any frame of coordinates since it has different values for different spins. Thus it is with respect to the instantaneous Larmor precession in the instantaneous local field alone that the fluctuation of the lattice must be fast. The frequency of the precession in the instantaneous local field is of the order of the rigid lattice *line* width expressed in cycles, that is, of the order of $(\overline{\Delta\omega_0^2})^{\frac{1}{2}}$ where $\overline{\Delta\omega_0^2}$ is the second moment calculated in Chapter IV. The criterion for motion narrowing would thus be

$$(\overline{\Delta\omega_0^2})^{\frac{1}{2}}\tau_c \ll 1. \qquad (1)$$

To find the real line width $\Delta\omega$ observable under those conditions, the following intuitive reasoning can be used (1). The local field is a stationary random function with a mean square value which is unaffected by the motion and is equal to $\overline{\Delta\omega_0^2} = \gamma^2 \overline{\Delta H^2}$. According to equation (17) of Chapter VIII, $\overline{\Delta\omega_0^2}$ is related to the power spectrum of the local field by the relation

$$\overline{\Delta\omega_0^2} = a \int_{-\infty}^{\infty} J(\omega)\,d\omega, \tag{2}$$

but this spectrum, which for a rigid lattice contained only the frequency zero, now extends over a wide band. Following an argument used several times before, only the quasi-adiabatic components of the local field will be effective for the broadening, that is, those in the vicinity of the frequency zero (in the rotating frame). Since the line has a width (unknown) $\Delta\omega$, Fourier components of the local field with frequencies between $\pm\Delta\omega$ can still be considered as adiabatic and taken as limits in the integral (2). The real line width is then defined implicitly by the relation

$$\Delta\omega^2 \simeq a \int_{-\Delta\omega}^{\Delta\omega} J(\omega)\,d\omega \simeq \overline{\Delta\omega_0^2}\, \frac{\int_{-\Delta\omega}^{\Delta\omega} J(\omega)\,d\omega}{\int_{-\infty}^{\infty} J(\omega)\,d\omega}. \tag{3}$$

It has already been mentioned that for a random function with a correlation time τ_c, $J(\omega)$ is approximately constant and equal to $J(0)$ for $\omega < 1/\tau_c$, and becomes very small as soon as ω is appreciably larger than $1/\tau_c$. Approximately, $\int_{-\infty}^{\infty} J(\omega)\,d\omega \simeq (2/\tau_c) J(0)$. If $\Delta\omega_0$ and thus also $\Delta\omega \ll 1/\tau_c$,

$$\left.\begin{array}{l} \Delta\omega^2 \simeq \overline{\Delta\omega_0^2}\,\dfrac{2\Delta\omega J(0)}{(2/\tau_c)J(0)} \simeq \overline{\Delta\omega_0^2}\,\Delta\omega\,\tau_c \\[6pt] \Delta\omega \simeq \overline{\Delta\omega_0^2}\,\tau_c \end{array}\right\} \tag{4}$$

If $(\overline{\Delta\omega_0^2})^{\frac{1}{2}}\tau_c \ll 1$ there will be appreciable narrowing.

Equation (4) can also be obtained by another intuitive argument. We can define qualitatively the line width $\Delta\omega$ as the inverse of the time t after which two spins precessing in their respective local fields, and initially in phase, are out of phase by an amount of the order of unity. Suppose that the difference between the two local fields (in cycles) keeps the same value $\Delta\omega_0$ as in the rigid lattice but changes its sign randomly with an average frequency $1/\tau_c \gg \Delta\omega_0$. It is clear that after a time t the average phase angle will be $\tau_c \Delta\omega_0 \sqrt{(t/\tau_c)}$ and this will be of order unity after a time $t \simeq 1/\Delta\omega \simeq 1/\overline{\Delta\omega_0^2}\tau_c$.

An interesting example of motion narrowing is the absence of Doppler width in nuclear resonance experiments on gases under pressure. The explanation of this result is similar to that given above: because of the collisions and of the subsequent changes of the velocity of the molecules, the Doppler frequency shifts undergo changes in sign and magnitude at a rate $1/\tau$ (where τ is the time between collisions), large compared with the absolute values $|\Delta\omega|$ of the instantaneous Doppler shifts. The contribution of the Doppler effect to the line width has the greatly reduced value $\overline{\Delta\omega^2}\,\tau$.

We shall now pass on to a more quantitative theory of motion narrowing.

II. The Adiabatic Case

A. General theory

The general formalism required for a quantitative theory of line width was outlined in Chapter IV. The shape $I(\omega)$ of the absorption line is given by the Fourier transform of the relaxation function of the magnetization:

$$G(t) = \operatorname{tr}\{\mathcal{M}_x(t)\mathcal{M}_x\} = \operatorname{tr}\{e^{i\mathcal{H}t}\mathcal{M}_x e^{-i\mathcal{H}t}\mathcal{M}_x\}. \tag{5}$$

It is further assumed that the Hamiltonian $\hbar\mathcal{H}$ is the sum of an unperturbed Hamiltonian $\hbar\mathcal{H}_T^0$ (the meaning of the index T will appear in a moment) and of a small perturbing Hamiltonian $\hbar\mathcal{H}_1$. In the absence of $\mathcal{H}_1$ the magnetic resonance spectrum of $\mathcal{H}_T^0$ contains one or several infinitely sharp lines. The simplest example for $\mathcal{H}_T^0$ is the Zeeman coupling of a system of spins with an applied d.c. field.

It should be realized that the existence of a discrete, infinitely sharp magnetic absorption spectrum for $\mathcal{H}_T^0$ does not imply that the energy spectrum of the Hamiltonian $\mathcal{H}_T^0$ itself has the same features. Clearly any Hamiltonian $\hbar\mathcal{F}$ with arbitrarily complicated spectrum could be added to $\hbar\mathcal{H}_T^0$ without changing the absorption $I(\omega)$ provided $\mathcal{F}$ commutes with $\mathcal{H}_T^0$ and $\mathcal{M}$.

The perturbing Hamiltonian $\hbar\mathcal{H}_1$, such as for instance the dipolar coupling between the spins, is responsible for the broadening of the infinitely sharp lines of the spectrum of $\hbar\mathcal{H}_T^0$. It is assumed to be sufficiently small for the magnetic absorption spectrum $I(\omega)$ of the total Hamiltonian $\hbar\mathcal{H}$ to remain a sharp line spectrum.

To understand the mathematical nature of motion narrowing it is best to consider a few examples. In most of the problems studied so far the main Hamiltonian $\hbar\mathcal{H}_T^0$ of the system has been composed of two parts (hence the index T for total), the Zeeman energy $\hbar\mathcal{H}_0$ of a system of spins I with a given Larmor frequency and a Hamiltonian $\hbar\mathcal{F}$, commuting with both $\mathcal{H}_0$ and the total magnetization $\mathcal{M}$ of the spins I.

Thus in liquids or in gases $\hbar\mathcal{F}$ is the kinetic energy of the Brownian motion of the molecules, in ionic crystals the energy of lattice vibrations, or in imperfect crystals, where diffusion is important, the kinetic energy of the diffusing nuclei. The Hamiltonian $\hbar\mathcal{F}$ need not, however, be a kinetic lattice energy. A scalar coupling $\hbar \sum_{i<k} J_{ik} \mathbf{I}_i \cdot \mathbf{I}_k$ between the spins I also satisfies the conditions

$$[\mathcal{F}, \mathcal{M}] = 0; \quad [\mathcal{F}, \mathcal{H}^0] = 0.$$

When there are two species of spins, the 'resonant' spins I and the non-resonant spins S, any couplings between the spins S can also be considered as a Hamiltonian $\mathcal{F}$. Commuting with both $\mathcal{H}_0$ and $\mathcal{M}$, $\mathcal{F}$ would be completely foreign to the magnetic resonance phenomenon but for the fact that it does *not* commute with the perturbation $\mathcal{H}_1$ and thus deeply affects the way in which $\mathcal{H}_1$ broadens the infinitely sharp spectrum of $\mathcal{H}^0$. To sum up:

$$\left. \begin{array}{c} \mathcal{H}_T^0 = \mathcal{H}^0 + \mathcal{F} \\ [\mathcal{H}^0, \mathcal{F}] = 0; \quad [\mathcal{M}, \mathcal{F}] = 0; \quad [\mathcal{H}_1, \mathcal{F}] \neq 0 \end{array} \right\}. \quad (6)$$

Although mathematically the separation of $\mathcal{H}_T^0$ into its two parts $\mathcal{H}^0$ and $\mathcal{F}$ according to (6) is not unique, the physical context will prevent any ambiguity.

The time-dependent magnetization $\mathcal{M}_x(t)$ of equation (5) can be obtained in the interaction representation by solving the differential equation obeyed by the operator

$$\left. \begin{array}{c} \mathcal{M}_x^*(t) = e^{-i\mathcal{H}_T^0 t} \mathcal{M}_x(t) e^{i\mathcal{H}_T^0 t} \\ \mathcal{M}_x(t) = e^{i(\mathcal{H}_T^0 + \mathcal{H}_1)t} \mathcal{M}_x e^{-i(\mathcal{H}_T^0 + \mathcal{H}_1)t} \end{array} \right\}. \quad (7)$$

where

The operator $\mathcal{M}_x^*(t)$ would be time-independent if $\mathcal{H}_1$ did not exist and, since $\mathcal{H}_1$ is a small perturbation, the rate of change of $\mathcal{M}_x^*(t)$ is expected to be slow. This rate of change is given by

$$\frac{1}{i}\frac{d\mathcal{M}_x^*}{dt} = [e^{-i\mathcal{H}_T^0 t}\mathcal{H}_1 e^{i\mathcal{H}_T^0 t}, \mathcal{M}_x^*] = [\mathcal{H}_1^*(t), \mathcal{M}_x^*] = [e^{-i\mathcal{H}^0 t}\mathcal{H}_1(t)e^{i\mathcal{H}^0 t}, \mathcal{M}_x^*], \quad (8)$$

where the following definitions are used:

$$\left. \begin{array}{c} \mathcal{H}_1^*(t) = e^{-i\mathcal{H}_T^0 t}\mathcal{H}_1 e^{i\mathcal{H}_T^0 t} = e^{-i(\mathcal{H}^0 + \mathcal{F})t}\mathcal{H}_1 e^{i(\mathcal{H}^0 + \mathcal{F})t} = e^{-i\mathcal{H}^0 t}\mathcal{H}_1(t)e^{i\mathcal{H}^0 t} \\ \mathcal{H}_1(t) = e^{-i\mathcal{F}t}\mathcal{H}_1 e^{i\mathcal{F}t} \end{array} \right\}. \quad (8')$$

where

In an explicit matrix representation equation (8) can be rewritten

$$-i\frac{d}{dt}(E_0 sf | \mathcal{M}^* | E_0' s' f')$$
$$= \sum_{E_0'' s'' f''} \{(E_0 sf | \mathcal{H}_1(t) | E_0'' s'' f'')(E_0'' s'' f'' | \mathcal{M}^* | E_0' s' f')e^{-i(E_0 - E_0'')t} -$$
$$-(E_0 sf | \mathcal{M}^* | E_0'' s'' f'')(E_0'' s'' f'' | \mathcal{H}_1(t) | E_0' s' f')e^{-i(E_0'' - E_0')t}\}. \quad (9)$$

This equation can also be written in a form intermediate between the operator expression (8) and the numerical matrix equations (9).

We can introduce symbols such as $(E_0 s | \mathcal{H}_1 | E_0' s')$ or $(E_0 s | \mathcal{H}_1(t) | E_0' s')$ which will still be operators with respect to the degrees of freedom of $\mathcal{F}$, with matrix elements defined by

$$(f | (E_0 s | \mathcal{H}_1 | E_0' s') | f') = (E_0 sf | \mathcal{H}_1 | E_0' s' f'). \quad (10)$$

With this convention the equations (8) or (9) can be rewritten in the semi-operator form

$$-i\frac{d}{dt}(E_0 s | \mathcal{M}^* | E_0' s')$$
$$= \sum_{E_0'' s''} (E_0 s | \mathcal{H}_1(t) | E_0'' s'')(E_0'' s'' | \mathcal{M}^* | E_0' s')e^{-i(E_0 - E_0'')t} -$$
$$-(E_0 s | \mathcal{M}^* | E_0'' s'')(E_0'' s'' | \mathcal{H}_1(t) | E_0' s')e^{-i(E_0'' - E_0')t}, \quad (11)$$

which in appearance closely resembles the equation (23) of Chapter IV except that:

(i) the matrix elements in (11) are still operators with respect to the degrees of freedom of $\mathcal{F}$;

(ii) $\mathcal{H}_1(t)$ is time-dependent through the definition

$$\mathcal{H}_1(t) = e^{-i\mathcal{F}t}\mathcal{H}_1 e^{i\mathcal{F}t}.$$

We saw in Chapter VIII that, given a spin system subject to a time-dependent perturbation $\mathcal{H}_1$, there were two alternative descriptions of this situation: the semi-classical one when matrix elements such as $(E_0 s | \mathcal{H}_1(t) | E_0' s')$ were random functions of time, and the quantum mechanical method where the time variation of the perturbing Hamiltonian $\hbar\mathcal{H}_1$ was introduced through a Hamiltonian of 'motion' $\hbar\mathcal{F}$ by the relation
$$\mathcal{H}_1(t) = e^{-i\mathcal{F}t}\mathcal{H}_1 e^{i\mathcal{F}t}.$$

The problem was greatly simplified by the assumption of short correlation times ($|\mathcal{H}_1|\tau_c \ll 1$) and it was possible to cast both approaches into a single formalism. On the other hand, to study incipient motion narrowing with relatively long correlation times, the semi-classical approach is much simpler and will be used henceforth.

We shall, then, assume that the various quantities of equation (11) are

c-numbers, random functions of time, rather than operators defined in equation (10).

Simple assumptions about the behaviour of the random functions will lead to detailed predictions about the shape of the resonance curves. On the other hand, at least in special cases, some rigorous statements about this shape can be derived from the quantum mechanical approach, thus permitting a test of some of the features of the random functions model.

We can now compare the rigid lattice case when the various matrix elements of $\mathcal{H}_1$ in (11) are time-independent, with that of motion narrowing where they vary randomly in time.

For the rigid lattice, it has already been stated in Chapter IV that on the right-hand side of (11), only the adiabatic terms with $E_0 = E_0''$ or $E_0 = E_0'$ should be retained, since all the others, because of their fast varying exponential coefficients, give a negligible contribution.

For the time varying matrix elements $(E_0 s \,|\, \mathcal{H}_1(t) \,|\, E_0'' s'')$ this argument is not necessarily valid since the variation contained in the exponential $e^{i(E_0-E_0'')t}$ may well be compensated by that of the matrix elements of $\mathcal{H}_1(t)$. We shall, however, assume for the time being that the latter variation is still sufficiently slow for the neglect of matrix elements with $E_0 - E_0'' \neq 0$ to be justified. This amounts to saying that the Fourier expansion of these matrix elements does not contain any appreciable frequency components in the vicinity of differences $E_0 - E_0''$, that is actually in the vicinity of the resonance frequency. Another way of putting it is to say that the corresponding correlation time τ_c is long compared with the Larmor period.

This will not be true in all cases, as for instance in liquids, but then the simplifications due to the shortness of the correlation time will come into play and results obtained in Chapter VIII will be used. The dropping in eqn. (11) of terms with $E_0 - E_0'' \neq 0$ amounts to the use of a truncated Hamiltonian $\mathcal{H}_1(t)$ that commutes with $\mathcal{H}^0$ (as has already been done for the rigid lattice in Chapter IV). It is possible to choose the second quantum number s, introduced to discriminate between eigenstates of $\hbar \mathcal{H}^0$ with the same unperturbed energy $\hbar E_0$, in such a way that matrix elements $(E_0 s \,|\, \mathcal{H}_1(t) \,|\, E_0' s')$ vanish unless $E_0 = E_0'$, $s = s'$. Equation (11) becomes

$$-i\frac{d}{dt}(E_0 s \,|\, \mathcal{M}^* \,|\, E_0' s')$$
$$= \{(E_0 s \,|\, \mathcal{H}_1(t) \,|\, E_0 s) - (E_0' s' \,|\, \mathcal{H}_1(t) \,|\, E_0' s')\}(E_0 s \,|\, \mathcal{M}^* \,|\, E_0' s'), \quad (12)$$

which integrates to

$$(E_0 s \,|\, \mathcal{M}^*(t) \,|\, E'_0 s') = (E_0 s \,|\, \mathcal{M}_0 \,|\, E'_0 s')\exp i\left\{\int_0^t [(E_0 s \,|\, \mathcal{H}_1(t') \,|\, E_0 s) - (E'_0 s' \,|\, \mathcal{H}_1(t') \,|\, E'_0 s')]\,dt'\right\},$$

and the relaxation function $G(t) = \mathrm{tr}\{\mathcal{M}_x(t)\mathcal{M}_x\}$ becomes

$$G(t) = \sum_{E_0 E'_0 ss'} |(E_0 s \,|\, \mathcal{M}_0 \,|\, E'_0 s')|^2 \exp\{i(E_0 - E'_0)t\} \times$$
$$\times \exp i\left\{\int_0^t [(E_0 s \,|\, \mathcal{H}_1(t') \,|\, E_0 s) - (E'_0 s' \,|\, \mathcal{H}_1(t') \,|\, E'_0 s')]\,dt'\right\}. \quad (13)$$

To avoid confusion, we have written $\mathcal{M}_0$ for the operator

$$\mathcal{M}^*(0) = \mathcal{M}(0)$$

that was previously called $\mathcal{M}$. Since a well-defined transition of the unperturbed Hamiltonian $\mathcal{H}^0$ is considered, the summation (13) must be restricted to terms such that $E_0 - E'_0 = \omega_0$. To proceed farther a simplified model must be chosen. We consider first the rigid lattice case where $\mathcal{H}_1$ and the quantity under the integral sign in equation (13) are time-independent. Each term of the sum (13) is defined by the three quantum numbers E_0, s, s' (since $E_0 - E'_0 = \omega_0$). It can be denoted instead by two indices ω and σ where

$$\omega = (E_0 s \,|\, \mathcal{H}_1 \,|\, E_0 s) - (E'_0 s' \,|\, \mathcal{H}_1 \,|\, E'_0 s')$$

and σ represents all the other indices necessary to specify this term when ω is given. $|(E_0 s \,|\, \mathcal{M}_0 \,|\, E'_0 s')|^2$ becomes a certain function $a(\sigma, \omega)$ and $G(t)$ becomes (for the rigid lattice)

$$G(t) = e^{i\omega_0 t} \sum_{\sigma, \omega} a(\sigma, \omega) e^{i\omega t},$$

where the frequencies ω form a quasi-continuous spectrum.

Let $f_\sigma(\omega)\,d\omega$ be the number of these frequencies between ω and $\omega + d\omega$ for a given value of σ. Then

$$G(t) = e^{i\omega_0 t} \int e^{i\omega t} \sum_\sigma a(\sigma, \omega) f_\sigma(\omega)\,d\omega.$$

If we define
$$P(\omega) = \sum_\sigma a(\sigma, \omega) f_\sigma(\omega),$$

$G(t)$ can be written
$$G(t) = e^{i\omega_0 t} \int e^{i\omega t} P(\omega)\,d\omega. \quad (14)$$

(14) can be interpreted by saying that the departure ω of the resonance frequency from its unperturbed value is a random variable with a probability distribution $P(\omega)$ for which we saw that the assumption

of a gaussian shape was not too bad an approximation. For brevity (14) can be rewritten

$$G(t) = e^{i\omega_0 t}\langle e^{i\omega t}\rangle, \tag{15}$$

where the symbol $\langle \rangle$ means average taken over the distribution $P(\omega)$. When we pass over to the time-dependent Hamiltonian $\mathscr{H}_1(t)$, (15) has to be replaced by

$$G(t) = e^{i\omega_0 t}\langle e^{iX(t)}\rangle, \tag{16}$$

where

$$X(t) = \int_0^t \omega(t')\,dt' = \int_0^t \{(E_0\,s\,|\,\mathscr{H}_1(t')\,|\,E_0\,s) - (E_0'\,s'\,|\,\mathscr{H}_1(t')\,|\,E_0'\,s')\}\,dt', \tag{17}$$

and the problem is to find the distribution law $P(X, t)$ for the random variable $X(t)$ and to calculate $\langle e^{iX(t)}\rangle$.

The model chosen, mainly for reasons of mathematical tractability, is the following (2): (i) the random function $\omega(t')$ is stationary and gaussian; (ii) its mean square value $\langle \omega^2\rangle$ has the same value as in the absence of motion. Physically this corresponds to the assumption that at each instant the microscopic distribution of local fields through the sample is the same as for the rigid lattice (for example, an instantaneous quasi-crystalline structure for a liquid), but that the local field at each point fluctuates at a rate described by a correlation function

$$G_\omega(\tau) = \langle \omega(t)\omega(t-\tau)\rangle = \langle \omega^2\rangle g_\omega(\tau), \tag{18}$$

where $\langle \omega^2\rangle$ is the second moment of the resonance line for a rigid lattice (written as $\overline{\Delta\omega_0^2}$ in Chapter IV). From the assumption of the gaussian character of $\omega(t)$ it follows that $X(t) = \int_0^t \omega(t')\,dt'$ is also gaussian, with $P(X, t) = [2\pi\langle X^2(t)\rangle]^{-\frac{1}{2}}\exp\{-\frac{1}{2}X^2/\langle X^2(t)\rangle\}$

$$\langle e^{-iX}\rangle = \int P(X, t)e^{-iX}\,dX$$
$$= [2\pi\langle X^2\rangle]^{-\frac{1}{2}}\int_{-\infty}^{\infty} \exp\{-\frac{1}{2}X^2/\langle X^2\rangle\}e^{-iX}\,dX = \exp\langle -\frac{1}{2}X^2\rangle.$$

The only remaining problem is to calculate

$$\langle X^2\rangle = \left\langle \left[\int_0^t \omega(t')\,dt'\right]^2\right\rangle = \left\langle \int_0^t dt' \int_0^t dt''\omega(t')\omega(t'')\right\rangle = 2\int_0^t (t-\tau)G_\omega(\tau)\,d\tau,$$

whence

$$G(t) = e^{i\omega_0 t}\exp\left\{-\int_0^t (t-\tau)G_\omega(\tau)\,d\tau\right\} = e^{i\omega_0 t}\exp\left\{-\langle \omega^2\rangle \int_0^t (t-\tau)g_\omega(\tau)\,d\tau\right\}, \tag{19}$$

where $g_\omega(\tau)$ is the reduced correlation function of $\omega(t)$ such that

$$g_\omega(0) = 1.$$

Before making any explicit assumptions about $g_\omega(\tau)$, consider two extreme cases.

(i) The correlation time τ_c is so long that $\langle\omega^2\rangle\tau_c^2 \gg 1$, which means that for $\tau \ll \tau_c$ we can replace $g_\omega(\tau)$ by unity in (19), whence

$$G(t) = e^{i\omega_0 t}\exp\{-\tfrac{1}{2}\langle\omega^2\rangle t^2\}. \tag{20}$$

This is the same expression as in the absence of motion. It is not true any more when t becomes of the order of τ_c but for these values of t, $\langle\omega^2\rangle t^2$ is very large, $G(t)$ is very small, and the corresponding contribution to the absorption,

$$I(\omega) \sim \int G(t)e^{-i\omega t}\,dt, \tag{20'}$$

is negligible. The shape of the resonance curve is the same as for the rigid lattice.

(ii) If, on the other hand, τ_c is so short that $\langle\omega^2\rangle\tau_c^2 \ll 1$, for $t \gg \tau_c$ it is permissible to write

$$\langle\omega^2\rangle\int_0^t (t-\tau)g_\omega(\tau)\,d\tau \cong \langle\omega^2\rangle t \int_0^\infty g_\omega(\tau)\,d\tau = \langle\omega^2\rangle t\tau_c', \tag{21}$$

where $\tau_c' = \int_0^\infty g_\omega(\tau)\,d\tau$ is of the order of τ_c. Then

$$G(t) \cong e^{-i\omega_0 t}e^{-\langle\omega^2\rangle|t|\tau_c'}. \tag{22}$$

The approximation (21) is not valid for $t \leqslant \tau_c$, but the contribution of these values of t to the integral (20'), giving $I(\omega)$, is negligible except for those values of ω for which $\omega\tau_c \cong 1$, that is, for $\omega \gg [\overline{\Delta\omega_0^2}]^{\frac{1}{2}}$ (i.e. very far in the wings of the curve). As a consequence the absorption curve, Fourier transform of (22), is a Lorentz curve with a width $\delta = \langle\omega^2\rangle\tau_c' \ll \langle\omega^2\rangle^{\frac{1}{2}}$. The intuitive consideration of the introduction has indeed predicted this order of magnitude for the width but not the Lorentz shape of the curve. Besides, the formula (19) provides, once $g_\omega(\tau)$ has been chosen, a detailed description of the resonance curve for the intermediate case when $\langle\omega^2\rangle\tau_c^2$ is neither large nor small.

The choice of the correlation function $g_\omega(\tau)$ is facilitated by the following two theorems (3).

(i) The second moment of the absorption curve $I(\omega)$ is unaffected by the narrowing 'motion'. This results most directly from the rigorous quantum mechanical approach to motion narrowing; the second moment (with respect to frequency zero) is given by the trace of the commutator

$$M_2 \propto \operatorname{tr}\{[\mathscr{H},\mathscr{M}_x]^2\},$$

where $\mathscr{H}$ is the total Hamiltonian. Since the 'motion' Hamiltonian $\mathscr{F}$ commutes with $\mathscr{M}_x$ it could not possibly affect the second moment.

This can be also demonstrated from the random function model used here. According to formula (31) of Chapter IV,

$$M_2 = -\left(\frac{d^2G}{dt^2}\right)_{t=0}$$

If we take $\quad G(t) = \left\langle \exp\left(i \int_0^t \omega(t')\, dt'\right)\right\rangle$

(where the unperturbed resonance frequency ω_0 is taken as origin),

$$\frac{dG}{dt} = \left\langle i\omega(t) \exp\left(i \int_0^t \omega(t')\, dt'\right)\right\rangle. \tag{23}$$

Since $\omega(t)$ is a stationary random function it is permissible to shift the time origin in (23) by any amount and write

$$\frac{dG}{dt} = \left\langle i\omega(0) \exp\left(i \int_{-t}^0 \omega(t')\, dt'\right)\right\rangle,$$

whence
$$\frac{d^2G}{dt^2} = -\left\langle \omega(0)\omega(-t) \exp\left(i \int_{-t}^0 \omega(t')\, dt'\right)\right\rangle \tag{24}$$

$$= -\left\langle \omega(t)\omega(0) \exp\left(i \int_0^t \omega(t')\, dt'\right)\right\rangle$$

and $\quad M_2 = -\left(\dfrac{d^2G}{dt^2}\right)_0 = \langle \omega^2(0)\rangle,$

which in our model was assumed to be the same as in the absence of motion.

This shows that a Lorentz shape predicted by the model for

$$\langle \omega^2\rangle \tau_c^2 \ll 1$$

could not possibly represent the absorption curve for all values of ω since it leads to an infinite second moment.

(ii) Although the quantities

$$\langle \omega^2\rangle = \int \omega^2 P(\omega)\, d\omega$$

and $\quad M_2 = \int \omega^2 I(\omega)\, d\omega \Big/ \int I(\omega)\, d\omega,$

where $P(\omega)$ is the instantaneous distribution of the frequencies $\omega(t)$ and $I(\omega)$ is the absorption curve, are identical, this is not true for higher moments:

$$M_n = \int \omega^n I(\omega)\, d\omega \Big/ \int I(\omega)\, d\omega \neq \langle \omega^n\rangle = \int P(\omega)\omega^n\, d\omega.$$

In particular, we shall show that
$$M_4 = \langle \omega^4 \rangle - \langle \omega^2 \rangle \left(\frac{d^2 g_\omega}{dt^2} \right)_{t=0}. \tag{25}$$

Taking the derivative, with respect to t, of equation (24), we get

$$\frac{d^3 G}{dt^3} = -\left\langle \omega'(t)\omega(0) \exp\left(i \int_0^t \omega(t')\, dt'\right) + i\omega^2(t)\omega(0) \exp\left(i \int_0^t \omega(t')\, dt'\right) \right\rangle$$

$$= -\left\langle \omega'(t)\omega(0) \exp\left(i \int_0^t \omega(t')\, dt'\right) + i\omega^2(0)\omega(-t) \exp\left(i \int_{-t}^0 \omega(t')\, dt'\right) \right\rangle.$$

Taking the derivative once more, we get

$$\left(\frac{d^4 G}{dt^4} \right)_{t=0} = M_4 = \langle \omega^4 \rangle - \langle \omega''(0)\omega(0) \rangle,$$

and the theorem is proved.

B. Exchange narrowing

We saw in Chapter VI that so-called indirect spin-spin couplings, distinct from the usual magnetic dipolar couplings and having a short range in non-metals, could exist between nuclear spins. Their importance was recognized particularly in high resolution studies of resonance in liquids where the dipolar interactions are wiped out, to the first order, by molecular Brownian motion, whilst the scalar part $\hbar J \mathbf{I}_1 \cdot \mathbf{I}_2$ of the indirect couplings remains unaffected. However, even in solids where the dipolar couplings have their full value, the indirect couplings may be comparable to, and for the heaviest atoms a good deal larger than, the dipolar ones. As previously mentioned, indirect couplings can be written as sums of traceless tensor couplings, which usually (but not necessarily) have the same form as the dipolar couplings (whence the name of pseudo-dipolar interactions) and scalar parts. Under certain circumstances (predominant s-character of the electronic wave function near the nucleus) the scalar part may be much more important than the traceless tensor part and then it has a very remarkable narrowing effect on the resonance line. Before we examine these effects it should be mentioned that exchange narrowing is much more important in electron spin resonance, where it was first suggested (4) and where the scalar couplings between electron spins are really due to electric rather than magnetic forces and have such spectacular effects as ferromagnetism or antiferromagnetism. As a criterion of the narrowing effect of the Hamiltonian of 'motion' we calculate the ratio

$M_4/(M_2)^2$. The scalar exchange interaction affords no contribution to the second moment but supplies to the fourth moment a term proportional to $\mathcal{H}_1^2 \mathcal{H}_{\text{exch}}^2$, where $\mathcal{H}_1$ is the dipolar broadening Hamiltonian. This contribution M_4' of exchange to the fourth moment can be calculated by the formula (34') of Chapter IV where $\mathcal{H}_1'$, the truncated dipolar Hamiltonian, is replaced by

$$\mathcal{H}_1' + \mathcal{H}_{\text{exch}} = \mathcal{H}_1' + \sum_{i<k} J_{ik} \mathbf{I}_i \cdot \mathbf{I}_k.$$

It is clear from this formula that, since $[\mathcal{H}_{\text{exch}}, I_x] = 0$, there are no terms of order higher than $\mathcal{H}_{\text{exch}}^2$ in M_4.

The result of a straightforward but cumbersome calculation, using the technique of traces, gives for this contribution (5)

$$M_4' = \gamma^4 \hbar^2 N^{-1} \sum_{j,k,l \neq} [2J_{jk}^2(b_{jl}-b_{kl})^2 + 2J_{jk}J_{kl}(b_{jl}-b_{jk})(b_{jl}-b_{kl})][\tfrac{1}{3}I(I+1)]^2 +$$
$$+ \gamma^4 \hbar^2 N^{-1} \sum_{k>j} b_{jk}^2 J_{jk}^2 [\tfrac{4}{5}I^2(I+1)^2 - \tfrac{3}{5}I(I+1)]. \quad (26)$$

In this expression terms that are linear with respect to J have been omitted as being much smaller than those proportional to J^2; b_{jk} is $\tfrac{3}{2}(1-3\cos^2\theta_{jk})r_{jk}^{-3}$ and a summation symbol such as $\sum_{jkl \neq}$ means that all three indices j, k, l must be different.

For a simple cubic lattice of lattice length d where exchange interactions between nearest neighbours only are considered and dipolar interactions between spins removed by more than $2d$ are disregarded, it is found (5) that

$$\frac{M_4'}{3(M_2)^2} \cong \frac{M_4}{3(M_2)^2} \cong \frac{0 \cdot 12 J^2}{\gamma^4 \hbar^2 d^{-6}} [\lambda_1^4 + \lambda_2^4 + \lambda_3^4 - 0 \cdot 187]^{-1}, \quad (27)$$

where $J\hbar$ is the exchange coupling between nearest neighbours and $\lambda_1, \lambda_2, \lambda_3$ are the cosines of the magnetic field with respect to the crystal axes.

It was suggested in Chapter IV that whenever the ratio $M_4/3(M_2)^2$ turned out to be a large number we might use as a trial model for the resonance curve a truncated Lorentz shape with a width

$$\delta = \frac{\pi}{2\sqrt{3}}(M_2)^{\frac{1}{2}} \left(\frac{(M_2)^2}{M_4}\right)^{\frac{1}{2}}. \quad (28)$$

The same problem of exchange narrowing can also be approached by using the random function model, where the effect of the exchange coupling is described as a random modulation of the dipolar local field (2, 3).

For that purpose we must choose a reduced correlation function:

$$g_\omega(\tau) = \langle \omega(t)\omega(t-\tau)\rangle/\langle\omega^2(t)\rangle.$$

The exponential $\exp(-|\tau|/\tau_c)$ cannot be used, for its second derivative is infinite for $\tau = 0$, thus leading according to equation (25) to an infinite fourth moment M_4, in contradiction to the finite value (26) obtained by a quantum mechanical calculation. On the other hand, the gaussian form $\exp(-\tfrac{1}{4}\pi\omega_e^2\tau^2)$, where the constant ω_e describing the rate of random modulation of the dipole field by the exchange can be determined from equation (25), yields (writing $\omega_p^2 = \langle\omega^2\rangle$)

$$M_4 = 3\omega_p^4 + \tfrac{1}{2}\pi\omega_p^2\omega_e^2, \tag{29}$$

where the second term represents the contribution M_4' of exchange.

For instance, for a simple cubic lattice, where for simplicity we perform an average over the angles in (27) and use for ω_p^2 the value of equation (39') of Chapter IV:

$$\omega_p^2 = 5\cdot 1\,\gamma^4\hbar^2 I(I+1)d^{-6}, \tag{29'}$$

it is found that
$$\omega_e^2 = 2\cdot 8 J_e^2 I(I+1). \tag{29''}$$

Once ω_e^2 has been determined, the detailed shape of the resonance curve can be determined from the formula (19), which now reads

$$G(t) = \exp\left\{-\omega_p^2 \int_0^t (t-\tau) e^{-\tfrac{1}{4}\pi\omega_e^2\tau^2}\,d\tau\right\}. \tag{30}$$

As previously stated, for $\omega_e \gg \omega_p$ (30) reduces to

$$\exp\left\{-\omega_p^2 t \int_0^\infty g_\omega(\tau)\,d\tau\right\} = \exp\left\{-\frac{\omega_p^2}{\omega_e}t\right\},$$

thus leading to a Lorentz shape with a width

$$\delta = \omega_p^2/\omega_e.$$

It is interesting to compare the line width $\delta = \omega_p^2/\omega_e$ obtained from the random function model, which through (29) (where we neglect $3\omega_p^4$ as compared with $\tfrac{1}{2}\pi\omega_p^2\omega_e^2$) can be written

$$\delta = \sqrt{(\tfrac{1}{2}\pi)}(M_2)^{\tfrac{3}{2}}(M_4)^{-\tfrac{1}{2}}, \tag{31}$$

with that predicted from the assumption of a truncated Lorentz shape by formula (28):

$$\delta' = \frac{\pi}{2\sqrt{3}}(M_2)^{\tfrac{3}{2}}(M_4)^{-\tfrac{1}{2}}. \tag{32}$$

The ratio of the numerical coefficients in front of (31) and (32) is $\delta'/\delta = \sqrt{(\tfrac{1}{6}\pi)} = 0\cdot 72$, exhibiting good agreement between two quite different approaches to the calculation of the line width.

The narrowing, as expressed by the ratio between the r.m.s. width $(M_2)^{\frac{1}{2}} = \omega_p$ and the half intensity width δ, is ω_p/ω_e which for a simple cubic lattice, by (29') and (29"), is

$$1 \cdot 35 \frac{\gamma^2 \hbar}{d^3 J}.$$

So far we have assumed scalar couplings between like spins only. Scalar couplings between unlike spins behave quite differently. A coupling $\hbar \sum_{i,k} J_{ik}(\mathbf{I}_i \cdot \mathbf{S}_k)$ does *not* commute with the x component, $I_x = \sum_i I_{ix}$, of the total spin of the 'resonant' nuclei and brings a contribution to the second moment which is easily computed to be

$$M'_2 = \frac{S(S+1)}{3} \sum_k J_{ik}^2. \tag{33}$$

The summation in (33) must be made over the sites of the spins S_k surrounding a spin I_i.

Thus, roughly speaking, scalar couplings between like spins narrow the resonance line, scalar couplings between unlike spins broaden it.

Finally, there is the influence of strong couplings (scalar or not) between the non-resonant spins S on the resonance of the spins I. It has already been seen in Chapter IV that since these couplings affect the fourth but not the second moment their effect is a narrowing one. More details on these points can be found in references (2) and (5).

In electron resonance, where many instances of exchange narrowing are known and where independent information on the strength of the exchange couplings can be obtained from measurements of susceptibility and specific heat at low temperatures, detailed comparison of the experimental results with theory is hampered by many complicating features: anisotropy, crystalline splittings, unresolved hyperfine structure, etc. (2).

In nuclear resonance there have so far been few instances of exchange narrowing or broadening. The most spectacular example is that of the resonance of the two isotopes of thallium (6). Both isotopes Tl[203] and Tl[205] have spin $\frac{1}{2}$, magnetic moments differing by less than 1 per cent, and isotopic abundances of 29·5 per cent and 70·5 per cent, respectively.

The very striking fact that the resonance of Tl[203] is much broader than that of Tl[205] has been explained (6) by the assumption of strong indirect scalar couplings between the various nuclear spins. Since such coupling between unlike neighbours broadens the line, but narrows it between like neighbours, it is conceivable that Tl[205] which has seven like neighbours for three unlike ones would have a narrower line than Tl[203] for which the situation is reversed. This surmise has been verified by using samples with various isotopic concentrations. It has been shown (6) that the higher the concentration of each isotope, the narrower is its line and that for equal concentrations both lines have equal widths (Fig. X, 1).

It should be added that the indirect coupling between the spins was certainly not purely scalar (as indeed, given the electronic structure of thallium, there was no reason to expect it to be). For a sample containing a single isotope, a purely scalar coupling would have led to a width that was much smaller than the dipolar width whilst the width observed was several times larger. A non-scalar traceless indirect coupling of strength about one-third of the strength of the scalar part had to be invoked to account for this. Finally, it should be mentioned that although these couplings appreciably exceed the dipolar ones, they are still small compared with the Zeeman energy. This justifies the use of the adiabatic approximation, for the time variation of the dipolar Hamiltonian $\mathcal{H}_1(t) = e^{-i\mathcal{F}t}\mathcal{H}_1 e^{i\mathcal{F}t}$ (where $\hbar\mathcal{F}$ is the indirect coupling), is slow compared with the Larmor precession.

C. Brownian motion narrowing

In many crystals, except at very low temperatures, self-diffusion exists. This is characterized by a jumping of the atoms from one crystal site to another. The effect on the resonance line width can be described in the frame of the previous theory by assuming a reduced correlation function $g_\omega(\tau)$ for the frequency $\omega(t)$ of the random local field that fluctuates because of the diffusion of the form

$$g_\omega(\tau) = e^{-|\tau|/\tau_c}. \tag{34}$$

This leads to the following expression for the Fourier transform $G(t)$ of the absorption curve $I(\omega)$:

$$e^{-i\omega_0 t}G(t) = \exp\left\{-\omega_p^2 \int_0^t (t-\tau)g_\omega(\tau)\,d\tau\right\}$$
$$= \exp\left\{-\omega_p^2 \tau_c^2\left[\exp\left(-\frac{t}{\tau_c}\right) - 1 + \frac{t}{\tau_c}\right]\right\}. \tag{35}$$

It is easily verified that, depending on whether $\omega_p\tau_c$ is very large or very small, (34) leads for $I(\omega)$ to a Gaussian shape with an r.m.s. width ω_p or to a Lorentzian shape with a half intensity width $\omega_p^2\tau_c$. However, (35) provides a detailed description of the resonance shape for intermediate values of $\omega_p\tau_c$.

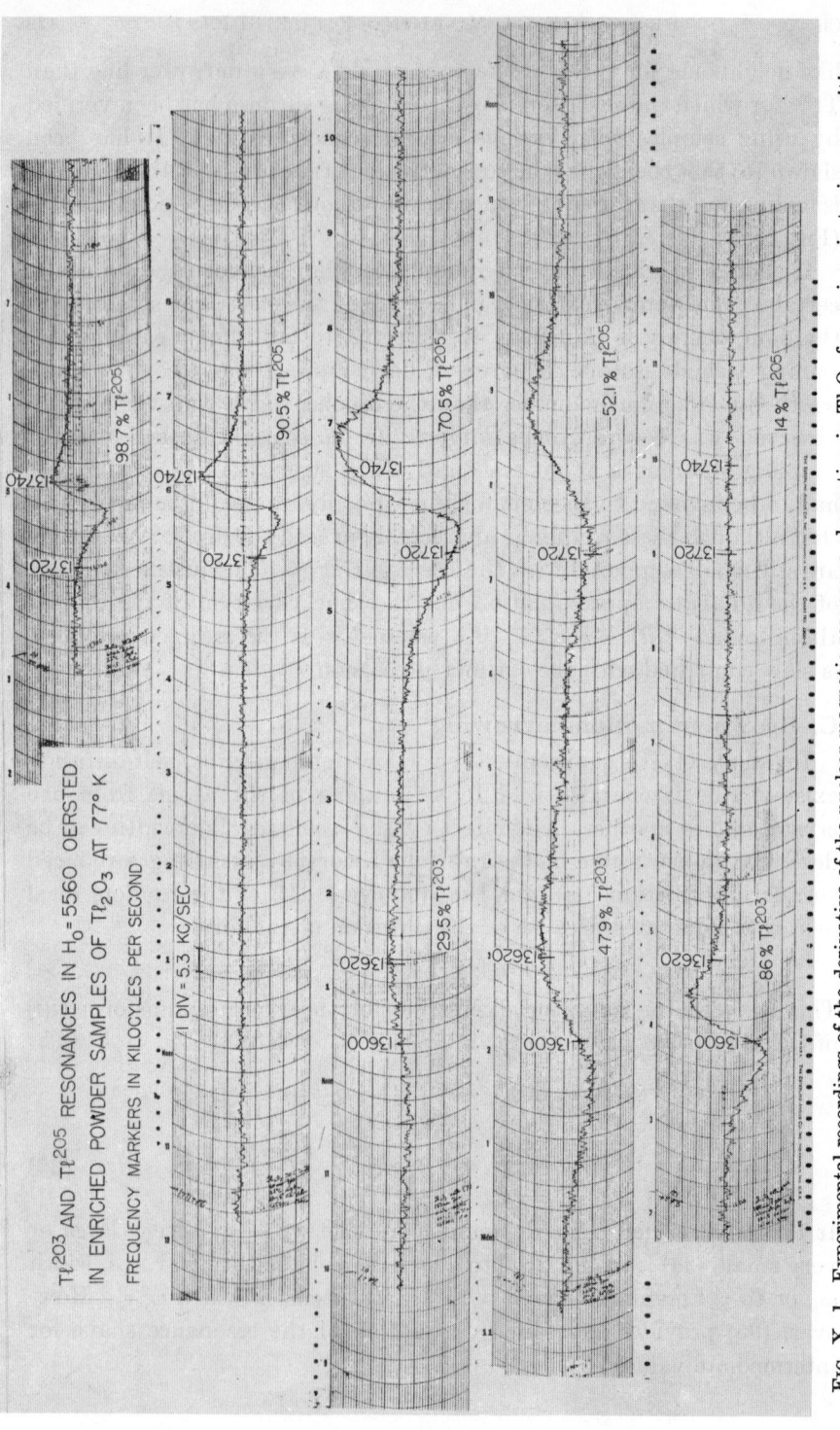

Fig. X, 1. Experimental recordings of the derivative of the nuclear magnetic resonance absorption in Tl_2O_3 for various isotopic compositions.

The choice of the reduced correlation function $g_\omega(\tau) = e^{-|\tau|/\tau_c}$ is dictated mainly by its simplicity. It was shown in Chapter VIII that if one assumes that the translational random motions of the atoms in the lattice are described correctly by the classical diffusion equation, the following reduced correlation function (given by eqn. (113) of Chapter VIII) should be used instead:

$$g_\omega(\tau) = 3 \int_0^\infty [J_{\frac{3}{2}}(u)]^2 \exp\left(-\frac{u^2 \tau}{\tau_c}\right) \frac{du}{u}, \qquad (36)$$

normalized to $g_\omega(0) = 1$, where $\tau_c = \frac{1}{2} d^2/D$, D being the diffusion coefficient and d the distance of closest approach between molecules.

More realistic descriptions of the diffusion in a crystal lattice (7, 8) lead to even more complicated expressions for the correlation functions or their Fourier transforms, that cannot be expressed in closed form and have to be computed numerically. They will be discussed briefly in Section V B.

III. THE NON-ADIABATIC LINE WIDTH

A. Line width and transverse relaxation time

When the fluctuation rate $1/\tau_c$ of the local field becomes comparable to the resonance frequency the adiabatic approximation breaks down and a more general treatment that takes into account the off-diagonal matrix elements of the perturbing Hamiltonian $\hbar \mathcal{H}_1(t)$ must be used. Fortunately in that case the shortness of the correlation time is, as was shown in Chapter VIII, a source of considerable simplification, for in contradistinction to the situation in a rigid lattice, it is legitimate to consider individual spins or at most groups of spins as separate systems with few degrees of freedom.

Indeed the problem of the line width was solved in Chapter VIII through the proof of the existence and the computation of a single transverse relaxation time T_2 for dipolar coupling between like or unlike spins and for some other couplings as well. The fundamental theorem proved in Chapter IV, which states that the time dependence of the signal of free precession is the Fourier transform of the non-saturated absorption curve $I(\omega)$, enables us to conclude that in all cases where a single T_2 exists the resonance curve has a Lorentz shape with a half intensity width $\Delta\omega = 1/T_2$.

As an example of the corrections to the adiabatic approximation required by the shortness of the correlation time, consider formula (79) of Chapter VIII, giving $1/T_2 = \Delta\omega$ for a system of two like spins I

coupled by a dipolar interaction. In the adiabatic approximation $J^{(1)}(\omega)$ and $J^2(2\omega)$ are vanishingly small and only the term $\frac{3}{8}J^{(0)}(0)$ contributes to $1/T_2$.

At the other end, when the narrowing is extreme,

$$J^{(2)}(2\omega) = J^{(2)}(0); \qquad J^{(1)}(\omega) = J^{(1)}(0)$$

and since, for an isotropic motion,

$$J^{(0)}(0) : J^{(1)}(0) : J^{(2)}(0) = 6 : 1 : 4,$$

the ratio of the full width to the adiabatic width is

$$(\tfrac{3}{8} \times 4 + \tfrac{15}{4} + \tfrac{3}{8} \times 6)/(\tfrac{3}{8} \times 6) = \tfrac{10}{3}.$$

The necessity of correcting the line width given by the adiabatic theory by this factor was first suggested in connexion with exchange narrowing in electron resonance, where exchange couplings between electron spins may be large compared to the Larmor frequency (2).

B. General case

The possibility of representing a strongly 'motion narrowed' resonance curve by a single Lorentz curve, and therefore the free decay of the transverse magnetization by a single exponential, was demonstrated in the adiabatic approximation using a specialized model (gaussian nature of the random functions involved), and outside this approximation for a certain number of relaxation mechanisms. However, for a quadrupole relaxation mechanism, it had already appeared impossible to fit the free decay of the magnetization by a single exponential (except for extreme narrowing).

We will now consider the general problem of predicting the line shape for a resonant transition ω_0 occurring in a system described by an unperturbed Hamiltonian $\hbar\mathcal{H}_0$ and subject to a perturbation $\hbar\mathcal{H}_1(t)$ with a correlation time τ_c not necessarily long compared with $1/\omega_0$.

For simplicity we shall use the random function description, equivalent to the quantum mechanical one, as shown in Section II D of Chapter VIII, for short correlation times and high temperatures.

The relaxation function $G(t)$ of the magnetization, that is the Fourier transform of $I(\omega)$, can be written

$$G(t) = \text{tr}\{\mathcal{M}_x(t)\mathcal{M}_x\} = \text{tr}\{e^{i\mathcal{H}_0 t}\mathcal{M}_x^*(t)e^{-i\mathcal{H}_0 t}\mathcal{M}_x\}, \tag{37}$$

where $\mathcal{M}_x^*(t) = e^{-i\mathcal{H}_0 t}\mathcal{M}_x(t)e^{i\mathcal{H}_0 t}$ obeys the differential equation

$$\frac{1}{i}\frac{d\mathcal{M}_x^*}{dt} = [e^{-i\mathcal{H}_0 t}\mathcal{H}_1(t)e^{i\mathcal{H}_0 t}, \mathcal{M}_x^*(t)],$$

which (apart from a trivial change of i into $-i$) is the same as the equation obeyed by the density matrix σ^* in the interaction representation. The equation for σ^*, and thus also that for $\mathscr{M}^*$, has been shown to reduce under certain assumptions (Chapter VIII, Section II C (c)) to a master equation with constant coefficients:

$$\frac{d\mathscr{M}^*_{\alpha\alpha'}}{dt} = \sum_{\beta\beta'} R_{\alpha\alpha',\beta\beta'} \mathscr{M}^*_{\beta\beta'} \qquad (38)$$

(eqn. (35) of Chapter VIII) where the summation is limited to the eigenstates of $\hbar\mathscr{H}_0$ such that their energies $\hbar\beta$, $\hbar\beta'$ satisfy

$$\beta - \beta' = \alpha - \alpha', \qquad (38')$$

(eqn. (36) of Chapter VIII). The equation (37) can be rewritten as

$$G(t) = \sum_{\alpha,\alpha'} e^{i(\alpha-\alpha')t}(\alpha|\mathscr{M}^*_x(t)|\alpha')(\alpha'|\mathscr{M}_x|\alpha).$$

In practice we are not interested in the whole of $G(t)$ but only in the part related to the transitions between levels $|\alpha\rangle$ and $|\alpha'\rangle$ of the unperturbed Hamiltonian $\hbar\mathscr{H}_0$, such that $\alpha - \alpha' = \omega_0$, whence

$$G(t) = e^{i\omega_0 t} \underset{\alpha-\alpha'=\omega_0}{\sum\nolimits'} (\alpha|\mathscr{M}^*_x(t)|\alpha')(\alpha'|\mathscr{M}_x|\alpha). \qquad (39)$$

Let N be the number of pairs of eigenstates $|\alpha\rangle$, $|\alpha'\rangle$ of $\mathscr{H}_0$ such that $\alpha - \alpha' = \omega_0$, and represent each pair of indices (α, α') by a single index ν.

The matrix elements $(\alpha|\mathscr{M}_x|\alpha')$ and $(\alpha|\mathscr{M}^*_x(t)|\alpha')$ can be considered as components X_ν and $X_\nu(t)$ of vectors $\mathbf{X}$ and $\mathbf{X}(t)$ in an N-dimensional space $\mathscr{E}$ and, because of the condition (38'), the matrix $\dot{R}_{\alpha\alpha',\beta\beta'}$ can be considered as a matrix $R_{\nu\nu'}$ representing a linear transformation $\mathscr{R}$ in that space.

The solution of (38) can be written formally:

$$\mathbf{X}(t) = e^{Rt}\mathbf{X} \quad \text{or} \quad X_\nu(t) = \sum_{\nu'} (e^{Rt})_{\nu\nu'} X_{\nu'},$$

$$G(t) = e^{i\omega_0 t} \sum_{\nu\nu'} X^*_\nu (e^{Rt})_{\nu\nu'} X_{\nu'} = e^{i\omega_0 t}(\mathbf{X}^* \cdot e^{Rt} \cdot \mathbf{X}). \qquad (40)$$

Let $\boldsymbol{\eta}_1, \boldsymbol{\eta}_2, \ldots, \boldsymbol{\eta}_N$ be the eigenvectors of the matrix R in space $\mathscr{E}$ with eigenvalues $r_1, r_2, \ldots, r_N$. Equation (40) can be rewritten

$$G(t) = e^{i\omega_0 t} \sum_{\lambda=1}^{N} |(\mathbf{X} \cdot \boldsymbol{\eta}_\lambda)|^2 e^{r_\lambda t}. \qquad (41)$$

It is clear from (41) that $G(t)$ is a sum of exponentials. The eigenvalues r_λ are not necessarily real but their real parts r'_λ are clearly negative for physical reasons (dissipation rather than exponential growth of the precessing magnetization). The imaginary parts r''_λ correspond to frequency shifts which will shortly be shown to be smaller than the line width and will be neglected after that. $I(\omega)$ is then a superposition of Lorentz curves. Their number is at most N, that is the number of couples of levels (α, α') between which the resonant transition of frequency ω_0 takes place, their relative weights in the superposition are $|(\mathbf{X} \cdot \boldsymbol{\eta}_\lambda)|^2$ and their widths are $|r'_\lambda|$, real parts of the eigenvalues of $R_{\nu\nu'}$ in the space $\mathscr{E}$. In particular, the condition for the resonance curve to reduce to a *single* Lorentz curve is that the vector $\mathbf{X}$ should itself be an eigenvector of R.

It is clear that a single constant T_2 cannot in general describe the width of the line.

One obvious exception to that statement occurs when there exists only a single pair of levels (a, b) of the Hamiltonian $\hbar \mathscr{H}_0$ such that $a - b = \omega_0$, for then the space $\mathscr{E}$ has only one dimension, and the resonance curve is a single Lorentz curve.

We are going to calculate its width and show that it admits of a simple physical interpretation.

The matrix $R_{\alpha\alpha',\beta\beta'}$ now reduces to a single coefficient $1/T_2 = R_{ab,ab}$. Its value can be obtained from the general equation (33) of Chapter VIII where we replace σ^* by $\mathscr{M}^*$ and calculate the rate of change of the matrix element $\mathscr{M}^*_{ab}$,

$$\frac{d\mathscr{M}^*_{ab}}{dt} = R_{ab,ab} \mathscr{M}^*_{ab}.$$

With a little algebra we obtain

$$-R_{ab,ab} = \int_0^\infty d\tau \Big\{ \sum_c \overline{(a|\mathscr{H}^*_1(t)|c)(c|\mathscr{H}^*_1(t-\tau)|a)} +$$
$$+ \overline{(b|\mathscr{H}^*_1(t-\tau)|c)(c|\mathscr{H}^*_1(t)|b)} -$$
$$- \overline{(a|\mathscr{H}^*_1(t)|a)(b|\mathscr{H}^*_1(t-\tau)|b)} -$$
$$- \overline{(a|\mathscr{H}^*_1(t-\tau)|a)(b|\mathscr{H}^*_1(t)|b)} \Big\} \quad (42)$$
$$= \int_0^\infty d\tau \Big\{ \overline{(a|\mathscr{H}_1(t)|a)(a|\mathscr{H}_1(t-\tau)|a)} -$$
$$- \overline{(a|\mathscr{H}_1(t)|a)(b|\mathscr{H}_1(t-\tau)|b)} +$$
$$+ \overline{(b|\mathscr{H}_1(t-\tau)|b)(b|\mathscr{H}_1(t)|b)} -$$
$$- \overline{(a|\mathscr{H}_1(t-\tau)|a)(b|\mathscr{H}_1(t)|b)} \Big\} +$$
$$+ \int_0^\infty d\tau \Big\{ \sum_{c \neq a} \overline{(a|\mathscr{H}_1(t)|c)(c|\mathscr{H}_1(t-\tau)|a)} e^{i\omega_{ac}\tau} +$$
$$+ \sum_{c \neq b} \overline{(b|\mathscr{H}_1(t-\tau)|c)(c|\mathscr{H}_1(t)|b)} e^{i\omega_{bc}\tau} \Big\} \quad (43)$$

where $|c)$ are all the eigenstates of $\hbar \mathscr{H}_0$. We now introduce the random function $\omega(t) = (a|\mathcal{H}_1(t)|a) - (b|\mathcal{H}_1(t)|b)$. The first integral in (43) can be written

$$\int_0^\infty \overline{\omega(t)\omega(t-\tau)}\,d\tau. \tag{43'}$$

This is precisely the line width given under strong narrowing by the adiabatic theory, and we shall call it $1/T_2' = (1/T_2)_{\text{adiab}}$. If we disregard for a moment the imaginary terms in the second integral of (43), it can be written

$$\tfrac{1}{2}\int_{-\infty}^\infty e^{i\omega_{ac}\tau}\,d\tau \sum_{c\neq a} \overline{(a|\mathcal{H}_1(t)|c)(c|\mathcal{H}_1(t-\tau)|a)} + $$
$$+ \tfrac{1}{2}\int_{-\infty}^\infty e^{i\omega_{bc}\tau}\,d\tau \sum_{c\neq b} \overline{(b|\mathcal{H}_1(t-\tau)|c)(c|\mathcal{H}_1(t)|b)}, \tag{44}$$

which, from equation (23) of Chapter VIII, is nothing but

$$\tfrac{1}{2}\Big\{\sum_{c\neq a} W_{ac} + \sum_{c\neq b} W_{bc}\Big\},$$

where W_{ac} and W_{bc} are the transition probabilities per unit time induced by the perturbing Hamiltonian $\hbar\mathcal{H}_1(t)$. If we define the lifetimes t_a and t_b of the states $|a)$ and $|b)$ through

$$\frac{1}{t_a} = \sum_{c\neq a} W_{ac}, \qquad \frac{1}{t_b} = \sum_{c\neq b} W_{bc},$$

we get
$$\frac{1}{T_2} = \frac{1}{T_2'} + \frac{1}{2}\Big(\frac{1}{t_a} + \frac{1}{t_b}\Big). \tag{45}$$

Equation (45) means that the adiabatic line width $1/T_2'$ has to be supplemented by the mean of the inverse lifetimes of either state. An even simpler situation occurs when $|a)$ and $|b)$ are not only the single pair of states separated in energy by $\hbar\omega_0$ but the only eigenstates of $\hbar\mathcal{H}_0$. Then it is clear that

$$\frac{1}{t_a} = \frac{1}{t_b} = \frac{1}{2T_1}$$

and
$$1/T_2 = 1/T_2' + 1/(2T_1). \tag{46}$$

If, furthermore, $T_1 = T_2$ (extreme narrowing), it follows that

$$T_2' = 2T_1 = 2T_2.$$

It should be remarked that the relation (46) has been established under very specialized assumptions. Even for such a simple case as that of two like spins relaxing through dipolar coupling it is not verified. In the extreme narrowing case it follows from formula (79) of Chapter VIII that
$$1\ T_2' = (1/T_2)_{\text{adiab}} = \tfrac{3}{10}(1/T_2)$$

and since then $T_1 = T_2$, we have instead of (46)

$$\frac{1}{T_2} = \frac{1}{T_2'} + \frac{7}{10}\frac{1}{T_1}. \qquad (47)$$

No such simple relation can be established between T_2, T_2', and T_1 outside the extreme narrowing case.

In replacing in (44) terms such as

$$\sum_c \int_0^\infty d\tau\, e^{i\omega_{ac}\tau}\overline{(a|\mathcal{H}_1(t)|c)(c|\mathcal{H}_1(t-\tau)|a)}$$

through $\frac{1}{2}\int_{-\infty}^\infty ...$, we have neglected an imaginary term

$$i\delta_a = i\sum_c \int_0^\infty d\tau \sin(\omega_{ac}\tau)\overline{(a|\mathcal{H}_1(t)|c)(c|\mathcal{H}_1(t-\tau)|a)} \qquad (48)$$

which corresponds to a frequency shift. In order to compare the shift δ_a of the level $|a)$ to its non-adiabatic width $\frac{1}{2}(1/t_a)$ let us assume for simplicity a reduced correlation function for the various matrix elements of $\mathcal{H}_1(t)$ equal to $\exp(-|\tau|/\tau_c)$. From (44) and (48) we find

$$2\delta_a t_a = \frac{\int_0^\infty \sin(\omega_0\tau)e^{-\tau/\tau_c}\,d\tau}{\int_0^\infty \cos(\omega_0\tau)e^{-\tau/\tau_c}\,d\tau} = \omega_0\tau_c.$$

Similarly, the ratio of δ_a to the adiabatic width is of the order of

$$\frac{\int_0^\infty \sin(\omega_0\tau)e^{-\tau/\tau_c}\,d\tau}{\int_0^\infty e^{-\tau/\tau_c}\,d\tau} \cong \frac{\omega_0\tau_c}{1+\omega_0^2\tau_c^2}. \qquad (49)$$

The frequency shift is thus much smaller than the total line width at both ends of the narrowing process, for $\omega_0\tau_c$ large or $\omega_0\tau_c$ small, and becomes comparable to it only in the vicinity of $\omega_0\tau_c \cong 1$. The general smallness of the second-order frequency shift and the corresponding lack of experimental evidence for its existence explain its frequent neglect in theories of relaxation processes.

On coming back to the general case of what we shall call a multiple line with a magnetization relaxation function described by equations (40) or (41), it may be worth while pointing out that $I(\omega)$ can be calculated explicitly without having to find the eigenvectors and eigenvalues of the relaxation matrix R:

$$I(\omega) = \int_{-\infty}^{\infty} G(t)e^{-i\omega t}\,dt.$$

Or, since $G(t)$ is an even function of t,

$$I(\omega) = \mathrm{re}\int_0^{\infty} G(t)e^{-i\omega t}\,dt = \mathrm{re}\int_0^{\infty} e^{i(\omega_0-\omega)t}(\mathbf{X}^*.e^{Rt}.\mathbf{X})\,dt$$

$$= \mathrm{re}\int_0^{\infty} (\mathbf{X}^*.e^{At}.\mathbf{X})\,dt \tag{50}$$

where, representing by the symbol I the unit matrix in space $\mathscr{E}$,

$$A = i(\omega_0-\omega)I + R, \tag{51}$$
$$I(\omega) = \mathrm{re}(\mathbf{X}^*.A^{-1}.\mathbf{X}), \tag{52}$$

and A^{-1} is the matrix inverse of (51) which can in principle be computed without solving the secular equation. In conclusion, it should be made clear that the foregoing theory provides only a schematic outline, based on somewhat restrictive hypotheses, of one of the key problems of nuclear magnetism and more generally of r.f. and microwave spectroscopy, the problem of line width. For each specific problem it will be necessary to appreciate whether it fits that frame and if this is so to perform explicitly the computation of the broadening caused by the interactions existing in that case.

IV. Destruction of Fine Structures through Motion

In the adiabatic theory of motion narrowing it was assumed that in the absence of motion the continuous distribution of frequencies $P(\omega)$, and thus the shape of the resonance curve $I(\omega)$ also, were gaussian. The effect of the 'motion' was to introduce a random jumping of the frequency ω from one value of the continuum to another, thus replacing the relaxation function for the rigid case $G_0(t) = \langle e^{i\omega t}\rangle$ by the 'motional' function $G(t) = \langle e^{i\int_0^t \omega(t')dt'}\rangle$. A somewhat different problem arises when in the absence of 'motion' the spin system has a finite number p of neighbouring frequencies $\omega_\alpha,\ldots,\omega_\beta$ corresponding to p discrete infinitely sharp resonance lines, and the effect of the 'motion' is to produce a jumping of the system at a random rate between these different frequencies. As an example we may consider a proton which through chemical exchange can belong to one or the other of two molecules and because of different chemical environments in its two positions has slightly different Larmor frequencies, owing to different chemical shifts. The problem is to evaluate $G(t) = \langle e^{i\int_0^t \omega(t')dt'} \rangle$ under those circumstances (3).

We assume that the jumping from one frequency to another is a stationary Markov process. We mean thereby that the probability that the frequency ω has a value ω_2 at time $t' = t+\Delta t$ when we *know* that it had a value ω_1 at time t is:

(i) independent of the values ω took prior to the time t;
(ii) dependent on the times t and t' only through the difference $t'-t = \Delta t$ and can accordingly be written as $W(\omega_1|\omega_2, \Delta t)$.

It is physically reasonable to assume that, for Δt sufficiently small,

$$W(\omega_1|\omega_2, \Delta t) = \delta_{\omega_1,\omega_2} + \pi(\omega_1, \omega_2)\Delta t \qquad (53)$$

and, since

$$\sum_{\omega_2} W(\omega_1|\omega_2, \Delta t) = 1,$$

$$\sum_{\omega_2} \pi(\omega_1, \omega_2) = 0.$$

At any given time t the probability of the system having a frequency ω_α is W_α, which is time-independent since the process is stationary.

We divide the time interval $(0, t)$ into n equal parts $\Delta t = t/n$. The relaxation function is

$$G(t) = \left\langle \exp\left\{i\int_0^t \omega(t')\,dt'\right\} \right\rangle = \lim_{n\to\infty} \sum_{(\omega_1,\ldots,\omega_n)} P(\omega_1, t_1; \ldots; \omega_n, t_n)e^{i(\omega_1+\ldots+\omega_n)t/n}, \qquad (54)$$

where $t_1 = \Delta t$, $t_2 = 2\Delta t$, $t_n = t = n\Delta t$; $P(\omega_1, t_1; \ldots; \omega_n, t_n)$ is the probability that the frequency has the values ω_1 at time t_1, ω_2 at time t_2, etc., and the summation has to be carried over all possible paths $(\omega_1, \ldots, \omega_n)$ for the frequency between 0 and t. For a Markov process,

$$P(\omega_1, t_1; \ldots; \omega_n, t_n) = W_1 W(\omega_1|\omega_2, \Delta t), \ldots, W(\omega_{n-1}|\omega_n, \Delta t). \qquad (55)$$

Let $G_\alpha(t)$ be the sum of the terms of (54) for which ω_n has a well-defined value ω_α at time t. Then

$$G(t) = \sum_{\omega_\alpha} G_\alpha(t), \qquad (56)$$

the summation being taken over the p possible frequencies of the system. From (54) and (55) it is clear that

$$G_\beta(t+\Delta t) = e^{i\omega_\beta \Delta t} \sum_{\omega_\alpha} G_\alpha(t) W(\omega_\alpha|\omega_\beta, \Delta t). \qquad (57)$$

Since Δt is very small we may replace $e^{i\omega_\beta \Delta t}$ throughout by $1+i\omega_\beta \Delta t$ and, using (53), write

$$G_\beta(t+\Delta t) = G_\beta(t)(1+i\omega_\beta \Delta t) + \Delta t \sum_{\omega_\alpha} \pi(\omega_\alpha, \omega_\beta) G_\alpha(t) \qquad (58)$$

or $$\frac{dG_\beta}{dt} = i\omega_\beta G_\beta + \sum_{\omega_\alpha} \pi(\omega_\alpha, \omega_\beta) G_\alpha(t). \tag{59}$$

Let us define in a p-dimensional space a vector **G** having components G_α, a matrix **π** with elements $(\pi)_{\alpha\beta} = \pi(\omega_\alpha, \omega_\beta)$, and a diagonal matrix **ω** with matrix elements $(\omega)_{\alpha\beta} = \omega_\alpha \delta_{\alpha\beta}$. (59) can be rewritten as

$$\frac{d\mathbf{G}}{dt} = \mathbf{G}(i\boldsymbol{\omega}+\boldsymbol{\pi}),$$

which integrates to
$$\mathbf{G}(t) = \mathbf{G}(0)\exp\{(i\boldsymbol{\omega}+\boldsymbol{\pi})t\}. \tag{60}$$

According to (54) and (55), $\mathbf{G}(0) = \mathbf{W}$, where **W** is a vector with components W_α. Equation (60) can thus be rewritten as

$$\mathbf{G}(t) = \mathbf{W}.\exp\{(i\boldsymbol{\omega}+\boldsymbol{\pi})t\}$$

or
$$G_\alpha(t) = \sum_{\omega_\beta} W_\beta [\exp\{(i\boldsymbol{\omega}+\boldsymbol{\pi})t\}]_{\beta\alpha},$$

$$G(t) = \sum_{\omega_\alpha} G_\alpha(t) = \sum_{\omega_\beta,\omega_\alpha} W_\beta [\exp\{(i\boldsymbol{\omega}+\boldsymbol{\pi})t\}]_{\beta\alpha}$$

or, introducing a vector **1** with all p components equal to unity,

$$G(t) = \mathbf{W}.\exp\{(i\boldsymbol{\omega}+\boldsymbol{\pi})t\}.\mathbf{1}.$$

The absorption

$$I(\omega) = \mathrm{re}\int_0^\infty G(t)e^{-i\omega t}\,dt = \mathrm{re}\int_0^\infty \mathbf{W}.\exp\{i(\boldsymbol{\omega}-\omega\mathsf{E})t+\boldsymbol{\pi}t\}.\mathbf{1}\,dt \tag{61}$$

$$= \mathrm{re}\{\mathbf{W}.\mathsf{A}^{-1}.\mathbf{1}\}. \tag{61'}$$

In (61) the matrix $\omega\mathsf{E}$, which is the unit matrix E times the constant ω, should not be confused with the diagonal matrix **ω** of which the eigenvalues are the ω_α. A^{-1} is the matrix inverse of the matrix

$$\mathsf{A} = i(\boldsymbol{\omega}-\omega\mathsf{E})+\boldsymbol{\pi}.$$

Before considering an example notice that the expression (53) for $W(\omega_1 | \omega_2, \Delta t)$ can be rewritten

$$W(\omega_1 | \omega_2, \Delta t) = \delta_{\omega_1\omega_2}(1-\Omega(\omega_1)\Delta t)+\Omega(\omega_1)P(\omega_1, \omega_2)\Delta t, \tag{62}$$

where $\quad \Omega(\omega_1) = \pi(\omega_1, \omega_1) \quad$ and $\quad P(\omega_1, \omega_2) = \dfrac{\pi(\omega_1, \omega_2)}{\pi(\omega_1, \omega_1)}.$

It is clear from equation (62) that $\Omega(\omega_1)\Delta t$ is the probability that the system will pass during the time Δt from the value ω_1 to a different value and $P(\omega_1, \omega_2)$ is the relative probability that this value will be ω_2. Clearly
$$\sum_{\omega_2 \neq \omega_1} P(\omega_1, \omega_2) = 1.$$

As an example let us consider a system with two frequencies $\pm\delta$ (we take their mean as the origin on the frequency scale), with equal a priori probabilities $W_+ = W_- = \frac{1}{2}$ and a probability per unit time Ω of jumping from one frequency to the other.

The matrices $\boldsymbol{\pi}$ and $\boldsymbol{\omega}$ can be rewritten:

$$i\boldsymbol{\omega} = \begin{pmatrix} i\delta & 0 \\ 0 & -i\delta \end{pmatrix}, \quad \boldsymbol{\pi} = \begin{pmatrix} -\Omega & \Omega \\ \Omega & -\Omega \end{pmatrix},$$

$$\mathsf{A} = \begin{pmatrix} i(-\omega+\delta)-\Omega & \Omega \\ \Omega & -i(\omega+\delta)-\Omega \end{pmatrix},$$

$$\mathsf{A}^{-1} = \begin{pmatrix} -i(\omega+\delta)-\Omega & -\Omega \\ -\Omega & i(-\omega+\delta)-\Omega \end{pmatrix} \Big/ \det\{\mathsf{A}\}, \tag{63}$$

whence, through (61'),

$$I(\omega) \propto \operatorname{re} \frac{2i\omega+4\Omega}{(\delta^2-\omega^2)+2i\omega\Omega} = \frac{4\delta^2\Omega}{\omega^4+2\omega^2(2\Omega^2-\delta^2)+\delta^4}. \tag{64}$$

Let us first consider the case where $\Omega^2 \ll \delta^2$. The denominator of (62) then has two sharp minima for $\omega \cong \pm\delta$. In the neighbourhood of $\pm\delta$,

$$I(\omega) \cong \frac{4\delta^2\Omega}{(\omega^2+\delta^2)^2+4\omega^2\Omega^2} \cong \frac{2\Omega}{[\omega-(\pm\delta)]^2+\Omega^2}$$

and the curve has a Lorentz shape with a width Ω. If, on the contrary, $\Omega^2 \gg \delta^2$ the denominator has a sharp minimum for $\omega = 0$ and in the neighbourhood of $\omega = 0$:

$$I(\omega) \cong \frac{4\delta^2\Omega}{4\omega^2\Omega^2+\delta^4} = \frac{\delta^2/\Omega}{\omega^2+(\delta^2/2\Omega)^2}.$$

The resonance curve is a single Lorentz curve centred on $\omega = 0$ and has a width $\frac{1}{2}\delta^2/\Omega = \delta(\frac{1}{2}\delta/\Omega)$. This is the fine-structure half spacing δ reduced through motion in a ratio $\frac{1}{2}\delta/\Omega$. Fig. X, 2 shows how the shape of $I(\omega)$ changes for various values of $x = 4\delta/\Omega$.

We have assumed that both lines $\pm\delta$ were infinitely sharp. If they have a Lorentzian shape with a width $1/\tau$ it is sufficient to replace $\pm i\delta$ in (63) by $\pm i\delta + 1/\tau$.

The previous formalism enables one to handle more complicated situations with more than two frequencies and different jumping probabilities. As such an example one may consider the resonance of a spin $I = \frac{1}{2}$ coupled through scalar coupling $\hbar J \mathbf{I}.\mathbf{S}$ to a spin $S > \frac{1}{2}$. Each time the spin S makes a transition from a state $S_z = m$ to a state $S_z = m'$ because of its quadrupole relaxation ($|m-m'| = 1, 2$), the frequency of the spin I makes a jump $J|m-m'|$. The various transition probabilities $W_{mm'}$ of the spin S enable us to construct the matrix $\boldsymbol{\pi}$

and to solve completely the problem of the shape of the spectrum for all values of (J, T_1), where T_1 is the relaxation time of the spin S. This will be done for some special cases in Chapter XI.

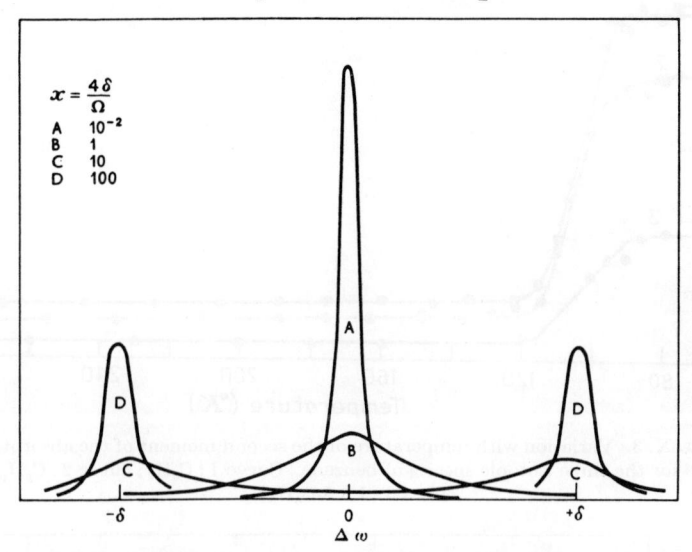

Fig. X, 2. Theoretical shape of the spectrum of a system with two frequencies $\pm\delta$, jumping from one to the other with an average frequency $1/\tau = \Omega$.

V. Influence of Internal Motions in Solids on the Width and Relaxation Properties of Zeeman Resonance Lines

A. Rotational motions

In many solids, even at temperatures well below the melting-point, the observed nuclear line width is much smaller than that predicted from the rigid lattice theory given in Chapter IV. As the temperature is lowered the width increases in step-like fashion, reaching the rigid lattice value at the lower temperatures.

Examples of such behaviour are exhibited in Figs. X, 3 (9) and X, 4 (10), representing the variation of the measured second moment of the proton line as a function of temperature in polycrystalline benzene and in a single crystal of ammonium chloride. This stepwise narrowing is generally considered to be caused by internal motions which have a temperature-dependence connected with that observed for the line width.

The statement that the second moment is reduced by lattice motions appears to be in contradiction to the earlier statement (eqn. (24)) of its invariance with respect to such motions and some explanation is needed in order to reconcile these statements.

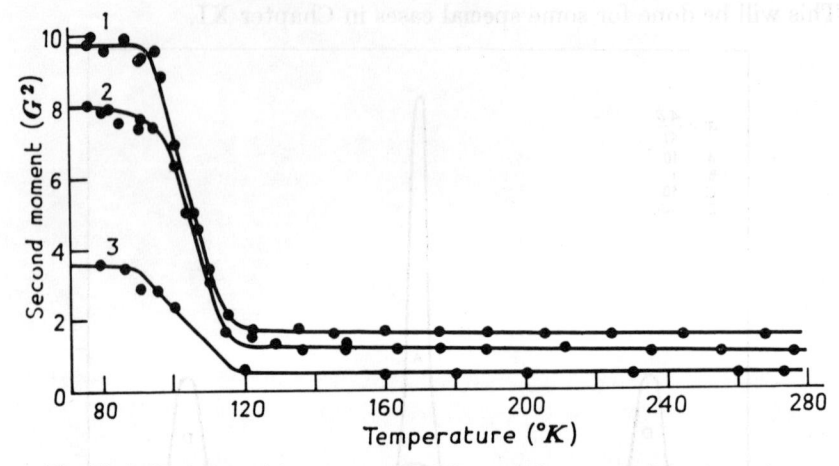

FIG. X, 3. Variation with temperature of the second moment of the absorption line for the three isotopic species of benzene. Curve 1: C_6H_6; curve 2: C_6H_5D; curve 3: 1,3,5-$C_6H_3D_3$.

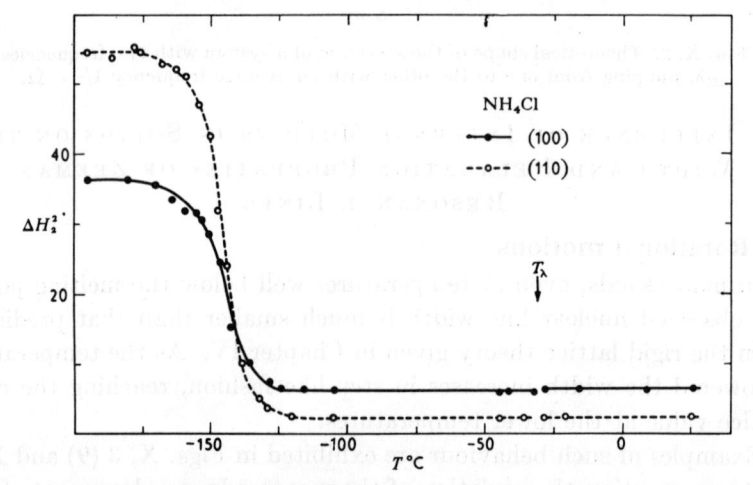

FIG. X, 4. The second moment of the proton magnetic resonance absorption line in a single crystal of ammonium chloride as a function of temperature, with the magnetic field applied in the (100) and (110) directions.

In the general equation (32) of Chapter VIII, which is satisfied by the density matrix σ^* (or, with a trivial change of i into $-i$, by the operator $\mathscr{M}_x^*(t)$), we assumed that the average value of the perturbing Hamiltonian $\overline{\mathscr{H}_1^*(t)}$ vanished and that the line width and relaxation were therefore determined by the second-order term in (32).

Although this assumption is generally correct in liquids it is not necessarily so for internal motions in solids.

Let us consider a molecule or a group of atoms which may rotate more or less freely in a solid and assume that the conditions for 'motion narrowing' are realized in the sense that the rotation rate is much higher than the strength of the dipolar interaction expressed in frequency units. The largest contribution to the dipolar energy comes generally (but not always) from couplings among nuclear spins inside a molecule ('intra' couplings), which we denote by $\mathcal{H}_1^i$. If the rotation takes place around a single axis, as for instance in benzene around the sixfold axis perpendicular to the plane of the molecule, the average Hamiltonian $\overline{\mathcal{H}_1^i}$ will not be zero.

Even if the rotation is isotropic in space, which implies $\overline{\mathcal{H}_1^i} = 0$, there still remains the contribution $\mathcal{H}_1^e$ from spins outside the molecule, which will not be averaged to zero by such a rotation.

The second moment of the resonance line is proportional to the trace of $\overline{[\mathcal{H}_1(t), I_x]^2}$, and, as was shown earlier, is unaffected by the lattice motion. It can be rewritten as the trace of $[\overline{\mathcal{H}}_1, I_x]^2 + \overline{[\mathcal{H}_1(t) - \overline{\mathcal{H}}_1, I_x]^2}$.

The two parts into which we have split the perturbing Hamiltonian, namely $\overline{\mathcal{H}}_1$ and $\mathcal{H}_1 - \overline{\mathcal{H}}_1$, although they may bring comparable contributions to the second moment, affect the shape of the resonance line very differently.

The effect of $\overline{\mathcal{H}}_1$, which is time-independent, is to give to the line a gaussian-like shape where the half intensity width and the r.m.s. width are comparable. The effect of $\mathcal{H}_1 - \overline{\mathcal{H}}_1$, which varies rapidly and has a vanishing average value, was shown earlier in this chapter to give to the line a Lorentzian shape. Then, practically all the contribution to the second moment came from regions very far in the wings where the absorption is so low as to be lost in the noise and unobservable. It follows that while, strictly speaking, the second moment is unaffected by the motion, the *observable* part of the second moment will be given by the trace of $[\overline{\mathcal{H}}_1, I_x]^2$ and will be smaller than in the absence of motion. The procedure of replacing $\mathcal{H}_1$ by $\overline{\mathcal{H}}_1$ in the calculation of the second moment is not unlike the truncation of the dipolar Hamiltonian described in Chapter IV. There we rejected the off-diagonal matrix elements of $\mathcal{H}_1$ because they brought into the second moment unobservable contributions from satellite lines; here we reject from $\mathcal{H}_1(t)$ the part $\mathcal{H}_1(t) - \overline{\mathcal{H}}_1$, responsible for the unobservable contributions of the wings.

The contributions to the second moment from the dipolar coupling between two nuclear spins $\mathbf{I}_1$ and $\mathbf{I}_2$ situated at points P_1 and P_2 was shown in Chapter IV to be proportional to $(1-3\cos^2\theta_{12})^2$, where θ_{12} is the angle of the vector $\mathbf{P}_1\mathbf{P}_2$ with the applied field. If the molecule to which these two spins belong is rotating around an axis OZ which makes an angle θ' with the applied field and an angle γ with $\mathbf{P}_1\mathbf{P}_2$, $(1-3\cos^2\theta_{12})^2$ must be replaced by $\overline{(1-3\cos^2\theta_{12})^2}$; the average must be taken over all the values of θ_{12} in the course of the rotation. The rotation need not be describable in classical terms; the molecule may have several equivalent positions with respect to the axis OZ, separated by potential barriers, and flip from one to the other at a rate which will in general be temperature-dependent. If there are more than two such equivalent positions, it is easy to show that averaging over a discrete number of possible orientations of the molecule or over a continuous set, as in classical rotation, gives the same result.

Using the addition theorem for spherical harmonics, we get

$$\overline{3\cos^2\theta_{12}-1} = \tfrac{1}{2}(3\cos^2\gamma-1)(3\cos^2\theta'-1). \qquad (65)$$

For a polycrystalline sample, where all orientations of the axis OZ with respect to the magnetic field are equally probable, the net effect of the reorientation is to multiply the second moment by the factor $\tfrac{1}{4}(1-3\cos^2\gamma)^2$. In particular, if the reorientation axis of each molecule is perpendicular to the spin-spin vector $\mathbf{P}_1\mathbf{P}_2$, this factor is $\tfrac{1}{4}$. This very simple reduction factor is only valid for the intramolecular contribution to the second moment. The intermolecular contribution resulting from interactions between spins belonging to different molecules is affected by the rotation in a more complicated way since the distances between the spins change as well as the orientations of the spin-spin vectors.

The change in the second moment observed in polycrystalline benzene C_6H_6 below $90°$ K and above $120°$ K and attributed to the rotation of the molecules around their hexad axes, can be read on Fig. X, 3 as approximately a factor 6 rather than 4, as is to be expected from the calculation above. This is due to the intermolecular contributions for which the ratio of the values below $90°$ K and above $120°$ K is different from 4. The inter- and intra-contributions could be separated by the method of isotopic substitution (9): replacing the 1, 3, 5 protons in C_6H_6 by deuterons affects these contributions differently, making their separation possible. It has been found that for the intra-contribution, isolated in this way, the ratio of the values below $90°$ K and above $120°$ K was indeed 4, as expected.

The passage from one plateau to the other as, say, in the curve of Fig. X, 3 for polycrystalline benzene between 90° K and 120° K, can be described as follows. Below 90° K the rate of molecular transitions from one equivalent position to another, characterized by a correlation time τ_c, is very low, and the lattice is practically rigid. It may be assumed that τ_c is of the form

$$\tau_c = \tau_0 e^{U/kT}, \qquad (66)$$

where U is an activation energy corresponding approximately to the height of the potential barrier between two equivalent molecular positions. As the temperature increases, τ_c decreases and the resonance likewise narrows.

A new plateau is reached when the rate $1/\tau_c$ is much faster than $\Delta\omega_0$, the rigid lattice width in frequency units.

The theoretical interpretation of the intermediate region between the two plateaux is difficult; when $\Delta\omega_0 \tau_c$ is neither large nor small, the experimental determination of the second moment becomes very uncertain if not meaningless. The measured value of the second moment will depend on how far in the wings it is possible to measure the nuclear absorption, that is, on the signal-to-noise ratio available for the measurement. It would thus appear more advisable, for the intermediate region, to measure and to plot against the temperature some unambiguous parameter of the observed resonance curve, such as the half width at half intensity or the distance between the maximum and the minimum of the absorption derivative. Then, if one chooses a model for the narrowing such as the one described in Section II C, it is possible to obtain from a formula such as (35), or rather from its Fourier transform, the detailed shape of the resonance curve, to correlate the observed line width parameter with the correlation time τ_c and obtain the variation of τ_c with temperature.

A formula such as (35) implies that in the course of the motion the average dipolar Hamiltonian $\overline{\mathscr{H}_1}$ is zero, and this implies that the height of the second plateau is zero also.

It is possible to adapt the formula (35) to the present situation by splitting the second moment ω_p^2 of formula (35) into two parts

$$\omega_p^2 = \omega_p'^2 + \omega_p''^2,$$

of which the first only is destroyed by the motion, and to write $G(t)$ as

$$G(t) = \exp(-\tfrac{1}{2}\omega_p''^2 t^2) \times \exp\left\{-\omega_p'^2 \int_0^t (t-\tau)e^{-\tau/\tau_c}\,d\tau\right\}. \qquad (67)$$

Instead of the simple correlation function $e^{-\tau/\tau_c}$ more elaborate expressions could be used, as explained in Section II C for the reduced correlation function $g_\omega(\tau)$.

The common practice (1) in the literature has been to correlate the narrowed line width $\delta\omega$ with the second moment $\delta\omega_0^2$ of the rigid lattice by the formula

$$\delta\omega^2 = \delta\omega_0^2 \frac{2}{\pi}\tan^{-1}[\alpha\,\delta\omega\,\tau_c], \qquad (68)$$

where the numerical factor α of order unity is rather ill-defined. This formula is an immediate consequence of the formula (3) where the spectral density $J(\omega)$ is assumed to be of the form $2\tau_c/[1+\omega^2\tau_c^2]$. The factor α in (68) stems from the idea that the integration limits in (3), of the order of $\pm\Delta\omega$, are not very well defined.

An *ad hoc* modification of (68), for the case where the dipolar interaction averaged over the motion does not vanish completely, is

$$\delta\omega^2 = \delta\omega_0''^2 + \delta\omega_0'^2 \frac{2}{\pi}\tan^{-1}[\alpha\,\delta\omega\,\tau_c]. \qquad (68')$$

Compared to the more elaborate procedure outlined in Sections II and III, the use of the formulae (68) or (68') has the great advantage of simplicity but suffers from the drawback that it gives no indication about the line shape.

Formulae such as (35) or (67) have the advantage of permitting far more detailed predictions and of being self-consistent within the framework of the model chosen to describe the motion. However, since there is some arbitrariness in the model chosen, and since the simple formulae (68) or (68') appear to fit the experimental results reasonably well, it remains to be seen whether the increased labour involved in the use of the formulae such as (35) or (67) is too high a price to pay for internal consistency. It would be interesting to translate some of the experimental results into the new language and to see if the relation between τ_c and the temperature T deduced from the formulae (68) or (68') is modified appreciably.

On the other hand, on either side of the transition region, and sufficiently far from it, the calculation and the measurement of the second moment provides unambiguous results. Thus, for instance, it was possible to conclude from the height of the high-temperature plateau that a complete reorientation of the ammonium ion occurs in ammonium halides, the remaining line width being due to intermolecular contributions.

A large number of publications have been devoted to the study of reorientation motions in solids and of their dependence on temperature, by means of nuclear resonance absorption. Some references to earlier work can be found in (8) and (9).

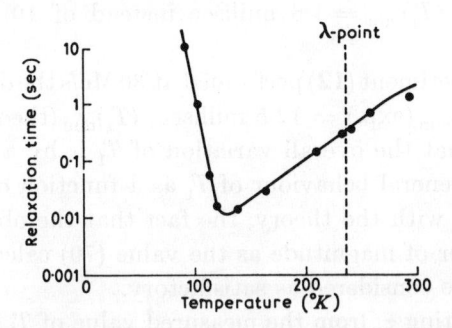

FIG. X, 5. Spin-lattice relaxation time for proton magnetic resonance in ammonium bromide plotted against temperature.

There have been rather fewer investigations of spin-lattice relaxation times. Fig. X, 5 (11) represents the spin-lattice relaxation time for the proton magnetic resonance in ammonium bromide, plotted against temperature. The main relaxation mechanism for the protons of the ammonium ion is the reorientation of that ion, already responsible for the narrowing of the resonance line.

It was stated at the end of Section II E (a) of Chapter VIII that the relaxation time of a system of four spins $\frac{1}{2}$, placed at the apices of a rotating tetrahedron could, to a very good approximation, be calculated by neglecting the correlations existing between the motions of the various spins, that is, by the formula (77) of Chapter VIII. If we assume that the reduced correlation function for the reorientation of the ammonium ion is of the form $e^{-\tau/\tau_c}$, this formula can be rewritten, using the formula (105) of Chapter VIII, as

$$\frac{1}{T_1} = \frac{6}{5}\frac{\gamma^4\hbar^2}{b^6}I(I+1)\left[\frac{\tau_c}{1+\omega^2\tau_c^2}+\frac{4\tau_c}{1+4\omega^2\tau_c^2}\right]$$

or, for $I = \frac{1}{2}$,

$$= \frac{9}{10}\frac{\gamma^4\hbar^2}{b^6}\left[\frac{\tau_c}{1+\omega^2\tau_c^2}+\frac{4\tau_c}{1+4\omega^2\tau_c^2}\right]$$

(69)

where b is the distance between two protons. $1/T_1$ is a maximum and T_1 a minimum for $\omega\tau_c \simeq 0{\cdot}6$.

$$\frac{1}{(T_1)_{\min}} = \frac{9}{10}\frac{\gamma^4\hbar^2}{b^6}\frac{1{\cdot}42}{\omega}.$$

(70)

The proton-proton distance in NH_4Br has been determined from second moment measurements. It is equal to $1\cdot 03\times\sqrt{\tfrac{8}{3}}$ Å (it is the distance N—H $=\sqrt{\tfrac{3}{8}}b$ which is given in the literature as 1·03 Å). For the frequency $\omega=2\pi\times 8\cdot 155$ Mc/s used in the measurements of reference (11), (70) gives $(T_1)_{\min}=1\cdot 6$ millisec instead of 10 millisec read on Fig. X, 5.

In another experiment (12) performed at 30 Mc/s the discrepancy is less severe since $(T_1)_{\min}$ (exp.) $= 12\cdot 5$ millisec; $(T_1)_{\min}$ (theor.) $= 6$ millisec. If one notices that the overall variation of T_1 is by a factor of 10^3 or more, that the general behaviour of T_1 as a function of a temperature is in accordance with the theory, the fact that the observed $(T_1)_{\min}$ is of the same order of magnitude as the value (70) calculated from first principles, can be considered as satisfactory.

If after extracting τ_c from the measured value of T_1 by means of the formula (69), $\log\tau_c$ is plotted against $1/T$, a straight line should be obtained, if the relation (66) is valid, and its slope should give the activation energy U. Within experimental error $\log\tau_c$ against $1/T$ is indeed a straight line and yields a value $U/k = 1800$ (11) (or 1870 (12)) for NH_4Br.

For the determination of τ_c, the measurement of T_1, which permits us to follow the variation of τ_c with the temperature, between say $\tau_c \sim 10^{-3}/\omega$ and $\tau_c \sim 10^3/\omega$, is clearly preferable to that of the line width.

In the examples considered so far, in the high-temperature range the line width reached a plateau, lower than the rigid lattice value, but still measurable. This was interpreted as due to the contribution of the intermolecular couplings which the rotation of the molecule could not average out. On the other hand, in many solids and in particular in metals there is no such plateau and the line becomes extremely narrow at high temperature, showing that the intermolecular contribution is averaged out as well. This can only occur if the molecules (or atoms) diffuse through the lattice, a process to be considered now.

B. Translational diffusion in solids

As a specific example we shall consider in some detail metallic lithium where both the line width and the spin-lattice relaxation time have been studied very thoroughly over a wide range of temperatures.

Fig. X, 6 (13) shows the variation of the line width of Li^7 (between peaks of the derivative) as a function of temperature between 150° K and 350° K.

At the lowest temperatures the experimental value of the second moment agrees to within experimental error with the theoretical value for a rigid lattice. As the temperature increases, motion narrowing sets in and around 20° C the line becomes sufficiently narrow for its real width to be masked by instrumental effects such as field inhomogeneity

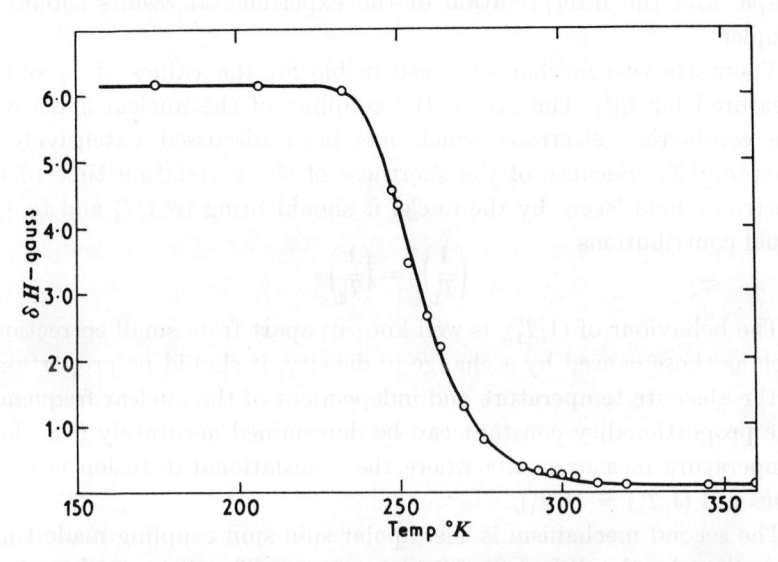

FIG. X, 6. The Li⁷ magnetic resonance line width transition in the metal.

and field modulation. Fortunately at that stage it becomes possible to pursue the investigation of the line width towards higher temperatures using spin echo techniques (14). Above 20° C there is a very considerable narrowing of the line and it follows from the theory given in Sections II and III of this chapter that the line has very nearly a Lorentzian shape. The decay of the free precession in a perfect magnet, or that of the echo envelope in an imperfect one, should be very nearly exponential with a time constant T_2, which is related to the distance δH between peaks of the derivative by

$$\gamma \delta H = \frac{2}{\sqrt{3}} \frac{1}{T_2}.$$

On the other hand, in the transition region between say 240° K and 300° K the shape of the line is expected to change, going progressively from the gaussian-like shape of the rigid lattice to a Lorentzian shape.

It is possible in principle, as already explained in Section V A, using the semi-empirical formula (68) or the more elaborate one (35), to analyse the behaviour of the line width in this region and to extract from it the dependence on the temperature of a correlation time τ_c for the diffusion process.

We prefer to discuss the behaviour of T_1 and T_2 in the high-temperature region from 20° C up, where the line has a well-defined Lorentzian shape, and the interpretation of the experimental results should be simpler.

There are two mechanisms responsible for the values of T_2 and T_1 measured for Li⁷. The first is the coupling of the nuclear spins with the conduction electrons which has been discussed extensively in Chapter IX. Because of the shortness of the correlation time of the electronic field 'seen' by the nuclei it should bring to $1/T_1$ and to $1/T_2$ equal contributions

$$\left(\frac{1}{T_1}\right)_e = \left(\frac{1}{T_2}\right)_e.$$

The behaviour of $(1/T_1)_e$ is well known: apart from small corrections, such as those caused by a change in density, it should be proportional to the absolute temperature and independent of the nuclear frequency. The proportionality constant can be determined accurately from low-temperature measurements where the translational diffusion is negligible and $(1/T_1) = (1/T_1)_e$.

The second mechanism is the dipolar spin-spin coupling made time-dependent by the diffusion of lithium atoms. The theoretical study of this coupling is relatively simple in the range of high temperatures and short correlation times, i.e. in regions such that the rigid lattice-line width $\Delta\omega_0$ is much greater than the narrowed width: $\Delta\omega \sim \Delta\omega_0^2 \tau_c$. These are the conditions where the general theory of Chapter VIII does apply and where we can define unambiguously relaxation times $(T_1)_d$ and $(T_2)_d$ given by the formulae (76) and (79) of Chapter VIII.

The observed relaxation times T_1 and T_2 are then given by

$$\frac{1}{T_1} = \left(\frac{1}{T_1}\right)_e + \left(\frac{1}{T_1}\right)_d, \quad \frac{1}{T_2} = \left(\frac{1}{T_2}\right)_d + \left(\frac{1}{T_2}\right)_e.$$

The outstanding theoretical problem is the calculation of the spectral densities $J_i(\omega)$ for diffusion in a crystal lattice, which we now discuss.

An approximate solution of this problem was given in Section II E (c) of Chapter VIII, equation (114). It was assumed there that the motion of the atoms was describable by a diffusion equation $\partial p/\partial t = D\,\Delta p$ and a distance of closest approach d between atoms had to be introduced in order to prevent the spectral densities $J_i(\omega)$ from being infinite. The actual process of diffusion in a crystal lattice can be visualized as a random walk where the atoms jump from a crystal site to a neighbouring one with an average frequency $1/\tau_r$. (We use the notation τ_r (r for random) to avoid confusion with other constants τ introduced elsewhere.) To appreciate whether the diffusion equation is adequate for calculation of the spectral densities encountered in the expressions of the relaxation times, one should remember that this equation can be obtained as a limit of a random-walk process, when the individual steps are very small; the diffusion constant D is then given by $\langle r^2 \rangle = 6D\tau_r$, where $\langle r^2 \rangle$ is the mean square value of one step. An intuitive physical agreement shows that the diffusion equation is a good approximation for the random-walk process in the calculation of $J(\omega)$, for $\omega\tau_r \ll 1$ but not for $\omega\tau_r \geqslant 1$. The main contribution to $J(\omega)$ comes from local fields that fluctuate at a rate comparable to ω. If $\omega\tau_r \ll 1$, the nearest neighbours of a spin will be relatively less efficient (because they are jumping too fast) than spins farther removed from the spin under consideration. A large number of spins will contribute to $J(\omega)$ and the discrete nature of a random walk will be relatively unimportant. On the other hand, for $\omega\tau_r \geqslant 1$, the influence of the nearest neighbours will be preponderant and the microscopic details of the diffusion process important.

The spectral densities $J_i(\omega)$ have been calculated both for f.c.c. and b.c.c. lattices using a model where a nucleus has a probability per unit time $1/\tau_r$ of jumping into a nearest neighbour position (7, 8). The following results are found for the spectral densities of correlation functions:

$$\left. \begin{array}{l} J_2(\omega) = 4J_1(\omega) \\[6pt] J_1(\omega) = \dfrac{8\pi}{15} \dfrac{n\tau_r}{k^3 l^3} G(k, \tfrac{1}{2}\omega\tau_r) \end{array} \right\} \quad (71)$$

where the different quantities have the following significance:

$$G(k, y) = \int_0^\infty [J_{\frac{3}{2}}(kx)]^2 \frac{1 - \sin x/x}{(1 - \sin x/x)^2 + y^2} \frac{dx}{x}. \quad (72)$$

The Bessel function $J_{\frac{3}{2}}$ should not be confused with the spectral densities J_1 and J_2.

For an f.c.c. lattice: $n = 4cb^{-3}$ where b is the lattice spacing, c the ratio of the number of nuclear spins to the number of lattice sites, l the distance between nearest neighbours is $b/\sqrt{2}$, and $k = 0.743$. The function $G(k, y)$ is tabulated in (7).

For a b.c.c. lattice: $n = 2cb^{-3}$, $l = b\sqrt{3}/2$, $k = 0.763$. The relaxation rate $(1/T_1)_d$ is given by

$$\left(\frac{1}{T_1}\right)_d = \tfrac{3}{2}\gamma^4\hbar^2 I(I+1)[J_1(\omega_0)+J_2(2\omega_0)] \tag{73}$$

or, by (71),
$$\left(\frac{1}{T_1}\right)_d = \frac{8\pi}{5}\gamma^4\hbar^2 I(I+1)\frac{n}{k^3 l^3 \omega}\Phi(k,y)$$

where $\quad y = \tfrac{1}{2}\omega_0 \tau_r, \quad \Phi(k,y) = yG(k,y)+4yG(k,2y). \tag{74}$

(Unfortunately, in references (7) and (8) $1/T_1$ is incorrectly given as being proportional to $J_1(\omega_0)+\tfrac{1}{2}J_2(2\omega_0)$ and for b.c.c. lattices tables are given for
$$\Psi(y) = yG(k,y)+2yG(k,2y) \tag{74'}$$
rather than for $\Phi(k,y)$ given by (74).)

It is interesting to compare the results of the formulae (71) and (72) with those obtained from the formula (114) of Chapter VIII based on the diffusion equation, for the two limiting cases $\omega\tau_r \ll 1$, $\omega\tau_r \gg 1$. We take as the distance of closest approach in the diffusion equation the distance l between two nearest neighbours in a crystal lattice (called d in the formula (114) of Chapter VIII). For $\omega\tau_r \ll 1$, the random-walk formula (71) yields for an f.c.c. lattice

$$J_1(\omega) = J_1(0) = \frac{8\pi}{45}\frac{n}{lD}\frac{G(0)}{2k^3} \simeq \frac{8\pi}{45}\frac{n}{lD}\times 0{\cdot}75, \tag{74''}$$

where for the sake of comparison with the results of the diffusion equation we introduce $D = l^2/6\tau_r$. The approximate formula (114) of Chapter VIII yields

$$J_1(\omega) = J_1(0) = \frac{8\pi}{45}\frac{n}{lD}\times 0{\cdot}4. \tag{74'''}$$

The discrepancy is not very serious.

As expected, the discrepancy is much more serious for $\omega\tau_r \gg 1$. The random walk formula gives for an f.c.c. lattice

$$J_1(\omega) = 98{\cdot}5b^{-6}\frac{8}{15\omega^2\tau_r} \simeq \frac{28Dn}{\omega^2 l^5}, \tag{75}$$

where we again take $D = l^2/6\tau_r$.

On the other hand, the diffusion equation gives a very different asymptotic behaviour of $J_1(\omega)$ since for large ω it varies as $\omega^{-3/2}$ which, as expected, is very different from (75).

A procedure, proposed at an early date by BPP (1) to estimate the translational spectral densities, is to assume arbitrarily that the correlation function for the interaction between two spins separated by a distance r is proportional to $e^{-t/\tau}$, where the correlation time τ is the time required for the spins to drift apart by a distance of the order of r: $\tau \simeq r^2/12D$ (12 rather than 6 because both spins drift). Adding the contributions from all spins outside a sphere of closest approach of radius l, it is found that

$$J_1(\omega) = \frac{16\pi}{15} n \int_l^\infty \frac{dr}{r^4} \frac{\tau}{1+\omega^2\tau^2} \tag{76}$$

with $\tau = r^2/12D$.

Taking $\omega\tau \ll 1$, that is, neglecting $\omega^2\tau^2$ in the denominator of (76), it is found that

$$J_1(\omega) = J_1(0) = \frac{8\pi n}{45Dl} \times 0\cdot 5, \tag{77}$$

in good agreement with both (74″) and (74‴).

For $\omega\tau \gg 1$, neglecting 1 in the denominator of (76), it is found that

$$J_1(\omega) = 2\cdot 56\pi \frac{Dn}{\omega^2 l^5}. \tag{77'}$$

Although the dependence on ω^2 is the same as in the random-walk model formula (75), the numerical coefficient is out by a factor

$$\frac{28}{2\cdot 56\pi} \simeq 3\cdot 5.$$

The conclusion of this comparison between the three methods used to compute the spectral density, namely the random walk method, the diffusion equation used in a consistent manner, and the crude BPP model, is that all agree for short correlation times, but that for long correlation times it is necessary to pay attention to the microscopic features of the diffusion process and to use the random-walk model of reference (7).

After this digression on the theory of translational diffusion in a crystal lattice we discuss the results of measurements of T_1 and T_2 in Li^7.

Fig. X, 7 shows the dependence of T_1 and T_2 as a function of temperature for a frequency of 9 Mc/s. For T_2, if we except the unexplained drop near the melting point, there is a continuous increase with temperature which is in accordance with the formula (79) of Chapter VIII. The behaviour of T_1 is different: since it results, as we said earlier, from a combination of two effects, coupling with conduction electrons and dipolar spin-spin coupling, T_1 exhibits a general decrease with increasing temperature for which the conduction electrons are responsible, on which is superimposed a minimum similar to that of Fig. X, 5, for BrNH$_4$, characteristic of a correlation time that varies very rapidly with temperature, and is caused by the diffusion.

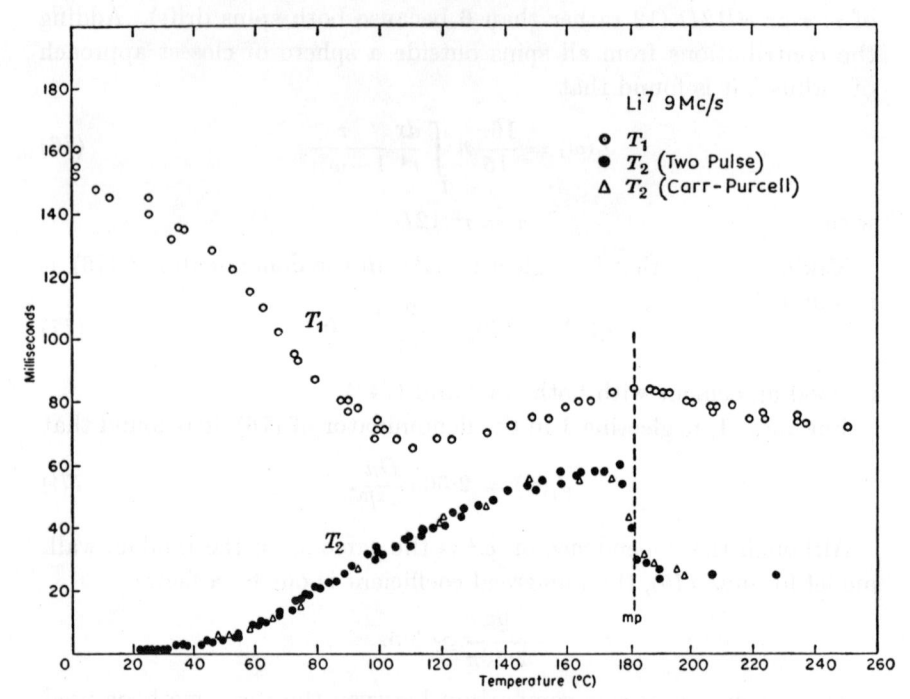

FIG. X, 7. T_1 and T_2 data for Li^7, measured in a sample of natural abundance, at a Larmor frequency of 9·0 Mc/s.

In order to separate the two mechanisms, T_1, measured at three different frequencies, is plotted against $1/T$ (Fig. X, 8). If the coupling with conduction electrons were the only relaxation mechanism this curve would be a straight line.

It may be seen that, for instance, for a frequency of 9 Mc/s the product $T_1 T$ is 42·3 sec degrees towards either end of the experimental range, namely near the melting point at 180° C, and near 20° C. The value 42·3 for $T_1 T$ is in good agreement with $T_1 T = 43 \pm 2$ resulting from measurements at the temperature of liquid helium (reference (4) of Chapter IX), where the coupling with conduction electrons is the only nuclear relaxation mechanism. This demonstrates that at both 20° C and 180° C the contribution of diffusion to $1/T_1$ is negligible, the diffusion motion being either too slow or too fast to be efficient for relaxation. In order to determine the contribution $(1/T_1)_d$ to $1/T_1$ it is sufficient to subtract from the observed $1/T_1$ the quantity

$T/42 \cdot 3$ contributed by conduction electrons. (This is for 9 Mc/s: there seems to be a slight and unexplained dependence on frequency of the

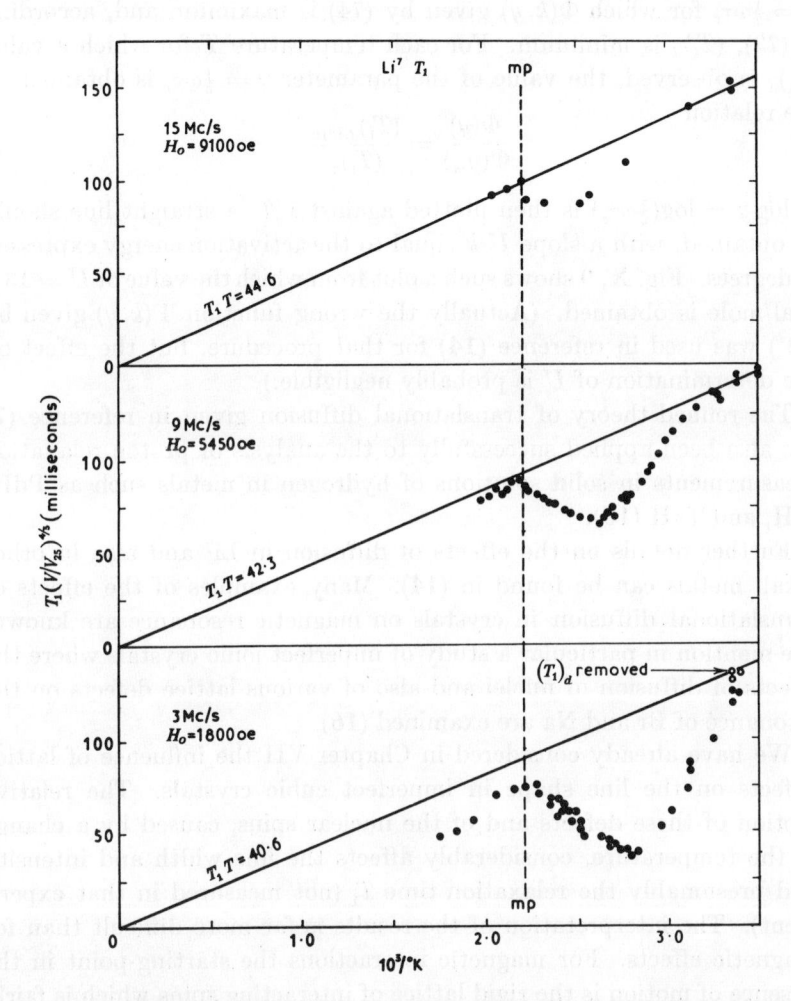

FIG. X, 8. T_1 data for Li^7, taken at three values of the Larmor frequency, plotted against the reciprocal of the absolute temperature in units $10^3/°K$. The data are corrected against a dependence of T_1 on the volume as $V^{-\frac{2}{3}}$, expected for the relaxation by conduction electrons in the free electrons approximation.

relaxation rate $(1/T_1)_e$ caused by the conduction electrons, as apparent from Fig. X, 8.)

From the curve $(1/T_1)_d$ against the temperature, thus obtained, the dependence of the correlation time τ_r on temperature can be extracted

using the formula (73) and numerical values of $\Phi(k,y)$. The procedure used for this extraction is the following. Let y_m be the value of $y = \frac{1}{2}\omega\tau_r$ for which $\Phi(k,y)$ given by (74) is maximum and, according to (73), $(T_1)_d$ is minimum. For each temperature T for which a value $(T_1)_d$ is observed, the value of the parameter $y = \frac{1}{2}\omega\tau_r$ is obtained by the relation
$$\frac{\Phi(y)}{\Phi(y_m)} = \frac{(T_1)_{d,\min}}{(T_1)_d}.$$

If $\log y = \log(\frac{1}{2}\omega\tau_r)$ is then plotted against $1/T$, a straight line should be obtained, with a slope U/k equal to the activation energy expressed in degrees. Fig. X, 9 shows such a plot from which the value of $U = 13\cdot 2$ kcal/mole is obtained. (Actually the wrong function $\Psi(k,y)$ given by (74') was used in reference (14) for that procedure, but the effect on the determination of U is probably negligible.)

The refined theory of translational diffusion given in reference (7) has also been applied successfully to the analysis of proton-relaxation measurements in solid solutions of hydrogen in metals such as PdH, TiH, and TaH (15).

Further details on the effects of diffusion in Li[7] and also in other alkali metals can be found in (14). Many examples of the effects of translational diffusion in crystals on magnetic resonance are known. We mention in particular a study of imperfect ionic crystals where the effects of diffusion of nuclei and also of various lattice defects on the resonance of Br and Na are examined (16).

We have already considered in Chapter VII the influence of lattice defects on the line shape in imperfect cubic crystals. The relative motion of these defects and of the nuclear spins, caused by a change of the temperature, considerably affects the line width and intensity and presumably the relaxation time T_1 (not measured in that experiment). The interpretation of the results is far more difficult than for magnetic effects. For magnetic interactions the starting-point in the absence of motion is the rigid lattice of interacting spins which is fairly well understood; the starting-point in the present study is the imperfect ionic crystal where the effects of imperfections on the resonance line, although qualitatively understood, as discussed in Section II A (b) (2) of Chapter VII, are difficult to calculate. The same applies to an even larger extent to moving defects. We shall not discuss the results of the study (16) beyond mentioning that in order to explain the effects observed for, say, bromine, anti-shielding factors γ of the order of at least 20 are required.

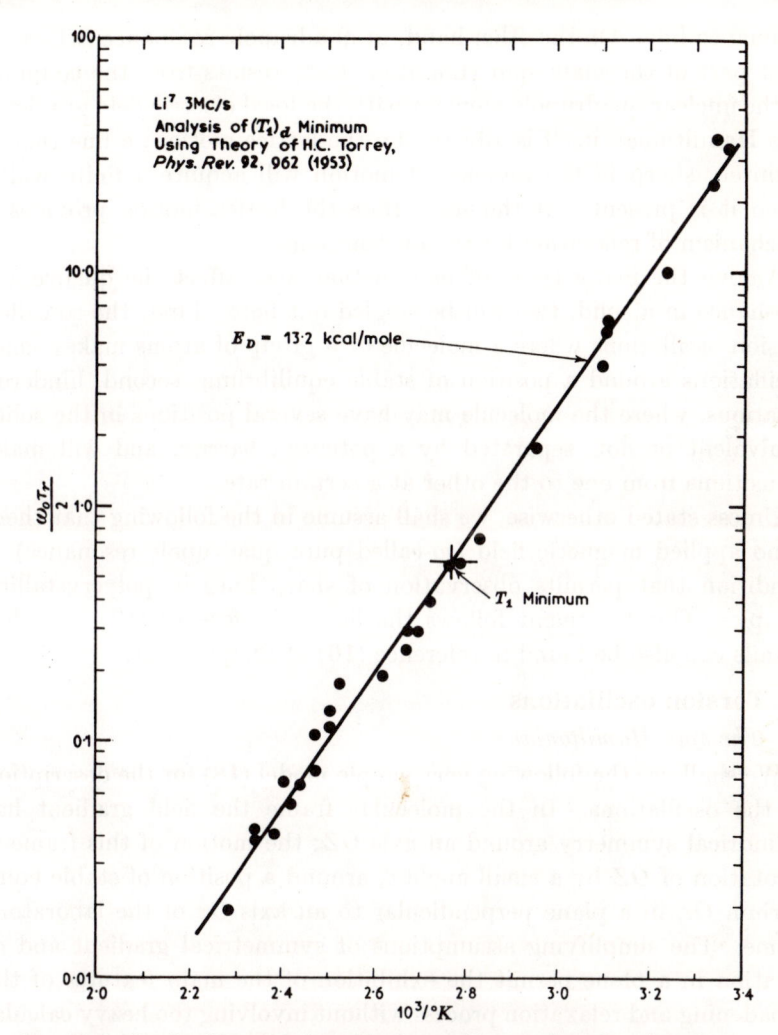

FIG. X, 9. Graph of $\tfrac{1}{2}\omega_0\tau_r$, in logarithmic coordinates, plotted against the reciprocal temperature, as explained in the text.

VI. Influence of Internal Motions in Solids on the Width and Relaxation of Quadrupole Resonance Lines

The effects of lattice motion on quadrupole line width differ from those considered for Zeeman resonance in one important respect. There the motion of the lattice did not affect the Zeeman Hamiltonian $\mathcal{H}_0$, the main part of the total Hamiltonian $\mathcal{H} = \mathcal{H}_0 + \mathcal{H}_1$, while its effect on the perturbing Hamiltonian $\mathcal{H}_1$ resulted in a narrowing of the resonance line. On the other hand, in quadrupole resonance, where at least part of the main spin Hamiltonian $\mathcal{H}_0$ results from the coupling of the nuclear quadrupole moment with the local electric field gradient, this Hamiltonian itself is affected by the lattice motion; a line that is infinitely sharp in the absence of motion will acquire a finite width when it is present. At the same time the lattice motion provides a mechanism of relaxation for the nuclear spins.

Among the many types of motion that may affect the quadrupole resonance in a solid, two will be singled out here. First, the so-called torsion oscillations where a molecule or a group of atoms makes small oscillations around a position of stable equilibrium; second, hindered rotations, where the molecule may have several positions in the solid, equivalent or not, separated by a potential barrier, and will make transitions from one to the other at a certain rate.

Unless stated otherwise, we shall assume in the following that there is no applied magnetic field (so-called pure quadrupole resonance), a condition that permits observation of sharp lines in polycrystalline samples. Our treatment follows the lines of reference (17). Further details can also be found in reference (10) of Chapter VII.

A. Torsion oscillations

(a) *The spin Hamiltonian*

We shall use the following very simple model (18) for the description of the oscillations. In the molecular frame the field gradient has cylindrical symmetry around an axis OZ; the motion of this frame is a rotation of OZ by a small angle θ, around a position of stable equilibrium Oz, in a plane perpendicular to an axis Ox of the laboratory frame. The simplifying assumptions of symmetrical gradient and of rotation in a plane permit the exhibition of the main features of the broadening and relaxation process without involving too heavy calculations. More realistic assumptions would have to be made for an actual comparison of theory with experiment in specific examples.

The components of the electric-field gradient in the molecular frame are
$$V_{ZZ} = eq, \qquad V_{XX} = V_{YY} = -\tfrac{1}{2}eq, \tag{78}$$
and related to those we have V_{xx}, etc. in the laboratory frame, by the usual tensor transformation coefficients, which for small θ yield

$$V_{zz} = eq\left(1 - \frac{3e\theta^2}{2}\right), \qquad V_{xx} = -\tfrac{1}{2}eq, \qquad V_{yy} = -\tfrac{1}{2}eq(1-3\theta^2),$$
$$V_{yz} = \frac{3eq\theta}{2}, \qquad V_{xz} = V_{xy} = 0. \tag{79}$$

In the laboratory frame the spin Hamiltonian can then be written, using the formula (22) of Chapter VII,

$$\mathscr{H} = \frac{e^2qQ}{4I(2I-1)}\left[\left(1-\frac{3\theta^2}{2}\right)(3I_z^2-I(I+1))-\tfrac{3}{4}\theta^2(I_+^2+I_-^2)+3\theta(I_zI_y+I_yI_z)\right]. \tag{80}$$

Because of the fluctuations of θ the Hamiltonian $\mathscr{H}$ is now time-dependent, this dependence being of a random nature, as in many similar examples of lattice motions encountered previously.

The Hamiltonian $\mathscr{H}$ of equation (80) can be rewritten as

$$\mathscr{H} = \overline{\mathscr{H}} + (\mathscr{H}-\overline{\mathscr{H}}),$$

where the bar means average taken over the oscillatory motion of θ. The average Hamiltonian $\overline{\mathscr{H}}$ will be the new zero-order Hamiltonian, instead of

$$\mathscr{H}_0 = \frac{e^2qQ}{4I(2I-1)}\{3I_z^2-I(I+1)\}$$

in the absence of motion, whereas $\mathscr{H}_1 = \mathscr{H}-\overline{\mathscr{H}}$ will be the time-dependent perturbation responsible for the broadening and the relaxation.

The average value $\bar{\theta}$ vanishes and the last term of the square bracket in equation (80) gives no contribution to $\overline{\mathscr{H}}$. The second term $-\tfrac{3}{4}\overline{\theta^2}(I_+^2+I_-^2)$, which represents a small departure from cylindrical symmetry, at least for half-integer spins, affects the energy levels of $\overline{\mathscr{H}}$ to second order only by a contribution proportional to $(\overline{\theta^2})^2$, and for small amplitudes of oscillation can be neglected.

There remains the first term, which corresponds to a change of the effective gradient q and thus of the resonance frequencies ω_m by a factor $(1-\tfrac{3}{2}\overline{\theta^2})$. Since $\overline{\theta^2}$ is expected to increase with temperature, the resonance frequencies are expected to and actually do go down with increasing temperature as was already noticed in the earliest observations of quadrupole resonance (19).

For instance, the resonance frequency of Cl^{35} in KClO$_3$ goes down from 28·55 Mc/s to 27·75 Mc/s as the temperature goes up from $-77°$ C to $85°$ C.

If we assume that the rotational motion is describable by the equation of an oscillator $I\ddot{\theta}+C\theta = 0$, so that the energy of motion is $\tfrac{1}{2}C\theta^2+\tfrac{1}{2}I\dot{\theta}^2$, in the high-temperature limit $C\overline{\theta^2} = kT$ and the change of frequency with temperature is given by

$$-\frac{1}{\omega_m}\frac{d\omega_m}{dt} = \frac{3k}{2C}.$$

(b) *The line width*

The line width results from the random Hamiltonian

$$\mathcal{H}_1 = \mathcal{H}-\overline{\mathcal{H}} = A\{-\tfrac{3}{2}(\theta^2-\overline{\theta^2})(3I_z^2-I(I+1))-\tfrac{3}{4}(\theta^2-\overline{\theta^2})(I_+^2+I_-^2)+ \\ +3\theta(I_zI_y+I_yI_z)\}, \quad (81)$$

where
$$A = \frac{e^2qQ}{4I(2I-1)}.$$

Consider a transition $|m) \to |m-1)$ with a frequency

$$\omega_m = \frac{3A}{\hbar}(2m-1).$$

There is another transition, $|-m) \to |-(m-1))$, that has the same frequency, and according to the general theory given in Section III of this chapter, one may expect the line shape to be represented by a superposition of two Lorentz curves. However, it is easy to verify that for half-integer spins, the two transitions $m \to (m-1)$ and $-m \to -(m-1)$ are actually uncoupled, all 'off-diagonal' matrix elements of the matrix R in equation (38), such as $R_{m,(m-1);(-m),(-m+1)}$, being zero for a Hamiltonian such as (81). Each transition $m \to (m-1)$ is thus a 'simple' line that has the shape of a single Lorentz curve and its width is given by the general formula (45)

$$\frac{1}{T_2} = \frac{1}{T_2'}+\frac{1}{2}\left(\frac{1}{t_m}+\frac{1}{t_{m-1}}\right),$$

where $1/T_2'$ is the so-called adiabatic width and t_m and t_{m-1} the lifetimes of the states $|m)$ and $|m-1)$. The adiabatic width $1/T_2'$ is given by (43') $1/T_2' = \int_0^\infty \overline{\omega(t)\omega(t-\tau)}\,d\tau$, where $\omega(t) = -\omega_m\tfrac{3}{2}(\theta^2-\overline{\theta^2})$ as given by (81) is the departure of the instantaneous resonance frequency from its average value $\omega_m(1-\tfrac{3}{2}\overline{\theta^2})$. We define $K_{II}(\tau)$ as the correlation function of the random variable $\theta^2(t)-\overline{\theta^2}$:

$$K_{II}(\tau) = \overline{(\theta^2(t)-\overline{\theta^2})(\theta^2(t-\tau)-\overline{\theta^2})}. \quad (82)$$

The spectral density $J_{II}(\omega)$ is defined as usual by the relation

$$J_{II}(\omega) = \int_{-\infty}^{\infty} K_{II}(\tau)e^{-i\omega\tau}\,d\tau. \quad (83)$$

(We write J_{II} rather than J_2 to avoid confusion with a different spectral density encountered earlier.) From these definitions we obtain

$$\frac{1}{T_2'} = \tfrac{9}{4}\omega_m^2 \int_0^\infty \overline{[\theta^2(t)-\overline{\theta^2}][\theta^2(t-\tau)-\overline{\theta^2}]}\,d\tau = \tfrac{9}{8}\omega_m^2\,J_{II}(0). \quad (84)$$

We also define
$$K_1(\tau) = \overline{\theta(t)\cdot\theta(t-\tau)}, \qquad J_I(\omega) = \int_{-\infty}^{\infty} K_I(\tau)e^{-i\omega\tau}\,d\tau. \qquad (85)$$

From the Hamiltonian (81) we obtain for the lifetime t_m of the state $|m\rangle$

$$\frac{1}{t_m} = \frac{9A^2}{16\hbar^2}[J_{II}(\omega'_m)|\langle m|I_+^2|m-2\rangle|^2 + J_{II}(\omega''_m)|\langle m|I_-^2|m+2\rangle|^2] +$$
$$+ \frac{9A^2}{4\hbar^2}[J_I(\omega_m)|\langle m|I_zI_+ + I_+I_z|m-1\rangle|^2 +$$
$$+ J_I(\omega_{(m+1)})|\langle m|I_zI_- + I_-I_z|m+1\rangle|^2], \quad (86)$$

where $\omega'_m = \dfrac{E_m - E_{m-2}}{\hbar}$, $\omega''_m = \dfrac{E_m - E_{m+2}}{\hbar}$.

A similar formula holds for $1/t_{(m-1)}$. We now apply these formulae in the special cases of $I = \tfrac{3}{2}$ and $I = \tfrac{5}{2}$.

$I = \tfrac{3}{2}$:
$$\frac{1}{T_2} = \frac{\omega^2}{2}[\tfrac{9}{4}J_{II}(0) + \tfrac{3}{8}J_{II}(\omega) + \tfrac{3}{2}J_I(\omega)], \qquad (87)$$

where $\omega = \omega_{\frac{3}{2}}$ is the resonance frequency.

$I = \tfrac{5}{2}$. There are two resonance frequencies: $\omega_{\frac{5}{2}}$ for the transition $\tfrac{5}{2} \to \tfrac{3}{2}$ and $\omega_{\frac{3}{2}}$ for the transition $\tfrac{3}{2} \to \tfrac{1}{2}$, with $\omega_{\frac{5}{2}} = 2\omega_{\frac{3}{2}} = 2\omega$. For the transition $\tfrac{5}{2} \to \tfrac{3}{2}$ we find

$$\frac{1}{T_2} = \frac{\omega^2}{2}[10J_I(2\omega) + \tfrac{9}{8}J_{II}(\omega) + 2J_I(\omega) + \tfrac{5}{8}J_{II}(3\omega) + 9J_{II}(0)]. \qquad (88)$$

For the transition $\tfrac{3}{2} \to \tfrac{1}{2}$,

$$\frac{1}{T_2} = \frac{\omega^2}{2}[4J_I(\omega) + \tfrac{9}{4}J_{II}(\omega) + \tfrac{5}{8}J_{II}(3\omega) + 5J_I(2\omega) + \tfrac{9}{4}J_{II}(0)]. \qquad (88')$$

Actually it is reasonable to expect that the quadrupole frequencies are small compared with the reciprocal correlation times of the random oscillation so that $J_I(\omega) \simeq J_I(0) = J_I$; $J_{II}(\omega) \simeq J_{II}(0) = J_{II}$. The equations (87), (88), (88') become:

$I = \tfrac{3}{2}$:
$$\frac{1}{T_2} = \frac{3\omega^2}{4}[J_I + \tfrac{7}{4}J_{II}]. \qquad (89)$$

$I = \tfrac{5}{2}$: transition $\tfrac{5}{2} \to \tfrac{3}{2}$,
$$\frac{1}{T_2} = \frac{\omega^2}{8}[48J_I + 43J_{II}], \qquad (90)$$

transition $\tfrac{3}{2} \to \tfrac{1}{2}$,
$$\frac{1}{T_2} = \frac{\omega^2}{16}[72J_I + 41J_{II}]. \qquad (90')$$

These formulae are easily extended to higher spins.

(c) Relaxation time

The non-adiabatic part of the random Hamiltonian (81) is also responsible for the spin-lattice relaxation. In quadrupole resonance, where for $I > \frac{3}{2}$ the levels are not equidistant, spin-spin couplings are unable rapidly to establish a spin temperature and the definition of a spin-lattice relaxation time is not unambiguous.

In general, the trend of the level populations towards their thermal equilibrium values will not be described by a single exponential and cannot be expressed by a single relaxation time. The rates of change of these populations can be obtained from the general equations (25) and (26") of Chapter VIII, where the transition probabilities $W_{\alpha\beta}$ are computed from the non-adiabatic part of the Hamiltonian (81).

As an example we consider the spin $I = \frac{3}{2}$ where, exceptionally, a single relaxation time T_1 can be defined.

Let $n'_{\frac{3}{2}},...,n'_{-\frac{3}{2}}$ be the departures of the populations from their thermal equilibrium values. They obey the equations

$$\left.\begin{aligned}\frac{dn'_{\frac{3}{2}}}{dt} &= -W_1(n'_{\frac{3}{2}}-n'_{\frac{1}{2}})-W_2(n'_{\frac{3}{2}}-n'_{-\frac{1}{2}}) \\ \frac{dn'_{\frac{1}{2}}}{dt} &= -W_1(n'_{\frac{1}{2}}-n'_{\frac{3}{2}})-W_2(n'_{\frac{1}{2}}-n'_{-\frac{3}{2}})\end{aligned}\right\}, \tag{91}$$

and two similar equations for $n'_{-\frac{1}{2}}$ and $n'_{-\frac{3}{2}}$.

Subtracting the second equation (91) from the first we get

$$\frac{d}{dt}(n'_{\frac{3}{2}}-n'_{\frac{1}{2}}) = -2W_1(n'_{\frac{3}{2}}-n'_{\frac{1}{2}})-W_2((n'_{\frac{3}{2}}-n'_{\frac{1}{2}})+(n'_{-\frac{3}{2}}-n'_{-\frac{1}{2}})).$$

Unless a truly rotating r.f. field which can make $(n'_{\frac{3}{2}}-n'_{\frac{1}{2}}) \neq (n'_{-\frac{3}{2}}-n'_{-\frac{1}{2}})$ has been used to disturb the populations, we have at all times

$$n'_{\frac{3}{2}}-n'_{\frac{1}{2}} = n'_{-\frac{3}{2}}-n'_{-\frac{1}{2}}$$

and the equation

$$\frac{d}{dt}(n'_{\frac{3}{2}}-n'_{\frac{1}{2}}) = -2(W_1+W_2)(n'_{\frac{3}{2}}-n'_{\frac{1}{2}}).$$

We define $1/T_1$ as a rate of change of $(n'_{\frac{3}{2}}-n'_{\frac{1}{2}})$ (or of $n_{\frac{3}{2}}-n_{\frac{1}{2}}$), by

$$\frac{1}{T_1} = 2(W_1+W_2). \tag{92}$$

From (81) we obtain immediately

$$\left.\begin{aligned}W_1 &= \tfrac{3}{4}\omega^2 J_I(\omega), \qquad W_2 = \tfrac{3}{16}\omega^2 J_{II}(\omega) \\ \frac{1}{T_1} &= \tfrac{3}{8}\omega^2[J_{II}(\omega)+4J_I(\omega)]\end{aligned}\right\}. \tag{93}$$

(d) The spectral densities

As a last step, in order to obtain some indications about the nature of the spectral densities J_I and J_{II}, a model must be chosen for the description of the rotational motion of the molecule. We can visualize this motion as that of a quantum mechanical oscillator which, because of its coupling with the rest of the lattice acting as a thermal bath, makes frequent transitions between its various energy levels $(n+\tfrac{1}{2})\hbar\Omega$, where Ω is the oscillation frequency. The correlation time of the random variable θ would thus be of the order of the lifetimes τ of the various oscillator states.

More generally (and more realistically) a system of a large number of such oscillators coupled with each other and forming a periodic lattice can be considered, and a relation can be established between the spectral densities J_I and J_{II} and the Raman spectrum of the sample (20).

We shall be content to consider the classical model of a damped harmonic oscillator where the influence of the thermal bath is represented by white gaussian noise. Its equation of motion (the Langevin equation) can be written

$$I(\ddot{\theta}-f\dot{\theta}+\Omega^2\theta) = F(t), \tag{94}$$

where $F(t)$ is a random gaussian function with a white spectrum.† According to (94) $\theta(t)$ is also a gaussian random function. This last result is actually far more general than the assumption (94) and can be demonstrated under assumptions as general as that of an assembly of interacting harmonic oscillators in thermal equilibrium (20).

For gaussian random functions the following relation can be shown to exist between $K_I(\tau)$ and $K_{II}(\tau)$:

$$K_{II}(\tau) = 2[K_I(\tau)]^2; \tag{95}$$

whence for the spectral densities, the corresponding folding relation:

$$\left.\begin{aligned} J_{II}(\omega) &= \frac{1}{\pi}\int_{-\infty}^{\infty} J_I(\omega')J_I(\omega-\omega')\,d\omega' \\ J_{II}(0) &= \frac{1}{\pi}\int_{-\infty}^{\infty} [J_I(\omega')]^2\,d\omega' \end{aligned}\right\}. \tag{96}$$

With a little algebra we obtain from (94)

$$J_I(\omega) = 2f\Omega^2 \frac{\overline{\theta^2}}{(\omega^2-\Omega^2)^2+f^2\omega^2} \tag{97}$$

† For a general discussion of random functions see reference (21).

which, using the second equation (96), yields

$$J_{II}(0) = \frac{2}{f}(\overline{\theta^2})^2, \qquad (98)$$

and, for the ratio $J_{II}(0)/J_I(0)$,

$$J_{II}(0)/J_I(0) = \left(\frac{\Omega}{f}\right)^2 \overline{\theta^2}. \qquad (99)$$

If the ratio $(\Omega/f)^2$ is a large number, that is, if the frequencies of the torsional oscillator are relatively sharp, J_{II} can be expected to be much larger than J_I unless $\overline{\theta^2}$ is vanishingly small. If such is the case the following interesting relations between the various relaxation times can be derived for the spins $\frac{3}{2}$ and $\frac{5}{2}$ by neglecting J_I compared with J_{II}.

Spin $I = \frac{3}{2}$. From (89) and (93) we obtain

$$\frac{1}{T_2} = \frac{7}{2}\frac{1}{T_1}. \qquad (100)$$

Spin $I = \frac{5}{2}$. From (90) and (90′) we obtain the ratio of the widths of the two resonances:

$$\left(\frac{1}{T_2}\right)_{\frac{5}{2}} \bigg/ \left(\frac{1}{T_2}\right)_{\frac{3}{2}} = \frac{86}{41} = 2 \cdot 1. \qquad (101)$$

For a large number of iodine compounds this ratio has indeed been found to be of the order of 2, in accordance with (101) (**22**).

We shall stop this theory of the effects of torsional oscillations here. Rather than to attempt an explanation of the known experimental results, its purpose was to show how a very simple model permitted a coherent description of motion broadening and relaxation in quadrupole resonance.

B. Hindered rotations

The consequences of hindered rotation for quadrupole resonance, whether in zero magnetic field or in a finite field, are far more drastic than for Zeeman resonance since, as explained earlier, it is the main Hamiltonian $\mathscr{H}_0$ itself that is affected by such motion. There are essentially two situations of interest where the resonance can be observed: very fast motion and very slow.

(a) *Fast motion*

Let us assume that a molecule or a group of atoms may occupy several equivalent positions in a crystal, and jump from one to another at a rate $1/\tau$.

In each of these positions the spin Hamiltonian $\mathcal{H}_0$ is a different operator $\mathcal{H}_0^a$, $\mathcal{H}_0^b$,.... These operators have the same spectrum in zero magnetic field but not in a finite field, and in any case have different eigenstates (unless the Zeeman part of the Hamiltonian is very much larger than the quadrupole coupling). If the jumping rate $1/\tau$ is much larger than the frequencies of the spin Hamiltonian, the nuclear spin will 'see' only an average Hamiltonian $\overline{\mathcal{H}_0}$ with frequencies different from those of $\mathcal{H}_a^0$, $\mathcal{H}_b^0$, etc., and a spectrum that is very different from that in the absence of motion. Assume for simplicity p positions where the spin 'sees' symmetrical gradients $V_{ZZ} = eq$, of axes $OZ_1, OZ_2, ..., OZ_p$, all making the same angle θ with the axis of rotation Oz and angles $\phi_1, ..., \phi_p$ with a plane xOz of the laboratory frame $Oxyz$.

The components of the field gradient 'seen' by the spin when the molecule is in a position P can be expressed in the laboratory frame $Oxyz$ by the usual formulae:

$$V_{xx} = \tfrac{1}{2}eq(3\sin^2\theta\cos^2\phi - 1),$$
$$V_{yy} = \tfrac{1}{2}eq(3\sin^2\theta\sin^2\phi - 1),$$
$$V_{zz} = \tfrac{1}{2}eq(3\cos^2\theta - 1),$$
$$V_{zx} = \frac{3eq}{2}\sin\theta\cos\theta\cos\phi, \tag{102}$$
$$V_{zy} = \frac{3eq}{2}\sin\theta\cos\theta\sin\phi,$$
$$V_{xy} = \frac{3eq}{2}\sin^2\theta\sin\phi\cos\phi.$$

As the molecule jumps to another position the angle ϕ in (102) changes and the values of the components of the gradients (102), averaged over the possible values of ϕ, give

$$\overline{V_{zz}} = V_{zz} = \tfrac{1}{2}eq(3\cos^2\theta - 1) = -2\overline{V_{xx}} = -2\overline{V_{yy}},$$
$$\overline{V_{zx}} = \overline{V_{zy}} = \overline{V_{xy}} = 0. \tag{103}$$

The average gradient 'seen' by the spin has axial symmetry around the axis of rotation and the quadrupole frequencies are reduced in the ratio $\tfrac{1}{2}(3\cos^2\theta - 1)$.

Just as in the problem of torsion oscillations the time-dependent parts $V_{ij} - \overline{V_{ij}}$ of the gradient are a cause of broadening and the corresponding line width can be computed by the same methods. It may be noticed that here the adiabatic width $1/T_2'$ due to fluctuations in the resonance frequency vanishes since $V_{zz} = \overline{V_{zz}}$.

The calculation of the line width is straightforward and we shall give only the very simple result that holds for a spin $I = \frac{3}{2}$ (**17**):

$$\frac{1}{T_2} = \omega^2 \tau \sin^2\theta. \tag{104}$$

Evidence of a change of quadrupole frequency caused by hindered rotation has been found in *trans*dichloroethane CH_2Cl—CH_2Cl, where

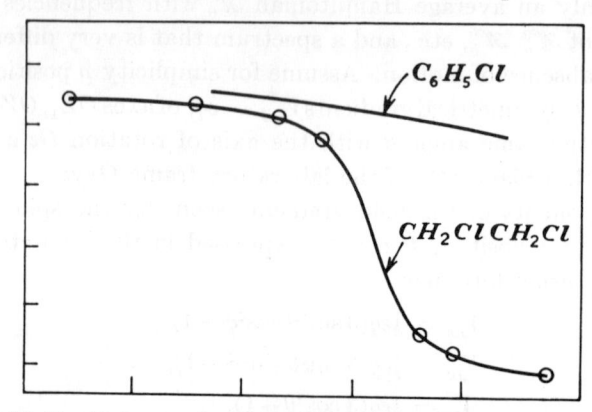

FIG. X, 10. Rapid change of a quadrupole frequency in CH_2Cl—CH_2Cl caused by hindered rotation occurring between 140° K and 180° K. The slow variation of the chlorine frequency in C_6H_5Cl is shown for contrast.

between 140° K and 180° K a change in the chlorine quadrupole frequency by more than 10 per cent is observed (**23**). This is shown in Fig. X, 10 where the much slower variation of the chlorine frequency in C_6H_5Cl is also shown for contrast.

If we assume that the rotation takes place around the Cl—Cl axis which makes an angle $\theta = 19° 16'$ with the directions of the two C—Cl bonds, the reduction factor $\frac{1}{2}(1 - 3\cos^2\theta)$ is in qualitative agreement with the observed reduction in frequency.

In the intermediate region where the transition takes place the line is unobservable because of excessive broadening by a rotation that is neither fast nor slow compared with the quadrupole frequency.

Another example of flipping of a molecule between two equivalent positions is provided by the observation of deuteron resonance in a single crystal of $Li_2SO_4 \cdot D_2O$ (**24**), where the experimental results can be explained by assuming that the D_2O molecule flips through 180° about the bisector of the DOD angle. Since the experiment was performed in a high magnetic field (6381 gauss) the two-spin Hamiltonians $\mathcal{H}_a$ and $\mathcal{H}_b$ that correspond to the two positions of the molecule will not in general have the same spectrum.

At low temperature ($-125°$ C), where the flipping does not exist, four pairs of lines are observed. This is in accordance with the fact that there are two heavy-water molecules in the unit cell, two deuterons per molecule, and per deuteron in a high magnetic field two lines corresponding to the two transitions $1 \to 0$ and $0 \to -1$ of $I_z = m$, quantized along the magnetic field. The spectra observed for various orientations of the crystal are in agreement with the assumption of quadrupole tensors having their maximum components along the OD bonds of the molecule, a quadrupole coupling constant e^2qQ/h of 237 ± 10 kc, and an asymmetry factor η of $0{\cdot}14\pm0{\cdot}04$.

At room temperature two pairs of lines only are observed (one for each water molecule). For each molecule the quadrupole coupling tensor has its maximum component perpendicular to the DOD plane and the direction of its minimum component is in the DOD plane and bisects the DOD angle. The values

$$e^2qQ/h = 123\pm3 \text{ kc}, \qquad \eta = 0{\cdot}80\pm0{\cdot}02$$

are in good agreement with the values deduced from the low-temperature results by tensor transformation, assuming at room temperature the fast 180° flipping described above. It is interesting to notice that such a flipping does *not* change the magnetic spin-spin coupling and thus cannot be observed in $Li_2SO_4 \cdot H_2O$.

(b) Slow motion

We have discussed in Section IV of this chapter the problem of a spin system that has several resonance frequencies, assumed for simplicity to be infinitely sharp, and passes from one to the other in a random fashion at an average rate $1/\tau$. If $1/\tau$ is much smaller than the differences between these frequencies, each line is broadened by an amount of the order of $1/\tau$.

For a molecule that makes jumps between equivalent positions where the quadrupole coupling constants and thus the resonance frequencies have the same values, the problem is somewhat different, the contrast with the situation of the Section IV being that the eigenkets rather than the eigenvalues of the spin Hamiltonian change (both change in the presence of a magnetic field). The actual problem of computing the line shape, which is more difficult here because the Hamiltonians $\mathcal{H}_0^a$, $\mathcal{H}_0^b$ that correspond to the different positions do not commute, can be formulated as follows.

The shape $I(\omega)$ of the absorption line is the Fourier transform of the function $G(t) = \langle I_x(t)I_x \rangle$, where $I_x(t)$ is expressed as follows. Assume that from $t = 0$ the molecule passes t_1 seconds in position a, t_2 in position b, t_3 in position a, etc., with $t = t_1+t_2+...+t_n$ (assuming for simplicity two positions only);

$$I_x(t) = \overline{e^{i\mathcal{H}_a t_n}... e^{i\mathcal{H}_b t_2}e^{i\mathcal{H}_a t_1}I_x e^{-i\mathcal{H}_a t_1}e^{-i\mathcal{H}_b t_2}... e^{-i\mathcal{H}_a t_n}}, \quad (105)$$

where the average is taken over all the distributions of the intervals $t_1,..., t_n$ within the interval $(0 \to t)$, with $\overline{t_1} = \overline{t_2} = ... \overline{t_n} = \tau$ (the Hamiltonian for the first interval t_1 and the last interval t_n being either $\mathcal{H}_a$ or $\mathcal{H}_b$). Although one feels intuitively that the resonance curve should be of the Lorentz type with a width of the order of $1/\tau$, the actual calculation of its shape is fairly difficult and has been performed rigorously only in special cases.

It may be instructive to consider an example that has no physical reality but where the result can be obtained very simply: a spin **I** has two positions a and b where it 'sees' magnetic fields **H**$_a$ and **H**$_b$ of equal magnitude but different directions, say perpendicular to each other, and jumps from one position to the other at an average rate $1/\tau$.

The problem is simple here because the motion of the nuclear magnetization is described rigorously by a classical equation.

We define $\mathbf{M}_a(t)$ and $\mathbf{M}_b(t)$ as the magnetizations of the spins that are at time t in position a and b respectively. The following equations are self evident:

$$\begin{aligned}\frac{d\mathbf{M}_a}{dt} &= \gamma(\mathbf{M}_a \wedge \mathbf{H}_a) - \frac{\mathbf{M}_a - \mathbf{M}_b}{\tau}, \\ \frac{d\mathbf{M}_b}{dt} &= \gamma(\mathbf{M}_b \wedge \mathbf{H}_b) - \frac{\mathbf{M}_b - \mathbf{M}_a}{\tau}.\end{aligned} \quad (106)$$

An elementary calculation shows that the component $(M_{ax}+M_{bx})$ of the nuclear magnetization along an axis Ox perpendicular to $\mathbf{H}_a$ and $\mathbf{H}_b$ is of the form $e^{-t/2\tau}\cos(\omega t+\phi)$, where $\omega = -\gamma|H_a| = -\gamma|H_b|$ and the line is a single Lorentz curve with a width $1/2\tau$.

We give the results of (17) for a problem that can be applied to a group of atoms such as CCl_3. The axial quadrupole gradient 'seen' by a spin $\frac{3}{2}$ takes at random three equivalent directions around an axis of hindered rotation (the directions of the three bonds C—Cl).

For a polycrystalline sample the line is a superposition of Lorentz curves proportional to

$$\frac{4}{(\omega-\omega_0)^2+(\tfrac{4}{3}\lambda)^2}+\frac{56}{(\omega-\omega_0)^2+(\tfrac{7}{6}\lambda)^2}+\frac{20}{(\omega-\omega_0)^2+(\tfrac{5}{6}\lambda)^2}+\frac{4}{(\omega-\omega_0)^2+(\tfrac{2}{3}\lambda)^2},$$

where $\lambda = 1/\tau$. \hfill (107)

The calculations are heavy and considerable mathematical ingenuity is displayed in obtaining the result (107) in a closed form.

REFERENCES

1. N. BLOEMBERGEN, E. M. PURCELL, and R. V. POUND, *Phys. Rev.* **73**, 679, 1948.
2. P. W. ANDERSON and P. R. WEISS, *Rev. Mod. Phys.* **25**, 269, 1953.
3. —— *J. Phys. Soc. Japan*, **9**, 316, 1954.
4. C. J. GORTER and J. H. VAN VLECK, *Phys. Rev.* **72**, 1128, 1947.
5. J. H. VAN VLECK, ibid. **74**, 1168, 1948.
6. N. BLOEMBERGEN and T. L. ROWLAND, ibid. **97**, 1679, 1955.
7. H. C. TORREY, ibid. **92**, 962, 1953.
8. —— ibid. **96**, 690, 1954.
9. E. R. ANDREW and R. G. EADES, *Proc. Roy. Soc.* A, **218**, 537, 1953.
10. R. BERSOHN and H. S. GUTOWSKY, *J. Chem. Phys.* **22**, 651, 1954.
11. A. H. COOKE and L. E. DRAIN, *Proc. Phys. Soc.* A, 1952, **65**, 894.
12. SACHS and TURNER quoted by E. M. PURCELL, *Physica*, **17**, 282, 1951.
13. H. S. GUTOWSKY and B. R. MCGARVEY, *J. Chem. Phys.* **20**, 1472, 1952.
14. D. F. HOLCOMB and R. E. NORBERG, *Phys. Rev.* **98**, 1074, 1955.
15. H. C. TORREY, *Suppl. Nuovo Cim.* **9**, 95, 1958.
16. F. REIF, *Phys. Rev.* **100**, 1597, 1955.
17. Y. AYANT, Thesis, published by Masson & Co., Paris, 1955.
18. H. BAYER, *Z. Phys.* **130**, 227, 1951.
19. H. G. DEHMELT and H. KRÜGER, ibid. **129**, 401, 1951.
20. Y. AYANT, *J. Phys. Rad.* **17**, 338, 1956.
21. M. C. WANG and G. E. UHLENBECK, *Rev. Mod. Phys.* **17**, 323, 1945.
22. H. G. DEHMELT, *Z. Phys.* **130**, 356, 1951.
23. H. W. DODGEN and J. L. RAGLE, *J. Chem. Phys.* **25**, 376, 1956.
24. S. KETUDAT and R. V. POUND, ibid. **26**, 708, 1957.

XI

MULTIPLET STRUCTURE OF RESONANCE LINES IN LIQUIDS

Les Grandes Familles

I. Energy Levels observed by Continuous Wave Methods

It has been shown in Chapter VI that in diamagnetic liquids the combined effects of electron nucleus couplings and fast molecular tumbling on nuclear spins was to introduce isotropic resonance frequency shifts δ (chemical shifts) proportional to the applied field, and scalar bilinear, field-independent, spin-spin couplings of the form $\hbar J \mathbf{I}_1 \cdot \mathbf{I}_2$. Owing to these effects, the otherwise single nuclear resonance line of a compound may be split into a spectrum of many lines with various spacings, intensities, and widths. It is the purpose of this chapter to show how these sometimes complex patterns can be unravelled, starting from a 'spin' Hamiltonian containing the constants δ and J defined above. First a few definitions are in order.

Consider nuclear spins in a molecule that have the same Larmor frequency in an applied field H_0, which requires not only that they belong to the same isotopic species but also that they have the same chemical shift. This will be the case if their positions in the molecule can be made to coincide through an operation of the symmetry group of the molecule. Such spins will be called 'isochronous'.

Consider a group G of isochronous spins. The spins of this group will be called 'equivalent' if every other spin of the molecule has the same coupling constant with all spins of the group. For instance, the two isochronous protons of difluoromethane CH_2F_2 are equivalent, but not those of difluoroethylene $C_2H_2F_2$ where each fluorine has different couplings with the two protons (Fig. XI, 1). It should be emphasized that the definition of equivalent spins does not imply that inside the group G of equivalent nuclei all spin couplings are equal. Thus in the molecule of BrF_5 which has the shape of a square-based pyramid with a fluorine at the apex and four fluorines at the corners of the square base, the latter which are clearly isochronous are also equivalent, according to our definition, although the coupling between two adjacent fluorines is different from that of two fluorines on a diagonal of the square.

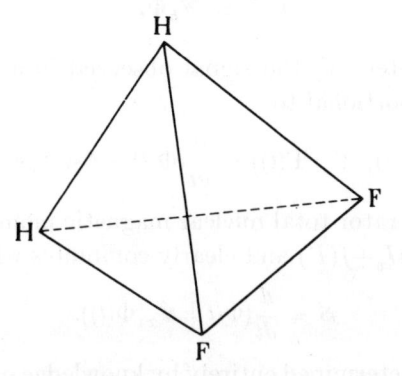

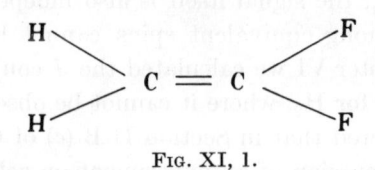

FIG. XI, 1.

The reason for introducing this definition of equivalent spins is based on the following theorem.

Scalar couplings between equivalent spins are unobservable in a nuclear resonance experiment (1).

Let $I_1,...,I_p,...,I_n$ be the equivalent spins of a group G, and $I'_1,...,I'_{q'},...,I'_{n'}$ all the other spins of the molecule. In the presence of a uniform magnetic field H (which may be any combination of d.c. and r.f. field) the Hamiltonian $\hbar\mathcal{H}$ of the spin system can be written

$$\hbar\mathcal{H} = -\gamma\hbar\mathbf{H}\cdot\mathbf{I} + \hbar\sum_{p<q}J_{pq}\mathbf{I}_p\cdot\mathbf{I}_q + \mathbf{I}\sum_{q'}\hbar J_{q'}\mathbf{I}'_{q'} + \hbar\mathcal{H}_1(I'), \qquad (1)$$

where $\mathbf{I} = \mathbf{I}_1+\mathbf{I}_2+...+\mathbf{I}_n$, and $\hbar\mathcal{H}_1(I')$ is the part of the Hamiltonian that does not depend on the spins I_p. The equivalence of the spins I_p manifests itself through the fact that a single constant $J_{q'}$ describes the couplings of a spin $I'_{q'}$ with all the spins I_p. The operator $\mathcal{H}_a = \sum_{p<q}J_{pq}\mathbf{I}_p\cdot\mathbf{I}_q$ commutes with all the other terms of $\mathcal{H}$. Let us define $\mathcal{H}_b = \mathcal{H}-\mathcal{H}_a$. In the Schrödinger equation

$$i\frac{\partial\Psi}{\partial t} = (\mathcal{H}_a+\mathcal{H}_b)\Psi$$

introduce Φ through $\Psi = \exp(-i\mathcal{H}_a t)\Phi$. The Schrödinger equation for Φ, thanks to the relation $[\mathcal{H}_a, \mathcal{H}_b] = 0$, is

$$i\frac{\partial \Phi}{\partial t} = \mathcal{H}_b \Phi. \tag{2}$$

As shown in Chapter IV, the signal observed in a nuclear resonance experiment is proportional to

$$S = \frac{d}{dt}(\Psi(t)|\mathcal{M}_x|\Psi(t)) = \frac{d}{dt}(\Phi(t)|e^{i\mathcal{H}_a t}\mathcal{M}_x e^{-i\mathcal{H}_a t}|\Phi(t)),$$

where $\mathcal{M}$ is the operator total nuclear magnetic moment of the sample, and since $\mathcal{M}_x = \gamma \hbar I_x + f(I')$ and clearly commutes with $\mathcal{H}_a$,

$$S = \frac{d}{dt}(\Phi(t)|\mathcal{M}_x|\Phi(t)). \tag{3}$$

Since the signal is determined entirely by knowledge of $\Phi(t)$ which by (2) is independent of $\mathcal{H}_a$, the signal itself is also independent of $\mathcal{H}_a$, and the couplings J_{pq} among equivalent spins cannot be observed. This explains why in Chapter VI we calculated the J coupling for the molecule HD rather than for H_2, where it cannot be observed.

It will be remembered that in Section II B (c) of Chapter VI it was mentioned in the discussion of an experiment on solid iodine that the experimental results were not inconsistent with an isotropic coupling $\hbar J \mathbf{I}_1 \cdot \mathbf{I}_2$ between the two iodine nuclear spins of the molecule. The observability of such a coupling does not contradict the present theorem, for the experiment discussed was performed in the solid state and the Hamiltonian included the quadrupole interaction $a(I_{1z}^2 + I_{2z}^2)$ that does not commute with the scalar coupling $\hbar J \mathbf{I}_1 \cdot \mathbf{I}_2$.

A few examples of multiplet structures will now be given.

A. $J \ll \delta$

Consider two groups G and G' of equivalent spins I_k and $I'_{k'}$ with a mutual coupling constant J much smaller than the difference $\delta = (\gamma - \gamma')H_0$ of their Larmor frequencies. This condition will always prevail when the two groups belong to different nuclear species (except possibly in exceedingly low fields such as the earth's field).

The Hamiltonian, from which irrelevant terms have been omitted in view of the previous theorem, is

$$\hbar \mathcal{H} = -(\gamma \hbar I_z + \gamma' \hbar I'_z)H_0 + \hbar J \mathbf{I} \cdot \mathbf{I}', \tag{4}$$

where we assume $H_0|\gamma - \gamma'| \gg J$ and $\mathbf{I} = \sum_k \mathbf{I}_k$, $\mathbf{I}' = \sum_{k'} \mathbf{I}'_{k'}$. The shielding constants σ and σ' are supposed to be included in γ and γ'. A first-order perturbation method can be used whereby the small coupling $\hbar J \mathbf{I} \cdot \mathbf{I}'$

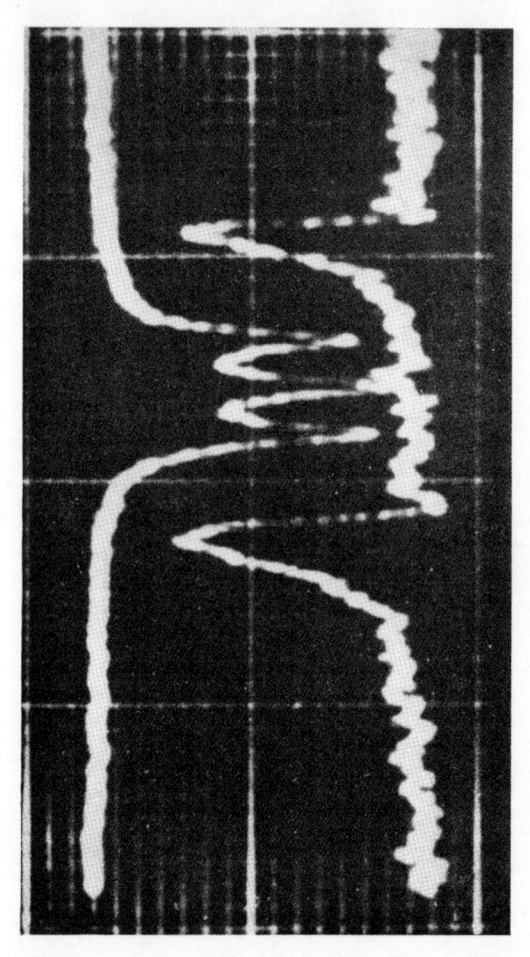

FIG. XI, 2. Simultaneous display of deuteron and proton resonances in HD. The proton trace is inverted.

is replaced by its part $\hbar J I_z I'_z$ that commutes with the main Hamiltonian $-(\gamma \hbar I_z + \gamma' \hbar I'_z) H_0$. The energy levels of the system are then given through
$$\hbar E_{MM'} = -(\gamma \hbar H_0 M + \gamma' \hbar H_0 M') + \hbar J M M', \qquad (5)$$
where $I_z = M$, $I'_z = M'$. The frequencies of the transitions $\Delta M = 1$, $\Delta M' = 0$, and $\Delta M = 0$, $\Delta M' = 1$ are given by
$$\omega = -\gamma H_0 + J M', \qquad \omega' = -\gamma' H_0 + J M. \qquad (6)$$
There are two multiplets, one for each group of spins, having respectively $(2p'i'+1)$ and $(2pi+1)$ components, where i is the spin of each of the p spins I_k of the group G, and the same for i' and p'. The relative intensity of a component $\omega = -\gamma H_0 + J M'$ is proportional to the number of ways in which a state $I'_z = M'$ can be formed from the p' spins $I'_{k'}$. It is proportional to the coefficient of $x^{(p'i'+M')}$ in the expansion of $P(x) = (1+x+...+x^{2i'})^{p'}$. If the experiment is performed at constant frequency, the spectrum being scanned by changing the field, the distances ΔH and $\Delta H'$ between two adjacent lines of the multiplets G or G', $|J/\gamma|$ and $|J/\gamma'|$ respectively, are independent of the applied field (in contradistinction to the chemical shifts) and in the ratio $|\gamma'/\gamma|$.

Consider as an example the molecule of BrF_5 (1). The spectrum includes two multiplets, a quintuplet, and a much stronger doublet separated by a distance which is 0·876 gauss in a field of 6365 gauss and varies linearly with the field. The intervals between the components are independent of the field and equal to 0·019 gauss. Finally, the relative intensities of the components of the quintuplet are 1:4:6:4:1, the two components of the doublet are equal, and the total intensity of the doublet is four times that of the quintuplet. All this is in perfect agreement with the assumption of a group G of four equivalent fluorines (those of the square base) coupled to a different fluorine at the apex of the pyramid. The doublet naturally corresponds to the resonance of the four fluorines G and the quintet to that of the solitary fluorine G'. The reason why the effect of the bromine nucleus is not felt will be discussed in a later section of this chapter.

As a second example we may quote the resonance of HD observed in the gas under pressure, Fig. XI, 2 (2). The triplet is due to the resonance of the proton split by the three possible orientations of the deuteron spin and similarly the deuteron resonance line has two components. Since the scanning of each multiplet is done through a change of the field it can be verified on Fig. XI, 2 that the intervals between components are inversely proportional to the γ of the nucleus. (In the previous example of BrF_5 this effect was negligible.) Both curves of Fig. XI, 2 were taken in the same field at two different frequencies. Furthermore, the two nuclei have different relaxation times and no meaningful comparison of the intensities of their resonances can be made.

The case when more than two groups of equivalent spins are interacting is a straightforward generalization of the above.

Because the difference in frequencies is much larger than J even in fields of a few gauss when the groups G and G' belong to different species, the J couplings between them can be observed at those fields, the loss in signal being compensated to a certain extent by the possibility of making very homogeneous fields over larger volumes and thus using larger samples.

B. J and δ comparable, for two spins $\frac{1}{2}$

The Hamiltonian $\hbar\mathcal{H}$ of the system can be written

$$\mathcal{H} = \omega_0(i_z+i'_z)+\tfrac{1}{2}\delta(i_z-i'_z)+J\mathbf{i}.\mathbf{i}', \tag{7}$$

where $\qquad \omega_0 = -\tfrac{1}{2}(\gamma+\gamma')H_0, \qquad \delta = -(\gamma-\gamma')H_0.$

Without loss of generality we will assume $\delta > 0$. The Hamiltonian (7) clearly commutes with $I_z = i_z+i'_z$, which is a good quantum number, and the four eigenstates of the system are

$$|a) = |+,+); \qquad |b) = p|+,-)+q|-,+);$$
$$|c) = -q|+,-)+p|-,+); \qquad |d) = |-,-), \tag{8}$$

where $p^2+q^2 = 1$ and the energies $\hbar E$ are given by

$$E_a = \omega_0+\tfrac{1}{4}J, \qquad E_d = -\omega_0+\tfrac{1}{4}J, \tag{9}$$

E_b and E_c being determined from the secular equation

$$\begin{vmatrix} \tfrac{1}{2}\delta-\tfrac{1}{4}J-E & \tfrac{1}{2}J \\ \tfrac{1}{2}J & -\tfrac{1}{2}\delta-\tfrac{1}{4}J-E \end{vmatrix} = 0,$$

whence $\qquad E_{\{^b_c\}} = -\tfrac{1}{4}J\pm\tfrac{1}{2}\sqrt{(J^2+\delta^2)}. \tag{10}$

Defining $J/\delta = \tan\phi$, $-\tfrac{1}{2}\pi \leqslant \phi \leqslant \tfrac{1}{2}\pi$, and assuming without loss of generality $p > 0$, we find

$$p = \cos\tfrac{1}{2}\phi, \qquad q = \sin\tfrac{1}{2}\phi. \tag{11}$$

There are four allowed transitions, with frequencies

$$\omega_{ab} = \omega_0+\tfrac{1}{2}J-\tfrac{1}{2}\sqrt{(J^2+\delta^2)},$$
$$\omega_{ac} = \omega_0+\tfrac{1}{2}J+\tfrac{1}{2}\sqrt{(J^2+\delta^2)},$$
$$\omega_{bd} = \omega_0-\tfrac{1}{2}J+\tfrac{1}{2}\sqrt{(J^2+\delta^2)},$$
$$\omega_{cd} = \omega_0-\tfrac{1}{2}J-\tfrac{1}{2}\sqrt{(J^2+\delta^2)}. \tag{12}$$

Their relative probabilities are

$$P_{ab} \propto |(a\,|\,i_x+i'_x\,|\,b)|^2 \propto (p+q)^2 = (\cos\tfrac{1}{2}\phi+\sin\tfrac{1}{2}\phi)^2,$$
$$P_{ac} \propto |(a\,|\,i_x+i'_x\,|\,c)|^2 \propto (p-q)^2 = (\cos\tfrac{1}{2}\phi-\sin\tfrac{1}{2}\phi)^2,$$
$$P_{bd} \propto |(b\,|\,i_x+i'_x\,|\,d)|^2 \propto (p+q)^2 = (\cos\tfrac{1}{2}\phi+\sin\tfrac{1}{2}\phi)^2,$$
$$P_{cd} \propto |(c\,|\,i_x+i'_x\,|\,d)|^2 \propto (p-q)^2 = (\cos\tfrac{1}{2}\phi-\sin\tfrac{1}{2}\phi)^2. \qquad (13)$$

There are four lines in a symmetrical pattern. Fig. XI, 3 represents the spectrum for $J > 0$. The two inner lines are always higher than

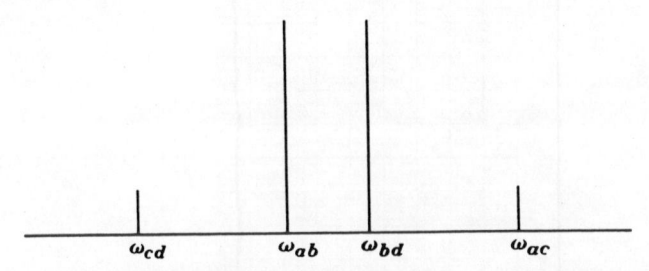

$\omega_{cd} \qquad \omega_{ab} \quad \omega_{bd} \qquad \omega_{ac}$

FIG. XI, 3. Theoretical spectrum for two spins $\tfrac{1}{2}$ with a scalar coupling J comparable to the difference δ of their Larmor frequencies.

the two outer ones and the distance between an inner line and the neighbouring outer line is exactly $|J|$. The formulae (9)–(13) are exact formulae, and it is interesting to see how the spectrum changes with the ratio J/δ. For $|J/\delta| \ll 1$, $\phi \to 0$, $p \to 1$, $q \to 0$,

$$P_{ab},\,P_{ac},\,P_{bd},\,P_{cd} \propto 1,$$

the four lines become equal and their frequencies become:

$$\omega_{ab} \to \omega_0+\tfrac{1}{2}J-\tfrac{1}{2}\delta,$$
$$\omega_{ac} \to \omega_0+\tfrac{1}{2}J+\tfrac{1}{2}\delta,$$
$$\omega_{bd} \to \omega_0-\tfrac{1}{2}J+\tfrac{1}{2}\delta,$$
$$\omega_{cd} \to \omega_0-\tfrac{1}{2}J-\tfrac{1}{2}\delta.$$

The pattern becomes that of two doublets of width J separated by a distance δ, as predicted by the first-order theory of Section A. On the other hand, for $J/\delta \gg 1$,

$$P_{ab},\,P_{bd} \propto 1, \qquad P_{ac},\,P_{cd} \to 0,$$

the two external lines grow weaker and eventually disappear. The two inner ones, with a separation

$$\omega_{bd}-\omega_{ab} = \sqrt{(J^2+\delta^2)}-J \to \frac{\delta^2}{2J},$$

tend to coalesce and give a single line in agreement with the general theorem that a J coupling between two equivalent spins is unobservable.

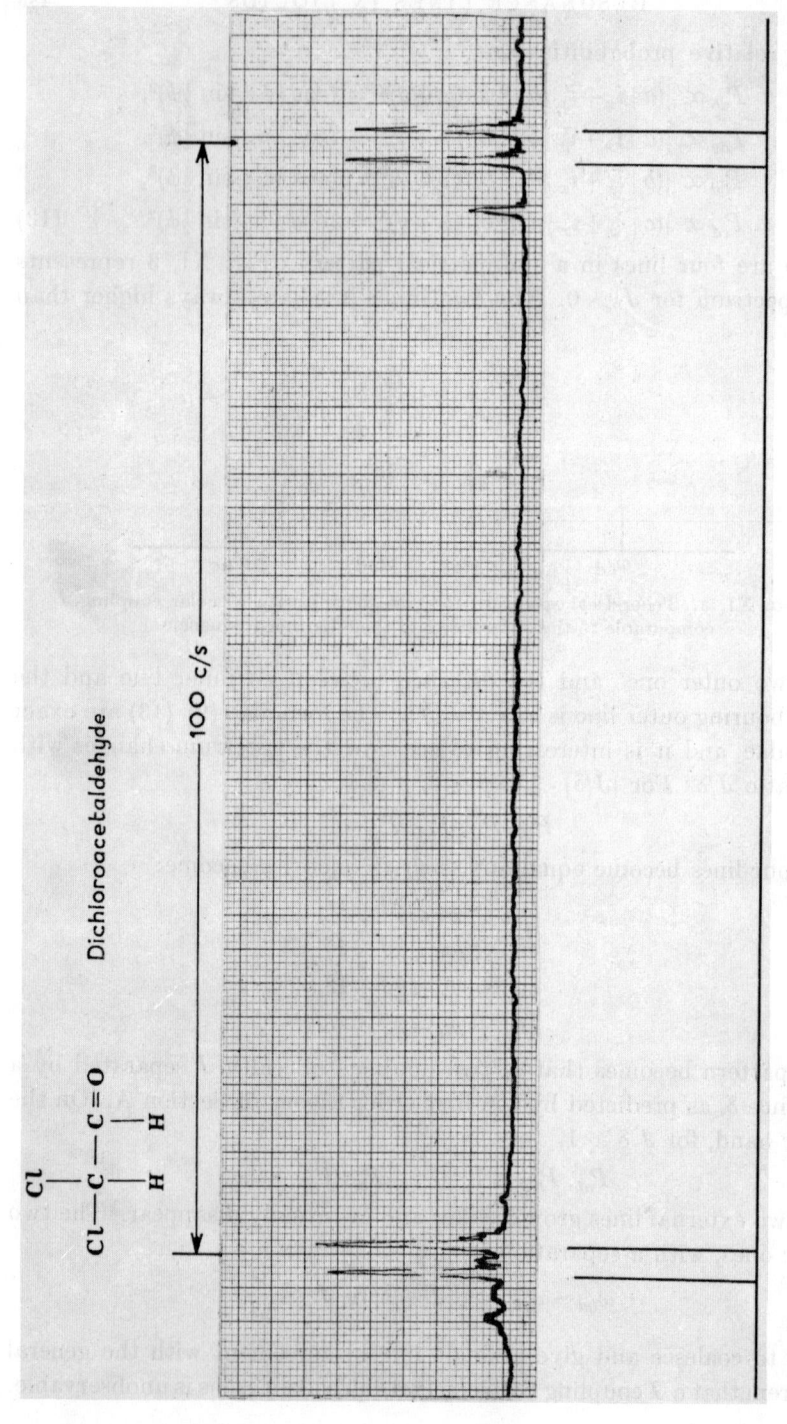

Fig. XI, 4. Spectrum of dichloroacetaldehyde. The small peaks are due to impurities. The theoretical spectrum was calculated assuming that $J/\delta = 0.029$.

An example of the case $J/\delta \ll 1$ is provided by the proton resonance in dichloroacetaldehyde (3) (Fig. XI, 4) where the analysis of the spectrum observed at a frequency of 30·5 Mc/s gives

$$\frac{\delta}{2\pi} = 100 \cdot 0 \pm 1 \text{ c/s}, \qquad \frac{J}{2\pi} = 2 \cdot 9 \pm 0 \cdot 3 \text{ c/s}.$$

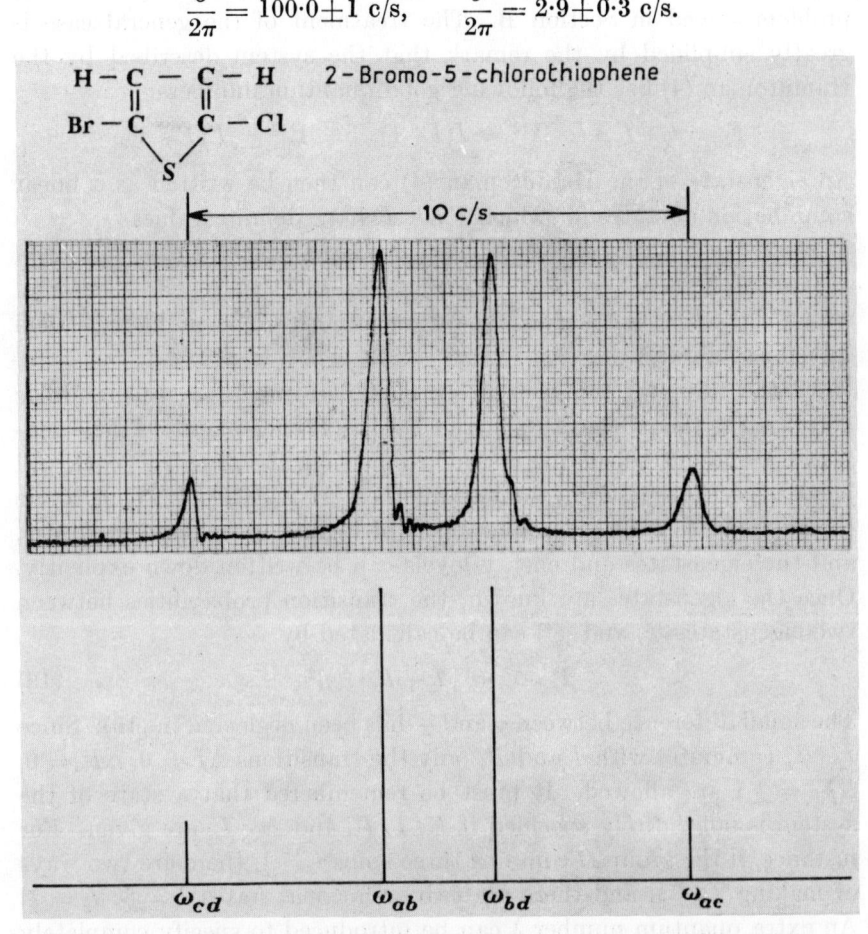

FIG. XI, 5. Spectrum of 2-bromo-5-chlorothiophene. The theoretical spectrum below was calculated assuming that $J/\delta = 0 \cdot 835$. The labelling of the lines is the same as in Fig. XI, 3.

An example where J and δ are comparable is that of 2-bromo-5-chlorothiophene (Fig. XI, 5) where it is found (3) that

$$\frac{\delta}{2\pi} = 4 \cdot 7 \text{ c/s}, \qquad \frac{J}{2\pi} = 3 \cdot 9 \text{ c/s}.$$

As in BrF_5, the lack of effect from the spins of Br and Cl in those molecules will be explained later.

C. J and δ comparable, for two groups G and G' of p equivalent spins i and p' equivalent spins i', respectively

The Hamiltonian (4) must now be diagonalized without as in Section A replacing $J\mathbf{I}.\mathbf{I'}$ by $JI_z I'_z$. If each group is a single spin $\tfrac{1}{2}$ this is the problem solved in Section B. The treatment of the general case is greatly simplified by the remark that the system described by the Hamiltonian (4) has the following good quantum numbers:

$$F_z = I_z + I'_z, \qquad |\mathbf{I}|^2 = I(I+1), \qquad |\mathbf{I'}|^2 = I'(I'+1).$$

An eigenstate of the Hamiltonian (4) can then be written as a linear combination of states for which F_z, I, I' have definite values

$$|\xi\rangle = \sum_M C_M \,|\, F_z, I, I', M\rangle, \tag{14}$$

whilst $I_z = M$ is not a good quantum number. The problem is then solved as follows.

All the states of the system are classified into separate manifolds according to the values of F_z, I, and I'. Inside each manifold the eigenstates are found by solving a secular equation, the order of which will be at most the smaller of the two numbers $2I+1$, $2I'+1$. Thus if one of the spins I, I' is $\tfrac{1}{2}$ no equation is of higher order than quadratic, and the eigenstates and energy levels can be written down explicitly. Once the eigenstates are known, the transition probabilities between two eigenstates $|\xi\rangle$ and $|\xi'\rangle$ can be calculated by

$$P_{\xi\xi'} \propto |\langle\xi\,|\,I_x + I'_x\,|\,\xi'\rangle|^2. \tag{15}$$

The small difference between γ and γ' has been neglected in (15). Since $I_x + I'_x$ commutes with I and I', only the transitions $\Delta I = 0$, $\Delta I' = 0$, $\Delta F_z = \pm 1$ are allowed. It must be remembered that a state of the system is not entirely specified if F_z, I, I', and say I_z, are given. For instance, if the group G contains three spins $i = \tfrac{1}{2}$, there are two ways of making $I = \tfrac{1}{2}$, and there are two orthogonal states $I = \tfrac{1}{2}$, $I_z = \tfrac{1}{2}$. An extra quantum number λ can be introduced to specify completely a state of the system as $|F_z, I, I', I_z, \lambda\rangle$. States with different values of λ have a different symmetry character with respect to a permutation of the spins i (or the spins i') among themselves. Since the Hamiltonian of the system (including the r.f. part) is a symmetrical function of the spins i (and of the spins i') it is independent of λ. The frequencies and the probabilities of the transitions can then be calculated disregarding the existence of λ, provided the intensity of a transition

$$\xi(I, I', F_z) \to \xi'(I, I', F_z - 1)$$

is weighted with $N(I, I')$, the number of independent ways of constructing a total spin I from the p spins i and I' from the p' spins i'. The generalization to more than two groups of equivalent spins is again straightforward.

As an example assume that G and G' contain two spins $\frac{1}{2}$ each, and seek the maximum number of lines in the spectrum. There are three independent groups of levels (and transitions): $I = 1$, $I' = 0$; $I = 0$, $I' = 1$; $I = 1$, $I' = 1$. The first and the second group clearly give a single line each. In the third group there is one state $F_z = 2$, two states $F_z = 1$, and three states $F_z = 0$. Thus there are *a priori* two transitions $2 \to 1$, six transitions $1 \to 0$, six transitions $0 \to -1$, and two transitions $-1 \to -2$, that is sixteen for the combination $I = 1$, $I' = 1$, and eighteen is the maximum number of lines in the spectrum. Before comparing this prediction with experiment, we briefly consider a perturbation method.

D. Perturbation method

If J/δ, without being so small as to justify the first-order treatment given in Section A, is still appreciably smaller than unity, a higher-order perturbation calculation is useful (3).

Assume that the molecule contains R groups of equivalent spins $G_A, ..., G_R$. As explained in Section C, it can be assumed that the total spin of each group G has a well-defined value I, since the contributions to the spectrum from the various manifolds of states $(I_1, I_2, ..., I_R)$ are additive.

The Hamiltonian $\hbar \mathscr{H}$ of the system can be written

$$\mathscr{H} = \mathscr{H}^{(0)} + \mathscr{H}^{(1)},$$

where

$$\mathscr{H}^{(0)} = \sum_R \omega_R I_z^R, \qquad \mathscr{H}^{(1)} = \tfrac{1}{2} \sum_{R \neq S} J_{RS} \{I_z^R I_z^S + I_+^R I_-^S\}, \qquad (16)$$

and $\mathscr{H}^{(1)}$ is considered as a perturbation of $\mathscr{H}^{(0)}$.

The zero-order wave functions of the system are

$$\psi_i^{(0)} = |m_A, ..., m_R), \quad \text{where } m_R = I_z^R,$$

and the zero-order energy values $\hbar E_i^{(0)}$ are given by

$$E_i^{(0)} = \sum_R \omega_R m_R.$$

The first-order corrections are (as shown in Section A)

$$E_i^{(1)} = \tfrac{1}{2} \sum_{R \neq S} J_{RS} m_R m_S$$

and the second-order corrections originating in the terms $I_+^R I_-^S$:

$$E_i^{(2)} = \sum_{j \neq i} \frac{(\psi_i^{(0)} | \mathcal{H}^{(1)} | \psi_j^{(0)})(\psi_j^{(0)} | \mathcal{H}^{(1)} | \psi_i^{(0)})}{E_i^{(0)} - E_j^{(0)}}$$

$$= \tfrac{1}{4} \sum_{R \neq S} \frac{J_{RS}^2}{\omega_R - \omega_S} \{m_R(I_S^2 + I_S - m_S^2) - m_S(I_R^2 + I_R - m_R^2)\}. \quad (17)$$

The frequency of a transition relative to the group G_A between two levels $E(m_A, m_B, \ldots, m_R)$ and $E(m_A - 1, m_B, \ldots, m_R)$ is given by

$$\omega = \omega_A + \sum_{R \neq A} J_{AR} m_R +$$

$$+ \tfrac{1}{2} \sum_{R \neq A} \frac{J_{AR}^2}{(\omega_A - \omega_R)} \{I_R(I_R + 1) - m_R(m_R + 1) + 2 m_A m_R\}. \quad (18)$$

For fixed values of the $m_{R \neq A}$, the various lines of the Zeeman transitions $m_A \to m_A - 1$ do not coincide as they do in the first-order approximation since, by (18), their frequencies depend on m_A. A single exception to this occurs when all the $m_{R \neq A}$ are zero. By calculating the first-order change in the wave function it is easily found that to the same order the transition probability $P_{m_A \to m_A - 1}$ is

$$P = (I_A - m_A + 1)(I_A + m_A)\left\{1 - \sum_{R \neq A} \frac{2 J_{AR} m_R}{\omega_A - \omega_R}\right\}. \quad (19)$$

If there are only two groups G_A and G_B, (19) shows that the components of the spectrum of G_A further removed from the spectrum of G_B are reduced in intensity (see, for instance, Figs. XI, 4 and XI, 5). Third-order energy terms have also been computed (3) and may be useful if the solution of high-order secular equations is necessary for an exact solution.

Apart from the magnitude of J, the question of determination of its sign may arise. If there are only two groups of spins G_A and G_B the sign of J can never be determined, as can be shown by an argument very similar to that used in Chapter VII to prove the impossibility of obtaining the sign of quadrupole interaction.

If there are more than two groups, and thus more than one coupling constant J, their relative signs may be determined provided that the ratios J/δ are not so small as to render the second and higher order terms negligible.

It is instructive to compute as an example on the basis of the perturbation theory the maximum number of lines to be expected from two groups G and G' of two equivalent spins $\tfrac{1}{2}$.

The combinations $I = 1$, $I' = 0$, and $I = 0$, $I' = 1$, clearly give one line each. For the combination $I = 1$, $I' = 1$, the perturbation theory predicts twelve lines: there are six transitions $M \to M-1$, namely $M = 1$, $M' = 0$, ± 1, and $M = 0$, $M' = 0$, ± 1, and similarly six transitions $M' \to M'-1$ (for $M' = 0$ the transitions $M = 1 \to 0$ and $M = 0 \to -1$ have the same frequency according to (18) but they are split in third order). This is to be contrasted with the prediction in Section C of eighteen transitions for the same system, and can be explained in the following way.

In the perturbation method, I_z and I'_z are approximately good quantum numbers. Thus the two eigenstates $F_z = 1$ are approximately the states $I_z = 1$, $I'_z = 0$, and $I_z = 0$, $I'_z = 1$. The three states $F_z = 0$ are approximately

$$I_z = 1,\ I'_z = -1;\quad I_z = -1,\ I'_z = 1;\quad I_z = 0,\ I'_z = 0.$$

In the exact theory we had, from the existence of two states $F_z = 1$ and three states $F_z = 0$, concluded the existence of six transitions $F_z = 1 \to F_z = 0$. However, the transitions

$$|1, 0\rangle \to |-1, 1\rangle \quad \text{and} \quad |0, 1\rangle \to |1, -1\rangle,$$

rigorously forbidden if I_z and I'_z are perfectly good quantum numbers, have very little intensity if they are approximately good quantum numbers, that is for J/δ small, which is the domain of validity of the perturbation theory.

Fig. XI, 6 shows the spectrum of β-propiolactone (3), together with the theoretical spectrum, computed for the value $J/\delta = 0.265$ adjusted for the best fit. This value is sufficiently small for the four forbidden lines predicted by the exact theory to escape detection. The agreement based on the assumption that the two protons bound to a carbon are equivalent is slightly surprising in view of the assumption generally made that the molecule has a plane of symmetry. There seems to be no reason why a proton of one carbon should have the same coupling with a *cis*-proton and a *trans*-proton of the other carbon. Still, any other assumption would introduce extra lines that, as will appear from the next paragraph, are not observed.

E. Isochronous non-equivalent spins

If a group G_A of such spins exists in a molecule it is easily seen that its total spin I_A is *not* a good quantum number and the problem is accordingly complicated. It has been suggested that when the molecule

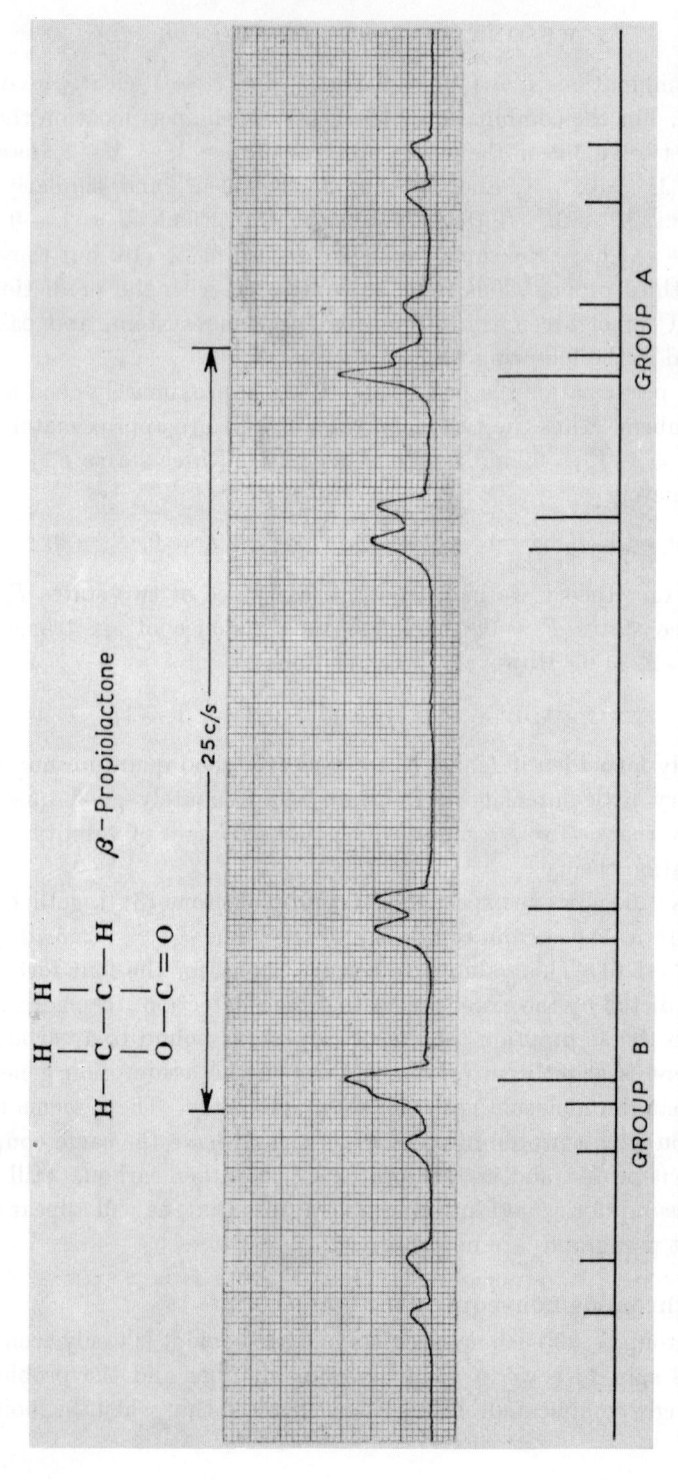

Fig. XI, 6. Spectrum of β-propiolactone. The theoretical spectrum below was calculated using perturbation theory and assuming for the best fit $J/\delta = 0.265$.

possesses any symmetry, group theory would be helpful in selecting the proper zero-order wave functions that would factorize the secular determinant, and in exhibiting selection rules, based on symmetry, that forbid otherwise possible transitions (**4, 5**).

We shall be content to consider the example of difluoroethylene (**4**) which is sufficiently simple for the simplifications brought about by the symmetry of the molecule to be evident without making use of the full mathematical apparatus of group theory. The J couplings in the molecule are as indicated below. Let s and s' be the proton spins, i and i' the fluorine spins.

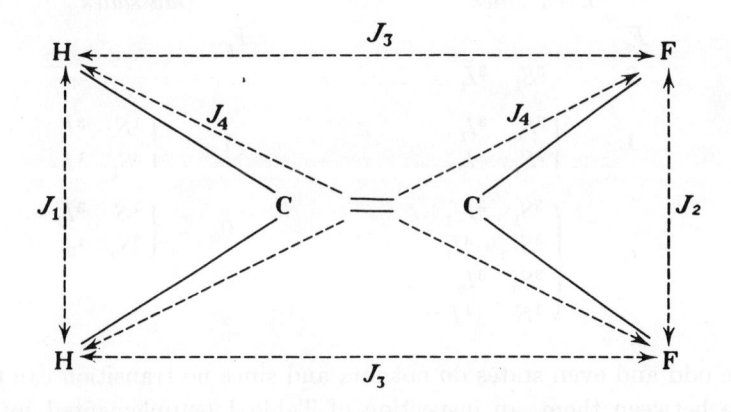

The Hamiltonian $\hbar \mathscr{H}$ of the system can be written

$$\mathscr{H} = \omega_S(s_z+s'_z)+\omega_I(i_z+i'_z)+$$
$$+J_1(\mathbf{s}.\mathbf{s}')+J_2(\mathbf{i}.\mathbf{i}')+J_3(\mathbf{i}.\mathbf{s}+\mathbf{i}'.\mathbf{s}')+J_4(\mathbf{i}.\mathbf{s}'+\mathbf{i}'.\mathbf{s}). \quad (20)$$

It is apparent from (20) that neither S, the total spin of the protons, nor I, the total spin of the fluorines, is a good quantum number. On the other hand, there is an important simplification due to the fact that $\delta = \omega_S - \omega_I$ is many orders of magnitude larger than the couplings J. It is instructive, however, not to take advantage of this straight away, in order to compare this problem with that of two groups of two equivalent spins ½, considered in Sections C and D. Our treatment would thus apply to a molecule containing two groups of isochronous but non-equivalent protons. The system still possesses a good quantum number which is the total spin $F_z = s_z+s'_z+i_z+i'_z$. Because of the symmetry of the molecule with respect to a plane P perpendicular to the plane of the molecule, there is a second good quantum number, the parity with respect to that operation. We may classify the states of the system as even or odd depending on whether the symmetry with respect to the plane P changes the sign of their wave functions, or not. The states are easily tabulated by making use of the fact that a triplet wave function of two spins $\tfrac{1}{2}$ is even with respect to these spins, while the singlet wave function is odd. We introduce the notations $^3S_0$ and $^3S_{\pm 1}$ to represent a triplet state with $S_z = 0$ or ± 1, and ^1S to represent the singlet state.

The states of the system are given in Table I for $F_z \geqslant 0$.

TABLE I

F_z	Even states		F_z	Odd states	
2	$^3S_1$	$^3I_1$			
1	$\begin{cases} ^3S_0 \\ ^3S_1 \end{cases}$	$\begin{matrix} ^3I_1 \\ ^3I_0 \end{matrix}$	1	$\begin{cases} ^1S \\ ^3S_1 \end{cases}$	$\begin{matrix} ^3I_1 \\ ^1I \end{matrix}$
0	$\begin{cases} ^3S_1 \\ ^3S_{-1} \\ ^3S_0 \\ ^1S \end{cases}$	$\begin{matrix} ^3I_{-1} \\ ^3I_1 \\ ^3I_0 \\ ^1I \end{matrix}$	0	$\begin{cases} ^1S \\ ^3S_0 \end{cases}$	$\begin{matrix} ^3I_0 \\ ^1I \end{matrix}$

Since odd and even states do not mix and since no transition can take place between them, an inspection of Table I (supplemented by the negative values of F_z) shows that the secular equation may be factorized into two equations of first order, five of second order, and one of fourth order, and that the maximum number of allowed transitions is twenty-eight. This must be contrasted with the problem treated in Section C of two groups of two equivalent spins where the maximum number of transitions was eighteen and the highest order of the secular equations was three.

If now we make use of the existence of a very large difference between the frequencies ω_S and ω_I, a drastic simplification occurs since first-order perturbation theory is applicable in calculating the energies of most states. The only exceptions are the two even states $|^3S_0\,^3I_0\rangle$ and $|^1S\,^1I\rangle$, with the same unperturbed energy, to be replaced as eigenstates by two linear combinations $|\xi\rangle$, $|\eta\rangle$, and the two odd states $|^3S_0\,^1I\rangle$ and $|^1S\,^3I_0\rangle$ to be replaced similarly by two linear combinations $|\xi'\rangle$ and $|\eta'\rangle$. These combinations and their energies are determined by solving secular equations of second order. The spectrum is separated into two widely distant parts, the proton spectrum and the fluorine spectrum.

The even part of the proton spectrum contains the following six lines:

$$|^3S_1, {}^3I_1) \to |^3S_0, {}^3I_1) \to |^3S_{-1}, {}^3I_1) \quad \text{(a single line)}$$
$$|^3S_1, {}^3I_{-1}) \to |^3S_0, {}^3I_{-1}) \to |^3S_{-1}, {}^3I_{-1}) \quad \text{(a single line)}$$
$$|^3S_1, {}^3I_0) \to |\xi); \quad |^3S_1, {}^3I_0) \to |\eta)$$
$$|\eta) \to |^3S_{-1}, {}^3I_0); \quad |\xi) \to |^3S_{-1}, {}^3I_0).$$

Similarly, the odd part contains the four lines

$$|^3S_1, {}^1I) \to |\xi'); \quad |^3S_1, {}^1I) \to |\eta');$$
$$|\xi') \to |^3S_{-1}, {}^1I); \quad |\eta') \to |^3S_{-1}, {}^1I).$$

Altogether there are ten lines in the proton spectrum and as many in the fluorine spectrum.

We shall not give the values of the different frequencies that can be easily computed with the indications given above (4). The important point is that the coupling constant J_1 and J_2 between isochronous spins can be and have been determined from an analysis of the observed spectrum. The distinction between isochronous and equivalent spins is essential.

We shall stop our survey of the analysis of the multiplet structures in liquids here. The technique of high resolution necessary for the detailed observation of these spectra has, during the last few years, developed greatly as a tool for the investigation of chemical structures, and a whole book could (and very likely will) be† devoted to all the problems connected with the analysis and interpretation of these spectra. It might be added that there is a certain trend of going to higher fields for the observation of complex spectra. This has the double advantage of simplifying the interpretation by increasing the validity of the simple perturbation theory and of increasing the strengths of the individual lines by reducing their number.

II. Multiplet Spectra observed by Transient Methods

A. The method of free precession

The fundamental theorem proved in Chapter IV stated that the free precession signal observed after a 90° pulse was the Fourier transform of the continuous wave spectrum, provided the field H_1 that produced the pulse was much larger than the width (in gauss) of the spectrum, and in particular much larger than the inhomogeneity ΔH of the applied field. Since, as was shown in Chapter III, a free precession signal is

† (Note added in proof) and indeed has been: *High Resolution Nuclear Magnetic Resonance*, J. A. Pople, W. G. Schneider and H. J. Bornstein.

always damped in a time $T^* \sim (\gamma \Delta H)^{-1}$ because of the destructive interference between signals from various parts of the sample, it is clear that a free precession method cannot, any more than the continuous wave method, resolve structures narrower than the inhomogeneity of the field. On the other hand, in a sufficiently homogeneous field the same information as in the C.W. method can be obtained by observing the free precession.

For instance, if the molecule contains two groups of spins I_k and $I'_{k'}$ with a relative chemical shift $\delta = (\gamma - \gamma')H_0$ (and no spin-spin couplings) such that $\gamma \Delta H \ll \delta \ll \gamma H_1$, the amplitude $A(t)$ of the signal exhibits beats at the frequency δ, described by $A(t) \propto [1 + \lambda \cos(\delta t)]^{\frac{1}{2}} \mathscr{F}(t)$, where $\mathscr{F}(t)$ would be the shape of the signal in the absence of chemical shift. It is easy to see that the coefficient $\lambda \leqslant 1$ is related to the ratio n/n' of the numbers of the spins I and I' in the molecule through

$$\lambda = \frac{2}{n/n' + n'/n}.$$

The existence of a J coupling can also manifest itself through beats in a free precession signal. Suppose, for instance, two spins I and S belong to different species so that $\delta = |\gamma_I - \gamma_S|H_0 \gg \gamma_{S,I} H_1 \gg J$, where the r.f. field has the frequency $\omega_I = -\gamma_I H_0$ of the spins I. Assume further that $J \gg |\gamma_I \Delta H|$ and (for simplicity) $S = \frac{1}{2}$. A 90° pulse at the frequency ω_I leaves the spins S unaffected and the spins I precess with the two frequencies $\omega_I \pm \frac{1}{2}J$ that correspond to the two possible orientations of the spin S, giving a signal with an amplitude

$$A(t) \propto (1 + \cos(Jt))^{\frac{1}{2}} \mathscr{F}(t).$$

If the spin S is not restricted to the value $\frac{1}{2}$ the amplitude of the signal becomes

$$A(t) \propto |1 + e^{iJt} + \ldots + e^{2iSJt}| \mathscr{F}(t) \propto \frac{\sin\{(S+\frac{1}{2})Jt\}}{\sin(\frac{1}{2}Jt)} \mathscr{F}(t).$$

The extension to situations where there is more than one spin S or I is straightforward. Fig. XI, 7, taken from reference (6), represents the free precession signals from the deuteron and the proton spins of the HD molecule, whence the first measurement of their J-coupling was obtained. This figure should be compared with Fig. XI, 2, which represents the same structure, observed by the C.W. method.

It has already been mentioned that continuous-wave measurements of multiplet spectra in low fields, while suffering from the weakness of the signal, benefited from the increase in resolution owing to a better absolute homogeneity of the applied field. The free-precession method

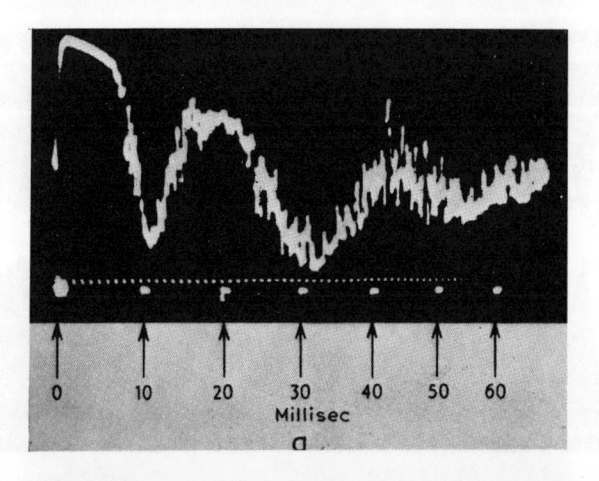

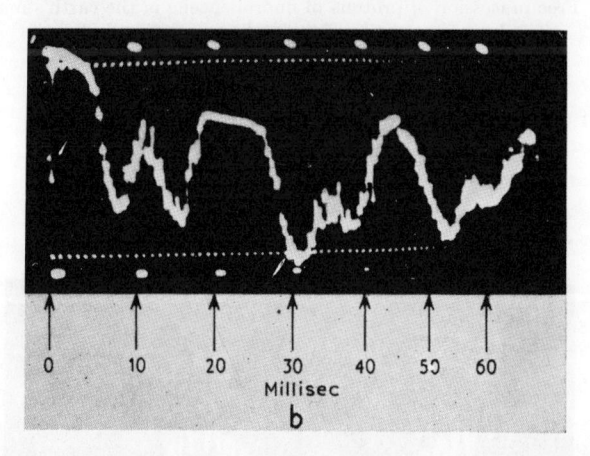

FIG. XI, 7. Free precession signals exhibiting beats caused by the I_1, I_2 coupling between the spins of the proton and of the deuteron in HD. (a) Deuteron signal. (b) Proton signal.

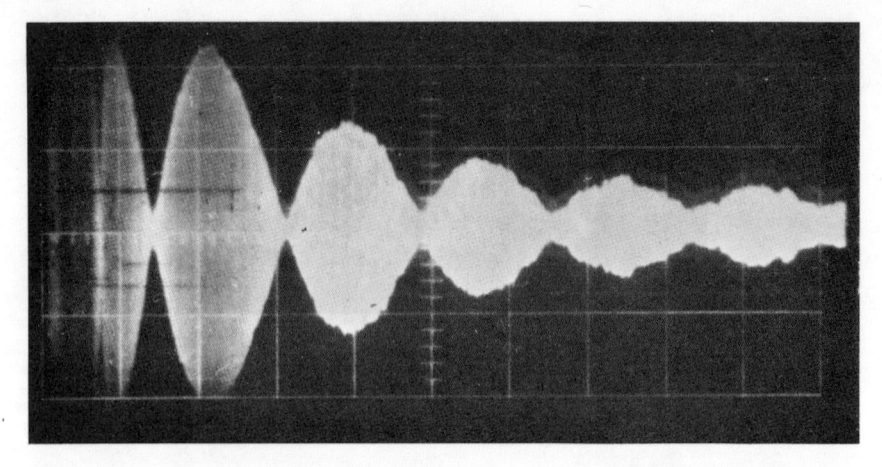

Fig. XI, 8. Free precession of protons in fluorobenzene in the earth's magnetic field. The beat frequency is 5·8 c/s, the sweep time is 1 sec.

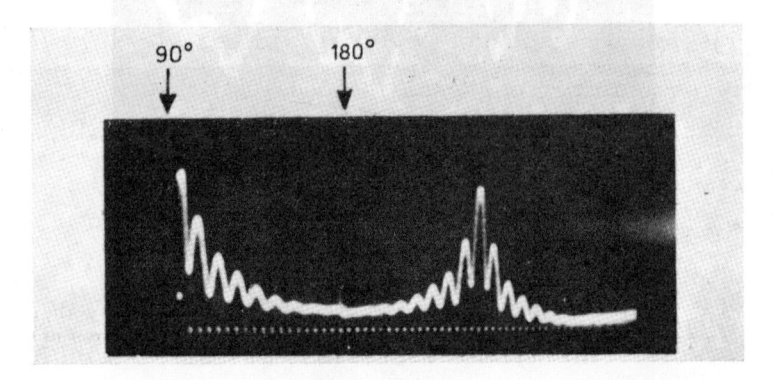

Fig. XI, 9. A tail and echo observed from a sample of acetic acid in a very homogeneous field.

permits the observation of multiplet structures in fields as low and as homogeneous as the earth's magnetic field without a prohibitive loss in signal. By an extension of a method described elsewhere (Chapter III), the sample is first polarized in a field of a few hundred gauss, and after a rapid cut-off of the polarizing field the decaying precession signal of the large magnetization acquired in the high field is observed in the earth's field. Fig. XI, 8, taken from reference (7), shows the free-precession proton signal obtained in this manner from fluorobenzene. The slow beats superimposed on the exponential decay of the signal have a frequency that corresponds to a coupling J between the protons and the fluorine of $2\pi \times 5 \cdot 8$ c/s. The existence of a single-beat frequency, consistent with the existence of a single doublet in the C.W. proton spectrum at high field (although with the slightly different coupling constant $J = 2\pi \times 6 \cdot 7$ c/s), would seem to require the equality of the couplings between the fluorine and the five protons of the molecule of fluorobenzene, a point for the chemists to explain.

B. The method of spin echoes

It has been shown in Chapter III that the spin echo technique made it possible to measure under certain conditions the relaxation time T_2, and thus to obtain the true width of a line unobtainable by the continuous-wave method, because it was much narrower than the inhomogeneity ΔH of the applied field. It was natural to apply the same method to the measurement of splittings J and δ much smaller than $|\gamma \Delta H|$ (8). It will be shown that the amplitude $E(2\tau)$ of the echo (where τ is the interval between the initial pulse starting the precession and the refocusing pulse) has, as a function of τ, an oscillatory behaviour from which the values of the shifts δ and the couplings J can be obtained. These oscillations should not be confused with those described in Section II A in connexion with the free-precession signal and represented in Figs. XI, 7 and XI, 8. It was assumed there that $|\gamma \Delta H| \ll \delta, J$ so that a few oscillations of the amplitude of the free-precession signal, of frequency J or δ, could be observed before the signal was damped in a time $\sim (\gamma \Delta H)^{-1}$ by the inhomogeneity of the field. The same oscillations appear in the echo itself, as is shown in Fig. XI, 9. The shape of the echo does not in that case provide any information not contained already in the free-precession signal or for that matter in the C.W. spectrum.

On the contrary, in the problem to be considered now, neither the observation of the free precession nor that of a *single* echo can give any information on J or δ, for their duration $|\gamma\Delta H|^{-1}$ is much smaller than the periods $1/\delta$ or $1/J$ that are being sought. It is from the comparison of the amplitude $E(2\tau)$ of a large number of echoes, for different values of τ, that information, unobtainable in the same magnet by the C.W. method, will be collected.

Already used for that purpose in Chapter VII, the formalism of the density matrix greatly simplifies the calculation of the echo amplitudes. A further simplification results from the use of refocusing pulses of 180° rather than 90° as in reference (8). Let the Hamiltonian $\hbar\mathcal{H}$ of the system be given by

$$\mathcal{H} = \omega_0 I_z + \sum_k \delta_k i_{kz} + \sum_{k<l} J_{kl} \mathbf{i}_k \cdot \mathbf{i}_l = \omega_0 I_z + \mathcal{H}_1, \tag{21}$$

where ω_0 is the average Larmor frequency of the spins of the molecule and δ_k the chemical shift of the spin i_k with respect to that frequency. For simplicity, we have assumed that all the spins of the molecule belonged to the same nuclear species, say protons, a restriction that can easily be lifted. The average Larmor frequency ω_0 does not have the same value through the sample because of the inhomogeneity of the applied field.

The contributions to the signal from the various parts of the sample are functions of ω_0 and interfere destructively except at the time 2τ of the echo, when they are independent of ω_0 and thus are all in phase. Strictly speaking, the δ_k also depend on the position inside the sample, but this dependence can be safely neglected. The amplitude of the signal $S(t)$ is proportional to $\mathrm{tr}\{\sigma(t)I_x\}$, where $\sigma(t)$ is the density matrix describing the spin system in the frame rotating at the frequency ω of the r.f. field applied in pulses along an axis of that frame.

Actually, unless special care is taken to maintain coherence between the pulses, the direction of the r.f. field in the rotating frame is not the same in general for the two pulses, a fact of no importance if the decay tail and the echo are well separated. In the rotating frame the Hamiltonian $\mathcal{H}$ of (21) should be replaced by

$$(\omega_0 - \omega)I_z + \mathcal{H}_1. \tag{22}$$

Before the first pulse the Boltzmann density matrix describing the spin system is proportional to $\exp\{-\hbar\mathcal{H}/kT\}$ or in the high-temperature approximation to $\mathcal{H}$, and since the term $\omega_0 I_z$ is much larger than $\mathcal{H}_1$, σ can be assumed to be proportional to I_z.

The amplitude H_1 of the rotating field is assumed to be much larger than ΔH and $(1/\gamma)(\delta, J)$ and the durations of 90° or 180° pulses are taken to be negligible. The first 90° pulse transforms I_z into I_x and at a time τ later, just before the second pulse, σ is given by

$$\sigma(\tau_-) = e^{-i[(\omega_0-\omega)I_z+\mathscr{H}_1]\tau}I_x e^{i[(\omega_0-\omega)I_z+\mathscr{H}_1]\tau}. \tag{23}$$

The effect of the refocusing 180° pulse is to change the signs of I_z, I_x, and $\sum_k \delta_k I_{kz}$ whilst leaving $\sum_{k<l} J_{kl}\mathbf{i}_k\cdot\mathbf{i}_l$ unchanged. If we define $\widetilde{\mathscr{H}}_1$ as

$$\widetilde{\mathscr{H}}_1 = -\sum_k \delta_k i_{kz} + \sum_{k<l} J_{kl}\mathbf{i}_k\cdot\mathbf{i}_l, \tag{23'}$$

immediately after the 180° pulse σ is given by

$$\sigma(\tau_+) = -e^{-i[-(\omega_0-\omega)I_z+\widetilde{\mathscr{H}}_1]\tau}I_x e^{i[-(\omega_0-\omega)I_z+\widetilde{\mathscr{H}}_1]\tau}.$$

Finally, τ seconds later σ becomes

$$\sigma(2\tau) = e^{-i[(\omega_0-\omega)I_z+\mathscr{H}_1]\tau}\sigma(\tau_+)e^{i[(\omega_0-\omega)I_z+\mathscr{H}_1]\tau}. \tag{24}$$

Since $(\omega_0-\omega)I_z$ clearly commutes with both $\mathscr{H}_1$ and $\widetilde{\mathscr{H}}_1$, this can be rewritten as

$$\sigma(2\tau) = -e^{-i\mathscr{H}_1\tau}e^{-i\widetilde{\mathscr{H}}_1\tau}I_x e^{i\widetilde{\mathscr{H}}_1\tau}e^{i\mathscr{H}_1\tau}, \tag{25}$$

an expression from which ω_0 has disappeared, and the amplitude of the echo is proportional to $E(2\tau)$:

$$E(2\tau) = \operatorname{tr}\{e^{-i\mathscr{H}_1\tau}e^{-i\widetilde{\mathscr{H}}_1\tau}I_x e^{i\widetilde{\mathscr{H}}_1\tau}e^{i\mathscr{H}_1\tau}I_x\}. \tag{26}$$

A very simple situation obtains if the ratios J/δ are sufficiently small for the first-order perturbation theory outlined in Section I A to apply. It is then permissible to replace $\sum_{k<l} J_{kl}\mathbf{i}_k\cdot\mathbf{i}_l$ by $\sum J_{kl} i_{kz} i_{lz}$, which commutes with $\sum_k \delta_k i_{kz}$, and (26) becomes simply

$$E(2\tau) = \operatorname{tr}\{e^{-2i\tau\sum_{k<l}J_{kl}i_{kz}i_{lz}}I_x e^{2i\tau\sum_{k<l}J_{kl}i_{kz}i_{lz}}I_x\}, \tag{27}$$

independent of the δ. This demonstrates that the chemical shifts δ cannot be obtained from the envelope of the echo in the absence of J couplings (in contradiction to the observation of the free-precession tail as explained in Section II A).

For instance, for two spins $\frac{1}{2}$ with a coupling J, it is easily found that (27) is simply proportional to $\cos(J\tau)$. As an example of the more general situation, where δ and J are comparable, consider the case of two spins $\frac{1}{2}$ for which the C.W. spectrum was calculated in Section I B.

The two Hamiltonians

$$\mathscr{H}_1 = \tfrac{1}{2}\delta(s_z-s'_z)+J\mathbf{s}\cdot\mathbf{s}',$$
$$\widetilde{\mathscr{H}}_1 = -\tfrac{1}{2}\delta(s_z-s'_z)+J\mathbf{s}\cdot\mathbf{s}'$$

have the same spectrum with energies (in frequency units)

$$a = d = \tfrac{1}{4}J, \quad b = -\tfrac{1}{4}J+\tfrac{1}{2}\Delta, \quad c = -\tfrac{1}{4}J-\tfrac{1}{2}\Delta.$$

where
$$\Delta = \sqrt{(J^2+\delta^2)}.$$

Their eigenstates, however, are different and $[\mathcal{H}_1, \tilde{\mathcal{H}}_1] \neq 0$. Those of $\mathcal{H}_1$ are
$$|a\rangle = |+,+\rangle; \quad |b\rangle = p|+,-\rangle + q|-,+\rangle,$$
$$|c\rangle = -q|+,-\rangle + p|-,+\rangle; \quad |d\rangle = |-,-\rangle$$
with
$$p = \cos\tfrac{1}{2}\phi, \quad q = \sin\tfrac{1}{2}\phi$$
and
$$\tan\phi = J/\delta, \quad -\tfrac{1}{2}\pi \leqslant \phi \leqslant \tfrac{1}{2}\pi.$$

The eigenstates of $\tilde{\mathcal{H}}_1$ are deduced from those of $\mathcal{H}_1$ by reversing all the spins:
$$|\tilde{a}\rangle = |-,-\rangle = |d\rangle; \quad |\tilde{b}\rangle = q|+,-\rangle + p|-,+\rangle;$$
$$|\tilde{c}\rangle = p|+,-\rangle - q|-,+\rangle; \quad |\tilde{d}\rangle = |+,+\rangle = |a\rangle.$$

The expression (26) is easily calculated, using a representation where one of the two Hamiltonians, say $\mathcal{H}_1$, is diagonal and noticing that the only products of eigenkets of $\mathcal{H}_1$ and $\tilde{\mathcal{H}}_1$ different from zero are:
$$\langle a|\tilde{d}\rangle = \langle \tilde{a}|d\rangle = 1,$$
$$\langle b|\tilde{b}\rangle = -\langle c|\tilde{c}\rangle = 2pq,$$
$$\langle b|\tilde{c}\rangle = \langle c|\tilde{b}\rangle = p^2 - q^2.$$

All are real as well as their complex conjugates. The result is
$$E(2\tau) = (p^2-q^2)^2 \cos\{(a+d-b-c)\tau\} + pq(p+q)^2 \cos\{(a+d-2b)\tau\} -$$
$$- pq(p-q)^2 \cos\{(a+d-2c)\tau\}$$
$$= \cos^2\phi \cos J\tau + \tfrac{1}{2}\sin\phi(1+\sin\phi)\cos(J-\Delta)\tau -$$
$$- \tfrac{1}{2}\sin\phi(1-\sin\phi)\cos(J+\Delta)\tau$$
$$= \frac{\delta^2}{\Delta^2}\cos J\tau + \frac{J^2}{\Delta^2}\cos\Delta\tau\cos J\tau + \frac{J}{\Delta}\sin\Delta\tau\sin J\tau. \qquad (28)$$

For $(J/\delta) \ll 1$, the expression (28) tends toward the simple expression $\cos(J\tau)$ as predicted.

A spin echo technique was described in Chapter III, where, in order to overcome the diffusion damping, rather than to observe as a function of τ the size of the first echo $E(2\tau)$ after a single refocusing pulse of 180° at time τ, one observed as a function of n the size of the nth echo at time $2n\tau'$ after n refocusing 180° pulses at times τ', $3\tau'$,..., $(2n-1)\tau'$. Through an obvious generalization of (26) the size of the nth echo is given by a product of alternating factors
$$E(2n\tau') = \text{tr}\{e^{-i\mathcal{H}_1\tau'}e^{-i\tilde{\mathcal{H}}_1\tau'}e^{-i\mathcal{H}_1\tau'}...I_x...e^{i\tilde{\mathcal{H}}_1\tau'}e^{i\mathcal{H}_1\tau'}I_x\}. \qquad (29)$$

(26) and (29) will not in general coincide for $n\tau' = \tau$ except in the approximation $J/\delta \ll 1$ when the non-commutativity of $\mathcal{H}_1$ and $\tilde{\mathcal{H}}_1$ can be neglected and when the second method should permit a precise measurement of J.

If the refocusing pulses are not of 180° the calculations are heavier, the results more complicated, and the interpretation of the patterns observed a good deal more difficult. Thus, for instance, for a refocusing pulse of 90° rather than 180° (28) must be replaced by (8):

$$|E(2\tau)| \propto \frac{\delta^2}{2(\delta^2+J^2)}\left\{1+\frac{J^2}{2\delta^2}-2\sin^2\tfrac{1}{2}J\tau\sin^2\left(\frac{\tau(J^2+\delta^2)^{\frac{1}{2}}}{2}\right)\right\} \quad (30)$$

which, in particular, for $J/\delta \ll 1$ is approximately

$$\tfrac{1}{2}\{1-2\sin^2\tfrac{1}{2}J\tau\sin^2\tfrac{1}{2}\tau\delta\},$$

an expression that is more complicated than the simple variation $\cos(J\tau)$ obtained after a 180° pulse. It should be mentioned, however, that in spite of these difficulties the authors of reference (8) were able to measure a good many constants J and δ with reasonable accuracy, as confirmed later by continuous wave measurements using high resolution techniques.

III. Line Width Problems in Multiplet Spectra

A. Effects of quadrupole relaxation and chemical exchange

In some of the previous examples of multiplet structures, such as those observed in BrF_5, in the dichloracetaldehyde $CHCl_2$—CHO, etc., there was no apparent effect from the nuclear spins of chlorine and bromine. This is due to the fact that those nuclei have spins I' larger than $\tfrac{1}{2}$ and quadrupole moments strongly coupled to the local electric fields by a relaxation mechanism as described in Chapter VIII. The lifetimes of the various states $I'_z = M'$ are accordingly short. It has been shown earlier that if two spins I and I' coupled by an interaction $\hbar J$, belong to different species, the resonance spectrum of the spin I is a multiplet with frequencies

$$\omega_I(M') = -\gamma_I H_0 + JM'. \quad (31)$$

If, because of the relaxation of the spin I', $\omega_I(M')$ jumps from one of the values (31) to another, at a rate fast compared with the frequency J, the various lines of the multiplet will coalesce and a single line will be observed at an average frequency $-\gamma_I H_0$. This problem was considered in some detail in Chapter X on motion narrowing, where a general formula (61), (61') was derived describing the shape of the spectrum as a function of the jumping rate. Alternative treatments have been given by various authors (1).

Similar effects can occur because of chemical exchange. If in a given molecule, a spin I' coupled to the spin I, the resonance of which is being observed, is replaced at frequent intervals by another spin I' of the same species, as far as the spectrum of I is concerned, the effect is the same as if the spin I', without leaving the molecule, made a transition.

The situation is still the same if it is the 'resonant' nucleus I, rather than the 'non-resonant' one I', that moves from one molecule to another. Then, if in its new molecule (chemically identical or not with the one it just left) the spin I has a different position and thus a different chemical shift, an averaging of the latter can clearly also take place.

Extensive use has been made by chemists of these effects of chemical exchange in order to study reaction rates, and a considerable and fast-growing literature exists on the subject.

As the jumping rate which takes the resonant frequency (31) from one value to another increases, the spectrum is changed as follows. First, each of the lines (31) is broadened by an amount which, as will appear shortly, is inversely proportional to the lifetime of the state $I'_z = M'$ of the spin I' (as 'seen' by the spin I if the jumping is caused by chemical exchange). As the jumping frequency Ω goes on increasing a complex broad pattern appears, which upon further increase of Ω narrows to a single line with a frequency weighted average, of the values (31). Finally, if Ω reaches values comparable to the Larmor frequency ω_0, the jumping becomes a mechanism of relaxation of the spin I. This last situation has already been dealt with in Chapter VIII, where examples had been given of a spin I coupled to another spin I' through an interaction $\hbar A \mathbf{I}.\mathbf{I}'$ and relaxing either through fast modulation of the constant A by the chemical exchange, or through very fast relaxation of the spin I'. An example of the former was provided by the relaxation of both nuclear spins in HF, and of the latter by the relaxation of protons in aqueous solutions of Mn^{++}.

To study the earlier stages of the distortion of a multiplet spectrum as a function of an increasing jumping frequency Ω, consider first a spin $I = \frac{1}{2}$ coupled to a spin I' with a quadrupole relaxation mechanism, by a coupling $\hbar J \mathbf{I}.\mathbf{I}'$. If $\Omega \ll J$, a transition of the spin I between two states $|a) = |+, M'\rangle$ and $|b) = |-, M'\rangle$ with a frequency

$$\omega_{ab} = \omega_0 + JM'$$

has a width given by formula (45) of Chapter X:

$$\frac{1}{T_2} = \frac{1}{T'_2} + \frac{1}{2}\left(\frac{1}{t_a} + \frac{1}{t_b}\right). \qquad (32)$$

In that formula, which gives the line width caused by a fluctuating perturbation, $1/T'_2$ is the so-called adiabatic width due to a modulation of the frequency ω_{ab} of the resonance by the perturbation, whereas t_a and t_b are the lifetimes of the initial and the final state. If the relaxation of the spin I' is considered as the only cause of broadening, it is clear that $1/T'_2 = 0$ since, the state $I'_z = M'$ of the spin I' being the same in both states $|a\rangle$ and $|b\rangle$ of the combined system, the relaxation of the spin I' cannot affect the resonance frequency ω_{ab}. On the other hand, $t_a = t_b = \tau_{M'}$ where $\tau_{M'}$ is the lifetime of the state $I'_z = M'$ and the width of the transition ω_{ab} is precisely $1/\tau_{M'}$. The inverse lifetime $1/\tau_{M'}$ is equal to

$$\frac{1}{\tau_{M'}} = \sum_{M''} P_{M'M''}, \tag{33}$$

where $P_{M'M''}$ is the transition probability per unit time from the state M' to the state M'', induced by the quadrupole coupling on the spin I' with the fluctuating local electric-field gradient. The description of this coupling was given in Chapter VIII, Section II F (c). If we assume for simplicity extreme narrowing (hardly a restriction, for in a liquid sufficiently viscous for this assumption to be invalid the lines would be too broad for any multiplet structure to be observed at all), a straightforward extension of the results given in Chapter VIII, Section II F (c), with the notations defined there, leads to

$$P_{M',M'\pm 2} = \frac{3}{80}\left[\frac{eQ}{\hbar}\frac{\partial^2 V}{\partial z'^2}\frac{1}{I(2I-1)}\right]^2 \left(1+\frac{\eta^2}{3}\right)|(M'\,|\,I'^2_{\mp}\,|\,M'\pm 2)|^2 \tau_c,$$

$$P_{M',M'\pm 1} = \frac{3}{80}\left[\frac{eQ}{\hbar}\frac{\partial^2 V}{\partial z'^2}\frac{1}{I(2I-1)}\right]^2 \left(1+\frac{\eta^2}{3}\right)|(M'\,|\,I'_{\mp}I'_z + I'_z I'_{\mp}\,|\,M'\pm 1)|^2 \tau_c. \tag{34}$$

Introducing the spin lattice relaxation time T of the nuclear spin I' (unambiguously defined by the formula (137) of Chapter VIII if extreme narrowing is assumed), (33) becomes

$$\frac{1}{\tau_{M'}} = \frac{1}{T}\frac{1}{2(4I(I+1)-3)}\{|(M'\,|\,I'^2_+\,|\,M'-2)|^2 + |(M'\,|\,I'^2_-\,|\,M'+2)|^2 + $$
$$+ |(M'\,|\,I'_z I'_+ + I'_+ I'_z\,|\,M'-1)|^2 + |(M'\,|\,I'_z I'_- + I'_- I'_z\,|\,M'+1)|^2\}; \tag{35}$$

for $I' = 1$ (35) gives

$$\frac{1}{\tau_1} = \frac{1}{\tau_{-1}} = \frac{3}{5}\frac{1}{T}, \qquad \frac{1}{\tau_0} = \frac{2}{5}\frac{1}{T};$$

for $I = \frac{3}{2}$,

$$\frac{1}{\tau_{\frac{3}{2}}} = \frac{1}{\tau_{\frac{1}{2}}} = \frac{1}{\tau_{-\frac{1}{2}}} = \frac{1}{\tau_{-\frac{3}{2}}} = \frac{1}{T}.$$

If a spin $I = \frac{1}{2}$ is coupled to a spin $I' = 1$, the spectrum of I is a triplet, the spectrum of I' a doublet. The two lines of the doublet I' have the same width $1/T$, the central line of the triplet I has a width $\frac{2}{5}(1/T)$ and the two lateral lines a width $\frac{3}{5}(1/T)$. For a spin $I = \frac{1}{2}$ coupled to a spin $I' = \frac{3}{2}$ the four lines of the quadruplet I and the two lines of the doublet I' should all have the same width $1/T$. These predictions are consistent with the observation of widths in the ratio $3:2:3$ for protons bound to N^{14} in NH_3, and of four lines of equal widths for protons bound to B^{11} ($I' = \frac{3}{2}$) in $NaBH_4$. If Ω becomes comparable to J the previous treatment breaks down and the general formula (61) of Chapter X must be used.

In this formula
$$I(\omega) \propto \mathrm{re}\{\mathbf{W} \cdot \mathbf{A}^{-1} \cdot \mathbf{1}\}, \tag{36}$$
the vector $\mathbf{W}$ with components proportional to the *a priori* probabilities of the various frequencies of the multiplet, that is, to the populations of the states $I'_z = M'$, is the vector

$$\begin{pmatrix} 1 \\ \vdots \\ 1 \end{pmatrix} = (\mathbf{1}).$$

The matrix $\mathbf{A}$ is given by

$$A_{M'M''} = \left[i(\omega_0 + JM' - \omega) - \frac{1}{\tau_{M'}}\right]\delta_{M'M''} + P_{M'M''}, \tag{37}$$

where $P_{M'M''}$ is given by (34). For a spin $I' = 1$,

$$\mathbf{A} = \begin{bmatrix} i(\omega_0 - \omega + J) - \dfrac{3}{5T} & \dfrac{1}{5T} & \dfrac{2}{5T} \\ \dfrac{1}{5T} & i(\omega_0 - \omega) - \dfrac{2}{5T} & \dfrac{1}{5T} \\ \dfrac{2}{5T} & \dfrac{1}{5T} & i(\omega_0 - \omega - J) - \dfrac{3}{5T} \end{bmatrix}$$

and (36) yields (9)

$$I(x) = \frac{45 + \eta^2(5x^2 + 1)}{225x^2 + \eta^2(34x^4 - 2x^2 + 4) + \eta^4(x^6 - 2x^4 + x^2)}, \tag{38}$$

where $x = (\omega_0 - \omega)/J$, $\eta = 5TJ$.

Fig. XI, 10, from reference (9), shows how the spectrum changes as the parameter η^2 varies from 1000 to 1.

The occurrence of similar distortions of a multiplet spectrum because of chemical exchange is best illustrated by the spectrum of the hydroxyl proton in ethyl alcohol CH_3-CH_2-OH (10). Unless special precautions are taken to keep the sample very pure, the hydroxyl proton gives

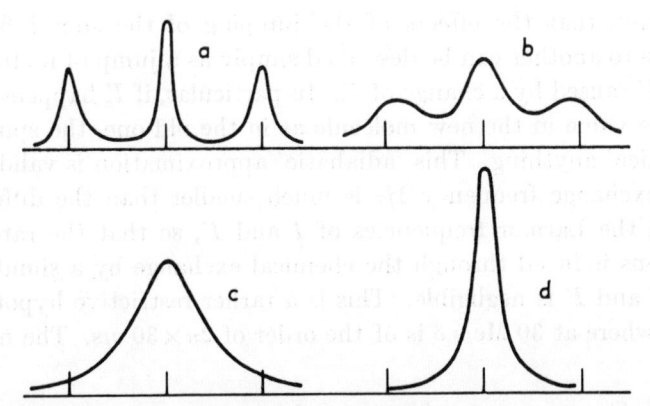

FIG. XI, 10. Theoretical line shapes for the spectra of nuclei of spin ½ coupled to a nucleus of spin I according to equation (38). (a) $\eta^2 = 10^3$, (b) $\eta^2 = 10^2$, (c) $\eta^2 = 10$, (d) $\eta^2 = 1$. The vertical lines indicate the positions of the triplet in the absence of quadrupole relaxation.

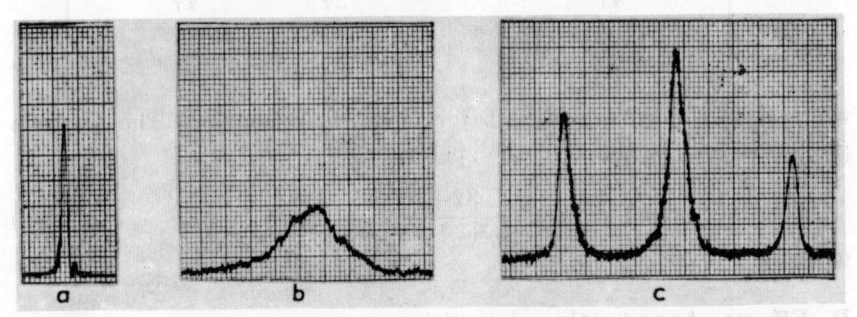

FIG. XI, 11. Signal from the hydroxyl proton in ethyl alcohol for varying concentrations of HCl. (a) 5×10^{-5} normal HCl, (b) 10^{-5} normal HCl, (c) 3.8×10^{-6} normal HCl.

a single line in spite of the fact that its coupling with the CH_2 group should (in the first-order approximation where

$$|\delta_H(CH_2) - \delta_{H'}(OH)| \gg |J_{HH'}|)$$

lead to a triplet with a central line twice the lateral ones. Fig. XI, 11, from reference (10), shows the spectrum of the hydroxyl proton for small varying concentrations of HCl which acts as a catalyst for exchange. The theoretical shape of the spectrum can be obtained from formula (36). The vector **W** should be taken as $\begin{pmatrix} 1 \\ 2 \\ 1 \end{pmatrix}$ to allow for the fact that the combination $I'_z = 0$ for the two protons of the CH_2 group is twice as probable as each of the combinations $I'_z = \pm 1$. We make the assumption that the effects of the jumping of the spin **I** from one molecule to another can be described simply as a jump of its frequency $\omega_0 + JM'$ caused by a change of I_z'. In particular, if I_z' happens to have the same value in the new molecule as in the old one, the spin I does not 'notice' anything. This 'adiabatic' approximation is valid as long as the exchange frequency $1/\tau$ is much smaller than the difference δ between the Larmor frequencies of I and I', so that the rate of the transitions induced through the chemical exchange by a simultaneous flip of I and I' is negligible. This is a rather restrictive hypothesis in alcohol where at 30 Mc/s δ is of the order of $2\pi \times 30$ c/s. The matrix **A** is then

$$\mathbf{A} = \begin{bmatrix} i(\omega_0-\omega+J)-\dfrac{3}{4\tau} & \dfrac{1}{2\tau} & \dfrac{1}{4\tau} \\ \dfrac{1}{4\tau} & i(\omega_0-\omega)-\dfrac{1}{2\tau} & \dfrac{1}{4\tau} \\ \dfrac{1}{4\tau} & \dfrac{1}{2\tau} & i(\omega_0-\omega-J)-\dfrac{3}{4\tau} \end{bmatrix}$$

where τ is the time constant for chemical exchange. The formula (36) gives for the shape of the spectrum

$$I(\omega) = \frac{32 + \eta^2(2x^2+1)}{64x^2 + \eta^2(1+8x^4) + \tfrac{1}{4}\eta^4 x^2(1-x^2)^2}, \tag{38'}$$

where $\eta = 4J\tau$.

B. Effects of magnetic relaxation

The fact that the width of a transition made by a spin I or by a group of equivalent spins, depends on the lifetimes of other spins I' coupled to it, can also manifest itself in the absence of chemical exchange and quadrupole relaxation when magnetic relaxation only is present.

A striking example is provided by a situation where the spins I' are two equivalent spins $i' = \tfrac{1}{2}$. As mentioned earlier, the spectrum of the spin I is then in first order a triplet, with the central line $I_z' = 0$ twice as high as the lateral lines. However, in higher order the two components of the central line which are $I_z' = 0$, $I' = 1$, and $I_z' = 0$, $I' = 0$ are split and can be resolved in practice. The singlet spin wave function $I' = 0$ is antisymmetrical with respect to the two spins and thus is insensitive to any symmetrical perturbation such as a bilinear coupling between the spins i' or a local field homogeneous over the distance between them. As a consequence this state has a much longer lifetime than the triplet state $I'_z = 0$, $I' = 1$, and the two central lines of the spectrum of the spin I should have unequal widths and, being of equal intensity, unequal heights.

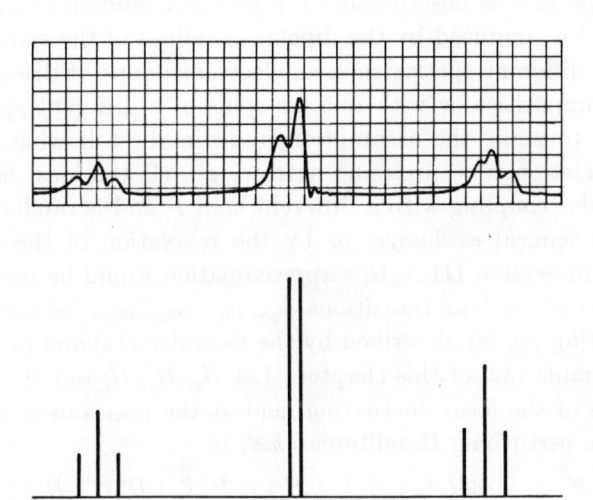

FIG. XI, 12. Spectrum of the methyl group CH_3 in ethyl alcohol. The three groups of lines correspond respectively to the values $I'_z = -1, 0, 1$ for the total spin I'_z of the two protons of the methylene group CH_2 to which the methyl group is coupled. In the central group of two lines, the narrower one corresponds to the singlet state $I'_z = 0$, $I' = 0$, of the methylene group, and the broader one to the triplet state $I'_z = 0$, $I' = 1$, as explained in the text.

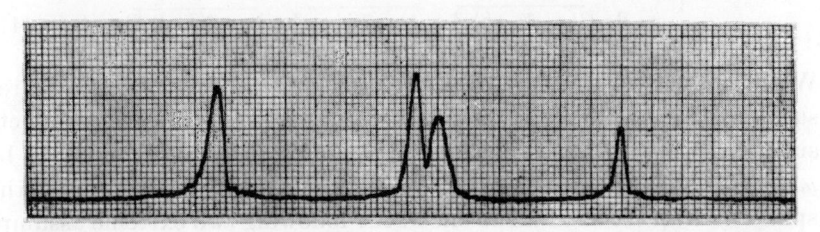

FIG. XI, 13. Spectrum of a single proton A in 1,1,2-trichloroethane, coupled to a group of two equivalent protons B. The narrowest (and highest) of the two central lines corresponds to the singlet state $I'_z = 0$, $I' = 0$ of the protons B, as explained in the text.

This is apparent in Figs. XI, 12, from reference (10), and XI, 13, from reference (3), taken from the spectrum of the CH_3 group of alcohol coupled to the CH_2 group, and the spectrum of a proton in 1,1,2-trichloroethane coupled to two equivalent protons, respectively.

To give a quantitative example of the theory of line widths in multiplet spectra, consider the simpler problem (Section I B) of two spins $i = i' = \frac{1}{2}$ with a coupling J and a relative shift δ of comparable magnitude.

We shall assume first that the broadening mechanism is of the 'external' type that is describable by a classical random magnetic field. If this field is produced by the dipolar couplings of the spins i, i' with spins of a different species such as electronic spins of dissolved paramagnetic impurities or with nuclear spins of a different species, it is legitimate to make the assumption, used in the following, of a very short correlation time (extreme narrowing). (If the local field results from a *scalar* coupling with a different spin I' and is randomly modulated by chemical exchange, or by the relaxation of the spin I' as described in Section III A, this approximation would be invalid.) The line widths of the four transitions ω_{ab}, ω_{ac}, ω_{bd}, ω_{cd}, between the four states $|a\rangle$, $|b\rangle$, $|c\rangle$, $|d\rangle$, described by the formulae (12) and (8), are given by the formula (32) of this chapter. Let H_x, H_y, H_z and H'_x, H'_y, H'_z be the values of the local fluctuating field at the positions of either spin so that the perturbing Hamiltonian $\hbar\mathcal{H}_1$ is

$$\mathcal{H}_1 = -\gamma\{H_x i_x + H_y i_y + H_z i_z + H'_x i'_x + H'_y i'_y + H'_z i'_z\}. \tag{39}$$

In the extreme narrowing approximation quantities such as, say, $1/T'_2(ab)$ and $1/\tau_a$ are given by

$$\frac{1}{T'_2(ab)} = \tau_c \overline{\{(a|\mathcal{H}_1|a) - (b|\mathcal{H}_1|b)\}^2}, \tag{40}$$

$$\frac{1}{\tau_a} = 2\tau_c\{\overline{|(a|\mathcal{H}_1|b)|^2} + \overline{|(a|\mathcal{H}_1|c)|^2} + \overline{|(a|\mathcal{H}_1|d)|^2}\}. \tag{41}$$

We make the usual assumption of complete isotropy so that squares such as $\overline{H_x^2}$ or $\overline{H'^2_x}$ are all equal ($\overline{H_x^2} = \overline{H'^2_x} = \frac{1}{3}\overline{H^2}$) and cross products such as $\overline{H_x H_y} = \overline{H'_x H'_y}$ vanish. A physical hypothesis remains to be *made* about the correlations $\overline{H_x H'_x}$ between the fields 'seen' by the spins i and i'. We shall make one of the following two extreme assumptions (a) or (b).

(a) The two fields are entirely uncorrelated (11) so that $\overline{H_x H'_x} = 0$. Such an assumption would not be too unreasonable if the distance of closest approach between either spin i or i' and say a paramagnetic ion, turned out to be appreciably smaller than the distance between i and i'. Then, using (40), (41), and (8), a very simple calculation yields the same width for the four transitions:

$$\frac{1}{T_2} = \gamma^2 \overline{H^2} \tau_c. \tag{42}$$

This is to be contrasted with the value $1/T_2 = \frac{2}{3}\gamma^2 \overline{H^2}\tau_c$ that is easily computed for a single spin $\frac{1}{2}$ under a similar relaxation mechanism. The four lines have a width that is $\frac{3}{2}$ times that obtained for a single spin.

(b) The two fields are completely correlated. A similar calculation gives

$$\frac{1}{T_2(ab)} = \frac{1}{T_2(bd)} = \gamma^2 \overline{H^2}\tau_c \left\{1 + \frac{2pq}{3}\right\},$$

$$\frac{1}{T_2(ac)} = \frac{1}{T_2(cd)} = \gamma^2 \overline{H^2}\tau_c \left\{1 - \frac{2pq}{3}\right\}. \quad (43)$$

The two outer (and weaker lines) are narrower than the two inner ones in the ratio $(1-\frac{1}{3}|\sin\phi|)/(1+\frac{1}{3}|\sin\phi|)$, where $\tan\phi = J/\delta$.

A measurement of the line widths in a sample sufficiently doped with paramagnetic ions for the external relaxation to be the main source of broadening, would enable one to discriminate between the assumptions (a) and (b).

The line widths of the four transitions can also be computed assuming a relaxation through dipolar coupling between the spins i and i'.

In comparing the line widths thus calculated to that in the absence of J coupling, it should be specified whether a frequency shift δ is assumed to exist or not. Indeed, the line width of two identical spins $\frac{1}{2}$ due to their dipolar coupling, is given under the extreme narrowing assumption by

$$\frac{1}{T_2} = \frac{1}{T_1} = \frac{3}{2}\frac{\gamma^4 \hbar^2}{b^6}\tau_c, \quad (44)$$

where b is the distance between the spins. On the other hand, if the two spins have sufficiently different Larmor frequencies for their lines to be well separated, their line width is given by the theory of the coupling between unlike spins and, as shown in Section II E (b) of Chapter VIII, is two-thirds of that for like spins and is thus equal to

$$\frac{1}{T_2} = \frac{\gamma^4 \hbar^2}{b^6}\tau_c. \quad (44')$$

The dipolar Hamiltonian is given by the formulae (67) to (69) of Chapter VIII and, calculating the line width by the formula (32) and making use of formulae such as (40) and (41), it is readily found that

$$\frac{1}{T_2(ab)} = \frac{1}{T_2(bd)} = \frac{17}{20}\frac{\gamma^4 \hbar^2}{b^6}\tau_c\{1 + \tfrac{7}{17}\sin\phi\},$$

$$\frac{1}{T_2(ac)} = \frac{1}{T_2(cd)} = \frac{17}{20}\frac{\gamma^4 \hbar^2}{b^6}\tau_c\{1 - \tfrac{7}{17}\sin\phi\}. \quad (45)$$

The ratio $\dfrac{1-\frac{7}{17}\sin\phi}{1+\frac{7}{17}\sin\phi}$ of the widths of the weaker to the stronger lines is not very different from that resulting from an external coherent relaxation, but the absolute values of the widths are smaller, not larger, than for two identical spins.

More complicated situations could be described using a treatment similar to that outlined above.

REFERENCES

1. H. S. GUTOWSKY, D. W. MCCALL, and C. P. SLICHTER, *J. Chem. Phys.* **21**, 279, 1953.
2. T. F. WIMETT, *Phys. Rev.* **91**, 499, 1953.
3. W. A. ANDERSON, ibid. **102**, 151, 1956.
4. H. M. MCCONNELL, C. A. REILLY, and A. D. MCLEAN, *J. Chem. Phys.* **24**, 479, 1955.
5. E. BRIGHT WILSON Jr., ibid. **27**, 60, 1957.
6. H. Y. CARR and E. M. PURCELL, *Phys. Rev.* **88**, 415, 1952.
7. D. F. ELLIOTT and R. T. SCHUMACHER, *J. Chem. Phys.* **26**, 1350, 1957.
8. E. L. HAHN and D. E. MAXWELL, *Phys. Rev.* **88**, 1070, 1952.
9. J. A. POPLE, *Molecular Physics*, **1**, 168, 1958.
10. J. T. ARNOLD, *Phys. Rev.* **102**, 136, 1956.
11. F. BLOCH, ibid., p. 104, 1956.

XII
THE EFFECTS OF STRONG RADIO-FREQUENCY FIELDS

The Power and the Glory

THROUGHOUT this book it has been made abundantly clear that the best, if not the only, means for the investigation of nuclear magnetism was the use of r.f. fields with frequencies in the neighbourhood of the Larmor frequency of the spins or more generally in the neighbourhood of a resonance frequency $\nu = \Delta E/h$ between two levels of the spin system.

It was shown in Chapter III that the phenomenological equations of Bloch could provide a description of the transient and the steady-state behaviour of a system of spins submitted to a rotating field of arbitrary strength.

This description was stated to be quantitatively correct in liquid samples. In solids, however, it is at best qualitatively correct and can even lead to completely wrong predictions. No justification from first principles of the validity of these equations was given. Elsewhere in this book the r.f. fields were generally assumed to be either sufficiently small as to exert a negligible perturbation on the state of the spin system, or on the contrary as in the pulse methods, so strong and applied during intervals so brief as to enable one to neglect during these intervals the effects of spin-spin and spin-lattice interactions and to justify the approximation of free spins, a problem dealt with in Chapter II.

We were also led to assume, in particular in the discussion of dynamic polarization experiments, that it was possible to reduce appreciably the polarization $\langle S_z \rangle$ of a spin S by applying a strong r.f. field at the Larmor frequency ω_S of that spin, or, more generally, to equalize the populations of two levels of a spin system.

We propose now to examine in a more fundamental way the behaviour of spin systems submitted to strong r.f. fields.

I. STRONG RADIO-FREQUENCY FIELDS IN LIQUIDS

A. 'Non-viscous liquids'

For a student of nuclear magnetism a sample can be said to be liquid if the internal motions average out the various spin-spin interactions described loosely as the local fields and if the correlation time τ_c associated with these motions is sufficiently short; since $\hbar\mathcal{H}_1$ is the Hamiltonian responsible for these interactions, $|\mathcal{H}_1|\tau_c$ must be a small number. Thus, some metals where diffusion narrows the line width appreciably might be called 'liquids' well below the melting-point. It was shown in Chapter VIII that in a 'liquid' a linear master equation could be written for the rate of change of the density matrix σ describing the statistical behaviour of the spin system. Using the master equation it was possible to show from first principles that for many relaxation mechanisms, such as for instance a dipolar coupling between like spins or a fluctuating quadrupole coupling (with the extra assumption of extreme narrowing), the macroscopic nuclear magnetization of the system did in the absence of applied r.f. fields obey Bloch equations. We propose to investigate now under what conditions spin systems which obey Bloch equations in the absence of r.f. fields still obey them when the r.f. field is present.

It was shown in Chapter VIII that the assumption of a high lattice temperature and a short lattice correlation time τ_c led to the following master equation in the interaction representation:

$$\frac{d\sigma^*}{dt} = -\int_0^t \overline{\left[\mathcal{H}_1^*(t), \left[\mathcal{H}_1^*(t-\tau), \sigma^* - \frac{1}{A}\frac{\hbar\mathcal{F}}{kT}\right]\right]} d\tau$$

$$\cong -\frac{1}{2}\int_{-\infty}^{\infty} \overline{\left[\mathcal{H}_1^*(t), \left[\mathcal{H}_1^*(t-\tau), \sigma^* - \frac{1}{A}\frac{\hbar\mathcal{F}}{kT}\right]\right]} d\tau, \qquad (1)$$

where
$$\mathcal{H}_1^*(t) = e^{i\mathcal{H}_0 t}e^{i\mathcal{F}t}\mathcal{H}_1 e^{-i\mathcal{F}t}e^{-i\mathcal{H}_0 t} = e^{i\mathcal{H}_0 t}\mathcal{H}_1(t)e^{-i\mathcal{H}_0 t},$$
$$\sigma^*(t) = e^{i\mathcal{H}_0 t}\sigma e^{-i\mathcal{H}_0 t},\qquad (1')$$

and the bar means trace taken over the degrees of freedom of the lattice.

A is the number of degrees of freedom of the spin system; $\hbar\mathcal{H}_0$, $\hbar\mathcal{F}$, and $\hbar\mathcal{H}_1$ are respectively the Hamiltonian for the spin system, the lattice, and their mutual coupling.

From the relation

$$\int_{-\infty}^{\infty} [\mathcal{H}_1^*(t-\tau), \mathcal{H}_0 + \mathcal{F}] d\tau = i\int_{-\infty}^{\infty} \frac{d}{d\tau}[\mathcal{H}_1^*(t-\tau)] d\tau = 0, \qquad (2)$$

which is a consequence of the definition (1'), we deduced in Chapter VIII

$$\frac{d\sigma^*}{dt} = -\tfrac{1}{2} \int_{-\infty}^{\infty} \overline{[\mathcal{H}_1^*(t), [\mathcal{H}_1^*(t-\tau), \sigma^* - \sigma_0]]} \, d\tau, \qquad (3)$$

where $\sigma_0 \simeq \dfrac{1}{A}\left\{1 - \dfrac{\hbar\mathcal{H}_0}{kT}\right\} \simeq e^{-\hbar\mathcal{H}_0/kT}/\mathrm{tr}\{e^{-\hbar\mathcal{H}_0/kT}\}$

is the Boltzmann density matrix for thermal equilibrium.

The equation (3) can be rewritten

$$\frac{d\sigma^*}{dt} = -\tfrac{1}{2} e^{i\mathcal{H}_0 t}\left\{ \int_{-\infty}^{\infty} \overline{[\mathcal{H}_1(t), [e^{-i\mathcal{H}_0 \tau}\mathcal{H}_1(t-\tau)e^{i\mathcal{H}_0 \tau}, \sigma - \sigma_0]]} \, d\tau \right\} e^{-i\mathcal{H}_0 t}, \qquad (4)$$

and the equation for the density matrix, $\sigma = e^{-i\mathcal{H}_0 t}\sigma^* e^{i\mathcal{H}_0 t}$, follows:

$$\begin{aligned}
\frac{d\sigma}{dt} &= -i[\mathcal{H}_0, \sigma] + e^{-i\mathcal{H}_0 t}\frac{d\sigma^*}{dt}e^{i\mathcal{H}_0 t} \\
&= -i[\mathcal{H}_0, \sigma] - \tfrac{1}{2} \int_{-\infty}^{\infty} \overline{[\mathcal{H}_1(t), [e^{-i\mathcal{H}_0 \tau}\mathcal{H}_1(t-\tau)e^{i\mathcal{H}_0 \tau}, \sigma - \sigma_0]]} \, d\tau \\
&= -i[\mathcal{H}_0, \sigma] + f(\sigma - \sigma_0). \qquad (5)
\end{aligned}$$

The equation (5) expresses two facts. First, for an infinite lattice temperature a master equation

$$\frac{d\sigma}{dt} = -i[\mathcal{H}_0, \sigma] + f(\sigma) \qquad (5')$$

is obtained, from which relaxation times can be calculated as shown in Chapter VIII. Secondly, the effect of a high but finite lattice temperature can be taken into account simply by replacing $f(\sigma)$ by $f(\sigma - \sigma_0)$ in (5'), where σ_0 is the Boltzmann density matrix for thermal equilibrium.

To demonstrate the relations (1) to (5), the unnecessarily broad assumption

$$\exp(-\hbar\mathcal{F}/kT) \simeq 1 - (\hbar\mathcal{F}/kT) \qquad (a)$$

was made. We saw in Chapter VIII, Section II D, that the less stringent assumption

$$\exp(-\hbar\mathcal{H}_0/kT) \simeq 1 - (\hbar\mathcal{H}_0/kT) \qquad (b)$$

was actually sufficient. In the following we keep for simplicity the expansion (a). The alternative, and more elaborate procedure used to demonstrate (3) under the more restrictive assumption (b), could also be applied in this chapter.

These results have now to be adapted to a situation where the spin Hamiltonian contains the time explicitly:

$$\hbar E(t) = \hbar\mathcal{H}_0 + \hbar E_1(t). \qquad (6)$$

In that case it is easy to show that the equation (1) is still valid with the following definitions:

$$\sigma^* = U(t)\sigma(t)U^{-1}(t),$$
$$\mathcal{H}_1^*(t) = U(t)e^{i\mathcal{F}t}\mathcal{H}_1 e^{-i\mathcal{F}t}U^{-1}(t) = U(t)\mathcal{H}_1(t)U^{-1}(t), \quad (7)$$

where the unitary operator $U(t)$ is a solution of the differential equation

$$\frac{1}{i}\frac{dU}{dt} = UE(t) \quad \text{with } U(0) = 1. \quad (8)$$

More generally, we define $U(t, t_0)$ as the solution of (8) such that

$$U(t_0, t_0) = 1, \quad \text{and thus} \quad U(t) = U(t, 0).$$

It follows from (8) that

$$U(t-\tau) = U(t) \cdot U(t-\tau, t). \quad (8')$$

If the time-dependent part $E_1(t)$ of the spin Hamiltonian vanishes, $U(t)$ is clearly $e^{i\mathcal{H}_0 t}$ and $U(t-\tau, t)$ is $e^{-i\mathcal{H}_0 \tau}$. Using (8'), the equation (1) can be rewritten in a form similar to (4):

$$\frac{d\sigma^*}{dt} = -U(t) \times$$

$$\times \left\{ \frac{1}{2}\int_{-\infty}^{\infty} \overline{\left[\mathcal{H}_1(t), \left[U(t-\tau,t)\mathcal{H}_1(t-\tau)U^{-1}(t-\tau,t), \sigma - \frac{1}{A}\frac{\hbar\mathcal{F}}{kT}\right]\right]} d\tau \right\} \times$$
$$\times U^{-1}(t), \quad (9)$$

and the corresponding equation for σ becomes

$$\frac{d\sigma}{dt} = -i[E(t), \sigma] -$$

$$-\frac{1}{2}\int_{-\infty}^{\infty} \overline{\left[\mathcal{H}_1(t), \left[U(t-\tau,t)\mathcal{H}_1(t-\tau)U^{-1}(t-\tau,t), \sigma - \frac{1}{A}\frac{\hbar\mathcal{F}}{kT}\right]\right]} d\tau.$$
$$(10)$$

When the spin Hamiltonian contains the time explicitly, two questions arise. First, for an infinite lattice temperature, under what conditions are the relaxation terms of the master equation unaffected by the addition to the static part $\hbar\mathcal{H}_0$ of (6), of a time-dependent part $\hbar E_1(t)$ (for instance are the relaxation times T_1 and T_2 affected by the presence of an r.f. field)? Secondly, is it possible to account for a finite lattice temperature by replacing σ in the relaxation terms by $\sigma - \sigma_0(t)$ where

$$\sigma_0(t) = \frac{e^{-\hbar E(t)/kT}}{\text{tr}\{e^{-\hbar E(t)/kT}\}} \quad (11)$$

is an instantaneous Boltzmann density matrix?

We shall answer the second question first: the term of equation (10)

$$\frac{1}{2A}\frac{\hbar}{kT}\int_{-\infty}^{\infty}\overline{[\mathscr{H}_1(t),[U(t-\tau,t)\mathscr{H}_1(t-\tau)U^{-1}(t-\tau,t),\mathscr{F}]]}\,d\tau$$

is equal to

$$-\frac{1}{2A}\frac{\hbar}{kT}\int_{-\infty}^{\infty}\overline{[\mathscr{H}_1(t),[U(t-\tau,t)\mathscr{H}_1(t-\tau)U^{-1}(t-\tau,t),\tilde{E}(t-\tau,t)]]}\,d\tau,$$

where $\qquad \tilde{E}(t-\tau,t) = U(t-\tau,t)E(t-\tau)U^{-1}(t-\tau,t).$ (12)

The result (12) is a consequence of the relation

$$\int_{-\infty}^{\infty}[U(t-\tau,t)\mathscr{H}_1(t-\tau)U^{-1}(t-\tau),\mathscr{F}+\tilde{E}(t-\tau,t)]\,d\tau$$

$$= i\int_{-\infty}^{\infty}\frac{d}{d\tau}[U(t-\tau,t)\mathscr{H}_1(t-\tau)U^{-1}(t-\tau,t)]\,d\tau = 0, \quad (13)$$

which follows from the relations (6), (7), (8), (8').

The master equation can then be rewritten as

$$\frac{d\sigma}{dt} = -i[E(t),\sigma]-$$

$$-\frac{1}{2}\int_{-\infty}^{\infty}\overline{\left[\mathscr{H}_1(t),\left[U(t-\tau,t)\mathscr{H}_1(t-\tau)U^{-1}(t-\tau,t),\sigma+\frac{1}{A}\frac{\hbar}{kT}\tilde{E}(t-\tau,t)\right]\right]}\,d\tau.$$

(13')

In this section which deals with 'non-viscous' liquids we assume that the correlation time τ_c is so short that the product $|E_1(t)|\tau_c$, where $E_1(t)$ is the time-dependent part of $E(t)$, is very small.

The constant part $\mathscr{H}_0$ of $E(t)$ may be much larger than $E_1(t)$ and the condition $|\mathscr{H}_0|\tau_c \ll 1$ may or may not be obeyed. If it is obeyed we have extreme narrowing.

In the integral (13') only the values of $|\tau| \lesssim \tau_c$ contribute appreciably. In that interval we may write

$$U(t-\tau,t) = e^{-i\mathscr{H}_0\tau}+O(|E_1|\tau_c),$$

$$\tilde{E}(t-\tau,t) = \mathscr{H}_0+e^{-i\mathscr{H}_0\tau}E_1(t-\tau)e^{i\mathscr{H}_0\tau}+O(|E_1|\tau_c). \quad (14)$$

Let us assume first that $|\mathscr{H}_0| \gg |E_1(t)|$ (without necessarily assuming extreme narrowing).

We may with good accuracy in (13') replace $\tilde{E}(t-\tau,t)$ by $\mathscr{H}_0$ and

$-(1/A)(\hbar/kT)\tilde{E}(t-\tau,t)$ by $-(\hbar\mathcal{H}_0/AkT)$, or, within the commutator (13'), by

$$\frac{1}{A}\left\{1-\frac{\hbar\mathcal{H}_0}{kT}\right\} \simeq e^{-\hbar\mathcal{H}_0/kT}/\text{tr}\{e^{-\hbar\mathcal{H}_0/kT}\} = \sigma_0.$$

The relaxation of σ is towards σ_0.

On the other hand, let us assume extreme narrowing (without necessarily having $|\mathcal{H}_0| \gg |E_1(t)|$).

Then $|\mathcal{H}_0|\tau_c \ll 1$ and, according to (14),

$$\tilde{E}(t-\tau,t) \simeq \mathcal{H}_0 + E_1(t-\tau) = E(t-\tau).$$

The relative change of $|E(t-\tau)|$ in the interval $|\tau| \lesssim \tau_c$ is

$$\sim |1/E||dE/dt|\tau_c.$$

If $E(t)$ is a periodic function of frequency ω, $|1/E||dE/dt|\tau_c \sim \omega\tau_c$ and, since in practice $\omega \sim \mathcal{H}_0$, $\omega\tau_c$ is very small and we may replace in (13')

$$\tilde{E}(t-\tau,t) \simeq E(t-\tau) \quad \text{by} \quad E(t).$$

It follows immediately that σ relaxes towards the instantaneous Boltzmann matrix $\sigma_0(t)$ defined by (11).

Consider as an example the motion of spins in a d.c. field $H_0 = |\omega_0/\gamma|$ and a rotating field $\mathbf{H}_1(t)$ of frequency $\omega \simeq \omega_0$ and amplitude $H_1 = |\omega_1/\gamma|$. The assumption of low viscosity of the thermal bath is $|\omega_1\tau_c| \ll 1$. If the relaxation mechanisms are such that the Bloch equations are valid for an infinite temperature of the bath, for a finite temperature the nuclear magnetization $\mathbf{M}$ will relax as follows:

(i) if $H_0 \gg H_1$, towards $\chi_0 \mathbf{H}_0$ (indistinguishable from
$$\chi_0[\mathbf{H}_0 + \mathbf{H}_1(t)]);$$

(ii) if $|\omega\tau_c| \ll 1$ (extreme narrowing) towards $\chi_0(\mathbf{H}_0 + \mathbf{H}_1(t))$. This modification of the Bloch equations, significant in low d.c. fields, has already been described without proof in Chapter III.

The dependence of the relaxation terms on the strength of the r.f. field is also governed by the magnitude of the product $|E_1(t)\tau_c|$.

For non-viscous liquids this product is very small. Then for $|\tau| \lesssim \tau_c$,

$$U(t-\tau,t) = e^{-i\mathcal{H}_0\tau} + O(|E_1(t)|\tau_c)$$

and, since the integrand of equation (10) is negligible for $|\tau| \gg \tau_c$, it is permissible to write, neglecting the small quantity $O(|E_1(t)|\tau_c)$,

$$\int_{-\infty}^{\infty} \overline{[\mathcal{H}_1(t),[U(t-\tau,t)\mathcal{H}_1(t-\tau)U^{-1}(t-\tau,t),\sigma]]}\,d\tau$$

$$\simeq \int_{-\infty}^{\infty} \overline{[\mathcal{H}_1(t),[e^{-i\mathcal{H}_0\tau}\mathcal{H}_1(t-\tau)e^{i\mathcal{H}_0\tau},\sigma]]}\,d\tau,$$

from which the time-dependent part of the Hamiltonian $E_1(t)$ has disappeared.

The relaxation terms of the master equation are thus unaffected by the presence of the time-dependent operator $E_1(t)$. For instance, if the physical situation is such that the Bloch equations are valid in the absence of an r.f. field, they will still be valid with the same relaxation times in the presence of such a field of amplitude H_1, provided

$$|\gamma H_1|\tau_c \ll 1.$$

B. 'Viscous liquids'

If the condition $|E_1(t)|\tau_c \ll 1$ is not satisfied, the problem is a good deal more complicated (1).

As an example we consider in some detail the behaviour of a system of nuclear spins in a d.c. field $H_0 = -\omega_0/\gamma$ and a rotating field of frequency $\omega \cong \omega_0$ and of amplitude $H_1 = -\omega_1/\gamma$ with $|\omega_1| \ll \omega_0$.

We assume that the correlation time τ_c of the lattice in which the spins are embedded is sufficiently long for $|\omega_1|\tau_c$ not to be small (and consequently for $|\omega_0|\tau_c$ to be very large). At the same time in order to deal with a 'liquid' we must assume that $\delta\omega . \tau_c$, where $\delta\omega$ expresses the strength of the instantaneous local field in frequency units, is a small number. An r.f. field at least an order of magnitude larger than the local field is required by the double inequality

$$\delta\omega . \tau_c \ll 1 \lesssim \omega_1 \tau_c. \qquad (15)$$

For instance, in metallic sodium at 200° K, that is 170° below the melting-point, the nuclear line width is of the order of 0·25 gauss or one-tenth of its value of 2·5 gauss at 77° K. The quantity $\delta\omega . \tau_c$ which expresses the motion narrowing is $\simeq \frac{1}{10}$ and the substance is thus a 'liquid'. A rotating r.f. field of 25 gauss would be required to make $|\omega_1|\tau_c \sim 1$, and exhibit significant departures from the simple results obtained in Section A.

The spin Hamiltonian for this problem is

$$E(t) = \omega_0 I_z + \omega_1 \{I_x \cos\omega t + I_y \sin\omega t\} \qquad (16)$$

and the operator U defined by equation (8) can be written

$$U(t) = R(t)S(t),$$

where $\quad S(t) = e^{i\omega I_z t}, \quad R(t) = e^{i\{(\omega_0-\omega)I_z + \omega_1 I_x\}t} = e^{ia\{I_z\cos\theta + I_x\sin\theta\}t} \qquad (17)$

with $\qquad a = [(\omega-\omega_0)^2 + \omega_1^2]^{\frac{1}{2}}, \quad \tan\theta = \dfrac{\omega_1}{\omega_0-\omega},$

and $R(t)$ can be rewritten

$$R(t) = e^{i\theta I_y} e^{iaI_z t} e^{-i\theta I_y}. \qquad (18)$$

$S(t)$ is the unitary operator for going from the laboratory frame into the frame rotating with velocity ω around the d.c. field H_0, already introduced in Chapter II (where it was called U).

$R(t)$ is the operator for going from that frame into the frame rotating with velocity a around the effective field. For brevity we shall call these the rotating frame and the doubly rotating frame, respectively. Let $\mathscr{H}_1$ be the Hamiltonian describing the coupling of the spin system with the lattice. We shall assume for simplicity that the lattice is at an infinite temperature and use the semi-classical description of $\mathscr{H}_1$ in terms of random functions as explained in Chapter VIII.

Besides the two density matrices σ and σ^* that describe the motion of the spins in the laboratory frame and the doubly rotating frame, it is convenient to introduce the matrix $\tilde{\sigma}$ that describes the motion in the rotating frame and is connected with σ and σ^* through

$$\tilde{\sigma} = S\sigma S^{-1} = R^{-1}\sigma^* R. \tag{19}$$

From (17), (19), and the equation (1) for σ^*, the following equation is easily obtained, for $\tilde{\sigma}$:

$$\frac{d\tilde{\sigma}}{dt} = -i[(\omega_0-\omega)I_z+\omega_1 I_x, \tilde{\sigma}] - \tfrac{1}{2}\int_{-\infty}^{\infty} [\tilde{\mathscr{H}}_1(t), [R^{-1}(\tau)\tilde{\mathscr{H}}_1(t-\tau)R(\tau), \tilde{\sigma}]]\, d\tau, \tag{20}$$

where $$\tilde{\mathscr{H}}_1(t) = S\mathscr{H}_1 S^{-1} = R^{-1}\mathscr{H}_1^* R. \tag{21}$$

The motion with respect to the rotating frame, of the macroscopic magnetization $\gamma\hbar\langle\mathbf{I}\rangle = \gamma\hbar\,\mathrm{tr}\{\tilde{\sigma}\mathbf{I}\}$ is given by the equation

$$\frac{d\langle\mathbf{I}\rangle}{dt} = \gamma\{\langle\mathbf{I}\rangle\wedge\mathbf{H}_{\text{eff}}\} - \tfrac{1}{2}\int_{-\infty}^{\infty} \langle[R^{-1}(\tau)\tilde{\mathscr{H}}_1(t-\tau)R(\tau), [\tilde{\mathscr{H}}_1(t), \mathbf{I}]]\rangle\, d\tau, \tag{22}$$

where $\mathbf{H}_{\text{eff}}$ is the effective field in the rotating frame and a symbol such as $\langle Q\rangle$, means $\mathrm{tr}\langle Q\tilde{\sigma}\rangle$.

To proceed further we must make an assumption about the relaxation mechanism described by the random Hamiltonian.

For simplicity we shall outline the calculation for exact resonance ($\omega = \omega_0$), the relaxation mechanism being a random local magnetic field.

The relaxation Hamiltonian $\hbar\mathscr{H}_1(t)$ is given by

$$\mathscr{H}_1(t) = -\gamma\mathbf{I}\cdot\mathbf{H}(t) = A_0 I_z + \tfrac{1}{2}A_+ I_- + \tfrac{1}{2}A_- I_+,$$

where $$A_0 = -\gamma H_z(t), \quad A_{\pm} = -\gamma\{H_x \pm iH_y\},$$

$$\overline{|A_0|^2} = \tfrac{1}{2}\overline{|A_+|^2} = \tfrac{1}{2}\overline{|A_-|^2} = \tfrac{1}{3}\gamma^2\overline{|\mathbf{H}(t)|^2}. \tag{23}$$

We assume as usual that the only non-vanishing correlation functions of the random functions A are the following:

$$\overline{A_0(t).A_0(t-\tau)} = \overline{|A_0|^2}\exp(-\tau/\tau_c),$$

$$\overline{A_+(t)A_-(t-\tau)} = \overline{|A_+|^2}e^{-\tau/\tau_c}. \qquad (23')$$

At resonance $R(t) = e^{i\omega_1 I_z t}$ and, using the definitions (17), (21), and (23), it is easily found that

$$\tilde{\mathscr{H}}_1(t) = A_0 I_z + \tfrac{1}{2}A_+ I_- e^{-i\omega t} + \tfrac{1}{2}A_- I_+ e^{i\omega t} = -\tfrac{1}{2}iA_0\{I'_+ - I'_-\} +$$
$$+ \tfrac{1}{2}A_+\{I_x - \tfrac{1}{2}i(I'_+ + I'_-)\}e^{-i\omega t} + \tfrac{1}{2}A_-\{I_x + \tfrac{1}{2}i(I'_+ + I'_-)\}e^{i\omega t}, \quad (24)$$

where $I'_\pm = I_y \pm i I_z$.

From (24) we get

$$R^{-1}(\tau)\tilde{\mathscr{H}}_1(t-\tau)R(\tau) = e^{-i\omega_1 I_z \tau}\tilde{\mathscr{H}}_1(t-\tau)e^{i\omega_1 I_z \tau}$$

$$= \tfrac{1}{2}I_x\{A_+(t-\tau)e^{-i\omega(t-\tau)} + A_-(t-\tau)e^{i\omega(t-\tau)}\} +$$

$$+ \tfrac{1}{2}iI'_+ e^{-i\omega_1\tau}\left\{-A_0(t-\tau) - \frac{A_+(t-\tau)}{2}e^{-i\omega(t-\tau)} + \frac{A_-(t-\tau)}{2}e^{i\omega(t-\tau)}\right\} -$$

$$- \tfrac{1}{2}iI'_- e^{i\omega_1\tau}\left\{-A_0(t-\tau) + \frac{A_+(t-\tau)}{2}e^{-i\omega(t-\tau)} - \frac{A_-(t-\tau)}{2}e^{i\omega(t-\tau)}\right\}. \quad (25)$$

Similarly,

$$[\tilde{\mathscr{H}}_1(t), I_x] = iA_0 I_y + I_z\{-\tfrac{1}{2}A_+ e^{-i\omega t} + \tfrac{1}{2}A_- e^{i\omega t}\},$$
$$[\tilde{\mathscr{H}}_1(t), I_y] = -iA_0 I_x + iI_z\{\tfrac{1}{2}A_+ e^{-i\omega t} + \tfrac{1}{2}A_- e^{i\omega t}\},$$
$$[\tilde{\mathscr{H}}_1(t), I_z] = \tfrac{1}{2}A_+ I_- e^{-i\omega t} - \tfrac{1}{2}A_- I_+ e^{i\omega t}. \quad (26)$$

We calculate according to (22) the commutators of (25) with the three expressions (26), and then take the averages of the resulting expressions over the ensemble of the random functions A, thus introducing their correlation functions given by (23').

Computing the integral (22), we find after a little algebra the following equation for the motion of the magnetization in the rotating frame:

$$\frac{d\langle \mathbf{I} \rangle}{dt} = \gamma[\langle \mathbf{I} \rangle \wedge \mathbf{H}_{\text{eff}}] - \frac{\mathbf{i}\langle I_x \rangle}{T_x} - \frac{\mathbf{j}\langle I_y \rangle}{T_y} - \frac{\mathbf{k}\langle I_z \rangle}{T_z}, \quad (27)$$

where $\mathbf{k}$, $\mathbf{i}$, $\mathbf{j}$ are unit vectors of the rotating frame along $\mathbf{H}_0$, $\mathbf{H}_1$, and $\mathbf{H}_0 \wedge \mathbf{H}_1$, respectively.

For T_x, T_y, T_z, the following values are found:

$$\frac{1}{T_x} = \frac{\gamma^2 \overline{|\mathbf{H}|^2} \tau_c}{6} \left\{ \frac{2}{1+\omega_1^2 \tau_c^2} + \frac{1}{1+(\omega+\omega_1)^2 \tau_c^2} + \frac{1}{1+(\omega-\omega_1)^2 \tau_c^2} \right\},$$

$$\frac{1}{T_y} = \frac{\gamma^2 \overline{|\mathbf{H}|^2} \tau_c}{6} \left\{ \frac{2}{1+\omega_1^2 \tau_c^2} + \frac{2}{1+\omega^2 \tau_c^2} \right\},$$

$$\frac{1}{T_z} = \frac{\gamma^2 \overline{|\mathbf{H}|^2} \tau_c}{6} \left\{ \frac{2}{1+\omega^2 \tau_c^2} + \frac{1}{1+(\omega+\omega_1)^2 \tau_c^2} + \frac{1}{1+(\omega-\omega_1)^2 \tau_c^2} \right\}. \quad (28)$$

Since ω_1/ω is assumed to be much smaller than unity, keeping in (28) the terms of lowest order in ω_1/ω, we get

$$\frac{1}{T_1} = \frac{1}{T_z} = \tfrac{2}{3} \gamma^2 \overline{|\mathbf{H}|^2} \frac{\tau_c}{1+\omega^2 \tau_c^2},$$

$$\frac{1}{T_2} = \frac{1}{T_x} = \frac{1}{T_y} = \frac{\gamma^2 \overline{|\mathbf{H}|^2}}{3} \frac{\tau_c}{1+\omega_1^2 \tau_c^2}. \quad (29)$$

The fact that $T_1 \gg T_2$ if $\omega \tau_c \gg 1$ has already been pointed out in Chapter VIII. On the other hand, a new and remarkable feature of the equations (29) is the dependence of the transverse relaxation time T_2 on the strength of the r.f. field. As a consequence, both the steady-state and the transient behaviour of the magnetization should be very different from that predicted by the Bloch equations of Chapter III. In our approximation of infinite lattice temperature we cannot calculate the steady-state magnetization. This calculation can be found, for instance, in (1). One should remark, however, that if the experimental conditions are such that $\omega_1 \tau_c$ is not small, $\omega \tau_c$, and thus also $T_z = T_1$, is very large and the coupling with the lattice very weak. In practice there are likely to be other relaxation mechanisms with shorter correlation times that will short circuit the $1/T_1$ of formula (29) (but not the $1/T_2$).

Using the equation (22), a phenomenological equation for $\langle \mathbf{I} \rangle = \text{tr}\{\mathbf{I}\tilde{\sigma}\}$ can be easily established for $\omega \neq \omega_0$. Instead of $R(t) = e^{-i\omega_1 I_x t}$, the more general formula (18) must be used. The calculations although quite straightforward are somewhat lengthy and will not be reproduced for that reason.

A further generalization would be to replace the local field relaxation mechanism expressed by the random Hamiltonian (23) with a dipolar Hamiltonian expressed by the formulae (67) to (69) of Chapter VIII. The results thus obtained are not qualitatively very different from those above, the main feature still being the dependence of the transverse relaxation times on the strength of the r.f. field.

The corresponding calculations can be found in reference (2).

In view of the present scarcity of experimental evidence on situations where the basic assumptions (15) are actually verified we shall be content to mention the following details derived in (2).

(i) At resonance T_x and T_y have a somewhat different dependence on ω_1 since
$$\frac{1}{T_x} \propto \frac{2}{1+(2\omega_1 \tau_c)^2},$$
$$\frac{1}{T_y} \propto 1 + \frac{1}{1+(2\omega_1 \tau_c)^2}. \tag{30}$$

(ii) Off resonance, for $\omega_1 \tau_c$ very large, T_x and T_y have pronounced maxima for $3\cos^2\theta - 1 = 0$ where
$$\tan\theta = \frac{\omega_1}{\omega_0 - \omega}.$$

To close the theoretical discussion of strong r.f. fields in 'viscous liquids' ($\omega_1 \tau_c \gtrsim 1$), the following remark can be made.

It has been known for some time (3) that in solids the saturation behaviour of the dispersion is very different from that predicted by the Bloch equations and this departure can be described formally by a dependence of T_2' on ω_1 as given by equations such as (29) or (30). It must be pointed out, however, that in real solids where the motion narrowing is negligible, no master equation can be written for the density matrix, and the free decay of the transverse magnetization is by no means exponential. The formalism outlined here does not apply to real solids and a different explanation of their behaviour must be sought.

Finally, it should be stressed that the continuous-wave method is ill adapted to the investigation of situations where inequalities such as (15) exist.

The relationship between the observed dispersion signal and the transverse relaxation time (or times) is far from direct and the interpretation of the results is difficult. In particular, the value of the steady-state signal is critically dependent on the value of T_1 which may be determined by causes, such as, for instance, paramagnetic impurities, quite alien to the mechanism investigated. It is much preferable then to use the method of forced transient precession (4) where the time-dependence of the magnetization in the presence of a suddenly introduced strong r.f. field is observed, and where the transverse relaxation times can be obtained much more directly.

Interesting examples of situations where weak local fields and long correlation times make it possible for the inequalities (15) to be satisfied and for the dependence of T_2 on ω_1 to be exhibited are provided by chemical shifts δ or spin-spin interactions $\hbar A\mathbf{I}\cdot\mathbf{S}$ modulated by chemical exchange, or by the fast relaxation of the spin S, as discussed in Chapter VIII, Section II F, and also in Chapter XI in connexion with the disappearance of fine structures.

The spin $\mathbf{I}$ 'sees' a fluctuating local field produced by the spin $\mathbf{S}$ and the contribution of this local field to the transverse relaxation rate of the spin $\mathbf{I}$ driven by an r.f. field $H_1 = |\omega_1/\gamma_I|$ at its Larmor frequency $\omega = -\gamma_I H_0$ is given by

$$\frac{1}{T_2} \simeq \tfrac{1}{3}A^2 S(S+1)\frac{\tau}{1+\omega_1^2\tau^2}, \tag{30'}$$

where τ is the relaxation time of the spin S. As the strength of the r.f. field is increased T_2 becomes longer and the decay of the transverse magnetization of the spin I becomes slower. This effect has been demonstrated in formamide

$$\text{H---C---NH}_2$$
$$\|$$
$$\text{O}$$

where the protons 'see' the fluctuating field of nitrogen N^{14} (5). The relaxation time of N^{14} is measured from its line width to be $\tau \simeq 0.9$ millisec.

Disregarding the complications caused by the fact that the various protons of the molecule have actually different chemical shifts and different couplings with the nitrogen spin, an average transverse relaxation time $T_2 \simeq 35$ millisec has been observed by standard spin echo methods.

This time becomes 80 millisec for $H_1 = 50$ milligauss, 0.5 sec for $H_1 = 200$ milligauss, and 3 sec for $H_1 = 600$ milligauss (see Fig. XII, 1), exhibiting a dependence of T_2 on H_1 in good agreement with the relation (30').

C. Bloch equations for a 'simple' line

Consider a spin system with a Hamiltonian $\hbar\mathcal{H}_0$, energy levels $\hbar\alpha$, $\hbar\beta$,..., and energy states $|\alpha)$, $|\beta)$,..., and assume that there exist among those, two states $|a)$ and $|b)$ with an energy difference

$$\hbar\omega_{ab} = \hbar(a-b) = \hbar\omega_0,$$

different from any other energy difference $\hbar(\alpha-\beta)$. Such a couple of

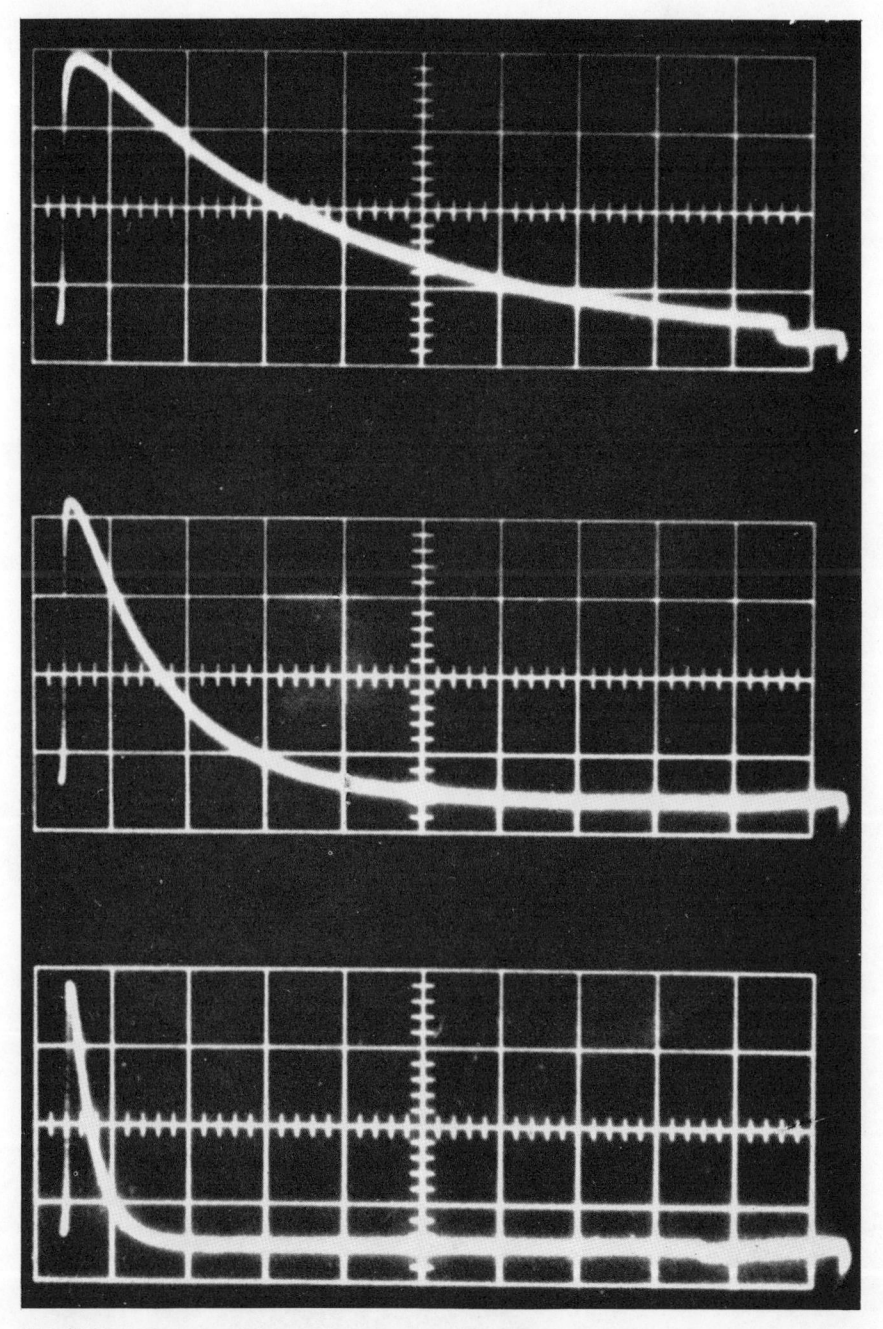

FIG. XII, 1. Decay of the nuclear magnetization of the protons of formamide, along the rotating r.f. field H_1 at exact resonance.

    Bottom $H_1 = 50$ milligauss
    Middle $H_1 = 200$ milligauss
    Top $H_1 = 600$ milligauss

states will be called for brevity a 'simple' line. Problems relative to simple lines have already been considered in this book.

Thus in Chapter II we considered a spin system with a simple line, free from interactions with other spin systems or with the lattice and submitted to a sinusoidal perturbation $\hbar E_1(t)$ of frequency ω in the neighbourhood of ω_0. A description of the behaviour of the system by a two-by-two density matrix $\sigma = \begin{pmatrix} \sigma_{aa} & \sigma_{ab} \\ \sigma_{ba} & \sigma_{bb} \end{pmatrix}$ was proposed, that disregarded all the other energy levels of the system.

Writing σ as $\tfrac{1}{2}+\mathbf{m}.\mathbf{s}$, where $\mathbf{s}$ was a fictitious spin $\tfrac{1}{2}$ and $\mathbf{m}$ a c-vector, an equation of the form

$$\frac{d\mathbf{m}}{dt} = \mathbf{m} \wedge \gamma \mathbf{H} \tag{31}$$

was written, where the components of the c-vector $\gamma\mathbf{H}$ could be calculated from the matrix elements of the Hamiltonian $\mathscr{H}_0 + E_1(t)$ of the spin system. Several examples of the usefulness of this formalism were given.

No such simple statements can in general be made for spin systems with a simple line, embedded in a 'liquid' and for which a master equation with relaxation terms has been established. The problem of the free motion or free decay of such systems was considered in Chapters VIII and IX. If the system has been removed from thermal equilibrium by, say, an r.f. pulse, its return towards it is described by the time-dependence of the density matrix elements $\sigma_{\alpha\beta}$ (or $\sigma^*_{\alpha\beta}$ in the interaction representation). It has been shown that the motion $\sigma^*_{ab}(t)$ of the off-diagonal matrix element for a single line was given by a single decreasing exponential

$$\sigma^*_{ab}(t) = \sigma^*_{ab}(0)e^{-t/T_2}, \tag{32}$$

where the constant T_2, given by formula (45) of Chapter X, is the inverse of the width $\delta\omega$ of the simple line observed with a vanishingly small r.f. field.

On the other hand, the time-dependence of the diagonal matrix elements σ_{aa} and σ_{bb} is more complex since, as was shown in Chapter VIII, their motion is coupled with that of all the other diagonal matrix elements $\sigma_{\alpha\alpha}$. No single decay constant T_1 analogous to T_2 can be defined in general for these elements, and their decay is described by a sum of several exponentials.

We consider now (6) the motion of a spin system with a simple line in the presence of an r.f. perturbation $\hbar E_1(t)$ with a frequency ω in the neighbourhood of $\omega_{ab} = \omega_0$. Both $\omega - \omega_0$ and the strength of the perturbation (measured in frequency units) are supposed to be small in comparison with all the differences $\omega_0 - \omega_{\alpha\beta}$.

We assume that the correlation time τ_c is sufficiently short to have $|E_1(t)|\tau_c \ll 1$, so that the master equation for the density matrix of the spin system can be written

$$\frac{d\sigma}{dt} = -i[\mathcal{H}_0 + E_1(t), \sigma] - \Gamma(\sigma - \sigma_0), \tag{33}$$

where
$$\sigma_0 = \exp(-\hbar\mathcal{H}_0/kT)/\mathrm{tr}\{\exp(-\hbar\mathcal{H}_0/kT)\}$$
$$\cong \frac{1}{A}\left[1 - \frac{\hbar\mathcal{H}_0}{kT}\right] = \frac{1}{A} - q\mathcal{H}_0, \tag{33'}$$

where A is the number of degrees of freedom of the spin system and $q = \hbar/kTA$. The relaxation term $-\Gamma(\sigma - \sigma_0)$ on the right-hand side of (33) is given by the formula (35) of Chapter VIII:

$$\{-\Gamma(\sigma - \sigma_0)\}_{\alpha\alpha'} = \sum_{\beta,\beta'} R_{\alpha\alpha',\beta\beta'}(\sigma_{\beta\beta'} - \sigma_{0,\beta\beta'}). \tag{33''}$$

In particular, for the simple line $|a\rangle \to |b\rangle$,

$$\{-\Gamma(\sigma - \sigma_0)\}_{ab} = \frac{-1}{(T_2)_{ab}}\sigma_{ab}. \tag{34}$$

For the diagonal terms $\sigma_{\alpha\alpha}$:

$$\{-\Gamma(\sigma - \sigma_0)\}_{\alpha\alpha} = -\sum_{\beta} W_{\alpha\beta}\{(\sigma_{\alpha\alpha} - \sigma_{0,\alpha\alpha}) - (\sigma_{\beta\beta} - \sigma_{0,\beta\beta})\}, \tag{34'}$$

where the $W_{\alpha\beta}$ are the transition probabilities per unit time, with $W_{\alpha\beta} = W_{\beta\alpha}$ in the high-temperature approximation, and the equation (34') is a transcription of equation (26'') of Chapter VIII.

The time-dependent perturbation $E_1(t)$ of frequency ω, being of necessity a hermitian operator, can be written

$$E_1(t) = De^{i\omega t} + D^+e^{-i\omega t}, \tag{35}$$

and we define
$$(b|D|a) = (a|D^+|b)^* = d. \tag{35'}$$

We introduce a reduced density matrix $\chi = \sigma - \sigma_0$ which satisfies the equation

$$\frac{d\chi}{dt} + i[\mathcal{H}_0, \chi] + i[D, \chi]e^{i\omega t} + i[D^+, \chi]e^{-i\omega t} + \Gamma(\chi)$$
$$= iq[D, \mathcal{H}_0]e^{i\omega t} + iq[D^+, \mathcal{H}_0]e^{-i\omega t}. \tag{36}$$

We look for a steady-state solution of (36), with constant diagonal elements $(\alpha|\chi|\alpha) = \chi_a$, and off-diagonal matrix elements $(\alpha|\chi|\beta)$ varying with time as $e^{\pm i\omega t}$. An inspection of this equation shows that if ω is in the neighbourhood of $\omega_0 = a-b$, the only off-diagonal matrix elements that are not very small are

$$(b\,|\,\chi\,|\,a) = ze^{i\omega t} \quad \text{and} \quad (a\,|\,\chi\,|\,b) = z^*e^{-i\omega t}.$$

Taking the matrix element $(b|\quad|a)$ of both sides of the operator equation (36) we find the following equation for z, using (34) and (35'),

$$\left\{(\omega-\omega_0) - \frac{i}{T_2}\right\}z + d(\chi_a - \chi_b) = q\omega_0 d, \tag{37}$$

where for brevity we drop the index ab of $(T_2)_{ab}$.

We assume first that the perturbing Hamiltonian $\hbar E_1(t)$, and thus also the quantity d, are vanishingly small. It is then permissible to consider that the populations $(a\,|\,\sigma\,|\,a)$ and $(b\,|\,\sigma\,|\,b)$ have their thermal equilibrium values, and that χ_a and χ_b, departures from these values, vanish. (37) yields

$$z = \frac{q\omega_0 d}{(\omega-\omega_0)-(i/T_2)} = \frac{qd(\omega_0 T_2)[(\omega-\omega_0)T_2+i]}{1+(\omega-\omega_0)^2 T_2^2}. \tag{37'}$$

If d is not vanishingly small the other matrix elements of σ must be calculated in order to obtain z. Taking the matrix elements $(a|\quad|a)$ and $(b|\quad|b)$ of (36) we find, using (34'),

$$\sum_{\alpha \neq a} W_{a\alpha}(\chi_a - \chi_\alpha) = \frac{1}{i}(zd^* - z^*d) = 2\,\text{im}(zd^*),$$

$$\sum_{\alpha \neq b} W_{b\alpha}(\chi_b - \chi_\alpha) = -\frac{1}{i}(zd^* - z^*d) = -2\,\text{im}(zd^*). \tag{38}$$

Finally, taking diagonal matrix elements $(\alpha|\quad|\alpha)$ of (36), with $\alpha \neq a, b$, we obtain a system of $(A-2)$ equations

$$\sum_\beta' W_{\alpha\beta}(\chi_\alpha - \chi_\beta) = 0, \tag{39}$$

where $\beta \neq \alpha$, $\alpha \neq a, b$. A last equation

$$\sum_\alpha \chi_\alpha = 0 \quad \text{expresses that} \quad \text{tr}\{\chi\} = \text{tr}\{\sigma - \sigma_0\} = 0. \tag{40}$$

From the system (37) to (40) the various matrix elements of χ can be obtained. One way of doing it is to extract z from (37) and carry it over into (38) which then becomes

$$\sum_{\alpha \neq a} W_{a\alpha}(\chi_a - \chi_\alpha) + V\{(\chi_a + (p_a)_0) - (\chi_b + (p_b)_0)\} = 0,$$

$$\sum_{\alpha \neq b} W_{b\alpha}(\chi_b - \chi_\alpha) + V\{(\chi_b + (p_b)_0) - (\chi_a + (p_a)_0)\} = 0, \tag{38'}$$

where

$$V = \frac{2|d|^2 T_2}{1+T_2^2(\omega-\omega_0)^2} = \frac{2|(b\,|\,D\,|\,a)|^2 T_2}{1+T_2^2(\omega-\omega_0)^2} \tag{41}$$

is the familiar transition probability per unit time induced by the r.f. perturbation (35) of frequency ω, the frequency of the transition being ω_0, and the shape of the line Lorentzian with a width $\delta\omega = 1/T_2$.

$$(p_a)_0 - (p_b)_0 = -q\omega_0 = -\frac{\hbar(a-b)}{kTA} \simeq \frac{e^{-\hbar a/kT} - e^{-\hbar b/kT}}{\sum_\alpha e^{-\hbar\alpha/kT}}$$

is the difference between the thermal equilibrium populations of the levels $|a\rangle$ and $|b\rangle$. The equations (38'), (39), and (40) are those usually written to calculate the steady-state populations of a spin system submitted to an r.f. perturbation.

The present calculation shows that the usual treatment, which seems to disregard the coherence of the r.f. field since this field appears only through a transition probability, is actually correct. Such a justification has already been given in Chapter II for the very special model of relaxation by strong collisions.

The coherence of the r.f. perturbation manifests itself through the existence of the off-diagonal element $(b|\chi|a) = ze^{i\omega t}$. In order to calculate it we rewrite the system (38), (39), (40) as

$$\sum_{\alpha \neq a} W_{a\alpha}(\chi_a - \chi_\alpha) = 2I,$$

$$\sum_{\alpha \neq b} W_{b\alpha}(\chi_b - \chi_\alpha) = -2I,$$

$$\sum_\beta{}' W_{\alpha\beta}(\chi_\alpha - \chi_\beta) = 0 \quad (\beta \neq \alpha, \alpha \neq a, b),$$

$$\sum_\alpha \chi_\alpha = 0, \quad (42)$$

where $I = \text{im}(zd^*)$.

This inhomogeneous system of $A+1$ linear equations with A unknown quantities χ_α has a unique solution which can be written $\chi_\alpha = T_\alpha I$, where the constants T_α, functions of the $W_{\alpha\beta}$, are independent of the r.f. field. Carrying $\chi_a = T_a I$, $\chi_b = T_b I$ into the equation (37) (and its complex conjugate), two linear equations are obtained for z and z^* from which we find

$$z = qd(\omega_0 T_2)\frac{(\omega-\omega_0)T_2 + i}{1+[(\omega-\omega_0)T_2]^2 + |d|^2 T_2 T_{ab}} \quad (43)$$

with $T_{ab} = T_a - T_b$.

For $d \to 0$, (43) reduces to (37') as it should.

An inspection of the system (42) shows that the quantity T_{ab} is positive. Indeed, the third relation (42) shows that neither the largest nor the smallest of the numbers χ_α can be other than χ_a or χ_b. We will assume for the sake of argument that $I > 0$. It is clear then from the first relation (42) that χ_a is the largest of the χ_α and $T_a > T_b$. The same result obtains if we assume $I < 0$ and this proves our point. For the difference $\chi_a - \chi_b$ we get

$$\chi_a - \chi_b = q|d|^2 \frac{\omega_0 T_2 T_{ab}}{1+[(\omega-\omega_0)T_2]^2+|d|^2 T_2 T_{ab}}$$

$$= \{(p_b)_0 - (p_a)_0\} \frac{|d|^2 T_2 T_{ab}}{1+[(\omega-\omega_0)T_2]^2+|d|^2 T_2 T_{ab}}. \quad (44)$$

The results (43) and (44) bear a strong resemblance to the steady-state solutions of Bloch equations given by the equations (15) of Chapter III.

Consider a system of Bloch equations

$$\frac{d\mathbf{M}}{dt} = \mathbf{M} \wedge \gamma \mathbf{H} - \frac{M_x}{T_2}\mathbf{i} - \frac{M_y}{T_2}\mathbf{j} - \frac{M_z - M_0}{T_1}\mathbf{k}. \quad (45)$$

We can establish the following correspondence between their steady-state solutions, given by the equations (15) of Chapter III, and the results (43) and (44):

$$2M_0 = q\omega_0 = \frac{\hbar\omega_0}{AkT} = (p_b)_0 - (p_a)_0,$$

$$T_2 = (T_2)_{ab}, \qquad T_1 = \frac{T_{ab}}{4} = \frac{T_a - T_b}{4},$$

$$\gamma \mathbf{H} = -\omega_0 \mathbf{k} + 2|d|\{\mathbf{j}\sin\omega t + \mathbf{i}\cos\omega t\},$$

$$z = \tilde{M}_x + i\tilde{M}_y, \qquad \chi_a - \chi_b = 2(M_0 - M_z). \quad (46)$$

This correspondence becomes trivial if the two levels $|a\rangle$ and $|b\rangle$ are those of a single spin $\tfrac{1}{2}$ in a rotating field.

The steady state behaviour of a simple line (but not the transient behaviour in general) can thus be predicted from a set of equations of the Bloch type where T_2 is the width of the simple line and the part of the longitudinal relaxation time T_1 is played by the quantity $\tfrac{1}{4}T_{ab}$. This method can be used, for instance, to investigate the saturation properties, described by generalized longitudinal relaxation times, of the various components of a multiplet. The calculation of their transverse relaxation times was described in Chapter XI, Section III B.

D. Decoupling of spins through 'stirring' by a radio-frequency field

(a) *Introduction*

We saw in Chapter XI that the resonance spectrum of a spin **I** coupled to a spin **S** by a scalar coupling $\hbar J \mathbf{I}.\mathbf{S}$ was a single line rather than a multiplet, when, because of a short relaxation time (or chemical exchange) the orientation of the spin S was undergoing rapid changes at a rate appreciably faster than the frequency J. It was natural to expect a similar effect in an experiment where the resonance of the spins I would be observed by means of a weak r.f. field at the Larmor frequency ω_I while rapid transitions of the spins S were induced by a strong r.f. field at a frequency ω_S. (In the following the stronger r.f. field will be called the 'stirring' field.) Indeed, if the 'stirring' field of frequency ω_S is sufficiently intense the multiplet structure of the spins I collapses into a single line.

In spite of a certain analogy this phenomenon is very different from the collapse of multiplets because of relaxation or chemical exchange, since the coherent motion of the spins S caused by the stirring r.f. field is very different from the random motion caused by relaxation. If the stirring r.f. field is increased progressively from a very small value where the multiplet I is unaffected, to a large value where its collapse is complete, the intermediate patterns are quite different from those observed for instance when the rate of chemical exchange is progressively increased by means of a catalyst. Before describing the theory of r.f. 'stirring' in liquids a few remarks are in order.

(i) In a frame rotating at the frequency $\omega = \omega_S$ of the stirring field, the Hamiltonian $\hbar \mathscr{H}$ can be written approximately, neglecting the direct coupling of the stirring field with the spins I, as

$$\mathscr{H} = (\omega_I - \omega_S)I_z + \omega_1 S_x + J\mathbf{I}\cdot\mathbf{S}. \qquad (47)$$

If both $|(\omega_I-\omega_S)|$ and $|\omega_1| \gg J$, the last term of (47) can be treated as a small perturbation. For the first two terms, $I_z = m$ and $S_x = m'$ are good quantum numbers and since $J\mathbf{I}\cdot\mathbf{S}$ then has no diagonal matrix elements in the representation (m, m') its effect vanishes in first order, whence the collapse of the multiplet.

(ii) The main interest of the r.f. stirring in liquids is to facilitate the analysis of complicated spectra by uncoupling the stirred spins S from the observed spins I. Fig. XII, 2 (7) shows the changes in the doublet spectrum of one of the protons of the molecule of dichloroacetaldehyde of Larmor frequency ω_I whilst the other proton of Larmor frequency $\omega_S = \omega_I + 100$ c/s, responsible for that doublet (see Fig. XI, 4), is being stirred by an r.f. field of frequency ω near $\omega_S = \omega_I + 100$ c/s. It is seen that for $\omega = \omega_I + 100$ c/s $= \omega_S$, the two components of the doublet ω_I collapse into a single line.

(iii) It is possible using double irradiation to measure the Larmor frequency of a nucleus without having any electromagnetic detection device for that frequency.

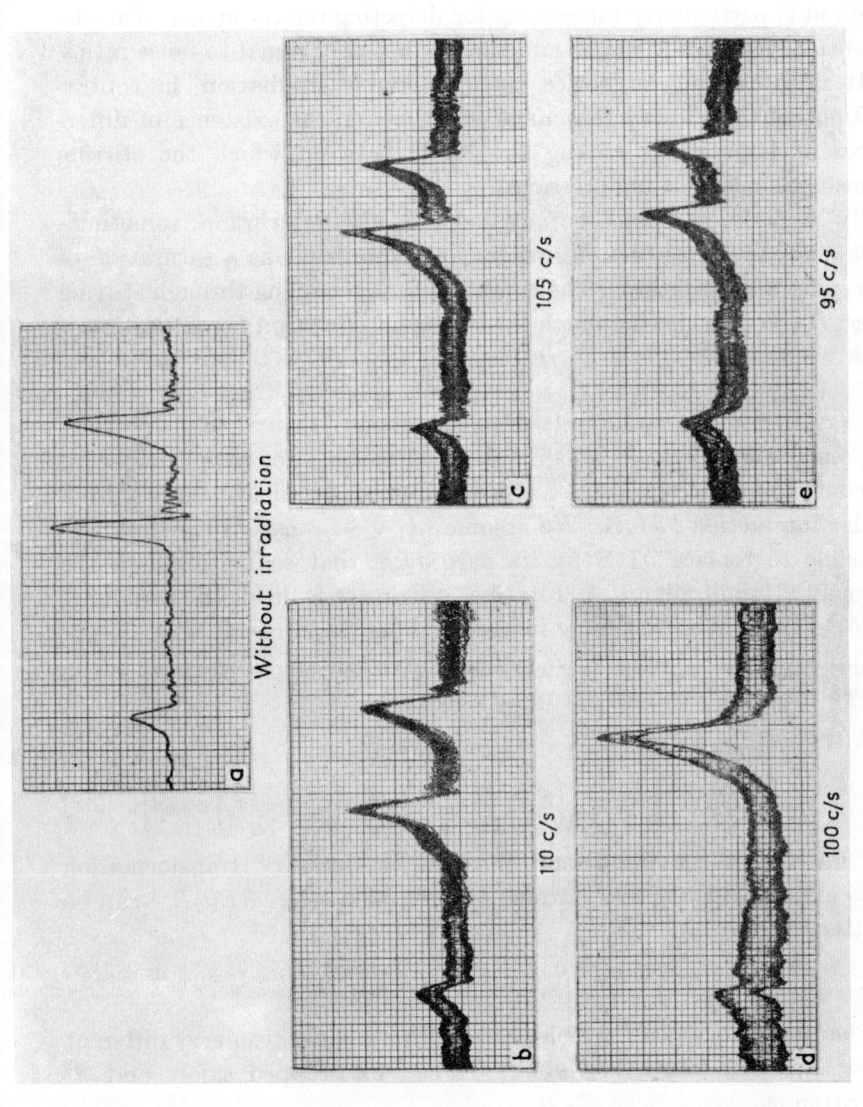

FIG. XII, 2. The doublet resulting from the coupling of a spin I (one of the two protons of dichloroacetaldehyde) with a spin S (the other proton), collapses when the spin S is being stirred at a frequency $\nu = \nu_S$. The difference $\nu - \nu_I$ is given for each trace. When $\nu - \nu_I = \nu_S - \nu_I = 100$ c/s, the doublet collapses. The small signal at the left of the doublet is due to an impurity.

Thus the Larmor frequency of C^{13} in CH_3I (enriched in C^{13}) was measured (8) by determining the stirring frequency $\omega(C^{13})$ at which the proton spectrum collapsed from a doublet into a single line. This method is particularly interesting for detecting the resonance of nuclei of small gyromagnetic ratio and thus with a poor signal-to-noise ratio.

It is interesting to notice that in double irradiation, in contradistinction to ordinary resonance experiments, the existence of differences of populations among the levels between which the stirring transitions occur, is unnecessary.

(iv) Finally, it should be stressed that the description, sometimes met with, of such a double irradiation experiment as a saturation of the spins S, is improper. The condition for decoupling through stirring is $|\gamma_S H_1| \gg J$, while the much less stringent condition for saturation is

$$\gamma_S^2 H_1^2 (T_2)_S (T_1)_S \gg 1.$$

(b) *The intermediate pattern (elementary theory)*

Consider two spins I and S (assumed to be $\tfrac{1}{2}$ for simplicity) with Larmor frequencies ω_I and ω_S in an applied d.c. field, coupled by a scalar interaction $\hbar J \mathbf{I} \cdot \mathbf{S}$. We assume $|J| \ll |\omega_I - \omega_S|$. It is then permissible to replace $J \mathbf{I} \cdot \mathbf{S}$ by its part $J I_z S_z$ that commutes with the Zeeman Hamiltonian. A rotating stirring r.f. field of amplitude $H_1 = -\omega_1/\gamma_S$ at a frequency in the neighbourhood of ω_S, $\omega = \omega_S - \Delta$, is applied to the system. If we neglect the relaxation for the time being, the spin Hamiltonian $\hbar E(t)$ can be written

$$E(t) = \omega_I I_z + \omega_S S_z + J I_z S_z +$$
$$+ \omega_1 \left\{ \left(S_x + \frac{\gamma_I}{\gamma_S} I_x \right) \cos \omega t + \left(S_y + \frac{\gamma_I}{\gamma_S} I_y \right) \sin \omega t \right\}. \quad (48)$$

By defining a moving frame through the unitary transformation $R = \exp\{it\omega_S S_z\}$, the transformed Hamiltonian $\tilde{\mathscr{H}} = R E(t) R^{-1}$ can be written

$$\tilde{\mathscr{H}} = \omega_I I_z + \Delta S_z + J I_z S_z + \omega_1 S_x + \omega_1 \frac{\gamma_I}{\gamma_S} \{ I_x \cos \omega t + I_y \sin \omega t \}. \quad (48')$$

The last term of (48'), which varies with a frequency ω very different from ω_I, has a negligible effect. It can be dropped safely and $\tilde{\mathscr{H}}$ rewritten as

$$\tilde{\mathscr{H}} = J I_z S_z + \Delta S_z + \omega_1 S_x + \omega_I I_z. \quad (49)$$

The four eigenstates and eigenvalues of $\tilde{\mathscr{H}}$ are then easily found since it is clear that $I_z = m_I$ is a good quantum number. For $I_z = +\tfrac{1}{2}$,

$$\tilde{\mathscr{H}}_+ = S_z(\tfrac{1}{2}J + \Delta) + \omega_1 S_x + \tfrac{1}{2}\omega_I$$

and S is quantized along a vector of components $Z = \frac{1}{2}J+\Delta$, $X = \omega_1$ and similarly along the vector $Z = -\frac{1}{2}J+\Delta$, $X = \omega_1$ for $I_z = -\frac{1}{2}$. The four eigenstates $|\alpha)$ and eigenvalues α of $\mathscr{H}$ are given by

$$|\xi_\pm) = \cos(\tfrac{1}{2}\theta_\pm)|\pm,+)+\sin(\tfrac{1}{2}\theta_\pm)|\pm,-),$$

$$\xi_+ = a_+ + \tfrac{1}{2}\omega_I, \qquad \xi_- = -a_- - \tfrac{1}{2}\omega_I,$$

$$|\eta_\pm) = -\sin(\tfrac{1}{2}\theta_\pm)|\pm,+)+\cos(\tfrac{1}{2}\theta_\pm)|\pm,-),$$

$$\eta_+ = -a_+ + \tfrac{1}{2}\omega_I, \qquad \eta_- = a_- - \tfrac{1}{2}\omega_I. \tag{49'}$$

In (49') a symbol such as, say, $|+,-)$ represents a state where $I_z = +\tfrac{1}{2}$, $S_z = -\tfrac{1}{2}$. The angles $\theta_\pm$ are given by $\tan(\theta_\pm) = \omega_1/(\Delta \pm \tfrac{1}{2}J)$ and the frequencies $a_\pm$ by

$$a_\pm = \tfrac{1}{2}[(\Delta \pm \tfrac{1}{2}J)^2 + \omega_1^2]^{\frac{1}{2}}. \tag{49''}$$

It may easily be seen that for $\omega_1 = \Delta = 0$: $\theta_\pm = 0$, $a_+ = a_- = \tfrac{1}{4}J$, and (49') reduces to the usual values:

$$|\xi_\pm) = |\pm,+), \qquad \xi_\pm = \pm(\tfrac{1}{4}J + \tfrac{1}{2}\omega_I);$$

$$|\eta_\pm) = |\pm,-), \qquad \eta_\pm = \pm(\tfrac{1}{2}\omega_I - \tfrac{1}{4}J). \tag{50}$$

If the resonance of the spin I flipping from $I_z = +\tfrac{1}{2}$ to $I_z = -\tfrac{1}{2}$ is observed, there are four transitions:

$$|\xi_+) \to |\xi_-), \quad |\xi_+) \to |\eta_-), \quad |\eta_+) \to |\xi_-), \quad |\eta_+) \to |\eta_-)$$

with frequencies (ω_I being the origin):

$$a_+ + a_-, \quad a_+ - a_-, \quad a_- - a_+, \quad -(a_+ + a_-). \tag{51}$$

Four lines should be observable in general in the spectrum of the spin I. If one assumes (without theoretical justification) that the intensity of each transition is proportional to the square of the matrix elements of I between the initial and the final state, the relative probabilities of the four transitions (51) are easily found by (49') to be, respectively,

$$\cos^2\tfrac{1}{2}(\theta_+-\theta_-), \quad \sin^2\tfrac{1}{2}(\theta_+-\theta_-), \quad \sin^2\tfrac{1}{2}(\theta_+-\theta_-), \quad \cos^2\tfrac{1}{2}(\theta_+-\theta_-). \tag{52}$$

Consider now a few special cases. Suppose first exact resonance $\omega = \omega_S$ for the stirring field. By (49''), $a_+ = a_- = a = \tfrac{1}{2}[\tfrac{1}{4}J^2+\omega_1^2]^{\frac{1}{2}}$ and $\theta_+ = -\theta_- = \theta$ with $\tan\theta = 2\omega_1/J$. By (51), only three lines at frequencies $2a$, 0, $-2a$ (with ω_I as origin) are observed, with, by (52), relative intensities

$$\cos^2\theta, \quad 2\sin^2\theta, \quad \cos^2\theta.$$

For a weak stirring field, θ is very small, $a \sim \frac{1}{4}J$. The spacing and the intensities of the two lateral lines are practically unaffected by the stirring, whereas the central line of relative intensity $2\sin^2\theta$ is very weak. As ω_1 and thus a and θ are made to grow, the lateral lines are pushed apart and become weaker, whilst the central line grows at their expense. For $\omega_1 \gg J$, only the central line remains. These effects are

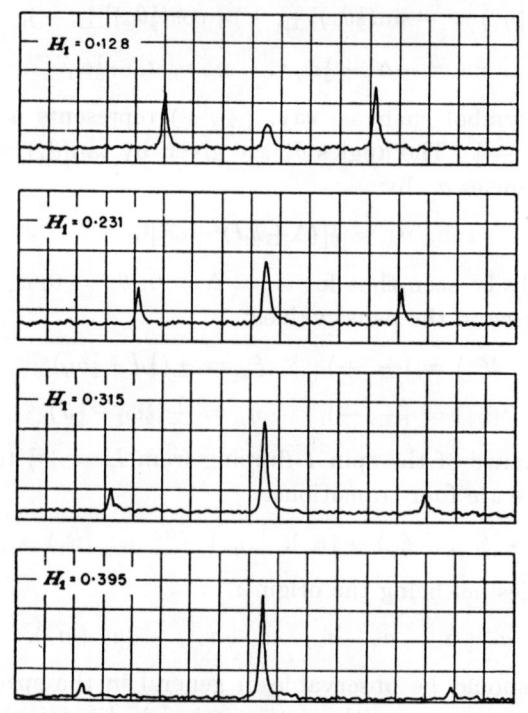

Fig. XII, 3. Effect on the fluorine spectrum of stirring the phosphorus at exact resonance with an r.f. field H_1 of varying magnitude. The top and bottom traces of Fig. XII, 4 are also part of this series.

illustrated in Fig. XII, 3 (9) showing the fluorine spectrum observed in an aqueous solution of Na_2PO_3F while stirring the phosphorus spins by an r.f. field $H_1 = -\omega_1/\gamma_P$ at exact resonance.

It is also possible, keeping the magnitude of the stirring field constant, to vary the stirring frequency. This is illustrated in Fig. XII, 4 (9) where four resonances of the fluorine spin I can be observed. The general agreement between theory and experiment is fair for the positions of the lines. No quantitative comparison with experiment is possible for the intensities predicted by a theory that neglects relaxation effects.

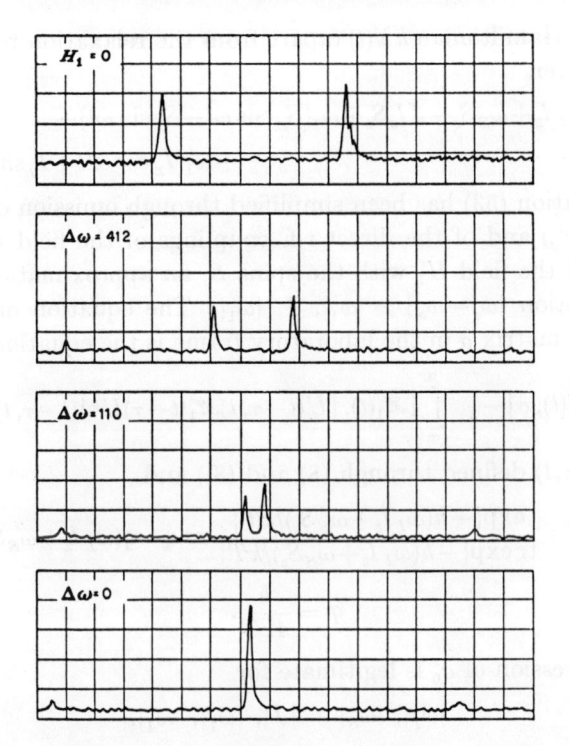

FIG. XII, 4. Effect on the fluorine spectrum of stirring the phosphorus with an r.f. field $H_1 = 0.464$ at a frequency ω near the phosphorus frequency $\omega_P = |\tfrac{1}{2}\gamma_P H_0/\pi|$. We define $\Delta\omega = \omega_P - \omega$.

(c) *The intermediate pattern (detailed theory)*

In the course of the previous calculation several assumptions were made, without justification, and the relaxation mechanisms were disregarded. A rigorous theory of the spin decoupling through r.f. stirring, which can be found in reference (6), although a straightforward consequence of the general formalism of relaxation, is fairly complicated. As an illustration of this theory we reconsider in more detail the problem of two spins $\tfrac{1}{2}$. The hypotheses and the notations are essentially the same as in the last section.

The relaxation mechanism is described by a random Hamiltonian $\hbar\mathcal{H}_1(t)$ which we shall assume for simplicity to be due to a fluctuating isotropic local field. Besides the stirring r.f. field of amplitude $H_1 = -\omega_1/\gamma_S$ at a frequency $\omega = \omega_S - \Delta$, we introduce explicitly in the spin Hamiltonian a vanishingly small 'observing' r.f. field of amplitude $H_1' = -\omega_1'/\gamma_I$ rotating with a frequency $\omega' = \omega_I - \delta$ in the neighbourhood of ω_I.

The spin Hamiltonian $\hbar E(t)$ (apart from the relaxation terms $\hbar \mathcal{H}_1(t)$) can be written

$$E(t) = \omega_I I_z + \omega_S S_z + J I_z S_z + \omega_1 \{S_x \cos \omega t + S_y \sin \omega t\} + \\ + \omega_1' \{I_x \cos \omega' t + I_y \sin \omega' t\}. \quad (53)$$

The equation (53) has been simplified through omission of the terms $J(I_x S_x + I_y S_y)$ and of the direct r.f. couplings of the field H_1 with the spin I, and the field H_1' with the spins S, an approximation based on the assumption $|\omega_I - \omega_S| \gg |\omega_1|, J, |\omega_1'|$. The equation of motion of the density matrix σ in the laboratory frame is the equation (13')

$$\frac{d\sigma}{dt} = -i[E(t), \sigma] - \tfrac{1}{2} \int_{-\infty}^{\infty} \overline{[\mathcal{H}_1(t), [U(t-\tau, t)\mathcal{H}_1(t-\tau)U^{-1}(t-\tau, t), \sigma-\sigma_0]]} \, d\tau \quad (53')$$

with $U(t-\tau, t)$ defined through (8) and (8') and

$$\sigma_0 = \frac{\exp\{-\hbar(\omega_I I_z + \omega_S S_z)/kT\}}{\operatorname{tr} \exp\{-\hbar(\omega_I I_z + \omega_S S_z)/kT\}} \simeq \tfrac{1}{4} - q(\omega_I I_z + \omega_S S_z), \quad (53'')$$

where
$$q = \frac{\hbar}{4kT}.$$

This expression of σ_0 is legitimate for

$$|\omega_I|, |\omega_S| \gg |J|, |\omega_1|, |\omega_1'|.$$

If we assume for simplicity, extreme narrowing for the relaxation Hamiltonian $\hbar \mathcal{H}_1(t)$, it is permissible to make $U(t-\tau, t) = 1$ in (53') and rewrite it as

$$\frac{d\sigma}{dt} = -i[E(t), \sigma] - \tau_c \overline{[\mathcal{H}_1(t), [\mathcal{H}_1(t), \sigma-\sigma_0]]}. \quad (54)$$

We define a moving frame through the unitary transformation

$$R = \exp\{it\omega S_z\}, \quad (55)$$

where the motion is described by a density matrix

$$\tilde{\sigma} = R\sigma R^{-1}.$$

We introduce $\tilde{E} = RE(t)R^{-1} = \omega S_z + \mathcal{H} + E_1(t)$ where, according to (53) and (55),

$$\mathcal{H} = \omega_I I_z + \Delta S_z + \omega_1 S_x + J I_z S_z,$$
$$E_1(t) = \tfrac{1}{2}\omega_1'\{I_+ e^{-i\omega't} + I_- e^{i\omega't}\}. \quad (56)$$

The equation of motion for $\tilde{\sigma}$ is

$$\frac{d\tilde{\sigma}}{dt} = -i[\mathcal{H}, \tilde{\sigma}] - i[E_1(t), \tilde{\sigma}] - \tau_c \overline{[\tilde{\mathcal{H}}_1(t), [\tilde{\mathcal{H}}_1(t), \tilde{\sigma}-\sigma_0]]} \quad (57)$$

with
$$\tilde{\mathcal{H}}_1(t) = R(t)\mathcal{H}_1(t)R^{-1}(t). \quad (57')$$

We write the relaxation Hamiltonian $\mathscr{H}_1(t)$ as

$$\mathscr{H}_1(t) = A_0^I(t)I_z + \tfrac{1}{2}\{A_+^I(t)\}I_- + \tfrac{1}{2}\{A_-^I(t)\}I_+ + \\ + A_0^S(t)S_z + \tfrac{1}{2}\{A_+^S(t)\}S_- + \tfrac{1}{2}\{A_-^S(t)\}S_+ \quad (58)$$

and assume that the random functions A^S and A^I, i.e. the random fields acting upon the spins S and the spins I, are entirely uncorrelated so that the only non-vanishing average products of the random functions A are
$$\overline{|A_0^I(t)|^2} = \tfrac{1}{2}\overline{|A_\pm^I(t)|^2}, \qquad \overline{|A_0^S(t)|^2} = \tfrac{1}{2}\overline{|A_\pm^S(t)|^2}.$$

From the definitions (55) of the operator R, (57′) of $\tilde{\mathscr{H}}_1(t)$, and (58) of $\mathscr{H}_1(t)$, it may easily be verified that on the right-hand side of (57), the product
$$\tau_c \overline{[\tilde{\mathscr{H}}_1(t), [\tilde{\mathscr{H}}_1(t), \tilde{\sigma} - \sigma_0]]}$$
can be replaced by $\tau_c \overline{[\mathscr{H}_1(t), [\mathscr{H}_1(t), \tilde{\sigma} - \sigma_0]]}$

and thus, since $\mathscr{H}_1(t)$ is a stationary function, this product can be written as $\Gamma(\tilde{\sigma} - \sigma_0)$, a linear function with constant coefficients, of the matrix elements of $\tilde{\sigma} - \sigma_0$. Since for spins $\tfrac{1}{2}$ and with our assumption of uncorrelated local fields, $\overline{|\mathscr{H}_1(t)|^2}$ is a c-number, $\Gamma(\tilde{\sigma} - \sigma_0)$ is simply

$$\Gamma(\tilde{\sigma} - \sigma_0) = 2\tau_c \{\overline{|\mathscr{H}_1(t)|^2}(\tilde{\sigma} - \sigma_0) - \overline{\mathscr{H}_1(t)(\tilde{\sigma} - \sigma_0)\mathscr{H}_1(t)}\}. \quad (59)$$

To solve the equation of motion (57) we make the substitution, $\tilde{\sigma} = \rho + \chi$ where ρ is a constant matrix, the steady-state solution of (57) for $E_1(t) = 0$, and thus obeying the equation

$$-i[\mathscr{H}, \rho] - \Gamma(\rho - \sigma_0) = 0. \quad (60)$$

The equation obeyed by χ is then

$$\frac{d\chi}{dt} + i[\mathscr{H}, \chi] + i[E_1(t), \chi] + \Gamma(\chi) = -i[E_1(t), \rho]. \quad (61)$$

To calculate ρ consider a matrix element $(\alpha| \quad |\alpha')$ of (60) where $|\alpha\rangle$ and $|\alpha'\rangle$ are eigenstates of $\mathscr{H}$. We get

$$-i(\alpha - \alpha')(\alpha|\rho|\alpha') - (\alpha|\Gamma(\rho - \sigma_0)|\alpha') = 0.$$

Since $\alpha - \alpha'$, the separation of two levels of $\mathscr{H}$, is much larger than the relaxation term Γ, which is of the order of the line width, off-diagonal matrix elements of ρ are very small.

The diagonal elements ρ_α are determined from the relation

$$(\alpha|\Gamma(\rho)|\alpha) = (\alpha|\Gamma(\sigma_0)|\alpha),$$

$$\Gamma(\sigma_0) = \tau_c \overline{[\mathscr{H}_1(t), [\mathscr{H}_1(t), \sigma_0]]}. \quad (62)$$

$\Gamma(\sigma_0)$ is easily calculated from the expressions (53″) and (58) of σ_0 and $\mathcal{H}_1(t)$, yielding

$$\Gamma(\sigma_0) = -q\tau_c \tfrac{2}{3}\{\overline{|\mathbf{A}_S|^2}\omega_S S_z + \overline{|\mathbf{A}_I|^2}\omega_I I_z\} = -q\left(\frac{\omega_S}{T_S}S_z + \frac{\omega_I}{T_I}I_z\right), \quad (63)$$

where T_S and T_I are the relaxation times of the spins S and I, in the absence of the coupling J and of the r.f. field ω_1, given by

$$\frac{1}{T_S} = \tfrac{2}{3}\tau_c \overline{|\mathbf{A}_S|^2}, \qquad \frac{1}{T_I} = \tfrac{2}{3}\tau_c \overline{|\mathbf{A}_I|^2}. \quad (63')$$

The matrix elements $(\alpha|\rho|\alpha) = \rho_\alpha$ are then calculated from (62) and (59),

$$2\tau_c\left\{\overline{\mathcal{H}_1^2}\rho_\alpha - \sum_\beta \overline{(\alpha|\mathcal{H}_1|\beta)^2}\rho_\beta\right\} = (\alpha|\Gamma(\sigma_0)|\alpha). \quad (64)$$

Passing on to the calculation of χ we notice that its equation (61) is very similar to the equation (36) of Section C, the only difference being the replacement on the left-hand side of (36) of $\mathcal{H}_0$ by $\mathcal{H}$ and on the right-hand side, that of the diagonal matrix $q\mathcal{H}_0$ by the diagonal matrix $-\rho$.

Then, as shown in that section, if the amplitude of the 'observing' r.f. Hamiltonian is vanishingly small and the 'observing' frequency ω' is much nearer to one frequency $\omega_{\alpha\alpha'} = \alpha' - \alpha$ than to all other such frequencies (which implies that no two such frequencies are equal), all diagonal matrix elements of χ are at most of the order of $\omega_1'^2$, and all the off-diagonal ones of the order of $\omega_1'/(\alpha-\alpha')$ and thus very small with the exception of

$$(\alpha|\chi|\alpha') = (\alpha'|\chi|\alpha)^* = z_{\alpha\alpha'} e^{i\omega' t}.$$

The value of $z_{\alpha\alpha'}$ is obtained by taking the $(\alpha|\ \ |\alpha')$ matrix element of (61) and yields

$$i(\omega' - (\alpha'-\alpha))z_{\alpha\alpha'} + \frac{z_{\alpha\alpha'}}{T_{\alpha\alpha'}} = -\tfrac{1}{2}i\omega_1'(\alpha|I_-|\alpha')(\rho_{\alpha'} - \rho_\alpha),$$

$$z_{\alpha\alpha'} = -\tfrac{1}{2}\omega_1'(\alpha|I_-|\alpha')(\rho_{\alpha'}-\rho_\alpha)\Big/\left[\omega' - (\alpha'-\alpha) - \frac{i}{T_{\alpha\alpha'}}\right], \quad (65)$$

where

$$\frac{1}{T_{\alpha\alpha'}} = 2\tau_c\{\overline{|\mathcal{H}_1|^2} - \overline{(\alpha|\mathcal{H}_1|\alpha)(\alpha'|\mathcal{H}_1|\alpha')}\}. \quad (66)$$

The signal observed at the frequency ω' and proportional to ω_1', due to a transition of the observed spin I, is then proportional to

$$\langle I_+\rangle = \mathrm{tr}\{I_+\chi\} = \sum_{\beta\beta'} (\beta|I_+|\beta')(\beta'|\chi|\beta).$$

For a transition $|\alpha\rangle \to |\alpha'\rangle$ where the spin I flips from $I_z = -\tfrac{1}{2}$ to $I_z = +\tfrac{1}{2}$ (there are four such transitions):

$$\langle I_+\rangle = (\alpha'|I_+|\alpha) z_{\alpha\alpha'} e^{i\omega' t} \propto |(\alpha'|I_+|\alpha)|^2. \tag{67}$$

The assumption of the elementary theory that the integrated intensity of each transition $|\alpha\rangle \to |\alpha'\rangle$ was proportional to $|(\alpha|I_x|\alpha')|^2$ is thus confirmed. That theory could not, however, foresee the extra factor $(\rho_{\alpha'}-\rho_\alpha)$ of (65) nor the values of the line widths, given by (66).

The foregoing treatment has to be modified when two frequencies $\alpha'-\alpha$ and $\beta'-\beta$ become equal. This is what happens according to (51) at exact resonance, when

$$\xi_+ - \eta_- = \eta_+ - \xi_-.$$

If ω' is in the neighbourhood of $\alpha'-\alpha$ it is also near $\beta'-\beta$ and the matrix element $(\beta|\chi|\beta') = z_{\beta\beta'} e^{i\omega' t}$ becomes appreciable at the same time as $z_{\alpha\alpha'}$. If we take the $(\alpha| \quad |\alpha')$ matrix element of (61) the first equation (65) has to be replaced by

$$i\{\omega'-(\alpha'-\alpha)\}z_{\alpha\alpha'} + \frac{z_{\alpha\alpha'}}{T_{\alpha\alpha'}} - z_{\beta\beta'} 2\tau_c \overline{(\alpha|\mathcal{H}_1|\beta)(\beta'|\mathcal{H}_1|\alpha')}$$
$$= -\tfrac{1}{2} i\omega_1'(\alpha|I_-|\alpha')(\rho_{\alpha'}-\rho_\alpha). \tag{67'}$$

A similar equation can be written for the $(\beta| \quad |\beta')$ matrix element of (61), whence $z_{\alpha\alpha'}$ and $z_{\beta\beta'}$ can be calculated. In the present situation the reader will easily convince himself that

$$z_{\eta_+,\xi_-} = -z_{\xi_+,\eta_-},$$

so that the only change is the replacement of the line width (66) by

$$\frac{1}{T_{(\eta_+,\xi_-)}} = \frac{1}{T_{(\xi_+,\eta_-)}}$$
$$= 2\tau_c\{\overline{|\mathcal{H}_1|^2} - \overline{(\xi_-|\mathcal{H}_1|\xi_-)(\eta_+|\mathcal{H}_1|\eta_+)} + \overline{(\xi_-|\mathcal{H}_1|\eta_-)(\xi_+|\mathcal{H}_1|\eta_+)}\}. \tag{68}$$

The problem is now entirely solved in principle. Using the expression (49') for the eigenvectors $|\alpha\rangle$, (58) for $\mathcal{H}_1(t)$, (63) and (64) for $\Gamma(\sigma_0)$ and the ρ_α, and (65), (66), (67), (68) for the intensities and the line widths, the last named can be calculated with some slightly tedious but straightforward algebra.

Off resonance for the stirring field, the four line widths obtained from (66) are given by

$$\left(\frac{1}{T}\right)_{\xi_+ \to \xi_-} = \left(\frac{1}{T}\right)_{\eta_+ \to \eta_-} = \frac{1}{T_I} + \frac{1}{4T_S}\{3-\cos(\theta_+-\theta_-)\},$$
$$\left(\frac{1}{T}\right)_{\xi_+ \to \eta_-} = \left(\frac{1}{T}\right)_{\eta_+ \to \xi_-} = \frac{1}{T_I} + \frac{1}{4T_S}\{3+\cos(\theta_+-\theta_-)\}. \tag{69}$$

If the stirring r.f. field vanishes, $\theta_+ = \theta_- = 0$ and only the two transitions $\xi_+ \to \xi_-$ and $\eta_+ \to \eta_-$ are observed. Their widths become $1/T_I + 1/2T_S$ in agreement with the formula (32) of Chapter XI for the widths of either line of a doublet.

At resonance for the stirring field, the width of the single central line $\xi_+ \to \eta_-$, $\xi_- \to \eta_+$ has to be calculated by (68) and yields

$$\left(\frac{1}{T}\right)_{\text{central}} = \frac{1}{T_I} + \frac{\cos^2\theta}{T_S}, \tag{70}$$

where $\theta = \theta_+ = -\theta_-$.

For a very large stirring field $\theta = \tfrac{1}{2}\pi$ and the width of the central line $1/T = 1/T_I$ becomes independent of T_S, as it is reasonable to expect from a complete 'decoupling'.

The calculation of the diagonal matrix elements ρ_α which determine the intensity of the various lines according to (65) and (67) offers little difficulty and will be omitted here. We shall be content to remark that at resonance, for $\theta_+ = \theta_-$ it is clear from reasons of symmetry that the two lateral lines $\xi_+ \to \xi_-$ and $\eta_+ \to \eta_-$ have equal intensities, so that

$$\rho_{\xi_+} - \rho_{\xi_-} = \rho_{\eta_+} - \rho_{\eta_-}.$$

The central line, the sum of transitions $\xi_+ \to \eta_-$ and $\eta_+ \to \xi_-$, has an intensity proportional to

$$(\rho_{\xi_+} - \rho_{\eta_-}) + (\rho_{\eta_+} - \rho_{\xi_-}) = 2(\rho_{\xi_+} - \rho_{\xi_-}).$$

It is thus seen that at resonance the population factors ρ do not modify the ratio (central line/lateral lines), predicted by the elementary theory to be $2\tan^2\theta$.

None of the simplifying assumptions made in the above theory of r.f. stirring (namely: spins $\tfrac{1}{2}$ with Larmor frequencies differing by an amount much larger than the coupling J and the stirring amplitude ω_1, relaxation by external uncorrelated fields and extreme narrowing) are necessary, and all could be given up at the cost of an increased complexity of calculations. They were made in order to exhibit the principle of the calculation, which under less simple assumptions might have been concealed by the complication of the formulae.

A problem related to the above but much simpler is that of a single species of spins I with a Larmor frequency ω_I in an applied d.c. field H_0 to which are applied a large 'stirring' rotating field of amplitude $H_1 = -\omega_1/\gamma$ and frequency ω, and a small 'observing' field of amplitude $H_1' = -\omega_1'/\gamma_I$ and frequency ω'. In the frame rotating at a frequency ω the effective spin Hamiltonian is $\hbar(\mathscr{H}+\tilde{E}_1(t))$, where

$$\mathscr{H} = (\omega_I-\omega)I_z+\omega_1 I_x,$$
$$\tilde{E}_1(t) = \omega_1'\{I_x\cos(\omega'-\omega)t+I_y\sin(\omega'-\omega)t\}. \tag{71}$$

It is clear from (71) that the resonance frequencies are given by

$$\omega' = \omega\pm[(\omega_I-\omega)^2+\omega_1^2]^{\frac{1}{2}} \tag{72}$$

and that signals proportional to ω_1' should be obtained from the observing field for the two frequencies (72). To calculate the intensities and widths of these signals, the general theory is fortunately unnecessary since, as shown generally in Section A, the magnetization **M** of the spins I obeys the Bloch equations

$$\frac{d\mathbf{M}}{dt} = \gamma_I \mathbf{M} \wedge \{\mathbf{H}_0+\mathbf{H}_1+\mathbf{H}_1'\} - \frac{M_x\mathbf{i}+M_y\mathbf{j}}{T_2} - \frac{M_z-\chi_0 H_0}{T_1}\mathbf{k}, \tag{73}$$

where writing $\chi_0 H_0 \mathbf{k}$, rather than $\chi_0(\mathbf{H}_0+\mathbf{H}_1+\mathbf{H}_1')$, is legitimate if the d.c. field is much larger than r.f. fields. The solutions of (73) can be found in reference (7). Fig. XII, 5, from that reference, shows the signals obtained at the observing frequency ω' for various values of the difference $(\omega'-\omega)/2\pi$ between stirring and observing frequency. (The relative magnitude of the signals is immaterial, for the observing receiver is overloaded by the stirring field.) According to (72), for each value of the difference $|\omega'-\omega| \geqslant \omega_1$ there are two values ω_I^a and ω_I^b of the Larmor frequency for which the resonance occurs. These are given by

$$\omega_I = \omega\pm[(\omega'-\omega)^2-\omega_1^2]^{\frac{1}{2}}. \tag{74}$$

Plotting $D^2 = (\omega_I^a-\omega_I^b)^2$ against $(\omega'-\omega)^2$, from the relation

$$D^2 = 4[(\omega'-\omega)^2-\omega_1^2],$$

should give a straight line, intersecting the axis of the $(\omega'-\omega)^2$ at the point $(\omega'-\omega)^2 = \omega_1^2$, thus affording a precise measurement of the strength of the stirring field.

II. Strong Radio-frequency Fields in Solids

A. Introduction

The theoretical description of the effects of strong r.f. fields in solids is a good deal more difficult than the one outlined in Section I for liquids, and at present no rigorous theory of such effects exists.

The reason for such a difference between the behaviour of liquids and solids as far as nuclear magnetism is concerned has already been stressed several times in this book; whereas in liquids, the rapid relative motion of nuclear spins uncouples them from each other, making it

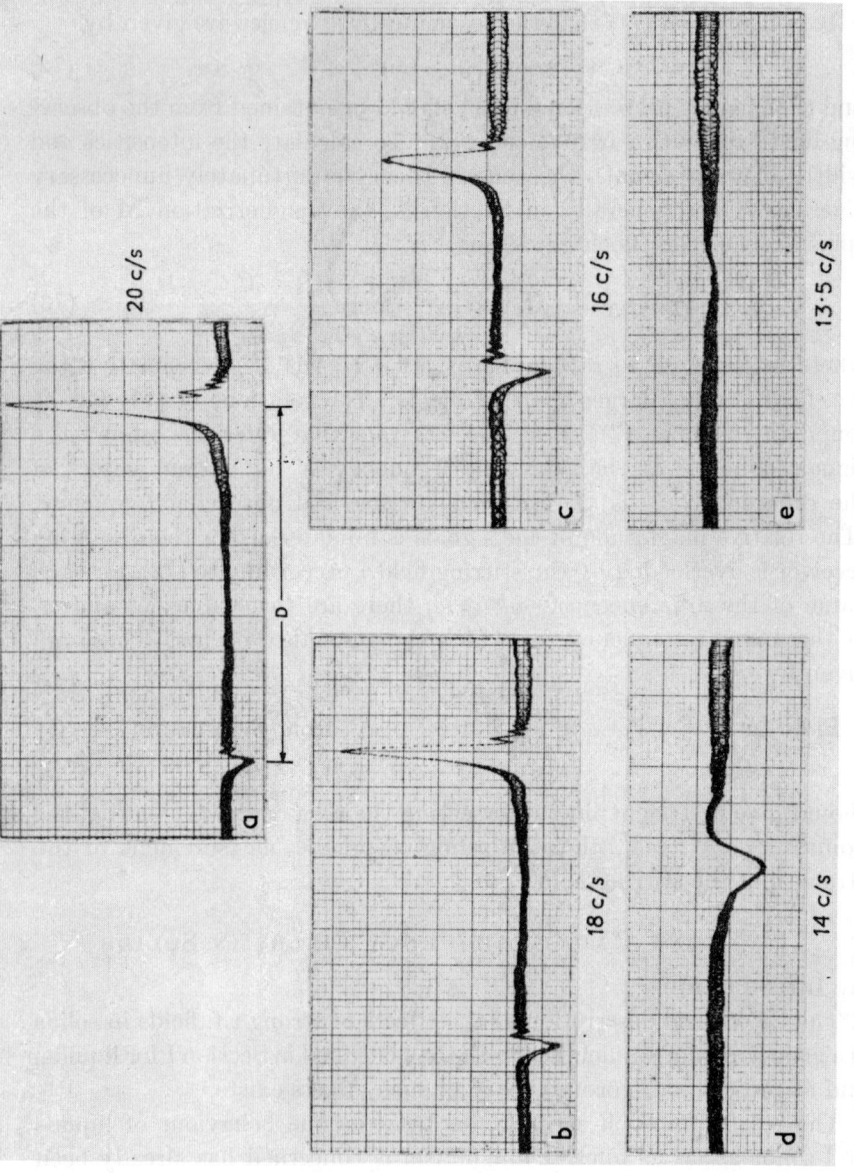

Fig. XII, 5. Signals observed with a small r.f. field at a frequency ω', while stirring the spins with a large r.f. field at a frequency ω. For each trace, the value of $(\omega'-\omega)/2\pi$ is given.

possible to consider individual spins (or groups of spins in a molecule) as simple systems with very few degrees of freedom, the tight coupling that exists between the nuclear spins in solids requires a collective description of all the spins of the sample as a single spin system with very many degrees of freedom.

On the other hand, as will appear shortly, the very strength of the spin-spin couplings and the ensuing complexity of the spin system make it possible further to generalize the concept of spin temperature, discussed in some detail in Chapter V, and to make quantitative predictions about the steady-state and transient behaviour of the spin system in the presence of strong r.f. fields.

In the following we shall assume explicitly that so-called 'inhomogeneous' broadening resulting from a distribution of Larmor frequencies of the individual spins, due to an inhomogeneous applied field or to small quadrupole splittings, is absent. The saturation behaviour of inhomogeneously broadened lines has already been discussed to some extent in Chapter III and there is nothing fundamental about the complications resulting from such a broadening, once the behaviour under saturation of the individual spin packets is known. Furthermore, when the inhomogeneous broadening takes place on an atomic scale as for quadrupole splittings in imperfect cubic crystals, it prevents the exchange of energy between neighbouring spins by quenching appreciably their mutual flips and making the assumption of spin temperature less reliable. Thus the only systems to be considered in this section are so-called Zeeman systems as defined in Chapter V, their spin Hamiltonian being the sum of their Zeeman coupling with the applied field and of the spin-spin couplings.

When an r.f. field is applied to such a system at or near its Larmor frequency, the two relevant parameters are the strength of the r.f. field and the time during which it is applied at a given frequency.

First we assume slow passage conditions where for each value of the applied frequency a steady state is reached for the combined system of spins-plus-lattice, and recall some elementary results established previously.

For very small amplitudes $H_1 = |\omega_1/\gamma|$ of the rotating r.f. fields it was stated in Chapter II that the relative departure of the longitudinal magnetization from its equilibrium value was given by

$$\frac{M_0 - M_z}{M_0} = 2WT_1 = \gamma^2 H_1^2 \pi f(\omega) T_1, \tag{75}$$

where $W = \gamma^2 H_1^2 (\tfrac{1}{2}\pi) f(\omega)$ was the transition probability per unit time induced by the r.f. field and $f(\omega)$, normalized to $\int f(\omega)\,d\omega = 1$, was the shape function of the resonance line. The calculation of the unsaturated shape $f(\omega)$ for Zeeman systems was described in Chapter IV.

The imaginary part $\chi''(\omega)$ of the r.f. susceptibility can then be calculated by energy considerations, writing that the energy $H_0(M_0-M_z)/T_1$ transferred per unit time to the lattice is equal to the r.f. energy $2\omega\chi''H_1^2$ absorbed by the spins, whence the relations

$$\chi''(\omega) = \tfrac{1}{2}\pi\chi_0\omega_0 f(\omega) \quad \text{and} \quad \chi''(\omega_0) = \tfrac{1}{2}\chi_0\omega_0 T_2 \tag{76}$$

(equation (8) of Chapter III) where, *by definition* of the constant T_2,

$$T_2 = \pi f(\omega_0).$$

Finally, the real part $\chi'(\omega)$ of the r.f. susceptibility is deduced from χ'' by the Kramers–Krönig relations.

These relations are valid in the absence of saturation, i.e. as long as

$$\left. \begin{array}{l} 2W \ll \dfrac{1}{T_1}, \quad \gamma^2 H_1^2 \pi f(\omega_0) \ll \dfrac{1}{T_1} \\ \text{or} \quad \gamma^2 H_1^2 T_1 T_2 \ll 1 \end{array} \right\} \tag{77}$$

If the free precession signal following a 90° pulse is observed, its shape is the Fourier transform of the shape function $f(\omega)$ and it is damped in a time of the order of the inverse width of the distribution $f(\omega)$ which is of the order of T_2.

All these results are independent of the nature of the broadening expressed by the shape function $f(\omega)$ and would be valid for inhomogeneous broadening as well. However, if spin echo techniques are used, a difference appears immediately between inhomogeneous and homogeneous broadening in solids. If the line width $\Delta H \sim 1/\gamma T_2$ is due to spin-spin coupling, after the free precession signal has decayed completely in a time of the order of, say, several T_2, a subsequent pulse will not restore it, as it would for a width caused by an inhomogeneous applied field or by small quadrupole splittings. The free decay is essentially irreversible and the line width cannot be visualized as caused by a spread in the Larmor frequencies of the individual spins.

As the strength of the r.f. field is increased so that the condition (77) is not fulfilled any longer, saturation occurs and χ' and χ'' change accordingly. Their behaviour could be predicted in liquids under the assumption of the validity of the Bloch equations, although even there, as indicated in Section I C of this chapter, attention must be paid in special cases to the possible dependence of the relaxation times T_1 and T_2 on the strength of the r.f. field. No such predictions are available in solids where the Bloch equations are known to be invalid, as shown, for instance, by the non-exponential character of the decay of the free precession.

We may tentatively, for in solids we have no justification for it, believe in the validity of the concept of r.f. induced transition probability per unit time and write a rate equation for the Zeeman energy of the spins:

$$\frac{d}{dt}(-M_z H_0) = \frac{M_z - M_0}{T_1} H_0 + 2WM_z H_0,$$

whence under steady-state conditions

$$\frac{M_0 - M_z}{M_0} = \frac{2WT_1}{1 + 2WT_1} = \frac{\gamma^2 H_1^2 \pi f(\omega) T_1}{1 + \gamma^2 H_1^2 \pi f(\omega) T_1}, \qquad (78)$$

of which (75) was a special case for $WT_1 \ll 1$.

The reader will easily convince himself that if the shape function were the Lorentzian curve $(T_2/\pi)\{1 + T_2^2(\omega - \omega_0)^2\}^{-1}$ the result (78) would be identical with that predicted for M_z by the Bloch equations.

$\chi''(\omega)$ can again be obtained from the energy relation:

$$2\omega \chi'' H_1^2 = H_0 \frac{M_0 - M_z}{T_1}, \qquad (79)$$

$$\chi'' = \tfrac{1}{2}\pi \frac{\chi_0 \omega_0 f(\omega)}{1 + \gamma^2 H_1^2 T_1 \pi f(\omega)}. \qquad (80)$$

In particular, at resonance,

$$\pi f(\omega_0) = T_2 \quad \text{and} \quad \chi''(\omega_0) = \frac{\chi_0 \omega_0 T_2}{2[1 + \gamma^2 H_1^2 T_1 T_2]}. \qquad (80')$$

Indeed, up to values of H_1 such that $\gamma^2 H_1^2 T_1 T_2$ is of the order of several units, that is for appreciable saturation, the formula (80') is well verified experimentally in copper and aluminium (**10**), giving some confidence in the measurements of spin-lattice relaxation times by saturation methods.

The values of T_1 obtained by these methods agree reasonably well with those obtained by rapid passage and quoted in Chapter IX. We emphasize again that T_2 is defined here as $\pi f(\omega_0)$ rather than the inverse unsaturated line width. The two quantities are, however, of comparable order of magnitude.

For very strong r.f. fields when $\gamma^2 H_1^2 T_1 T_2 \gg 1$

$$\chi''(\omega_0) \to \frac{\chi_0 \omega_0}{2\gamma^2 H_1^2 T_1}. \qquad (81)$$

We notice that all indication of line shape has disappeared from the asymptotic relation (81), which follows directly from the energy relation (79) if one assumes that in the presence of r.f. field $M_z \ll M_0$, and can thus be expected to hold under fairly general conditions.

The previous considerations of energy balance, whatever their validity, provide no clue for the behaviour of χ' which in presence of saturation cannot be deduced from χ'' by the K.-K. relations.

Since the results thus obtained for M_z and χ'' happen to coincide with those predicted by the Bloch equations if the shape function $f(\omega)$ is Lorentzian, there is a strong temptation to believe (and it was actually generally believed prior to the publication (3)) that the predictions of the Bloch equations for the saturation of χ' would at least be qualitatively correct.

At resonance $\chi'(\omega_0)$ vanishes but its derivative with respect to H_0, which is the quantity directly measured, is given by (first equation (15) of Chapter III)

$$\frac{\partial \chi'}{\partial H_0} = \frac{\chi_0 \gamma^2 M_0 T_2^2}{1+\gamma^2 H_1^2 T_1 T_2}$$

and is expected to saturate with increasing H_1 in the same way as $\chi''(\omega_0)$.

This prediction is in *complete disagreement* with *experiment* (3). For instance, in copper for values of H_1 of the order of 0·6 gauss where $\chi''(\omega_0)$ has decreased from its unsaturated value by a factor of the order of 7, $\partial \chi'/\partial H_0$ is practically unchanged and has decreased by a factor of 2 only for an H_1 as large as 3 gauss. Similar results are observed in other solids. It is clear that an entirely new approach to the problem of strong r.f. fields in solids is required.

Before describing this approach we discuss briefly some aspects of the transient effects of r.f. fields in solids. It is convenient for that purpose to consider the effective field $\mathbf{H}_e = \{H_0+(\omega/\gamma)\}\mathbf{k}+H_1\mathbf{i}$ in the rotating frame. This field can be changed either by changing H_1, generally by pulse methods, or by sweeping H_0 through the resonant value $H^* = -\omega/\gamma$. In pulse methods one strives to make H_1 several times the local field and it is applied during times τ of the order of $|1/\gamma H_1|$. Under such conditions, during the pulse not only the effects of T_1 but also those of the spin-spin couplings, expressed by the local field, can be neglected and the behaviour observed is essentially that of free spins. Between pulses one has the free precession, correlated by the Fourier relation to the unsaturated line width, a problem discussed at some length in Chapter IV.

In the rapid passage problem, where H_0 is varied continuously, the significant parameter is the time τ required to cross the resonance line.

In order for such a passage to be adiabatic and therefore reversible, according to the Bloch equations the following conditions have to be fulfilled:
$$\tau \gg (\gamma H_1)^{-1}, \qquad \tau \ll T_1, T_2.$$

However, in solids where T_2 is often a fraction of a millisecond, and such conditions are very difficult to satisfy, reversible rapid passages have been observed under conditions where $\tau \gg T_2$ (3). Indeed, since 1955 rapid passage has become a standard tool for the study of nuclear magnetism in solids. The new approach will have to explain this feature as well.

B. Spin temperature in the rotating frame, reversible fast passage

In order to explain the experimental results mentioned in Section II A, such as the lack of saturation of the real r.f. susceptibility and the possibility of performing a reversible fast passage in solids, an important hypothesis has been put forward: the existence of a spin temperature in the rotating frame (3). While the calculation of the steady-state r.f. susceptibility necessarily involves the consideration of the spin-lattice coupling, this coupling can be disregarded in the fast passage which takes place in a time short compared with T_1. We shall therefore discuss this problem first and neglect the spin-lattice relaxation altogether.

It has been shown in Chapter V that for Zeeman spin systems, in the absence of r.f. fields, the assumption of a spin temperature T_S, that is, of a statistical description of a spin system with a Hamiltonian $\hbar \mathcal{H}$, by a statistical Boltzmann operator

$$\sigma_B = \exp(-\hbar\mathcal{H}/kT_S)/\text{tr}\{\exp(-\hbar\mathcal{H}/kT_S)\} \tag{81'}$$

(where the index B stands for Boltzmann) was under certain conditions well verified experimentally. If by some means such a spin system is so prepared as to be described by a density matrix σ' that does not have the form (81') and is then left free, in a time of the order of T_2 it reaches a spin temperature T_S. The actual value of the temperature T_S reached is obtained by writing that the energy of the spin system is conserved during the process:

$$\text{tr}\{\mathcal{H}\sigma'\} = \text{tr}\{\mathcal{H}\sigma_B\} \cong -\frac{\hbar}{kT_S}\frac{\text{tr}\{\mathcal{H}^2\}}{\text{tr}\{1\}}. \tag{82}$$

The reader is referred back to Chapter V for a detailed discussion of the various conditions for the validity of (81'), of the way in which two different spin species may come into thermal equilibrium with each other, and of the various features of isentropic demagnetization.

Consider now a Zeeman spin system to which we apply a rotating field $\mathbf{H}_1$ with components

$$H_x = H_1 \cos \omega t, \qquad H_y = H_1 \sin \omega t, \qquad H_z = 0.$$

If we assume first for simplicity that a single-spin species is present in the sample, its Hamiltonian can be written

$$\hbar \mathscr{H} = -\gamma \hbar \mathbf{I} \cdot (\mathbf{H}_0 + \mathbf{H}_1) + \hbar \mathscr{H}_{ss} = \hbar(Z + \mathscr{H}_{ss}). \tag{83}$$

The spin-spin Hamiltonian $\hbar \mathscr{H}_{ss}$ can be written most generally

$$\hbar \mathscr{H}_{ss} = \sum_{j<k} \tilde{A}_{jk} \mathbf{I}_j \cdot \mathbf{I}_k + \tilde{B}_{jk} \left\{ \mathbf{I}_j \cdot \mathbf{I}_k - 3 \frac{(\mathbf{I}_j \cdot \mathbf{r}_{jk})(\mathbf{I}_k \cdot \mathbf{r}_{jk})}{r_{jk}^2} \right\}. \tag{84}$$

We have allowed in (84) for the possible existence of indirect spin-spin interactions, both scalar and pseudo-dipolar. If they are absent,

$$\tilde{A}_{jk} = 0, \qquad \tilde{B}_{jk} = \gamma^2 \hbar^2 r_{jk}^{-3}.$$

(This notation is different from that of reference (11), where $\tilde{B}_{jk}$ stands for the coefficient of the pseudo-dipolar coupling only and vanishes if indirect interactions are absent.)

The Zeeman part $\hbar Z$ of the Hamiltonian (83) contains the time explicitly and it is not clear how a spin temperature could be defined for such a system. However, if we perform the transformation to rotating coordinates, $e^{i\omega I_z t} \mathscr{H} e^{-i\omega I_z t}$ we know that in the rotating frame the coupling with the applied field is described by an effective time-independent Zeeman Hamiltonian $\hbar Z^*$, given by

$$\hbar Z^* = -\gamma \hbar \mathbf{I} \cdot \mathbf{H}_e = \hbar[(\omega_0 - \omega) I_z + \omega_1 I_x],$$

where
$$\omega_0 = -\gamma H_0, \qquad \omega_1 = -\gamma H_1, \qquad |\gamma H_e| = [(\omega_0 - \omega)^2 + \omega_1^2]^{\frac{1}{2}}. \tag{85}$$

On the other hand, some parts of $\mathscr{H}_{ss}^* = e^{i\omega I_z t} \mathscr{H}_{ss} e^{-i\omega I_z t}$ will now contain the time explicitly. According to an expansion of the dipolar (or pseudo-dipolar) interaction, performed several times before, $\mathscr{H}_{ss}^*$ will contain time-independent terms, terms with factors $e^{\pm i\omega t}$, and terms with factors $e^{\pm 2i\omega t}$. If we assume that the d.c. field $H_0 = -\omega_0/\gamma \sim -\omega/\gamma$ is much larger than the local field, the time-dependent terms of $\mathscr{H}_{ss}^*$ have a negligible influence and can be discarded. (The argument is the same as that for discarding the counter-rotating component of an oscillating r.f. field.) The time-independent part of $\mathscr{H}_{ss}^*$ is that which commutes with $-\gamma\hbar H_0 I_z$, that is, the truncated spin-spin Hamiltonian $\mathscr{H}'_{ss}$ used in the computation of various moments of the resonance line in Chapter IV.

We are thus left with an effective static Hamiltonian in the rotating frame,

$$\hbar\mathscr{H}^* = \hbar(Z^* + \mathscr{H}'_{ss}) = -\gamma\hbar\mathbf{I}\cdot\mathbf{H}_e + \sum_{j<k} A_{jk}\mathbf{I}_j\cdot\mathbf{I}_k + B_{jk}I_{jz}I_{kz},$$

with

$$A_{jk} = \tilde{A}_{jk} + \tfrac{1}{2}\tilde{B}_{jk}(3\cos^2\theta_{jk}-1), \quad B_{jk} = -\tfrac{3}{2}\tilde{B}_{jk}(3\cos^2\theta_{jk}-1). \tag{86}$$

We are now ready to discuss the rapid passage problem. As the d.c. field is swept from an initial value, H'_0 say, far above the resonant value $H^* = -\omega/\gamma$ to another H''_0 far below, the magnitude of the effective field starts with a large initial value $H_{ei} = [(H'_0 - H^*)^2 + H_1^2]^{\frac{1}{2}}$, goes down to a minimum value H_1, then up again to a large final value $H_{ef} = [(H^* - H''_0)^2 + H_1^2]^{\frac{1}{2}}$. If we postulate the existence of a spin temperature in the rotating frame, which is expressed mathematically by the fact that the density matrix of the spin system in the rotating frame is of the form $\tilde{\sigma}_B \propto \exp(-\hbar\mathscr{H}^*/kT_S)$, where $\mathscr{H}^*$ is given by (86), this passage is formally identical with the process of isentropic demagnetization (or magnetization) described in Chapter V and the results established there can be taken over immediately. The spin temperature varies as

$$T_S \propto [H_e^2 + H_L'^2]^{\frac{1}{2}}$$

and the magnetic moment, constantly aligned along $\mathbf{H}_e$, varies as H_e/T_S, that is, as $H_e/[H_e^2 + H_L'^2]^{\frac{1}{2}}$. The quantity $H_L'^2$ is defined here as

$$H_L'^2 = \frac{\mathrm{tr}\{\mathscr{H}'^2_{ss}\}}{\mathrm{tr}\{\gamma^2 I_z^2\}}. \tag{87}$$

The only difference from the treatment of Chapter V is that the spin-spin Hamiltonian $\mathscr{H}'_{ss}$ is here the truncated spin-spin Hamiltonian rather than the full one $\mathscr{H}_{ss}$, and $H_L'^2$ is accordingly smaller than H_L^2 introduced in Chapter V.

It is easy to verify by comparing (87) with the formula (34) of Chapter IV giving the second moment of the unsaturated resonance line, that

$$H_L'^2 = \tfrac{1}{3}\overline{\Delta H^2} + \frac{I(I+1)}{\hbar^2\gamma^2 N}\sum_{k>j}\tilde{A}_{jk}^2, \tag{88}$$

where N is the number of spins and $\overline{\Delta H^2}$ is the second moment expressed in gauss^2.

The magnitude of the magnetic moment along the effective field starts from an initial value M_0, which is the thermal equilibrium value if one has waited off-resonance for a time appreciably larger than T_1, passes through a minimum equal to

$$M_t = M_0 H_1 / [H_1^2 + H_L'^2]^{\frac{1}{2}} \tag{89}$$

at resonance, and returns to the value M_0 along the effective field far below resonance. Since $\mathbf{H}_e$ is then antiparallel to $\mathbf{H}_0$, so also is the magnetic moment.

The reversibility of the rapid passage is exhibited very clearly in Fig. III, 16. The signal of Si^{29} shown there was swept in more than a minute, which, with a modulation frequency of 20 c/s, corresponded to more than a thousand passages back and forth through resonance, and thus evidently to an exceedingly small (if any) loss of magnetic moment per passage.

If the sweep is stopped at any time during the passage when the magnetic moment along the effective field has the value

$$M_t = M_0 H_e / [H_e^2 + H_L'^2]^{\frac{1}{2}}$$

(in particular at resonance $M_t = M_0 H_1 / [H_1^2 + H_L'^2]^{\frac{1}{2}}$), the magnetization along the effective field will keep this value until the effects of the spin-lattice relaxation are felt. In that sense one might say that in the presence of a rotating r.f. field the relaxation time of the magnetization along the effective field (in particular along $\mathbf{H}_1$ if one stops at resonance) is comparable to T_1.

This statement requires some qualification, as is shown by the following example.

Assume that at a time $t = 0$ when the spin system is in equilibrium with the lattice at a temperature T_L and the magnetization $M_0 = \chi_0 H_0$ is along the d.c. field, a rotating r.f. field of amplitude H_1 and frequency ω in the neighbourhood of ω_0 is introduced suddenly, that is, in a time much shorter than T_2. The effective field in the rotating frame is $(H_0 - H^*)\mathbf{k} + H_1 \mathbf{i}$ and the effective Zeeman Hamiltonian can be written

$$\hbar Z^* = \hbar \omega_e [I_z \cos \Theta + I_x \sin \Theta], \tag{90}$$

where $\omega_e = -\gamma H_e$ and Θ is the angle of the effective field with Oz given by

$$\tan \Theta = \frac{\omega_1}{\omega_0 - \omega} = \frac{H_1}{H_0 - H^*}.$$

The spin system in the rotating frame is not in thermal equilibrium immediately after the pulse introducing the r.f. fields. Once this equilibrium has been reached, in a time of the order of T_2, the component of the magnetization perpendicular to $\mathbf{H}_e$ will have disappeared. We propose to show that during that period the component along the effective field will not keep its initial value $M_t^0 = M_0 \cos \Theta$ but will decrease irreversibly to a lower value (unless $H_1 \gg H'_L$). Before the introduction of the r.f. field the density matrix σ' of the spin system which is in equilibrium with the lattice at a temperature T_L, can be written, omitting normalization factors and assuming the temperatures are sufficiently high to allow an expansion of the Boltzmann exponential,

$$\sigma' \propto \left(1 - \frac{\hbar \omega_0 I_z}{kT_L}\right). \tag{91}$$

Since σ' commutes with I_z this density matrix is the same in the laboratory and the rotating frame and since H_0 is high the neglect of $\mathscr{H}_{ss}$ in (91) is legitimate. After the introduction of the r.f. field and once thermal equilibrium among spins is established in the rotating frame, the density matrix is

$$\tilde{\sigma}_B \propto 1 - \frac{\hbar \mathscr{H}^*}{kT_S} = 1 - \frac{\hbar(Z^* + \mathscr{H}'_{ss})}{kT_S}.$$

The value of T_S is obtained by writing that the expectation value $\langle \mathscr{H}^* \rangle$ has not changed during the establishment of equilibrium:

$$\operatorname{tr}\{\sigma' \mathscr{H}^*\} = \operatorname{tr}\{\tilde{\sigma}_B \mathscr{H}^*\},$$

which yields immediately

$$\frac{1}{T_S} = \frac{1}{T_L} \frac{\operatorname{tr}\{\omega_0 I_z(Z^* + \mathscr{H}'_{ss})\}}{\operatorname{tr}\{(Z^* + \mathscr{H}'_{ss})^2\}} = \frac{1}{T_L} \frac{H_e H_0 \cos \Theta}{H_e^2 + H'^2_L}.$$

The equilibrium magnetization $(M_t)_{eq}$ along the effective field $\mathbf{H}_e$ is related to its initial value $M_t^i = M_0 \cos \Theta$ by

$$\frac{(M_t)_{eq}}{M_t^i} = \frac{(M_t)_{eq}}{M_0 \cos \Theta} = \frac{1}{\cos \Theta} \frac{T_L}{T_S} \frac{H_e}{H_0} = \frac{1}{1 + (H'_L/H_e)^2}. \tag{92}$$

Thus, after the sudden introduction of the r.f. field H_1, the component of the magnetization perpendicular to the effective field goes in a time of the order of T_2 from $M_0 \sin \Theta$ to zero, whilst the component along $\mathbf{H}_e$ decreases from $M_0 \cos \Theta$ towards

$$\frac{M_0 \cos \Theta}{1 + (H'_L/H_e)^2} = \frac{M_0 \cos \Theta}{1 + \sin^2\Theta(H'_L/H_1)^2} \leqslant M_0 \cos \Theta.$$

Similarly, if we assume that by a combination of pulses the magnetization is brought along a transverse field $\mathbf{H}_1$ rotating at a frequency $\omega = \omega_0$ in a time short compared to T_2, a calculation analogous to the one above shows that the transverse magnetization will reach irreversibly an equilibrium value $M_0 H_1^2/[H_1^2 + H'^2_L]$, which is smaller than the value $M_0 H_1/[H_1^2+H_L'^2]^{\frac{1}{2}}$ obtainable by adiabatic passage. Once this value is reached, any further decrease can only come from the spin-lattice relaxation.

At this stage of the discussion the following remarks can be made:

(i) Just as in the absence of r.f. fields, the existence of a spin temperature is a hypothesis for which no justification has been given from first principles; it is, however, fairly plausible and in good agreement with known experimental facts.

(ii) All the previous results about rapid passage become very simple if $H_1 \gg H_L'$. The fact that the magnetization along the effective field remains constant until affected by the spin lattice coupling can be expressed simply by saying that in the rotating frame the spins are quantized along that field and that the spin-spin coupling is unable to provide the large quanta $\gamma \hbar H_e$ required for the reversal of a spin quantized along $\mathbf{H}_e$. This argument is, however, inadequate to explain, as does the assumption of spin temperature, why a reversible passage can be produced by an r.f. field which is a small fraction of the local field.

(iii) The results obtained for the magnitude of the transverse magnetization at resonance bear a strong resemblance to those that would obtain in the absence of any spin-spin interaction in a strongly inhomogeneous field. If this inhomogeneity has a value $\delta H \gg H_1$, the maximum transverse magnetization obtainable by rapid passage in that case is of the order of $M_0 H_1/\delta H$, which is comparable with the result $M_0 H_1/H_L'$ that follows from (89). This analogy should not be taken too literally, for the two situations are really very different. For instance, if there is a strong scalar coupling between the spins the resonance line becomes very narrow (exchange narrowing). Yet according to (88), the strong scalar coupling will increase $H_L'^2$ and decrease the maximum transverse magnetization given by (89). The connexion between rapid passage signal and line width is less simple than that provided by the model of the inhomogeneous field.

Now consider a sample with two spin species $\mathbf{I}_k$ and $\mathbf{S}_{k'}$, with Larmor frequencies $\omega_I = -\gamma_I H_0$ and $\omega_S = -\gamma_S H_0$. Assume between the spins $\mathbf{I}$ scalar couplings $\tilde{A}_{jk}^I \mathbf{I}_j \cdot \mathbf{I}_k$ and tensor couplings

$$\tilde{B}_{jk}^I \left\{ \mathbf{I}_j \cdot \mathbf{I}_k - 3\frac{(\mathbf{I}_j \cdot \mathbf{r}_{jk})(\mathbf{I}_k \cdot \mathbf{r}_{jk})}{r_{jk}^2} \right\}.$$

If indirect couplings are absent,

$$\tilde{A}_{jk}^I = 0, \qquad \tilde{B}_{jk}^I = \frac{\gamma_I^2 \hbar^2}{r_{jk}^3}.$$

Similar couplings $\tilde{A}^S_{j'k'}$ and $\tilde{B}^S_{j'k'}$ are defined for the spins S, and couplings $\tilde{A}^{IS}_{jk'}$ and $\tilde{B}^{IS}_{jk'}$ between the spins I and the spins S. If a rotating r.f. field $\mathbf{H}_1$ at a frequency ω in the neighbourhood of ω_I is applied to the system, the total Hamiltonian can be written immediately as the sum of the Zeeman couplings of either spin system with the applied field and of the various spin-spin couplings. If we transform $\mathscr{H}$ by the relation $e^{i(\omega I_z+\omega S_z)t}\mathscr{H}e^{-i(\omega I_z+\omega S_z)t}$ and discard from the transformed spin-spin interaction time-dependent terms as explained previously, we obtain an effective time-independent Hamiltonian

$$\begin{aligned}\hbar\mathscr{H}^* &= \hbar(Z^*_S+Z^*_I+\mathscr{H}'_{ss}),\\ \text{where}\quad \hbar Z^*_S &= -\gamma_S\hbar\mathbf{S}.\mathbf{H}^S_e, \quad \hbar Z^*_I = -\gamma_I\hbar\mathbf{I}.\mathbf{H}^I_e\\ \hbar\mathscr{H}'_{ss} &= \sum_{k>j}\{A^I_{jk}\mathbf{I}_j.\mathbf{I}_k+B^I_{jk}I_{jz}I_{kz}\}+\sum_{j,k'}C^{IS}_{jk'}I_{jz}S_{k'z}+\\ &\quad +\sum_{k'>j'}\{A^S_{j'k'}\mathbf{S}_{j'}.\mathbf{S}_{k'}+B^S_{j'k'}S_{j'z}S_{k'z}\}\end{aligned}\Bigg\}, \quad (93)$$

where the coefficients A, B, C are related to the $\tilde{A}$ and $\tilde{B}$ by

$$\left.\begin{aligned}A^I_{jk} &= \tilde{A}^I_{jk}+\tfrac{1}{2}\tilde{B}_{jk}(3\cos^2\theta_{jk}-1)\\ B^I_{jk} &= -\tfrac{3}{2}\tilde{B}^I_{jk}(3\cos^2\theta_{jk}-1),\\ \text{similar relations exist for } A^S_{j'k'} \text{ and } B^S_{j'k'} \text{ and}\\ C^{IS}_{jk'} &= \tilde{A}^{IS}_{jk'}+(1-3\cos^2\theta_{jk'})\tilde{B}^{IS}_{jk}\end{aligned}\right\}. \quad (94)$$

The effective fields $\mathbf{H}^I_e$ and $\mathbf{H}^S_e$ are given by

$$\begin{aligned}\mathbf{H}^I_e &= \left(H_0+\frac{\omega}{\gamma_I}\right)\mathbf{k}+H_1\mathbf{i},\\ \mathbf{H}^S_e &= \left(H_0+\frac{\omega}{\gamma_S}\right)\mathbf{k}+H_1\mathbf{i}.\end{aligned} \quad (95)$$

Since the applied frequency ω is in the neighbourhood of $\omega_I=-\gamma_I H_0$, and is thus very different from $\omega_S=-\gamma_S H_0$, the z-component $(H_0+\omega/\gamma_S)$ of the effective field $\mathbf{H}^S_e$ seen by the non-resonant spins S is much larger than its transverse component H_1 which can be safely neglected and the first term of (93) can be rewritten as

$$\hbar Z^*_S = -\gamma_S\hbar\mathbf{S}.\mathbf{H}^S_e \cong -\gamma_S\hbar S_z\left(H_0+\frac{\omega}{\gamma_S}\right). \quad (96)$$

It is then clear that it commutes with all the other terms of (93) and that no energy exchange can take place between this term and all the others. On the other hand, the effective Zeeman Hamiltonian

$$\hbar Z^*_I = -\gamma_I\hbar\mathbf{I}.\mathbf{H}^I_e$$

of the spins I does not commute with the other terms of (93) (unless

$H_0+\omega/\gamma_I = (\omega-\omega_0)/\gamma_I$ is very much larger than H_1 so that $\mathbf{H}_e^I$ is also practically aligned along Oz). We can then expect a spin temperature to be established between all the degrees of freedom of $\mathcal{H}^*$ except those corresponding to the first term $Z_S^* \cong -\gamma_S S_z(H_0+\omega/\gamma_S)$.

The situation is similar to that of a hypothetical experiment where, given two spin systems **S** and **I** interacting with each other, one would be able to keep the spins **I** in a low field while maintaining the spins S in a high field.

Nothing is changed in the results established for the adiabatic fast passage of the spins I when no spins S are present, except $H_L'^2$ which now becomes

$$H_L'^2 = \frac{\text{tr}\{\mathcal{H}_{ss}'^2\}}{\text{tr}\{\gamma_I^2 I_z^2\}},\qquad(97)$$

where $\mathcal{H}_{ss}'$ is given by (93).

A straightforward but somewhat tedious calculation of traces yields for two ingredients:

$$H_L'^2 = \tfrac{1}{3}\overline{(\Delta H^2)}_{II} + \tfrac{1}{3}[f_S\gamma_S^2 S(S+1)][f_I\gamma_I^2 I(I+1)]^{-1}\overline{(\Delta H^2)}_{SS} + \overline{(\Delta H^2)}_{IS} +$$
$$+ I(I+1)(f_I N\hbar^2\gamma_I^2)^{-1}\Big\{\sum_{j>k}\tilde{A}_{jk}^{I^2} + S^2(S+1)^2 I^{-2}(I+1)^{-2}\sum_{j'>k'}\tilde{A}_{j'k'}^{S^2}\Big\}.\qquad(98)$$

In (98), N is the total number of all spins, $f_I N$ and $f_S N$ the numbers of spins I and S respectively, $\overline{(\Delta H^2)}_{II}$ the contribution to the second moment of the resonance I from the spins I, $\overline{(\Delta H^2)}_{IS}$ the contribution to that moment from the spins S, and $\overline{(\Delta H^2)}_{SS}$ the contribution to the second moment of the resonance S from the spins S.

A striking feature of (98) is the appearance of the term $\tfrac{1}{3}(f_S/f_I)\overline{(\Delta H^2)}_{SS}$. If the spins I are much less numerous than the spins S, f_S/f_I and $H_L'^2$ may become very large and the signal, proportional to $M_0 H_1/[H_1^2+H_L'^2]^{\frac{1}{2}}$, very small.

Results in qualitative agreement with (98) were observed for rapid passage in LiF. The ratio of the equilibrium magnetization of Li[7] and Li[6], because of the small isotopic proportion (and also smaller magnetic moment) of the latter, is of the order of 160. With an r.f. field of 0·5 gauss, signal-to-noise ratios of the order of several hundred were observed for Li[7] in nearly reversible fast passages. Signals of Li[6] with signal-to-noise ratios of a few units were also observed, in qualitative agreement with the ratio 160 of their static magnetizations (see Fig. IX, 5). However, these passages were irreversible and upon sweeping the field back through resonance no signal of Li[6] could be seen. On the other hand, if one passed through resonance very slowly in an attempt to produce a reversible passage, no signal at all was seen. These results are consistent with the assumption that for a reversible passage the ratios of the signals of Li^7 and Li^6 are

$$\frac{S(7)}{S(6)} \cong \frac{M_0(7)}{M_0(6)} \times \frac{H'_L(6)}{H'_L(7)}.$$

Since, according to (98), $H'_L(6) \gg H'_L(7)$, this is consistent with the signal of Li^6 being well below the noise level if the passage is reversible. If, however, the resonance is crossed very rapidly, the effective Zeeman energy $Z^*(\text{Li}^6)$ has no time to come into equilibrium with the large spin-spin couplings among the other spins, its transverse magnetization is accordingly larger, and the signal can be seen. The price that has to be paid for it is irreversibility.

The attempt to cross the resonance reversibly, accepting the situation that no signal would be seen during this passage, also failed, for after crossing the resonance very slowly (without signal) one way, no signal was seen upon crossing it rapidly on the way back. This point will be discussed shortly.

It is to be hoped that more quantitative experiments will be performed in the future to test this part of the theory.

Dynamical problems connected with the rapid passage, such as the rapidity with which thermal equilibrium is established between the effective Zeeman Hamiltonian and the spin-spin coupling, and the permissible rate of change dH_0/dt of the d.c. field, are similar to those encountered in the process of isentropic demagnetization and discussed in Chapter V. In view of their difficulty and the scarcity of experimental data only a few brief indications will be given.

It is clear that energy will only be exchanged between Z^* and $\mathcal{H}'_{ss}$ if these operators do not commute, and it is useful for that purpose and also for later discussion to rewrite the Hamiltonian $\mathcal{H}^* = Z^* + \mathcal{H}'_{ss}$ in a manner which exhibits explicitly the part of $\mathcal{H}'_{ss}$ that does not commute with Z^*. We introduce (assuming for simplicity a single spin species) the components I_Z, I_X, I_Y of the angular momentum, along the effective field and at right angles to it respectively. We define

$$I_\pm = I_X \pm i I_Y.$$

We may for instance take

$$\begin{aligned} I_Z &= I_z \cos\Theta + I_x \sin\Theta, \\ I_X &= -I_z \sin\Theta + I_x \cos\Theta, \\ I_Y &= I_y. \end{aligned} \qquad (99)$$

With a little algebra we obtain

$$\hbar \mathcal{H}^* = -\gamma \hbar H_e I_Z + \sum_{k>j} \{A_{jk\rho} \mathbf{I}_j \cdot \mathbf{I}_k + B_{jk\rho} I_{jZ} I_{kZ} +$$
$$+ D_{jk\rho}[I_{j+} I_{k+} + I_{j-} I_{k-}] + E_{jk\rho}[(I_{j+} + I_{j-})I_{kZ} + I_{jZ}(I_{k+} + I_{k-})]\}, \quad (100)$$

where, since A_{jk} and B_{jk} are being defined by (86), and $\tilde{A}_{jk}$ and $\tilde{B}_{jk}$ by (84), we have

$$\left.\begin{array}{l} A_{jk\rho} = \tilde{A}_{jk} + \tfrac{1}{2}(3\cos^2\Theta - 1)(A_{jk} - \tilde{A}_{jk}) \\ B_{jk\rho} = \tfrac{1}{2}(3\cos^2\Theta - 1)B_{jk} \\ D_{jk\rho} = \tfrac{1}{4}\sin^2\Theta\, B_{jk} \\ E_{jk\rho} = \tfrac{1}{2}\sin\Theta\cos\Theta\, B_{jk}. \end{array}\right\} \quad (101)$$

Far from resonance, the angle Θ is very small, and so are the terms D and E of (100) that do not commute with I_Z and are responsible for thermal mixing, the term D being proportional to Θ^2 and thus smaller than E which is proportional to Θ only.

Let us first assume $H_1 \ll H'_L$. In an order of magnitude calculation, we may assume the matrix elements of A and B in (101) to be of the order of $\gamma\hbar H'_L$ and those of E of the order of $\Theta\gamma\hbar H'_L$. The important region for the thermal mixing is that where H_e becomes comparable to H'_L and thus Θ is of the order of H_1/H'_L. Finally, the width of the transition induced by E is also of the order of $\gamma\hbar H'_L$. The inverse of the time constant τ for the transfer of energy between $\hbar Z^*$ and $\hbar \mathcal{H}'_{ss}$ is of the order of

$$\frac{1}{\tau} \sim \frac{1}{\hbar} E^2 \frac{1}{\hbar\gamma H'_L} \simeq \left(\frac{H_1}{H'_L}\right)^2 \gamma^2 H'^2_L \frac{1}{\gamma H'_L} \simeq \frac{\gamma H_1^2}{H'_L}. \quad (102)$$

The reversibility would require that the relative change of H_e, which is in the neighbourhood of H'_L, be small during the time τ:

$$\tau \frac{dH_0}{dt} \frac{1}{H'_L} \ll 1, \qquad \frac{dH_0}{dt} \ll \frac{H'_L}{\tau}, \quad \text{or} \quad \frac{dH_0}{dt} \ll \gamma H_1^2, \quad (103)$$

which is the same condition as for free spins. Since this is also the condition if $H_1 \gg H'_L$ it is not unreasonable to expect it to be valid for all values of H_1/H'_L.

These considerations may break down completely in some special cases. Thus, for instance, for Li⁶ in LiF, off-diagonal matrix elements coupling $Z^*(\text{Li}^6)$ to spin-spin interactions will be one or two orders of magnitude smaller than H'_L, which in that case is of the order of 200 gauss, and the mixing time τ will be increased by several orders of magnitude, in accordance with failure to produce a reversible passage for Li⁶ in LiF even without signal, as mentioned earlier.

C. Spin temperature in the rotating frame, steady state solutions

In order to calculate the steady-state value of the nuclear magnetization it is necessary to introduce the coupling of the spin system with the lattice. We shall assume that this coupling is weak in comparison with the interactions within the spin system. More specifically we assume that the latter are sufficiently strong to maintain a temperature within the spin system at all times, in the rotating frame, and that the effect of the spin-lattice interaction is to change this temperature slowly until it reaches an equilibrium value. A similar approach has already been used in Chapter IX to calculate the spin-lattice relaxation time in low fields. There is, however, an important difference between the two problems. In the laboratory frame and in the absence of r.f. fields, the spins and the lattice are two ordinary thermodynamic systems and it is clear that an equilibrium between them will exist when their temperatures are equal. No such simple prediction can be made in the present problem since spin and lattice temperatures are defined in two different frames of reference and there is a danger of being led astray by intuitive physical arguments in this unfamiliar situation. For this reason we shall first obtain the result by using the general formalism of spin-lattice relaxation developed in Section I, and only then attempt to rederive it by more elementary arguments.

If for simplicity we assume a single spin species, the total Hamiltonian of the spin system is

$$\hbar E(t) = \hbar\{\omega_0 I_z + \omega_1(I_x \cos \omega t + I_y \sin \omega t) + \mathscr{H}_{ss}\}, \tag{104}$$

where the first term is much larger than the others. According to equation (13′), the rate of change of the density matrix σ, which now describes in the laboratory frame the whole spin system, is given by

$$\frac{d\sigma}{dt} = -i[E(t), \sigma] -$$

$$-\frac{1}{2} \int_{-\infty}^{\infty} \overline{\left[\mathscr{H}_1(t), \left[U(t-\tau, t)\mathscr{H}_1(t-\tau)U^{-1}(t-\tau, t), \sigma + \frac{\hbar}{AkT_L}\tilde{E}(t-\tau, t)\right]\right]} d\tau \tag{105}$$

where $U(t-\tau, t)$ is defined by (8′), $\tilde{E}(t-\tau, t)$ by (12), and T_L is the lattice temperature.

We shall assume that the correlation time for the spin-lattice interaction is very short (extreme narrowing) as would be the case for instance in nuclear relaxation by conduction electrons in metals. The equation (105) becomes much simpler, for we may then replace $U(t-\tau, t)$

by unity, $\tilde{E}(t-\tau,t)$ by $E(t-\tau) \cong \mathcal{H}_0$, and the integral by a simple product, and write

$$\frac{d\sigma}{dt} \cong -i[E(t),\sigma] - \tau_c\left[\mathcal{H}_1(t), \left[\mathcal{H}_1(t), \sigma + \frac{\hbar\omega_0 I_z}{AkT_L}\right]\right]. \tag{106}$$

Going into the rotating frame, that is, introducing

$$\tilde{\sigma} = e^{i\omega I_z t}\sigma e^{-i\omega I_z t}, \qquad \tilde{\mathcal{H}}_1(t) = e^{i\omega I_z t}\mathcal{H}_1(t)e^{-i\omega I_z t}$$

we get $\quad \dfrac{d\tilde{\sigma}}{dt} = -i[\mathcal{H}^*,\tilde{\sigma}] - \tau_c\left[\tilde{\mathcal{H}}_1(t), \left[\tilde{\mathcal{H}}_1(t), \tilde{\sigma} + \dfrac{\hbar\omega_0 I_z}{AkT_L}\right]\right], \qquad (107)$

where $\mathcal{H}^*$ is the effective Hamiltonian in the rotating frame given by (86).

We now introduce the assumption that $\tilde{\sigma}$ is actually a Boltzmann matrix $\tilde{\sigma}_B$ in the rotating frame

$$\tilde{\sigma} \propto e^{-\hbar\mathcal{H}^*/kT_S} \quad \text{or} \quad \tilde{\sigma} \cong \frac{1}{A}\left\{1 - \frac{\hbar\mathcal{H}^*}{kT_S}\right\}.$$

The first commutator of (107) then vanishes and, writing

$$\frac{\hbar}{AkT_S} = \beta_S, \qquad \frac{\hbar}{AkT_L} = \beta_L,$$

this equation can be rewritten as

$$\mathcal{H}^*\frac{d\beta_S}{dt} = -\tau_c[\tilde{\mathcal{H}}_1(t), [\tilde{\mathcal{H}}_1(t), \mathcal{H}^*\beta_S - \omega_0 I_z\beta_L]]. \tag{108}$$

Multiplying both sides of (108) by $\mathcal{H}^*$ and taking the trace with respect to spin variables, we get

$$\frac{d\beta_S}{dt} = \tau_c\frac{\text{tr}\{[\tilde{\mathcal{H}}_1(t),\mathcal{H}^*]\cdot[\tilde{\mathcal{H}}_1(t),\mathcal{H}^*\beta_S - \omega_0 I_z\beta_L]\}}{\text{tr}\{\mathcal{H}^{*2}\}} \tag{109}$$

which bears some resemblance to equation (17) of Chapter IX. By writing that (109) is zero, we obtain the steady-state value of

$$\frac{\beta_S}{\beta_L} = \frac{T_L}{T_S},$$

$$\frac{\beta_S}{\beta_L} = \frac{\text{tr}\{[\tilde{\mathcal{H}}_1(t),\mathcal{H}^*]\cdot[\tilde{\mathcal{H}}_1(t),\omega_0 I_z]\}}{\text{tr}\{[\tilde{\mathcal{H}}_1(t),\mathcal{H}^*]\cdot[\tilde{\mathcal{H}}_1(t),\mathcal{H}^*]\}}. \tag{110}$$

We assume that the spin-lattice relaxation mechanism can be represented by the coupling of the nuclear spins with a local random field of the form $\mathcal{H}_1(t) = \sum \mathbf{A}_p(t)\cdot\mathbf{I}_p$, where the random fields $\mathbf{A}_p$ and $\mathbf{A}_q$ at two different spins are incoherent. We recall that

$$\mathcal{H}^* = Z^* + \mathcal{H}'_{ss} = -\gamma H_e\{I_z\cos\Theta + I_x\sin\Theta\} + \mathcal{H}'_{ss},$$

and so a simple calculation similar to that leading to formula (19) of Chapter IX gives

$$\frac{\beta_S}{\beta_L} = \frac{T_L}{T_S} = \frac{H_0 H_e \cos\Theta}{H_e^2 + 2H_L'^2}, \tag{111}$$

where $H_L'^2$ is given by (87).

The factor 2 in the denominator of (111) has an origin similar to that of the ratio between the nuclear relaxation times in high and low fields and could be expressed in simple terms by saying that the relative (logarithmic) rate of relaxation of an expression that is bilinear with respect to the spins such as $\mathcal{H}_{ss}'$, is twice as fast as that of a linear expression such as a Zeeman coupling. For more complicated relaxation mechanisms one may expect (111) to be approximately valid with $H_e^2 + \alpha H_L'^2$ in the denominator, where α is a number of the order of unity.

The magnetic moment M_Z in the rotating frame is aligned along the effective field and related to the static equilibrium value $M_0 = \chi_0 H_0$ in the absence of r.f. field by

$$M_Z = M_0 \frac{H_e}{H_0} \frac{\beta_S}{\beta_L} = \frac{M_0 \cos\Theta}{1 + 2H_L'^2/H_e^2}, \tag{112}$$

which expresses the Curie law in either frame, laboratory or rotating.

The elementary argument that leads to (112) is as follows (3). We assume that for each spin $\mathbf{I}_j$, the coupling with the lattice causes a change in its expectation value given by

$$\left[\frac{\partial}{\partial t}\right]_{SL} \langle \mathbf{I}_j \rangle = -\frac{1}{T_1}\{\langle \mathbf{I}_j \rangle - \mathbf{I}_0\} \tag{113}$$

(the index SL stands for spin lattice), where

$$\mathbf{I}_0 = \frac{\chi_0 \mathbf{H}_0}{\gamma \hbar N} = \frac{\mathbf{M}_0}{\gamma \hbar N} = \frac{\gamma \hbar \mathbf{H}_0 I(I+1)}{3kT_L}.$$

The rate of change of $\langle \mathbf{I}_j \rangle$ as given by (113) can be considered as a sum of two processes. The first, which would be the only one if the lattice temperature T_L were infinite, corresponds for each spin I_j to a change during a time δt equal to $\delta_R \langle \mathbf{I}_j \rangle = -(1/T_1)\langle \mathbf{I}_j \rangle \delta t$. There is no preferred direction for that change (whence the subscript R for random). The corresponding change in the effective Zeeman energy, $\langle \hbar Z^* \rangle = -M_Z H_e$, is

$$\delta_R(\hbar Z^*) = \delta_R(-M_Z H_e) = \frac{M_Z H_e}{T_1} \delta t,$$

and the change in the spin-spin effective energy is

$$\delta_R \langle \hbar \mathcal{H}'_{ss} \rangle = -2 \frac{\langle \mathcal{H}'_{ss} \rangle}{T_1} \delta t,$$

the factor 2 being justified by the bilinear dependence of $\mathcal{H}'_{ss}$ on the spins.

The second term of (113) corresponds to the same change

$$\delta_B \langle \mathbf{I}_j \rangle = \frac{\mathbf{I}_0}{T_1} \delta t$$

for all the spins (the subscript B stands for Boltzmann). The corresponding change of $-\mathbf{M}.\mathbf{H}_e = -M_Z H_e$ is $(-M_0 H_e \cos \Theta)/T_1$ since $\mathbf{M}_0 = N\hbar \gamma \mathbf{I}_0$ makes an angle Θ with the effective field $\mathbf{H}_e$. On the other hand, a change $\delta_B \langle \mathbf{I}_j \rangle$, the same for all spins, brings no contribution to $\langle \mathcal{H}'_{ss} \rangle$. This can be seen as follows: the effective spin-spin energy $\langle \mathcal{H}'_{ss} \rangle$ is the sum (or rather the half sum) of the Zeeman energies of each spin in the local field produced by its neighbours. These local fields have different values and orientations for different spins, their sum is zero and a uniform change $\delta_B \langle \mathbf{I}_j \rangle = (\mathbf{I}_0/T_1) \delta t$ for all spins, does not change $\langle \mathcal{H}'_{ss} \rangle$.

The steady-state condition is obtained by writing that $\delta_{SL} \langle \mathcal{H}^* \rangle = 0$, or

$$\frac{M_Z H_e}{T_1} - \frac{2 \langle \hbar \mathcal{H}'_{ss} \rangle}{T_1} - \frac{M_0 H_e \cos \Theta}{T_1} = 0. \tag{114}$$

From the assumption of a spin temperature it follows that

$$\frac{\langle \hbar \mathcal{H}'_{ss} \rangle}{-M_Z H_e} = \frac{\langle \mathcal{H}'_{ss} \rangle}{\langle Z^* \rangle} = \frac{\mathrm{tr}\{\tilde{\sigma}_B \mathcal{H}'_{ss}\}}{\mathrm{tr}\{\tilde{\sigma}_B Z^*\}} = \frac{H_L'^2}{H_e^2}, \tag{115}$$

and, combining (114) and (115), the result (112) follows.

If there are two spin species in the sample, the 'resonant' spins I and the non-resonant spins S, with relaxation times T_1^I and T_1^S, in the spin-spin Hamiltonian $\mathcal{H}'_{ss}$ there will be terms $\mathcal{H}'^I_{ss}$, $\mathcal{H}'^S_{ss}$, and $\mathcal{H}'^{IS}_{ss}$ representing interactions among the spins I, the spins S, and between the spins I and S. The contribution of these various terms to $[\partial/\partial t]_{SL} \langle \mathcal{H}'_{ss} \rangle$ will be

$$-\frac{2}{T_1^I} \langle \mathcal{H}'^I_{ss} \rangle, \quad -\frac{2}{T_1^S} \langle \mathcal{H}'^S_{ss} \rangle, \quad -\left(\frac{1}{T_1^I} + \frac{1}{T_1^S}\right) \langle \mathcal{H}'^{IS}_{ss} \rangle, \tag{116}$$

respectively. The formula (112) will still be valid with a value of $H_L'^2$ given by (98) if the two relaxation times T_1^I and T_1^S are equal. If they are different it follows from (116) that $H_L'^2$ should be replaced in (112) by

$$H_L''^2 = \mathrm{tr}\left\{\mathcal{H}'^{I2}_{ss} + \frac{T_1^I}{T_1^S} \mathcal{H}'^{S2}_{ss} + \frac{1}{2}\left(1 + \frac{T_1^I}{T_1^S}\right) \mathcal{H}'^{IS2}_{ss}\right\}$$

which, according to (98), gives

$$H_L''^2 = \tfrac{1}{3}\overline{(\Delta H^2)}_{II} + I(I+1)(f_I N\hbar^2\gamma_I^2)^{-1}\left(\sum_{j>k} \tilde{A}_{jk}^{I2} + \frac{S^2(S+1)^2}{I^2(I+1)^2}\frac{T_1^I}{T_1^S}\sum_{j'>k'}\tilde{A}_{j'k'}^{S2}\right)$$
$$+ \frac{1}{3}\frac{f_S}{f_I}\frac{\gamma_S^2}{\gamma_I^2}\frac{S(S+1)}{I(I+1)}\frac{T_1^I}{T_1^S}\overline{(\Delta H^2)}_{SS} + \frac{1}{2}\left(1 + \frac{T_1^I}{T_1^S}\right)\overline{(\Delta H^2)}_{IS}. \quad (117)$$

If $T_1^S < T_1^I$, (117) is larger than (98) and the steady-state magnetization is reduced.

The real r.f. susceptibility χ' is immediately obtainable from (112) since it is simply

$$\chi' = \frac{M_x}{2H_1} = \frac{M_z \sin\Theta}{2H_1} = \frac{-M_0\gamma(\omega_0-\omega)}{2[(\omega_0-\omega)^2 + \gamma^2(H_1^2 + 2H_L'^2)]}, \quad (118)$$

and at resonance its derivative is

$$\left(\frac{\partial\chi'}{\partial H_0}\right)_{\omega=\omega_0} = \frac{M_0}{2[H_1^2 + 2H_L'^2]}. \quad (119)$$

The assumption of a magnetization aligned along the effective field leads to a vanishing value for the imaginary susceptibility χ''. However, this assumption is only approximately true (it becomes the more true as the spin-lattice relaxation time T_1 is increased) and the value of χ'' is obtained from the energy relation

$$2\omega\chi''H_1^2 = \frac{M_0-M_z}{T_1}H_0 = \frac{M_0-2H_1\chi'\cot\Theta}{T_1}H_0, \quad (120)$$

whence
$$\chi'' = \frac{\gamma^2(H_1^2+2H_L'^2)M_0 H_0}{2[(\omega-\omega_0)^2+\gamma^2(H_1^2+2H_L'^2)]T_1 H_1^2 \omega}, \quad (121)$$

which is a Lorentzian line with a width $\gamma[H_1^2+2H_L'^2]^{\frac{1}{2}}$.

It is noticeable that at resonance the value of χ'' coincides with the asymptotic value $M_0/2\gamma T_1 H_1^2$ predicted by the Bloch equations, as soon as $\gamma^2 H_1^2 T_1 T_2 \gg 1$, which explains the relative success of these equations in describing the saturation of the absorption in solids.

On the other hand, the asymptotic value of $\partial\chi'/\partial H_0$ at resonance predicted by the Bloch equations is

$$\left(\frac{\partial\chi'}{\partial H_0}\right)_{\text{Bloch}} \to \frac{M_0}{2H_1^2}\frac{T_2}{T_1}$$

which in the limit of very large H_1 is smaller than the value (119) by a factor of the order of T_2/T_1, that is, in most solids by several orders of magnitude. The above theory is only valid if the spin-lattice relaxation can be considered as a small perturbation of the motion of the spins in the rotating frame, that is, in the range of H_1 well above the level where the absorption begins to saturate.

For values of H_1 smaller than these but sufficiently large to saturate the absorption appreciably, no theoretical description is available at present.

For a comparison with experiment of the formulae (119) and (121) giving $\partial \chi'/\partial H_0$ and χ'' the reader is referred to the reference (3). The agreement is quite good and has to be contrasted with the glaring discrepancy with experiment of the predictions of the equations of Bloch. A particular point of interest is the possibility of obtaining from high level measurements using the formula (88) which gives $H_L'^2$, more information about the scalar spin-spin interactions $\tilde{A}_{jk}$ than is obtainable from the observation of the unsaturated absorption shape. There the scalar couplings do not contribute to the second moment of the resonance line and, unless they are much larger than the dipolar interactions, they affect its shape in a complicated way.

D. Spin-lattice relaxation in the rotating frame

(a) *Relaxation for a single spin species*

It has already been mentioned in Chapter IX that if a spin system could be described by a temperature T_S, the spin-lattice relaxation time would be defined unambiguously by the relation

$$\frac{d}{dt}\left(\frac{1}{T_S}\right) = -\frac{1}{T_1}\left[\frac{1}{T_S} - \left(\frac{1}{T_S}\right)_0\right]. \tag{122}$$

The only peculiarity of a spin temperature defined in the rotating frame is that its equilibrium value $(T_S)_0$ is different from the lattice temperature T_L, to which it is related by equation (111).

The expectation value $\langle A \rangle$ of any operator A with a vanishing trace is proportional to $\mathrm{tr}[\{1-(\mathscr{H}^*/kT_S)\}A] \propto 1/T_S$ and thus also obeys the equation (122) and relaxes towards its equilibrium value $\langle A \rangle_0$ with the same time constant. This constant is easily computed using the same assumptions as in Section II C, namely a nuclear relaxation by random uncorrelated fields with a very short correlation time. Using the basic equation (113) we get, for a sample with a single spin species,

$$\left[\frac{\partial}{\partial t}\right]_{SL} \langle Z^* \rangle = -\frac{1}{T_1}\{\langle Z^* \rangle - \langle Z^* \rangle_0\},$$

$$\left[\frac{\partial}{\partial t}\right]_{SL} \langle \mathscr{H}'_{ss} \rangle = -\frac{2}{T_1} \langle \mathscr{H}'_{ss} \rangle, \tag{123}$$

where T_1 is the spin-lattice relaxation time in a high d.c. field and in the absence of r.f. field. Adding the two equations (123) and making use of equation (115)

$$\frac{\langle Z^*\rangle}{H_e^2} = \frac{\langle \mathscr{H}'_{ss}\rangle}{H_L'^2} = \frac{\langle \mathscr{H}^*\rangle}{H_e^2+H_L'^2},$$

we get $\qquad \left[\dfrac{\partial}{\partial t}\right]_{SL}\langle \mathscr{H}^*\rangle = -\dfrac{1}{T_1}\dfrac{H_e^2+2H_L'^2}{H_e^2+H_L'^2}\{\langle\mathscr{H}^*\rangle - \langle\mathscr{H}^*\rangle_0\}.$ (124)

The relaxation time in the rotating frame, which will be, for instance, that of the magnetization along the effective field, and which we shall denote by $T_{1\rho}$, is related to T_1, the spin-lattice relaxation time in high d.c. field and zero r.f. field, by

$$\frac{1}{T_{1\rho}} = \frac{1}{T_1}\frac{H_e^2+2H_L'^2}{H_e^2+H_L'^2}, \qquad (125)$$

which bears a strong resemblance to the equation (19) of Chapter IX, in so far as it depends on the relative magnitude of H_e and H_L'; $1/T_{1\rho}$ varies between $1/T_1$ and $2/T_1$.

If H_1 and thus H_e become appreciably larger than H_L', the exchange of energy between the effective Zeeman Hamiltonian Z^* and the effective spin-spin interaction $\mathscr{H}'_{ss}$ is slowed down. The time constant τ for this exchange, sometimes called cross-relaxation time, can be expected to grow as $\exp\{+\alpha(H_e^2/H_L'^2)\}$, where α is a factor of order unity, and for values of $H_e^2 \gg H_L'^2$ to become much longer than the spin-lattice relaxation time T_1. The hypothesis of a unique temperature for Z^* and $\mathscr{H}'_{ss}$ breaks down and both energies relax independently. This distinction is of little practical importance since, as has been pointed out in Chapter V, at the point where Z^* and $\mathscr{H}'_{ss}$ become uncoupled, Z^* is the main part of the effective Hamiltonian anyway. The relaxation time $T_{1\rho}$ found for $\langle Z^*\rangle$ is then clearly T_1, which is also what is found from (125) in the limit $H_e^2 \gg H_L'^2$, and there is no discontinuity in the dependence of $T_{1\rho}$ on H_1 as the latter is made to grow. We shall see shortly that the situation may be quite different if there are two spin species in the sample.

These results have to be modified if the assumption of a very short correlation time τ_c for the spin-lattice interaction is not valid. If τ_c is so long that $|\omega_0 \tau_c| = |\gamma H_0 \tau_c|$ is not small (but still sufficiently short to have $|\gamma H_1 \tau_c| \ll 1$), the basic equation (113) has to be replaced by

$$\left[\frac{\partial}{\partial t}\right]_{SL}\langle I_{jz}\rangle = -\frac{1}{T_{1z}}\{\langle I_{jz}\rangle - I_0\},$$

$$\left[\frac{\partial}{\partial t}\right]_{SL}\langle I_{jx}\rangle = -\frac{1}{T_{1x}}\langle I_{jx}\rangle. \qquad (126)$$

In (126) we write T_{1z} and T_{1x} rather than T_1 and T_2 to avoid a confusion between T_{1x} and the much shorter constant T_2 which is independent of the spin-lattice coupling.

For simplicity we assume $H_1 \gg H'_L$ so that the effective spin-spin couplings and the Zeeman energy are uncoupled from each other. The existence of a spin temperature for Z^* implies that the component of the magnetization perpendicular to the effective field $\mathbf{H}_e$ vanishes:

$$\langle I_X \rangle = -\langle I_z \rangle \sin\Theta + \langle I_x \rangle \cos\Theta = 0,$$

and consequently

$$\langle I_z \rangle = \langle I_Z \rangle \cos\Theta, \qquad \langle I_x \rangle = \langle I_Z \rangle \sin\Theta. \tag{127}$$

Combining (126) and (127), we find

$$\frac{d}{dt}\langle M_Z \rangle = -\frac{1}{T_{1\rho}}\{\langle M_Z \rangle - M_\rho\},$$

where

$$\frac{1}{T_{1\rho}} = \frac{\cos^2\Theta}{T_{1z}} + \frac{\sin^2\Theta}{T_{1x}}, \qquad M_\rho = M_0 \cos\Theta\left(\frac{T_{1\rho}}{T_{1z}}\right). \tag{128}$$

(b) *Relaxation in the presence of two spin species*

We assume again that the correlation time of the spin-lattice coupling is very short.

We consider first values of H_1 comparable with the local field so that a single temperature exists in the rotating frame for the whole effective Hamiltonian

$$\mathscr{H}^* = Z^* + \mathscr{H}'_{ss} = Z^* + \mathscr{H}'^I_{ss} + \mathscr{H}'^S_{ss} + \mathscr{H}'^{IS}_{ss}. \tag{129}$$

From the relation

$$\left[\frac{\partial}{\partial t}\right]_{SL} \langle \mathscr{H}^* \rangle = -\frac{1}{T_1^I}\{\langle Z^* \rangle - \langle Z^* \rangle_0\} - \frac{2}{T_1^I}\langle \mathscr{H}'^I_{ss} \rangle - \frac{2}{T_1^S}\langle \mathscr{H}'^S_{ss} \rangle -$$
$$-\left(\frac{1}{T_1^I} + \frac{1}{T_1^S}\right)\langle \mathscr{H}'^{IS}_{ss} \rangle, \tag{130}$$

we get by the same method as for a single species

$$\frac{1}{T_{1\rho}} = \frac{1}{T_1^I} \frac{\text{tr}\{Z^{*2} + 2\mathscr{H}'^{I^2}_{ss} + 2(T_1^I/T_1^S)\mathscr{H}'^{S^2}_{ss} + (1+T_1^I/T_1^S)\mathscr{H}'^{IS^2}_{ss}\}}{\text{tr}\{\mathscr{H}^{*2}\}}$$

$$= \frac{1}{T_1^I} \frac{H_e^2 + 2H''^2_L}{H_e^2 + H'^2_L}, \tag{131}$$

where

$$H'^2_L = \frac{\text{tr}\{\mathscr{H}'^2_{ss}\}}{\text{tr}\{\gamma^2 I_Z^2\}},$$

$$H''^2_L = \frac{\text{tr}\{\mathscr{H}'^{I^2}_{ss} + (T_1^I/T_1^S)\mathscr{H}'^{S^2}_{ss} + \tfrac{1}{2}(1+T_1^I/T_1^S)\mathscr{H}'^{SI^2}_{ss}\}}{\text{tr}\{\gamma^2 I_Z^2\}} \tag{132}$$

are given by (98) and (117), respectively.

If T_1^S is shorter than T_1^I, $H_L''^2$ will be larger than $H_L'^2$, and according to (131) and (125) the relaxation time $T_{1\rho}$ will be shorter than for a single spin species. The physical reason for the shortening of $T_{1\rho}$ is clear, the spins S are closely coupled to the lattice; since at the same time in the rotating frame their effective spin-spin energy is even more tightly bound to that of the spins I, the magnetization of the spins I is also closely coupled to the lattice. In the limiting case when $T_1^S \ll T_1^I$ and consequently $H_L''^2 \gg H_L'^2$, we get from (131)

$$\frac{1}{T_{1\rho}} \simeq \frac{1}{T_1^S} \frac{\mathrm{tr}\{2\mathcal{H}_{ss}'^{S^2}+\mathcal{H}_{ss}'^{IS^2}\}+(T_1^S/T_1^I)\mathrm{tr}\{Z^{*2}\}}{\mathrm{tr}\{\mathcal{H}^{*2}\}} \tag{133}$$

or, using (117) and (132),

$$\frac{1}{T_{1\rho}} \simeq \frac{1}{T_1^S} \frac{2H_L'''^2+(T_1^S/T_1^I)H_e^2}{H_L'^2+H_e^2}, \tag{134}$$

where

$$H_L'''^2 = \frac{1}{3}\frac{f_S}{f_I}\frac{\gamma_S^2\,S(S+1)}{\gamma_I^2\,I(I+1)}(\overline{\Delta H^2})_{SS}+\tfrac{1}{2}(\overline{\Delta H^2})_{IS}+$$
$$+S^2(S+1)^2\{f_I N\hbar^2\gamma_I^2 I(I+1)\}^{-1}\sum_{j'>k'}\tilde{A}_{j'k'}^{S^2} \tag{135}$$

and $H_L'^2$ is given by (98). Unless the proportion f_S of the spins S, or their magnetic moments $\gamma_S\hbar S$, are very small, $1/T_{1\rho}$ is comparable to $1/T_1^S$.

If the r.f. field H_1 increases up to values a few times the local field, the cross-relaxation time for the transfer of energy between Z^* and $\mathcal{H}_{ss}'$, which increases as $\exp\{\alpha(H_e^2/H_L'^2)\}$, becomes very long and the assumption of a temperature common to Z^* and H_{ss}' breaks down.

The relaxation time $T_{1\rho}$ for the spins I becomes simply T_1^I, which is much longer than the value predicted by the formula (134).

An interesting study of the behaviour of a system with two spin species in a strong rotating field has been performed on a single crystal of CsBr where the nuclear spins of caesium are the spins I and those of bromine the spins S (11). The relaxation time T_1^I of caesium, because of its anomalously small quadrupole moment, is

$$(T_1)_{\mathrm{Cs}} \simeq 10^3 \text{ sec} \quad \text{whereas} \quad T_1^S = (T_1)_{\mathrm{Br}} \simeq 0{\cdot}1 \text{ sec}.$$

In this experiment the decay of the nuclear magnetization M_ρ of Cs^{133} along a strong r.f. field rotating at a frequency $\omega \simeq \omega^I = \omega(\mathrm{Cs})$ was studied for various values of H_1.

Before describing the experimental results let us consider briefly what the expected behaviour of that decay should be on the basis of the foregoing theory.

For vanishingly small values of H_1, the decay should be essentially that of the free precession and occur in a time of the order of T_2, that is, of the order of a millisecond. As H_1 is increased a complicated situation arises for which no theory is available. For values of H_1 well above those where the absorption begins to saturate, but possibly still a good deal smaller than the local field, the assumption of a spin temperature for the effective Hamiltonian $\mathcal{H}^*$ becomes valid. The decay of M_ρ is then a spin-lattice phenomenon. It should be exponential and described by a time constant $T_{1\rho}$ comparable to T_1^S, according to equation (133), that is in the present example of the order of 0·1 sec. As H_1 grows further this behaviour should persist until H_1 becomes appreciably larger than the local field and the cross-relaxation time τ for the exchange of energy between Z^* and $\mathcal{H}'_{ss}$ becomes longer than T_1^S. From then on the decay rate should decrease very rapidly (as $\exp\{-\alpha|H_1/H'_L|^2\}$) towards the very small value $1/T_1^I$ and then remain constant.

The investigation (11) was performed in the region of large H_1 when the thermal contact between Z^* and $\mathcal{H}'_{ss}$ breaks down. Surprisingly enough it turns out that over a wide range of values of H_1, between say 3 gauss and up to 20 gauss, the observed $T_{1\rho}$ is very nearly proportional to H_1^2. Its absolute value, which depends on the orientation of the field with respect to the crystal, is, for instance, equal to one minute (that is, it is much shorter than T_1^I), for $H_1 = 11$ gauss, when $\mathbf{H}_0$ is parallel to the [111] crystalline axis, and 1·52 times longer for the [100] direction of the field.

The explanation of this remarkable behaviour is as follows. In the range of values of H_1 considered, the coupling of the spins S with the lattice is stronger than their coupling with the spins I. It is therefore reasonable to consider the spins S as part of the lattice, and their coupling with the spins I as a part of the Hamiltonian $\mathcal{H}_1$ that couples the spins I to the lattice. A similar approach has been used several times in this book and in particular in the study of the mechanism of nuclear relaxation by paramagnetic impurities where the electronic field 'seen' by the nucleus was considered as a random field with a correlation time τ_c equal to the electron relaxation time. In the absence of the rotating field H_1, the relaxation flips of the spins S are a negligible relaxation mechanism for the spins I: the inverse of the relaxation time $1/T_1^S$, which is the inverse of the correlation time for the field 'seen' by the spins I, is negligibly small in comparison with their Larmor frequency ω^I. On the other hand, in the rotating frame, for a study of the decay of the magnetization of the spins I along H_1, the relevant frequency is $\omega_1^I = -\gamma_I H_1$ which is three orders of magnitude smaller, making this relaxation mechanism 10^6 times more effective than in the laboratory frame.

To calculate the relaxation time due to this process, in the coupling between the spins I and S given by (93), namely $\sum_{jk'} C_{jk'}^{IS} I_{jz} S_{k'z}$, we take into account the fact that the spins I are quantized along the direction OZ of the effective field, which makes an angle Θ with Oz, by replacing I_z by

$$I_Z \cos\Theta + I_X \sin\Theta = I_Z \cos\Theta + \tfrac{1}{2}(I_+ + I_-)\sin\Theta.$$

The part of the I–S coupling relevant for the relaxation of the spins I can be rewritten as

$$\hbar \mathcal{H}_1 = \tfrac{1}{2}\sin\Theta \sum_{j,k'} C_{jk'}^{IS} S_{zk'}(I_{j+} + I_{j-}), \tag{136}$$

where the $C_{jk'}^{IS}$ are given by (94). If we assume for each spin S a correlation function $\overline{S_{zk'} \cdot S_{zk'}(t+\tau)} = \tfrac{1}{3} S(S+1)\exp(-\tau/T_1^S)$ and a spectral density $\tfrac{1}{3} S(S+1) 2T_1^S/\{1+\omega^2(T_1^S)^2\}$, the relaxation time of the spins I caused by the coupling (136) is given by the usual formula

$$\frac{1}{T_{1\rho}} = \frac{2}{\hbar^2} \frac{\sin^2\Theta}{4} \frac{2T_1^S}{1+\omega_e^2 T_1^{S^2}} \frac{S(S+1)}{3} \sum_{k'} (C_{j,k'}^{IS})^2, \tag{137}$$

where $\omega_e = -\gamma_I H_e = -\gamma_I H_1/\sin\Theta$. Since $(\omega_e T_1^S) \gg 1$, (137) can be rewritten as

$$\frac{1}{T_{1\rho}} = \frac{1}{\hbar^2} \sin^4\Theta \frac{S(S+1)}{3} \frac{1}{\gamma_I^2 H_1^2 T_1^S} \sum_{k'} (C_{j,k'}^{IS})^2 \tag{138}$$

or

$$T_{1\rho} = T_1^S \frac{1}{\sin^4\Theta} \frac{H_1^2}{(\overline{\Delta H^2})^{IS}}, \tag{139}$$

where

$$(\overline{\Delta H^2})^{IS} = \frac{S(S+1)}{3\gamma_I^2} \sum_{k'} (C_{jk'}^{IS})^2$$

is the contribution from the couplings with the spins S to the second moment of the resonance line of the spins I, and where the proportionality of $T_{1\rho}$ to H_1^2 is exhibited. The measurement of $T_{1\rho}$ for various orientations of the crystal showed that dipolar couplings alone could not account for the observed values of $T_{1\rho}$, and that the existence of indirect interactions, scalar and pseudo-dipolar, between caesium and bromine, had to be assumed as well.

Indeed, the measurement of $T_{1\rho}$, by allowing measurement of $(\overline{\Delta H^2})^{IS}$ alone, provided much more accurate values for the indirect couplings $\tilde{A}_{jk'}$ and $\tilde{B}_{jk'} - \gamma_I \gamma_S \hbar^2/r_{jk'}^3$ than it would have been possible to obtain from second moment measurements where contributions from the I–S couplings and the I–I couplings are present simultaneously.

Another effect that is also based on the fact that the spins S have a much faster relaxation time than the spins I was observed in caesium bromide. This was a dynamic polarization of the caesium spins along the effective field (11).

The large r.f. field driving the caesium nuclei was at exact resonance. After the caesium magnetization had reached its steady state value along H_1, which was zero, an r.f. field H_1' was applied near the Larmor frequency ω_S of bromine, at a frequency $\omega = \omega_S \pm \omega_1^I$ where $\omega_1^I = -\gamma_I H_1$ is the Larmor frequency of the spins I in the field H_1. A nuclear magnetization of the spins I equal to $\pm(\gamma_S/\gamma_I)M_0^I$ could be built up along H_1 in a time much shorter than $T_{1\rho}$ for sufficiently strong r.f. fields H_1'. This effect, referred to in (11) as 'transverse Overhauser effect' or 'Overhauser induction' is formally identical to the 'solid-state effect' described in Section III B of Chapter IX. The theory given there assumed two spin species S and I with Larmor frequencies ω_S and ω_I and a much shorter relaxation time for the spins S than for the spins I. Finally, in order to permit the driving of forbidden transitions $\Delta m_S = 1$, $\Delta m_I = \pm 1$, a spin-spin coupling containing the operators $S_z I_+$ was required.

Under these conditions the saturation of the forbidden transitions at frequencies $\omega_S \pm \omega_I$ was shown to lead for the spins I to a polarization $\langle I_z \rangle$ equal or opposite to that $\langle S_z \rangle$ of the spins S, apart from a factor $I(I+1)/S(S+1)$,

$$\frac{\langle I_z \rangle}{I(I+1)} = \frac{\langle S_z \rangle}{S(S+1)}.$$

All these conditions are realized in the present problem, and the fact that the spins S are quantized along the d.c. field and the spins I along a different field H_1 (also d.c. in the rotating frame) is unimportant.

E. Double irradiation methods

In this section we discuss the principles of some experiments of double irradiation where at least one of the applied r.f. fields is large. The difference from experiments of dynamic polarization described in Chapter IX is that in the phenomena considered here, the coherent character of the motion of the spins submitted to a strong r.f. field plays an essential part.

(a) Rotary saturation

(1) *Rotary saturation in liquids*

When a strong r.f. field is applied to a system of nuclear spins, except at exact resonance where it is very small, the steady state precessing magnetization is very nearly parallel to the effective field. This is true in solids according to the assumption of spin temperature in the rotating frame; it is also true in liquids: for $\gamma^2 H_1^2 T_1 T_2 \gg 1$ and at a distance $H_0 - H^* = H_1 \cot\Theta$ from resonance, the steady-state solutions of the Bloch equations given by equation (15) of Chapter III can be rewritten

$$\frac{\tilde{M}_x}{M_0} \simeq \tan\Theta \Big/ \Big[1 + \frac{T_1}{T_2}\tan^2\Theta\Big],$$

$$\frac{\tilde{M}_y}{M_0} \simeq \frac{\tan^2\Theta}{\gamma H_1 T_2} \frac{1}{[1+(T_1/T_2)\tan^2\Theta]} \ll 1 \quad \text{if } \gamma H_1 T_2 \gg 1,$$

$$\frac{\tilde{M}_z}{M_0} \simeq 1 \Big/ \Big[1 + \frac{T_1}{T_2}\tan^2\Theta\Big], \qquad (140)$$

from which it is apparent that the steady-state magnetization $\mathbf{M}_\rho$, with its three components in the rotating frame given by (140), is very nearly parallel to the effective field $\mathbf{H}_e$. In the rotating frame R the magnetization $\mathbf{M}_\rho$ appears as an equilibrium d.c. magnetization aligned along a d.c. field $\mathbf{H}_e$ and possessing a Larmor frequency $\omega_e = -\gamma \mathbf{H}_e$.

A second field $\mathbf{H}_a$ rotating in a plane perpendicular to $\mathbf{H}_e$, with an angular velocity ω_a with respect to the frame R, will strongly affect the magnetization and make its component M_Z along the effective field depart from its steady-state value M_ρ, if ω_a is in the neighbourhood of ω_e. Such a field is easily produced by means of an oscillating field parallel to the d.c. field $\mathbf{H}_0$. This field is decomposed into two oscillating fields, the first parallel to $\mathbf{H}_e$ and the second in the plane perpendicular to $\mathbf{H}_e$. The latter is decomposed in the usual way into two rotating components. If $\tilde{H}_a$ is the amplitude of the oscillating field, the rotating component of interest has an amplitude $H_a = \tfrac{1}{2}\tilde{H}_a \sin\Theta$.

The correspondence between such a rotary resonance experiment and an ordinary one is as follows:

$$H_e \to H_0, \quad M_\rho \to M_0, \quad H_a \to H_1, \quad M_\rho - M_Z \to M_0 - M_z.$$

Just as in an ordinary resonance experiment the saturation makes the longitudinal magnetization depart from its equilibrium value M_0, so in rotary saturation the component M_Z along the effective field H_e departs from its equilibrium value M_ρ.

The distinction between ordinary and rotary saturations is in the detection. In ordinary resonance the absorption of r.f. energy by the spin system is detected directly. In rotary resonance, for the fields H_1 obtainable in practice, the frequency $\omega_e = -\gamma H_e$ falls in the acoustic range and the corresponding absorption of energy is exceedingly small.

The occurrence of rotary resonance for $\omega_a = \omega_e$ is detected by the change it induces in M_Z, which causes a change of the dispersion signal at the high frequency $\omega \sim -\gamma H_0$. The analogue of that kind of detection in an ordinary resonance experiment would be the observation of a change in the static magnetization, observed, for instance, by conventional susceptibility measurements, when the spin resonance is saturated. The feasibility of such an experiment was discussed briefly at the end of Chapter I.

At exact resonance the magnetization $M_\rho = 2\chi' H_1$ vanishes. However, its derivative with respect to H_0 is a maximum and equal to $M_0 T_2/H_1 T_1$. If the d.c. field H_0 is modulated to either side of the resonant value $H^* = -\omega/\gamma$ under slow-passage conditions, that is, at a frequency $\Omega \ll 1/T_1, 1/T_2, \gamma H_1$, for each $H_0 \neq H^*$, the value $M_\rho(H_0)$ of the quasi-steady state magnetization will be

$$M_\rho(H_0) \cong (H_0 - H^*) \frac{M_0}{H_1} \frac{T_2}{T_1}. \qquad (141)$$

The audio-frequency field $\mathbf{H}_a$ at a frequency

$$\omega_a = -\gamma H_1 \cong -\gamma[(H_0-H^*)^2 + H_1^2]^{\frac{1}{2}}$$

will reduce $M_\rho(H_0)$ to a smaller value $M_Z(H_0) < M_\rho(H_0)$, and the dispersion derivative signal will be decreased.

This experiment provides a method for measuring H_1 that is useful for its calibration (3).

If the amplitude of the audio field and that of the modulation of the d.c. field are not both much smaller than H_1, the rotary saturation curve will be asymmetrical, and its maximum will be shifted towards values of $|\omega_a|$ *higher* than $|\gamma H_1|$. If H_a/H_1 is not very small this is caused by the Bloch–Siegert shift (see Chapter II), and if $|(H_0-H^*)/H_1|$ is not very small, by the fact that during the modulation cycle $H_e > H_1$, both effects being quadratic with respect to $|H_a/H_1|$ or $|(H_0-H^*)/H_1|$.

A more quantitative justification of rotary saturation is as follows. In the rotating frame and in the presence of an audio field $\mathbf{H}_a(t)$ of amplitude H_a, rotating at a frequency $\omega_a \cong \omega_e = -\gamma H_e$, the equation of motion of the magnetization can be written (assuming $T_1 = T_2$ for simplicity)

$$\frac{d\mathbf{M}}{dt} = \gamma \mathbf{M} \wedge \{\mathbf{H}_e + \mathbf{H}_a(t)\} - \frac{\mathbf{M}-\mathbf{M}_0}{T_1}$$

$$= \gamma \mathbf{M} \wedge \{\mathbf{H}_e + \mathbf{H}_a(t)\} - \frac{\mathbf{M}-\mathbf{M}_\rho}{T_1} - \frac{\mathbf{M}_\rho - \mathbf{M}_0}{T_1}. \qquad (142)$$

Since $\mathbf{M}_\rho$ is a solution of $\gamma\mathbf{M}_\rho \wedge \mathbf{H}_e - (\mathbf{M}_\rho - \mathbf{M}_0)/T_1 = 0$, the vector $(\mathbf{M}_\rho - \mathbf{M}_0)/T_1$ is perpendicular to $\mathbf{H}_e$.

If we transform (142) to the doubly rotating frame, rotating at the frequency $\omega_a \simeq \omega_e = -\gamma H_e$ around the vector $\mathbf{H}_e$ very nearly parallel to $\mathbf{M}_\rho$, it becomes

$$\frac{\partial \mathbf{M}}{\partial t} = \gamma \mathbf{M} \wedge \left[\mathbf{H}_e\left(1 - \frac{\omega_a}{\omega_e}\right) + \mathbf{H}_a\right] - \frac{\mathbf{M} - \mathbf{M}_\rho}{T_1} + \text{time-dependent terms}.$$

(143)

The time-dependent terms result from the vector $(\mathbf{M}_\rho - \mathbf{M}_0)/T_1$, which in the doubly rotating frame is seen as rotating at the frequency ω_a. If one assumes that the effect of these terms can be neglected, the equation (143) is the exact analogue of an ordinary Bloch equation in a simply rotating frame, with a d.c. field $\mathbf{H}_e$ and an r.f. field $\mathbf{H}_a$, the relaxation taking place towards the equilibrium d.c. magnetization $\mathbf{M}_\rho$. The treatment of rotary resonance then becomes formally identical with that of ordinary resonance.

If $T_1 \neq T_2$ but if the high frequency ω is very nearly ω_0, that is, if $|H_0 - H^*| \ll 1$, the reader will easily convince himself that the above conclusions are still valid with the following correspondence:

$$H_e \to H_0, \quad H_a \to H_1, \quad M_\rho \to M_0, \quad \frac{1}{T_1} \to \frac{1}{T_2}, \quad \frac{1}{T_2} \to \frac{1}{2}\left(\frac{1}{T_1} + \frac{1}{T_2}\right).$$

(2) *Rotary saturation in solids*

Since according to the assumption of spin temperature in the rotating frame the r.f. magnetization $\mathbf{M}_\rho$ is aligned along the effective field $\mathbf{H}_e$. a change of M_ρ to a smaller value M_Z by means of a field $\mathbf{H}_a$ of a frequency $\omega_a \sim \omega_e = -\gamma H_e$ can be expected to occur, and has been observed (3), in the same manner as in liquids. A quantitative description of the phenomenon is, however, difficult since it requires a theory of saturation in solids for values of the d.c. field (which here is $\mathbf{H}_e$, or at resonance $\mathbf{H}_1$) that may not be large compared to the local field, and by means of an r.f. field (represented here by $\mathbf{H}_a$) that may either be comparable to the applied field, or else fall into the intermediate range of saturation where the theory of Section II is not applicable.

Referring the reader to reference (3) for a discussion of this complex problem, we consider briefly a simplified situation where:

(i) the r.f. field $\mathbf{H}_1$, and *a fortiori* $\mathbf{H}_e$, are much larger than the local field;

(ii) H_a is so small that the relative change $(M_\rho - M_Z)/M_\rho$ is given by a formula similar to (75);

$$\frac{M_\rho - M_Z}{M_\rho} = \gamma^2 H_a^2 \pi f(\omega_a) T_{1\rho}. \tag{144}$$

By $T_{1\rho}$ we mean the relaxation time along the effective field, which is not necessarily equal to T_1, as was shown in the last section.

A rotary saturation experiment then gives the line shape $f(\omega)$.

If, as we assumed, H_e is much larger than the local field, in the spin-spin couplings given by the formulae (100) and (101) only the terms A_ρ and B_ρ should be retained. At resonance, for $\Theta = \frac{1}{2}\pi$, if $\tilde{A}_{jk} = 0$ the terms A_ρ and B_ρ are exactly one-half of the terms A and B, which appear in the calculation of the line shape in ordinary resonance. It follows then that under the assumptions (i) and (ii) quoted above, the rotary saturation line should have exactly the same shape but be half as wide as the unsaturated absorption line in high fields.

It will also be noticed in (101) that for an angle Θ such that $3\cos^2\Theta - 1 = 0$, B_ρ vanishes and an extremely narrow line should be obtained.

(b) *Line narrowing by double frequency irradiation*

We consider again a system with two spin species I and S in a strong rotating field of frequency ω in the neighbourhood of ω_I, but this time we are interested in the influence of this irradiation on the spins S.

We shall begin by giving a description of this effect, which is definitely incorrect but may be illuminating for certain problems where the correct treatment is too difficult.

Assume for simplicity that the broadening of the resonance of the spins S is primarily caused by their coupling with the spins I, as will be the case if γ_I is appreciably larger than γ_S. If the spins I did not flip among themselves, the width Δ_{SI} of the resonance S would be of an inhomogeneous character and one might argue that the r.f. field, by making the spins I flip at a fast rate, would average out the local field 'seen' by the spins S. The criterion for such an averaging is that the probability per unit time W, of the flip of a spin I induced by the r.f. field should be large compared with the unperturbed width expressed in frequency units. Actually, the assumption that the spins I do not flip with each other is incorrect and such flips may have a considerable influence on the shape of the resonance S, as we saw in Chapter IV for the resonance of silver and potassium which are the spins S in AgF and KF, respectively, the spins I being those of fluorine. If the rate of flips among the spins I, which is comparable to the strength Δ_{II} (expressed in frequency units) of the interaction between two neighbouring spins I, is much larger than the strength Δ_{IS} of the interaction between a spin I and a spin S, the resonance line will have a width of the order of $\Delta_{IS}(\Delta_{IS}/\Delta_{II})$ and it is clear that in order to narrow it appreciably any further the rate W of the r.f. induced flips will have to be fast compared with the 'natural' rate, $\Delta_{II} \gg \Delta_{IS}$.

If we assume that this condition is fulfilled, the remaining width is that caused by the couplings of the spins S among themselves, and the measured second moment should then be that calculated from these interactions only. This statement does not really conflict with the one already discussed in Section V of Chapter X, namely that a coupling of the spins I with an r.f. field, which evidently commutes with all the components of the spins S, could not possibly affect the second moment of S. It was argued there that the contributions to the second moment of S from rapidly fluctuating couplings came from regions spread over a large distance in the wings of the resonance S and were for that reason lost in the noise and unobserved.

The approach just described is naturally incorrect in so far as it disregards the coherent nature of the motion of the spins I in the rotating field. A similar remark has already been made in Section I D of this chapter in connexion with the collapse of multiplets induced by r.f. field stirring. One of the aspects of this distinction between coherent and incoherent motion of the spins S is that in the former case, as will be shown shortly, the contributions to the second moment of S from the spins I, instead of spreading more or less uniformly in the wings of the resonance S, are localized in well-defined side bands.

The elementary (but qualitatively correct) argument for predicting a narrowing of the resonance S, caused by the r.f. stirring of the spins I, is as follows (12).

Assume a rotating field $H_1 = -\omega_1/\gamma_I$ applied at a frequency

$$\omega = \omega_I - \omega_1 \cot \Theta,$$

where $\omega_I = -\gamma_I H_0$ is the Larmor frequency of the spins I. If H_1 is much larger than the local field, in the rotating frame the spins I will precess around $\mathbf{H}_e$ with a velocity $\omega_e = -\gamma_I H_e$. It follows that the component of a spin $\mathbf{I}$ along $\mathbf{H}_0$, the only one instrumental in broadening the resonance of S, will have a static part reduced from its value in the absence of r.f. by the factor $\cos \Theta$, and a part varying with the frequency ω_e.

The local field originating in the spins I and 'seen' by the spins S contains then a static part reduced by $\cos\Theta$ and vanishing at exact resonance, and a time-dependent part which corresponds to a modulation of the effective Larmor frequency of the spins S at a frequency ω_e. This gives rise to side-bands in the resonance of S. For sufficiently high values of the frequency ω_e the intensity of the side-bands with separation $\pm\omega_e$ from the central line is much larger than that of all the other side-bands and in turn small compared to the central line. These are the results we now propose to rederive more rigorously.

As shown in Chapter IV, the absorption curve of the spins S is the Fourier transform of the function

$$G(t) = \mathrm{tr}\{S_+(t)S_-\}. \tag{145}$$

where $S_+(t)$ is given by $S_+(t) = V(t)S_+ V^{-1}(t)$ and $V(t)$ is a unitary operator solution of the differential equation

$$\frac{1}{i}\frac{dV}{dt} = \mathscr{H}(t)V \quad \text{with} \quad V(0) = 1.$$

$\hbar\mathscr{H}(t)$ is the total Hamiltonian of the spins including their interactions in the presence of the d.c. field and of the r.f. field rotating at the frequency $\omega \approx \omega_I$. The integrated intensity of the absorption line is proportional to $G(0)$ and the second moment is

$$-\left(\frac{d^2G}{dt^2}\right)_{t=0}\bigg/ G(0).$$

We apply to all the operators A inside the trace symbol in (145) the canonical transformation $e^{-i\omega(I_z+S_z)t}Ae^{i\omega(I_z+S_z)t}$ which does not change the value of the trace. This operation which is simply the transformation to the rotating frame, discussed in Section II B, yields

$$G(t) = e^{i\omega_s t}\mathrm{tr}\{e^{i[-\gamma_I(\mathbf{I}\cdot\mathbf{H}_e)+\mathscr{H}'_{ss}]t}S_+ e^{-i[-\gamma_I(\mathbf{I}\cdot\mathbf{H}_e)+\mathscr{H}'_{ss}]t}S_-\}, \tag{146}$$

where $\mathscr{H}'_{ss}$ is the truncated spin Hamiltonian given by (93) and (94). If $|\gamma_I H_e|$ is much larger than $|\mathscr{H}'_{ss}|$ it is legitimate to consider terms of $\mathscr{H}'_{ss}$ that do not commute with $-\gamma_I(\mathbf{I}\cdot\mathbf{H}_e)$ as non-secular and to disregard them.

In order to select the secular terms of $\hbar\mathscr{H}'_{ss}$ it is convenient to perform on the components of the spins I the transformation (99), which amounts to quantizing them along the direction OZ of the effective field, the spins S remaining quantized along the direction Oz of the d.c.

field. An inspection of (93) shows that the secular terms are of the form

$$\sum_{j<k} A^I_{jk\rho}\mathbf{I}_j\cdot\mathbf{I}_k + B^I_{jk\rho}I_{jZ}I_{kZ} + \sum_{j,k'} C^{IS}_{jk'}\cos\Theta I_{jZ}S_{k'z} +$$
$$+ \sum_{j'<k'} A^S_{j'k'}\mathbf{S}_{j'}\cdot\mathbf{S}_{k'} + B^S_{j'k'}S_{j'z}S_{k'z}, \quad (147)$$

which we rewrite as

$$\hbar[f_0(\mathbf{I}) + g_0(\mathbf{I},\mathbf{S}) + k_0(\mathbf{S})], \quad (148)$$

each of these three terms commuting with I_Z. Carrying (148) over into (146), we get

$$G_0(t) = e^{i\omega_s t}\operatorname{tr}\{e^{it[f_0(\mathbf{I})+g_0(\mathbf{I},\mathbf{S})+k_0(\mathbf{S})]}S_+ e^{-it[f_0(\mathbf{I})+g_0(\mathbf{I},\mathbf{S})+k_0(\mathbf{S})]}S_-\}. \quad (149)$$

From the expression of $\hbar g_0(\mathbf{I},\mathbf{S}) = \cos\Theta \sum_{j,k'} C_{jk'}I_{Zj}S_{zk'}$ it is clear that the contribution of the spins I to the second moment of S has been reduced by $\cos^2\Theta$, and vanishes at exact resonance, as predicted by the classical argument.

The existence of side-bands is demonstrated as follows (12). The effective Hamiltonian $-\gamma_I(\mathbf{I}\cdot\mathbf{H}_e) + \mathscr{H}'_{ss}$, where we quantize the spins I along OZ parallel to $\mathbf{H}_e$ and the spins S along Oz parallel to $\mathbf{H}_0$, can be rewritten as $\omega_e I_Z + \sum_{|m|=0,1,2} a^m$. The various terms a^m are classified according to their transformation under a rotation of angle ωt of the spins I around OZ:

$$e^{i\omega I_z t}a^m e^{-i\omega I_z t} = a^m e^{-im\omega t}. \quad (150)$$

The term a^0 is simply $f_0(\mathbf{I}) + g_0(\mathbf{I},\mathbf{S}) + k_0(\mathbf{S})$ defined in (147) and (148); the terms a^m with $m \neq 0$ can be obtained from the expressions (93) and (94) of $\mathscr{H}'_{ss}$, using the transform (99), but we shall not write them out explicitly. The expression of $G(t)$ is now

$$G(t) = e^{i\omega_s t}\operatorname{tr}\{e^{i(\omega_e I_Z + \Sigma a^m)t}S_+ e^{-i(\omega_e I_Z + \Sigma a^m)t}S_-\}. \quad (151)$$

Its exact calculation is very difficult and we shall use a perturbation method valid for large r.f. fields. Since in the limit of very large H_1, $G(t)$, according to (149), is simply $e^{i\omega_s t}\operatorname{tr}\{e^{ia^0 t}S_+ e^{-ia^0 t}S_-\}$, we rewrite (151) as

$$G(t) = e^{i\omega_s t}\operatorname{tr}\{e^{ia^0 t}U(t)e^{-ia^0 t}S_-\} \quad (152)$$

with $\quad U(t) = e^{-i[\omega_e I_Z + a^0]t}e^{i[\omega_e I_Z + \Sigma a^m]t}S_+ e^{-i[\omega_e I_Z + \Sigma a^m]t}e^{i[\omega_e I_Z + a^0]t}. \quad (153)$

The operator U satisfies the equation

$$\frac{1}{i}\frac{dU}{dt} = \sum_m{}'[a^m(t), U]e^{im\omega_e t} \quad \text{with} \quad U(0) = S_+, \quad (154)$$

where $a^m(t) = e^{-ia^0 t}a^m e^{ia^0 t}$ and the symbol $\sum_m{}'$ means that the value

$m = 0$ is excluded from the summation. The equation (154) can be integrated by successive approximations.

$$U(t) = S_+ - i \int_0^t {\sum}' [a_m(t'), S_+] e^{im\omega_e t'} \, dt' -$$
$$- {\sum_{m,n}}' \int_0^t dt' \int_0^{t'} dt'' [a^m(t'), [a^n(t''), S_+]] e^{i\omega_e(mt' + nt'')} + \ldots \quad (155)$$

This expression can be written as

$$U(t) = \sum_m U^{(m)}(t), \qquad (156)$$

where the $U^{(m)}$ are classified according to the same criterion (150) as the a^m. From the invariance of a trace with respect to a canonical transformation $e^{i\omega I_z t} A e^{-i\omega I_z t}$, it follows that only the term $U^0(t)$ of the expansion (156) brings a non-vanishing contribution to $G(t)$ as given in (152).

The only terms of (155) that contribute to $U^{(0)}(t)$ are thus:

$$U^{(0)}(t) = S_+ - {\sum_m}' \int_0^t dt' \int_0^{t'} dt'' [a^m(t'), [a^{-m}(t''), S_+]] e^{im\omega_e(t'-t'')} + \ldots \quad (157)$$

The first term of this expansion, carried over into (152), will simply give the Fourier transform (149) of the narrowed central line. In the second term we neglect the variation of $a^m(t) = e^{-ia^0 t} a^m e^{ia^0 t}$ in comparison with the much faster variation of the exponential $e^{i\omega_e m(t'-t'')}$ and replace $a^m(t')$ and $a^{-m}(t'')$ by a^m and a^{-m} obtaining
$-S_+ + U^{(0)}(t)$
$$= - {\sum_m}' \frac{1}{(m\omega_e)^2} [a^m, [a^{-m}, S_+]] - i\omega_e t {\sum_m}' \frac{1}{m\omega_e^2} [a^m, [a^{-m}, S_+]] +$$
$$+ {\sum_m}' \frac{1}{(m\omega_e)^2} [a^m, [a^{-m}, S_+]] e^{im\omega_e t}. \quad (158)$$

Since in the integration which led to (158) we neglected the slow time dependence of $a^m(t) = e^{-ia^0 t} a^m e^{ia^0 t}$, to be consistent we also neglect in carrying the right-hand side of (158) into (152), the factors $e^{\pm ia^0 t}$ present in that formula. By doing so we give up any attempt to obtain the shape of the side bands, and get only their positions and intensities.

The first term on the right-hand side of (158), when carried over into (152), can be interpreted as a relative decrease in the intensity of the central line equal to

$$\delta = {\sum_m}' \frac{1}{(m\omega_e)^2} \operatorname{tr}\{[S_+, a^{-m}][a^m, S_-]\} / \operatorname{tr}\{S_+ S_-\}. \quad (159)$$

It is a very small effect very difficult to observe. The second term of (158) corresponds to a small frequency shift of the central line, also unobservable.

The last term of (158) corresponds to side bands shifted with respect to the main line of frequency ω_S by $\pm m\omega_e$. It yields a contribution $G_1(t)$ to $G(t)$:

$$G_1(t) = e^{i\omega_S t} \sum_m{}' \frac{e^{im\omega_e t}}{(m\omega_e)^2} \mathrm{tr}\{[S_+, a^{-m}][a^m, S_-]\}. \tag{160}$$

It is apparent from (160) that the total intensity of the side bands comes from the small loss in the intensity of the central line given by (159). In the dipolar Hamiltonian rewritten as $\sum_m a^m$, the only terms with $m \neq 0$ that do not commute with S_+ or S_- are the terms originating in the I, S coupling, namely:

$$\frac{\sin\Theta}{2\hbar} \sum_{j,k} C^{IS}_{jk'}(I_{j+}+I_{j-})S_{zk'}.$$

Thus the only side bands are those at frequencies $\omega_S \pm \omega_e$. The intensity of each of them, taking that of the central line as unity, is

$$A = \frac{1}{\hbar^2 \omega_e^2} \frac{\sin^2\Theta}{4} \mathrm{tr}\Big\{\Big[S_+, \sum_{j,k'} C^{IS}_{jk'} I_{j-} S_{k'z}\Big]\Big[\sum_{j,k'} C^{IS}_{jk'} I_{j+} S_{k'z}, S_-\Big]\Big\}, \tag{161}$$

related to the contribution $(\overline{\Delta\omega^2})_{SI}$ to the second moment of the resonance S, from the spins I, by

$$A = \frac{1}{\omega_e^2}(\overline{\Delta\omega^2})_{SI}\frac{\sin^2\Theta}{2}.$$

The contribution of the two side-bands to the second moment of the resonance S is $2A \times \omega_e^2 = (\overline{\Delta\omega^2})_{SI} \sin^2\Theta$ which, added to the reduced second moment $(\overline{\Delta\omega^2})_{SI} \cos^2\Theta$ of the narrowed central line, verifies the invariance of the second moment of S with respect to r.f. stirring of the spins I.

In an attempt to verify the above theory an experiment of r.f. stirring was performed on a polycrystalline sample of sodium fluoride, the 'stirred' spin I being the spin of fluorine and the 'observed' spin S that of sodium (13).

The contributions to the second moment of sodium from couplings with sodium and with fluorine are

$$(\Delta\nu^2)_{\mathrm{Na-Na}} = 1\cdot 89 \ (\mathrm{kc/sec})^2, \qquad (\Delta\nu^2)_{\mathrm{Na-F}} = 7\cdot 75 \ (\mathrm{kc/sec})^2,$$

respectively.

For sufficiently large values of the stirring field (denoted in (13) as H_2) an appreciable narrowing of the sodium line should thus be expected. On the other hand, the second moment fluorine-fluorine, $(\Delta\nu^2)_{F-F}$, is equal to 99 (kc/sec)^2 and, as explained earlier, in order to eliminate the broadening of sodium by the fluorines the quantity

$$\left(\frac{\omega_e}{2\pi}\right)^2 = \left(\frac{\gamma_I}{2\pi}\right)^2 \frac{H_2^2}{\sin^2\Theta}$$

must be large not only compared to $(\Delta\nu^2)_{Na-F}$ but also to the much larger quantity $(\Delta\nu^2)_{F-F}$. For the largest value of the stirring amplitude $H_2 = 6$ gauss, used in this experiment, the ratio

$$\frac{\gamma_F H_2}{2\pi[(\Delta\nu^2)_{F-F}]^{\frac{1}{2}}}$$

was only 2·4, which was not sufficient to average out completely the contribution of fluorine to the second moment of sodium. Fig. XII, 6 shows absorption curves of sodium for various strengths of the r.f. field H_2 stirring the fluorines at exact resonance. Fig. XII, 7 is a plot of the second moment determined from these curves, against the strength H_2 of the stirring field. It demonstrates clearly the inadequacy of H_2 to average out completely $(\Delta\nu^2)_{Na-F}$. The side-bands were not observed in that experiment. The difficulty inherent in their observation is that they are not well defined unless $|\omega_e|$ is much larger than the strength of the dipolar couplings (in frequency units) but then their intensity is down by a factor $1/\omega_e^2$. Their detection is thus essentially a question of signal-to-noise ratio.

Another way in which a resonance line can be narrowed is by rotating the sample itself. For a line broadened by an inhomogeneous applied field this method was described in Section IV of Chapter III and an example of the effect was shown in Fig. III, 19. It is also possible to use this method to narrow lines broadened by dipolar interactions. Indeed, this is what happens when molecules get reoriented by hindered rotation in a crystal. Because of the random character of this rotation the contribution to the second moment, missing from the central part of the narrowed line, is spread in the wings. It is, however, possible by producing a coherent rotation of the sample to localize this contribution in well-defined side-bands.

In an experiment where a single crystal of sodium chloride was rotated around the [001] axis normal to the applied magnetic field, with angular velocities ω_R up to $2\pi \times 800$ sec^{-1}, a narrowing of the resonance line and the appearance of side-bands could be observed (14).

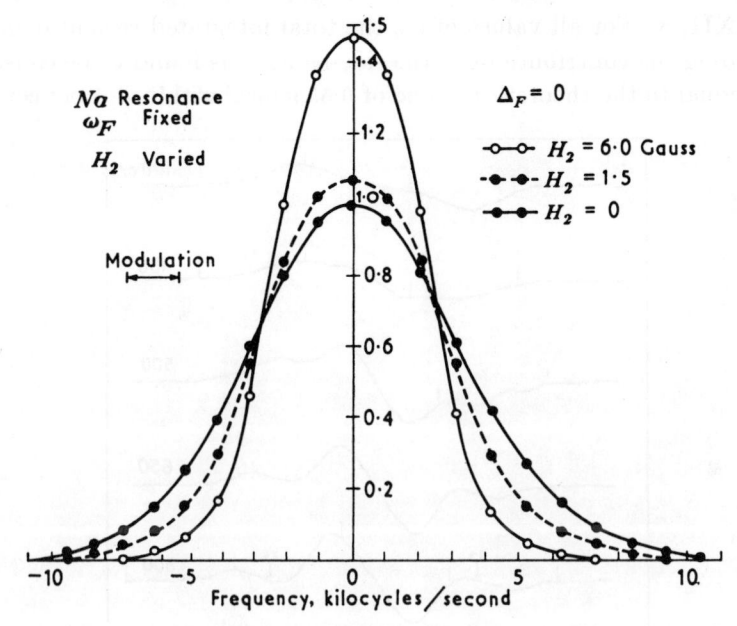

FIG. XII, 6. Absorption curves of sodium for various strengths of the r.f. field H_2 stirring the fluorines.

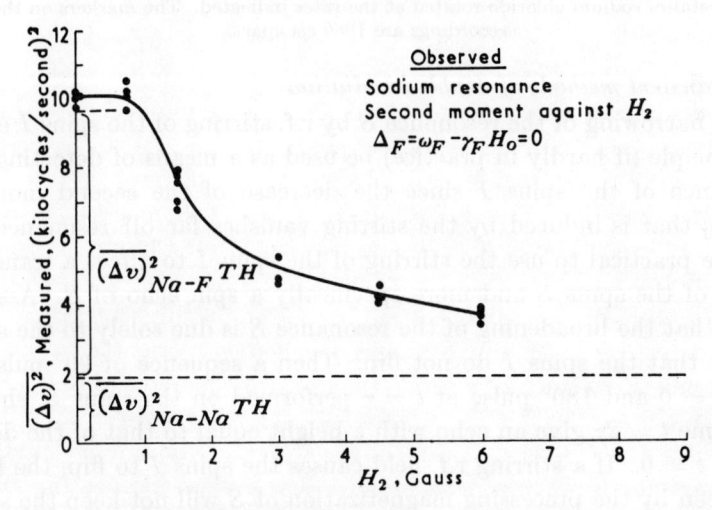

FIG. XII, 7. Measured second moment of the Na resonance against the amplitude of the r.f. field stirring the fluorines.

Since a rotation of the crystal by 180° does not change the dipolar interactions, the positions of the first side-bands are at $\pm 2\omega_R$ from the central line rather than at $\pm \omega_R$. This effect is clearly exhibited in

Fig. XII, 8. For all values of ω_R the total integrated second moment including the contributions of the side-bands was found to be constant and equal to the theoretical value of 0·55 gauss^2 within ± 5 per cent.

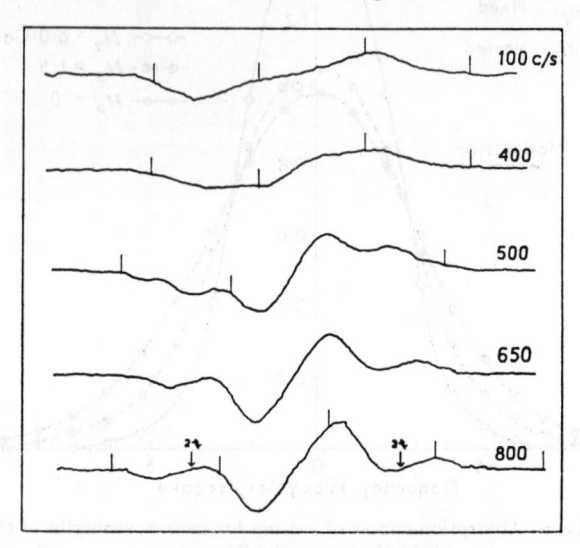

Fig. XII, 8. Na^{23} nuclear magnetic resonance derivative spectra of a single crystal of sodium chloride rotated at the rates indicated. The markers on the recordings are 1660 c/s apart.

(c) *Transient methods of double irradiation*

The narrowing of the resonance S by r.f. stirring of the spins I could in principle (if hardly in practice) be used as a means of detecting the resonance of the spins I since the decrease of the second moment $\overline{(\Delta\omega^2)}_{SI}$ that is induced by the stirring vanishes far off resonance. It is more practical to use the stirring of the spins I to affect a transient signal of the spins S and more specifically a spin echo of S. Assume again that the broadening of the resonance S is due solely to the spins I, and that the spins I do not flip. Then a sequence of 90° pulse at time $t = 0$ and 180° pulse at $t = \tau$ performed on the spins S, should at a time $t = 2\tau$ give an echo with a height equal to that of the decay tail at $t = 0$. If a stirring r.f. field causes the spins I to flip, the local field seen by the precessing magnetization of S will not keep the same value during the intervals 0–τ and τ–2τ, and the amplitude of the echo will be affected. Actually the assumption that the spins I do not flip among themselves is unrealistic. In the absence of the stirring r.f. field the statistical model of a random modulation of the local field seen by the spins S, described in Section II A of Chapter X, can be used to account for the presence of such flips and to calculate the amplitude of the echo.

We assume that the spins S 'see' a fluctuating local field which results in a distribution $\omega(t)$ of their Larmor frequencies around the central frequency ω_0. For each value of t, $\omega(t)$ has a gaussian distribution with a mean square value $\langle \omega^2 \rangle = \omega_p^2$ and a correlation function $G(\tau) = \langle \omega(t)\omega(t+\tau) \rangle$ which we shall assume for simplicity to be of the form $\omega_p^2 e^{-Rt}$, although a gaussian form proportional to $e^{-\alpha t^2}$ which leads to a finite fourth moment is certainly less unrealistic. Under these assumptions the free precession signal given by

$$S(t) = \left\langle \exp\left(-i \int_0^t \omega(t')\,dt'\right) \right\rangle$$

is, by equation (35) of Chapter X, equal to

$$S(t) = \exp\left(-\int_0^t (t-\tau)G(\tau)\,d\tau\right)$$
$$= \exp\left(-\frac{\omega_p^2}{R^2}(e^{-Rt}-1+Rt)\right).$$

Assume now that at time τ we produce a 180° pulse. The amplitude of the signal, t seconds after the pulse, will be

$$\left\langle \exp\left[-i\left(\int_\tau^{\tau+t} \omega(t')\,dt' - \int_0^\tau \omega(t')\,dt'\right)\right] \right\rangle. \quad (162)$$

If we write $A = \int_0^\tau \omega(t')\,dt'$, $B = \int_\tau^{\tau+t} \omega(t')\,dt'$, (162) is equal to

$$\exp\{-\langle \tfrac{1}{2}(A-B)^2 \rangle\},$$

where $\quad Y = \langle \tfrac{1}{2}(A-B)^2 \rangle = \langle A^2 + B^2 - \tfrac{1}{2}(A+B)^2 \rangle$
is clearly

$$Y = \frac{2\omega_p^2}{R^2}\{[e^{-R\tau}-1+R\tau]+[e^{-Rt}-1+Rt]-\tfrac{1}{2}[e^{-R(t+\tau)}-1+R(t+\tau)]\}. \quad (163)$$

For $t = \tau$, which is where the echo would occur in the absence of spin flips,

$$Y = \frac{\omega_p^2}{R^2}\{4e^{-R\tau}-e^{-2R\tau}+2R\tau-3\}.$$

If the constant R, which is a measure of the rate of flip of the spins I, is very small, an expansion of (163) gives

$$Y = \omega_p^2\{\tfrac{1}{2}(t-\tau)^2 + \tfrac{1}{6}R[(t+\tau)^3 - 2t^3 - 2\tau^3] - \tfrac{1}{24}R^2[(t+\tau)^4 - 2t^4 - 2\tau^4]\}. \quad (164)$$

The equation (164) shows that for R small, Y is a minimum and $\exp(-Y)$, which is the signal following a 180° pulse, is a maximum, for approximately $t = \tau$, and the position of the echo is not shifted. If R is increased the amplitude $\exp\{-Y(R)\}$ of the echo signal changes and may either increase or decrease depending on the relative values of the various parameters. Assuming that it is possible to describe the effect of an r.f. field stirring the spins I as an increase in the value of R which is a measure of the rate of spin-flip, the occurrence of the resonance I could be detected by a change of the amplitude of the echo of the spins S.

This method has been applied (15) to a measurement of the small quadrupole splitting of sodium Na^{23} in sodium chlorate, using as indicator the signal from the large quadrupole splitting of chlorine. A peculiarity of the situation existing in the presence of quadrupole splittings is that the spin-spin flips are quenched to a large extent and this corresponds to a small value of the constant R in the above description. The main advantage of the very crude model described is its simplicity and more elaborate and more realistic descriptions could certainly be attempted.

An improvement of the experimental technique is to use pulsed r.f. fields acting on either spin species. Since all that is required to affect an echo of S is to make the values of the local field 'seen' by S different in the intervals $0-\tau$ and $\tau-2\tau$, a 180° pulse performed on the spins I at the same time as the 180° pulse refocusing the spins S will affect appreciably the amplitude of the echo S. It is not at all necessary for the success of the method that the coupling with the spins I be the only or even the main broadening agent for the spins S. Very small quadrupole splittings of K^{39} and K^{41} could be measured by that method in $KClO_3$ observing the effect on the quadrupole echo of chlorine (16). A 180° pulse on K^{41} could be detected by its effect on the echo of Cl^{35} in $KClO_3$ with a signal-to-noise ratio of 10, in spite of the fact that the contribution of K^{41} to the local field 'seen' by Cl^{35} is smaller than 1 per cent. The signal-to-noise ratio for the direct observation of the splitting of K^{41} is 5×10^{-5}.

REFERENCES

1. F. Bloch, *Phys. Rev.* **105**, 1206, 1957.
2. K. Tomita, *Progr. Theor. Phys.* **19**, 541, 1958.
3. A. G. Redfield, *Phys. Rev.* **98**, 1787, 1955.
4. I. Solomon, *C.R. Acad. Sci.* **248**, 92, 1959.
5. —— ibid. **249**, 1631, 1959.

6. F. BLOCH, *Phys. Rev.* **102**, 104, 1956.
7. W. A. ANDERSON, ibid., p. 151, 1956.
8. V. ROYDEN, ibid. **96**, 543, 1954.
9. A. L. BLOOM and J. N. SHOOLERY, ibid. **97**, 1261, 1955.
10. A. G. REDFIELD, ibid. **101**, 67, 1956.
11. N. BLOEMBERGEN and P. P. SOROKIN, ibid. **110**, 865, 1958.
12. F. BLOCH, ibid. **111**, 841, 1958.
13. L. R. SARLES and R. M. COTTS, ibid., p. 853, 1958.
14. E. R. ANDREW, A. BRADBURY, and R. G. EADES, *Nature*, **182**, 1659, 1958.
15. B. HERZOG and E. L. HAHN, *Phys. Rev.* **103**, 148, 1956.
16. M. E. EMSHWILLER, E. L. HAHN, and D. KAPLAN, ibid. **118**, 414, 1960.

INDEX OF NUCLEAR SPECIES

Nuclear species	Material in which studied	Property mentioned	Page no.
μ Meson		magnetic moment	9
n¹ Neutron		magnetic moment	6, 7
H¹ Proton	H₂ solid	static susceptibility	2
		resonance	223–32
	liquid		350–2
	gas		349, 350
		relaxation theory	223–32, 316–21
	HD	resonance	483, 484, 496
	gas	relaxation	321, 350
	liquid	,,	352
	solid	,,	231
	H₂O	static susceptibility	2
		formation and detection of resonance signal	68–84
	H₂O containing dissolved impurities:		
	Mn⁺⁺ ions		330, 331, 379, 380, 502
	Fe⁺⁺⁺ ions		328–30
	Gd⁺⁺⁺ ions		328–30
	D₂O		328–30
	H₂O as water of crystallization in:		
	CuSO₄:5H₂O		197, 198
	CaSO₄:2H₂O		218–20, 224
	CuCl₂:2H₂O		198, 211
	Li₂SO₄:H₂O		476, 477
	KAl(SO₄)₂:12H₂O doped with Cr⁺⁺⁺ ions		381, 386
	HF hydrogen fluoride	relaxation	333–8, 502
	NH₃ ammonia	line width	504
	NH₄Cl ammonium chloride	,,	451, 452
	NH₄Br ammonium bromide		458
	NaBH₄		504
	PdH	relaxation	466
	TiH	,,	,,
	TaH	,,	,,
	CH₄ methane, dissolved in CS₂	relaxation	327
	CHCl₃ trichloromethane	,,	331, 332
	CH₂F₂ difluoromethane	equivalence of protons	480, 481
	CH₃I methyl iodide	double irradiation	530
	C₂H₂F₂ difluoroethylene	multiplet structure	480, 481, 493–5

INDEX OF NUCLEAR SPECIES

Nuclear species	Material in which studied	Property mentioned	Page no.
H¹ Proton (cont.)	$C_2H_4Cl_2$ 1, 2-dichlorethane	fine structure of resonance	220, 221
	$C_2H_3Cl_3$ trichlorethane	,, ,,	220–3, 507
	C_2H_5OH ethyl alcohol	relaxation	324, 325
		multiplet structure	504–8
	$C_3H_4O_2$ β-propiolactone	,, ,,	491, 492
	$(CH_2-CH_3O)_2$ 1, 2-dimethoxyethane doped with ionized naphthalene	polarization	338, 339
	$C_4H_4O_2$ diketene	fine structure of resonance	222
	CCl_3-COOH trichloracetic acid	,, ,,	221, 222
	$CHCl_2-CHO$ dichloracetaldehyde	multiplet structure and stirring of nuclei	486, 487, 501, 528, 529
	$CH_2OH-CHOH-CH_2OH$ glycerin	relaxation	327
	C_6H_6 benzene and C_6H_5D; $1,3,5\text{-}C_6H_3D_3$	line width	451–4
	C_6H_5F fluorobenzene	free precession	497
	$C_6H_4Cl_2$ para-dichlorobenzene	polarization by thermal mixing	143
	polystyrene doped with DPPH	dynamic polarization	397, 398
	$HCO-NH_2$ formamide	relaxation in strong r.f. fields	522
	C_4H_2BrClS 2-bromo, 5-chlorothiophene	multiplet structure	487
	Various solvents, doped with free radical peroxylamine disulphonate $ON(SO_3)_2$	dynamic polarization	339–44
H² or D Deuteron	D_2 solid	resonance	231, 232
	HD	,,	231, 483, 484, 496
	D_2O (doped)	signal size and detection	84
		quadrupole relaxation	347, 348
	$Li_2SO_4:D_2O$	quadrupole coupling and resonance	476, 477
He³ Helium		resonance and relaxation	352
Li⁶ Lithium	metallic	relaxation and hyperfine coupling with conduction electrons	370–2
	LiF lithium fluoride	fast-passage experiment	552–4
		spin temperature	142, 143, 150–4
		'solid-state' polarization	394–8, 422

INDEX OF NUCLEAR SPECIES

Nuclear species	Material in which studied	Property mentioned	Page no.
Li^7	metallic	susceptibility and Knight shift	203, 204
		relaxation and hyperfine coupling with conduction electrons	370–3
		Overhauser effect	373–8
		motion narrowing of line width	458–68
	LiF	fast-passage experiments	149, 552, 553
		spin temperature	144–54
		cross relaxation with fluorine nuclei	155–7
		relaxation	386–9
		'solid-state' polarization	398
	LiF in solution doped with free radical peroxylamine disulphonate $ON(SO_3)_2$	dynamic polarization	341
C^{13} Carbon	element, graphite	Overhauser polarization	392
	CH_3I methyl iodide (enriched in C^{13})	resonance by stirring of protons	530
	CCl_4 carbon tetrachloride	relaxation	348, 349
	CS_2 carbon disulphide	,,	,,
N^{14} Nitrogen	$HCO—NH_2$ formamide	relaxation	522
	$ON(SO_3)_2$ peroxylamine disulphonate, free radical	hyperfine interaction	339–44
F^{19} Fluorine	HF hydrogen fluoride	relaxation	333–8, 502
	LiF lithium fluoride	spin temperature and relaxation	142–54
		cross-relaxation	155–7
		relaxation	386–9
		'solid-state' polarization	397
	LiF in solution doped with free radical proxylamine disulphonate $ON(SO_3)_2$	dynamic polarization	341
	CaF_2 calcium fluoride	dipolar line width	115–22
		relaxation time	386–9
	MnF_2 manganese fluoride	resonance and hyperfine coupling	198, 199, 212, 213
	BrF_5 bromine pentafluoride	multiplet structure	480, 483, 501
	Na_2PO_3F sodium fluoro-orthophosphite	multiplet structure and double irradiation	532, 533
	$C_2H_2F_2$ difluoroethylene	resonance	480, 481, 493–5
	CH_2F_2 difluoromethane	resonance	480, 481
Na^{23} Sodium	metallic	Knight shift	203, 204
		Overhauser effect	373–8

INDEX OF NUCLEAR SPECIES

Nuclear species	Material in which studied	Property mentioned	Page no.
Na^{23} Sodium (cont.)	metallic	relaxation and hyperfine interaction with conduction electrons	370–3
		line width	517
	ionic	antishielding factor	168, 416
		quadrupole moment	416
	NaF sodium fluoride	line width in presence of stirring of fluorine nuclei	575–7
	NaCl sodium chloride	quadrupole relaxation	416
		ultrasonic saturation of quadrupole transitions	139, 140, 421
		line width with rotating sample	576–8
	NaI sodium iodide	ultrasonic saturation of quadrupole transitions	421
	NaClO$_3$ sodium chlorate	quadrupole splitting	580
	NaNO$_3$ sodium nitrate	quadrupole relaxation	411–15
		quadrupole splitting	235
Al^{27} Aluminium	metallic	relaxation	370–3, 543, 544
	Al—Zn alloy	second-order quadrupole shifts of resonance frequency	248
	Al$_2$O$_3$ alumina	second-order quadrupole shifts of resonance frequency	235, 236
Si^{29} Silicon	element	relaxation	390, 391
		Overhauser effect	391, 392
	element doped with phosphorus	'solid-state' polarization	398–400
	SiO$_2$ silica	fast-passage signal	86, 548
P^{31} Phosphorus	as impurity in silicon lattice	relaxation	144
	PCl$_3$ phosphorus trichloride	relaxation	332
	PBr$_3$ phosphorus tribromide	relaxation	332, 333
	Na$_2$PO$_3$F solution sodium fluoro-ortho-phosphite	stirring of phosphorus nuclei	532, 533
P^{32} (radioactive)	as impurity in silicon lattice		16
Cl^{35} Chlorine	NaCl sodium chloride	quadrupole relaxation	416
		ultrasonic saturation of quadrupole transitions	139, 420, 421
	CsCl caesium chloride	polarization of caesium nuclei by thermal mixing	154

INDEX OF NUCLEAR SPECIES

Nuclear species	Material in which studied	Property mentioned	Page no.
Cl^{35} and Cl^{37}	CCl_4 carbon tetrachloride	relaxation	348, 349
	$TiCl_4$, $VOCl_3$, CrO_2Cl_2, $SiCl_4$	quadrupole relaxation	348
Cl^{35}	$NaClO_3$ sodium chlorate	quadrupole resonance	255–7, 260
		ultrasonic saturation of quadrupole transitions	420
	$KClO_3$ potassium chlorate	quadrupole resonance	469
		double irradiation	580
	$CHCl_3$ trichloromethane	relaxation	331, 332
	$C_2H_4Cl_2$ trans dichloroethane	line width with hindered rotations	476
	C_6H_5Cl chlorobenzene	line width	476
Cl^{35} and Cl^{37}	$C_6H_4Cl_2$ para dichlorobenzene	thermal mixing, 'cross over'	143
K^{39} Potassium	KF potassium fluoride	line width	123, 124, 570
K^{39} and K^{41}	$KClO_3$ potassium chlorate	quadrupole splitting	580
Mn^{55} Manganese	MnF_2 manganese fluoride	hyperfine coupling	193, 212, 213
Fe^{57} Iron	metallic	resonance	213
Co^{59} Cobalt		chemical shifts	181
	metallic	quadrupole coupling	213
	CoO cobalt oxide	resonance	195
	$KCoF_3$ potassium cobalt fluoride	,,	195, 196
	CoF_2 cobalt fluoride	,,	213
Co^{60} (radioactive)		dynamical polarization	15
		γ-radiation	72
Cu^{63} and Cu^{65} Copper	metallic	hyperfine coupling	188, 189
		relaxation and hyperfine coupling with conduction electrons	370–3
		quadrupole broadening	239
		relaxation	543, 544
	alloys Zn–Cu	second-order quadrupole effects	247–9
As^{76} Arsenic (radioactive)	as impurity in silicon lattice	dynamical polarization and resonance	16
Br^{79} and Br^{81} Bromine		quadrupole relaxation	347
	BrF_5 bromine pentafluoride	quadrupole broadening	501
	KBr potassium bromide	resonance	239

INDEX OF NUCLEAR SPECIES

Nuclear species	Material in which studied	Property mentioned	Page no.
Br^{79} and Br^{81} Bromine (cont.)	KBr potassium bromide	quadrupole broadening	416
	LiBr, AgBr, TlBr	quadrupole relaxation and chemical shifts	417
	CsBr caesium bromide	relaxation in strong r.f. fields	563–6
	NaBrO$_3$ sodium bromate	quadrupole spin echoes	260
Rb^{85} and Rb^{87} Rubidium	ionic	antishielding factor	168
	metallic	Knight shift	204
		relaxation and hyperfine coupling with conduction electrons	373
	RbCl rubidium chloride	quadrupole relaxation	415
Ag^{109} Silver	AgF silver fluoride	line width, motion narrowing	123, 124, 570
Cd^{111} in excited state		magnetic moment	11
In^{115} Indium	InSb indium antimonide	ultrasonic saturation of quadrupole transition	419
Sn Tin	white tin, crystalline metal	Knight shift	206
Sb^{121} and Sb^{123} Antimony	AlSb aluminium antimonide	quadrupole relaxation	415
Sb^{122} (radioactive)	as impurity in silicon lattice	resonance and dynamic polarization	16
I^{127} Iodine	element (solid)	spin-spin coupling	482
		quadrupole resonance	189, 190
	KI potassium iodide	quadrupole relaxation and broadening	239, 243, 416
	NaI sodium iodide	ultrasonic saturation of quadrupole transitions	421
	ICl iodine chloride	quadrupole resonance	252
	various compounds	quadrupole resonance and line width	474
Xe^{129} and Xe^{131} Xenon	gaseous element	relaxation	352, 353
Cs^{133} Caesium	ionic	antishielding factor	168
	metallic	Knight shift	204
	CsCl caesium chloride	polarization by thermal mixing	154

INDEX OF NUCLEAR SPECIES

Nuclear species	Material in which studied	Property mentioned	Page no.
Cs^{133} Caesium (cont.)	CsBr	relaxation in strong r.f. fields, dynamic polarization and cross relaxation	563–6
Hg^{201} Mercury		resonance	11
Tl^{203} and Tl^{205} Thallium	metallic Tl_2O_3 thallium oxide	Knight shift ,, line width, exchange narrowing	206 206 438–40
Pb^{204} Lead (excited state)		magnetic moment	11
U^{233} Uranium	uranyl compounds	quadrupole coupling	252

SUBJECT INDEX

Absorption function, v, 46; saturation behaviour, 47, 48; in inhomogeneous fields, 50, 51.
— of energy from r.f. fields, 4, 13, 40–48, 264, 265; in zero field, 55; detection, 76–79; by system of interacting spins, 98–103, 572.
— — negative, 89–92.
— of phonons: direct process, 405, 406; Raman process, 405–9; multiple phonon process, 408, 409.
— of ultrasonic waves in crystals, 419.
Acoustic irradiation, 17, 417 ff.
— lattice vibrations, 402 ff.
Adiabatic approximation, for dipolar broadening, 105 ff., 427–41.
— change: contrast quantum mechanical and thermodynamic meanings, 135, 136. *See also* Isentropic.
— fast passage, 66; *see* Adiabatic passage.
— passage, 35, 36, 64–68, 146, 149, 154; detection, 68, 86–87; with strong r.f. fields, 548–53.
— theorem, 34, 35.
Amplifiers, for detection of nuclear magnetism, 78–80.
Angular correlation of radioactive emission, 11.
Antiferromagnetic state, 199, 210–12, 435.
Antishielding factor, γ, 168, 240, 241, 416, 420, 421.
Atoms, free: electric field gradient at nucleus, 165; quadrupole coupling, 169; magnetic coupling with nucleus, 173.
Atomic beam experiments, 170, 317.
Audio-frequency irradiation, 151.
— modulation (of fields), 78, 79.
Autocorrelation function: of magnetization, 114–22; series expansion for, 119, 120; of a random function, 271, 272; quantum mechanical analogue, 284. *See also* Correlation function, Relaxation function.
Auto-oscillation: conditions for, 90.

Beam methods, 5–13. *See also* Atomic beams; Molecular beams.
Beats in free precession signal, 496, 497. *See also* 'Wiggles'.
Bilinear coupling: electron–nucleus, 379, 380.

Bilinear coupling: spin-spin, 183 ff., 305–7, 480 ff.
— — spins with applied field, 305 ff.
Bloch equation, 44–53; steady-state solutions, 46, 47, 567; including diffusion term, 60; with strong r.f. fields, 67–70, 512 ff.; for two-level system, 522–7; for colliding gas molecules, 318; applicability for solids and liquids, 84, 520, 542, 543, 559.
— — modified for low fields, 53–57 ff., 516.
— — electron paramagnetism, 374–8.
Bloch–Siegert shift of resonance frequency, 22, 567, 568.
Bloch wave functions, 199, 207, 358 ff.
Boltzmann distribution: energies of spins, 2, 266–8; defining spin temperature, 133–5; establishment and maintenance, 139–43; system of many spin species, 151.
— — energies of conduction electrons, metals, 364; semiconductors, 389, 390.
— statistical operator, 145 ff., 513 ff., 545.
Bose statistics (of phonons), 404.
Breit–Rabi formula (energy levels with hyperfine interaction), 142, 144, 340, 341.
Bridge methods for detection of nuclear magnetism, 76.
Broadening of resonance line width, *see* Dipolar, Inhomogeneity, and Quadrupole broadening.
Brownian motion: in liquids, 59, 268, 273, 298 ff.; in solids, 439, 441.

Central limit theorem (line shape in presence of defects), 239, 240.
Chemical bonding, in molecular crystals, 169; theoretical description, 186, 190.
— exchange: relaxation effects, 268, 308, 309, 334, 335; effect of catalyst, 337, 338; line width and resolution of multiplet structure, 501–6, 522, 527, 528.
— shifts of n.m.r. frequency: theory, 175–83; in metals, 203; anisotropy and relaxation mechanisms, 315, 316, 348, 349; relation to covalency in ionic crystals, 417; contributions to multiplet structure of n.m.r. in liquids, 480–501.
— structure: inference from n.m.r. data, 495.
Coherence of radiation field, 4, 528 ff.
— of r.f. pulses, 62, 87, 89.
— of motion of spins, 571.

Coils, r.f., 71, 74, 75.
Collisions, in gases: in hydrogen, 317–21, 349–50; in monatomic gases, 322, 323, 352, 353; effect on line width, 427. *See also* Strong collisions.
Conduction electrons, coupling with nucleus, 199–206, 355–63, (in Li metal) 460 ff.; paramagnetism, 200, 363, 365; e.s. resonance, 366 ff., 391, 392.
Contact hyperfine interaction (via s-electrons), 172, 173, 192, 355 ff.
Continuous wave (c.w.) detection of n.m.r., 74–82, 521.
Correlation function: (theoretical introduction), 270 ff.; magnetization of spins in rigid lattice, 101, 102; rotation of molecules (in liquids), 297–300, (in solids), 470 ff.; translation of molecule (in liquids), 300–2; (in solids), 439–41, 460; random local fields, 432, 433; random exchange couplings, 308.
— time: 271, 279, 280; compared with Larmor period, 297, 298, 442; for molecular motion in liquids, 300, 302, 324–7, 515, 517; relation to collision time in gases, 349, 350; random local fields, 357, 425, 426; for rotational states in solids, 455–8, 473.
Coupled equations, for magnetization of unlike spins, 294–6, 308, 311; electron-nucleus, 368 ff., 460 ff.
Coupling of spins: types of coupling, 58, 59.
— — — with lattice, 267, 268; semiclassical description, 272–83; quantum mechanical description, 283–9.
— — — with radiation field, 264–6.
Covalent bonding in crystals, 169, 402, 416–17, 421.
Crossed-coil, detection of nuclear magnetism, 76, 77.
'Cross-over' contact between spin species, 143, 144.
Cross-relaxation: spin-spin mechanism, 155–7; spin-lattice mechanism, 562–4.
— time, 157. *See also* Mixing time.
Crystal defects, 239 ff., 354.
Crystalline potential (in solid hydrogen), 227–9.
Crystals: ionic, 166 ff., 416; molecular, 169, 170.
Curie law, 2, 39, 303; effective field in paramagnetic sample, 197; in terms of spin temperature, 135; in rotating frame, 557.

Damping, of transverse magnetization, 51, 58, 59, 73; of free precession, 72, 73; of 'wiggles', 86.

Damping, radiation, 73, 74, 91.
Debye formula (for dielectric dispersion), 56.
— temperature, 403, 406 ff.
Decoupling of spins by strong r.f. fields, 527 ff.
Degeneracy, of electronic energy levels, 164, 192.
— of quadrupole energy levels, 249–53; effect of perturbing field, 253–7.
Demagnetization, reversibility, in relation to spin temperature, 144–53, 546 ff.
Density matrix, introduced, 26; time variation, 27; for two-level system, 36, 523–7; for spin system in lattice, 39–41, (with dipolar interactions), 99–101, 114; in relation to establishment of a spin temperature, 134, 145.
— — general master equation for spin system with random perturbations (semiclassical), 276 ff.; (quantum mechanical), 285–9, 361, 512 ff.
Detection of n.m.r. by optical methods, 10; by thermal methods, 15; by electromagnetic methods, 13; by anisotropy of radioactive emission, 15; by double resonance methods, 194; by 'stirring' of other nuclei, 528, 530.
— — fast passage, 68, 86, 87; free precession signal, 71, 72; spin echoes 87–89; c.w. detection, 74–82.
— of resonance: 'trigger' methods, 5, 6, 10, 12, 224.
— of r.f. absorption and dispersion, 48, 76 ff.
— of 'rotary resonance', 568.
— of ultrasonic (quadrupole) transitions, 419.
Diamagnetic shielding, 177, 178, 183.
— state: electron-nucleus coupling, 173–91.
Diamagnetism, nuclear, 1.
Diffusion constant: determination by spin-echo methods, 62, 325; dependence on viscosity and temperature, 298, 324, 325.
— equation: random rotations, 298; translations, 301.
— of conduction electrons, 377, 378.
— narrowing of line width (in metals), 512.
— of magnetization beyond spin depth, 377, 378.
— of spins, in liquid, 59–62, 88; in crystal lattice, 439–41, 460–7. *See also* Spin diffusion.
Dilute magnetic systems (line width), 125–8.

SUBJECT INDEX

Dipolar broadening of line width: rigid lattice, 97, 98, 103–6; (like spins), 115 ff., 123, 129, 130; (unlike spins), 123, 124, 130; with fine structure, 220–3; with quadrupole interactions, 128 ff., 237–41; in dilute systems, 125–8; in liquids, 289 ff.; in solids, including motions of spins, 427–41, 453–4, 460.
— relaxation: establishment of spin temperature, 136, 137, 143, 144, (in strong r.f. fields) 520, 521; in liquids, 289, (rigid molecule) 294–7, 298–305, 324–7, 509, 510.
Dipole-dipole coupling: pair of spins in rigid lattice, 97 ff., 424; fine structure of n.m.r., 216–32; in molecules, 453, 454; random coupling in liquids, 289 ff.
— — electron-nucleus, in solids, 380, 381.
Dipole moment (electrical), 159, 161; (magnetic), 1, 170.
Dispersion function, u, 46; saturation behaviour, 47–52; (in solids) 521; detection, 76, 79.
Distortion of electron shells, 167–9.
Doping (with paramagnetic impurities), 53, 70, 79, 82, 85, (gases) 352, (liquids) 302, 303, 324, 326, 502, 509, (solids) 381 ff.
Doppler broadening of n.m.r. (in gases), 427.
Double irradiation, 15, 333 ff., 528–30, 566–78; transient methods, 578–80.
— resonance (ENDOR), 194.
Driving of electronic transitions, 15, 16, 364.
— of nuclear transitions, 333 ff.; see Dynamic polarization.
Dynamic polarization of nuclei, 15, 16, (negative) 89; in liquids (coupling nuclei), 333–8, (coupling electron spins with nuclei), 338–46, 364; in metals, 367 ff.; in semiconductors, 391, 392; in (insulating) crystals with paramagnetic impurities, 392–401, 422. See also Overhauser effect and 'Solid-state' polarization.

Earth's field, 64, 65.
Echoes, 58 ff.; see Spin echoes.
Effective field, 19 ff.; see Field.
Electric dipole transitions, 6.
— field gradient tensor: 160; interaction with quadrupole moment, 163–4; operator, 164–5; polarization of electron shells, 167–70, 416; location of principal axes, 232, 236, 253 ff.
— field gradient: random fluctuation, 269, 270; lattice vibration, 414 ff.; molecular rotation, 468, 469, 475.

Electric moments (dipole) 159, 161, (multipole) 160, 162.
— quadrupole —, see under Quadrupole moment.
Electromagnetic detection of n.m.r., 13, 71 ff.
— radiation field (coherence), 4; see Radiation.
Electron, in magnetic field, Hamiltonian, 171.
— polarization and distortion of closed shells, 167–70; (collisions in gas), 323.
Electron-nucleus (magnetic) coupling, 159; in diamagnetic material, 173–91; in paramagnetic non-metals, 191–9; in metals, 199–210; in antiferromagnetic or ferromagnetic state, 210–13. See also Hyperfine coupling.
— relaxation: dipolar, 296, 303, 304, 328–30; scalar coupling, 312, 330, 331; coupled equations (metal), 368, 369, 460 ff.; (liquid), 364, 379, 380.
— simultaneous transitions, 356–61, 394–400.
Electron-spin, coupling with phonons, 410.
— (paramagnetic) relaxation, 303, 304, 331, 342–6; conduction electrons, 357, 366; paramagnetic impurities in solids, 379, 380.
— (paramagnetic; e.s.r.) resonance, 3, 15; (saturation), 333 ff.; conduction electron, 366 ff., 376–8; line width, 396–8; (exchange narrowing), 435, 438.
— states, lifetime, 195.
Electronic (magnetic) fields at nucleus, 172, 195–9; (correlation), 357; in antiferromagnetic or ferromagnetic state, 210, 211; at different nuclei (coherence), 362, 363.
— paramagnetism, 3; free electrons, 200, 203; conduction electrons, 363–70; Van Vleck (temperature independent) paramagnetism, 303, 304.
— wave functions, 164; perturbations, 167, 169; in metals, 199, 202–4, 207, 358 ff.
Electrostatic coupling electrons with nuclei, 159, 160. See also Quadrupole coupling.
Ellipsoid, representing solutions of Bloch's equation, 46, 47.
E.M.F. (induced by transverse magnetization), 57, 72 ff.
Emission of phonons by lattice, 405–9.
ENDOR (electron nucleus double resonance), 194.
Energy levels: pair of spins in rigid lattice, 217; (with quadrupole coupling), 233–

SUBJECT INDEX

Energy levels (*cont.*):
6; quadrupole in low field, 249 ff.; pair of spins with *J* coupling, 484.
Ensemble of free spins, quantum mechanical description, 24 ff.
Equivalence of spins in molecule, 480 ff.
Ergodicity (existence of temperature), 141, 142, 150.
Exchange, *see* Chemical exchange.
— coupling of electrons, 188, 195, 435.
— narrowing of line width, 435–40, 550.
Excited states of nuclei: magnetic moments, 11.
Extreme narrowing approximation, conditions, 279.

F-centres (as 'paramagnetic impurities'), 386–9, 397.
Fast passage, *see* Adiabatic (fast) passage.
Fast-relaxing spins, 309–12, 331, 522 ff.
Fermi distribution (conduction electrons), 199 ff., 356 ff.
Ferromagnetic state, 210–14, 435.
Ferromagnetism, nuclear, 1.
Fictitious field (in rotating coordinates), 19 ff.
— spin (description of two-level system), 36, 37, 192, 257–61, 522, 523.
Field—*see also* Electric field.
— (magnetic), at electrons, 376–8.
— effective, (in rotating frame) 19–21 ff., 518, 544, 551.
— homogeneity, criterion, 58.
— local, due to dipolar coupling, 97 ff.; *see* Local field.
—— modulation of, 36, 78, 79.
— stability, 93.
Filling factor, for r.f. coils, 74, 75, 83.
Fine structure, of hydrogen atom, 7; of positronium, 8.
—— of n.m.r. (dipolar coupling), 217 ff.; (quadrupole coupling), 234–41; effect of motion of spins, 447 ff. *See also* Multiplet structure.
Flip-flop interaction of spins, 105, 112, 122, 123, 137, 138, 155–7, 269; electron-nucleus, 356–61, 379–82.
—— of electron spins, 195.
Folding of Lorentz distributions, 50.
Forced precession of nuclear magnetism, 14, 521.
Fourth moment of (dipolar) line shape, 113, 125.
Free precession of nuclear magnetization, 32, 33; (relation to line shape), 33, 39, 63–65, 114–21, 441, 442, 543, 544; (with quadrupole coupling), 242, 257–61; transient nutation, 68–70; beats, 496,

497; inhomogeneity damping, 495, 497.
Free precession decay (relation to T_2), 51, 85, 86, 114–21, 429.
— radicals, electron-nucleus coupling, 338, 339.
— spins in field (classical), 19–22; (quantum mechanical), 22 ff.
Frequency dependence of r.f. susceptibility, 42–43, 48, 49.

g-tensor, 192.
Gases, n.m.r., 316 ff., 349–52, 427.
Gaussian line shape (moments), 107, 108; with dipolar coupling, 113, 424, 425.
Gerlach and Stern experiment, 5, 8.
Gibbs ensemble (for spin system), 99, 137, 145; (conduction electrons), 364, 365.

Hamiltonian, electron in magnetic field, 171.
— electron-nucleus coupling (electrostatic), 159, 161; (magnetic), 171–8; random scalar coupling, 307 ff.
— pair of spins with dipolar coupling, 99 ff., 146 ff., 216 ff.; (random dipolar), 289 ff.; (in rotating coordinates), 546, 547; 'truncated' dipolar, 106 ff., 363, 430.
— system of spins: Zeeman, 40; interacting, 145 ff., 427 ff.; with quadrupole coupling, 163 ff., 232 ff.; (with Zeeman perturbation), 253, 469 ff.; (random quadrupole), 313, 314; equivalent spins, 481; (with scalar coupling), 482.
— paramagnetic ion with hyperfine coupling, 144, 342.
Heitler–London approximation, 186–90.
Higher-order resonance, linearly polarized r.f., 21, 22, 56; dipolar interaction, 105, 108.
Hindered rotation in solids, 424, 468; (fast), 474–7; (slow), 477–9.
Homogeneity of fields (criterion), 58.
Hydrogen, atom, fine structure, 7, 8.
— deuteride, 188 ff., 318–21, 349, 352.
— fluoride (relaxation), 333 ff.
— molecule, 183, 223–32; relaxation (scalar), 318, 319; (dipolar), 319–21; (measurement), 349–52.
Hyperfine coupling (free atom), 208, 187 ff., 192–6; conduction electrons in metal, 200–2; effect on polarization of nuclei, 341-3, 355–62; relaxation mechanism, 296, 339–44.
— structure of e.s.r., 159.
— tensor, 187 ff.

Imaginary component of susceptibility (define), 41.
Impurities in samples, see 'Doping'.
Incoherence, of r.f. pulses, 62.
Indirect coupling of spins, 183 ff.; see Spin-spin, indirect coupling.
Infinite temperature (of spins), 134 ff; (of lattice), 513 ff.
Inhomogeneity (of fields), 27; broadening of n.m.r. line width, 52, 53, 97, 98; effect on absorption, 49–53; effect on adiabatic passage, 36, 67 ff.; damping of transverse magnetization, 33, 51, 58, 59, 69, 72, 74, (of 'wiggles') 86, 495–7.
Inhomogeneous fields, refocusing, 63.
Interaction representation for magnetization, 105 ff., 428 ff.
Intermolecular forces (solid hydrogen), 225–7.
Ionic bonding in molecule, 169.
— crystals, 166–9, 417.
Irradiation of crystals (to produce paramagnetic centres), 386.
Isentropic change (define), 135, 136.
— demagnetization, 144–9, 546–53.
Isochronous spins, 480, 491–5.

J coupling, fine structure, 480–95, (transient) 495–501; effect on line width, 435–40; relaxation, 527 ff. See also Spin-spin indirect coupling; Scalar coupling.
Jahn–Teller theorem, 191.
Jumping rate, between states (chemical exchange), 448–51, 501–6, (rotational states) 468, 474–8.

Knight shift of n.m.r. frequency, 201 ff.; s-electron contribution, 202, 203; temperature dependence, 204; anisotropy, 204, 206, 355, 356, (relaxation effects) 363, 364, 373, 374.
Korringa relation, 363, 364, 373.
Kramers–Krönig relation, 42, 43, 49, 90, 99, 100, 542 ff.; proof, 93–96.
Kramers' theorem, 192, 251.

Lamb shift, 7–9.
Larmor precession, 1; frequency introduced, 20.
Lattice (classical) 268, (quantum mechanical) 267.
— defect (effect on line shape), 411, 466.
— temperature, 133.
— vibration, 401 ff. See also Phonons.
Lifetime of electron spin states, 195.
Line shape, with dipolar interaction, 107 ff., 124 ff., 220–3, 229, 237–41; with quadrupole broadening, 237, 238, 246–9; anisotropic Knight shift, 205, 206; with spin in motion, 424 ff., 541 ff.; of multiplet components, 501–10; of rotary resonance, 570.
Line width of n.m.r.: effect on signal to noise ratio, 85, (motion narrowing) 123, 124, 425 ff.; multiplet components, 501–10.
— — of quadrupole resonance, torsional oscillations, 470, 471, 474; hindered rotation, 475, 476, (spin-echo method) 459.
Linearly polarized r.f., 21, 22, 568; susceptibility components, 41; solutions of modified Bloch equation, 55–57.
Local fields, dipole interaction, 97, 98, 425, (correlation) 432, 433, 517 ff., 547, 554 ff.
Lock-in detection of n.m.r., 79–82, 86, 87.
Longitudinal magnetization, 17, 24 ff.
— relaxation time (define), 44, 57–59, (measurement), 64–67.
Lorentz shape function, 32, 33, (susceptibility components) 43–50; dipolar line shape, 107 ff., 124 ff., (including motion of spins) 422, 425, 433 ff.
Low fields, measurement, 64, 65.

Magnetic dipole transitions, 6 ff.; induced by ultrasonic irradiation, 421, 422.
— fields, production of, 93. See also Field.
— moments of nuclei, 1 ff., (dipole) 170; determination of sign, 77.
— relaxation, see Spin-spin relaxation.
— resonance, 20 ff. See also Nuclear magnetic resonance (n.m.r.).
Magnetization of a system of spins, 2, 24 ff.; in zero field, 55; interacting spins, 98 ff.; in terms of spin temperature, 135 ff.; in rotating coordinates, 519.
Magnetometer (proton), 65, 91.
Maser oscillation, 89–93, 339, 340.
Marginal oscillator, for detection of n.m.r., 77, 78, 90.
Markov process (jumping between resonance frequencies), 448.
Metals, adiabatic passage, 67; electron-nucleus coupling, 199 ff.; relaxation, 355 ff., 378; strong r.f., 512, 517.
Mixing, 150 ff., (mixing field) 150 ff., (mixing time) 153 ff.; in strong r.f. fields, 553–4; also Thermal mixing.
Modified Bloch equations, 53 ff., 516.
Modulation, of fields, 36, 78, 79.
Molecular beams, 178, 224, 228.
— crystals, 169, 170.
— motions, 268 ff., 424, 439 ff., 453, 454, 468 ff.
— orbitals, 190 ff.
— symmetry, 480, 493–6.

SUBJECT INDEX

Moments (of line shape), theory, 106 ff.; dipolar broadening, 111 ff., (with spins in motion) 425, 433 ff. *See also* Second moment *and* Fourth moment.
Mössbauer effect, 213.
Motion of spins, 268 ff., 424 ff., (coherent) 527 ff.
Multiple echoes (in solids), 89, 241 ff.
— quantum absorption, 56.
Multiplet structure of n.m.r. spin-spin coupling, 183, 184, (protons in water) 197–9, 216 ff.; J coupling and chemical shift, 308, 480 ff., (groups of lines) 490 ff.; effect of chemical exchange, 501–6, 527, 528; effects of 'stirring', 531–41, 571 ff.
Multipole (electric) 160–2, (magnetic) 170.

Narrowing of line width, by motion of spins, 123 ff., 424 ff., 451–3, 501, 570 ff.; by rotating sample, 93, 576.
Néel transition, 199, 210–12.
Negative absorption, 89–92.
— spin temperature, 134 ff.
Noise, 78 ff.
Normal modes of lattice vibration, 402 ff.
Nuclear ferromagnetism, 1.
— magnetic resonance (n.m.r.), 3, 13–17; conditions for observation, 40, (in paramagnetic materials) 193 ff.
— magnetization, 2, 17, 39 ff.; comparison of solids and liquids, 269.
— orbital motion, 2, 174.
— paramagnetism, 1–3 ff.
— quadrupole coupling, *see* Quadrupole coupling.

Octopole (magnetic) moment, 170.
Optical pumping, 10, 11, 15.
Orbital coupling of nuclear spins, 184–6.
— momentum, nuclear, 2, 174; electronic, 173, 174, 228 ff.
ortho-Hydrogen, 226 ff., 317 ff.
Oscillating r.f. fields, 21; *see* Linearly polarized r.f. fields.
Oscillation, maintained (auto-oscillation), 90–92.
Overhauser effect: in metals, 367–78, 389; in semiconductors, 391, 392; compare with 'solid state' effect, 399–401. *See also* Dynamic polarization.
Overlap factor for neighbouring spins, 155, 156.
Oxygen dissolved in liquids, 324, 326, (in gas) 352.

'Pancake' analogy for spin echoes, 63.
para-Hydrogen, 226 ff., 317 ff.

Paramagnetic centres, formed by irradiation, 386.
— frequency shift (of n.m.r.), 169–99, 211, 312, 331.
— impurities, *see* Doping.
— relaxation, *see* Electron spin relaxation.
— resonance, *see* Electron spin resonance.
— shielding, 179–83.
Paramagnetism of nuclei, 1–3 ff.
Penetration of electron inside nucleus, 160, 165.
Perturbation method for transition probabilities, 27–29 ff.
Perturbations, *see* Random perturbations.
Phonon, properties, 401–4; emission and absorption operators, 404, 405; relaxation mechanism, 406–9.
Photon, density in r.f. field, 3, 4, 265.
Polarization of nuclei, in beam, 6–7; preparation for low field observation, 64, 65, 497; by thermal mixing, 143, 154; by coupling with electron spins, 338–46, 364. *See also* Dynamic polarization.
— of conduction electrons, 366–9.
— of electron shells, 167–70, 240, 402, 416, 417, (collisions in gas), 323.
Populations of states, 2, 13, 14; relation to spin temperature, 133 ff.; master equation for time dependence, 274 ff., 360 ff.; conditions for dynamic polarization, 345, 346.
Positronium, fine structure, 8.
'Pound box' oscillator, 77, 78.
Precession forced, 14, 521, Larmor, 20; free, 32 ff. *See also* Free precession.
Probability distribution of resonance frequencies, 27 ff., 431, 432.
— of transition, *see* Transition probability.
Proton magnetometer, 65, 91.
— maser, 91, 339, 340.
— relaxation (experimental), 324 ff.
— resonance in water, 68 ff., 324–8; (of crystallization), 216 ff.; in hydrogen, 223 ff.
Pulse (of r.f.) method, 32 ff., 58, 63; detection of nuclear magnetism, 87–89. *See also* Spin echoes.

Q factor r.f. coils, 71 ff.
Q meter, detection of n.m.r., 75, 76.
Quadrupole broadening of line width, 237–41, (spin echo) 241–6; second order effects, 246–9.
— coupling, 17, 39, 40, (with dipolar broadening) 98, 128–32, 139 ff. (electron-nuclear coupling), 163 ff., 213; in zero field, 249–52; in low fields, 253–61; high fields, 233–49; relaxation process,

SUBJECT INDEX

Quadrupole coupling (cont.)
305, 313–15, 346 ff.; random coupling, 411–15, (sign of coupling) 261, 262.
— moment (electric) of nucleus, 16, 159–67, (sign) 261.
— relaxation, 305; liquids, 313–15, 321, 331, 332, 346–9, 450, 451, 468 ff., 501–6; gas, 321; crystals, 333, 411–17.
— resonance, 16, 189, 190, 216, 249–52; line width, 470–9.
— transitions, stimulated by ultrasonic irradiation, 419–21.
Quenching of orbital angular momentum, 174, 229.
— of spin flips, 142.

Rabi experiment, 5, 8, 12.
Rabi, formula for transition probability, 23.
'Race track' analogy for spin echoes, 62, 63.
Radiation damping, 73, 74, 91.
— field, 3, 4; coupling with spins, 264, 265.
Radioactive nuclei, 15, 16.
Radio-frequency (r.f.), absorption, 13, 40 ff., 264, 265; coupling of levels, 36, 37; strong fields, 511 ff., (measurement) 568; stirring by field, 527–39, 571 ff.
— spectroscopy, 3 ff.
— susceptibility, see Susceptibility.
Raman process in crystals, 405–9, (spectra) 473.
Random functions, (mathematical) properties, 270–2.
— jumping between states: 447–51, 501–6, (rotational) 468, 474–8.
— magnetic fields: (in rigid lattice) 97 ff., (conduction electrons) 357, 432–3, 508–10, (in presence of strong r.f. fields) 518–22, 554.
— molecular motion, liquids, 297, (rotations) 298–300, (translations) 300–2; solids, 354, 428, 429, (translation) 439–41, 460 ff., (oscillation) 470 ff.; see Brownian motion.
— perturbations (mathematical theory), 272 ff.; including motions of spins, 427 ff.
— spin-spin coupling, 292 ff., 302 ff., 436 ff., 527 ff.
— voltage (noise), 79, 80.
— walk: model for diffusion in solids, 460–3.
Real part of r.f. susceptibility (define), 41.
Refocusing of transverse magnetization, 58, 63. See also Spin echoes.
Relaxation, electronic, see Electron spin relaxation.
— function, 101, 108–11, 427 ff. See also Autocorrelation function and Correlation function.
Relaxation mechanisms: gases, 316–22, 337, 338, 349–52.
— — liquids; anisotropic chemical shift, 315, 316, 348, 349; chemical exchange, 308 ff., 501–6; dipolar coupling, 290 ff., 506–10 ff.; quadrupole, 313–15, 502–6.
— — solids: paramagnetic impurities, 139, 378–86; dipolar coupling, 97 ff., 427 ff.; spin-phonon coupling, 404 ff.; quadrupole, 333, 411 ff., (torsional oscillations) 424, 468 ff.; hyperfine coupling (metals), 173, 356–64, 370–3; ultrasonic irradiation, 418, 419.
— mechanism: strong collision model, 30–32, 318, 526.
— spin-lattice; (introduced) 14, (thermodynamic representation), 133 ff., 355 ff., 401 ff., 429; in rotating frame, 555 ff.
— spin-spin: 14–15; flip-flop mechanism, 105 ff.; in relation to establishment of spin temperature, 133 ff.
— in solids, 355 ff.; in rotating frame with strong r.f. fields, 546 ff.
— time, T_1, 14, 44, 64 ff., 85, 275, (dipolar coupling) 324 ff., (time dependence of spin temperature) 359–64, (low field measurements) 372, 373; in rotating fields, 561 ff.
— — T_2, 15, 44, 51, 67 ff., 85; dependence on correlation time of random perturbation, 297 ff., 422 ff.; experimental (protons in water), 324 ff., (HF) 333–8; in metals, 359–64; in solid, 457 ff.; dependence on strength of r.f. field, 520–2.
— toward rotating field, 53, 516.
Reorientation of molecules in solid, 451–8.
Resonance, see nuclear magnetic resonance, etc.
— methods (contrasted with static methods), 4–13.
Reversal of magnetization, adiabatic, 36 ff., 89; negative spin temperature, 146. See also Adiabatic passage.
Reversibility of demagnetization, 145, 150–2, 546 ff.
Rigid lattice: dipolar fine structure, 216 ff.; dipolar broadening, 97 ff.
— molecule, 293, 294, 480 ff.
Rotating coordinates, classical theory, 19–22, Schrödinger equation, 23; solutions of Bloch equations, 45, 46; use with strong r.f. fields, 518 ff., 545 ff.; spin temperature, 555–60.
— field, susceptibility components (define), 42.
Rotation of sample, 93, 576.

Rotational motion of molecules, 177, 178, (liquids) 298–300, (solids) 451 ff., 474–9.
Rotary echoes, 70.
— resonance, 567–70.

s-electrons, 165, 172, 173, 202 ff.; contribution to hyperfine coupling, 192, 355 ff.
Satellite lines: interacting dipoles, 105, 108, 155, 156; quadrupole, 233 ff., 411 ff.
Saturation of electron spin resonance, 333–46, (metals) 366 ff., (semiconductors) 391, 392.
— of n.m.r., 40, 44–50, 85; in terms of spin temperature, 133 ff.; by ultrasonic irradiation, 140; of quadrupole satellites, 411–15; (of forbidden transitions) by 'stirring', 566; in strong r.f. fields, 542–4, 559, 560; of 'rotary resonance', 567–70.
— parameter (define), 367 ff.
Scalar coupling, electron-nucleus, 435.
— — of nuclear spins, 306–12, 318, 324 ff., (unlike spins) 330–3, 438–9; in strong r.f. fields, ('stirring') 527 ff., 550. See also J coupling.
'Scanning' of field, 79.
Scattering amplitude (paramagnetic impurities), 382–5.
Schrödinger equation, in rotating coordinates, 23.
Second moment (of resonance line shape), 111–13, 125, 221, 363, 432–6, 453–4, (stirring of coupled spins) 575 ff.
Self-diffusion, see Diffusion.
Semiconductors (relaxation), 389–92.
Shape function, calculation of transition probability, 27, 32 ff., 41 ff., 50, 99; moments, 106 ff.; representing interaction of other spins, 138, 427 ff. See also Line shape.
Shielding, diamagnetic, 177, 178, 482; paramagnetic, 179, 180.
Side bands (rotary resonance), 571 ff.
Signal shape and size, 85; see Detection.
— to noise ratio (in detection of n.m.r.), 78–86.
'Simple line'; Bloch equation, 522–7. See also Two-level system.
Skin depth, 375–8.
'Solid state' polarization, 392–401; by ultrasonic irradiation, 422; in strong r.f., 566.
Spectral densities of random perturbations, 272 ff.
Spin diffusion, relaxation mechanism, 379 ff.; dynamic polarization, 395, 396. See also Diffusion.

Spin echo, 33, 34, 58–63; detection, 87–89; with quadrupole interaction, 241–6, 260; measurement of diffusion constant, 325; measurement of line width, 459; multiplet structure, 497–501; with 'stirring' of spins, 578–80.
— — rotary echoes, 70; multiple echoes, 89, 241 ff.
— fictitious, 36, 37, 192, 257–61, 523.
Spin-lattice coupling, 401 ff., 429, 555, 556. See also Relaxation, spin lattice.
Spin-orbit coupling, (nuclear) 174; electronic, 184 ff., 192.
Spin-phonon coupling, direct, 405, 406; Raman, 405–9; multiple phonon, 408–9; magnetic, 409–11; quadrupole, 411–17.
Spin-spin coupling, 39, 40; indirect processes, 175, 177, 183 ff., (in metals) 206–10, 305 ff.; relation to spin temperature, 135 ff. See Relaxation, spin-spin, Dipole-dipole coupling, J coupling, scalar coupling. See also Electron-nucleus coupling.
Spin system, macroscopic view, 99, 137, 145.
— temperature, 14, 15, 133 ff.; two spin species, 142 ff., 151, 152, 269; time to establish, 355; time dependence, 359–64, 372; in rotating frame, 541, 545 ff.
— temperature, infinite, 134 ff., 283 ff.; negative, 134.
Spinning of sample, 93, 576.
Spins, decoupled by 'stirring', 527 ff.
— equivalent, 480 ff., isochronous, 480, 491 ff.
Spontaneous transitions, 4, 264.
Stability of fields, 93.
Statistical operator, 39–41. See Density matrix.
'Stirring' of spins, 527 ff.; measurement of stirring field, 539; saturation of forbidden transitions, 566; collapse of multiplet structure, 571 ff. (transient methods) 578–80.
Strong collision (model for relaxation), 30–32, 318, 526.
— r.f. fields, (liquids) 511 ff., (solids) 539 ff.
Superconducting state, 363, 372, 373.
Susceptibility, electronic, see Electronic paramagnetism.
— r.f., components, 41, 57; in rotating fields, 42 ff.; system of interacting dipoles, 99–102, (solid) 542, 543, 559, 560.
— — tensor, 49, 57.
— static, 2, 39, 40.
Symmetry of molecules, 480 ff., 493–5.

SUBJECT INDEX

Temperature, criterion for existence, 141 ff.; contrast spin temperature and thermodynamic temperature, 146–9. See also Spin temperature.
— dependence of n.m.r., 197, 230, (Knight shift) 204; of quadrupole resonance, 469, 476; of relaxation time, 324–7, (spin-phonon) 406–9.
Tensor operators (transformation properties), 161; electric field gradient, 164, 165; hyperfine coupling, 187 ff., (with conduction electrons) 200 ff.; Zeeman coupling (g), 192; shielding, 177–9; spin-spin coupling, 305–7.
— susceptibility, 49–57.
Thermal detection of resonance, 15.
— equilibrium, within spin system, 14, 15; between lattice and spins, 133 ff., 548; of lattice, between spins and radiation field, 265, 266; of conduction electrons, 364 ff.
— mixing, 151–4, (with strong r.f.) 553, 554.
— relaxation liquids, 264 ff., solids, 354 ff.
Three-spin flip-flop process, 155–7.
Time, see Correlation time; Relaxation time.
— for driving resonant transition, 12.
Torsional oscillations (of molecules in solids), 424, 468–74.
Transient behaviour of magnetization, effect on steady-state detection, 85, 86; methods of detection of nuclear magnetization, 86–89; free spins, 32–35; with quadrupole coupling, 241 ff.; coupled unlike spin, 335–8; with strong r.f. fields, 542 ff.; with double irradiation, 578–80.
— nutation, 68–70.
Transitions, induced, probability: Rabi formula, 23, 24; perturbation theory, 27–29 ff.; dipolar flip-flop, 137 ff., 382, (electron-nucleus) 358–61, random lattice perturbation (classical), 272–4, (quantum mechanical) 284, 285; spin-phonon process, 405–9; between rotational states, 455, 468, 474–8.
—— (forbidden) by 'stirring', 566; by ultrasonic irradiation, 139–41, 418 ff.
— spontaneous, 4, 264.
Translational diffusion, in liquids, 59–62, 300–2; in solids, 424, 439, 441, 458–67.

Transverse magnetization, equilibrium, 25; decay, 33, 34, 521; detection, 57; inhomogeneity damping, 51, 58, 59; refocusing, 58, 63 ff.; spins with dipolar interaction, 291 ff.; existence of spin temperature, 134 ff.
— relaxation time T_2, 44, 51; physical meaning, 57, 58; measurement, 59, 67, 70, 72; for system of two levels, 523–7; dependence on strength of r.f. field, 520–2.
'Trigger' method for detection of resonance, 5, 6, 10, 12, 224.
Two-level system (coupled by r.f.), 36, 37, 444, 445, 522–7.
Two-spin species, thermal contact, 138 ff.; in molecule, 493–6; fast passage experiments, 550–3, 558 ff. See also Dynamic polarization.

Ultrasonic irradiation, 139–41, 417–22; amplitude measurement, 420, 421.

Valence electrons (polarization by), 169, 170.
Van Vleck formula for second moment of dipole line width, 112, 138.
—— temperature independent paramagnetism, 303, 304.
Vibrational spectrum of lattice, 403 ff.
Viscosity, relation to diffusion and relaxation, 298 ff., 324, 325, 512 ff.
Voltages induced by precessing magnetization, 57, 72 ff.

Water molecules, diffusion, 62; random motion, 298–305; proton resonance, 68–83, (water of crystallization) 216 ff.
Wave function, electronic, 164, 167, 169; in metals, 199, 202–4, 207, 358 ff.
Width of resonance, 21. See also Line width.
'Wiggles', 85, 86.

Zeeman coupling with field, 150, 151, 216, 217 ff., 228, 541 ff.; as a perturbation of quadrupole splitting, 253–61; conditions for establishment of spin temperature, 139 ff.; conditions for 'cross-over', 143.
Zero field magnetization, 55, 56.